# Reagents for Organic Synthesis

# Reagents for Organic Synthesis

VOLUME 4

**Mary Fieser**

Research Fellow in Chemistry
Harvard University

**Louis F. Fieser**

Sheldon Emery Professor of Organic Chemistry, Emeritus
Harvard University

A WILEY–INTERSCIENCE PUBLICATION

JOHN WILEY & SONS

NEW YORK · LONDON · SYDNEY · TORONTO

77150

# PREFACE

This volume covers literature on reagents published for the most part in 1970–1972, with some references to literature published in the early months of 1973. It includes references to 297 reagents reviewed by us for the first time as well as new references to 350 reagents previously discussed. We are surprised that so many new reagents have been introduced for organic synthesis. It is also interesting that new procedures have been developed for old reagents, for example, aluminum chloride and potassium permanganate.

We are indebted to our colleagues who have sent us additional information or suggested suitable topics for inclusion. We are pleased to acknowledge the help of Professor P. L. Fuchs, Professor J. Secrist, Dr. M. Wuonola, Dr. R. Wingard, R. L. Danheiser, and R. H. Wollenberg who have agreed to read proof.

Miss Theodora S. Lytle and Miss Lenor Kirkeby typed the manuscript and drew the formulas.

Dr. S. Brachwitz and Dr. A. Vasella have also been most helpful in proof-reading. The photograph on the dust cover was taken by E. M. Bellott.

We thank Research Corporation for continued financial support.

This is the first of our books, except for unauthorized translations, that does not include a picture of a cat. It is dedicated, however, to the many cats that we have had the pleasure of knowing.

*Cambridge, Massachusetts*　　　　　　　　　　MARY FIESER
*June 3, 1973*　　　　　　　　　　　　　　　LOUIS F. FIESER


We sincerely regret the delay in publication of this book. It has been caused by several factors, one of which has been beyond the control of the publisher. The manuscript was set in type in England during the energy crisis there.

# CONTENTS

# Reagents for Organic Synthesis

# Introduction

*Arrangement.* For enhanced usefulness the book is provided not only with a subject and an author index but also with an index of types, that is, types of reactions or types of compounds, for example: acetylation, bromination, cycloaddition, decarboxylation, or: acetonides, benzyne precursors, carbene precursors. Listed alphabetically under each such entry are all the reagents which figure in the operation or group cited, whether as prime reactant, catalyst, solvent, scavenger, etc. A given reagent may fit appropriately in two or more categories. When a reagent does not fit easily into a reasonable category, we leave it unclassified rather than make a forced assignment. With so many reagents available as oxidants and for use as reducing reagents, it seems out of the question to attempt to indicate in the index of types further details about these general reactions.

*Names and spelling.* One guideline we have followed is the rule recently adopted by *Organic Syntheses* that when an ester, ether, or peroxide contains two or more alkyl, aryl, or acyl groups the name must indicate the number of such groups:

| Formula | Correct | Incorrect |
|---------|---------|-----------|
| $(CH_3)_2O$ | Dimethyl ether | Methyl ether |
| $(C_2H_5O)_2SO_2$ | Diethyl sulfate | Ethyl sulfate |
| $(C_6H_5)_2O$ | Diphenyl ether | Phenyl ether |
| $(CO_2CH_3)_2$ | Dimethyl oxalate | Methyl oxalate |
| $CH_2(CO_2C_2H_5)_2$ | Diethyl malonate | Ethyl malonate |
| $(C_6H_5COO)_2$ | Dibenzoyl peroxide | Benzoyl peroxide |
| $HC(OC_2H_5)_3$ | Triethyl orthoformate | Ethyl orthoformate |
| $(C_2H_5O)_4C$ | Tetraethyl orthocarbonate | Ethyl orthocarbonate |

That the situation previously was highly confused is evident from the following entries in the index of *Org. Syn., Coll. Vol.*, **4**: "Diethyl oxalate" and "Diethyl malonate" (both correct), but "Ethyl orthoformate" and "Ethyl orthocarbonate" (both incorrect). The following entry is describable as a double error: "Triethyl orthoformate, *see* Ethyl orthoformate." To locate all references to a given ester, it is thus necessary to search under two names. We urge suppliers to revise their catalogs in accordance with the rule cited. In this book we do not even list, with cross references, names which we consider to be incorrect.

Similar reform in the nomenclature of polyhalogen compounds may come some day, but for the present we consider it imprudent to do more than make a start. Thus the correct names for $BF_3$ and for $ClCH_2CH_2Cl$ surely are boron trifluoride and ethylene dichloride, and we feel no restraint from using them. However, although the names methylene chloride for $CH_2Cl_2$ and aluminum chloride for $AlCl_3$ seem incorrect, we cannot bring ourselves to break with tradition and employ other names.

*Abbreviations.* Short forms of abbreviations of journal titles are as follows:
Accounts of Chemical Research
Journal of the American Chemical Society
Analytical Letters
Angewandte Chemie
Angewandte Chemie, international Edition in English
Annalen der Chemie
Annales de chimie (Paris)
Australian Journal of Chemistry
Chemische Berichte (formerly Berichte der deutschen chemischen Gesellschaft)
Bulletin de la société chimique de France
Canadian Journal of Chemistry
Carbohydrate Research
Chemical Communications
Chemical and Pharmaceutical Bulletin Japan
Acta Chemica Scandinavica
Chemistry and Industry
Chemical Reviews
Collection of Czechoslovak Chemical Communications
Comptes rendus hebdomadaires des séances de l'académie des sciences
Gazzetta Chimica Italiana
Helvetica Chimica Acta
Inorganic Synthesis
Journal of Chemical Education
Journal of the Chemical Society (London)
J.C.S. Chemical Communications
Journal of Heterocyclic Chemistry
Journal of Medicinal Chemistry
Journal of Organic Chemistry
Journal of Organometallic Chemistry
Journal für praktische Chemie
Monatschefte für Chemie
Organic Syntheses
Organic Syntheses, Collective Volume
Proceedings of the Chemical Society
Records of Chemical Progress
Receuil des travaux chimique des Pays-Bas (The Netherlands)
   The book by one of us, **Organic Experiments**, 2nd Ed., D. C. Heath and Co., Boston (1968), is referred to as **Org. Expts.**

### *Abbreviations*

| | |
|---|---|
| Ac | Acetyl |
| AcOH | Acetic acid |
| BuOH | Butanol |
| Bz | Benzoyl |
| CAN | Ceric ammonium nitrate |
| Cathyl | Carboethoxy |
| Cb | Carbobenzoxy |
| DABCO | 1,4-Diazabicyclo[2,2,2]octane |
| DCC | Dicyclohexylcarbodiimide |
| DDQ | 2,3-Dichloro-5,6-dicyano-1,4-benzoquinone |
| Diglyme | Diethylene glycol dimethyl ether |
| Dimsyl sodium | Sodium methylsulfinylmethide |
| DMA | Dimethylacetamide |
| DME | Dimethoxyethane |
| DMF | Dimethylformamide |
| DMSO | Dimethyl sulfoxide |
| DNF | 2,4-Dinitrofluorobenzene |
| DNP | 2,4-Dinitrophenylhydrazine |
| EtOH | Ethanol |
| Glyme | 1,2-Dimethoxyethane |
| HMPT | Hexamethylphosphoric triamide |
| MeOH | Methanol |
| DME | Dimethoxyethane |
| MMC | Magnesium methyl carbonate |
| Ms | Mesyl, $CH_3SO_2$ |
| NBA | N-Bromoacetamide |
| NBS | N-Bromosuccinimide |
| Ph | Phenyl |
| Phth | Phthaloyl |
| PPA | Polyphosphoric acid |
| PPE | Polyphosphate ester |
| Py | Pyridine |
| THF | Tetrahydrofurane |
| TMEDA | N,N,N,N-Tetramethylethylenediamine |
| Triglyme | Triethylene glycol dimethyl ether |
| Trityl | $(C_6H_5)_3C-$ |
| Ts | Tosyl, $p\text{-}CH_3C_6H_4SO_2-$ |
| TsCl | Tosyl chloride |
| TsOH | Tosic acid, $p\text{-}CH_3C_6H_4SO_3H$ |
| TTFA | Thallium(III) trifluoroacetate |

# A

**Acetic anhydride–Pyridine hydrochloride.**

**Ether cleavage.**[1] Treatment of the keto ether (1, 9-oxatricyclo[4.3.3.0]dodecane-3-one) with acetic anhydride and pyridine hydrochloride (reflux 5.5 hr.) yields the diacetate (2, 4-acetoxy-1-(2-acetoxyethyl)bicyclo[4.3.0]nonadiene-4,6) in 93% yield.

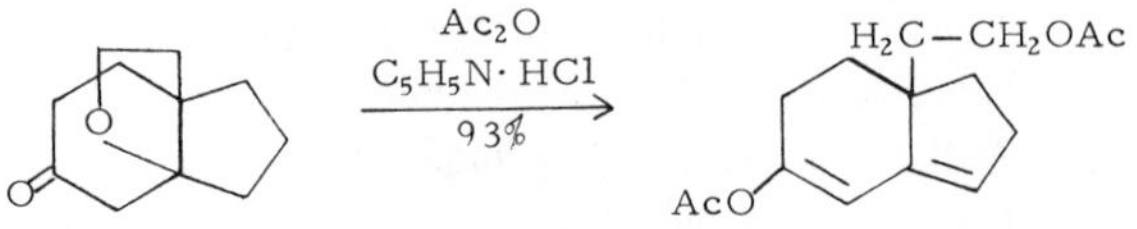

$$\underset{(1)}{\qquad} \xrightarrow[\substack{C_5H_5N \cdot HCl \\ 93\%}]{Ac_2O} \underset{(2)}{\qquad}$$

[1]N. P. Peet and R. L. Cargill, *J. Org.*, **38**, 1215 (1973).

**Acetic anhydride–Zinc chloride.**

**cis-*Hydrindane-1-ol*.**[1] Reaction of the dibromide (1)[2] with acetic anhydride–zinc chloride in methylene chloride gives the ketone (2) in 30% yield together with the dibromide (3), 21% yield. Acetic anhydride is essential for the transannular cyclization reaction; treatment of (1) with zinc chloride alone in methylene chloride merely isomerizes (1) to the *trans*-isomer of (3). Baeyer–Villiger oxidation of (2) affords (4),

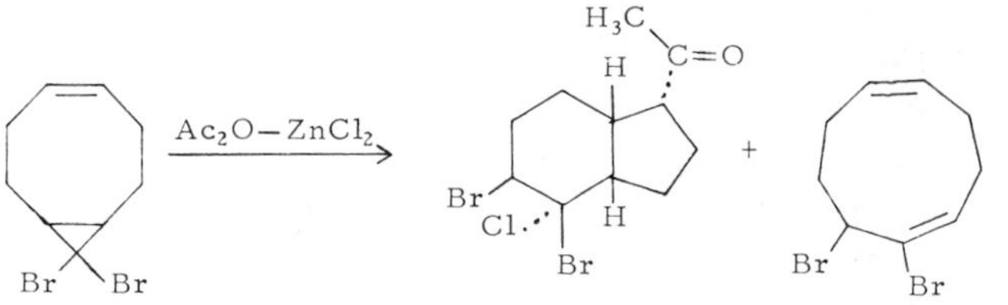

(1)          (2, 30%)          (3, 21% yield)

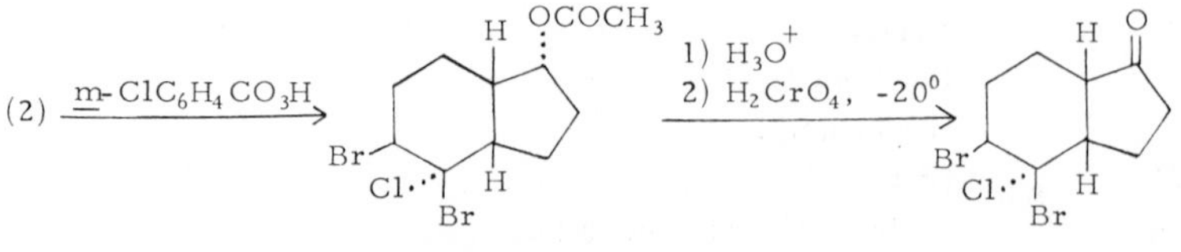

(4)          (5)

4

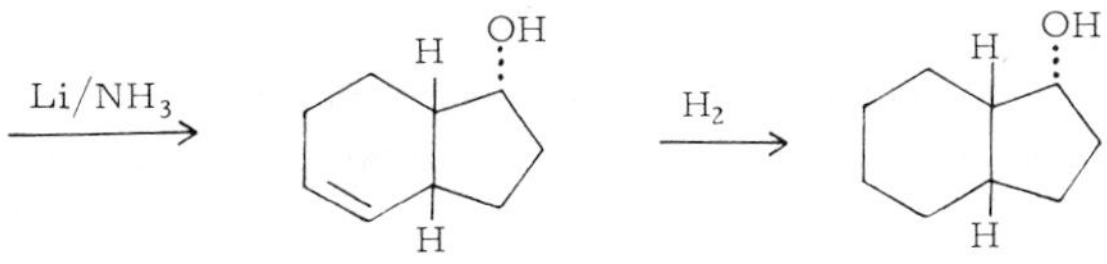

(6)                              (7)

which is hydrolyzed to the alcohol and oxidized with Jones reagent. Reduction of (5) with lithium in ammonia yields a single alcohol (6), which is hydrogenated to *cis*-hydrindane-1-ol (7).

[1]L. W. Boyle and J. K. Sutherland, *Tetrahedron Letters*, 839 (1973).
[2]L. Skattebøl, *Chem. Scand.*, **17**, 1683 (1963).

**Acetic-formic anhydride, 1,** 4; **2,** 10–12; **3,** 4.

*Diazoacetaldehyde.*[1] Diazoacetaldehyde is prepared conveniently in 46% yield by the reaction of acetic-formic anhydride with diazomethane. The coproduct, methyl

$$CH_3\overset{O}{\overset{||}{C}}-O-\overset{O}{\overset{||}{C}}-H \ + \ CH_2N_2 \ \xrightarrow[46\%]{ether} \ N_2CHCHO \ + \ CH_3COOCH_3 \ + \ N_2$$

acetate, is readily removed by rotary evaporation. Diazoacetaldehyde is a precursor of formylcarbene (**2,** 101–102) and has been used for ethanalation of olefins (**3,** 73).

[1]J. Hooz and G. F. Morrison, *Org. Prep. Proc. Int.*, **3**, 227 (1971).

**Acetophenone,** $C_6H_5COCH_3$. B.p. 202°/749 mm.

Photoisomerization of bicyclo[2.2.1]hepta-2,5-diene (1) in the presence of aceto-phenone as sensitizer provides a method for the preparation of quadricyclane (2). The

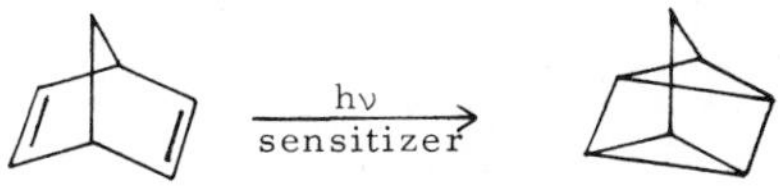

(1)                              (2)

reaction is carried out with 180 g. of (1) in 1 l. of ether in a commercial 550-W immersion photochemical reactor equipped with a stirrer and a condenser under nitrogen with 8 g. of acetophenone as photosensitizer for 36–48 hr.; yield 70–80%.[1]

[1]C. D. Smith, *Org. Syn.*, **51**, 133 (1971).

**Acetyl chloride, 1,** 11; **2,** 383; **3,** 8.

*Acetylation of γ-(naphthyl-2)-butyric acid* ethyl ester with aluminum chloride in nitrobenzene.[1]

*Reduction of sulfoxides.*[2] When an alkyl *o*-carboxyphenyl sulfoxide (1) is treated with acetyl chloride in methylene chloride at room temperature an exothermic reaction takes place with evolution of chlorine to give the sulfide (2) in nearly quantitative yield.

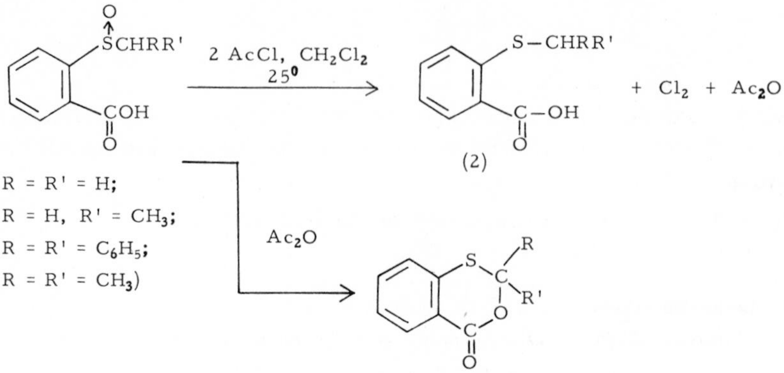

(1, R = R' = H;
    R = H, R' = $CH_3$;
    R = R' = $C_6H_5$;
    R = R' = $CH_3$)

(2)

(3)

Note that treatment of (1) with acetic anhydride gives a 3,1-benzoxathiane-4-one (3) via a Pummerer rearrangement in good yield.[3]

The reduction of sulfoxides by acetyl chloride is a general reaction and yields are generally high. However, the reduction of di-*n*-butyl sulfoxide with acetyl chloride gives di-*n*-butyl sulfide in only 70% yield.

The mechanism formulated is suggested for the reduction.

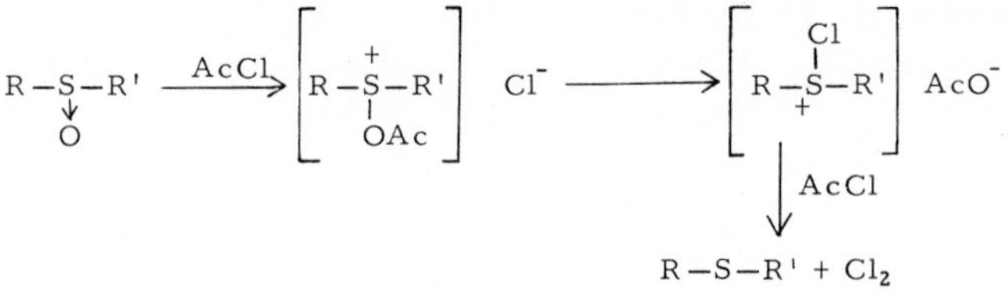

Sulfilimines (4) are also reduced to sulfides (5) by acetyl chloride in nearly quantitative yield.

$C_6H_5S-R \longrightarrow C_6H_5S-R$

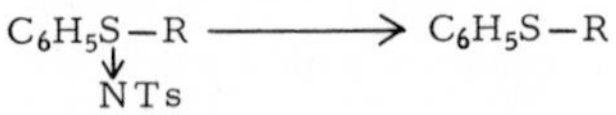

(5)

(4, R=$CH_3$, $CH_2C_6H_5$, $C_6H_5$)

[1] A. U. Rahman and C. Perl, *Ann.*, **718**, 127 (1968).
[2] T. Numata and S. Oae, *Chem. Ind.*, 277 (1973).
[3] *Idem, ibid.*, 726 (1972).

**1-Acetyl-1-methylhydrazine,** $CH_3N(Ac)NH_2$. Mol. wt. 88.11, b.p. 103°/8 torr, m.p. 16°.
The reagent is best prepared by acetylation of methylhydrazine (Matheson, Coleman and Bell) with acetic anhydride in pyridine (76% yield).[1]

*1-Alkyl-2-methylhydrazines.*[2] 1-Acetyl-1-methylhydrazine reacts with aldehydes or ketones, usually in quantitative yield, to give acetylmethylhydrazones, which are

$$CH_3N(Ac)NH_2 \;+\; \underset{R_2}{\overset{R_1}{>}}C=O \longrightarrow CH_3N(Ac)N=C\underset{R_2}{\overset{R_1}{<}} \xrightarrow{NaBH_4}$$

$$CH_3N(Ac)NHCH\underset{R_2}{\overset{R_1}{<}} \xrightarrow{H_3O^+} CH_3NHNHCH\underset{R_2}{\overset{R_1}{<}}$$

converted into 1-alkyl-2-methylhydrazines by reduction ($NaBH_4$) and hydrolysis (yields 40–80%).

[1] F. E. Condon, *J. Org.*, **37**, 3608 (1972).
[2] *Idem, ibid.*, **37**, 3615 (1972).

**Acetyl sulfuric acid.**
The statement in Vol. **2**, 389 that the correct formula for sulfoacetic acid is $HOOCCH_2SO_2OH$ needs to be elaborated by the further statement that the compound mentioned in **1**, 1117 is acetyl sulfuric acid.

Sulfoacetic acid is best prepared by the action of sodium sulfite on sodium chloroacetate.[1]

[1] E. E. Gilbert, *Sulfonation and Related Reactions*, Interscience, New York, 277 (1965).

**1-Acetyl-2-thiourea,** $CH_3\overset{O}{\overset{\|}{C}}NH\overset{S}{\overset{\|}{C}}-NH_2$. Mol. wt. 118.16, m.p. 165–169°. Supplier: Aldrich.

The reagent (1) is prepared by the reaction of sodium thiocyanate (**1**, 1105–1106) with acetyl chloride followed by treatment with aqueous ammonia[1]:

$$NaSCN \;+\; CH_3COCl \xrightarrow[-HCl]{NH_3} CH_3\overset{O}{\overset{\|}{C}}NH\overset{S}{\overset{\|}{C}}-NH_2$$

$$(1)$$

*Synthesis of mercaptans.*[1] Mercaptans can be obtained in 30–75% yield by heating (1) with an alkyl halide in ethanol for 24 hr. Acetylurea (3) is the other product. An intermediate 1-acetyl-2-alkyl-2-thiopseudourea hydrohalide (a) is involved. These thiopseudoureas can be isolated if prepared in acetonitrile solution. Yields of mercaptans are high when primary halides are used but rather low in the case of secondary

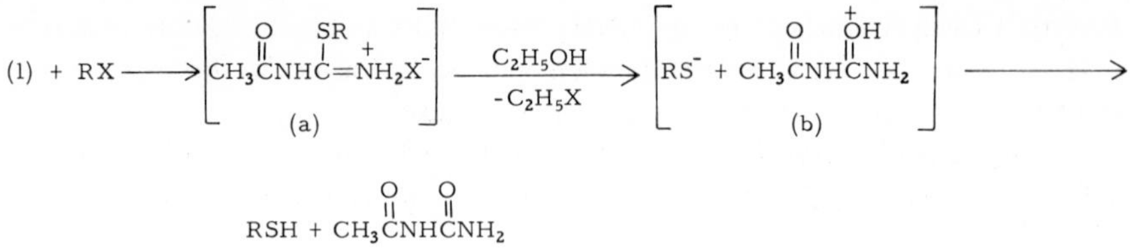

$$(1) \ + \ RX \longrightarrow \left[ \begin{array}{c} O \quad SR \\ \parallel \quad \mid \quad + \\ CH_3CNHC=NH_2X^- \\ (a) \end{array} \right] \xrightarrow[-C_2H_5X]{C_2H_5OH} \left[ \begin{array}{c} O \quad \overset{+}{O}H \\ \parallel \quad \parallel \\ RS^- + CH_3CNHCNH_2 \\ (b) \end{array} \right] \longrightarrow$$

$$\begin{array}{cc} & O \quad O \\ & \parallel \quad \parallel \\ RSH + & CH_3CNHCNH_2 \\ (2) & (3) \end{array}$$

halides. The reaction of (1) with *t*-butyl bromide did not give a mercaptan. Benzoyl-thiourea can be used in place of (1), but yields of mercaptans are lower.

[1]D. L. Klayman, R. J. Shine, and J. D. Bower, *J. Org.*, **37**, 1532 (1972).

**Acetyl *p*-toluenesulfonate, 2**, 14–15. Supplier: Aldrich.
  Definitive papers on the preparation and reactions have now been published.[1]

[1]M. H. Karger and Y. Mazur, *J. Org.*, **36**, 528, 532, 540 (1971).

**Alumina, 1**, 19–20; **2**, 17; **3**, 6.
  *Sulfoxide dehydration.* When a mixture of the sulfoxide 1,3-dihydrobenzo[*c*]-thiophene 2-oxide and grade I neutral alumina (Woelm) is heated under 25-mm. pressure at 120–130° in a sublimer, almost pure benzo[*c*]thiophene (2) condenses on the

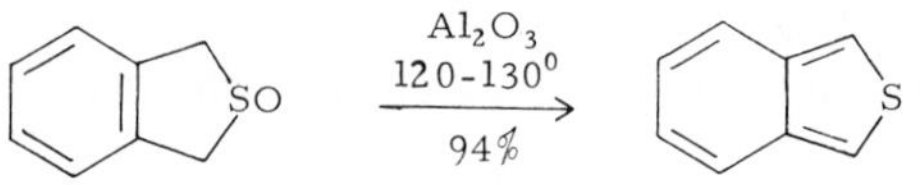

(1)        (2)

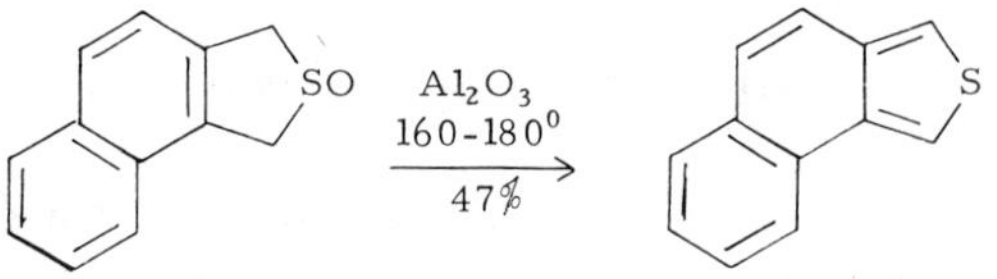

(3)        (4)

cold finger in 94% yield. Naphtho[1,2-*c*]thiophene (4) was obtained from 1,3-dihydro-naphtho[1,2-*c*]thiophene 2-oxide (3) in 47% yield.[1]
  *Tropone.*[2] Tropone (2) can be prepared from tropylium fluoroborate (**1**, 1261) in

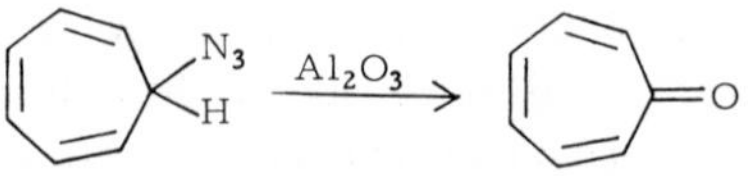

(1)        (2)

52% yield by treatment with sodium azide in water to give tropylium azide (1). Treatment of the azide with alumina (Fisher A-540) with vigorous stirring overnight yields tropone (2).

[1]M. P. Cava, N. M. Pollack, O. A. Mamer, and M. J. Mitchell, *J. Org.*, **36**, 3932 (1971).
[2]L. N. McCullagh and D. W. Wulfman, *Synthesis*, 422 (1972).

**Aluminum, 2,** 19.

*Dehalogenation.*[1] A solution of 1 g. of *cis*-3,4-dibromohexachloro-1,2-dimethylenecyclobutane (1) in 50 ml. of absolute ether is refluxed with 2 g. of aluminum foil with exclusion of moisture for 3 hr. The solution is filtered, washed with water, dried, and evaporated to give colorless needles of (2) from acetone, m.p. 145–146°.

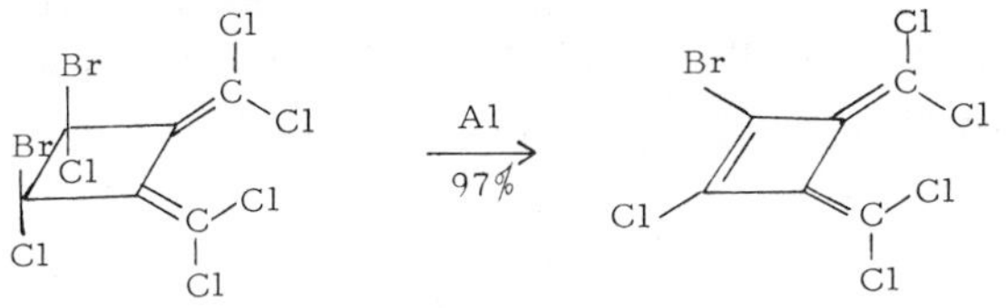

(1)                    (2)

Similarly:

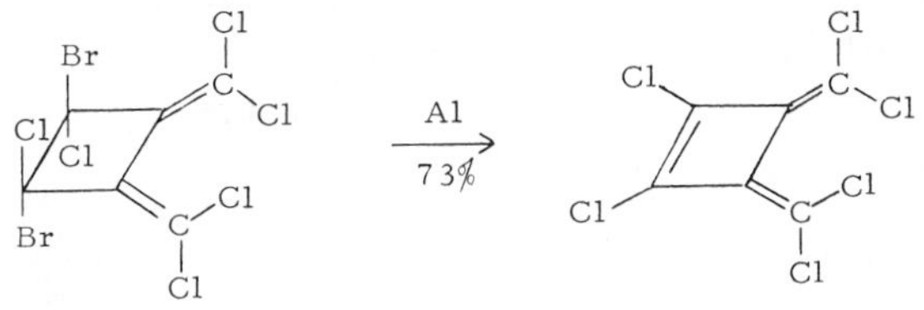

(3)                    (4)

*Reduction of nitro groups*[2]*:*

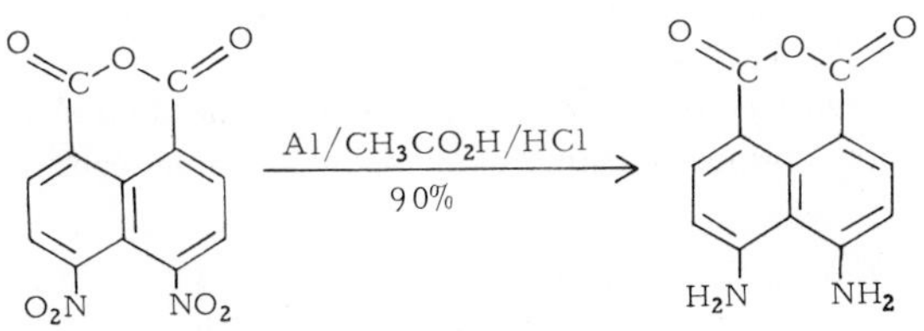

[1]A. Roedig, N. Detzer, and G. Bonse, *Ann.*, **752**, 60 (1971).
[2]O. Christmann, *ibid.*, **716**, 147 (1968).

**Aluminum bromide, 1,** 22–23; **2,** 19–21; **3,** 7.

*Rearrangement of α-bromoethyldiethylborane.*[1] α-Bromoethyldiethylborane (1) is isomerized almost instantaneously at 25° to s-butylethylboron bromide (2) by treatment

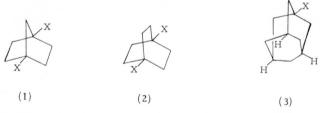

with aluminum bromide in carbon disulfide. Some other Lewis acids are about as effective ($AlCl_3$, $ZnCl_2$, $AgBF_4$); $HgCl_2$, $SnCl_4$, $SbCl_3$, and $TiCl_4$ are less effective.

[1]H. C. Brown and Y. Yamamoto, *J.C.S. Chem. Comm.*, 71 (1972).

**Aluminum bromide prepared *in situ*.**

Aluminum bromide can be prepared *in situ* from aluminum foil and bromine (or iodine for aluminum iodide).

*Bridgehead halogen exchange.*[1] Exchange of bridgehead halogen is usually diffi-cult. It can be performed easily and rapidly using aluminum bromide prepared *in situ* and the solvents $CH_3I$ or $CH_2I_2$ for iodine exchange, $CH_2Br_2$ or $CHBr_3$ for bromine transfer, and $CHCl_3$ or $CCl_4$ for chlorine transfer. Yields are in the range 50–90%. The method has been used for exchange at the bridgeheads of (1), (2), and (3).

(1)            (2)            (3)

[1]J. W. McKinley, R. E. Pincock, and W. B. Scott, *Am. Soc.*, **95,** 2030 (1973).

**Aluminum chloride, 1,** 24–34; **2,** 21–23; **3,** 7–9.

*Diels–Alder catalyst* (**1,** 31–32; **2,** 21–22; **3,** 8–9).[1] The Diels–Alder adduct (1) of 1,3-pentadiene with 3-bromo-4-methylpentene-3-one-2 (bromomesityl oxide) can be

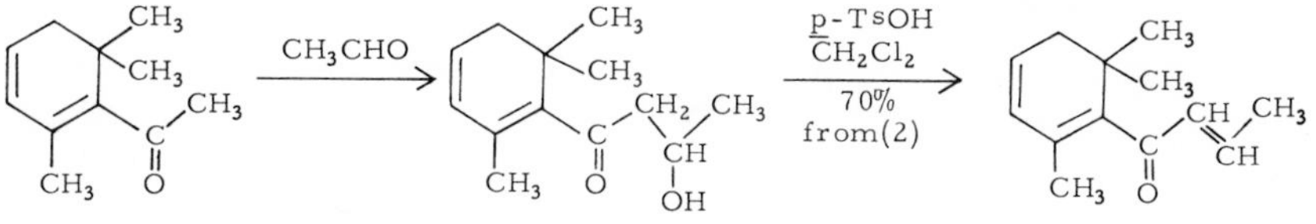

(2)          (3)          (4)

obtained in 87% yield (glc) by use of aluminum chloride as catalyst. The adduct (1) was converted into $\beta$-damascenone (4) by dehydrobromination (2), condensation with acetaldehyde catalyzed by N-methylanilinomagnesium bromide[2] (3), and dehydration.

$\delta$-Damascone (6) was synthesized by a similar sequence starting with the condensation of 1,3-pentadiene with mesityl oxide to give the adduct (5).

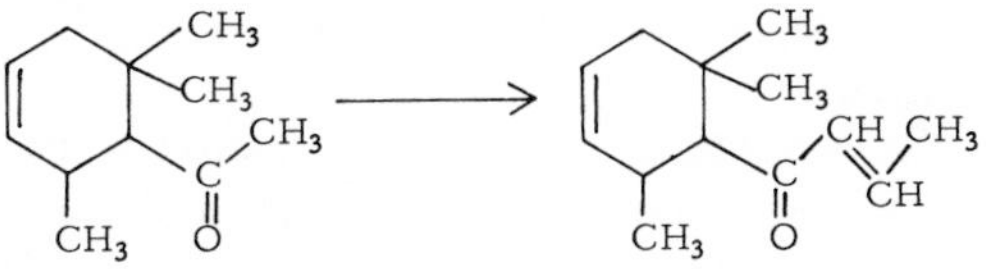

(5)                    (6)

*Addition reactions* (**1**, 29). 1,5-Dichloropentane-3-one can be prepared by reaction of 3-chloropropionyl chloride with ethylene in methylene chloride in the presence of anhydrous aluminum chloride.[3] The yield of product as a dark brown oil is 93–96%.

$$ClCH_2CH_2COCl + C_2H_4 \xrightarrow{AlCl_3} (ClCH_2CH_2)_2CO$$

*Acid chloride—olefin addition and Friedel-Crafts cyclization.*[4] A previous procedure was improved by use of methylene chloride as solvent rather than carbon disulfide. To check the progress of the reaction, one can quench a 2–3-ml. aliquot with water in a test tube, separate and dry the organic phase, and evaporate. The infrared spectrum will show disappearance of the acid chloride carbonyl band at 5.60 $\mu$ and appearance of the

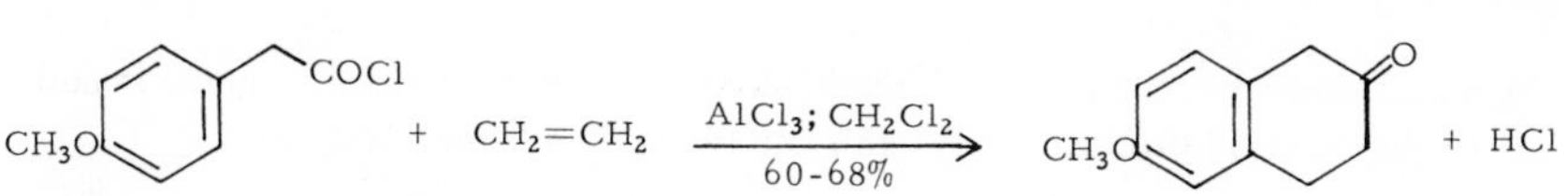

6-methoxy-$\beta$-tetralone carbonyl at 5.88 $\mu$; in addition, peaks at 5.78 and 6.02 $\mu$ finally disappear. Workup and distillation gives 21–24 g. (60–68%) of 6-methoxy-2-tetralone, b.p. 114–116° (0.2 mm.). On standing in a refrigerator the product solidifies to a white solid, m.p. 33.5–35°.

*Cleavage of benzyl groups.*[5] On refluxing 5,5-dibenzyldithiohydantoin (1) and 4,4-dibenzyl-2,5-bis(methylthio)-4H-imidazole (2) with an aromatic hydrocarbon and

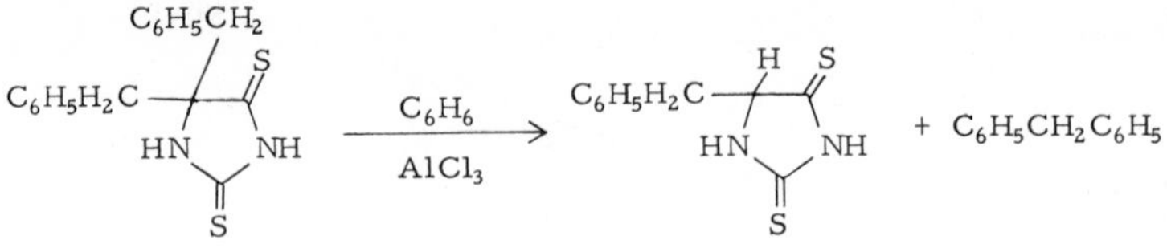

(1)

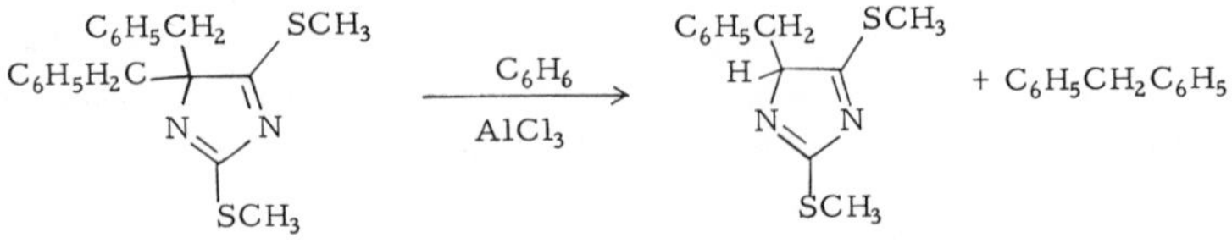

(2)

aluminum chloride, one of the C-benzyl groups is cleaved off the heterocycle and transferred to the solvent.

**Rearrangement of N,N-dihaloamines.**[6] N,N-Dichloro-tri-n-butylcarbinamine (2), prepared by treatment of tri-n-butylcarbinamine (1) with calcium hypochlorite,[7] when treated with aluminum chloride in methylene chloride at $-30°$, followed by acid hydrolysis, gives di-n-butyl ketone (3) and n-butylamine (4) in high yield. The reaction is considered to involve alkyl migration to electron-deficient nitrogen.

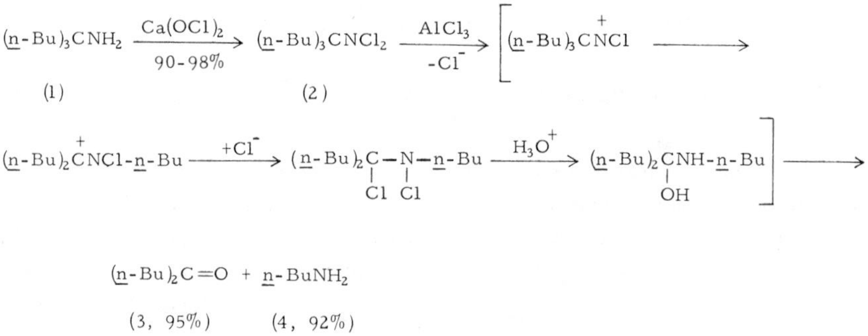

$$(\underline{n}\text{-}Bu)_3CNH_2 \xrightarrow[90-98\%]{Ca(OCl)_2} (\underline{n}\text{-}Bu)_3CNCl_2 \xrightarrow[-Cl^-]{AlCl_3} \left[ (\underline{n}\text{-}Bu)_3\overset{+}{C}NCl \longrightarrow \right.$$

(1)                              (2)

$$(\underline{n}\text{-}Bu)_2\overset{+}{C}NCl\text{-}\underline{n}\text{-}Bu \xrightarrow{+Cl^-} (\underline{n}\text{-}Bu)_2\underset{\underset{Cl}{|}}{C}\text{-}\underset{\underset{Cl}{|}}{N}\text{-}\underline{n}\text{-}Bu \xrightarrow{H_3\overset{+}{O}} \left. (\underline{n}\text{-}Bu)_2\underset{\underset{OH}{|}}{C}NH\text{-}\underline{n}\text{-}Bu \right] \longrightarrow$$

$$(\underline{n}\text{-}Bu)_2C{=}O + \underline{n}\text{-}BuNH_2$$

(3, 95%)          (4, 92%)

Application of the reaction to 1-N,N-dichloroaminoapocamphane (5) gives, in addition to the expected product (6), (7), (8), and (9), arising from $\beta$-scission.[8]

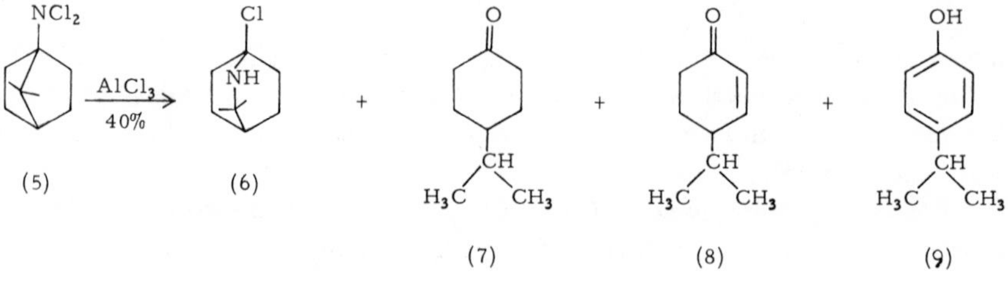

(5)         (6)                    (7)              (8)              (9)

*Preparation of diamantane* (3, 7).[9] The isomerization of tetrahydro Bisnor-S to diamantane can be effected in 82% yield (isolation) by use of aluminum chloride in refluxing methylene chloride for 12 hr. 1- and 4-Chlorodiamantane are obtained as minor products.

*Polystyrene–Aluminum chloride.*[10] Aluminum chloride forms a water-stable complex with polystyrene-divinylbenzene (1.8%) copolymer[11] denoted as Ⓟ–AlCl$_3$. In a typical preparation, the copolymer beads (59–100 mesh, 31.0 g.) are treated with carbon disulfide and then anhydrous aluminum chloride (7.5 g.). The mixture is stirred at reflux for 40 min. and then excess AlCl$_3$ is destroyed by cautious addition of water. The mixture is stirred until the original orange color fades to light yellow. The Ⓟ–AlCl$_3$ is filtered, washed with water, then successively with ether, acetone, hot isopropanol, and ether, and finally dried in a vacuum oven for 18 hr. The complex is stable to the atmosphere for over one year. The anhydrous aluminum chloride can be released from the polymer by swelling with various solvents: benzene, hexane, carbon disulfide.

The polymer-protected aluminum chloride has been used for the synthesis of ethers in moderate to high yields. Thus treatment of dicyclopropylcarbinol (1) with Ⓟ–AlCl$_3$

$$2 \ H-\underset{\triangle}{\overset{\triangledown}{C}}-OH \quad \xrightarrow[\text{81\%}]{\text{Ⓟ}-\text{AlCl}_3} \quad H-\underset{\triangle}{\overset{\triangledown}{C}}-O-\underset{\triangle}{\overset{\triangledown}{C}}-H$$

(1)                          (2)

produces di(dicyclopropylcarbinyl) ether (2) in yields as high as 81%. Mixed ethers can also be prepared by using two different carbinols. The new method is particularly valuable in the case of carbinols that are sensitive to aluminum chloride.

*Ether cleavage* (1, 30; 3, 7–8). Another example: when phloroacetophenone trimethyl ether (1) is heated with aluminum chloride at 110°, it is converted into the dimethyl ether (2).[12] Likewise, the 5-methoxyl group in a polymethoxyflavone under-

(1)                          (2)

goes preferential demethylation to the corresponding 5-hydroxyflavone.[13] This observation has been used for the synthesis of several naturally occurring flavones such as genkwanin (5).[14]

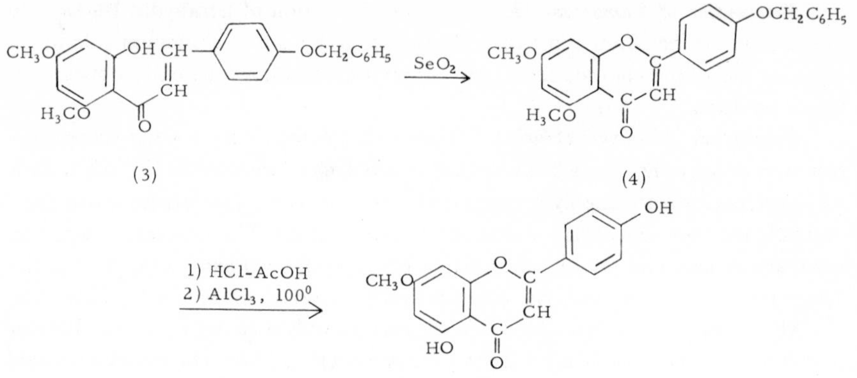

(3)     (4)

(5)

In a total synthesis of daunomycinone (8, the aglycone of an antitumor antibiotic), Ho *et al.*[15] effected selective ether cleavage of (6, daunomycinone trimethyl ether) to (7, 4-O-demethyl-7-O-methyldaunomycinone) by treatment with aluminum chloride at room temperature.

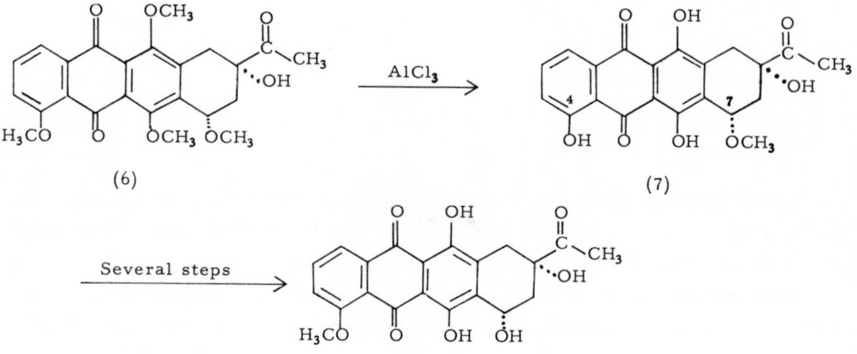

(6)     (7)

(8)

[1] K. S. Ayyar, R. C. Cookson, and D. A. Kagi, *J.C.S. Chem. Commun.*, 161 (1973).
[2] A. T. Nielsen, C. Gibbons, and C. A. Zimmerman, *Am. Soc.*, **73**, 4696 (1951).
[3] G. R. Owen and C. B. Reese, *Org. Syn.*, submitted 1972.
[4] J. J. Sims, L. H. Selman, and M. Cadogan, *ibid.*, **51**, 109 (1971).
[5] R. Markovits-Kornis, J. Nyitrai, and K. Lempert, *Ber.*, **104**, 3080 (1971).
[6] T. A. Kling, R. E. White, and P. Kovacic, *Am. Soc.*, **94**, 7416 (1972).
[7] Calcium hypochlorite is available as "HTH" from Olin Mathieson Chemical Corp. in about 70% purity; see P. Kovacic and S. S. Chaudhary, *Org. Syn.*, **48**, 4 (1968).
[8] R. D. Fisher, T. D. Bogard, and P. Kovacic, *Am. Soc.*, **94**, 7599 (1972).
[9] T. Courtney, D. E. Johnston, M. A. McKervey, and J. J. Rooney, *J.C.S. Perkin I*, 2691 (1972).
[10] D. C. Neckers, D. A. Kooistra, and G. W. Green, *Am. Soc.*, **94**, 9284 (1972).

[11]Appropriately cross-linked polystyrene beads are available from Bio-rad. Dr. Neckers mainly used beads obtained from Dow Chemical Co.

[12]St. von Kostanecki and J. Tambor, *Ber.*, **32**, 2260 (1899).

[13]K. Venkataraman and G. K. Bharadwaj, *Curr. Sci.*, **2**, 50 (1933); K. C. Gulati and K. Venkataraman, *J. Chem. Soc.*, 267 (1936).

[14]H. S. Mahal and K. Venkataraman, *ibid.*, 569 (1936).

[15]C. M. Wong, R. Schwenk, D. Popien, and T.-L. Ho, *Canad. J. Chem.*, **51**, 46 (1973).

**Aluminum cyclohexoxide, 1, 34.** Supplier: Eastman supplies the reagent in cyclohexanol.

**Reduction.** Hinshaw[1] effected reduction of the tetramethylanthraquinone (1) to 2,3,6,7-tetramethylanthracene (2) in 85% yield by treatment with aluminum cyclohexoxide in boiling cyclohexanol.

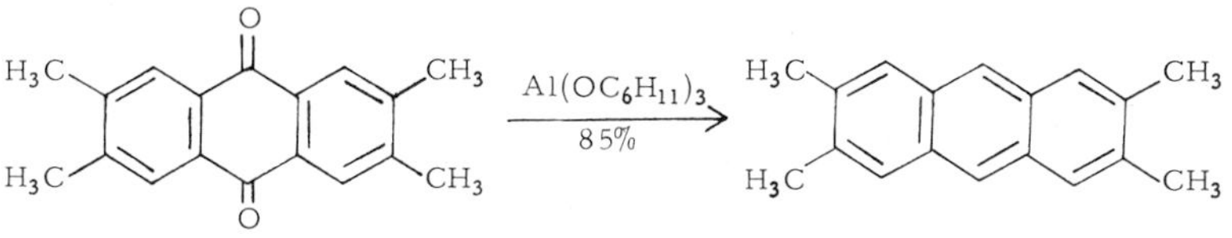

(1)                    (2)

[1]J. C. Hinshaw, *Org. Prep. Proc. Int.*, **4**, 211 (1972).

**Aluminum isopropoxide, 1, 35–37; 3, 10.**

**Meerwein–Ponndorf reduction (1, 35–36).** A recent study[1] of the Meerwein–Ponndorf reduction of mono- and bicyclic ketones shows that, contrary to commonly held views, the reduction of such ketones proceeds at a relatively high rate. The reduction of cyclohexanone and of 2-methylcyclohexanone is immeasurably rapid. Even

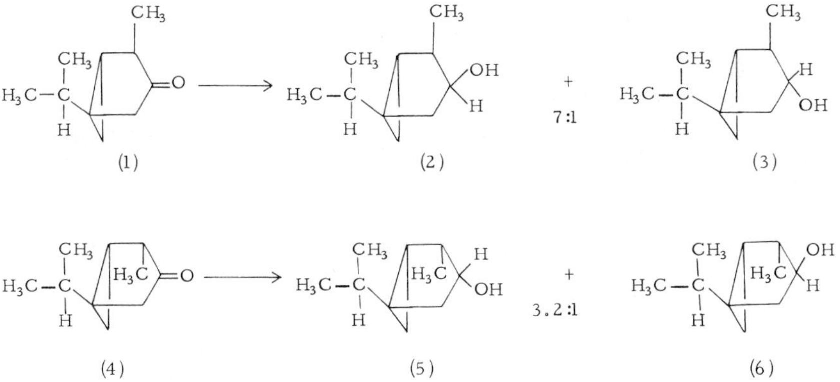

(1)                    (2)                    (3)

(4)                    (5)                    (6)

menthone is reduced almost completely in 2 hr. Hach also examined the stereochemistry of the reduction of 3-isothujone (1) and of 3-thujone (4). The former ketone (1) produces a preponderance of the *cis*-alcohol (2). The stereoselectivity is less pronounced in the case of 3-thujone (4), although again the *cis*-alcohol (5) predominates. The preponderance of the *cis*-alcohols can be increased by decreasing the concentration of ketone and alkoxide.

[1] V. Hach, *J. Org.*, **38**, 293 (1973).

## Amidines, bicyclic.

Review: "Bicyclic amidines as reagents in organic synthesis."[1]

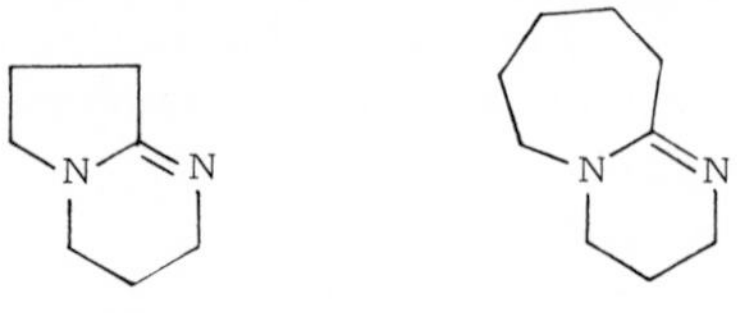

DBN

**1**, 189–190
**2**,  98–99
This volume

DBU

**2**, 101

This volume

These reagents are attractive for use in the dehydrohalogenation of sensitive organic compounds because of the mild conditions under which they are effective. Thus treatment of the halogenated compound (1) with DBN leads to smooth elimination of

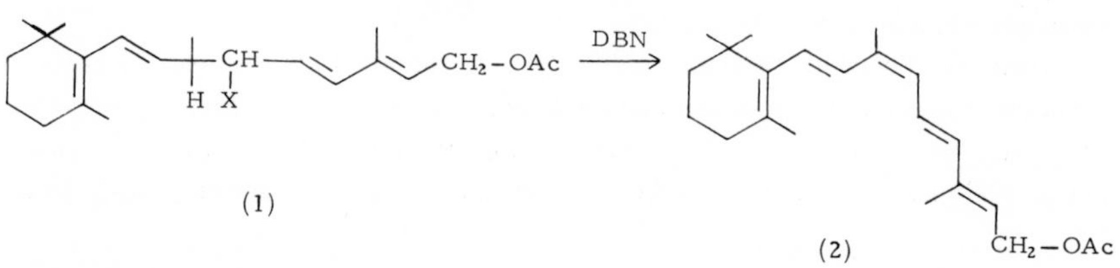

(1)

(2)

hydrogen halide and formation of vitamin A acetate (2) without rearrangements, which occur frequently in the vitamin A series. Use of the UV extinction coefficient as a parameter for the degree of conversion to (2) shows that eight other bases are inferior to DBN.

The use of DBN made possible for the first time the synthesis of naphthalene-1,2-oxide (4) from 4-bromo-1,2-epoxy-1,2,3,4-tetrahydronaphthalene (3):

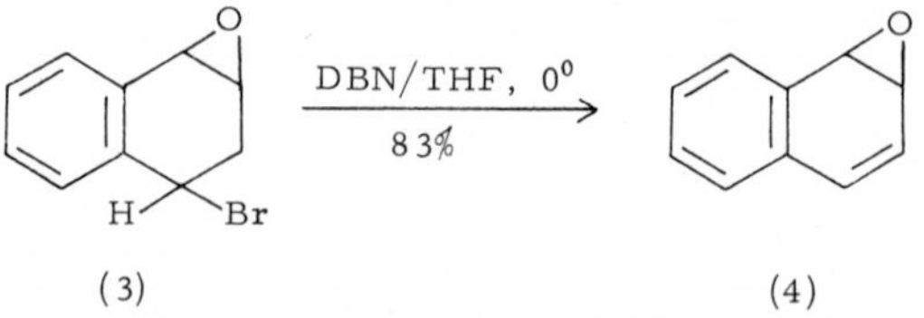

(3)

DBN/THF, 0°
83%

(4)

The product crystallizes as colorless needles in 83 % yield. The melting point cannot be determined exactly because of fast rearrangement to α-naphthol.

The first successful synthesis of the extremely unstable antimonin (stibiabenzene, 6) was performed by reaction of (5) with DBN:

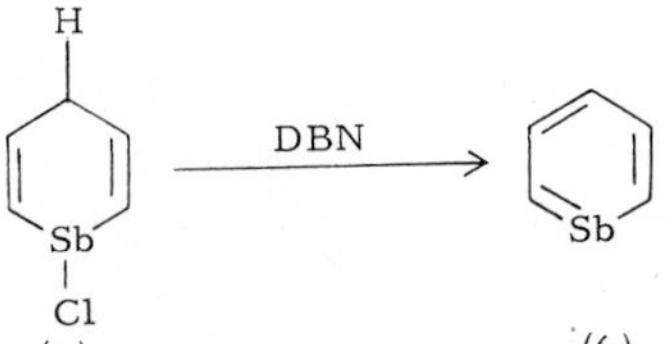

(5)                                      (6)

The selectivity of DBN is illustrated by the dehydrobromination of the dibromide (7).
Use of potassium *t*-butoxide leads to (8), whereas the product obtained with DBN
is (9).

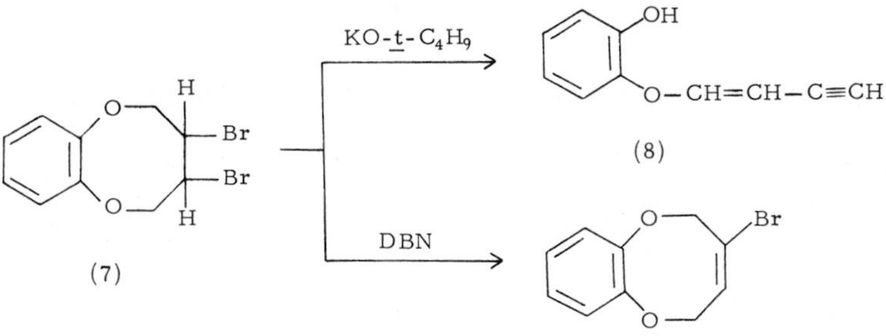

In a series of simple bromoalkanes (Table I), the yields of alkenes obtained with DBU
are in some cases more than twice those afforded by DBN.

**Table I**   Comparative dehydrohalogenation reactions with DBN and DBU

| Reaction | Yield (%) | |
|---|---|---|
| | with DBN | with DBU |
| H₃C–CH₂–CH₂–CH–CH₂–CH₂–CH₃ → \| Br | | |
| $\quad$ H₃C–CH₂–CH=CH–CH₂–CH₂–CH₃ | 60 | 91 |
| H₃C–CH–CH₂–(CH₂)₃–CH₃ → \| Br | | |
| H₃C–CH=CH–(CH₂)₃–CH₃ + H₂C=CH–CH₂–(CH₂)₃–CH₃ 4:1 | 36 | 78 |
| H₃C–CH–CH₂–(CH₂)₄–CH₃ → \| Br | | |
| H₃C–CH=CH–(CH₂)₄–CH₃ + H₂C=CH–CH₂–(CH₂)₄–CH₃ 4:1 | 40 | 84 |

*Elimination of sulfonic acids.* Double bonds can be introduced into organic molecules by elimination of sulfonic acids from the corresponding sulfonic acid esters. The reaction proceeds particularly smoothly using DBN or DBU as reagent. Thus, treatment of 3-tosyloxyhexa-1,5-diyne (10) with excess DBN in ether at room temperature (1 hr.) affords a 40:60 mixture of *cis*- and *trans*-hex-3-ene-1,5-diyne (11; 70%).[2]

$$HC\equiv C-CH_2-\underset{\underset{OTos}{|}}{CH}-C\equiv CH \xrightarrow{DBN} HC\equiv C-CH=CH-C\equiv CH$$

(10)                                              (11)

Other reactions for which the two cyclic amidines are useful catalysts are as follows:

Preparation of phosphorus ylides
Elimination of nitrous acid
Steric control of elimination reactions
Aldol condensation
Rearrangements
Reactions with aryl isocyanates
Preparation of carboxylic acid chlorides
Synthesis of macromolecules

[1]H. Oediger, F. Möller, and K. Eiter, *Synthesis*, 591 (1972).
[2]T. J. Barton, M. D. Martz, and R. G. Zika, *J. Org.*, **37**, 552 (1972).

*p*-**Aminoacetophenone**, *p*-$CH_3COC_6H_4NH_2$. Mol. wt. 135.16, m.p. 104–106°. Suppliers: Aldrich, Fisher, Matheson, Coleman and Bell, Pfalz and Bauer, Sargent, Fluka, Schuchardt.

*Meerwein arylation.* Detailed directions are given[1] for diazotization of *p*-aminoacetophenone with 48% hydrobromic acid and aqueous sodium nitrite in acetone and Meerwein reaction[2] of the diazonium salt in the presence of copper(I) bromide (Fisher, reagent grade, washed with acetone until the washings are colorless and dried) with acrylic acid to produce *p*-acetyl-α-bromohydrocinnamic acid. The yield of white needles, m.p. 159–160°, is 56–59%.

$$\underline{p}\text{-}CH_3COC_6H_4NH_2 \xrightarrow[HBr]{NaNO_2} \underline{p}\text{-}CH_3COC_6H_4\overset{+}{N}\equiv NBr^- \xrightarrow[CuBr,\ HBr]{CH_2=CHCO_2H}$$

$$\underline{p}\text{-}CH_3COC_6H_4CH_2\underset{\underset{Br}{|}}{CH}CO_2H \ +\ N_2$$

[1]G. H. Cleland, *Org. Syn.*, **51**, 1 (1971).
[2]C. S. Rondestvedt, Jr., *Org. React.*, **11**, 189 (1960).

**4-Amino-3-hydrazino-5-mercapto-1,2,4-triazole** (1). Mol. wt. 146.19, m.p. 231–233° dec. Supplier: Aldrich ("Purpald").

*Preparation.* A detailed procedure for the preparation of this triazole has been submitted to *Organic Syntheses.*[1]

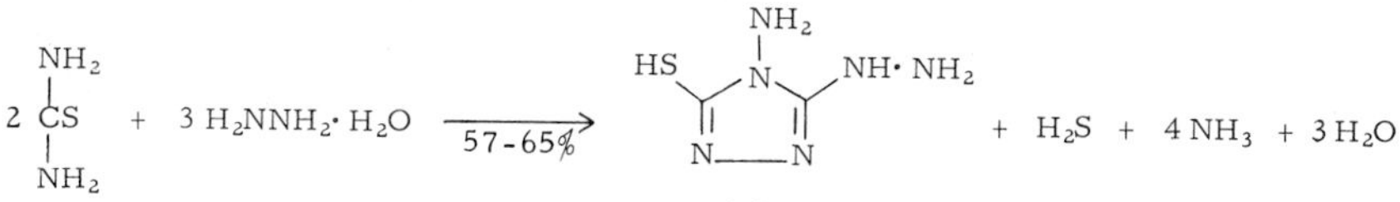

(1)

*Aldehydes.* The reagent (1) reacts with aldehydes only to give purple to magenta 6-mercapto-5-triazolo(4.3-*b*)-*s*-tetrazines (2). One drop of the aldehyde is added to 100–200 mg. of (1) dissolved in *ca.* 2 ml. of 1 *N* sodium hydroxide. Aeration produces the intense color in a few minutes.[2]

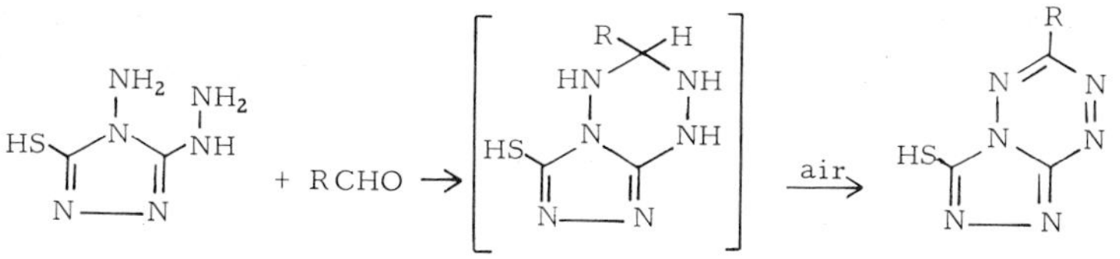

(1)                          (2)

[1]N. W. Jacobsen and R. G. Dickinson, *Org. Syn.*, submitted 1971.
[2]R. G. Dickinson and N. W. Jacobsen, *Chem. Commun.*, 1719 (1970).

**2-Aminothiazole** (1). Mol. wt. 100.14, m.p. 90–92°. Supplier: Eastman.

2-Aminothiazole (1) reacts with propiolic ester to give 2-oxo-7H-thiazo[3.2-*a*]pyrimidine (2a) and small amounts of the 1:1 and 1:2 adducts (3, 4). With acetylenedicarboxylic ester the 5-methoxycarbonyl derivative (2b) is formed.[1]

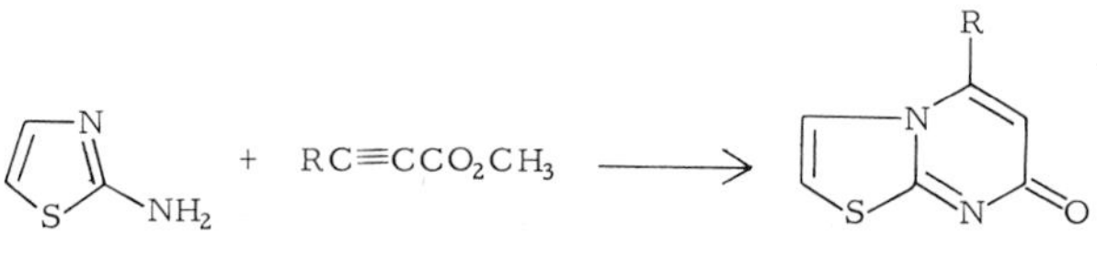

|   | R |
|---|---|
| a | H |
| b | $CO_2CH_3$ |
| c | $CO_2H$ |

(1)                          (2)

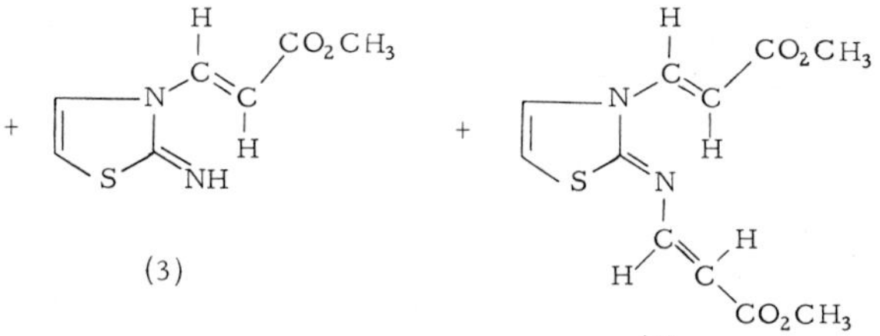

(3)                          (4)

[1]H. Reimlinger, *Ber.*, **104**, 2232 (1971).

**t-Amyl hydroperoxide (HPTA)**, $CH_3CH_2C(CH_3)_2$. Mol. wt. 104.15, b.p. 26°/3.5 mm.
$\underset{\displaystyle OOH}{|}$

*Preparation.*[1] The hydroperoxide can be prepared by the reaction of t-amyl hydrogen sulfate with concentrated hydrogen peroxide at 0°.

*Oxaziridines.*[2] Oxidation of Schiff bases (1) with HPTA catalyzed by molybdenum hexacarbonyl (**2**, 287, **3**, 206–207; this volume) or by $MoCl_5$ in benzene solution gives oxaziridines (2) generally in high yield.

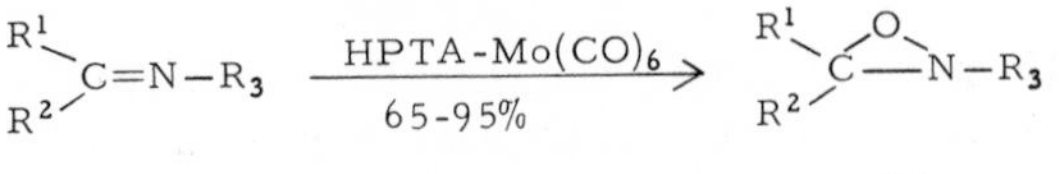

$$\underset{(1)}{\overset{R^1}{\underset{R^2}{>}}C=N-R_3} \quad \xrightarrow[65-95\%]{HPTA-Mo(CO)_6} \quad \underset{(2)}{\overset{R^1}{\underset{R^2}{>}}\overset{O}{\overset{|}{C}}-N-R_3}$$

*N-Oxides.*[2] The reagent also reacts with aromatic nitrogen-containing heterocycles to give high yields of N-oxides (90–100%). However, 2,2'-bipyridyl and o-phenanthroline are not oxidized.

*Nitramines.*[2] Prolonged oxidation of nitrosamines (1) with HPTA–MoCl$_5$ gives nitramines (2) in about 80% yield.

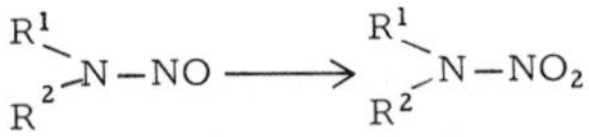

$$\underset{(1)}{\overset{R^1}{\underset{R^2}{>}}N-NO} \longrightarrow \underset{(2)}{\overset{R^1}{\underset{R^2}{>}}N-NO_2}$$

[1] N. A. Milas and D. M. Surgenor, *Am. Soc.*, **68**, 643 (1946).
[2] G. A. Tolstikev, U. M. Jemilev, V. P. Jurjev, F. B. Gershanov, and S. R. Rafikov, *Tetrahedron Letters*, 2807 (1971).

**t-Amylquinolyl-8-carbonate**. Mol. wt. 259.3, m.p. 46–46.5°.

*See* **t-Butylquinolyl-8-carbonate**.

**Anthraquinone,** 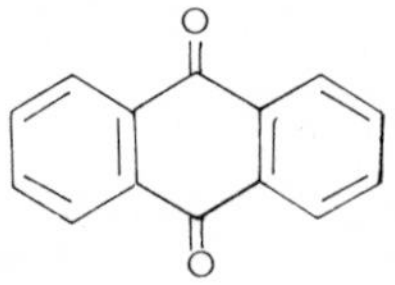 . Mol. wt. 208.20, m.p.286°.

Suppliers: Aldrich, J. T. Baker, Eastman, Fisher, Howe and French, Koch-Light, MCB, P and B, Sargent, Schuchardt.

*Preparation:*[1]

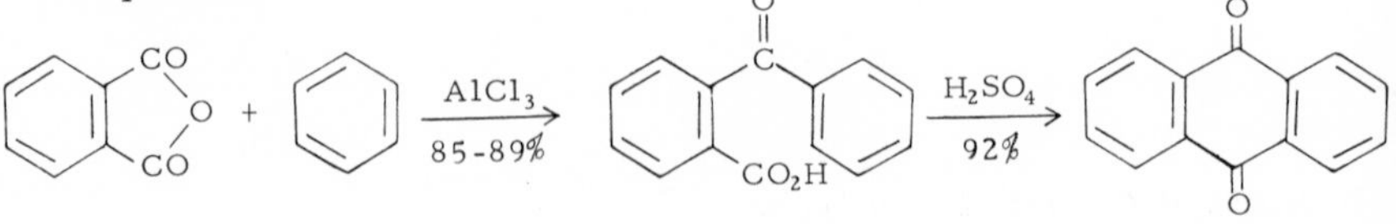

*Dehydrogenation.*[2] Treatment of the Diels–Alder adducts of 1,3-dienes and citracononitrile and mesacononitrile (1) with potassium *t*-butoxide and anthraquinone (3 eq.) in benzene at room temperature gives the substituted benzonitrile (2). This is apparently the first recorded instance of quinone dehydrogenation of carbanions.

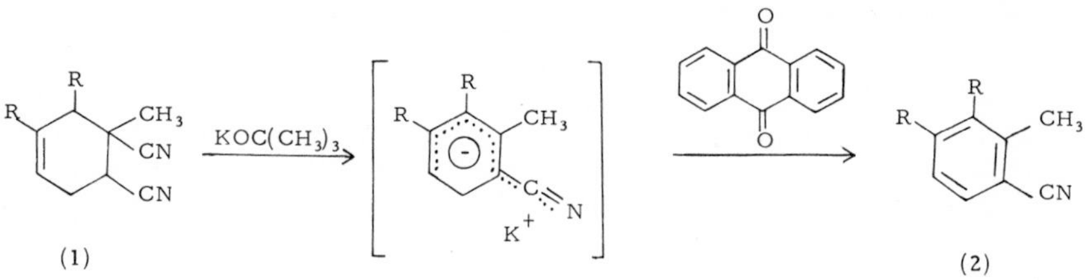

Anthraquinone is apparently converted into the potassium salt of anthracene quinhydrone.

[1]L. F. Fieser, *Org. Expts.*, 2nd Ed., D. C. Heath and Co., Boston, 199–203 (1968).
[2]W. R. Vaughan and D. R. Simonson, *J. Org.*, **38**, 566 (1973).

## Antimony(V) chloride, 1, 42.

*Selective* **cis**-*chlorination of olefins.*[1] Antimony(V) chloride reacts smoothly with olefins in $CCl_4$ to give predominantly *cis,vic*-dichloroalkanes. Thus the reaction of cyclohexene gives the three products shown as the only organic products.

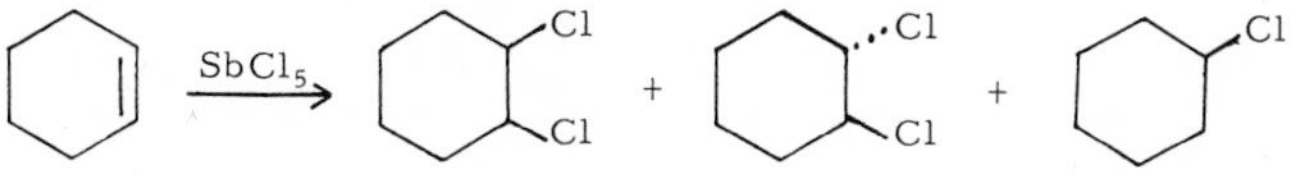

[1]S. Uemura, O. Sasaki, and M. Okano, *Chem. Commun.*, 1064 (1971).

## Ascarite (90 % NaOH on asbestos). Suppliers: Fisher, Sargent.

*Dehydrocyanation of dinitriles.*[1] 1,2-Cyclobutanedicarbonitrile (1) is converted into 1-cyclobutenecarbonitrile (2) by treatment with sodium hydroxide at 190–225° at

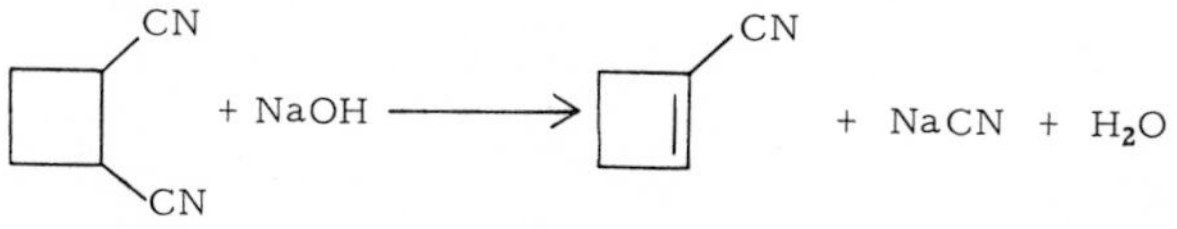

1 mm. Highest yields (59 %) are obtained with Ascarite. The method apparently is only useful for dehydrocyanation of *vic*-dinitriles.

[1]D. M. Gale and S. C. Cherkofsky, *J. Org.*, **38**, 475 (1973).

**Aziridine,** $\begin{array}{c} CH_2 \\ | \quad \diagdown NH \\ CH_2 \diagup \end{array}$ . Mol. wt. 43.07.

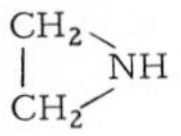

***Preparation.***[1] Manufactured in Europe by BASF and in the United States by Dow Chemical Company.

***Reaction with chloroform***[2] :

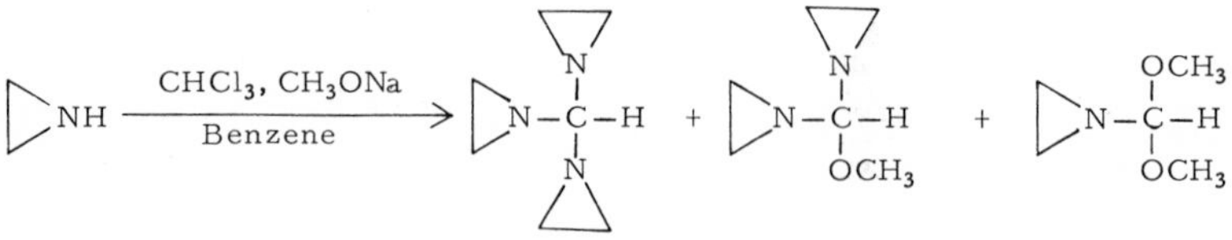

[1] H. Bestian, *Methoden zur Herstellung und Umwandlung von 1, 2- und 1, 3-Alkyleniminen,* Houben Weyl, Methoden der Organischen Chemie, Band XI/2.
[2] W. Funke, *Ann.,* **725,** 15 (1969).

# B

**Barium hydroxide**, $Ba(OH)_2$. Mol. wt. 171.38. Supplier: Fisher.

*α-Bromo acids and α-bromo esters*. Stotter and Hill[1] have reported a new synthesis of α-bromo esters and α-bromo acids from acetoacetic esters (1). The ester is converted into (2) by alkylation of the sodium enolate in THF with a primary alkyl iodide or bromide. Solutions of (2) in THF can be brominated exclusively in the α-position by

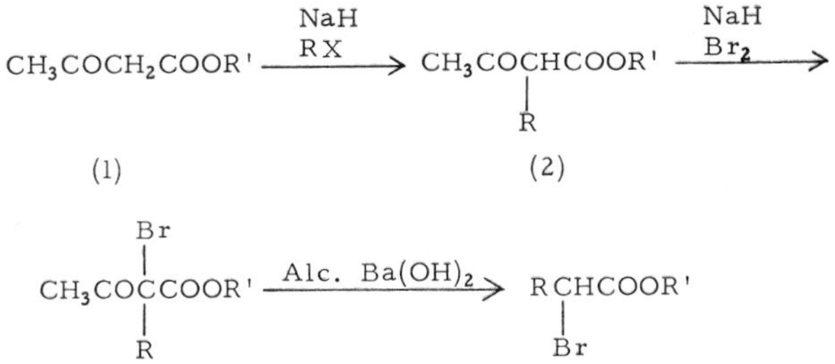

$$CH_3COCH_2COOR' \xrightarrow[\text{RX}]{\text{NaH}} CH_3COCHCOOR' \xrightarrow[\text{Br}_2]{\text{NaH}}$$

with R below the CH in (2).

(1)　　　　　　　　　　　(2)

$$CH_3COCCOOR' \xrightarrow{\text{Alc. Ba(OH)}_2} RCHCOOR'$$

with Br above and R below the central carbon in (3); Br below the CH in (4).

(3)　　　　　　　　　　　(4)

conversion into the sodium enolate (NaH) followed by addition of bromine in methylene chloride at 0°. Deacetylation of (3) is then accomplished by treatment with 1 eq. of barium hydroxide (dried to constant weight under vacuum at 125°) in absolute ethanol at 0° or below for 30 min. Use of excess barium hydroxide in wet alcohol gives poor results.

In the case of *t*-butyl esters deacetylation can be effected by treatment of (3) in refluxing benzene containing a catalytic amount of *p*-toluenesulfonic acid.

Johnson et al.[2] effected deacetylation of α-alkyl-α-chloro-β-diketones by treatment with barium hydroxide in 95% ethanol at 0°. For example, (5) was converted "smoothly" into the chloroketone (6).

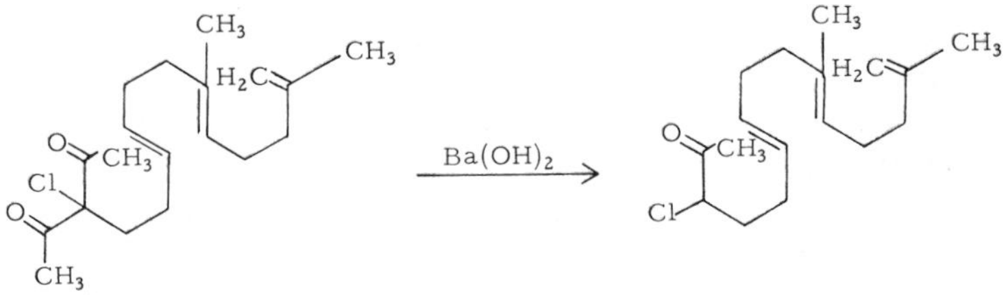

(5)　　　　　　　　　　　　　　　　(6)

[1]P. L. Stotter and K. A. Hill, *Tetrahedron Letters*, 4067 (1972).

[2]W. S. Johnson, K. Wiedhaup, S. F. Brady, and G. L. Olson, *Am. Soc.*, **90**, 5277 (1968); W. S. Johnson, T. Li, D. J. Faulkner, and S. F. Campbell, *ibid.*, **90**, 6225 (1968).

## Benzenediazonium chloride.

*Coupling with an aldehyde phenylhydrazone.*[1] The coupling of the reagent with trimethylacetaldehyde phenylhydrazone in pyridine–DMF gives the bisazo compound (1), which is rearranged by a strong base to the formazane (2). Irradiation does not convert (2) into an open, yellow form.

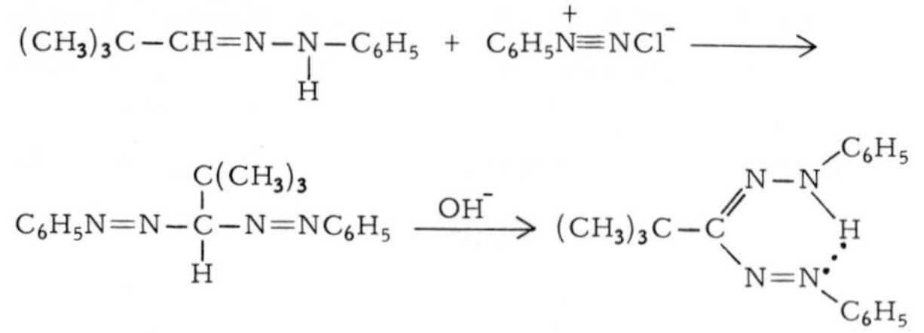

$$(CH_3)_3C-CH=N-\underset{\underset{H}{|}}{N}-C_6H_5 \quad + \quad C_6H_5\overset{+}{N}\equiv NCl^- \longrightarrow$$

(1)          (2)

Nitric acid oxidizes (2) to 5-*tert*-butyl-2,3-diphenyltetrazolium nitrate (3), which on irradiation yields 5-*tert*-butyl-2,3-[2,2'-biphenylene]tetrazolium nitrate (4). Sodium hydrosulfite reduces (4) to the deep blue radical (5).

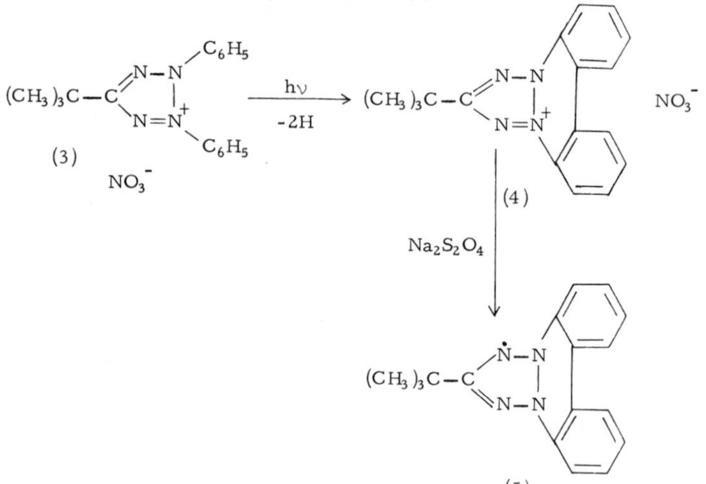

(3)          (4)

Na$_2$S$_2$O$_4$

(5)

[1]F. A. Neugebauer and H. Trischmann, *Ann.*, **706**, 107 (1967).

## Benzenesulfonyl azide, 3, 17–18.

*Reactions.* Correction to p. 17, lines 11–13. The reagent reacts with norbornene (1)

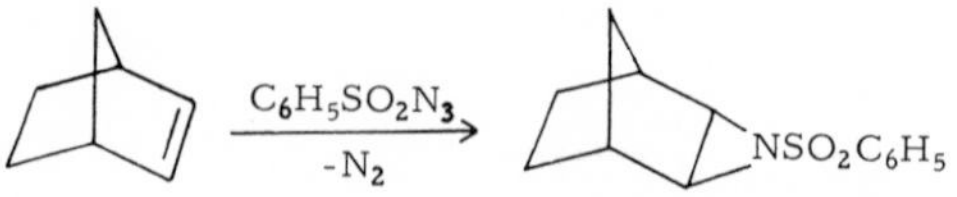

$$\xrightarrow[-N_2]{C_6H_5SO_2N_3}$$

(1)          (2)

to give an *exo*-aziridine (2).[1] Norbornadiene (3) reacts anomalously to give a nitrogen insertion product tentatively assigned structure (4).[2]

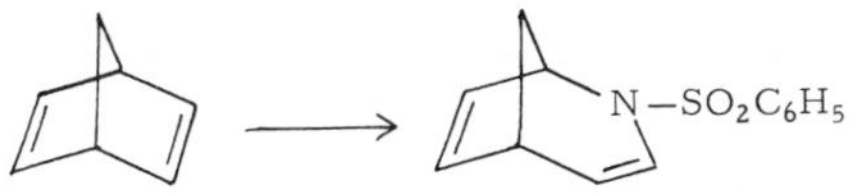

(3)                    (4)

[1]J. E. Franz, C. Osuch, and M. W. Dietrich, *J. Org.*, **29**, 2922 (1964).
[2]J. E. Franz and C. Osuch, *Chem. Ind.*, 2058 (1964).

**1,3,2-Benzodioxaborole** (1). Mol. wt. 119.92, b.p. 76–77°/100 mm. See also p. 69.

The reagent (1) is prepared in 80% yield by the reaction of catechol with borane in THF.[1] The borole reacts with olefins at 100° to give the corresponding 2-alkyl-

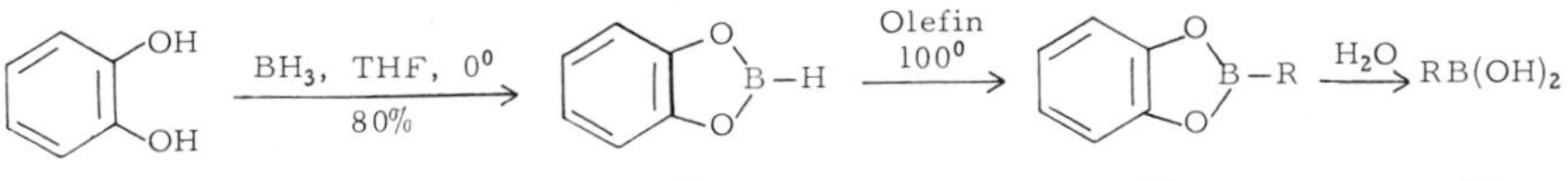

(1)                    (2)                    (3)

1,3,2-benzodioxaboroles (2) in nearly quantitative yield. The products are rapidly hydrolyzed to the corresponding alkaneboronic acids (3). One disadvantage is that a sealed ampoule is required in the case of low-boiling olefins.[1] In this case Brown recommends the following alternative route to 2-alkyl-1,3,2-benzodioxaboroles (2), based on the ready redistribution of trialkylboranes with *o*-phenylene borate[2] to give the desired boroles (2) in essentially quantitative yield:

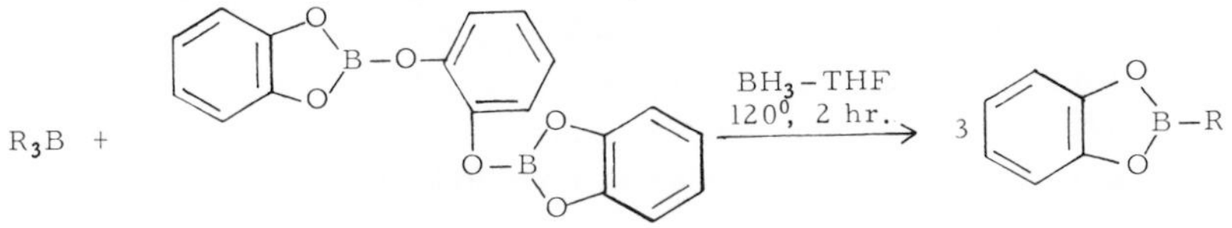

(2)

[1]H. C. Brown and S. K. Gupta, *Am. Soc.*, **93**, 1816 (1971).
[2]L. H. Thomas, *J. Chem. Soc.*, 820 (1946).

**Benzoin, cyclic carbonate (Diphenylethylene carbonate)**, (1),

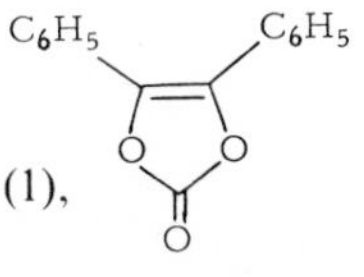

(1)

Mol wt. 238.23, m.p. 75–76°.

*Preparation.*[1] The carbonate (1) is prepared in 65–70% yield by treatment of benzoin (1.0 eq.) and phosgene (1.1 eq.) in benzene with distilled N,N-dimethylaniline at 5°. After stirring overnight at room temperature the amine hydrochloride is filtered, and the filtrate refluxed for 3 hr. to cyclize the initially formed chloroformate.

*Protection for primary amines.* Sheehan and Guziec[1] have introduced this reagent for replacement of both hydrogens of a primary amine function by incorporation into the very stable 4,5-diphenyl-4-oxazoline-2-one ring system (3).[2] These Ox-protected derivatives are obtained by treatment of a primary amine or amino acid tetramethyl-ammonium salt with (1) in DMF. After stirring for 0.5 hr., the mixture is acidified (HCl) and extracted with ethyl acetate to obtain a diastereomeric mixture of hydroxy-oxazolidinones (2). The mixture is dehydrated to the Ox derivative (3) by trifluoroacetic acid (TFA). Ox derivatives are stable to aqueous base, refluxing ethanolic hydrazine,

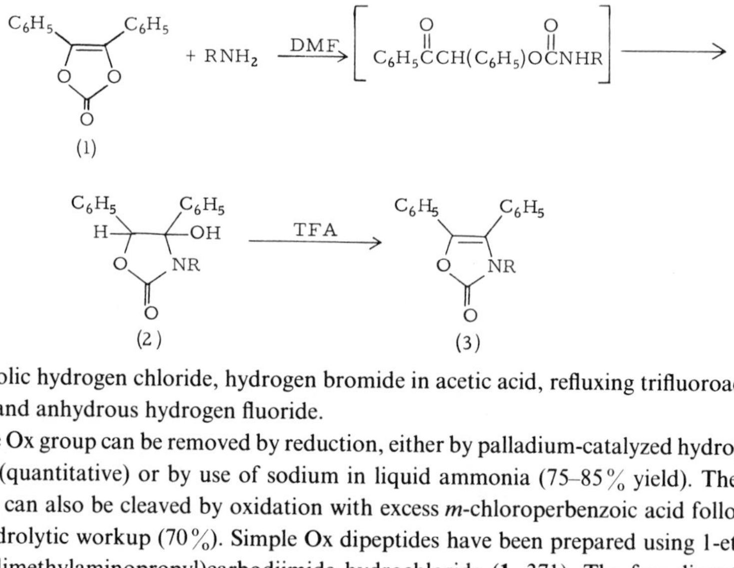

(1)

(2)                    (3)

ethanolic hydrogen chloride, hydrogen bromide in acetic acid, refluxing trifluoroacetic acid, and anhydrous hydrogen fluoride.

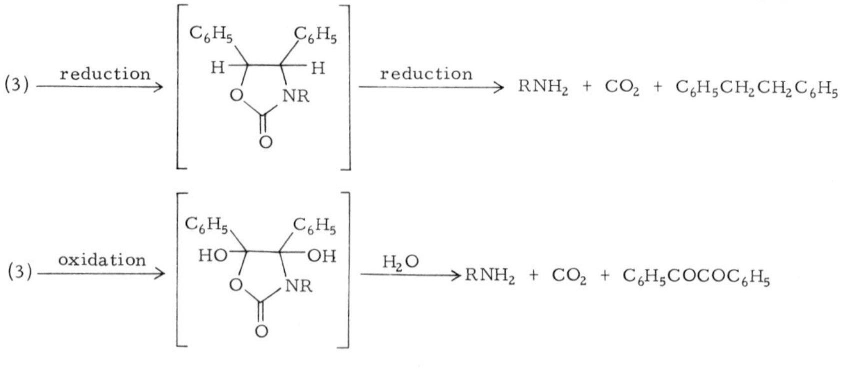

The Ox group can be removed by reduction, either by palladium-catalyzed hydrogenation (quantitative) or by use of sodium in liquid ammonia (75–85% yield). The Ox group can also be cleaved by oxidation with excess *m*-chloroperbenzoic acid followed by hydrolytic workup (70%). Simple Ox dipeptides have been prepared using 1-ethyl-3-(3′-dimethylaminopropyl)carbodiimide hydrochloride (1, 371). The free dipeptides are obtained by hydrolysis. No racemization has been observed either in the preparation of the Ox derivatives or in coupling or deblocking reactions.

[1] J. C. Sheehan and F. S. Guziec, Jr., *Am. Soc.*, **94**, 6561 (1972).
[2] R. Filler, *Advan. Heterocyclic Chem.*, **4**, 103 (1965).

**Benzyltriethylammonium chloride, 3, 19.** Additional suppliers: Aldrich, Columbia, Fluka, K and K, MCB, Pfaltz and Bauer, Sargent, Schuchardt.

**Dichlorocarbene.** Polish chemists[1] have reported a new method for generation of dichlorocarbene (or a carbenoid species); it involves the reaction of an olefin with chloroform in the presence of a 50 % aqueous NaOH solution and a catalytic amount of benzyltriethylammonium chloride. For example, dichloronorcarane was obtained from cyclohexene in this way in 72 % yield. In the absence of the catalyst, a yield of only 0.5 % has been reported.[2]

Benzyltriethylammonium chloride functions as a phase-transfer catalyst. The chloride is soluble in the basic aqueous phase and is converted into benzyltriethylammonium hydroxide, soluble in the organic phase. The hydroxide ion reacts with chloroform to give dichlorocarbene with regeneration of benzyltriethylammonium chloride.

German chemists[3] have used the method successfully for preparation of dichlorocyclopropanes from olefins which yield little or no products when the dichlorocarbene is generated from chloroform and potassium *t*-butoxide. They also generated dibromocarbene in the same way. Cyclopropenes are obtained in only low yields from acetylenes owing to side reactions.

The method was used by Japanese chemists[4] for generation of dichlorocarbene for reaction with adamantane (1) to give 1-dichloromethyladamantane (2); yield 54%,

$$CHCl_2$$

$$\xrightarrow[\ [(C_2H_5)_3\overset{+}{N}CH_2C_6H_5]Cl^-\ ]{CHCl_3,\ NaOH}$$

(1)                                (2)

91 % based on adamantane consumed. This reaction is the first example of dichlorocarbene insertion into a saturated hydrocarbon. The high selectivity for a bridgehead position is also noteworthy. The exclusive monosubstitution is probably a result of a retarding effect of the dichloromethyl substituent.

The same Japanese chemists[5] found that alcohols are converted in high yield into chlorides by reaction with dichlorocarbene generated in this way. Thus 1-adamantyl alcohol is converted into 1-adamantyl chloride in 94 % yield, and benzyl alcohol into benzyl chloride in 90 % yield. The reaction proceeds with predominant retention of configuration ($S_Ni$).

The reaction of dichlorocarbene with diamantane (3) is not so selective as in the case of adamantane. In this reaction a mixture of 1- and 4-dichloromethyldiamantane, (4) and (5), is obtained in almost quantitative yield in 1.7:1 ratio. Note that the 4-position of diamantane is less hindered than the 1-position.[6]

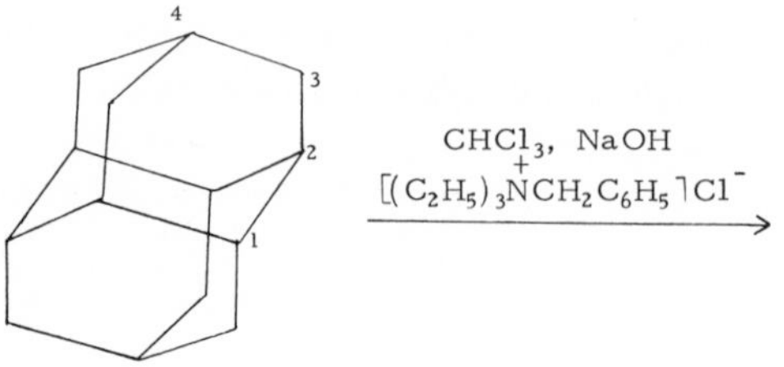

$$\begin{array}{c} CHCl_3, \ NaOH \\ + \\ [(C_2H_5)_3NCH_2C_6H_5]Cl^- \end{array} \longrightarrow$$

(3)

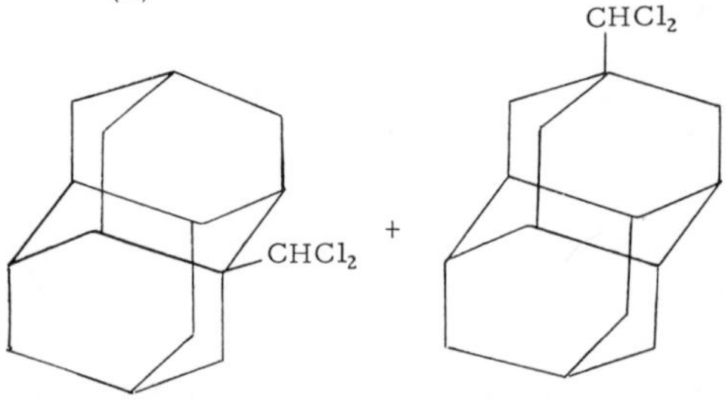

1.7:1

(4)    (5)

Many years ago, Hofmann[7] showed that isonitriles are formed on reaction of primary amines with chloroform and a strong base. Yields by this method are low and the reaction has been little used. Recently, Weber and Gokel[8] reported that isonitriles can

$$R-NH_2 + :CCl_2 \longrightarrow [R-\overset{+}{N}H_2-\overset{-}{C}Cl_2 \longrightarrow R-N=CHCl] \longrightarrow R-NC$$

be obtained in 40–60% yield if the dichlorocarbene is generated by the procedure of Mąkosza and Wawrzyniewicz.[1]

The Mąkosza reagent also is superior for reaction of dichlorocarbene with dienes; bis-adducts can be obtained in reasonable yield.[9]

Indian chemists[10] report that other cationic agents are even more efficient for generation of dichlorocarbene, for example, cetyltrimethylammonium chloride and an Indian detergent sold under the trademark Cetrimide. Starks[11] has used in the same way a "tricaprylylmethylammonium chloride,"[12] in which the alkyl groups are a mixture of $C_8$–$C_{12}$ straight chains.

Israeli chemists[13] have synthesized some sterols having a cyclopropane system in the side chain. Some naturally occurring sterols have recently been found to possess this structural feature.[14] Thus the reaction of desmosteryl acetate (6) with dichlorocarbene generated by the Mąkosza and Wawrzyniewicz technique gives the adduct (7) in 50% yield. This was reduced to $\Delta^5$-24,25-methylenecholestene-3$\beta$-ol (8) in 70% yield by lithium–*t*-butanol–THF (**1**, 604–606).

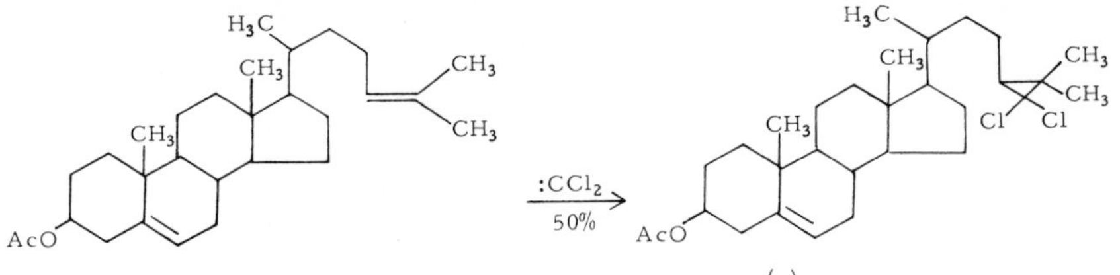

(6)　　　　:CCl$_2$ / 50% →　　　　(7)

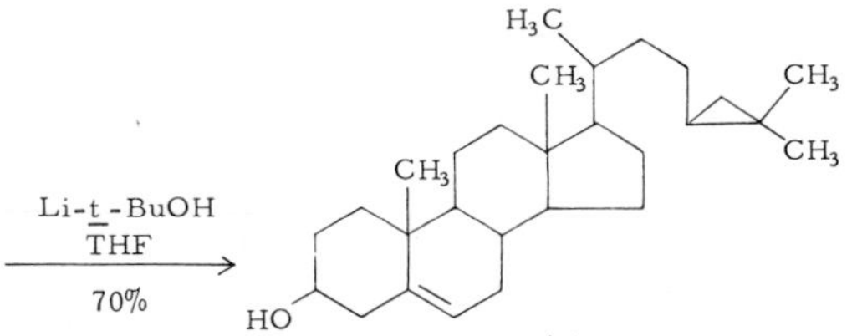

Li-*t*-BuOH / THF / 70% →　　(8)

Joshi *et al.*[15] have now been able to isolate the adduct (9) of phenanthrene with dichlorocarbene, generated in the presence of cetyltrimethylammonium chloride. The adduct rearranges at its melting point (140°) to 6-chlorodibenzo[*a*,*c*]tropylium chloride (10)

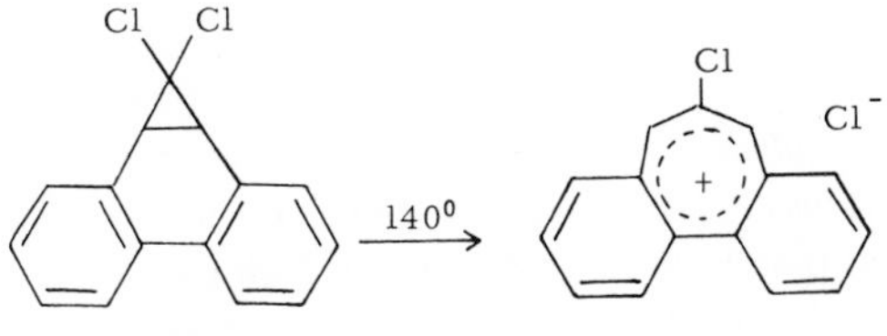

(9)　　140°→　　(10)

***Fluoroiodocarbene.*** German chemists[16] have generated fluoroiodocarbene in the following way:

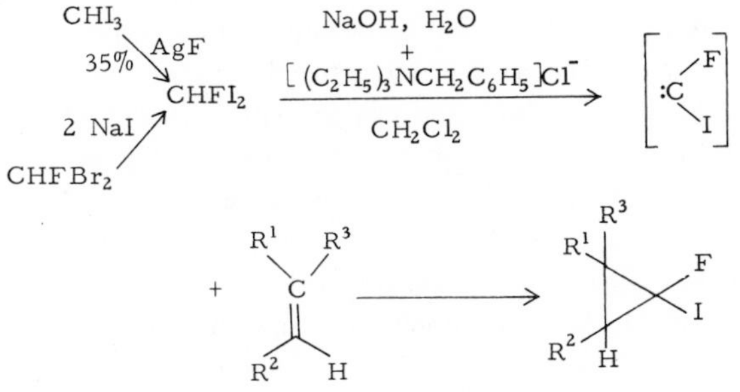

$$\text{CHI}_3 \xrightarrow[\text{AgF}]{35\%} \text{CHFI}_2 \xleftarrow[\text{CHFBr}_2]{2\ \text{NaI}}$$

$$\xrightarrow[\text{CH}_2\text{Cl}_2]{\substack{\text{NaOH, H}_2\text{O} \\ + \\ [(C_2H_5)_3NCH_2C_6H_5]Cl^-}} \quad \left[ :C\!\!\underset{I}{\overset{F}{<}} \right]$$

The reaction of the carbene with olefins yields 1-fluoro-1-iodocyclopropanes in 20–60% yield.

*Cyclohexylidene carbene.*[17] Treatment of a solution at −10° to −5° of 1-(N-nitrosoacetylaminomethyl)cyclohexanol (1) in pentane containing cyclohexene with a solution of 50% NaOH containing a tetraalkylammonium chloride[18] affords (2) in 80% yield.

(1)    (2)

*Alkylation of ketones.* Ketones are readily alkylated in the α-position on reaction with alkyl halides in the presence of 50% aqueous sodium hydroxide and catalytic amounts of benzyltriethylammonium chloride.[19] The catalytic effect of the salt is particularly marked in the case of weakly active alkyl halides. Thus the reaction of phenylacetone with *n*-butyl bromide in the absence of catalyst gives 3-phenyl-2-heptanone in 5% yield; the yield is 90% in the presence of the catalyst. Highest yields are obtained with ketones bearing an aromatic substituent at the α-CH$_2$ group.

*Glycidic nitriles.*[20] Glycidic nitriles can be obtained in good yield by the reaction of ketones with chloroacetonitrile in aqueous sodium hydroxide with benzyltriethylammonium chloride as catalyst.

*Permanganate oxidations by phase-transfer catalysis.* Starks[11] reported that oxidation of terminal alkenes to the one-carbon shorter carboxylic acids can be carried out by aqueous neutral potassium permanganate in high yield in the presence of a little quaternary ammonium salt. The oxidation of 1-decene to nonanoic acid is carried out as follows. The olefin (0.2 mole) is added to a stirred mixture of benzene (50 ml.), "tricaprylylmethylammonium chloride"[12] (0.01 mole), and potassium permanganate (0.8 mole) in water (100 ml.) at such a rate that the temperature is maintained at 40–50° (exothermic reaction). After addition is complete, the mixture is stirred for an additional 0.5 hr. Excess permanganate is destroyed with sodium sulfite; the reaction mixture is filtered to remove MnO$_2$ and acidified with dilute HCl. The benzene solution is shaken with 10% NaOH solution and the aqueous alkaline phase washed with ether and then acidified with HCl. The carboxylic acid is isolated by extraction with ether. The yield of nonanoic acid is 91%. 1-Octene is oxidized by the same procedure to heptanoic acid in essentially quantitative yield.

Weber and Shepherd[21] have now applied this new procedure to oxidation of internal alkenes to *cis*-1,2-glycols and obtained yields of about 50%. Ordinarily this reaction proceeds in low yield; the transformation for that reason has been generally effected with osmium tetroxide (toxic and expensive, **1**, 759–760) or with silver acetate and iodine in moist acetic acid (**1**, 1003).

The oxidation of *cis*-cyclooctene with basic permanganate illustrates the new method. The octene (0.1 mole) in methylene chloride is treated with a 40% aqueous solution of sodium hydroxide and 1 g. of benzyltriethylammonium chloride. The mixture is cooled to 0° and $KMnO_4$ (0.1 mole) is added in small portions over 2 hr. with vigorous stirring; after standing overnight at 0°, the precipitated $MnO_2$ is dissolved by $SO_2$. The *cis*-1,2-cyclooctanediol is isolated by ether extraction and crystallization in 50% yield. The yield of diol obtained without a phase-transfer catalyst is 7%.[22]

The oxidation of *trans*-cyclododecene to *trans*-1,2-cyclododecanediol by the new procedure also gives a yield of 50%.

Lower yields are obtained, however, if the glycol is highly soluble in the aqueous phase. Thus oxidation of cyclohexene gives *cis*-1,2-cyclohexanediol in only 15% yield together with significant amounts of adipic acid.

[1] M. Mąkosza and M. Wawrzyniewicz, *Tetrahedron Letters*, 4659 (1969): M. Mąkosza, *Org. Syn.*, submitted 1971.

[2] W. v. E. Doering and A. K. Hoffmann, *Am. Soc.*, **76**, 6162 (1954).

[3] E. V. Dehmlow and J. Schönefeld, *Ann.*, **744**, 42 (1971).

[4] I. Tabushi, Z. Yoshida, and N. Takahashi, *Am. Soc.*, **92**, 6670 (1970).

[5] *Idem, ibid.*, **93**, 1820 (1971).

[6] I. Tabushi, Y. Aoyama, N. Takahashi, T. M. Gund, and P. v. R. Schleyer, *Tetrahedron Letters*, 107 (1973).

[7] A. W. Hofmann, *Ann.*, **144**, 114 (1867).

[8] W. P. Weber and G. W. Gokel, *Tetrahedron Letters*, 1637 (1972).

[9] E. V. Dehmlow, *Tetrahedron*, **28**, 175 (1972).

[10] G. C. Joshi, N. Singh, and L. M. Pande, *Tetrahedron Letters*, 1461 (1972).

[11] C. M. Starks, *Am. Soc.*, **93**, 195 (1971).

[12] Available from General Mills Co., Chemical Division, Kankakee, Illinois 60901.

[13] R. Ikan, A. Markus, and Z. Goldschmidt, *J.C.S. Perkin I*, 2423 (1972).

[14] R. L. Hale, J. Leclereq, B. Tursch, C. Djerassi, R. A. Gross, A. J. Weinheimer, K. Gupta, and P. J. Scheuer, *Am. Soc.*, **92**, 2179 (1970); N. C. Ling, R. L. Hale, and C. Djerassi, *ibid.*, **92**, 5281 (1970); F. J. Schmitz and T. Pattabhiraman, *ibid.*, **92**, 6073 (1970).

[15] G. C. Joshi, N. Singh, and L. M. Pande, *Synthesis*, 317 (1972).

[16] P. Weyerstahl, R. Mathias, and G. Blume, *Tetrahedron Letters*, 611 (1973).

[17] M. S. Newman and Z. Din, *J. Org.*, **38**, 547 (1973).

[18] Methyltricaprylammonium chloride (Aliquat 336) was used.

[19] A. Jończyk, B. Serafin, and M. Mąkosza, *Tetrahedron Letters*, 1351 (1971).

[20] A. Jończyk, M. Fedoryński, and M. Mąkosza, *ibid.*, 2395 (1972).

[21] W. P. Weber and J. P. Shepherd, *ibid.*, 4907 (1972).

[22] A. C. Cope, S. W. Fenton, and C. F. Spencer, *Am. Soc.*, **74**, 5884 (1952).

## Birch reduction, 1, 54–56; 2, 27–29; 3, 19–20.

*Reviews.* Birch and Subba Rao[1] have reviewed reductions by metal–ammonia solutions.

Kaiser[2] has published a review which discusses the similarities and differences between reductions by the Birch procedure and by lithium in low molecular weight amines (Benkeser reduction). He concludes that the Benkeser reduction is more powerful but less selective than the Birch reduction.

**Reduction of styrenoid double bond.** The final step in a total synthesis of $(\pm)$-2,3,4-trimethoxyestra-1,3,5(10)-triene-17$\beta$-ol (2) involved stereoselective reduction of the styrenoid bond of the tetraenol (1) by sodium–liquid ammonia. When this was first

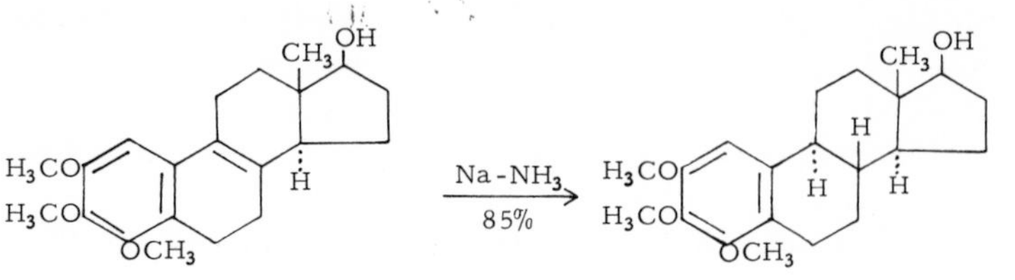

(1)                                           (2)

conducted with 4 g.-atom eq. of sodium in liquid ammonia–aniline for 30 min., the major product obtained was shown to have lost the 3-methoxy group. It was then found that the desired reduction could be effected in 85% yield by use of 2.5 g.-atom eq. of sodium and by limiting the reaction time to 10 min.[3]

**Metal reduction of malonates.** Dimethyl dimethylmalonate (1) is reduced by sodium dispersed in xylene containing trimethylchlorosilane (TMCS) to dimethylketene methyl trimethylsilyl acetal (2)[4]:

$$(CH_3)_2C(COOCH_3)_2 \xrightarrow[87\%]{\substack{Na-xylene \\ TMCS}} (CH_3)_2C=C(OCH_3)OSi(CH_3)_3 + CO + CH_3OSi(CH_3)_3$$

(1)                                           (2)

Since (2) is readily hydrolyzed to the ester (3), the two-step process provides a convenient decarboxylation procedure.

$$(2) + CH_3OH \longrightarrow (CH_3)_2CHCOOCH_3 + CH_3OSi(CH_3)_3$$

(3)

Reduction of (1) with 4 eq. of sodium in liquid ammonia followed by treatment with TMCS gives five products, the most interesting of which is 3,3-dimethyl-cis-1,2-bistrimethylsilyloxycyclopropane (4).[5]

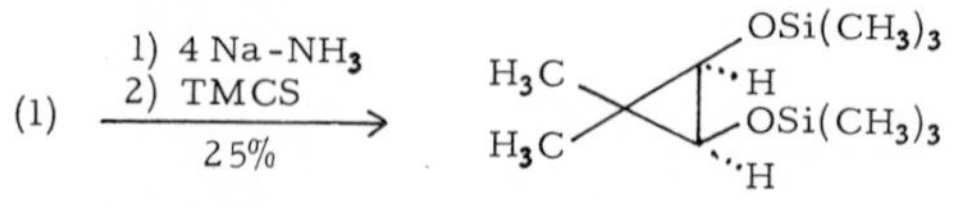

(4)

[1]A. J. Birch and G. Subba Rao, *Advan. Org. Chem., Methods and Results*, **8**, 1 (1972).
[2]E. M. Kaiser, *Synthesis*, 391 (1972).
[3]P. N. Rao, E. J. Jacob, and L. R. Axelrod, *J. Chem. Soc.* (*C*), 2855 (1971).
[4]Y.–N. Kuo, F. Chen, C. Ainsworth, and J. J. Bloomfield, *Chem. Commun.*, 136 (1971).
[5]F. Chen and C. Ainsworth, *Am. Soc.* **94**, 4037 (1972).

**N,N-Bisbromomagnesiumaniline,** $C_6H_5N(MgBr)_2$.

*Preparation.*[1] Aniline is added dropwise with stirring to an ethereal solution of ethylmagnesium bromide and refluxing is continued for 30 min.

$$2\ C_2H_5MgBr\ +\ C_6H_5NH_2\ \longrightarrow\ C_6H_5N(MgBr)_2\ +\ 2\ C_2H_6$$

*Reaction with 2-acetylamino-3-ethoxycarbonyl-5-ethylthiophene*[1] :

[1]W. Ried and E. Kahr, *Ann.*, **716**, 219 (1968).

**Bis-(1,5-cyclooctadiene)nickel(0),** $Ni(COD)_2$,

Mol. wt. 230.27, yellow crystals, extremely sensitive to oxygen in solution.

*Preparation.* The reagent is best prepared[1] by a modification of the original procedure of Wilke.[2] The preparation involves reduction of nickel acetylacetonate (Alfa Inorganics) with triethylaluminum (Texas Alkyls) in the presence of 1,5-cyclooctadiene. Because of the sensitivity of $Ni(COD)_2$ to oxygen, special inert-atmosphere techniques must be used. The yield is 81 %.

$\pi$-*Allylnickel halides* (**2**, 291). Walter and Wilke[3] reported briefly that $Ni(COD)_2$ is more reactive than nickel carbonyl in the reaction of allylic halides to form $\pi$-allylnickel halides. The reaction proceeds below 0° and in quantitative yields. The method has not been used extensively since nickel carbonyl is commercially available. However, Semmelhack recommends use of $Ni(COD)_2$ for the preparation of thermally sensitive $\pi$-allylnickel halides such as $\pi$-(2-carboethoxyallyl)nickel bromide,[4] which was prepared from ethyl 2-bromomethylacrylate with this nickel reagent in 76 % yield.

*Coupling of aryl halides.*[5] $Ni(COD)_2$ reacts with a variety of aryl halides at temperatures of 25–40° in DMF to form diaryls, nickel dihalides, and COD:

77150

$$\text{Ni(COD)}_2 + 2\,\text{ArX} \xrightarrow[25\text{-}40^0]{\text{DMF}} \text{Ar}-\text{Ar} + \text{NiX}_2 + 2\,\text{COD}$$

Yields are generally high (80–90%), but an *ortho*-substituent slows or even inhibits the reaction. Aryl halides containing acidic functional groups (OH, COOH) undergo reduction of the carbon–halogen bond instead of coupling. DMF is apparently the only satisfactory solvent; no reaction occurs in THF or toluene. DMSO and HMPT cannot be used owing to rapid decomposition of the nickel reagents in these solvents.

*Coupling of alkenyl halides.*[6] Alkenyl halides react with Ni(COD)$_2$ in DMF at 25° to give 1,3-dienes:

$$2\,\text{RCH}{=}\text{CHX} + \text{Ni(COD)}_2 \longrightarrow \text{RCH}{=}\text{CHCH}{=}\text{CHR} + \text{NiX}_2 + 2\,\text{COD}$$

The reaction can also be carried out in ether if a donor ligand is added (triphenylphosphine or tri-*n*-butylphosphine). Yields of dimer are high if the alkenyl halide carries an electron-withdrawing group. Thus *trans*-2-bromostyrene is converted into *trans,trans*-1,4-diphenyl-1,3-butadiene in 70% yield.

With simple alkenyl halides, the yields of coupling products are only moderate, 34–60%, and a mixture of geometric isomers results, with some preference for the more stable isomer. The low yields are probably the result of oligomerization of the initially formed 1,3-dienes by Ni(COD)$_2$.

2-Halo- and 3-haloacrylates undergo the coupling reaction in high yield and with complete retention of configuration as shown in the examples (a) and (b).

(a)  $\text{BrCH}\overset{c}{=}\text{CHCOOCH}_3 \xrightarrow[99\%]{} \text{CH}_3\text{OCOCH}\overset{c}{=}\text{CHCH}\overset{c}{=}\text{CHCOOCH}_3$

(b)  $\text{BrCH}\overset{t}{=}\text{CHCOOCH}_3 \xrightarrow[92\%]{} \text{CH}_3\text{OCOCH}\overset{t}{=}\text{CHCH}\overset{t}{=}\text{CHCOOCH}_3$

Semmelhack[1] has reported one example of intramolecular coupling with Ni(COD)$_2$ in DMF. Thus *cis,cis*-1,6-diiodo-1,5-hexadiene (1) is converted into 1,3-cyclohexadiene (2) in 67% yield (isolated).

(1)                    (2)

*[2 + 2] Cross-addition of olefins.*[7] The reaction of norbornadiene (1, 4.0 mmole)

and methylenecyclopropane (2, 4.0 mmole) catalyzed by Ni(COD)$_2$ (0.2 mmole) and triphenylphosphine (0.22 mmole) in benzene at 20° for 24 hr. gives the 1:1 cycloadduct (3) in 86% yield.

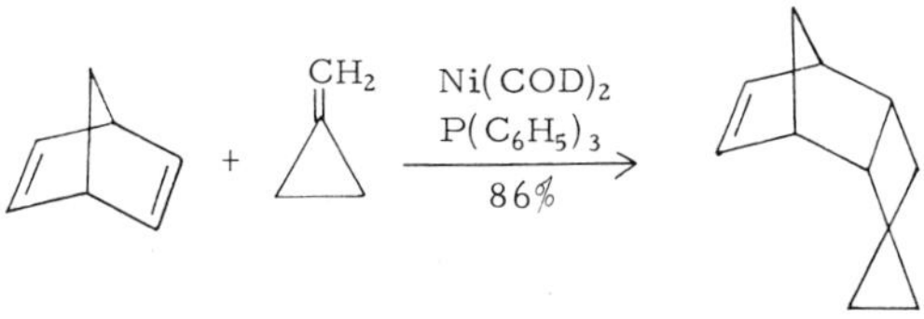

(1)          (2)                          (3)

If the reaction between (1) and (2) is carried out in the presence of Ni(COD)$_2$ and (−)-benzylmethylphenylphosphine,[7] (3) is obtained in an optically active form, $\alpha_D$ −0.8°.[8]

[1] M. F. Semmelhack, *Org. React.*, **19**, 115 (1972).
[2] B. Bogdanović, M. Kröner, and G. Wilke, *Ann.*, **699**, 1 (1966).
[3] D. Walter and G. Wilke, *Angew. Chem., internat. Ed.*, **5**, 897 (1966).
[4] M. F. Semmelhack, *Org. React.*, **19**, 179 (1972).
[5] M. F. Semmelhack, P. M. Helquist, and L. D. Jones, *Am. Soc.*, **93**, 5908 (1971).
[6] M. F. Semmelhack, P. M. Helquist, and J. D. Gorzynski, *ibid.*, **94**, 9234 (1972).
[7] R. Noyori, T. Ishigami, N. Hayashi, and H. Takaya, *ibid.*, **95**, 1674 (1973).
[8] K. Naumann, G. Zon, and K. Mislow, *ibid.*, **91**, 7012 (1969).

**1,8-Bis(dimethylamino)naphthalene, 3, 22.**

*Improved preparation.*[1] This base can be obtained in 87% yield by treatment of 1,8-diaminonaphthalene in refluxing THF with excess dimethyl sulfate in the presence of sodium hydride.

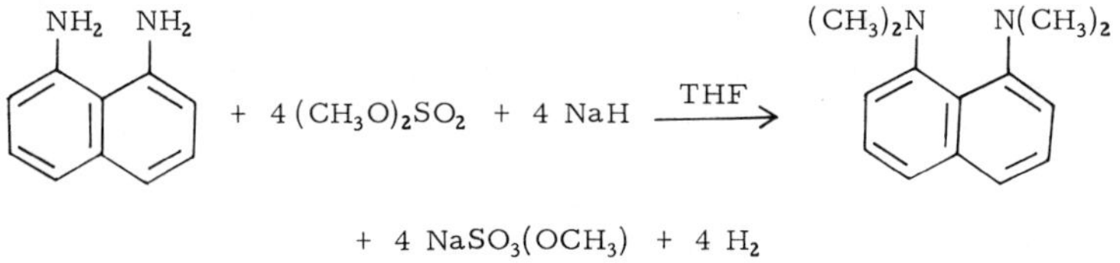

+ 4 NaSO$_3$(OCH$_3$)  +  4 H$_2$

[1] H. Quast, W. Risler, and G. Döllscher, *Synthesis*, 558 (1972).

**Bis(3,6-dimethyl)borepane (3,6-DMB-7)** (1). Mol. wt. 274.12.

*Preparation.*[1] This dialkylborane (1) is prepared in 84% yield by the reaction of 2,5-dimethyl-1,5-hexadiene (Chemical Samples Company, Columbus, Ohio 43221)

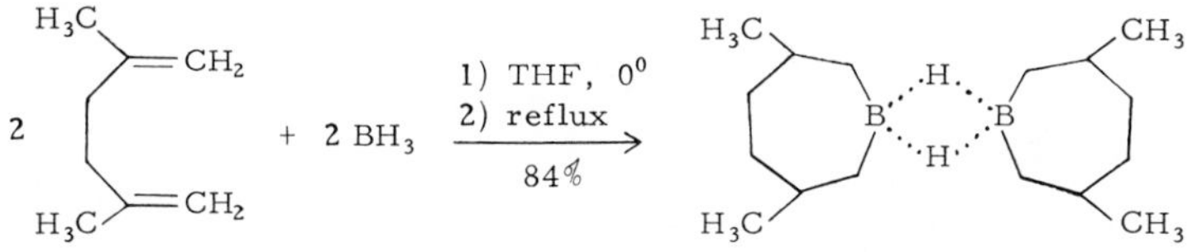

(1)

with borane in refluxing THF for 1 hr. Bis(3,5-dimethyl)borinane (2), 3,5-DMB-6, was prepared in the same way from 2,4-dimethyl-1,4-pentadiene (Chemical Samples Company).

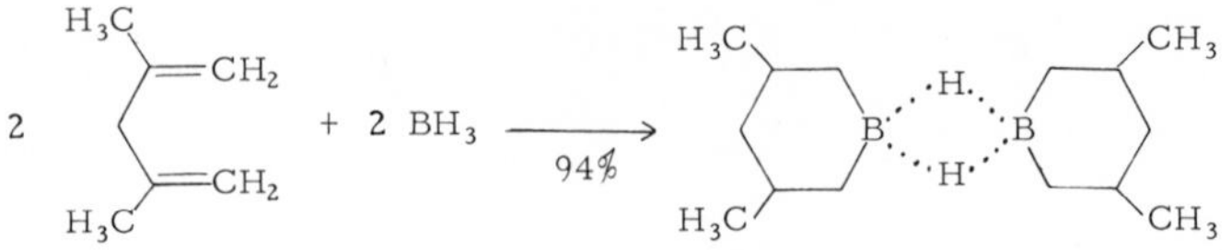

(2)

*Hydroboration.*[1] Both bisboracyclanes (1) and (2) are highly selective hydroborating agents, allowing quantitative conversion of olefins into the anti-Markownikoff alcohols. These are comparable to 9-borabicyclo[3.3.1]nonane, 9-BBN (**2**, 31; **3**, 24–29; this volume).

*Synthesis of secondary alcohols from 1-alkynes.*[2] Dihydroboration at room temperature of a terminal alkyne with either (1) or 9-BBN gives a 1,1-diborylalkane (3); this is treated at 0–5° with 1 eq. of methyllithium in ether. The product (4) rearranges to (5). An alkyl halide (100% excess) is then added, and the resultant secondary organoborane (6) is oxidized with alkaline hydrogen peroxide. Secondary alcohols (7) are obtained in 70–85% yield.

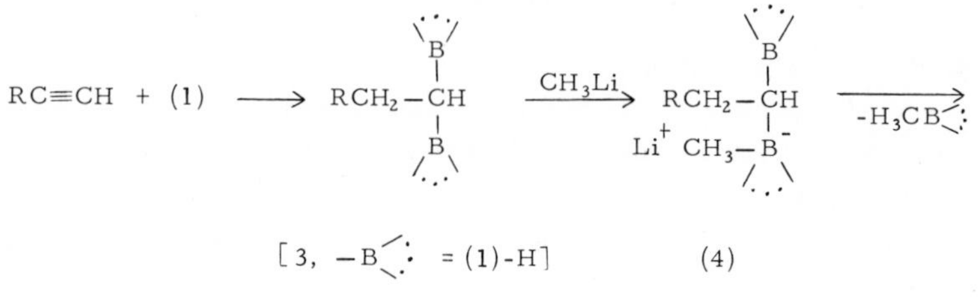

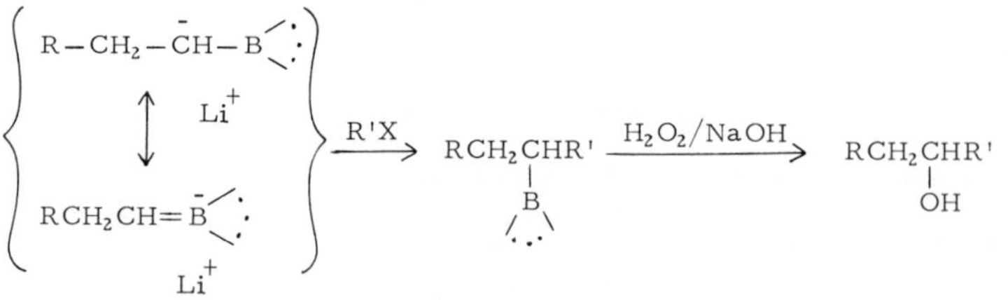

$$[3, \ -B\overset{\cdot\cdot}{\diagup} \ = (1)\text{-}H]$$  (4)

(5)          (6)          (7)

[1] H. C. Brown and E. Negishi, *J. Organometal. Chem.*, **28**, C1 (1971).
[2] G. Zweifel, R. P. Fisher, and A. Horng, *Synthesis*, 37 (1973).

**Bis(fluoroxy)difluoromethane,** $CF_2(OF)_2$. Mol. wt. 101.01, colorless liquid at $-184°$.
  The reagent is prepared by fluorination of $CO_2$ in the presence of CsF.[1]

*Electrophilic fluorination.* Barton *et al.*[2] have examined a number of fluoroxy reagents as electrophilic fluorinating agents, and conclude that bis(fluoroxy)difluoromethane is as useful as fluoroxytrifluoromethane (**2**, 200; **3**, 146–147) and is potentially cheaper.

[1]F. A. Hohorst and J. M. Shreeve, *Am. Soc.*, **89**, 1809 (1967); P. G. Thompson, *ibid.*, **89**, 1811 (1967); R. L. Cauble and G. H. Cady, *ibid.*, **89**, 1962 (1967).
[2]D. H. R. Barton, R. H. Hesse, M. M. Pechet, G. Tarzia, H. T. Toh, and N. D. Westcott, *Chem. Commun.*, 122 (1972).

## Bis-3-methyl-2-butylborane (Disiamylborane), $Sia_2BH$, 1, 57–59; 3, 22.

*Reduction of acetylenes* (**1**, 58).[1] The last step in a synthesis of the insect sex attractive *cis*-8-dodecene-1-ol acetate (2)[2] involved reduction of the acetylene (1). This was

$$CH_3(CH_2)_2C\equiv C(CH_2)_6CH_2OCOCH_3 \xrightarrow[\substack{\text{diglyme} \\ 2)\ HOAc \\ 3)\ H_2O_2 \\ 66\%}]{1)\ (Sia)_2BH,} CH_3(CH_2)_2\overset{H}{\underset{|}{C}}=\overset{H}{\underset{|}{C}}(CH_2)_6CH_2OCOCH_3$$

(1)                                        (2)

carried out with disiamylborane followed by hydrogen peroxide treatment to remove boron-containing impurities; (2), which assayed > 98 % *cis*, was obtained in 66 % yield.

[1]G. Holan and D. F. O'Keefe, *Tetrahedron Letters*, 673 (1973).
[2]W. L. Roelofs, A. Comeau, and R. Selle, *Nature* **224**, 723 (1969).

## Bis(*trans*-2-methylcyclohexyl)borane,  . Mol. wt. 204.16.

The reagent is prepared by the reaction of 1-methylcyclohexene with diborane in THF.[1]

*Hydroboration–iodination of alkynes.*[2] Olefinic side chains can be introduced stereoselectively into cyclic systems by hydroboration–iodination of alkynes. Thus the reaction of 1-hexyne with bis(*trans*-2-methylcyclohexyl)borane (1) gives the vinylborane (2). Treatment of (2) successively with 6 *N* sodium hydroxide and a solution of iodine in THF[3] yields *trans*-1-methyl-2-(*cis*-1-hexenyl)cyclohexane (3) in 85 % yield. Migration of the 2-methylcyclohexyl group thus occurs with retention of configuration.

The reaction can be applied to any substituted cyclic olefin, provided that hydroboration stops at the dialkylborane stage.

[1]H. C. Brown and G. Zweifel, *Am. Soc.*, **83**, 2544 (1961).
[2]G. Zweifel, R. P. Fisher, J. T. Snow, and C. C. Whitney, *ibid.*, **93**, 6309 (1971).
[3]See also, G. Zweifel, H. Arzoumanian, and C. C. Whitney, *ibid.*, **89**, 3652 (1967).

**1,3-Bis(methylthio)allyllithium,** $CH_3S \overset{\_}{\frown}\overset{\_}{\frown} SCH_3Li.^+$    Mol. wt. 142.21.

*Preparation.*[1] The reagent is prepared in THF from epichlorohydrin as illustrated. The deep-purple solution is stable for 12 hr. at 0°.

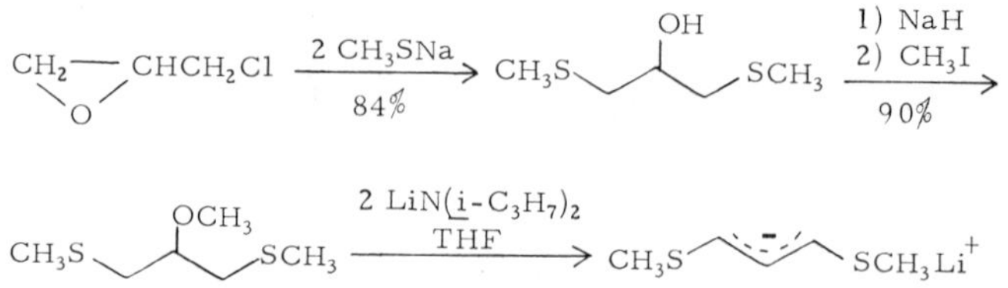

(1)

*α,β-Unsaturated aldehydes.*[1] The reagent (1) functions as the nucleophilic equivalent of $-CH=CHCHO$. Thus reaction of (1) with 1-bromopentane for 2 hr. at $-75°$ furnishes 1,3-bis(methylthio)-1-octene (2) in 90% yield. The bisthio ether is hydrolyzed by mercuric chloride in aqueous acetonitrile to give *trans*-2-octenal (3) in 84% yield.

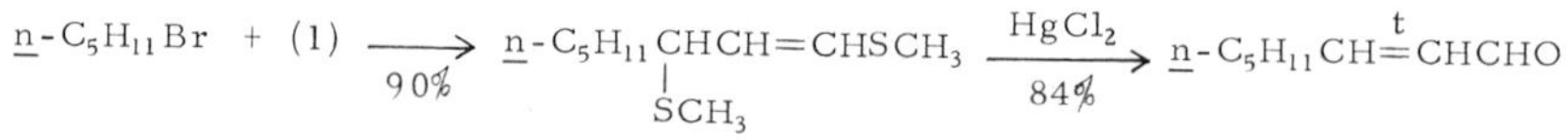

(2)            (3)

Reaction of (1) with an aldehyde or ketone followed by hydrolysis furnishes a γ-hydroxy-α,β-unsaturated aldehyde. For example, addition of (1) to propanal forms 1,3-bis(methylthio)-1-hexene-4-ol (4) in 89% yield; hydrolysis with mercuric chloride gives *trans*-4-hydroxy-2-hexenal (5) in 48% yield.

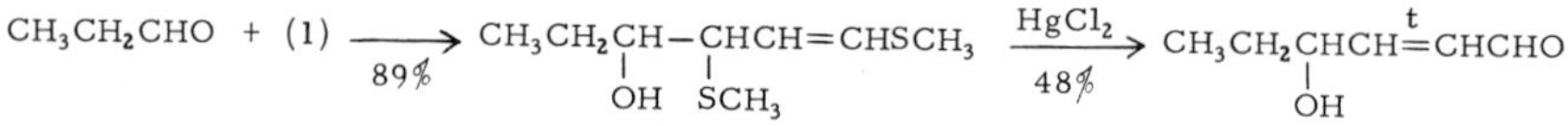

(4)            (5)

The reaction of (1) with cyclohexene oxide produces the *trans*-cyclohexanol derivative (6). Acetylation of (6) followed by mercuric ion-promoted hydrolysis afforded the acetoxy aldehyde (7) in 82% overall from cyclohexene oxide.

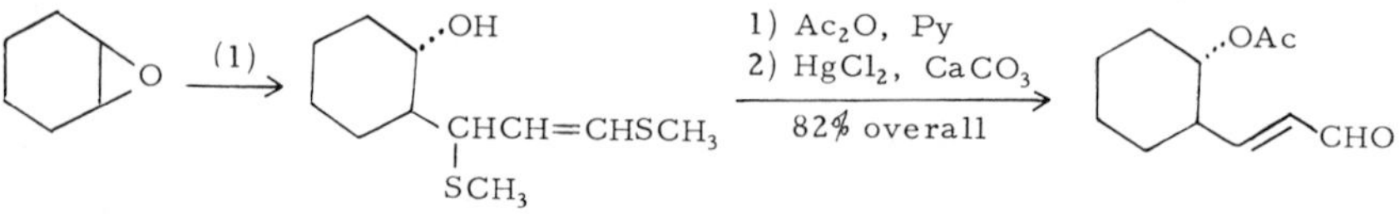

(6)            (7)

The reagent (1) has also been used in a key step in a total synthesis of prostaglandin $F_{2\alpha}$.[2] Thus treatment of the oxidoacetal (8) with 1,3-bis(methylthio)allyllithium (1) under argon at $-78°$ gave a mixture of the desired coupling product (9) and the isomer (10). These were not separated, but hydrolyzed with aqueous mercuric chloride–calcium carbonate at $50°$ under argon. The resulting two unsaturated aldehydes (11) and (12)

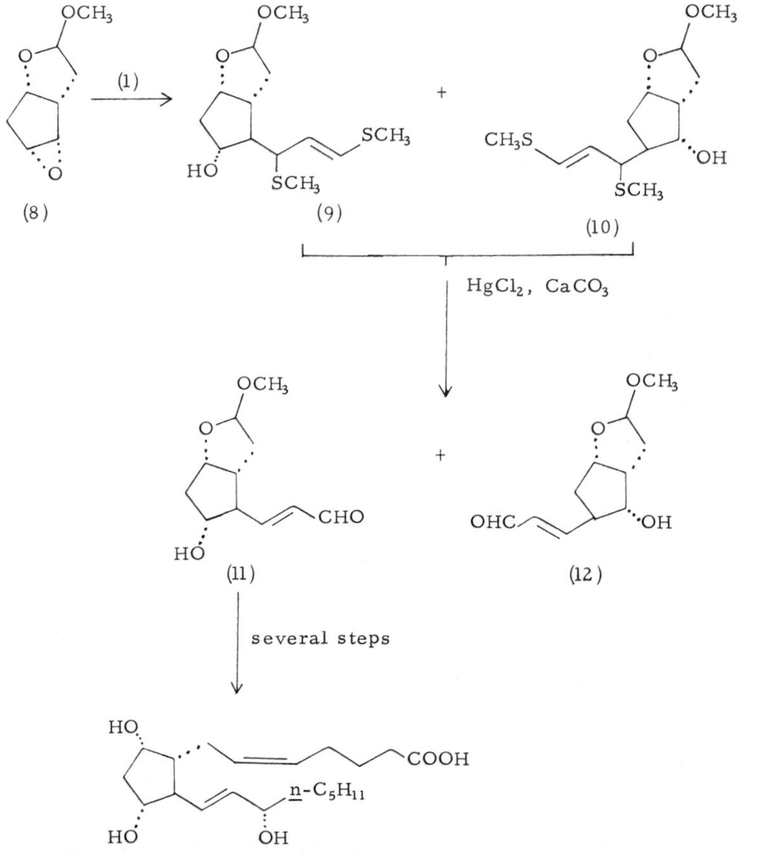

were separated by preparative layer chromatography. The desired hydroxyaldehyde (11) was obtained in 30% yield; the isomeric aldehyde in 40% yield. The hydroxyaldehyde (11) was transformed in several steps into dl-prostaglandin $F_{2\alpha}$ (13).

This synthetic route to (13) is attractive because of its directness; unfortunately its usefulness is diminished by the occurrence of ring opening in both possible directions in the reaction of the oxide (8) with 1,3-bis(methylthio)allyllithium (1).

[1] E. J. Corey, B. W. Erickson, and R. Noyori, *Am. Soc.*, **93**, 1724 (1971).
[2] E. J. Corey and R. Noyori, *Tetrahedron Letters*, 311 (1970).

**Bismuth triacetate,** $Bi(OCOCH_3)_3$. Mol. wt. 322.13.

*Preparation.*[1] Bismuth oxide (50 g.) is heated at 140–150° with acetic acid (300 ml.) and acetic anhydride (35 ml.) until it dissolves (1.5–2 hr.). The triacetate separates on cooling as glistening, colorless tablets (yield 90%).

*Reaction with amines and alcohols.*[2] Bismuth triacetate reacts with amines (equation 1) and with alcohols (equation 2) at ∼150° to give N- and O-acetylated derivatives, respectively. Yields in both cases are in the range 30–80%. The inorganic residue in both cases is bismuthyl acetate.

$$
1. \quad Bi(OAc)_3 + HN\underset{R^1}{\overset{R}{\diagup}} \longrightarrow CH_3CON\underset{R^1}{\overset{R}{\diagup}} + Bi(O)OCOCH_3 + CH_3COOH
$$

$$
2. \quad Bi(OAc)_3 + HOR \longrightarrow CH_3COOR + Bi(O)OCOCH_3 + CH_3COOH
$$

[1]W. Rigby, *J. Chem. Soc.*, 794 (1951).
[2]A. L. Reese, K. McMartin, D. Miller, and P. P. Wickham, *J. Org.*, **38**, 764 (1973).

**Bis(pyridine)dimethylformamidodichlororhodium borohydride,** $(Py)_2(DMF)(RhCl_2)BH_4$. Mol. wt. 419.96, dark red crystals.

The catalyst is prepared from $(Py)_3RhCl_3$ and sodium borohydride in DMF.

*Homogeneous hydrogenation.* The complex is a highly active catalyst for homogeneous hydrogenation.[1]

[1]P. Abley, I. Jardine, and F. J. McQuillin, *J. Chem. Soc. (C)*, 840 (1971).

**Bis(triphenylphosphine)dicyanonickel(0),** $[(C_6H_5)_3P]_2Ni(CN)_2$ (1). Mol. wt. 635.29, m.p. 212° dec.

*Preparation.*[1] The complex is prepared in 62% yield from the reaction of nickel cyanide and triphenylphosphine in ethanol (48-hr. reflux). It is similar to bis(triphenylphosphine)dicarbonylnickel(0) (**1**, 61) but more stable to air and safer to handle.

*Diels–Alder catalyst.*[1] The complex (1) catalyzes the homo-Diels–Alder condensation of vinyl compounds to norbornadiene. Thus the reaction of acrylonitrile with norbornadiene in the presence of this complex gives 8-cyanotetracyclo[4.3.0.0$^{2,4}$.0$^{3,7}$]-nonane (2)[2] as a mixture of epimers in 93% yield. The yield of (2) using bis(triphenylphosphine)dicarbonylnickel(0) as catalyst is 86%.

$$
+ \ CH_2{=}CHCN \xrightarrow[\text{93\%}]{\overset{(1)}{120°, \ 15 \ hrs.}}
$$

(2)

The complex also catalyzes the addition of alkynes to norbornadiene; yields are somewhat lower than those obtained with vinyl compounds.

$$\text{(norbornadiene)} + CH_3OOCC\equiv CCOOCH_3 \xrightarrow[\substack{32\%}]{\substack{(1)\\120^0}}$$

(3)

[1]G. N. Schrauzer and P. Glockner, *Ber.*, **97**, 2451 (1964).
[2]The name deltacyclane has been proposed for the parent ring system of (2); *see* P. K. Freeman, D. M. Balls, and J. N. Blazevich, *Am. Soc.*, **92**, 2051 (1970).

**Bis(triphenylphosphine)nickel(0),** $[(C_6H_5)_3P]_2Ni(0)$. Mol. wt. 295.00; red-brown, crystalline, air sensitive.

The reagent is prepared by sodium amalgam reduction of bis(triphenylphosphine)-nickel dibromide in acetonitrile or benzene.

*Allene cyclooligomerization.*[1] Allene is catalytically converted into a mixture of isomeric trimers, a tetramer, a pentamer, isomeric hexamers, and polymers by this complex. The reaction involves the initial formation of $[(C_6H_5)_3P]_2NiC_3H_4$, which can be isolated.

*Alkene carbonates.*[2] Ethylene carbonate (1) is formed when ethylene oxide and carbon dioxide are heated at 100° in benzene solution containing this nickel(0) complex. Under similar conditions, 2-methyl-1,2-epoxypropane is converted into 1,1-dimethyl-ethylene carbonate and 2,3-epoxybutane into 1,2-dimethylethylene carbonate. Several other catalysts of the type $L_2Ni(0)$ are effective.

(1)

[1]R. J. De Pasquale, *J. Organometal. Chem.*, **32**, 381 (1971).
[2]*Idem, J.C.S. Chem. Comm.*, 157 (1973).

**9-Borabicyclo[3.3.1]nonane (9-BBN), 2,** 31; **3,** 24–29.
Aldrich supplies 9-BBN as a 0.5 $M$ solution in THF. The company also supplies the reagent as a white, crystalline powder, m.p. 150–152°. This material can be stored under dry $N_2$ indefinitely at room temperature. Contact with moisture or oxygen should be minimized in handling the reagent.

*Caution:* Alkylboranes are potentially pyrophoric.

**Boric acid, 1,** 63–66; **2,** 32; **3,** 29–30.
*Esterification of phenols.*[1] The direct esterification of phenols is catalyzed by a combination of boric acid and sulfuric acid. Thus phenyl benzoate (1) can be prepared

in almost quantitative yield by azeotropic distillation of a refluxing toluene solution of phenol, benzoic acid, and catalytic amounts of boric acid and sulfuric acid.

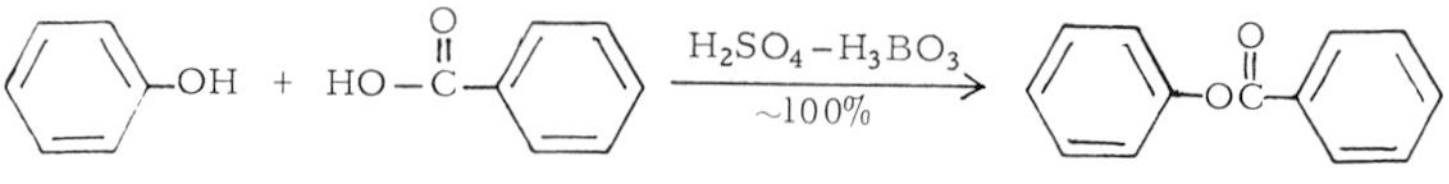

(1)

[1]W. W. Lawrence, Jr., *Tetrahedron Letters*, 3453 (1971).

**Boron tribromide**, **1**, 66–67; **2**, 33–34; **3**, 30–31. Additional supplier: Trona Division, American Potash and Chemical Corporation.

*Hexabromocyclopentadiene* (**2**, 90). An improved procedure for the West preparation of hexabromocyclopentadiene (**2**) by an exchange reaction of hexachlorocyclopentadiene (**1**, Hooker Chemical Company) with boron tribromide in the presence of aluminum bromide and bromine has been published.[1]

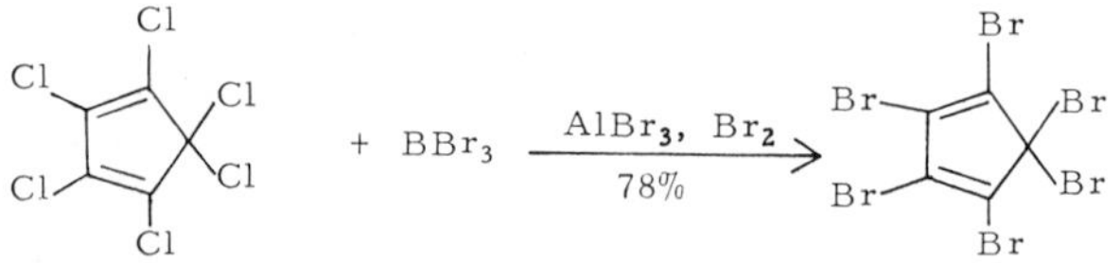

(1)                                    (2)

[1]G. A. Ungefug and C. W. Roberts, *J. Org.*, **38**, 153 (1973).

**Boron trichloride**, **1**, 67–68; **2**, 34–35; **3**, 31–32.

*Cleavage of hindered esters.*[1] Hindered esters can be cleaved efficiently by treatment with boron trichloride in methylene chloride. Thus methyl O-methylpodocarpate (1) is cleaved by the reagent at 0° to the corresponding acid, O-methylpodocarpic acid,

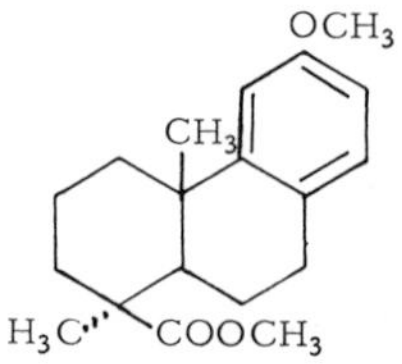

(1) 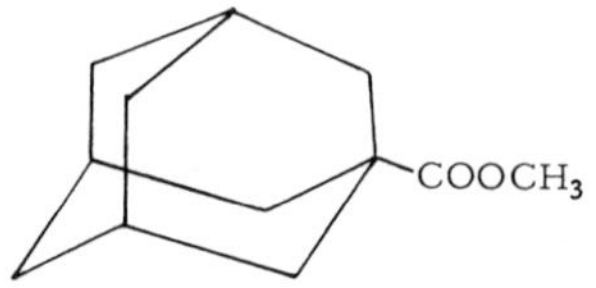                                    (2)

in 90% yield. No cleavage of the methyl ether function is observed. Similarly, the methyl esters of adamantane-1-carboxylic acid (2) and of 2,4,6-trimethylbenzoic acid gave the corresponding acids in high yield.

[1]P. S. Manchand, *Chem. Commun.*, 667 (1971).

*Demethylation* (**2**, 34–35). The selective demethylation of a methoxy group *ortho* to a carbonyl group has been reported. However, Barton *et al.*[2] have observed selective demethylation *para* to a carbonyl group in certain polyoxygenated benzenes. For example, treatment of (1) with boron trichloride in dichloromethane at −80° gives (2) as the only isolated product.

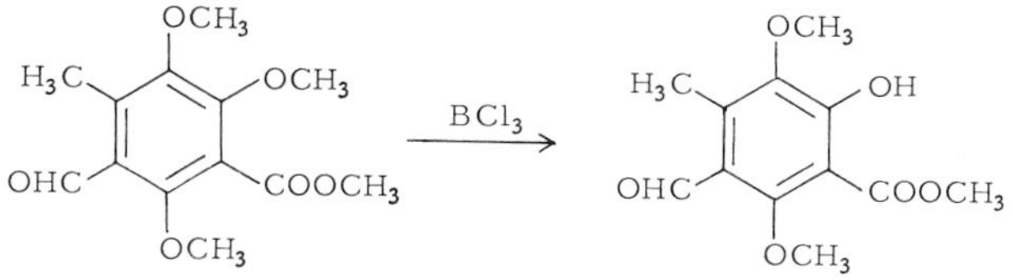

(1, 702 mg.)                    (2, 507 mg.)

*O-Demethylenation.*[3] By proper selection of conditions, cleavage of a methylene-dioxy group in preference to aromatic methoxyls can be achieved with boron trichloride. Thus the isoquinoline derivative (1) can be cleaved to (2) in 78% yield by treatment with boron trichloride at 4° in methylene chloride. Boron tribromide under the same conditions is not selective.

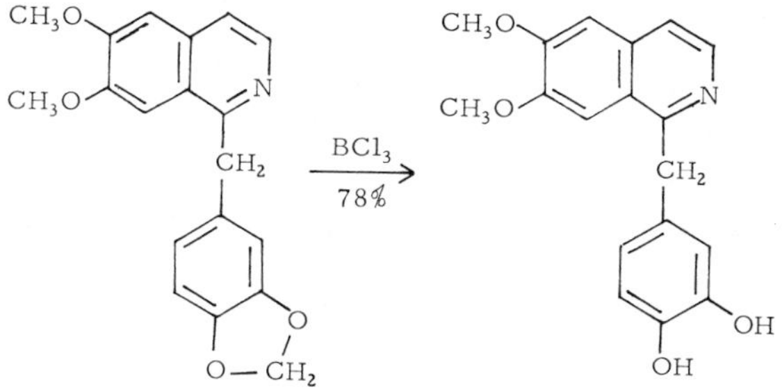

(1)                    (2)

[2]D. H. R. Barton, L. Bould, D. L. J. Clive, P. D. Magnus, and T. Hase, *J. Chem. Soc.* (*C*), 2204 (1971).

[3]S. Teitel, J. O'Brien, and A. Brossi, *J. Org.*, **37**, 3368 (1972).

**Boron trifluoride *n*-butyl etherate**, $BF_3 \cdot O(CH_2CH_2CH_2CH_3)_2$. Mol. wt. 198.04. Supplier: Fluka.

*Cyclopropanecarboxaldehyde.* Dehydration of a mixture of *cis*- and *trans*-cyclo-butane-1,2-diol (1) with boron trifluoride *n*-butyl etherate at 230° yields cyclopropane-carboxaldehyde (2) in 65–80% yield.[1] Dehydration of (1) with *p*-toluenesulfonic acid gives (2) in somewhat lower yields (66%).[2]

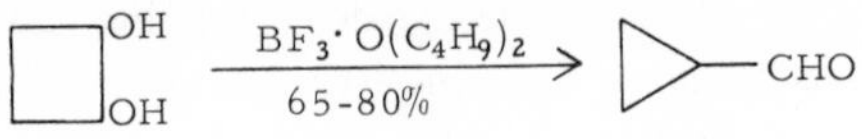

[1]J. P. Barnier, J. Champion, and J. M. Conia, *Org. Syn.*, submitted 1972.
[2]J. M. Conia and J. P. Barnier, *Tetrahedron Letters*, 4981 (1971).

**Boron trifluoride etherate, 1**, 70–72; **2**, 35–36; **3**, 33.

*β-Glucopyranosides.*[1] β-Glucopyranosides can be obtained in good yield by treatment of an aglycone with 2,3,4,6-tetra-O-acetyl-β-D-glucopyranose (1) in the presence of boron trifluoride etherate in 1,2-dichloroethane at a low temperature ($-20°$). The original β-configuration in the glucose moiety is preserved in the glucoside, but the

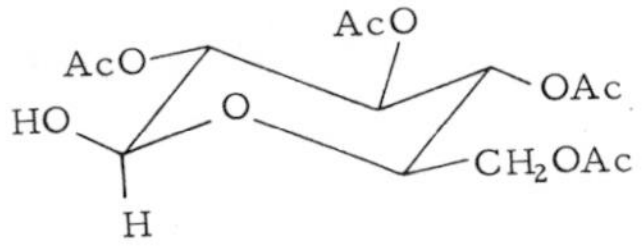

(1)

configuration at the aglycone hydroxyl group often is not, because the reaction proceeds through a carbonium ion.

*Cleavage of* t-*BOC peptide derivatives.*[2] The *t*-butyloxycarbonyl protective group is usually cleaved by trifluoroacetic acid or hydrogen chloride. It has been reported recently that cleavage can also be effected by boron trifluoride etherate in either glacial acetic acid or acetic acid–chloroform mixtures. Exclusion of moisture is essential. The reaction proceeds readily at room temperature. The new procedure is advantageous when a strongly acidic solvent is a disadvantage.

*Intramolecular addition of diazoketones to olefins.* Erman and Stone[3] have reported the first example of an acid-catalyzed addition of a diazoketone to an olefin. The reaction has useful applications for the synthesis of bicyclic ketones, particularly in the sesquiterpene series. Thus treatment of the diazoketone (1) with boron trifluoride etherate for 3 hr. at 0–27° yields the ketones (2) and (3) in yields of about 30 and 3%,

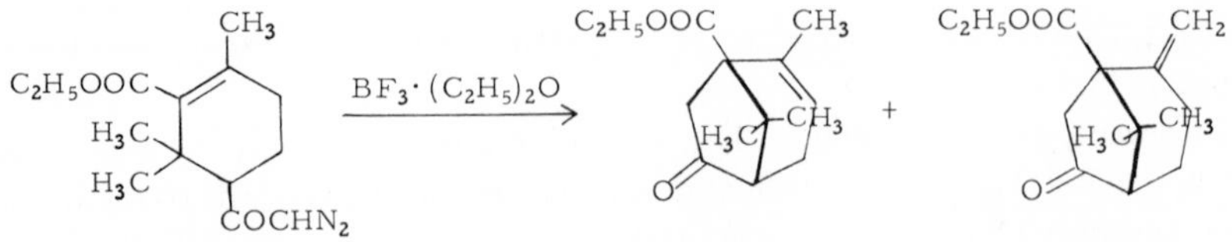

(1)                    (2, 30%)                    (3, 3%)

respectively. The new reaction is simpler, therefore, than the more usual copper-catalyzed thermal cyclization to a cyclopropyltricyclic ketone followed by acid-catalyzed cleavage to a bicyclic ketone.

*Esterification.* Boron trifluoride etherate in combination with a large excess of an alcohol is an effective reagent for the esterification of 4-aminobenzoic acid.[4] The combination is particularly useful for esterification of unsaturated carboxylic acids.[5]

The boron trifluoride etherate–alcohol reaction is particularly useful for esterification of heterocyclic carboxylic acids.[6]

*Cyclization of costunolide.*[7] Treatment of costunolide (1) with $BF_3 \cdot (C_2H_5)_2O$ at room temperature for 1–10 min. affords three products, (2), (3), and (4), in yields of 28,

(1)                (2, *exo*-isomer)              (4)

(3, *endo*-isomer)

37, and 1.5%, respectively. The formation of (4), 4α-hydroxycyclocostunolide, is interesting because this sesquiterpene could conceivably arise by enzymatic addition of water to (1).

*Steroidal O-tetrahydropyran-2-yl derivatives.* Ott *et al.*[8] prepared the O-tetrahydropyran-2-yl derivative of testosterone by the reaction of testosterone with 2,3-dihydropyrane catalyzed by *p*-toluenesulfonic acid monohydrate. The reaction required three weeks and the yield was 59%. Use of boron trifluoride etherate as catalyst raises the yield to 67% and lowers the reaction time to 10 hr.[9]

[1]M. Kuhn and A. von Wartburg, *Helv.*, **51**, 1631 (1968).
[2]R. G. Hiskey, L. M. Beacham (III), V. G. Matl, J. N. Smith, E. B. Williams, Jr., A. M. Thomas, and E. T. Wolters, *J. Org.*, **36**, 488 (1971).
[3]W. F. Erman and L. C. Stone, *Am. Soc.*, **93**, 2821 (1971).
[4]P. K. Kadaba, M. Carr, M. Tribo, J. Triplett, and A. C. Glasser, *J. Pharm. Sci.*, **58**, 1422 (1969).
[5]P. K. Kadaba, *Synthesis*, 316 (1971).
[6]*Idem, ibid.*, 628 (1972).
[7]T. C. Jain and J. E. McCloskey, *Tetrahedron Letters*, 1415 (1971).
[8]A. C. Ott, M. F. Murray, and R. L. Pederson, *Am. Soc.*, **74**, 1239 (1952).
[9]H. Alper and L. Dinkes, *Synthesis*, 81 (1972).

## Boron trifluoride–Trifluoroacetic anhydride.

*Pomeranz–Fritsch reaction.* In the Pomeranz–Fritsch reaction,[1] isoquinolines are formed by the cyclization of benzalaminoacetals (1) with concentrated sulfuric acid as

(1)

catalyst. In recent work, improved yields have been obtained with use of boron trifluoride in trifluoroacetic anhydride.[2] The reagent has been used in the same way for formation of N-substituted indoles and benzthiophene.

[1]W. J. Gensler, *Org. React.*, **6**, 191 (1951).
[2]M. J. Bevis, E. J. Forbes, N. N. Naik, and B. C. Uff, *Tetrahedron*, **27**, 1253 (1971).

**Boron tris(trifluoroacetate) (BTFA)**, $B(OOCCF_3)_3$. Mol. wt. 349.88.
   *Preparation.*[1,2] The reagent is prepared by the reaction of $BBr_3$ and $CF_3COH$

$$3 \; CF_3COOH + BBr_3 \longrightarrow B(OOCCF_3)_3 + 3HBr$$

in $CH_2Cl_2$ at 0°. A precipitate of the reagent is formed; the solvent is evaporated under vacuum at 20°.

   **Removal of protective groups in peptide synthesis.**[2] The reagent, dissolved in trifluoroacetic acid, removes various acid-labile N-protecting groups (Cb, BOC, etc.) at 0°. It also removes nitro, tosyl, or *p*-methoxybenzyl groups used for protection of side-chain groups.

   The new reagent thus is similar to liquid HF.

[1]W. Gerrard, M. F. Lappert, and R. Schafferman, *J. Chem. Soc.*, 3648 (1958).
[2]J. Pless and W. Bauer, *Angew. Chem., internat. Ed.*, **12**, 147 (1973).

**Bromine**, 3, 34. Suppliers: Alfa, J. T. Baker, Columbia, Fisher, Howe and French, Koch-Light, MCB, E. Merck, Riedel-de Haën, Sargent, Schuchardt.

   *Cyclobutane-1,2-dione.*[1] Attempts to prepare cyclobutane-1,2-dione by oxidation of the readily accessible α-hydroxycyclobutanone have all failed. Heine now has prepared the dione by bromination of 1,2-bistrimethylsiloxycyclobutene-1 (1).

   To 46 g. of (1) and 50 ml. of absolute isopentane, a solution of 32 g. of bromine in 50 ml. of isopentane was added with stirring at $-70°$ in 20 min. On warming the weakly yellow solution to $-10°$ yellow crystals began to separate with the eventual yield of 11.7 g. (70%), m.p. 67–68°.

The preparation was reported simultaneously by French chemists.[2]

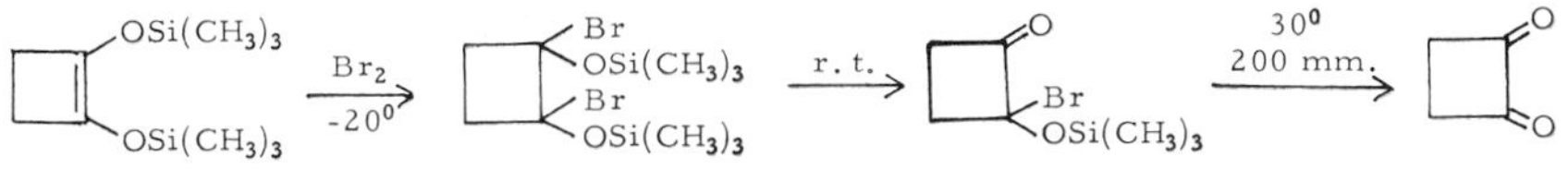

(1)          (a)          (b)          (2)

The reaction is carried out first at $-20°$, then at room temperature, and finally with vacuum distillation. The course of the reaction can be followed by IR and NMR examination. The yield of (2) is essentially quantitative.

[1]H.-G. Heine, *Ber.*, **104**, 2869 (1971).
[2]J. M. Conia and J. M. Denis, *Tetrahedron Letters*, 2845 (1971).

**N-Bromoacetamide (NBA)**, **1**, 74; **2**, 39.

*Reaction with olefins.* Wohl[1] in 1919 briefly examined the reaction of tetra-methylethylene with NBA and tentatively suggested that the reaction involved allylic bromination. Wolfe and Awang[2] have examined the reaction of NBA with olefins and find that the reaction does not involve allylic bromination but furnishes 2-bromo-N-bromoacetimidates, a new class of compounds. Thus the reaction of cyclohexene (1)

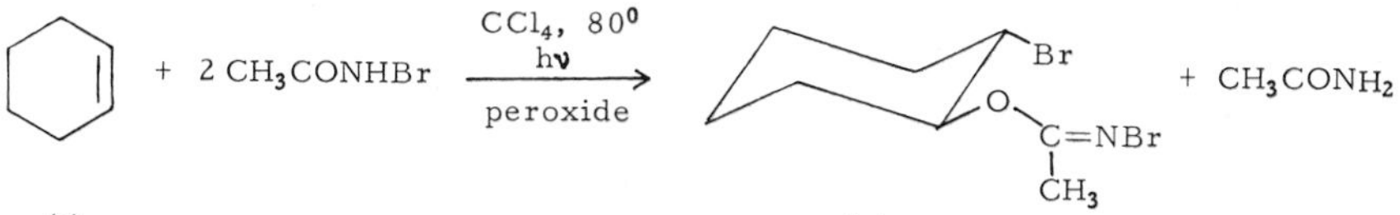

(1)                              (2)

with NBA in refluxing carbon tetrachloride under illumination gives as the major product (36 % yield) 2-bromocyclohexyl-N-bromoacetimidate (2). In the same way, the reaction of tetramethylethylene (3) gives (4).

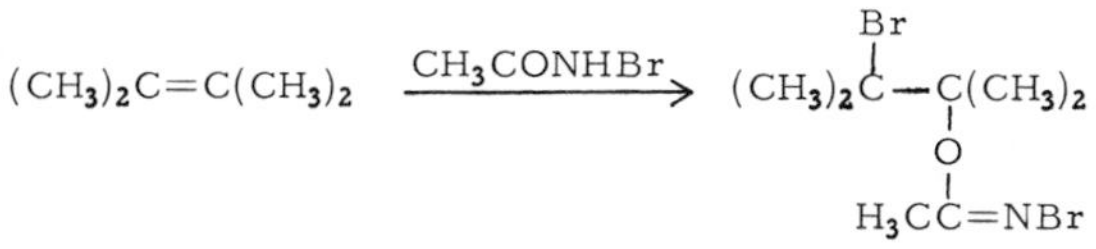

(3)                              (4)

The process appears to proceed in two steps: a free radical reaction of NBA with itself to form N,N-dibromoacetamide (NDBA), followed by an ionic addition of NDBA to the double bond.

*Bufalin.* The last steps in the synthesis of bufalin acetate (3) involve conversion of 14-dehydrobufalin acetate (1) into (3). This is accomplished conveniently by treatment of (1) with NBA in aqueous acetone followed by reduction of the resulting bromohydrin

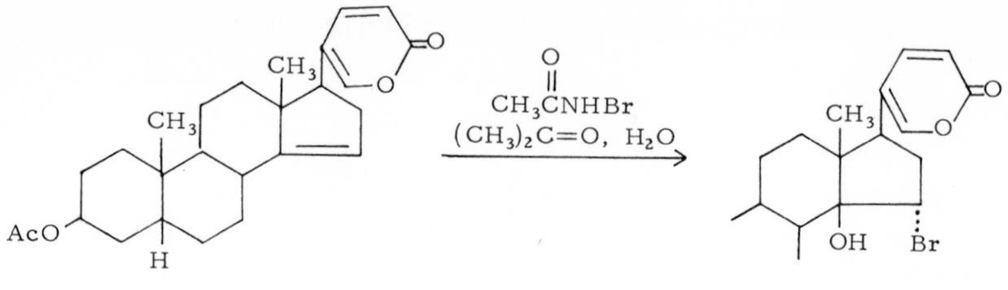

(1)                                                    (2)

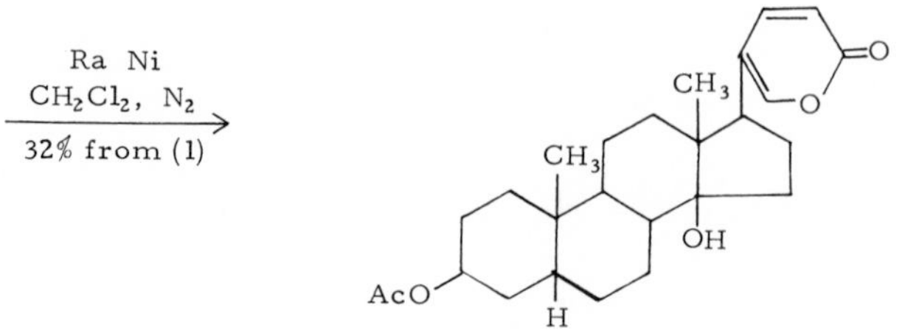

(3)

(2) with Raney nickel in methylene chloride. The acetyl group of (3) can be hydrolyzed with hydrochloric acid in aqueous methanol (20°, 48 hr., 68% yield).[3]

[1] A. Wohl, *Ber.*, **52**, 51 (1919).
[2] S. Wolfe and D. V. C. Awang, *Canad. J. Chem.*, 1384 (1971).
[3] F. Sondheimer and R. L. Wife, *Tetrahedron Letters*, 765 (1973).

**N-Bromoacetamide–Hydrogen fluoride, 1, 75; 2, 39–40.**

*Addition of bromine fluoride to acetylenes.*[1] Bromine fluoride reacts with terminal alkynes by Markownikoff addition:

$$CH_3(CH_2)_3C{\equiv}CH \xrightarrow{\ BrF\ } CH_3(CH_2)_3CF{=}CHBr$$

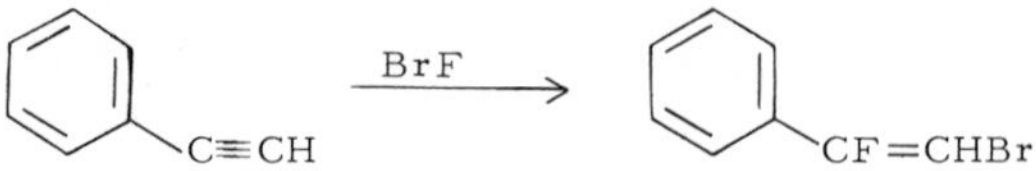

The mode of addition is predominantly *trans*. Electron-withdrawing groups inhibit the reaction; thus $CH_3OC(CF_3)_2C{\equiv}CH$ and $CH_3OC(CF_3)_2C{\equiv}CCl$ fail to react. Dimethyl acetylenedicarboxylate gives intractable products.

[1] R. E. A. Dear, *J. Org.*, **35**, 1703 (1970).

**N-Bromopolymaleimide (PNBS),**

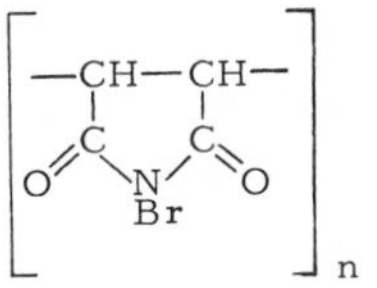

$$\left[\begin{array}{c} -CH-CH- \\ O{=}C \quad C{=}O \\ N \\ Br \end{array}\right]_n$$

Polymaleimide is prepared by free-radical polymerization of maleimide in the presence of divinylbenzene (2.5–5 %) as cross-linking agent. PNBS is obtained by bromination of the polymer in aqueous sodium hydroxide solution.

*Bromination.*[1] Bromination of cumene (1) with PNBS in boiling carbon tetrachloride (dibenzoyl peroxide) resulted in the three products shown. The results contrast with those obtained with NBS, which leads to (5) and (6), the latter being the main

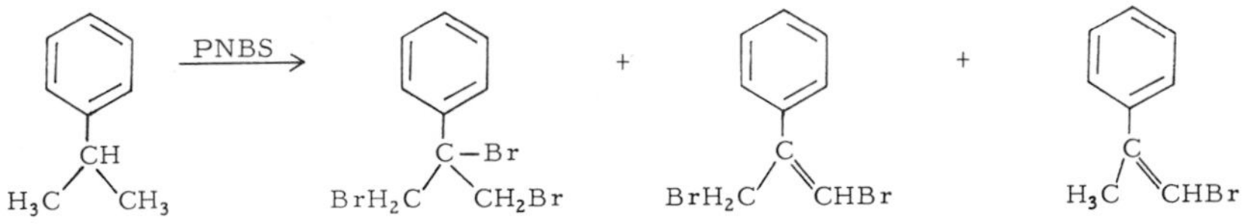

| (1) | (2, 48%) | (3, 15%) | (4, 13%) |

| (5) | (6) |

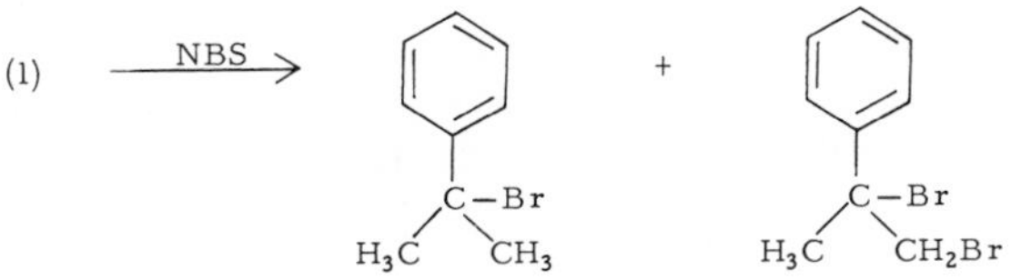

product when a large excess of NBS is used. Differences between PNBS and NBS were also observed in the case of other alkylbenzenes. However, bromination with NBS in a polar medium resembles that of PNBS in carbon tetrachloride. Katchalski attributes the unusual reactions of PNBS to the polymeric structure. *Compare* **N-Chloropolymaleimide,** this volume.

[1]C. Yaroslavsky, A. Patchornik, and E. Katchalski, *Tetrahedron Letters,* 3629 (1970).

**N-Bromosuccinimide, 1,** 78–80; **2,** 40–42; **3,** 34–36.

*Hypobromous acid,* HOBr **(1,** 80). Erickson and Kim[1] generated hypobromous acid from NBS and water and showed that it reacts with methylenecyclobutane (1, Aldrich) to give 1-(bromomethyl)cyclobutanol (2) in 78 % yield of about 90 % purity. The reaction thus is regioselective.

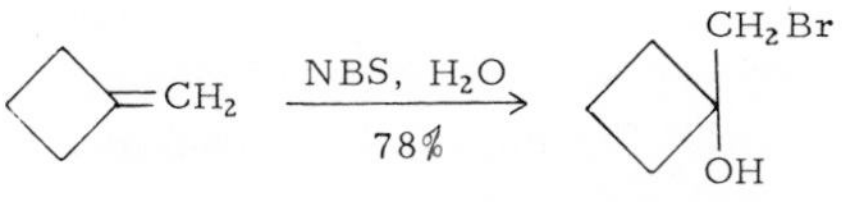

| (1) | (2) |

Japanese chemists[2] have reported that hypobromous acid generated from NBS and water reacts with the 4′,5′-didehydroadenosine derivative (3) to give (4) as the major product. The formation of (4) represents the first instance of N,4′-cyclization with concomitant ring cleavage of the base moiety in a purine nucleoside.

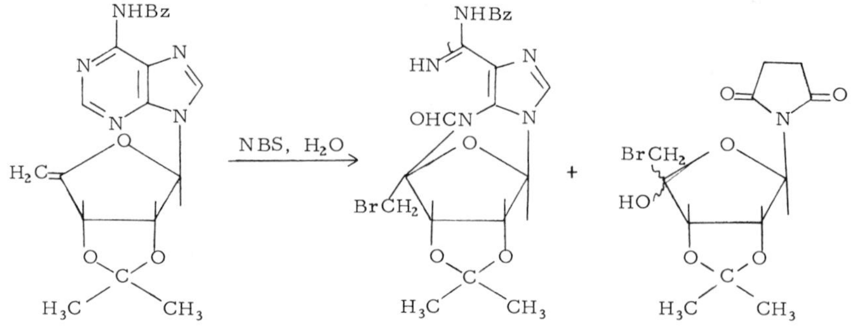

(3)                    (4, 20%)                    (5, 12%)

*Selective epoxidation.* van Tamelen and Curphey[3] noted that NBS in an aqueous polar solvent (1,2-dimethoxyethane) generated HOBr, and that this reagent showed some selectivity in reaction with double bonds. The method was used for selective epoxidation of the terminal double bond of farnesyl acetate (1) to give 10,11-epoxyfarnesyl acetate (4).[4] Complete details of this transformation have been described by Hanzlik.[5]

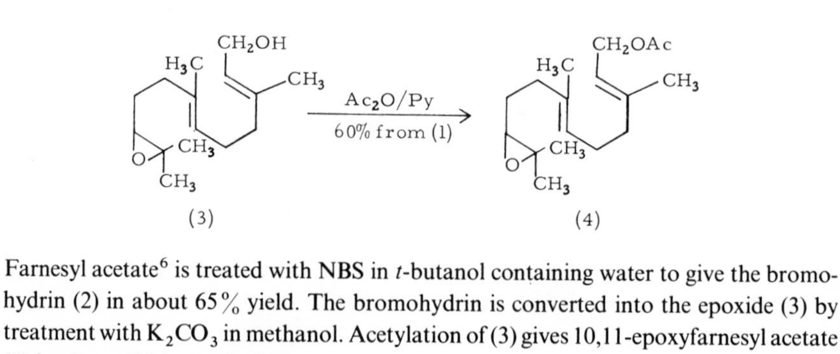

(1)                    (2)

(3)                    (4)

Farnesyl acetate[6] is treated with NBS in *t*-butanol containing water to give the bromohydrin (2) in about 65% yield. The bromohydrin is converted into the epoxide (3) by treatment with $K_2CO_3$ in methanol. Acetylation of (3) gives 10,11-epoxyfarnesyl acetate (4) in about 60% overall yield.

The method has been used in biogenetic-type synthesis of di- and triterpenes[7]; it has also been used in the synthesis of *Cecropia* juvenile hormone.[8]

*Allylic oxidation* (**3**, 35). Dr. Thomson[9] has informed us that dioxane is the only suitable solvent for allylic oxidation and that high yields are obtainable only when the double bond is highly hindered. For example, 4,4-dimethylcholesteryl acetate gives a quantitative yield of the 7-keto derivative, and cholesteryl acetate gives 7-ketocholesteryl acetate in 80% yield. However, cholestene-1, -2, and -3 give no products of allylic oxidation, but rather bromohydrins, dibromides, and transformation products therefrom. The use of calcium carbonate is not necessary.

*Oxidation of acetylenes.*[10] The addition of 2 mole eq. of NBS to a solution of diphenylacetylene in DMSO leads to the formation of benzil in 98% yield. DMSO is

required as solvent and NBS is unique in its ability to catalyze the reaction [molecular bromine and pyrrolidone-2-hydrotribromide (**3**, 240–241) are ineffective].

*Synthesis of karahanaenone.*[11] The ionic reaction of NBS in $CCl_4$ at room temperature with a $\gamma$-ethylenic tertiary alcohol leads to an $\alpha$-bromotetrahydrofurane. The reaction has been used in a convenient synthesis of karahanaenone (**4**), a constituent of hop oil, from linalool (**1**). Thus reaction of (**1**) with NBS affords 2-methyl-2-vinyl-5-(1-bromo-1-methylethyl)tetrahydrofurane (**2**) in 85% yield. Dehydrohalogenation of (**2**) with collidine at 110° leads to the allyl vinyl ether (**3**), which immediately

rearranges through a [3,3]-sigmatropic process to karahanaenone (**4**), which is obtained in 62% overall yield from (**1**).

*Reaction with trialkylboranes.* Lane and Brown[12] in 1971 reported a simple procedure for synthesis of highly substituted alcohols from trialkylboranes. For example, bromination of triethylborane (**1**) under irradiation in the presence of water followed by oxidation with alkaline hydrogen peroxide gives 3-methyl-3-pentanol (**2**) in 88% yield. The reaction involves as the first step free-radical bromination in the $\alpha$-position

$$(C_2H_5)_3B \xrightarrow[\text{H}_2\text{O}]{2 \text{ Br}_2,\text{ h}\nu} \underset{CH_3CH_2}{\overset{CH_3CH_2}{CH_3\overset{|}{\underset{|}{C}}B(OH)_2}} \xrightarrow[88\%]{H_2O_2,\text{ OH}^-} \underset{CH_3CH_2}{\overset{CH_3CH_2}{CH_3\overset{|}{\underset{|}{C}}-OH}}$$

(1)                                                              (2)

to give an α-bromoorganoborane (a), which undergoes facile rearrangement in the presence of water to give the borinic acid (b). This is then α-brominated (c) and in the presence of water rearranges to (d). Oxidation then yields the alcohol (2).

$$(C_2H_5)_3B \xrightarrow{Br_2,\text{ h}\nu} \left[ \underset{Br}{\overset{}{CH_3\overset{|}{\underset{|}{CH}}B(C_2H_5)_2}} \right. \xrightarrow[-HBr]{H_2O} \underset{OH}{\overset{C_2H_5}{CH_3\overset{|}{\underset{|}{CH}}BC_2H_5}} \xrightarrow{Br_2,\text{ h}\nu}$$

(1)                        (a)                          (b)

$$\left. \underset{Br\ \ OH}{\overset{C_2H_5}{CH_3\overset{|}{\underset{|\ \ |}{C}}-B-C_2H_5}} \right] \xrightarrow[-HBr]{H_2O} \underset{C_2H_5}{\overset{C_2H_5}{H_3C-\overset{|}{\underset{|}{C}}-B(OH)_2}} \xrightarrow{H_2O_2,\text{ OH}^-} (2)$$

(c)                          (d)

In order to effect successful α-bromination–migration, it is important to add the bromine slowly in order to minimize polybromination. Brown and Yamamoto[13] have reported recently that use of NBS in the presence of water in the reaction above leads to increased yields: (1) → (2), 97%. Bromine is known to be produced in small concentrations in the reaction of NBS with hydrogen bromide (which is produced in the

reaction process). Sodium bromate (**1**, 1055–1056) also serves satisfactorily as a bromine source.

Other examples:

$$(\underline{n}\text{-}C_4H_9)_3B \xrightarrow[78\%]{2 \text{ NBS}} \underset{OH}{\overset{CH_2CH_2CH_3}{CH_3(CH_2)_3\overset{|}{\underset{|}{C}}(CH_2)_3CH_3}}$$

5-Propyl-5-nonanol

1-Cyclohexylcyclohexanol

[1]K. L. Erickson and K. Kim, *J. Org.*, **36**, 2915 (1971).
[2]T. Sasaki, K. Minamoto, and K. Hattori, *Am. Soc.*, **95**, 1350 (1973).

[3]E. E. van Tamelen and T. J. Curphey, *Tetrahedron Letters*, 121 (1962).

[4]E. E. van Tamelen, A. Storni, E. J. Hessler, and M. Schwartz, *Am. Soc.*, **85**, 3295 (1963).

[5]R. P. Hanzlik, *Org. Syn.*, submitted 1973.

[6]A mixture of *trans, trans-* and *cis, trans-*isomers, or pure *trans, trans-*farnesyl acetate.

[7]E. E. van Tamelen, *Accts. Chem. Res.*, **1**, 111 (1968); E. E. van Tamelen *et al.*, *Am. Soc.*, **94**, 8225, 8228, 8229 (1972).

[8]E. J. Corey, J. A. Katzenellenbogen, N. W. Gilman, S. A. Roman, and B. W. Erickson, *ibid.*, **90**, 5618 (1968); E. E. van Tamelen and J. P. McCormick, *ibid.*, **92**, 737 (1970).

[9]Dr. J. B. Thomson, personal communication.

[10]S. Wolfe, W. R. Pilgrim, T. F. Garrard, and P. Chamberlain, *Canad. J. Chem.*, **49**, 1099 (1971).

[11]E. Demole and P. Enggist, *Helv.*, **54**, 456 (1971).

[12]C. F. Lane and H. C. Brown, *Am. Soc.*, **93**, 1025 (1971).

[13]H. C. Brown and Y. Yamamoto, *Synthesis*, 699 (1972).

**1,3-Butadiene monoxide**, $CH_2{=}CHCH\underset{\displaystyle O}{\diagdown \diagup}CH_2$. Mol. wt. 70.09; b.p. 65–66°.

Suppliers: Aldrich, K and K, Pfaltz and Bauer, ROC/RIC.

*Homologation of trialkylboranes.*[1] Trialkylboranes, in the presence of oxygen or other free-radical initiators, react with 1,3-butadiene monoxide to give 4-alkyl-2-butene-1-ols resulting from four-carbon-atom homologation, for example:

$$(C_2H_5)_3B \ + \ CH_2{=}CHCH\underset{\diagdown O \diagup}{-}CH_2 \ \xrightarrow[\substack{68\%}]{\substack{1)\ O_2 \\ 2)\ H_2O}} \ C_2H_5CH_2CH{=}CHCH_2OH \ + \ (C_2H_5)_2BOH$$

(excess)                                            (89% trans)

Benzene and ether are the preferred solvents.

[1]A. Suzuki, N. Miyaura, M. Itoh, H. C. Brown, G. W. Holland, and E. Negishi, *Am. Soc.*, **93**, 2792 (1971).

**3-Butenyltri-*n*-butyltin**, $(n\text{-}C_4H_9)_3SnCH_2CH_2CH{=}CH_2$. Mol. wt. 345.13, b.p. 91–93°/0.35 mm.

*Preparation.*[1] The reagent is prepared in 81% yield by the reaction of tri-*n*-butyltin chloride (Metal and Thermit Corporation) with $CH_2{=}CHCH_2CH_2MgBr$.

*Cyclopropylcarbinyl compounds.*[2] Electrophilic reagents ($Cl_2$, $SO_3$, RSCl, $HgCl_2$) add to 3-butenyltri-*n*-butyltin to give cyclopropylcarbinyl compounds in 72–86% yield.

$$(\underline{n}\text{-}C_4H_9)_3SnCH_2CH_2CH{=}CH_2 \ + \ E^+ \ \longrightarrow$$

$$\triangleright\!\!-CH_2E \ + \ (\underline{n}\text{-}C_4H_9)_3Sn^+$$

[1]D. Seyferth and M. A. Weiner, *Am. Soc.*, **84**, 361 (1962).

[2]D. J. Peterson and M. D. Robbins, *Tetrahedron Letters*, 2135 (1972).

*n*-**Butylamine**, $CH_3(CH_2)_3NH_2$. Mol. wt. 73.14, b.p. 78°, $n_D^{20}$. 4015. Suppliers: Aldrich, J. T. Baker, Columbia, Eastman, Fisher, Howe and French, K and K, Koch-Light, MCB, P and B, Riedel-de Haën, Sargent, Schuchardt.

*Arylacetylenes.* Treatment of 5,5-disubstituted 3-nitroso-2-oxazolidones (1)[1] with *n*-butylamine in ether at room temperature gives arylacetylenes in almost quantitative yield, when at least one of the groups of $C_5$ is an aryl group.[2,3] The method has also been used to prepare 2-ethynylthiophene.[4]

$$R_2-\overset{R_1}{\underset{\underset{NO}{\overset{|}{N}}}{\underset{|}{\overset{5}{C}}}}\overset{O}{\underset{3}{\diagdown}}C=O \quad \xrightarrow[99-100\%]{CH_3(CH_2)_3NH_2} \quad R_1C\equiv CR_2 + CO_2 + H_2O + N_2$$

(1, $R_1$ or $R_2$ = $C_6H_5$)

For other useful transformations of (1) see **3**, 186–187.

[1]Preparation: **1**, 748–749.
[2]H. P. Hogan and J. Seehafer, *J. Org.*, **73**, 4466 (1972).
[3]M. S. Newman and L. F. Lee, *ibid.*, **37**, 4468 (1972).
[4]T. B. Patrick, J. M. Disher, and W. J. Probst, *ibid.*, **37**, 4467 (1972).

*t*-**Butyl azidoformate, 1**, 84–85; **2**, 44–45; **3**, 36.

*Preparation.* Sakai and Anselme[1] recommend that the reagent be prepared by the reaction of *t*-butyl chloroformate and tetramethylguanidinium azide (**2**, 403–404):

$$COCl_2 + HOC(CH_3)_3 \longrightarrow ClCOOC(CH_3)_3$$

$$ClCOOC(CH_3)_3 + \left[(CH_3)_2N\right]_2C\overset{+}{=}\overset{-}{NH_2N_3} \xrightarrow[95-97\%]{} N_3COOC(CH_3)_3 + \left[(CH_3)_2N\right]_2C\overset{+}{=}NH_2Cl^-$$

*Caution*: All operations should be conducted in a good hood and behind a safety shield.

*N-Protection of amino acids* (**3**, 36). *t*-Butyloxycarbonyl (BOC) amino acids (2) can be prepared in 80–100% yield by the reaction of 1,1,3,3-tetramethylguanidinium salts of amino acids (1) with *t*-butyl azidoformate in DMF.[2]

$$(CH_3)_2N-\underset{\overset{\parallel}{\underset{NH_2}{+}}}{C}-N(CH_3)_2 \quad R\underset{\overset{|}{NH_2}}{CHCOO^-} \quad \xrightarrow{(CH_3)_3COCON_3} \quad R\underset{\overset{|}{\underset{COOC(CH_3)_3}{NH}}}{CHCOOH}$$

(1)                                    (2)

[1]K. Sakai and J.-P. Anselme, *J. Org.*, **36**, 2387 (1971); *Org. Syn.*, submitted 1972.
[2]A. Ali, F. Fahrenholz, and B. Weinstein, *Angew. Chem.*, *internat. Ed.*, **11**, 289 (1972).

*t*-**Butyl chromate, 1**, 86–87; **2**, 48; **3**, 37.

*Oxidation of* (+)-*car-3-ene.*[1] *t*-Butyl chromate is superior to permanganate or chromium trioxide for oxidation of (+)-car-3-ene (1) to (−)-car-3-ene-5-one (2).

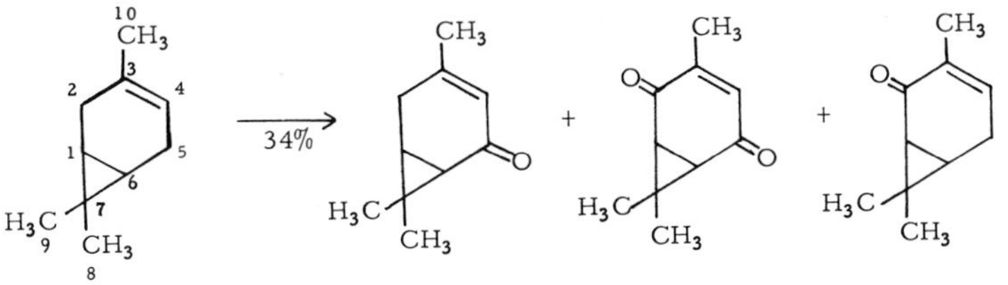

(1)          (2, 50%)        (3, 11. 5%)      (4, 11. 7%)

Among other products (3) and (4) are formed in significant amounts. The major products are readily separated by column chromatography.

[1]P. H. Boyle, W. Cocker, and D. H. Grayson, *J. Chem. Soc.* (*C*), 1073 (1971).

*t*-Butylcyanoketene,    $(CH_3)_3C$ $C=C=O$. Mol. wt. 123.15.    $NC$

*Preparation.*[1]  This ketene (2) is obtained in nearly quantitative yield by thermal cleavage of 2,5-diazido-3,6-di-*t*-butyl-1,4-benzoquinone (1)[2] in refluxing benzene. This

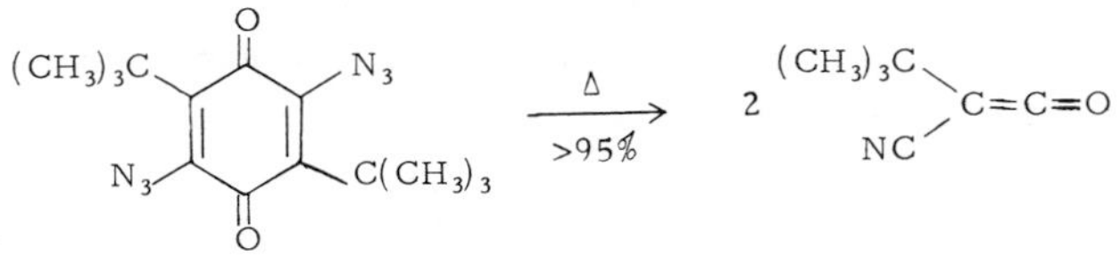

(1)                          (2)

ketene is unusually stable; only minor decomposition is noted when the ketene is kept in benzene solution at room temperature for eight days.

[2 + 2] *Cycloadditions.* Moore and Weyler[1] noted that *t*-butylcyanoketene (2) is unusually reactive in cycloaddition reactions. Thus it reacts with cyclohexene (3) to form the cyclobutanone (4) in 63 % isolated yield. Note that phenylcyanoketene does not react with cyclohexene.[3]

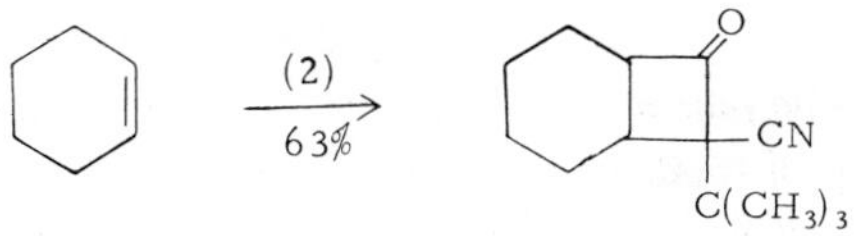

(3)                          (4)

The ketene reacts with *cis*- and *trans*-cyclooctene stereospecifically to give the cyclobutanones (5) and (6), respectively.[4]

(5)                              (6)

The ketene (2) reacts rapidly at room temperature with the cyclic allene 1,2-cyclo-nonadiene (7) to give a mixture of the diastereomeric cyclobutanones (8) and (9) in a ratio of 3:2.

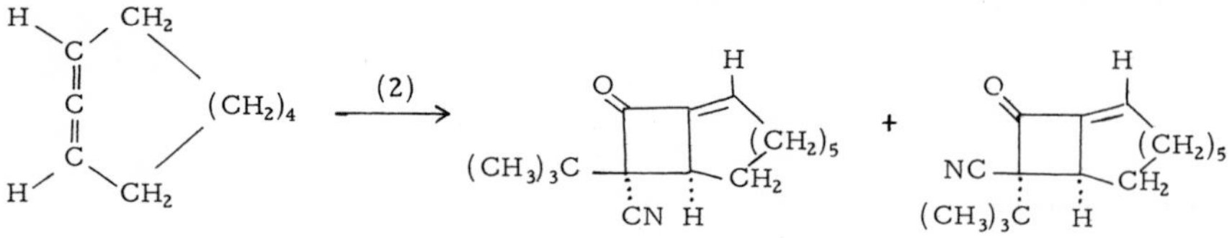

(7)                              (8)        3:2        (9)

When partially resolved (7) was used, both (8) and (9) showed appreciable optical activity.

The ketene (2) also undergoes cycloaddition to acetylenes to give 2-cyclobutene-1-one derivatives in 40–80% yield.[5] Thus *t*-butylcyanoketene reacts with phenylacetylene in benzene at room temperature to give the cyclobutenone (10). Only one of the two possible products of cycloaddition is formed; thus the reaction is regiospecific. The structure

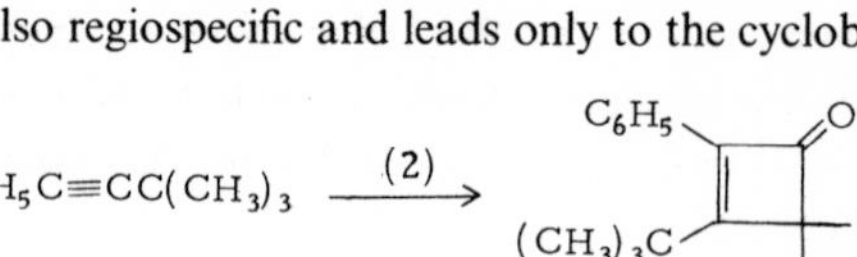

(10)

of (10) was established by spectroscopic data. The reaction of the ketene with *t*-butyl-phenylacetylene is also regiospecific and leads only to the cyclobutenone (11).

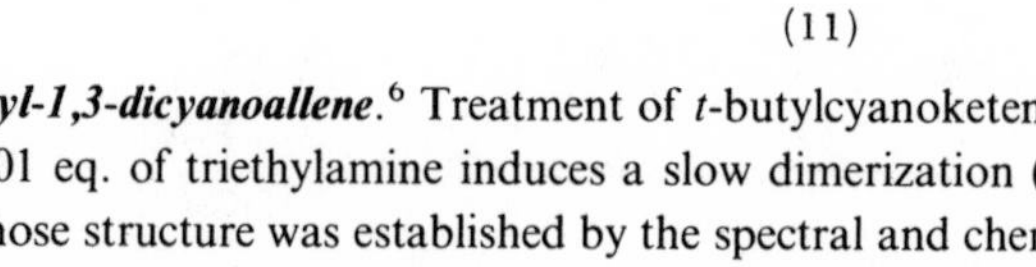

(11)

**1,3-Di-t-butyl-1,3-dicyanoallene.**[6] Treatment of *t*-butylcyanoketene (1) in benzene solution with 0.01 eq. of triethylamine induces a slow dimerization (8–12 hr.) to the β-lactone (2), whose structure was established by the spectral and chemical properties. On treatment of (2) with 0.1 eq. of triethylamine at room temperature, carbon dioxide is evolved, with formation of 1,3-di-*t*-butyl-1,3-dicyanoallene (3, 68% yield). The allene (3) can be obtained directly from (1) by treatment with 0.1 eq. of triethylamine (50% yield).

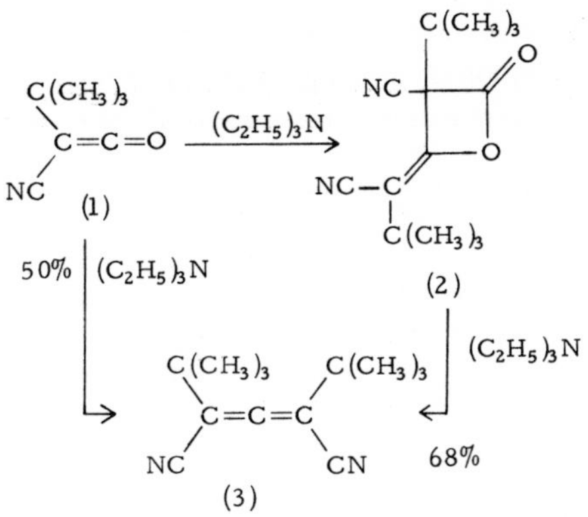

[1]H. W. Moore and W. Weyler, Jr., *Am. Soc.*, **92**, 4132 (1970).
[2]*Idem, ibid.*, **93**, 2812 (1971).
[3]R. C. DeSelms, *Tetrahedron Letters*, 1179 (1969).
[4]W. Weyler, Jr., L. R. Byrd, M. C. Caserio, and H. W. Moore, *Am. Soc.*, **94**, 1027 (1972).
[5]M. D. Gheorghiu, C. Drăghici, L. Stănescu, and M. Avram, *Tetrahedron Letters*, 9 (1973).
[6]H. W. Moore and W. G. Duncan, *J. Org.*, **38**, 156 (1973).

*t*-**Butyldimethylchlorosilane**, $(CH_3)_3CSiCl$.    Mol. wt. 150.73, bp. 125°/733 mm., m.p. 91.5°. Supplier: WBL.

*Preparation.*[1,2] The reagent is prepared by the reaction of *t*-butyllithium with dimethyldichlorosilane in pentane (reflux 2 hr.).

*Protection of hydroxyl groups.*[1,2] The *t*-butyldimethylsilyl (TBDMS) group is recommended for the protection of the 5′-primary hydroxyl groups of nucleosides. For example, the reaction of thymidine (1) with *t*-butyldimethylchlorosilane in DMF containing imidazole (2.5 molar eq., **1**, 492–494; **2**, 220) gives the ether (2) in 60% yield.

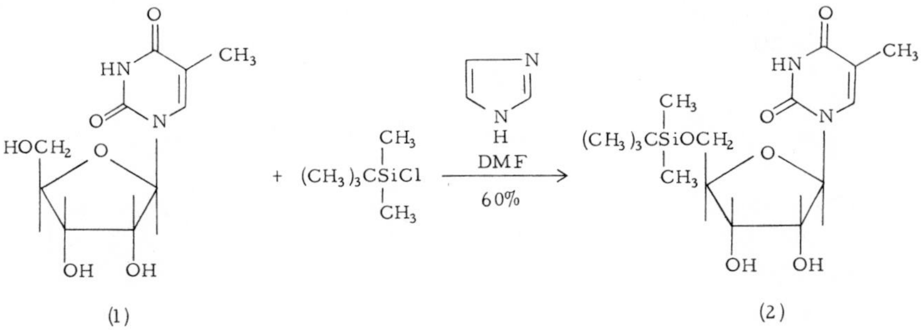

The TBDMS group is stable to weak bases (15% $NH_4OH$ in ethanol) and to normal conditions of phosphorylation. It is readily removed by 80% acetic acid (steam bath, 15 min.) or by tetra-*n*-butylammonium fluoride (22°, 30 min., this volume).

[1]L. H. Sommer and L. J. Tyler, *Am. Soc.*, **76**, 1030 (1954).
[2]K. K. Ogilvie and D. J. Iwacha, *Tetrahedron Letters*, 317 (1973).

**t-Butyl hypochlorite**, **1**, 90–94, **2**, 50; **3**, 38.

   *Chlorination of aminoquinones.*[1] 2-Amino-1,4-quinones (1) are selectively chlorinated in an ethereal or a benzene solution by 1 eq. of *t*-butyl hypochlorite at room

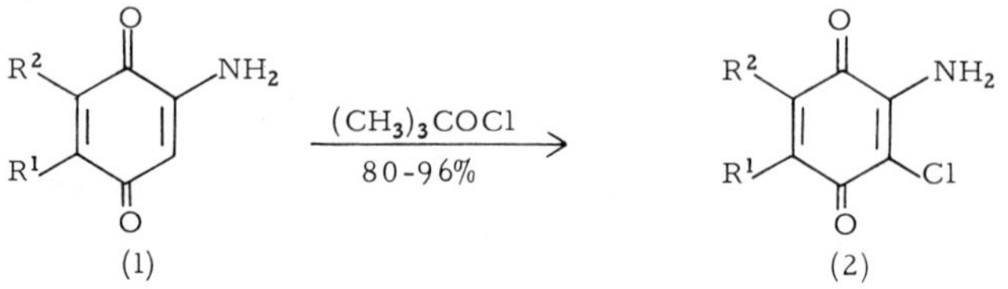

(1)                                        (2)

temperature within minutes. Chlorination occurs specifically at the position adjacent to the amino function. Quinones lacking an amino group fail to react.

   *Azoalkanes.* Timberlake *et al.*[2] recommend the following procedure for synthesis of azoalkanes (4) from N,N'-dialkylsulfamides (1).[3] Overall yields are 35–85%.

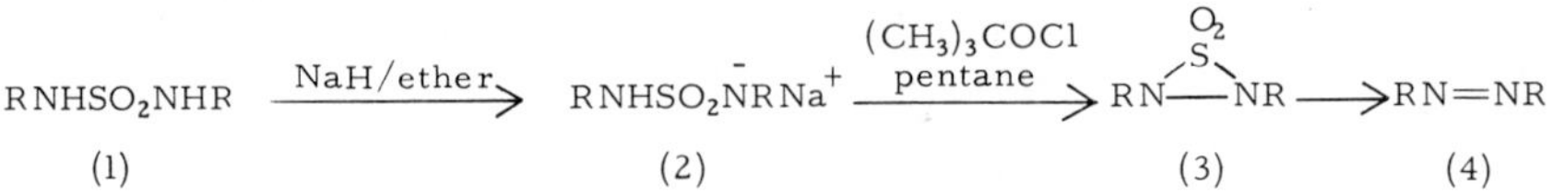

$$RNHSO_2NHR \xrightarrow{\text{NaH/ether}} RNHSO_2\overset{-}{N}RNa^+ \xrightarrow[\text{pentane}]{(CH_3)_3COCl} RN\overset{O_2}{\underset{|}{S}}NR \longrightarrow RN{=}NR$$

(1)                    (2)                    (3)        (4)

**Reaction with 1,3-dioxo-2-diazo-1,2-dihydrophenalene**[4] :

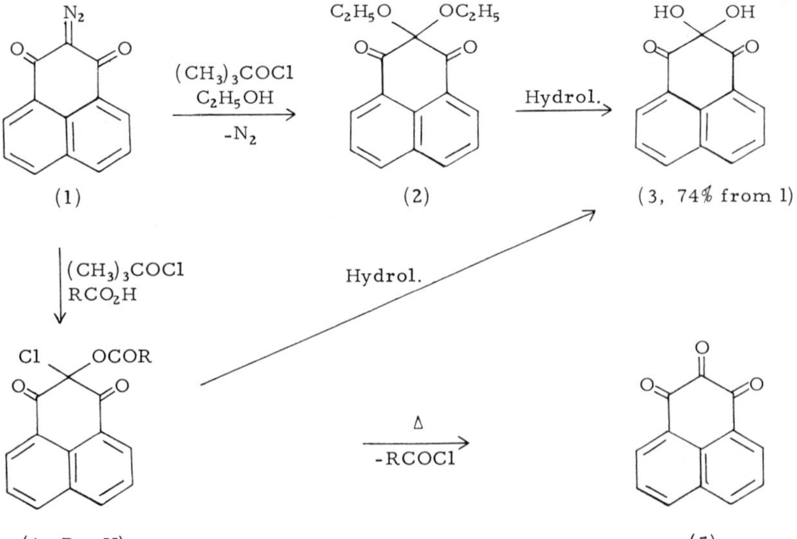

(1)                         (2)                    (3, 74% from 1)

(4, R = H)                                        (5)

*6-Methoxylation of penicillin derivatives*. Oxidation of the anhydropenicillin (1) with *t*-butyl hypochlorite in methanol containing sodium perborate (1, 1102) as buffer gives the 6-methoxy derivative (2) in nearly quantitative yield.[5] The thiazolidine sulfur atom can be protected either as the sulfone or sulfoxide.

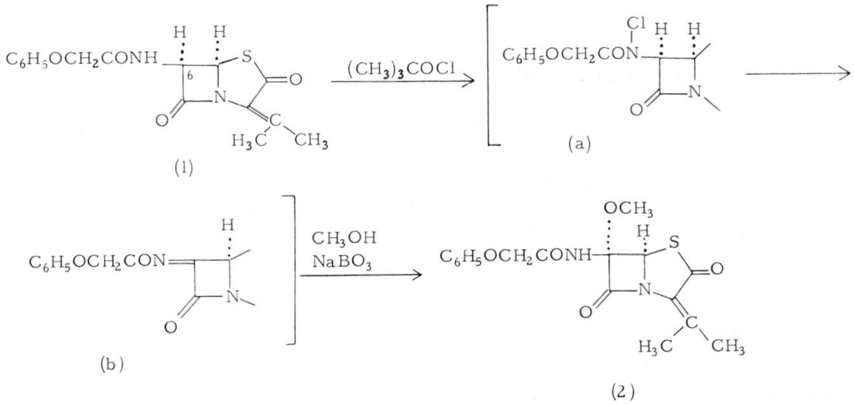

$C_6H_5OCH_2CONH$ ... (1) $\xrightarrow{(CH_3)_3COCl}$ $C_6H_5OCH_2CON$ ... (a) $\longrightarrow$

$C_6H_5OCH_2CON$ (b) $\xrightarrow[NaBO_3]{CH_3OH}$ $C_6H_5OCH_2CONH$ ... (2)

6-Methoxypenicillins can also be prepared by another procedure.[6] Thus treatment of (3) with lithium methoxide in THF at $-80°$ generates the amide anion (c); this on treatment with *t*-butyl hypochlorite and then with acetic acid gives (4) in 70% yield.

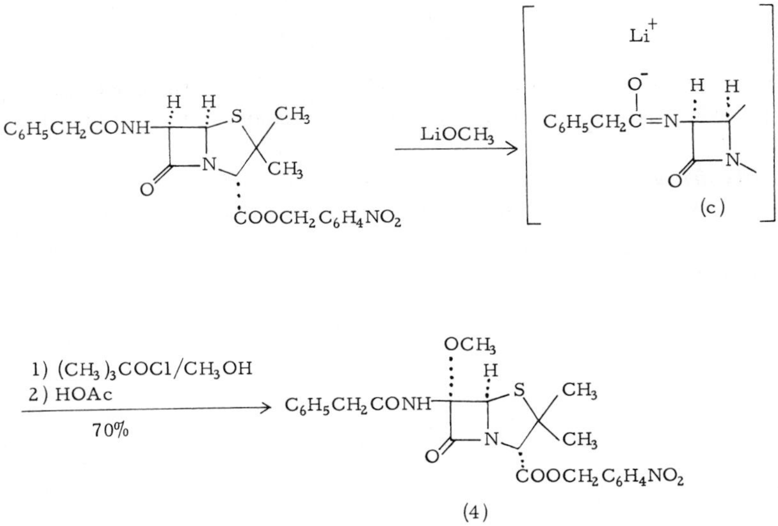

$C_6H_5CH_2CONH$ ... $COOCH_2C_6H_4NO_2$ $\xrightarrow{LiOCH_3}$ $C_6H_5CH_2C=N$ ... Li$^+$ (c)

1) $(CH_3)_3COCl/CH_3OH$
2) HOAc
$\xrightarrow{70\%}$ $C_6H_5CH_2CONH$ ... $COOCH_2C_6H_4NO_2$ (4)

The same procedure can be used for 7α-methoxylation of cephalosporins. Thus (5) is converted into (6) in 73% yield.

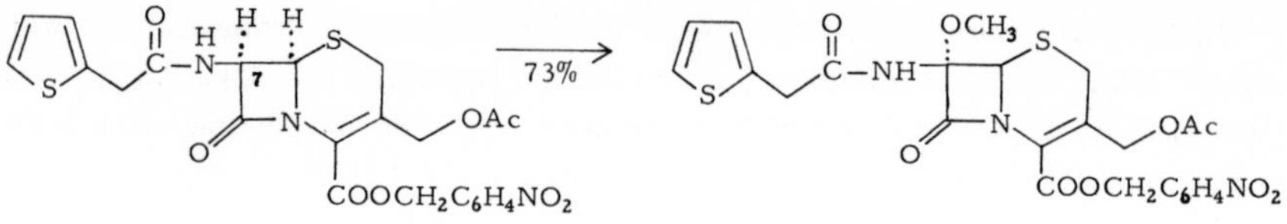

(5)                                      (6)

[1]H. W. Moore and G. Cajipe, *Synthesis*, 49 (1973).
[2]J. W. Timberlake, M. L. Hodges, and K. Betterton, *ibid.*, 632 (1972).
[3]For preparation see R. Ohme and H. Preuschhoff, *Ann.*, **713**, 74 (1968); J. C. Stowell, *J. Org.*, **32**, 2360 (1967).
[4]M. Regitz and H.-G. Adolph, *Ann.*, **723**, 47 (1969).
[5]J. E. Baldwin, F. J. Urban, R. D. G. Cooper, and F. L. Jose, *Am. Soc.*, **95**, 2401 (1973); see also, R. A. Firestone and B. G. Christensen, *J. Org.*, **28**, 1437 (1973).
[6]G. A. Koppel and R. E. Koehler, *Am. Soc.*, **95**, 2403 (1973).

**n-Butyllithium, 1**, 95–96; **2**, 51–53.

*Addition to aldehydes and ketones.*[1] Maximum yields of addition products of aldehydes and ketones with *n*-butyllithium are obtained by addition of the carbonyl compound in hexane or ether to the lithium reagent at $-78°$. Under these conditions only traces of products resulting from reduction or enolization are formed. Quantitative

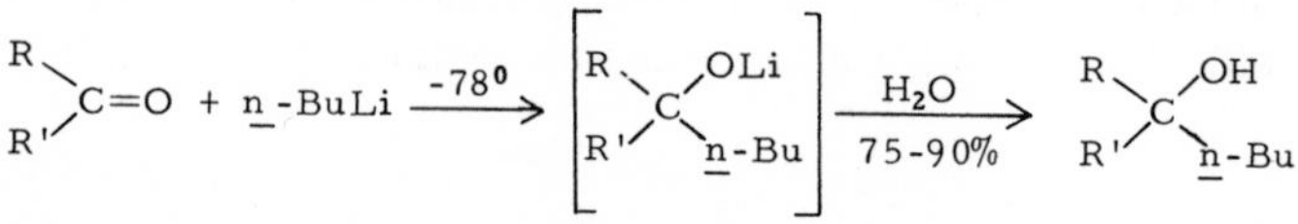

yields of alcohols are obtained from nonenolizable aldehydes and ketones. Somewhat lower yields of alcohols are obtained under these conditions with use of *t*-butyllithium (**1**, 16–97).

*Desulfurization of episulfides.* Desulfurization of *cis*- and *trans*-2-butene episulfides by *n*-butyllithium proceeds with complete stereospecificity to give *cis*- and *trans*-2-butene, respectively. Trost[2] suggests that the reaction may proceed through a sulfurane intermediate (a).

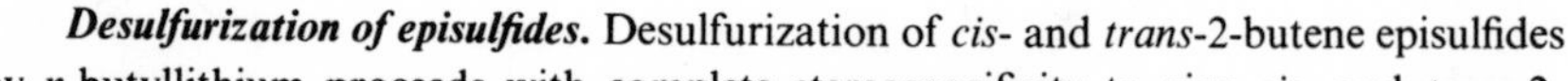

(a)

Desulfurization of episulfides with diiron nonocarbonyl (**1**, 259–260; **2**, 139–140; **3**, 101) is not so stereospecific; the loss (2–6 %) of stereospecificity is attributed to subsequent olefin isomerization.

$\alpha,\beta$-*Ethylenic sulfones.*[3] The reaction of a sulfonomethylphosphonate ester (1) with *n*-butyllithium in THF at $-78°$ generates the anion, which reacts with aliphatic and aromatic aldehydes or ketones to give alkylidene sulfones (2) in 70–97% yield.

$$(C_2H_5O)_2POCH_2SO_2R \xrightarrow[\substack{70\text{-}97\%}]{\substack{1)\ n\text{-BuLi, THF, }-78^0 \\ 2)\ \overline{R}^1R^2C=O}} R^1R^2C=CHSO_2R$$

$(1,\ R = CH_3,\ \underline{p}\text{-}ClC_6H_4)$                     (2)

Sodium hydride or sodium methoxide has been used previously for this Horner-Wittig reaction[4]; in this case the reaction is limited to the preparation of arylidene sulfones.

For the reaction of $\alpha,\beta$-ethylenic sulfones with dialkylcopperlithium reagents, *see* **Dimethylcopperlithium**, this volume.

*1,3-Dehydroadamantane.*[5] Reaction of 1,3-dibromoadamantane (1) with *n*-butyllithium in HMPT gives 1,3-dehydroadamantane (2), which can be isolated by glpc or by sublimation. The two bridgehead carbon atoms of the cyclopropyl ring are

(1)                     (2)

"inverted"; that is, all attached atoms lie within a hemisphere around each bridgehead carbon atom. The carbon–carbon bond length of the 1,3-cyclopropane system is remarkably long (1.64 Å). Even so, the strain energy in 1,3-dehydroadamantane is not particularly great.

Various 5-substituted 1,3-dehydroadamantanes have been prepared by the same method starting with 1,3,5-tribromoadamantane.[6]

*Selective metalation of methylated pyridines and quinolines.*[7] Treatment of 2,4,6-trimethylpyridine (1) with *n*-butyllithium in ether–hexane gives the 2-lithiomethyl derivative (2), which on methylation with methyl iodide affords the dimethylethylpyridine (3). On the other hand, metalation with sodium or potassium amide in liquid ammonia or with lithium diisopropylamide in ether–hexane gives the 4-alkali methyl derivatives (4), since methylation with methyl iodide leads to the dimethylethylpyridine (5). Since the 4-methyl group of pyridine is more acidic than one at the 2-position, metalation of (1) to give (4) is not unexpected. Kaiser ascribes the unexpected reaction of (1) with *n*-butyllithium to give (2) to complexation of the lithium cation with the ring nitrogen, thereby locking in the basic butyl group near the 2-methyl group.

The same paper also reports selective metalation of methylated quinolines.

The salts (2) and (4) also react with electrophiles other than alkyl halides, such as aldehydes and ketones.

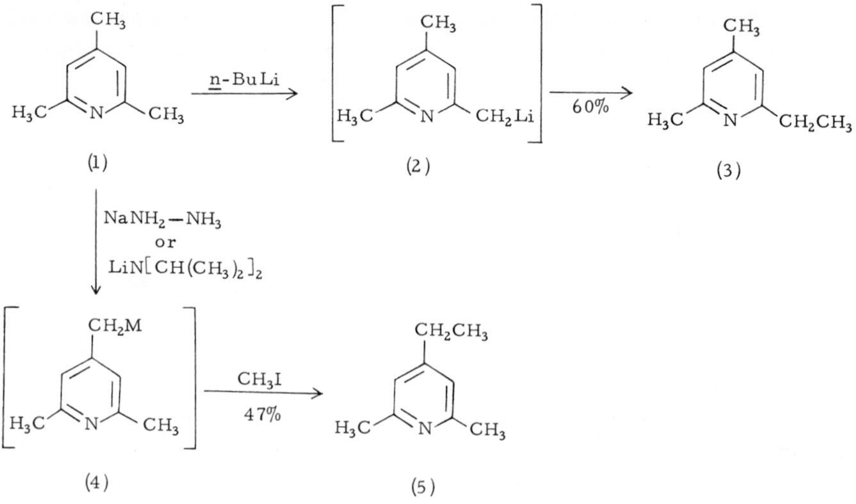

α,β-*Unsaturated aldehydes.* Corey and Terashima[8] have developed a simple route to α,β-unsaturated aldehydes from propargylic alcohols. For example, 2-octyn-1-ol[9] is converted into the tetrahydropyranyl derivative (1) in the usual way.This is metalated to form a propargylic lithium compound (2) by treatment with *n*-butyllithium (1.05 eq.) in THF at −25° under argon (2.5 hr.). On quenching with methanol–ice containing a little potassium carbonate this intermediate gives a mixture of the allene (3) and the

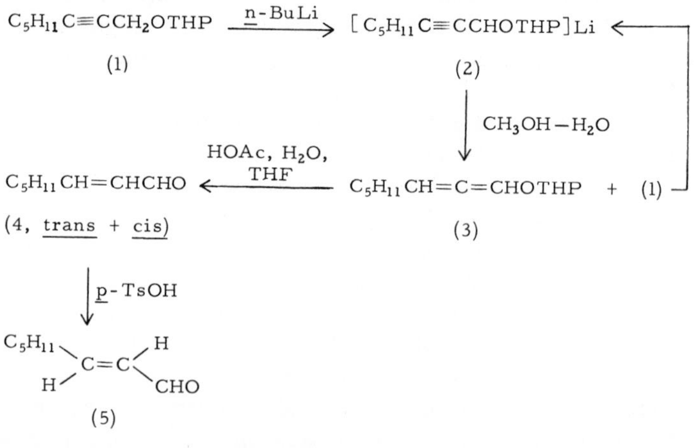

starting acetylene (1) in a ratio of 70:30. Selective hydrolysis of the enol ether (3) is readily accomplished by treatment of the mixture of (3) and (1) with acetic acid–water–THF (ratio by volume 1 : 1 : 2) at 40–45° for 4 hr. A mixture of *trans*- and *cis*-2-octene-1-al (4, ratio 76:24) is obtained in this way. This material can be isomerized quantitatively to the *trans*-aldehyde (5) by treatment with 0.02 eq. of *p*-toluenesulfonic acid in methylene chloride.

The starting acetylene (1) can be readily recovered during isolation of the octenals and can be recycled. Thus the only appreciable loss is mechanical.

[1] J. D. Buhler, *J. Org.*, **38**, 904 (1973).
[2] B. M. Trost and S. D. Ziman, *ibid.*, **38**, 932 (1973).
[3] G. H. Posner and D. J. Brunelle, *ibid.*, **37**, 3547 (1972).
[4] I. C. Popoff, J. L. Dever, and G. R. Leader, *ibid.*, **34**, 1128 (1969): I. Shahak and J. Almog, *Synthesis*, 170 (1969); 145 (1970).
[5] R. E. Pincock, J. Schmidt, W. B. Scott, and E. T. Torupka, *Canad. J. Chem.*, **50**, 3958 (1972).
[6] W. B. Scott and R. E. Pincock, *Am. Soc.*, **95**, 2040 (1973).
[7] E. M. Kaiser, G. J. Bartling, W. R. Thomas, S. B. Nichols, and D. R. Nash, *J. Org.*, **38**, 71 (1973).
[8] E. J. Corey and S. Terashima, *Tetrahedron Letters*, 1815 (1972).
[9] Farchan Research Laboratories.

**n-Butyllithium-N,N,N′,N′-tetramethylethylenediamine complex.**

*See* N,N,N′,N′-**Tetramethylethylenediamine**, **2**, 403; **3**, 284; this volume.

*t*-**Butylmagnesium chloride**, $(CH_3)_3CMgCl$, see **1**, 415–424.

*Intramolecular aldol condensation.* In a synthesis of a precursor to epiallogibberic acid, House et al.[1] effected an unfavorable intramolecular aldol condensation by use of a covalent magnesium alkoxide intermediate. Thus treatment of (1) with 2 eq. of

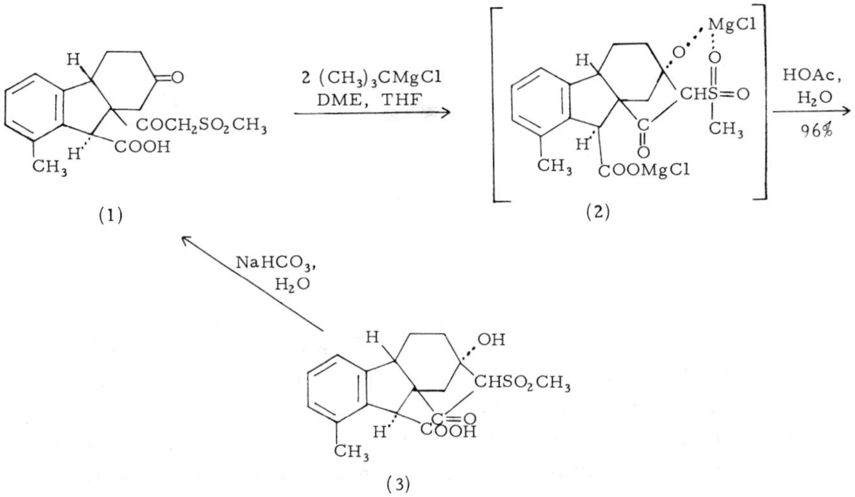

(1)    (2)

(3)

*t*-butylmagnesium chloride in THF and DME followed by hydrolysis with aqueous acetic acid afforded (3) in 96% yield. Use of methylmagnesium bromide reduced the yield of (3) to 45%. The aldol product (3) is hydrolyzed to (1) by aqueous sodium bicarbonate.

[1] H. O. House, D. G. Melillo, and F. J. Sauter, *J. Org.*, **38**, 741 (1973).

**n-Butylmercaptan, 2,** 53–54.

*α-Methylation of a ketone.* In a total synthesis of loganin (1) Büchi *et al.*[1] methylated the ketone (2) by the method of Ireland and Marshall (**2,** 53–54). Treatment of (2)

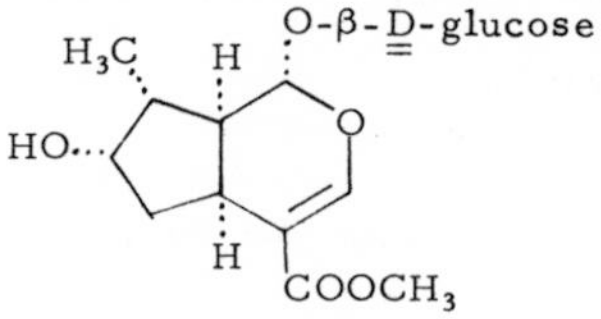

(1)

with methyl formate and base and then with tosyl chloride and *n*-butylmercaptan gave three *n*-butylthiomethylene ketones, (3), (4), and (5). Fortunately the desired isomer (5)

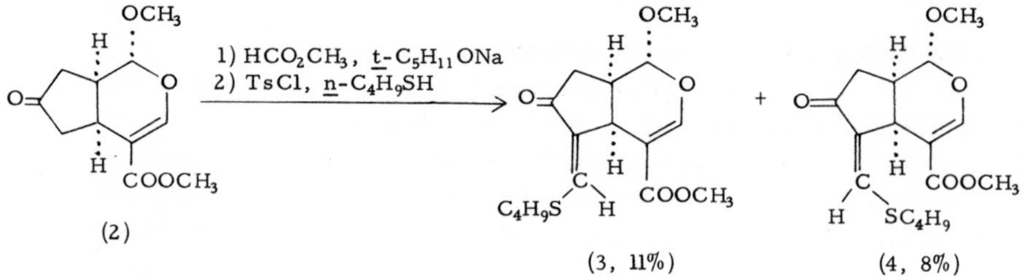

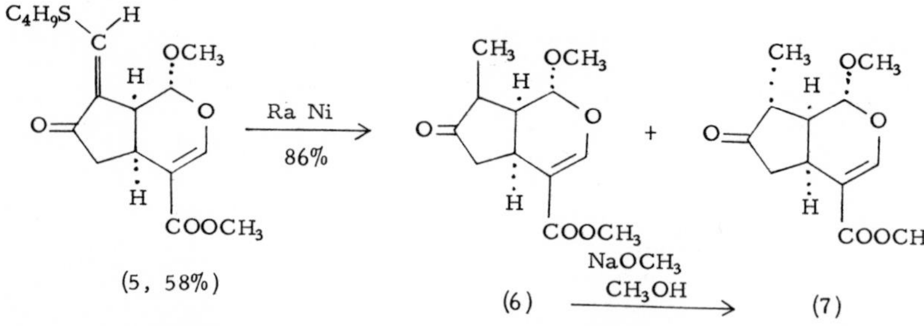

predominated. Raney nickel hydrogenolysis of (5) gave a mixture of epimeric methyl ketones in which (6) usually predominated. Treatment of (6) with base gave the desired thermodynamically more stable methyl ketone (7) exclusively.

*Geminal alkylation of ketones.* Coates and Sowerby[2] have reported a new method for site-selective geminal alkylation of ketones which involves reduction of the *n*-butyl-thiomethylene derivative of the ketone by lithium–ammonia to give a methyl-substituted enolate anion which can be alkylated *in situ.* The ketone, for example cyclohexanone (1), is condensed with ethyl formate and then transformed into the *n*-butyl-thiomethylene derivative (2) by reaction with *n*-butyl mercaptan (**2,** 53–54). This is then reduced with excess lithium in liquid ammonia at −33° with 2 eq. of a proton donor (water is usually used to avoid overalkylation). The lithium enolate is then

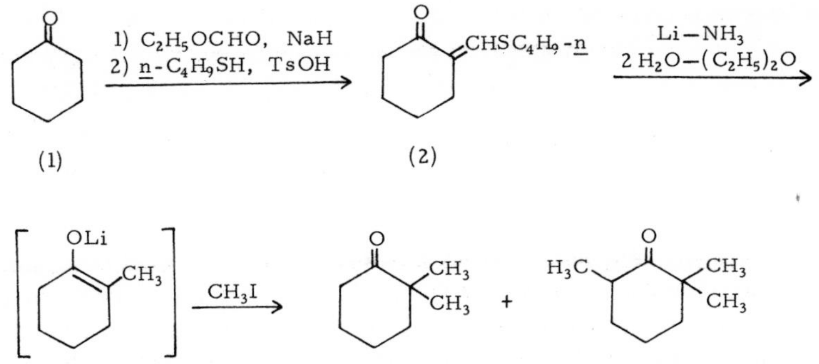

(3, 83%)     (4, 2%)

alkylated with methyl iodide in large excess. 2,2-Dimethylcyclohexanone (3) is obtained in 83% yield; a trace of 2,2,6-trimethylcyclohexanone (4) is formed as by-product. Alkylation is most satisfactory with primary reactants.

The same chemists report that 2-*n*-butylthiomethylenecyclohexanone (2) undergoes double conjugate addition with dimethylcopperlithium in ether at 0° to give, after hydrolysis, 2-isopropylcyclohexanone (5) in over 95% yield.

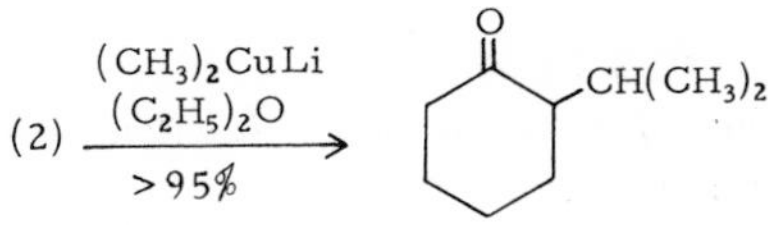

(5)

[1]G. Büchi, J. A. Carlson, J. E. Powell, Jr., and L.-F. Tietze, *Am. Soc.*, **95**, 540 (1973).
[2]R. M. Coates and R. L. Sowerby, *ibid.*, **93**, 1027 (1971).

### *tert*-Butyloxycarbonylfluoride.

**Preparation at − 25°:**

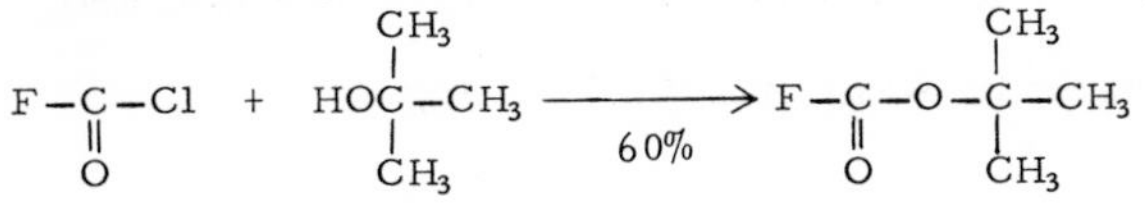

b.p. 4°/15 mm.

Used for the synthesis of BOC-amino acids.[1]

[1]E. Schnabel, H. Herzog, P. Hoffmann, E. Klauke, and I. Ugi, *Ann.*, **716**, 175 (1968).

**1-*t*-Butyloxycarbonyltriazole-1,2,4,**

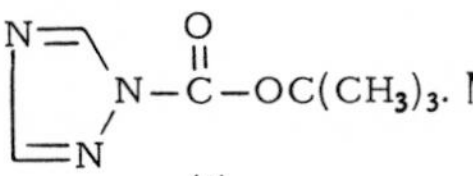

Mol. wt. 169.18, m.p. 40°.

(1)

The reagent is prepared[1] in 80% yield by the reaction of 1-trimethylsilyltriazole-1,2,4[2] with *t*-butyl azidoformate (1, 84–85; 2, 44–45; this volume).

$$\underset{\substack{\text{N}=\text{N}}}{\boxed{\phantom{x}}}\text{N}-\text{Si}(\text{CH}_3)_3 \; + \; (\text{CH}_3)_3\text{CO}\overset{\text{O}}{\overset{\|}{\text{C}}}\text{N}_3 \; \xrightarrow[80\%]{} \; \underset{\substack{\text{N}=\text{N}}}{\boxed{\phantom{x}}}\text{N}-\overset{\text{O}}{\overset{\|}{\text{C}}}-\text{OC}(\text{CH}_3)_3 \; + \; \text{N}_3\text{Si}(\text{CH}_3)_3$$

(1)

*N-t-Butyloxycarbonylamino acids.*[1] Amino acids, in the form of their salts with benzyltrimethylammonium hydroxide (Triton B), react with (1) in DMSO or DMF within a few hours to give BOC-amino acids in 80–90% yield.

BOC-amino acids can also be prepared, in somewhat lower yields, by the reaction of the benzyltrimethylammonium salts of amino acids with *t*-butyl phenylcarbonate (2, 1, 85) in the presence of 1 molar eq. of triazole-1,2,4.

$$(\text{CH}_3)_3\text{CO}\overset{\text{O}}{\overset{\|}{\text{C}}}\text{OC}_6\text{H}_5$$

(2)

[1]G. Bram, *Tetrahedron Letters*, 469 (1973).
[2]L. Birkoffer, P. Richter, and A. Ritter, *Ber.*, **93**, 2804 (1960).

*t*-Butyl perbenzoate, 1, 98–101; 2, 54–55.

*Acyloxylation reaction.*[1] The acyloxylation reaction (Kharasch-Sosnovsky reaction) has been reviewed in detail.

[1]D. J. Rawlinson and G. Sosnovsky, *Synthesis*, 1 (1972).

**t-Butylquinolyl-8-carbonate.** Mol. wt. 245.3, m.p. 69.5–72°.

*Preparation* from 8-hydroxyquinoline, *t*-butyl chloroformate, and triethylamine in tetrahydrofurane–ether at −60° to −25° (yield 72%). With this reagent the preparation of N$^\varepsilon$-protected lysine is particularly easy.

The preparation of several other BOC-L-amino acids is described.[1]

[1]B. Rzeszotarska and S. Wiejak, *Ann.*, **716**, 216 (1968).

# C

**Cadmium carbonate**, $CdCO_3$. Mol. wt. 172.42. Supplier: Baker.

*Koenigs–Knorr synthesis of aryl glucuronides.* Steroid glucuronides of estrone, $17\beta$-estradiol, estriol, and equilenin can be prepared in excellent yield by the reaction of the phenolic steroid with the bromo sugar (1) in toluene using cadmium carbonate as

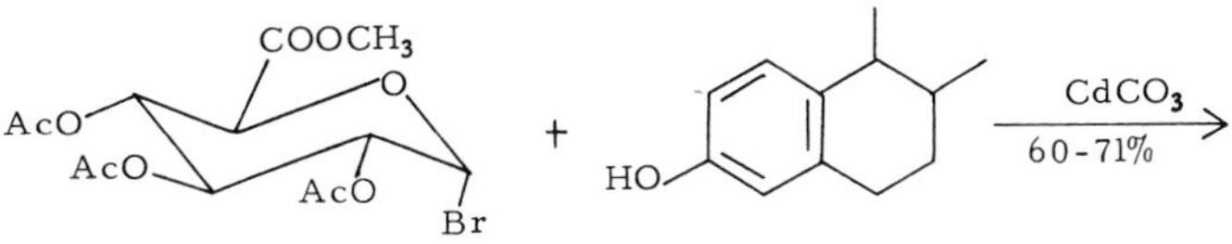

(1)

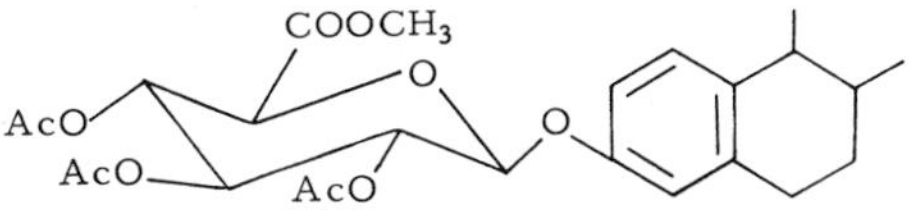

(2)

catalyst.[1] Use of silver carbonate, the usual catalyst for the Koenigs–Knorr reaction,[2] leads to yields of (2) of approximately 7%.

The catalyst also gives good yields of some steroidal alicyclic glucuronides.

[1] R. B. Conrow and S. Bernstein, *J. Org.*, **36**, 863 (1971).
[2] W. Koenigs and E. Knorr, *Ber.*, **34**, 957 (1901); see W. Pigman, *The Carbohydrates*, Academic Press, New York, 150, 191 (1957).

**Calcium carbonate, 1**, 103–104; **2**, 57–58.

*Pinacolic rearrangement.* 7,7-Dimethylbicyclo[4.1.1]octane-3-one (2) can be prepared[1] conveniently from the tosylate (1) by the procedure of Corey[2] for pinacolic

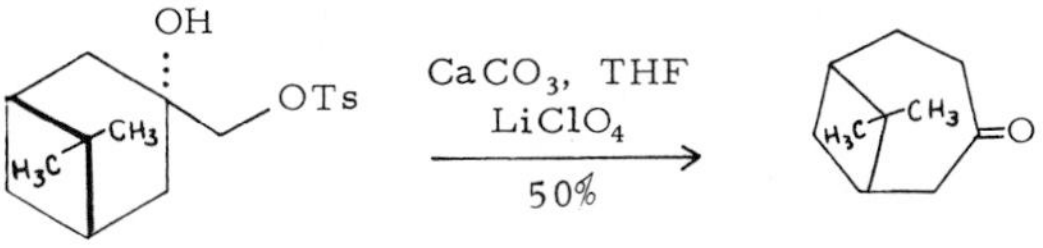

(1)                                    (2)

rearrangement. The tosylate (4.0 mmole) is treated with calcium carbonate (4.0 mmole) and a catalytic amount of lithium perchlorate in THF (72 hr., 60°).

[1] W. Tubiani and B. Waegell, *Angew. Chem., internat. Ed.*, **11**, 640 (1972).
[2] E. J. Corey, *Am. Soc.*, **83**, 1251 (1961); **86**, 478 (1964).

***d*-( + )-Camphor-10-sulfonic acid**, 1, 108–109; 2, 58–59.

*Preparation of chiral triarylphosphines.*[1] Stepwise treatment of phenyldichloro-phosphine (1) with organometallic reagents of different reactivity gives chiral triaryl-phosphines (2) in good yield. Treatment of (2) with paraformaldehyde and 0.5 eq. of

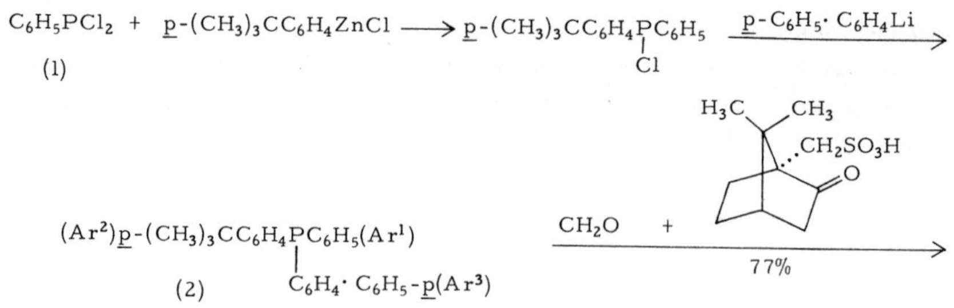

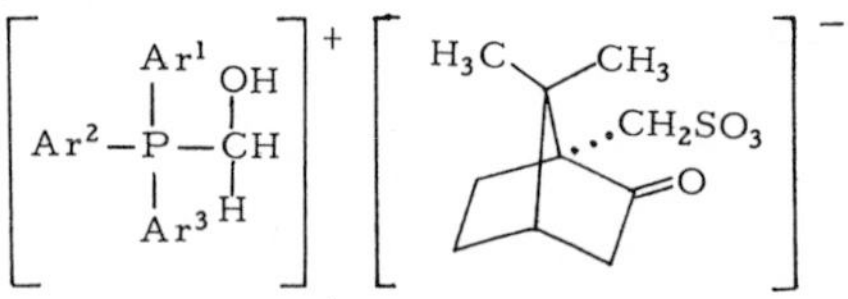

$$C_6H_5PCl_2 \ + \ \underline{p}\text{-}(CH_3)_3CC_6H_4ZnCl \longrightarrow \underline{p}\text{-}(CH_3)_3CC_6H_4\underset{\underset{Cl}{|}}{P}C_6H_5 \quad \underline{p}\text{-}C_6H_5\cdot C_6H_4Li$$

(1)

$$(Ar^2)\underline{p}\text{-}(CH_3)_3CC_6H_4\underset{\underset{C_6H_4\cdot\,C_6H_5\text{-}\underline{p}(Ar^3)}{|}}{P}C_6H_5(Ar^1)$$

(2)

$$CH_2O \ + \quad \xrightarrow{\ 77\%\ }$$

(3)

*d*-( + )-camphor-10-sulfonic acid gives the crystalline triarylhydroxymethylphos-phonium ( + )-camphor-10-sulfonate. The antipode can be isolated from the mother liquor. The salts (3) are split to optically active triarylphosphines (2) by treatment with base (NaOH, KOH, NH₃).

[1]G. Wittig, H. Braun, and H.-J. Cristau, *Ann.*, **751**, 17 (1971).

***p*-Carbomethoxyperbenzoic acid (*p*-Methoxycarbonylperbenzoic acid) (2).** Mol. wt. 196.15, begins to decompose at 125°.

This peracid is prepared[1] in 80–95% yield by photooxidation of methyl *p*-formyl-benzoate (1)[2] in carbon tetrachloride. It is as stable as *m*-chloroperbenzoic acid; the

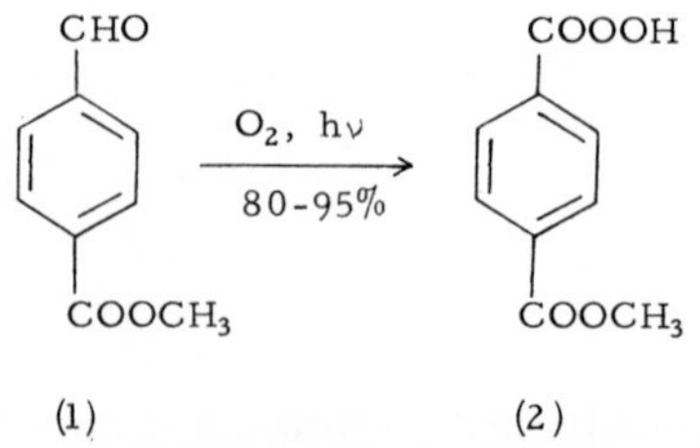

decomposition after storage for one year at 10° is less than 5%. The peracid is fairly soluble in dioxane, ethanol, acetone, and acetonitrile; it is less soluble in chloroform, benzene, and ether.

The peracid has been used for epoxidation and for Baeyer-Villiger oxidations; yields in the former case range from 60 to 90%, and from 64 to 85% in the latter case. It corresponds generally to the reactivity of perbenzoic acid and perphthalic acid.

[1]N. Kawabe, K. Okada, and M. Ohno, *J. Org.*, **37**, 4210 (1972).
[2]Prepared by oxidation of methyl *p*-methylbenzoate according to the method of S. V. Lieberman and R. Connor, *Org. Syn.*, *Coll.* Vol. II, 441 (1955).

**Carbon monoxide, 2,** 60, 204; **3,** 41–43.

*Trialkylcarbinols.*[1]  Mixed trialkylboranes, such as (1), react with carbon monoxide in the presence of ethylene glycol at 150° and 1000 psi to form 2-*t*-alkyl-1,3,2-dioxaborolanes (2), which are not isolated but oxidized with hydrogen peroxide in aqueous sodium hydroxide to give trialkylcarbinols (3). Yields of isolated products are 60–90%.

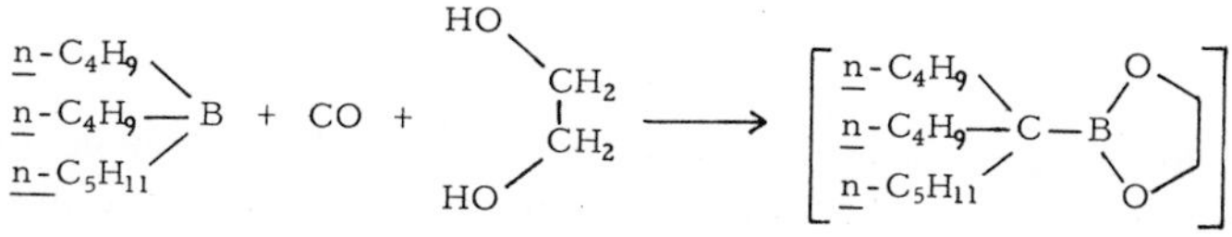

(1)                                                         (2)

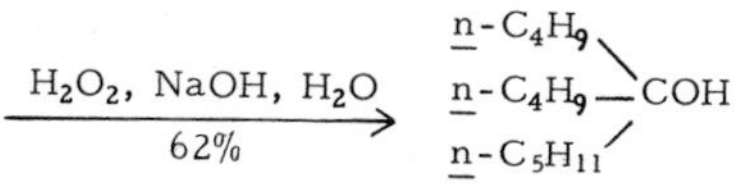

(3)

[1]H. C. Brown, E. Negishi, and S. K. Gupta, *Am. Soc.*, **92**, 6648 (1970).

**Carbon tetrachloride, 3,** 43.

*Telomerization of monochlorotrifluoroethylene.*[1] Carbon tetrachloride, in combination with ferric chloride, triethylamine hydrochloride, and benzoin (as reducing agent), and with acetonitrile as solvent, reacts with monochlorotrifluoroethylene to give telomers of the formula (1).

$$CCl_4 + nCF_2{=}CFCl \longrightarrow CCl_3(CF_2CFCl)_nCl$$

(1, n = 1 to 15)

[1]B. Boutevin and Y. Pietrasanta, *Tetrahedron Letters*, 887 (1973).

**Catecholborane (1,3,2-Benzodioxaborole),**       Mol. wt. 119.92.

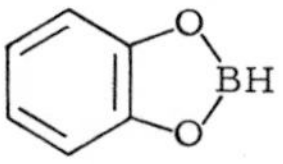

See also p. 25.

(1)

Catecholborane is prepared by the reaction of catechol with diborane in THF.[1]

*Hydroboration of alkynes.* The reagent (1) reacts stereospecifically and regioselectively with alkynes to give monohydroboration products, 2-alkenyl-1,3,2-benzodioxaboroles (2), in nearly quantitative yield. The hydroboration proceeds in a *cis*

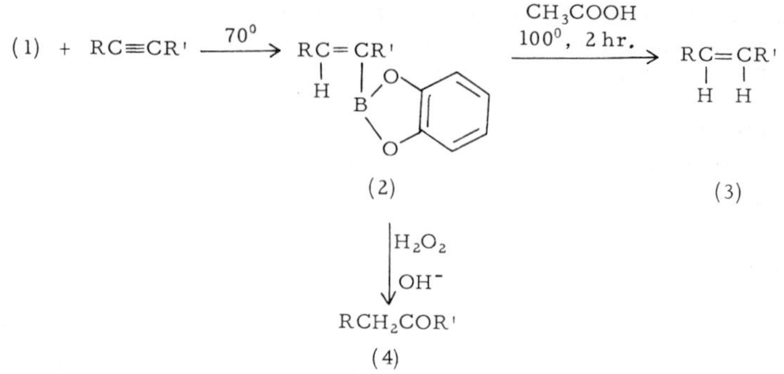

$$(1) \ + \ RC{\equiv}CR' \ \xrightarrow{70^0} \ \underset{\underset{B}{\overset{|}{\underset{}{}}}{\overset{R}{\underset{H}{}}C{=}C}R' \ \xrightarrow[100^0, \ 2 \ hr.]{CH_3COOH} \ \underset{H \ \ H}{\overset{| \ \ |}{RC{=}CR'}}$$

(2)                    (3)

$$\downarrow {\small H_2O_2}$$
$$\downarrow {\small OH^-}$$
$$RCH_2COR'$$
(4)

manner, with the boron being attached to the less hindered carbon atom of the triple bond. The esters (2) are readily hydrolyzed to the corresponding alkeneboronic acids, RC=CR'. Protonolysis of the esters (2) gives the corresponding *cis*-olefins (3); oxida-
$\ \ |\ \ \ |$
H  B(OH)₂
tion with alkaline hydrogen peroxide at 25–30° for 2 hr. gives aldehydes or ketones (4).[2]

2-Alkenyl-1,3,2-benzodioxaboroles (2) react with mercuric acetate at 0° to give alkenylmercuric acetates in nearly quantitative yield.[3]

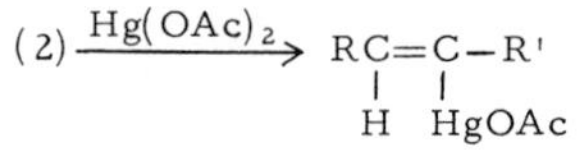

$$(2) \xrightarrow{Hg(OAc)_2} \underset{H \ \ \ HgOAc}{\overset{| \ \ \ \ \ |}{RC{=}C{-}R'}}$$

[1]H. C. Brown and S. K. Gupta, *Am. Soc.*, **93**, 1816 (1971).
[2]*Idem, ibid.*, **94**, 4370 (1972).
[3]R. C. Larock, S. K. Gupta, and H. C. Brown, *ibid.*, **94**, 4371 (1972).

**Catechol dichloromethylene ether**, **1**, 119–120.

*Carboxylation.* The reagent has been used in a preparation of dimethyl 2,4-thiophenedicarboxylate (3) from methyl 3-thiophenecarboxylate (1).[1]

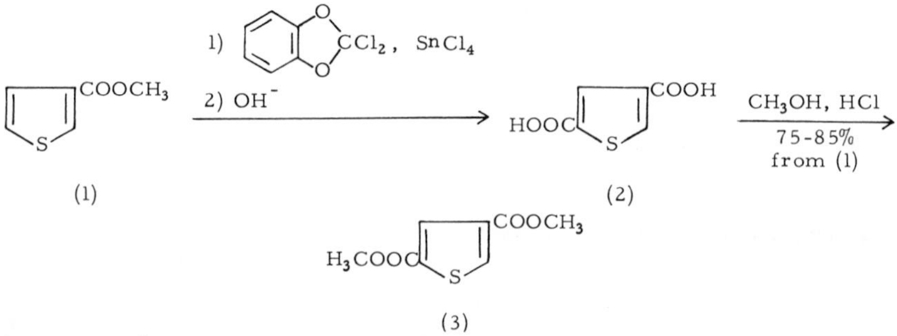

(1)                    (2)

(3)

[1]M. Janda, J. Šrogl, M. Němec, and I. Stibor, *Org. Prep. Proc. Int.*, **3**, 295 (1971).

**Ceric acetate,** $Ce(OAc)_4$. Mol. wt. 376.31.

*Preparation.* Ceric acetate is prepared by ozonolysis of cerous acetate in the presence of nitrate ion.[1] Stable yellow solid.

*Reaction with olefins.*[2] When ceric acetate is heated with styrene in glacial acetic acid containing 10% potassium acetate at 110° for 20 hr. the major product is the lactone (1), 70% yield. For the synthetic preparation of lactones, the more available ceric ammonium nitrate can be used in place of ceric acetate.

$$C_6H_5CH=CH_2 \quad \xrightarrow[\substack{CH_3COOH \\ 70\%}]{Ce(OAc)_4} \quad \underset{\underset{O}{\underset{\|}{\overset{|}{\underset{C}{O}}}\underset{}{\diagdown}\underset{}{\diagup}CH_2}}{C_6H_5CH-CH_2}$$

(1)

In contrast, the photochemical reaction of ceric acetate with *trans-β*-methylstyrene produces mainly the ester (2) and only minor amounts of lactone (3).

$$C_6H_5CH=CHCH_3 \quad \xrightarrow[\substack{h\nu \\ 95\%}]{Ce(OAc)_4} \quad \underset{\underset{OAc}{|}\,\underset{CH_3}{|}}{C_6H_5CH-CH-CH_3} \quad + \quad \underset{\underset{O}{\underset{\|}{C}}\diagdown\diagup CH_2}{\overset{C_6H_5-CH-CH-CH_3}{\diagdown\diagup O}}$$

(2)                                   (3)

main product                    minor product

The reaction of ceric acetate with aromatic hydrocarbons can be used for synthesis of arylacetic acids. Thus the reaction of ceric acetate with toluene in acetic anhydride–acetic acid gives as the major products benzyl acetate and tolylacetic acid (29% yield).

[1]N. Hay and J. K. Kochi, *J. Inorg. Nucl. Chem.*, **30**, 884 (1968).
[2]E. I. Heiba and R. M. Dessau, *Am. Soc.*, **93**, 995 (1971).

**Ceric ammonium nitrate (CAN), 1,** 120–121; **2,** 63–65; **3,** 44–45. Additional supplier: Baker.

*Oxidative cleavage of 1,2-glycols.* Trahanovsky *et al.*[1] have studied the relative rates of oxidative cleavage of 1,2-glycols with CAN and with lead tetraacetate and have concluded that the mechanism in the case of CAN involves formation of a monodentate complex followed by a one-electron cleavage to give an intermediate radical that is oxidized further (scheme I). The cleavage with lead tetraacetate (**1,** 554–

557; **2**, 235–237; **3**, 168–171), on the other hand, involves formation of a bidentate complex (scheme II).

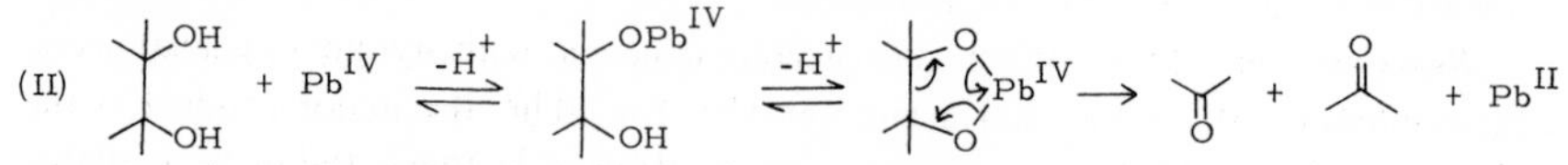

(II)

*Aromatic substitution with peroxydicarbonates.*[2] The reaction of toluene with peroxydicarbonates and CAN in acetonitrile at 60° gives tolyl alkyl carbonates in high yield:

$$2\ C_6H_5CH_3\ +\ (ROCO_2)_2\ +\ 2\ Ce(NH_4)_2(NO_3)_6\ \xrightarrow[\ 60^0\ ]{CH_3CN}$$

$$2\ ROCO_2C_6H_4CH_3\ +\ 4\ NH_4NO_3\ +\ 2\ Ce(NO_3)_3\ +\ 2\ HNO_3$$

Ceric potassium nitrate is about as efficient as CAN, but ceric hydroxide and nitrate are ineffective in this reaction.

*Cyclobutadiene.* Cyclobutadiene itself is unknown but the complex cyclobutadieneiron tricarbonyl (1, **2**, 140; **3**, 101) is readily available. The diene can be liberated from the complex as a transient intermediate for use in synthesis by oxidative degradation with ceric ammonium nitrate. If the reactants or products are sensitive to the acidic solutions of CAN, lead tetraacetate in pyridine can be used as oxidant. Thus the reaction of cyclobutadieneiron tricarbonyl with *p*-benzoquinone in the presence of CAN leads to formation of *endo*-tricyclo[4.4.0.0$^{2.5}$]deca-3,8-diene-7,10-dione (2) in about 40% yield.[3]

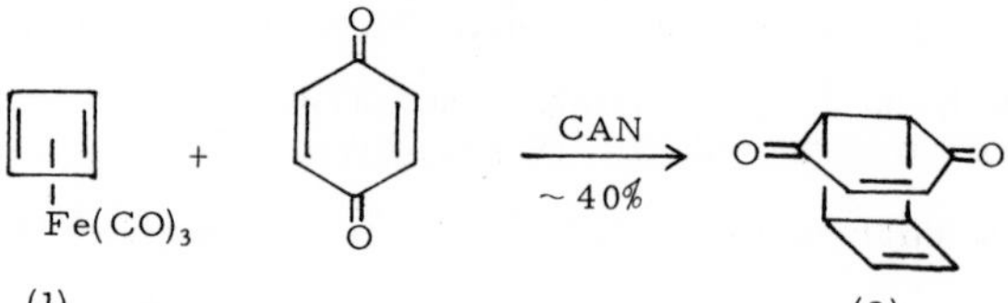

(1)                    (2)

*Oxidation of hydroquinones.*[4] Hydroquinones are oxidized to quinones in high yield by treatment with CAN for a few minutes in aqueous acetonitrile. The method is applicable to the production of *o*-, *p*-, and diquinones.

*Oxidation of benzoins.*[5] Benzoins are oxidized to aldehydes and acids in good yield by treatment with 2 mole eq. of CAN in aqueous acetonitrile at 60° for 10 min.

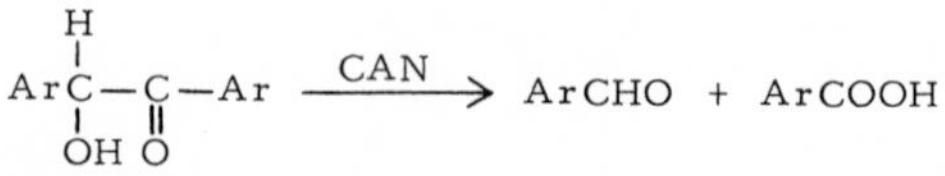

*Oxidation of diaryl sulfides.*[6] Ceric ammonium nitrate oxidizes diaryl sulfides in high yield to the corresponding sulfoxides without overoxidation to sulfones. The reaction is carried out in aqueous acetonitrile at room temperature and is generally complete within 3 min. Unfortunately the method is not suitable for oxidation of dialkyl

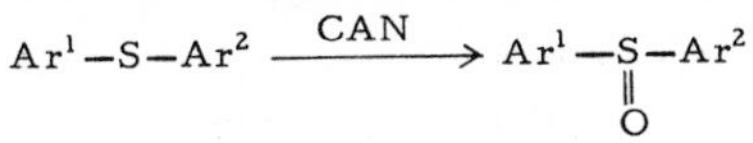

$$Ar^1-S-Ar^2 \xrightarrow{\ CAN\ } Ar^1-S-Ar^2$$
$$\underset{O}{\overset{\parallel}{}}$$

sulfides possessing α-hydrogen atoms, probably owing to Pummerer rearrangement of the resulting sulfoxides.

*Oxidation of carboxylic acid hydrazides.*[7] Both aliphatic and aromatic carboxylic acid hydrazides are rapidly oxidized in good yield by CAN in aqueous acetonitrile to the parent acids.

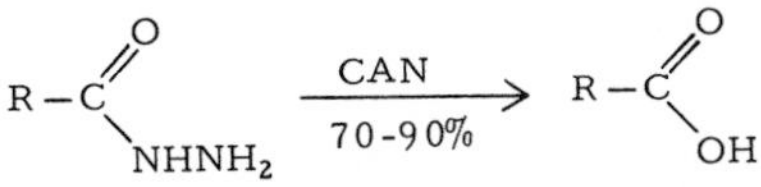

*Oxidation of* $\Delta^{1,3,5(10)}$*-4-methylestratrienes.* Oxidation of the ring A aromatic steroid (1) with CAN gives the corresponding 6-acetate (2) in 60–65% yield.[8]

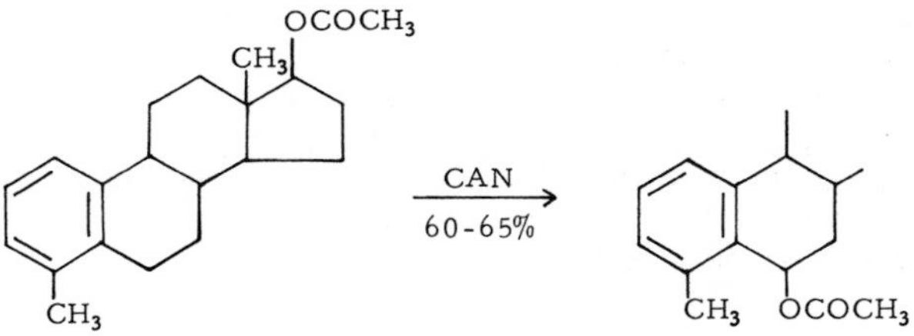

(1)                                        (2)

*Oxidation of tropilidene.* Cycloheptatriene-$d_8$ is oxidized by CAN at the same rate as cycloheptatriene itself. It can be argued therefore that CAN reacts with a non-aromatic carbon–carbon double bond in preference to a very reactive carbon–hydrogen bond.[9]

*cis- and trans-Stilbene from phenyldiazomethane.*[10] Phenyldiazomethane in pentane at −5° rapidly decomposes when treated with catalytic amounts of ceric ammonium nitrate (0.03 mole eq.) to give *cis-* and *trans*-stilbene, the former being the predominant

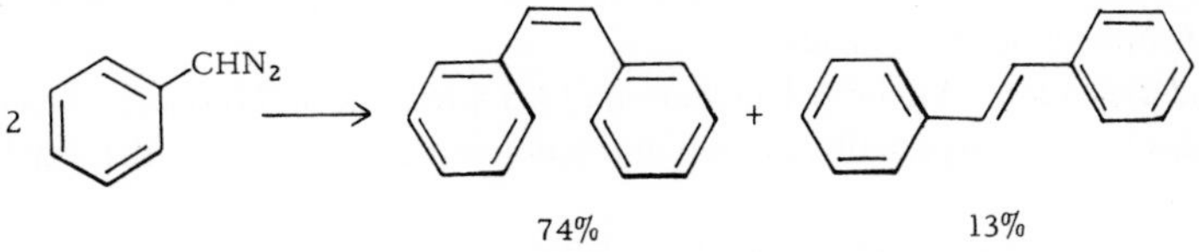

74%                    13%

product. An oxidative chain reaction is proposed, but an explanation for the high yield of the *cis*-isomer is lacking. Strong electron-withdrawing substituents inhibit the reaction.

*Regeneration of ketones from 1,3-dithiolanes and 1,3-dithianes.* Treatment of 1,3-dithiolanes or 1,3-dithianes in 75% aqueous acetonitrile with CAN at room temperature for 3 min. gives the parent carbonyl compounds in 70–87% yield.[11]

This method was used in a synthesis of *cis*-jasmone to effect removal of a dithioketal group.[12] Thus oxidation of (1) with ceric ammonium nitrate gave undecane-2,5-dione (2) in 80% yield. In this case use of chloramine-T (this volume) gave (2) in 67.3% yield.

(1)     CAN 80% →     (2)

[1]W. S. Trahanovsky, J. R. Gilmore, and P. C. Heaton, *J. Org.*, **38**, 760 (1973).
[2]M. E. Kurz, E. S. Steele, and R. L. Vechio, *J.C.S. Chem. Commun.*, 109 (1973).
[3]L. Brener, J. S. McKennis, and R. Pettit, *Org. Syn.*, submitted 1972.
[4]T.-L. Ho, T. W. Hall, and C. M. Wong, *Chem. Ind.*, 729 (1972).
[5]T.-L. Ho, *Synthesis*, 560 (1972).
[6]T.-L. Ho and C. M. Wong, *ibid.*, 561 (1972).
[7]T.-L. Ho, H. C. Ho, and C. M. Wong, *ibid.*, 562 (1972).
[8]D. M. Piatak and L. S. Eichmeier, *Chem. Commun.*, 772 (1971).
[9]P. Müller, E. Katten, and J. Roček, *Am. Soc.*, **93**, 7114 (1971).
[10]W. S. Trahanovsky, M. D. Robbins, and D. Smick, *ibid.*, **93**, 2086 (1971).
[11]T.-L. Ho, H. C. Ho, and C. M. Wong, *Chem. Commun.*, 791 (1972).
[12]*Idem, Canad. J. Chem.*, **50**, 2718 (1972).

**Chloramine**, **1**, 122–125; **2**, 65–66; **3**, 45.

*Oxidation.* Aniline is oxidized by chloramine (room temperature, 12 hr.) to *trans*-azobenzene (50% yield) together with a trace of *cis*-azobenzene and to 4-anilinoazobenzene (12%). In certain cases this method offers a convenient route to *cis*-azobenzenes,

$$2 \ C_6H_5NH_2 \ + \ 2 \ NH_2Cl \longrightarrow C_6H_5N=NC_6H_5 \ + \ 2 \ NH_4Cl$$

which otherwise are obtainable only in low yield by irradiation of the *trans*-isomer. Alcohols are oxidized to carbonyl compounds in high yields, comparable with those obtained with 1-chlorobenzotriazole (**2**, 67; **3**, 46–47).

Benzyl alcohol is oxidized to benzaldehyde (74% yield), and diphenylmethanol gives benzophenone (87% yield).[1]

*1,2,3-Triazoles.*[2] Attempted amination of 1,2,4-triazine-3(2H)-ones[3] with ethereal chloramine leads, unexpectedly, to ring contraction to 1,2,3-triazoles, with formal extru-

(1)     NH₂Cl 94% →     (2)     + NH₄Cl + CO

sion of carbon monoxide. Thus 5,6-diphenyl-1,2,4-triazine-3(2H)-one (1) is converted into 4,5-diphenyltriazole (2) in 94% yield. Chloramine is thus acting as an oxidant, probably with initial N-chlorination.

[1]G. A. Jaffari and A. J. Nunn, *J. Chem. Soc. (C)*, 823 (1971).
[2]C. W. Rees and A. A. Sale, *Chem. Commun.*, 532 (1971).
[3]Triazinones are readily available by the reaction of benzils with semicarbazide: R. C. Elderfield, *Heterocyclic Compounds*, Wiley, New York, **7**, 759 (1961).

**Chloramine-T (Sodium N-chloro-*p*-toluenesulfonamide)**, $p\text{-}CH_3C_6H_4SO_2NClNa\cdot3H_2O$. Mol. wt. 281.68. Suppliers: Eastman, Fluka.

*Removal of ethylenethioketal protecting group.*[1] A variety of 1,3-dithiolanes (ethylenethioketals) react smoothly with chloramine-T in aqueous methanol or ethanol to give the corresponding carbonyl compounds in good to excellent yield. For example, treatment of (1) with chloramine-T in aqueous methanol/ethanol yields fluorenone (2)

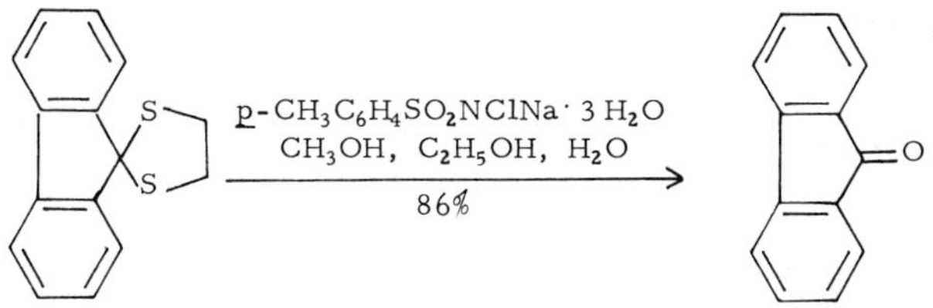

(1)

in 86% yield. Optimum yields of the carbonyl compound are obtained with 2 eq. of reagent, but workup is facilitated by use of 4 mole of chloramine-T.

[1]W. F. J. Huurdeman, H. Wynberg, and D. W. Emerson, *Tetrahedron Letters*, 3449 (1971).

**Chloranil**, **1**, 125–127; **2**, 66–67; **3**, 46.

*Dehydrogenation* of (3) to the blue hydrocarbon octazethrene (4)[1]:

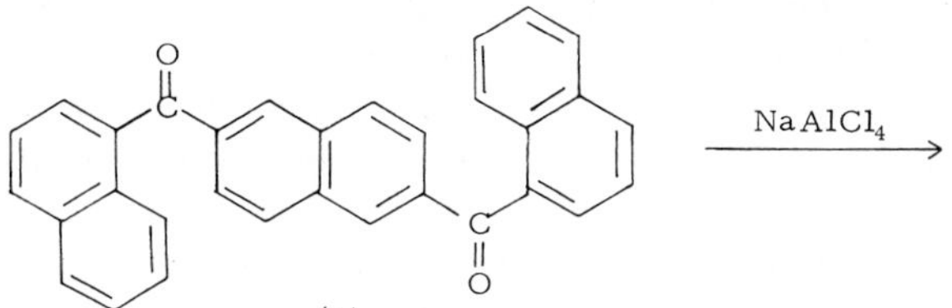

(1)

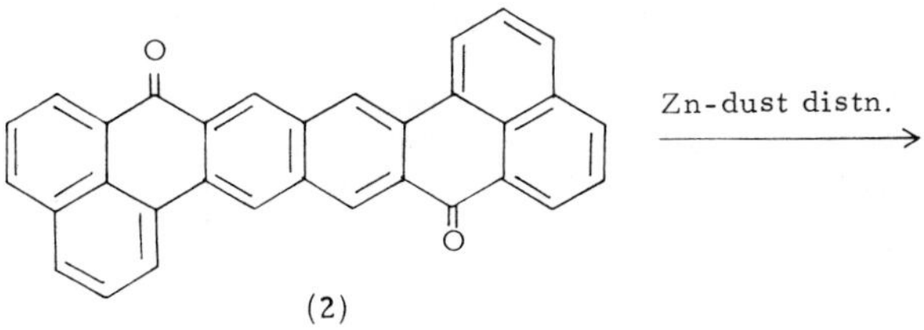

(2)

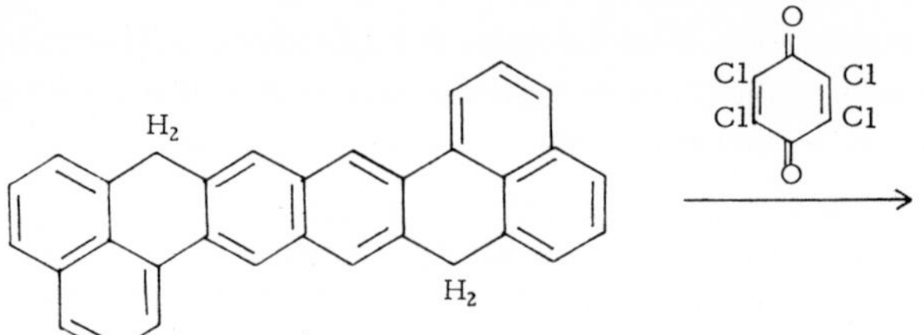

(3)

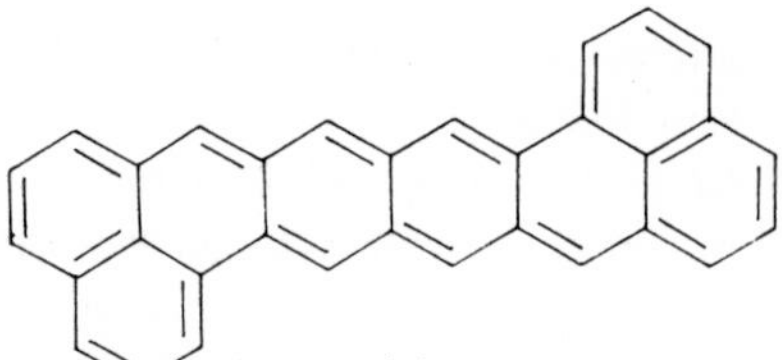

(4)

[1]R. K. Erünlü, *Ann.*, **721**, 43 (1969).

*o*-**Chloranil, 1**, 128–129; **2**, 345.

**Oxidation** of 4-methylcatechol in ether to 4-methyl-*o*-benzoquinone[1]:

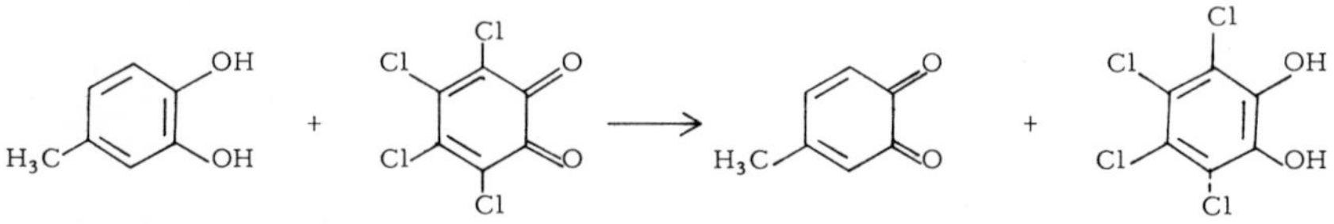

9 g.              20 g.                      6.3 g.

[1]L. Horner and T. Burger, *Ann.*, **708**, 105 (1967).

**2-Chloroacrylonitrile,** $CH_2$=CClCN. Mol. wt. 87.51, $n_D^{20}$ 1.4323. Supplier: WBL.

*Diels-Alder reaction.*[1] Both 2-acetoxyacrylonitrile (**1**, 7; **2**, 13–14) and 2-chloro-acrylonitrile (**3**, 66–67) have been used to some extent as ketene equivalents in Diels-Alder reactions. Evans *et al.*[1] have now made a direct comparison between the two reagents in the reaction with dihydroanisole (**1**) to give the diene adducts (**2**) and (**3**). 2-Chloroacrylonitrile is not only more reactive but also more regioselective. Thus 2-chloroacrylonitrile reacts with (**1**) to give the adduct (**2**) almost exclusively rather than (**3**). 2-Acetoxyacrylonitrile reacts to give mainly the product corresponding to (**2**), but in addition gives some of the isomer corresponding to (**3**).

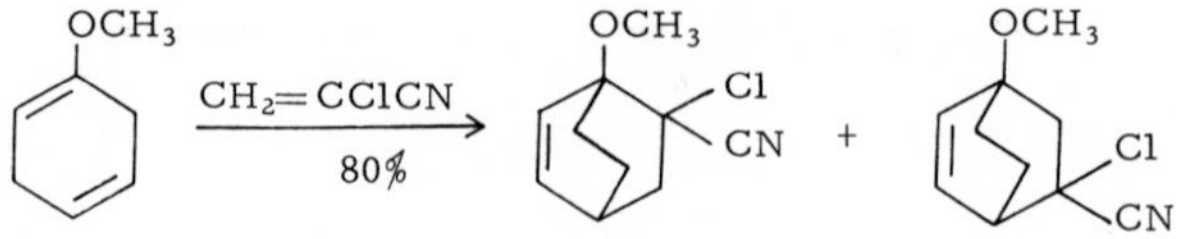

(1)              (2, >99.9%)       (3, <0.1%)

The same chemists also report an improved procedure for conversion of chloro nitrile adducts into the corresponding ketones. This reaction can be carried out in high yield (> 80%) by use of sodium sulfide in refluxing ethanol.

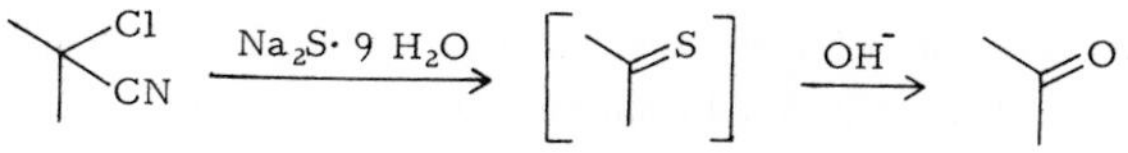

[1]D. A. Evans, W. L. Scott, and L. K. Truesdale, *Tetrahedron Letters*, 121 (1972).

**2-Chloroacrylyl chloride**, $CH_2=CClCOCl$. Mol. wt. 124.96, b.p. 45–48°/78–80 mm. Supplier: Willow Brook Laboratories, Inc.

*Preparation.*[1] The reagent is obtained in 37% yield by dehydrochlorination of α,β-dichloropropionyl chloride with diethylaniline.

*Diels-Alder reaction.*[2] This substance exhibits high dienophilic reactivity comparable to that of maleic anhydride. It reacts with 5-substituted cyclopentadienes to form adducts in which the 7-substituent is exclusively *anti* to the bridge bearing the chloro and chloroformyl groups. Thus it reacts with 5-benzyloxymethylcyclopentadiene (1) to give the adduct (2) as a 2:1 mixture of *exo*- and *endo*-acid chlorides in about 99%

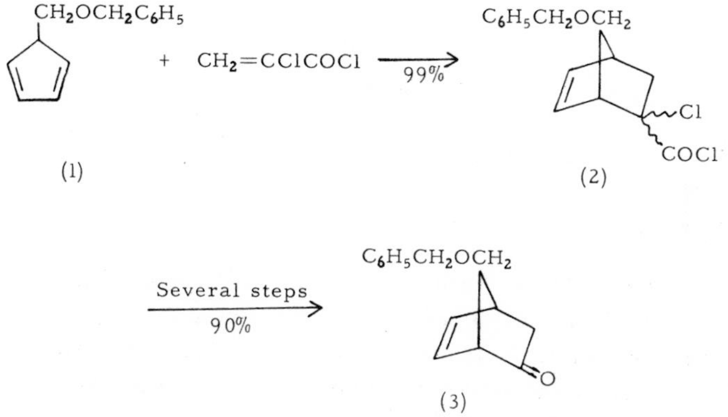

yield. Treatment of (2) with sodium azide gives the corresponding acyl azide, which on heating undergoes Curtius rearrangement to the isocyanate, leading after hydrolysis to the bicyclic ketone (3) in about 90% overall yield. The series of reactions thus provides a method for the 1,4-addition of the $-CH_2CO-$ unit to 1,3-dienes.

[1]C. S. Marvel, J. Dec, H. G. Cooke, Jr., and J. C. Cowan, *Am. Soc.*, **62**, 3495 (1940).
[2]E. J. Corey, T. Ravindranathan, and S. Terashima, *ibid.* **93**, 4326 (1971).

**1-Chlorobenzotriazole, 2**, 67; **3**, 46–47.

A report from Aldrich Chemical Company states that a batch of this chemical burst into flames while being packaged and that the material has been removed from their listing.[1]

*See*
*P190*

    *Chlorination of indole alkaloids.*[2] Indole alkaloids of the type (1) on reaction with a 5–10% molar excess of 1-chlorobenzotriazole in dry methylene chloride or benzene are converted in high yield into chloroindolenines (2). *t*-Butyl hypochlorite has been used previously for such conversions. Reported yields are not so high as those obtained

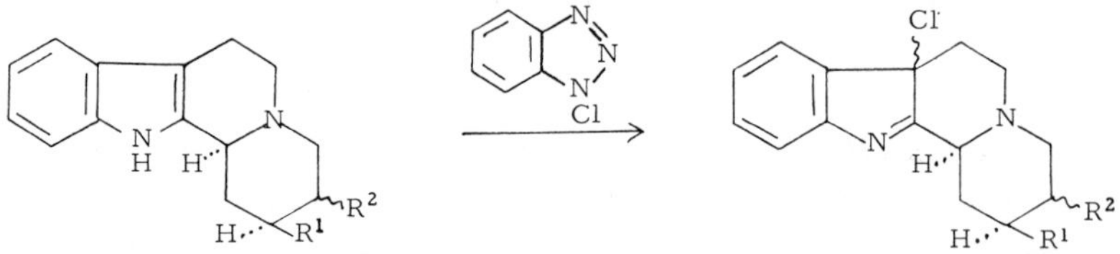

(1)                                           (2)

with 1-chlorobenzotriazole, which also has the advantage that it is less hazardous than *t*-butyl hypochlorite. The reaction was applied successfully to yohimbine without oxidation of the secondary hydroxyl group.

    *Chlorination of carbazole.*[3] Carbazole (1) is converted into 3-chlorocarbazole (2) in 79% yield when treated with 1 molar eq. of the reagent in methylene chloride at room temperature. 3,6-Dichlorocarbazole is obtained (64% yield) by use of 2 molar eq. of

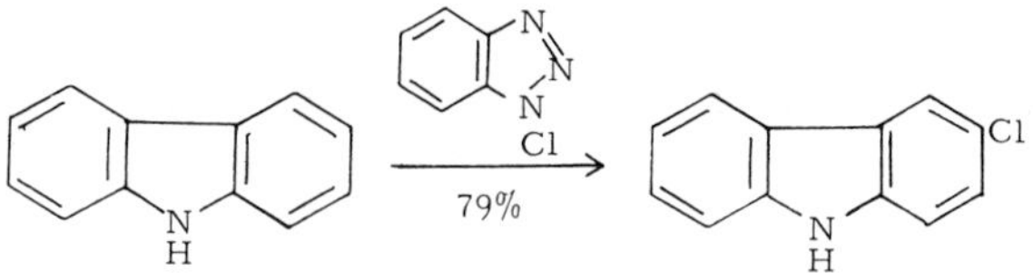

(1)                                           (2)

reagent; 1,3,6,8-tetrachlorocarbazole is obtained in 61% yield by use of 4 molar eq. of reagent. The chlorination was also applied to N-alkylated carbazoles.

    *Oxidation of sulfides.*[4] Sulfides are oxidized by the reagent to an adduct (1), which on reaction with an alcohol followed by addition of silver fluoroborate forms an alkoxy-sulfonium fluoroborate (2) in 30–40% yield.

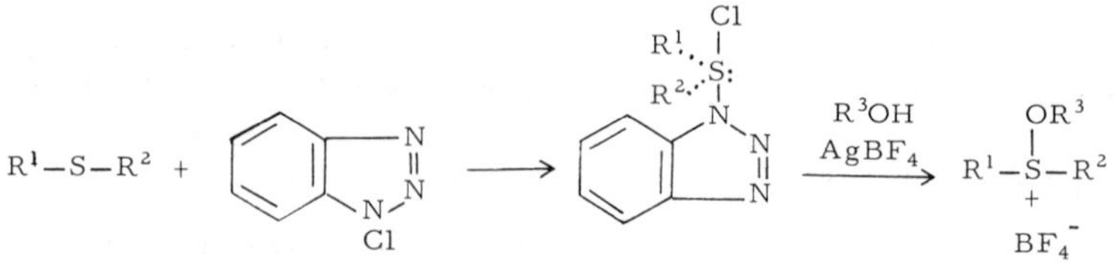

(1)                                           (2)

Aminosulfonium salts are obtained by the reaction of (1) with primary or secondary amines (40–85 % yield).

$$\text{(1)} \quad \xrightarrow[\text{AgBF}_4]{\text{R}^3\text{R}^4\text{NH}} \quad \begin{array}{c} \text{R}^3-\text{N}-\text{R}^4 \\ | \\ \text{R}^1-\text{S}-\text{R}^2 \\ + \\ \text{BF}_4^- \end{array}$$

***Steroidal thioacetals → ketones.***[5] Steroidal ethylene and trimethylene thioacetals (1) are acid-stable protective groups for ketones, but heretofore regeneration of the carbonyl function has presented difficulties. A simple method is now available for

(1, n = 2, 3)                    (2)                    (3)

regeneration. The thioacetal is oxidized by 1-chlorobenzotriazole to the disulfoxide (2), which is not isolated but decomposed to the ketone (3) directly by treatment with excess sodium hydroxide.

[1] H. B. Hopps, *Chem. Eng. News*, July 26, 1971, p. 3.
[2] K. V. Lichman, *J. Chem. Soc. (C)*, 2539 (1971).
[3] P. M. Bowyer, D. H. Iles, and A. Ledwith, *ibid.*, 2775 (1971).
[4] C. R. Johnson, C. C. Bacon, and W. D. Kingsbury, *Tetrahedron Letters*, 501 (1972).
[5] P. R. Heaton, J. M. Midgley, and W. B. Whalley, *Chem. Commun.*, 750 (1971).

**_p_-Chlorobenzoyl nitrite (PCBN)**, $p\text{-ClC}_6\text{H}_4\text{CO·ONO}$. Mol. wt. 185.57, b.p. 50°/0.2 mm.
    The reagent is synthesized from silver *p*-chlorobenzoate and nitrosyl chloride and stored as a solution in benzene.[1]

$$\underline{p}\text{-ClC}_6\text{H}_4\text{COOAg} + \text{NOCl} \longrightarrow \underline{p}\text{-ClC}_6\text{H}_4\text{CO·ONO} + \text{AgCl}$$

***Benzyne.*** Reactions of acetanilide in dry benzene with PCBN in the presence of tetraphenylcyclopentadienone, anthracene, 9,10-dimethoxyanthracene, and methyl methacrylate as trapping agents give the corresponding benzyne adducts, 1,2,3,4-

$$\text{C}_6\text{H}_5\text{NHAc} \xrightarrow{\text{PCBN}} \text{C}_6\text{H}_5\text{N(NO)Ac} + \underline{p}\text{-ClC}_6\text{H}_4\text{COOH}$$

$$\downarrow$$

$$\text{C}_6\text{H}_5\text{N}_2^+\text{AcO}^-$$

$$\downarrow$$

tetraphenylnaphthalene (70% yield), triptycene (16% yield), 9,10-dimethoxytriptycene (14% yield), and ethyl 2-methylene-3-phenylpropionate (31% yield), respectively.

The last example[2] is an ene reaction[3]:

$$
\begin{array}{c}
CH_3 \\
| \\
\text{+ } CH_2 = C - COOCH_3
\end{array}
\xrightarrow[31\%]{}
C_6H_5CH_2\overset{\overset{\displaystyle CH_2}{\|}}{C} - COOCH_3
$$

Somewhat lower yields are obtained with use of aniline plus acetic anhydride followed by PCBN. Substituted anilines can also be used; far higher yields are obtained from *meta*-substituted anilines than the corresponding *para*-isomers: many *ortho*-derivatives give no aryne adduct.

The *p*-chlorobenzoate system was selected because N-nitroso-*p*-chlorobenzoylanilines give high yields of arynes as compared with the corresponding acetyl derivatives.

[1] B. Baigrie, J. I. G. Cadogan, J. R. Mitchell, A. K. Robertson, and J. T. Sharp, *J.C.S. Perkin 1*, 2563 (1972).
[2] I. Tabushi, K. Okazaki, and R. Oda, *Tetrahedron*, **25**, 4401 (1969).
[3] Ene reactions have been reviewed by H. M. R. Hoffman, *Angew. Chem., internat. Ed.*, **8**, 556 (1969).

**α-Chloro-N-cyclohexylpropanaldonitrone**, Mol. wt. 189.69, m.p. 73–75° dec.

**Preparation.**[1] The α-chloronitrone is prepared by reaction of α-chloropropanal[2] with N-cyclohexylhydroxylamine (Fluka) in dry ether at 0°; after the second hour additional ether and methylene chloride are added. The solution is allowed to stand for 15 hr. over MgSO₄ at 0°. After filtration of the drying agent, the solvent is removed and the product crystallized from ethane–pentane. Yield is 79%. The reagent is stable

at −20°, but becomes yellow when stored at room temperature. α-Chloro-N-cyclohexylacetaldonitrone has been prepared in the same way. The cyclohexyl group is used because it improves the stability of the nitrone.

***1,4-Cycloadditions.***[1] Ordinarily, unactivated olefinic double bonds do not undergo Diels-Alder or 1,3-cycloaddition reactions. The outstanding property of α-chloronitrones is that, in the presence of silver ions, they undergo cycloaddition reactions with unactivated olefinic double bonds. Hence they can be described as enophiles. The actual reagent is considered to be an N-alkyl-N-vinylnitrosonium ion (a). For example,

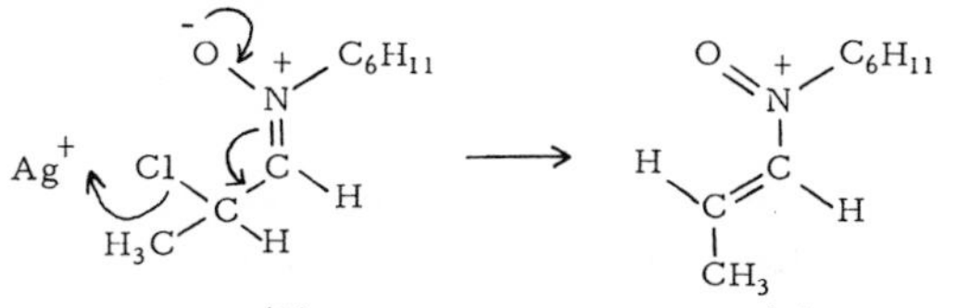

(1)                    (a)

the α-dichloronitrone (1) in the presence of silver tetrafluoroborate reacts with cyclohexene in 1,2-dichloroethane (a satisfactory solvent for AgBF$_4$) in the dark with efficient stirring; AgCl separates and the cycloaddition product is isolated as the cyano adduct (2, two epimers) in 85% yield. The iminium ion (3) can be obtained in 94% yield by

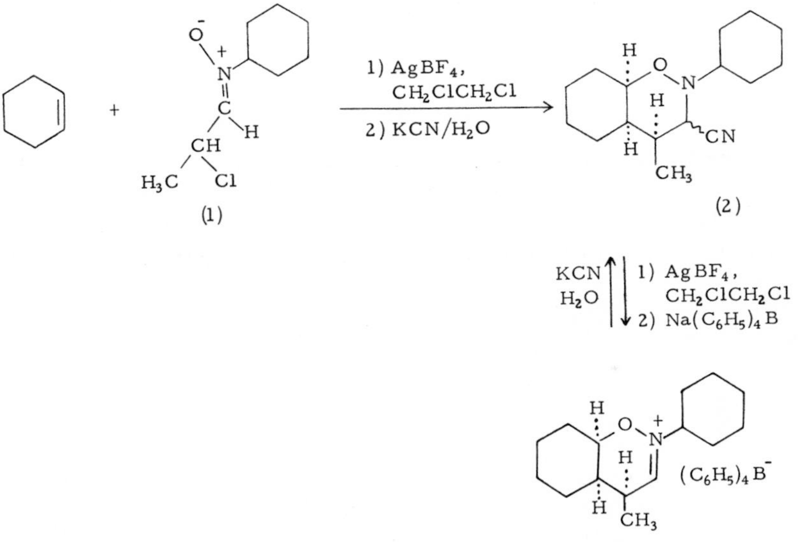

(1)                                                                      (2)

(3)

decyanation of (2) with AgBF$_4$ followed by ion exchange with sodium tetraphenylborate.

Eschenmoser et al.[3] have described two useful reactions of these addition products for synthesis. Thus (2) can be converted efficiently into a γ-lactone. Treatment of the mixture of epimeric nitriles (2) with 1.1 molar eq. of potassium t-butoxide in t-butanol at 80° under nitrogen gives the imino lactone (4) in 85% yield. This on acid hydrolysis is converted into the γ-lactone (5) in 82% yield. The two isomers (5α, 5β) are separable by gas chromatography.

$$
(2) \xrightarrow[85\%]{t\text{-}BuOK} \quad (4) \quad \xrightarrow[82\%]{H^+/H_2O} \quad (5)
$$

(4)                              (5)

A second interesting synthetic use of the cycloaddition reaction involves fragmentation of the original olefinic double bond; it is called an indirect "carboxolytic" cleavage.[4] Experimental conditions involve first deprotonation of the olefin–α-chloronitrone adduct (3) with potassium carbonate in methylene chloride to give an enaminoid derivative (6); this undergoes clean cycloreversion at slightly elevated temperatures to give an aldimine (7). The final step involves hydrolysis by passage through wet silica gel to give an α,β-unsaturated aldehyde (8).

$$
(3) \xrightarrow[0°]{\substack{K_2CO_3 \\ CH_2Cl_2}} \quad (6) \quad \xrightarrow[92\text{-}97\%]{80°} \quad (7) \quad \xrightarrow[78\%]{\substack{H_2O \\ SiO_2}} \quad (8)
$$

(6)                    (7)                    (8)

**Selective hydrolysis of an amide.** The analogous reagent α-chloro-N-cyclohexylacetaldonitrone was used in a total synthesis of vitamin $B_{12}$[5] to effect selective hydrolysis of an amide function in the presence of six methyl ester groups. Thus treatment of the α-chloronitrone with silver tetrafluoroborate generates the electrophilic cation (1). This

$$
R-C\!\!\begin{smallmatrix}O\\ \\NH_2\end{smallmatrix} + CH_2=CH-N\!\!\begin{smallmatrix}+ \\ \\ C_6H_{11}\end{smallmatrix}\!\!\!O \longrightarrow R-C\!\!\begin{smallmatrix}O \\ \\ NH\end{smallmatrix}\!\!\begin{smallmatrix}C_6H_{11} \\ +\!\!=N \\ O^-\end{smallmatrix} \xrightarrow{H_2O} R-C\!\!\begin{smallmatrix}O \\ \\ O H\end{smallmatrix}\!\!=O
$$

(2)              (1)                              (3)

$$
\xrightarrow{(CH_3)_2NH} \quad (4) \quad \longrightarrow R-C\!\!\begin{smallmatrix}O \\ \\ O\end{smallmatrix}\!\!\begin{smallmatrix}OH \\ \\ N(CH_3)_2\end{smallmatrix} \xrightarrow{H_2O} R-C\!\!\begin{smallmatrix}O \\ \\ OH\end{smallmatrix}
$$

(4)                              (5)

reacts with an amide (2) to give a 2-keto ethyl ester (3), which is rearranged to a labile ester (4) by treatment with dimethylamine in isopropanol. Hydrolysis of (4) gives the acid (5).

[1] U. M. Kempe, T. K. DasGupta, K. Blatt, P. Gygax, D. Felix, and A. Eschenmoser, *Helv.*, **55**, 2187 (1972).

[2] Prepared according to the procedure of C. R. Dick, *J. Org.*, **27**, 272 (1962).

[3] T. K. DasGupta, D. Felix, U. M. Kempe, and A. Eschenmoser, *Helv.*, **55**, 2198 (1972).

[4] P. Gygax, T. K. DasGupta, and A. Eschenmoser, *ibid.*, **55**, 2205 (1972).

[5] R. B. Woodward, *Pure Appl. Chem.*, **33**, 145 (1973).

**Chlorofluoromethylenetriphenylphosphorane**, $(C_6H_5)_3P{=}CFCl$. Mol. wt. 328.75.
The reagent is generated[1] *in situ* from sodium dichlorofluoroacetate[2] and triphenyl-phosphine.

*3β-Substituted 1-chloroperfluoro olefins.*[1] The reagent reacts with polyfluorinated ketones to give mixtures of *cis*- and *trans*-β-substituted 1-chloroperfluoro olefins in 34–70 % yield. The reagent reacts with nonfluorinated aldehydes and ketones to give

$$Ar(R)COR_f + (C_6H_5)_3P{=}CFCl \xrightarrow[\quad 90^0 \quad]{Triglyme} Ar(R)C(R_f){=}CFCl + (C_6H_5)_3PO$$

$$(R_f = CF_3, \quad C_2F_5, \quad CF_2Cl, \quad etc.)$$

the corresponding chlorofluoro olefins in rather low yields (9–49 %). Triglyme is the most satisfactory solvent for this Wittig reaction.

[1]D. J. Burton and H. C. Krutzsch, *J. Org.*, **35**, 2125 (1970).
[2]Sodium dichlorofluoroacetate is prepared by neutralization of dichlorofluoroacetic acid in ether with sodium carbonate, with subsequent removal of ether and water under reduced pressure followed by heating at 50° (4 mm.) overnight. Lithium dichlorofluoroacetate can be prepared in the same way.

**Chloroiridic acid, 1**, 131–132; **2**, 67–68; **3**, 47–48.
*Henbest reduction.* Reduction of the ketone (1) with chloroiridic acid and tri-methyl phosphite resulted in a 73 % yield of the axial 3β-alcohol (2). Sodium borohydride

(1)          (2)

reduction of (1) gave as the major product (74 %) the equatorial 3α-alcohol and only 17 % of the desired alcohol (2).[1]

[1]G. R. Pettit and J. R. Dias, *J. Org.*, **36**, 3207 (1971).

**Chloromethyl methyl ether, 1**, 132–135.
*Caution*: The reagent is carcinogenic. Bis(chloromethyl) ether is even more potent.[1]

*Chloromethylation of nitrophenols.*[2] *o*-Nitrophenols and, in some instances, *p*-nitrophenols can be chloromethylated with the reagent. The reaction fails with dinitro-phenols. A catalyst (zinc chloride) is required for the reaction.

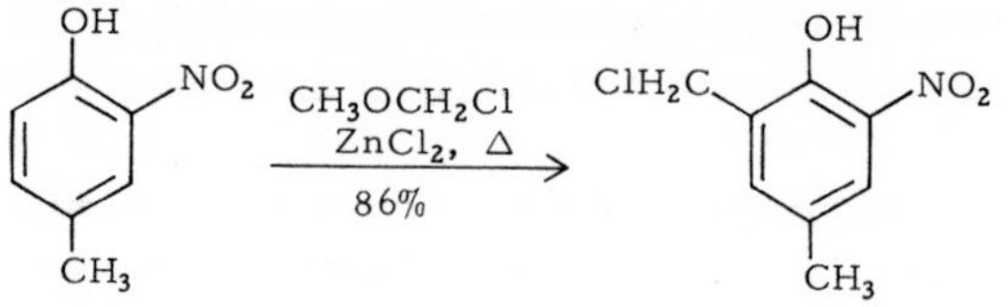

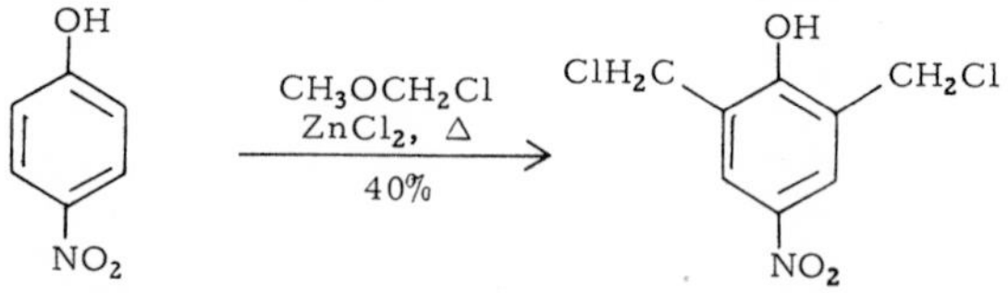

***Bischloromethylation.*** Electron-poor tetrasubstituted aromatic compounds such as nitromesitylene (1) are bischloromethylated when treated with chloromethyl methyl ether and 60% fuming sulfuric acid.[3]

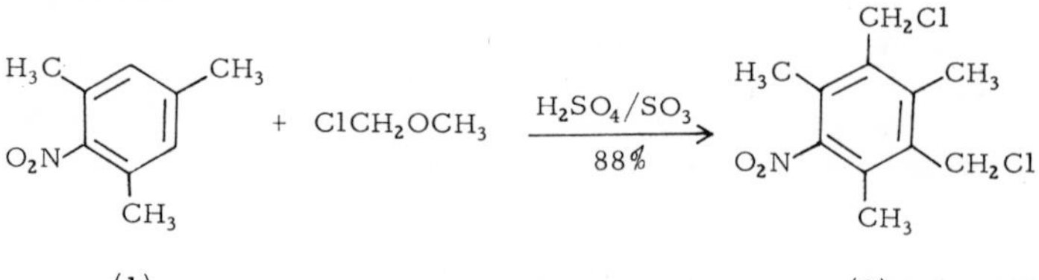

(1)                                                    (2)

***Methoxymethylation of β-keto esters.***[4]  The sodium salts of β-keto esters on alkylation with chloromethyl methyl ether in HMPT give almost exclusively the product of O-alkylation (96–100%). Yields of O-alkylated products are lower in less polar aprotic solvents. These enol ethers are reduced with lithium in liquid ammonia to the corresponding saturated esters (23–61% yield). The method is most efficient for reduction of a relatively hindered keto group of a β-keto ester.

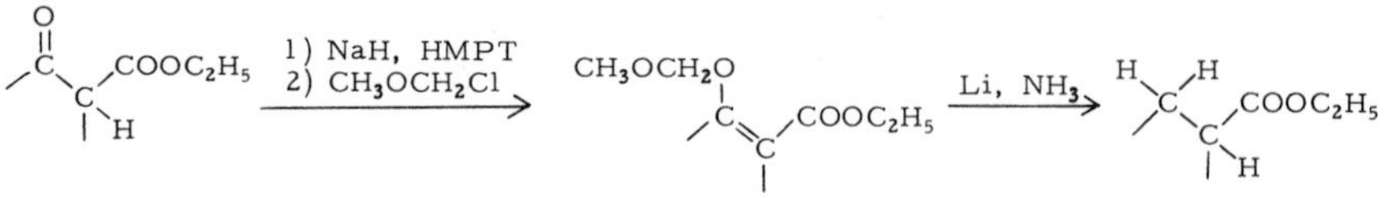

[1]B. L. Van Duisen, A. Sivak, B. M. Goldschmidt, C. Katz, and S. Melchionne, *J. Nat. Cancer Inst.*, **43**, 481 (1969); R. T. Drew, V. Cappiello, S. Larkin, and M. Kuschner, 1970 Conference American Industrial Hygiene Association, Abstract 164.
[2]V. Böhmer and J. Deveaux, *Org. Prep. Proc. Int.*, **4**, 283 (1972).
[3]H. Suzuki, *Bull. Chem. Soc. Japan*, **43**, 3299 (1970).
[4]R. M. Coates and J. E. Shaw, *J. Org.*, **35**, 2597, 2601 (1970).

**Chloromethyl methyl sulfide**, $ClCH_2SCH_3$. Mol. wt. 96.58, b.p. 105°. Suppliers: Aldrich, Columbia, Fluka, K and K, P and B, Schuchardt.

***Methylthiomethyl esters.***[1]  Methylthiomethyl esters are useful as protective groups for carboxylic acids. They are prepared by treatment of the acid with the reagent and

$$ClCH_2SCH_3 + RCOOH \xrightarrow[72-78\%]{CH_3CN} RCOOCH_2SCH_3$$

triethylamine in refluxing acetonitrile for 24 hr.[2] The esters are fairly stable toward aqueous alkali and are stable to mild reducing reagents ($NaBH_4$, $Zn-CH_3OH$). They can be hydrolyzed under acidic or neutral conditions. In one method, the ester is hydrolyzed by treatment with $CF_3COOH$ (15 min. at room temperature, 80–98% yield). In the other method, the ester is refluxed with methyl iodide and water in acetone for 17 hr. (yields 62–90%). In this procedure the first step consists in S-methylation:

$$RCOOCH_2SCH_3 \xrightarrow{CH_3I} RCOOCH_2\overset{+}{S}(CH_3)_2I^-$$

[1]T.-L. Ho and C. M. Wong, *J.C.S. Chem. Commun.*, 224 (1973).
[2]Procedure of R. H. Mills, M. W. Farrar, and O. J. Weinkauff, *Chem. Ind.*, 2144 (1962); see also J. H. Wagenkneckt, M. M. Baizer, and J. L. Chruma, *Syn. Commun.*, **2**, 215 (1972).

*m*-**Chloroperbenzoic acid**, **1**, 135–139; **2**, 68–69; **3**, 49–50.

**Baeyer–Villiger oxidation of carbohydrates.** Heyns *et al.*[1] have effected Baeyer–Villiger oxidation of carbohydrates with *m*-chloroperbenzoic acid in ethanol-free chloroform at room temperature.

Example:

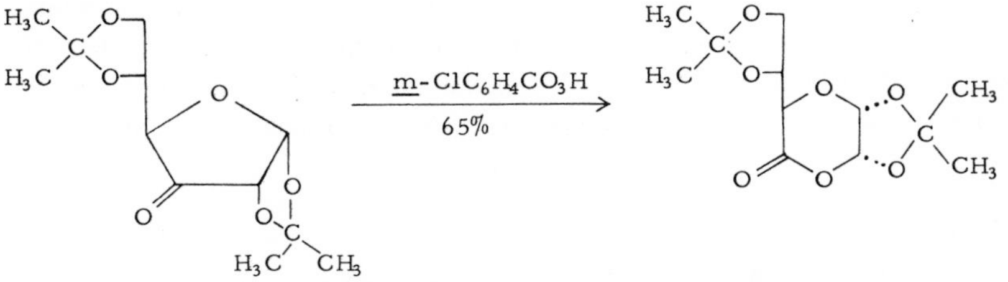

**Epoxidation at elevated temperatures.** Japanese chemists,[2,3] in an investigation directed toward a synthesis of tetrodotoxin, wanted to epoxidize the double bond of the olefin (1). However, this olefin is unreactive to *m*-chloroperbenzoic acid, peracetic

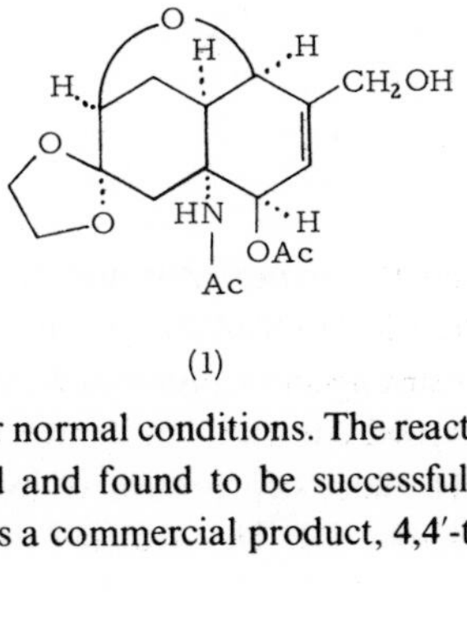

(1)

acid, and performic acid under normal conditions. The reaction at an elevated temperature (90°) was then examined and found to be successful if a radical inhibitor was present. The best inhibitor was a commercial product, 4,4'-thiobis-(6-*t*-butyl-3-methyl-phenol). Actually (1) could be converted into the desired epoxide in greater than 95 % yield by reaction with *m*-chloroperbenzoic acid in ethylene dichloride containing a small amount of the inhibitor at 90° for 2 hr.

The new procedure was found to be applicable to octene-1, dodecene-1, and methyl methacrylate, olefins which previously have been epoxidized only with the potent pertrifluoroacetic acid.

[1]P. Köll, R. Dürrfeld, U. Wolfmeier, and K. Heyns, *Tetrahedron Letters*, 5081 (1972).
[2]Y. Kishi, M. Aratani, H. Tanino, T. Fukuyama, T. Goto, S. Inoue, S. Sugiura, and H. Kakoi, *Chem. Commun.*, 64 (1972).
[3]Y. Kishi, M. Aratani, T. Fukuyama, F. Nakatsubo, T. Goto, S. Inoue, H. Tanino, S. Sugiura, and H. Kakoi, *Am. Soc.*, **94**, 9217 (1972).

**4-Chloro-2-phenylquinazoline,**

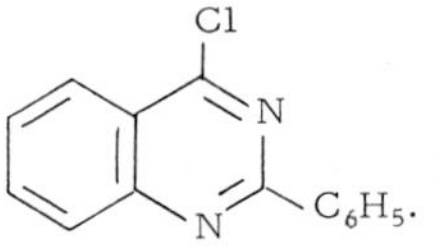

Mol. wt. 240.69,

(1)

m.p. 124–124.5°. Supplier: Aldrich (trade name AM-ex-OL).

*Preparation.*[1] The reagent can be prepared from benzoylanthranilamide.

*Phenols → anilines.*[2] A new general method for conversion of phenols into anilines has been described by Scherrer and Beatty.[2] A sodium phenoxide (prepared conveniently with sodium hydride) is condensed with 4-chloro-2-phenylquinazoline (1) at about 100° for 10 min. to give a 4-aryloxy-2-phenylquinazoline (2) in 70–85 % yield.

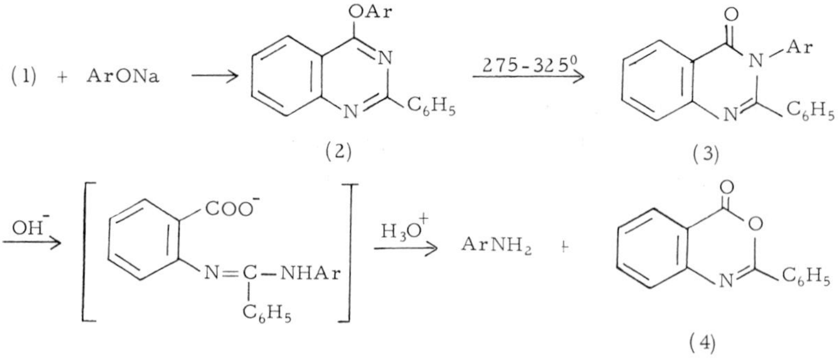

The key step in the process is the thermal rearrangement of (2) to a 3-aryl-2-phenyl-4(3H)-quinazolinone (3). This 1,3-O to N aryl migration was first observed by Tschitschi-babin and Jeletzky.[3] This rearrangement proceeds at useful rates in the temperature range 275–325°; it can be carried out neat, but the reaction is generally cleaner when run in heavy mineral oil. The final step consists of hydrolysis to the aniline and 2-phenyl-4H-3,1-benzoxazine-4-one (4). This reaction can be carried out by alkaline hydrolysis to an amidine intermediate, which on acidification yields the aniline and (4). Alternatively, (3) can be heated with potassium hydroxide in ethylene glycol. Overall yields are in the range 40–70%.

[1]M. Endicott, E. Wick, M. L. Mercury, and M. L. Sherrill, *Am. Soc.*, **68**, 1299 (1946).
[2]R. A. Scherrer and H. R. Beatty, *J. Org.*, **37**, 1681 (1972).
[3]A. E. Tschitschibabin and N. P. Jeletzky, *Ber.*, **57**, 1158 (1924).

**Chloroplatinic acid–Stannous chloride.**

*Carbonylation of terminal olefins.*[1] Terminal olefins can be carbonylated to linear esters in high yield by use of a chloroplatinic acid–stannous chloride couple at 90° and

$$RCH{=}CH_2 + CO + CH_3OH \xrightarrow{\ H_2PtCl_6\text{-}SnCl_2\ } RCH_2CH_2COOCH_3 + RCH(CH_3)COOCH_3$$

3000 psi of CO in acetone, methyl isobutyl ketone, or 1,2-dimethoxyethane. Acids are obtained by substitution of water for methanol.

[1]L. J. Kehoe and R. A. Schell, *J. Org.*, **35**, 2846 (1970).

**N-Chloropolymaleimide (PNCS),**

Polymaleimide is prepared by free-radical polymerization of maleimide in the presence of divinylbenzene (5%) or N,N′-methylenebisacrylamide (5–10%) as cross-linking agents. PNCS is obtained by addition of chlorine in carbon tetrachloride to a suspension of the polymer in aqueous sodium hydroxide solution.

*Chlorination of alkylbenzenes.*[1] Chlorination of alkylbenzenes with PNCS at 100–140° for 16–20 hr. yields exclusively products of aromatic substitutions. In contrast chlorination of alkylbenzenes under the same conditions with N-chlorosuccinimide (NCS) yields a mixture of products resulting from chlorination of the benzene ring and of the alkyl group(s). The difference in specificity between PNCS and NCS is attributed to the influence of the polymeric backbone in PNCS. Indeed, chlorination of alkylbenzenes with NCS in the presence of succinimide results mainly in chlorination of the benzene ring. *Cf.* **N-Bromopolymaleimide**, this volume.

[1]C. Yaroslavsky and E. Katchalski, *Tetrahedron Letters*, 5173 (1972).

**N-Chlorosuccinimide–Dimethyl sulfide,**       Mol. wt. 163.67,

m.p. 70–72°.

Vilsmaier and Sprügel[1] obtained this complex in 85% yield by the reaction of N-chlorosuccinimide with dimethyl sulfide in methylene chloride at 0°. It was shown to be an intermediate in the reaction of NCS with dimethyl sulfide to give chloromethyl methyl thioether:

*Oxidation of primary and secondary alcohols to carbonyl compounds.* Corey and Kim[2] report that the complex of N-chlorosuccinimide and dimethyl sulfide is somewhat superior to the complex of dimethyl sulfide and chlorine (this volume) for oxidation of primary and secondary alcohols; the formation of hydrogen chloride is avoided and yields are generally higher. The procedure is illustrated for the oxidation of 4-*t*-butylcyclohexanol (2) to 4-*t*-butylcyclohexanone (4). The complex (1) is prepared by addition of dimethyl sulfide (4.1 mmole) to a stirred solution of NCS (3.0 mmole) in toluene at 0° under argon. The mixture is cooled to −25° and a solution of 4-*t*-butylcyclohexanol (2.0 mmole, mixture of *cis* and *trans*) in toluene is added dropwise. The stirring is continued for 2 hr. at −25° and then triethylamine (3.0 mmole) in toluene is added dropwise. The ketone (4) is obtained in almost quantitative yield. As in oxidation with the complex of dimethyl sulfide and chlorine, an intermediate sulfoxonium complex (3) is involved.

1-Octanol was oxidized to 1-octanal in 96% yield by a similar procedure.

The new oxidation process has one important limitation. Allylic or benzylic alcohols are not oxidized but instead replacement of hydroxyl by chlorine is observed. Still another reaction may occur in polar media; thus, methylthiomethyl ether formation becomes pronounced when methylene chloride–dimethyl sulfoxide is used as solvent.

The reagent was used by Corey and Kim[3] in an improved synthesis of prostaglandins. Thus oxidation of the hydroxy acid (5) to the keto acid (6) was effected in > 90% yield with use of N-chlorosuccinimide–dimethyl sulfide. The oxidation had been carried out previously with Jones reagent at −20° in about 70% yield. The carboxyl function of (5) was protected by *in situ* conversion to the isopropyldimethylsilyl ester by reaction of (5) with isopropyldimethylsilyl chloride and triethylamine in toluene. The oxidizing reagent was then added to give the silyl ester of (6), which was hydrolyzed by dilute hydrochloric acid to (6).

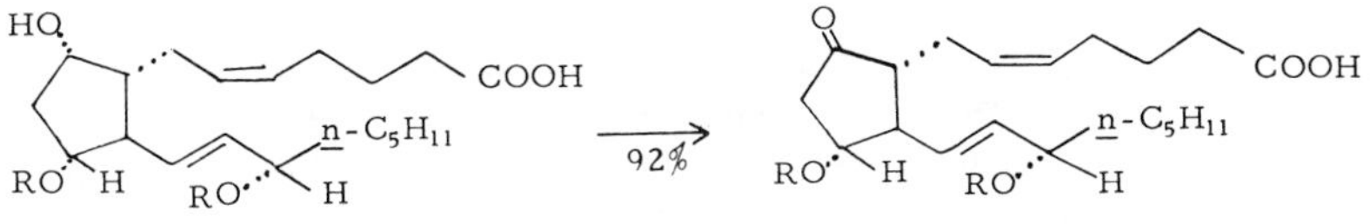

(5, R = THP)                                    (6)

The same paper reports oxidation of the hydroxy lactone (7) to the aldehyde (8) using the 1:1 complex of chlorine and methyl phenyl sulfide (thioanisole) in $CCl_4$–$CH_2Cl_2$.

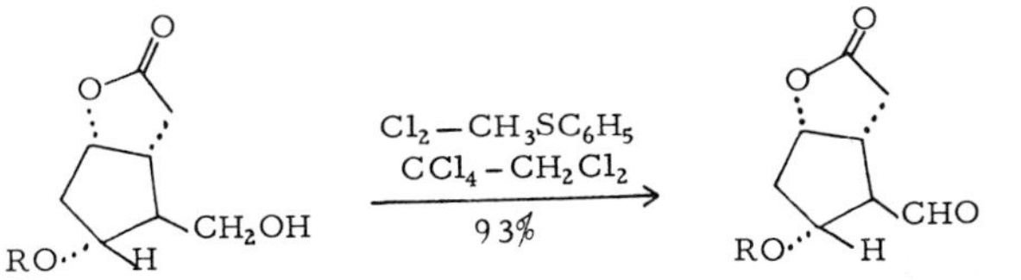

$$Cl_2 - CH_3SC_6H_5$$
$$CCl_4 - CH_2Cl_2$$
$$93\%$$

(7, R = p-$C_6H_5C_6H_4CO$)                     (8)

*Selective conversion of allylic and benzylic alcohols into halides.*[4] It was noted above that allylic or benzylic alcohols are not oxidized by the N-chlorosuccinimide–dimethyl sulfide reagent but instead are converted into chlorides. Corey et al.[4] now report that this procedure can be used to form allylic and benzylic chlorides in high yield if the sulfoxonium intermediate corresponding to (3, above) is allowed to decompose in methylene chloride without addition of a tertiary amine. For example, the allylic or benzylic alcohol is treated with the complex of N-chlorosuccinimide–dimethyl sulfide as above in methylene chloride and then allowed to stand for 4 hr. at −25°. In the case of benzhydrol [$(C_6H_5)_2$CHOH], benzhydryl chloride can be obtained in > 95% yield.

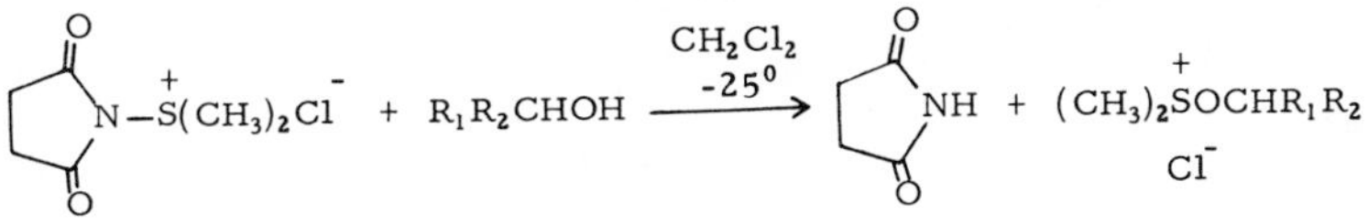

$$R_1R_2CHCl + (CH_3)_2SO$$
(3)

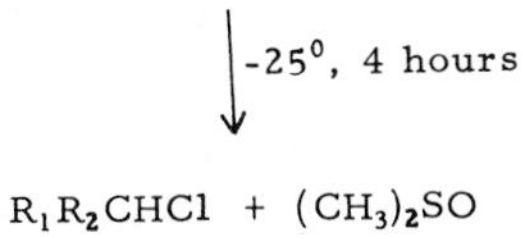

Under the same conditions 2-cyclohexene-1-ol is converted into 3-chlorocyclohexene in > 95% yield.

Use of the 1:1 complex of N-bromosuccinimide and dimethyl sulfide in place of (1) similarly converts allylic and benzylic alcohols into the corresponding bromides in 80–90% yield. For example, geraniol was converted into geranyl bromide in 82% yield.

Under these conditions saturated primary and secondary aliphatic and alicyclic alcohols are recovered essentially unchanged. This selectivity was illustrated impressively by conversion of Z-3-methyl-2-pentene-1,5-diol (4) by treatment with the complex (1) in methylene chloride briefly at −20° and then at 0° for 1 hr. into the allylic monochloride (5), isolated in 87% yield.

$$
\begin{array}{ccc}
\underset{\text{HOCH}_2\text{CH}_2}{\overset{\text{H}_3\text{C}}{>}}\text{C}{=}\text{C}\underset{\text{CH}_2\text{OH}}{\overset{\text{H}}{<}} & \xrightarrow[87\%]{(1)} & \underset{\text{HOCH}_2\text{CH}_2}{\overset{\text{H}_3\text{C}}{>}}\text{C}{=}\text{C}\underset{\text{CH}_2\text{Cl}}{\overset{\text{H}}{<}} \\
(4) & & (5)
\end{array}
$$

[1]E. Vilsmaier and W. Sprügel, *Ann.*, **747**, 151 (1971).
[2]E. J. Corey and C. U. Kim, *Am. Soc.*, **94**, 7586 (1972).
[3]*Idem, J. Org.*, **38**, 1233 (1973).
[4]E. J. Corey, C. U. Kim, and M. Takeda, *Tetrahedron Letters*, 4339 (1972).

## Chlorosulfonyl isocyanate (CSI), 1, 117–118; 2, 70; 3, 51–53.

[2+2] *Cycloadditions with olefins* (3, 51–52). Olefins with an internal, unbranched double bond react with the reagent by [2+2] cycloaddition to give α,β-disubstituted β-lactam-N-sulfochlorides (1) and (2) as the major products.[1]

$$
\begin{array}{l}
\text{R}^1\text{CH}_2\text{CH}{=}\text{CHCH}_2\text{R}^2 \\
+ \\
\text{O}{=}\text{C}{=}\text{NSO}_2\text{Cl}
\end{array}
\left.
\begin{array}{l}
\\
\\
\end{array}
\right\}
\begin{array}{l}
\nearrow \\
\\
\xrightarrow{<5\%}
\end{array}
$$

$$
\begin{array}{cc}
\underset{\overset{|}{\text{SO}_2\text{Cl}}}{\overset{\overset{\text{R}^1\text{CH}_2\text{CHCHCH}_2\text{R}^2}{| \quad |}}{\text{N}{-}\text{C}{=}\text{O}}} & + \quad \underset{\overset{|}{\text{SO}_2\text{Cl}}}{\overset{\overset{\text{R}^1\text{CH}_2\text{CHCHCH}_2\text{R}^2}{| \quad |}}{\text{O}{=}\text{C}{-}\text{N}}} \\
(1) & (2)
\end{array}
$$

$$
\begin{array}{cc}
\underset{\text{O}{=}\text{C}{-}\text{NHSO}_2\text{Cl}}{\text{R}^1\text{CH}{=}\text{CHCHCH}_2\text{R}^2} & + \quad \underset{\text{O}{=}\text{C}{-}\text{NHSO}_2\text{Cl}}{\text{R}^1\text{CH}_2\text{CHCH}{=}\text{CHR}^2} \\
(3) & (4)
\end{array}
$$

Malpass and Tweddle[2] have isolated a β-lactam derivative (6) by the reaction of CSI with camphene (5) in $CH_2Cl_2$ at −60°. At higher temperatures, (6) rearranges to thermodynamically preferred products (7) and (8).

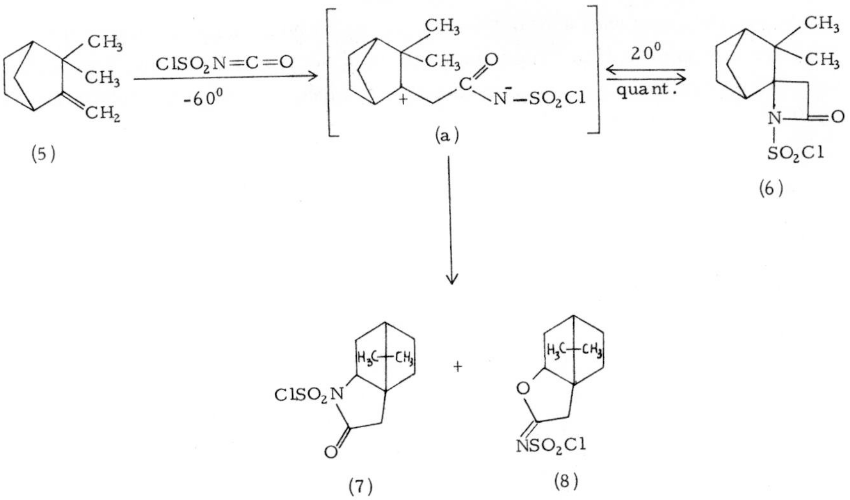

The scheme shows compound (5) reacting with $ClSO_2N=C=O$ at $-60°$ to give intermediate (a), which at $20°$ (quant.) gives (6), and branches down to give (7) + (8).

***Addition to bicyclic monoterpene olefins.***[3] The reagent reacts regiospecifically with
α-pinene (1) at $-73°$ in ether to give the unstable 1:1 adduct (2; 3-chlorosulfonyl-2,8,8-
trimethyl-3-azatricyclo[5.1.1.0$^{2,5}$]nonane-4-one) in 65% yield. This adduct rearranges
either thermally or on contact with silicic acid via a Wagner–Meerwein rearrangement
to (3), yield 40%. The adduct (2) is reduced by sodium sulfite to the β-lactam (4; 2,8,8-
trimethyl-3-azatricyclo[5.1.1.0$^{2,5}$]nonane-4-one) in 61% yield.

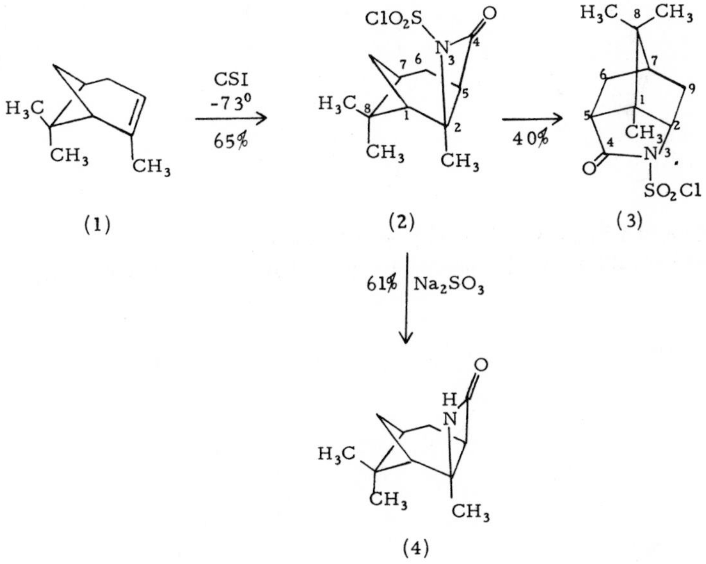

β-Pinene (5) reacts with CSI at −73° to give the very unstable 1:1 adduct (6); the structure is established by reduction to (9; 2-azetidinone-4-spiro-2′-(6′,6′-dimethylbicyclo[3.1.1]heptane). The unstable adduct rearranges on a silica gel column to (7) and (8).

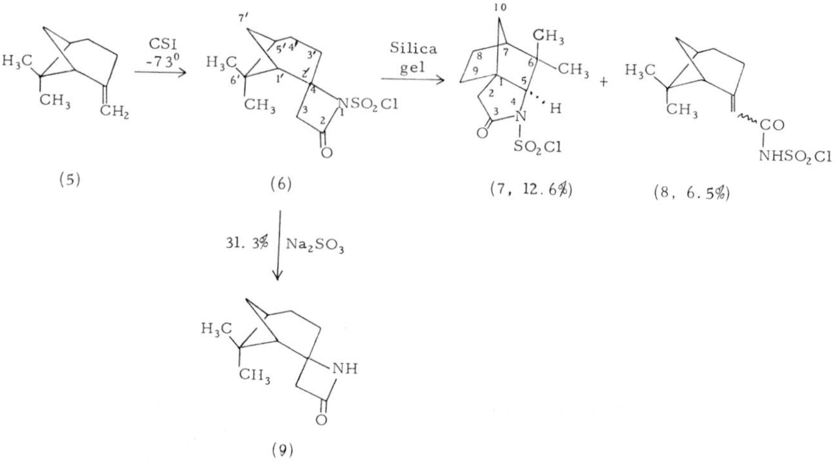
CSI
-73°
Silica
gel
31.3%
$Na_2SO_3$
(5)
(6)
(7, 12.6%)
(8, 6.5%)
(9)

Camphene (10) reacts with CSI in ether at −3 to 0° to give the adduct (12; 4-chlorosulfonyl-10,10-dimethyl-4-azatricyclo[5.2.1.$0^{1,5}$]decane-3-one) in 77% yield.

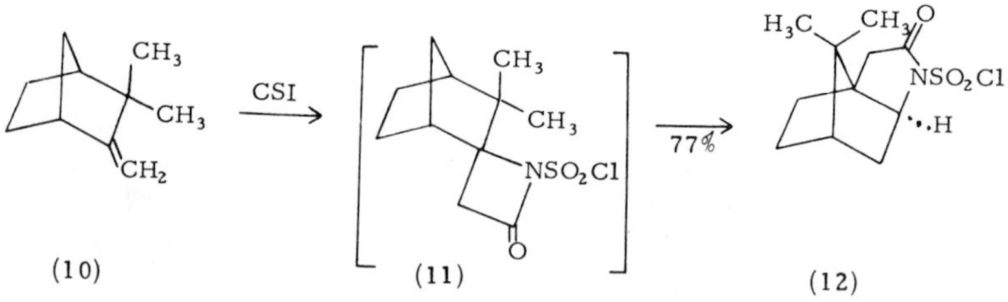
CSI
77%
(10)
(11)
(12)

Addition of CSI to $\Delta^3$-carene (13) in refluxing ether (23 hr) gives the 1:1 adduct (14; 4-chlorosulfonyl-3,9,9-trimethyl-4-azatricyclo[6.1.0.$0^{3,6}$]nonane-5-one) in 72% yield.

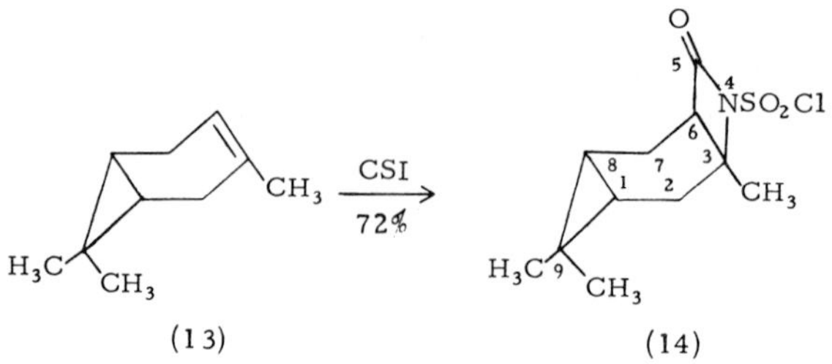
CSI
72%
(13)
(14)

*Cycloadditions to strained bicyclic hydrocarbons.*[4] CSI reacts with simple alkenes by [2 + 2] cycloadditions. However, as the complexity of the system is increased [1+4], [1+5], and [1+6] cycloadditions can be realized. Thus CSI reacts with tricyclo[4.1.0.0$^{2,7}$]heptane (1) to give, after hydrolysis, the lactam (2) as the major product. Under the same conditions the lactam (4) is obtained as the major product

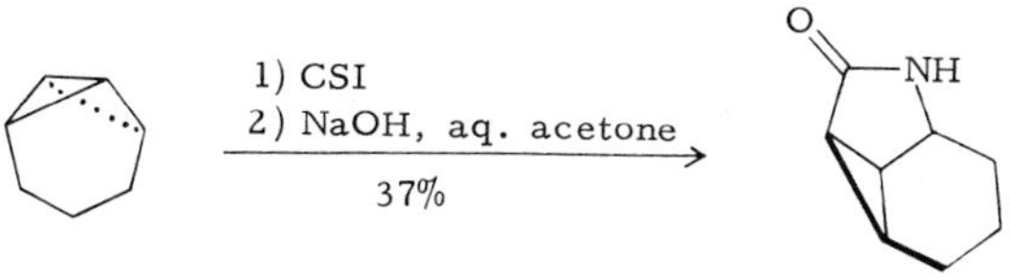

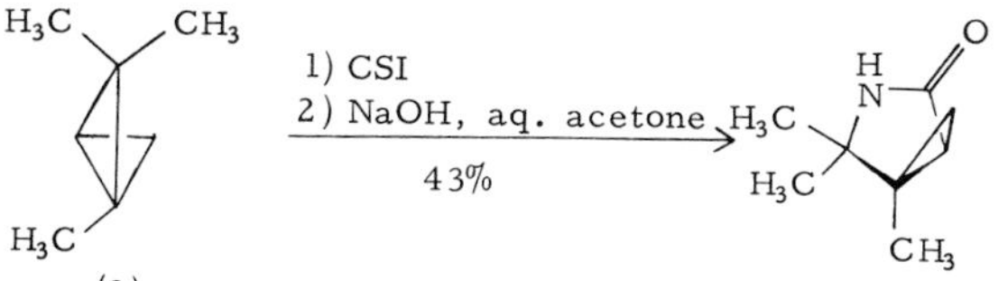

from 1,2,2-trimethylbicyclo[1.1.0]butane (3). 1,2,2,3-Tetramethylbicyclo[1.1.0]butane (5) affords the lactam (6). Under similar conditions, 1,3-dimethylbicyclo[1.1.0]butane (7) gives the bicyclic lactam (8).[4,5]

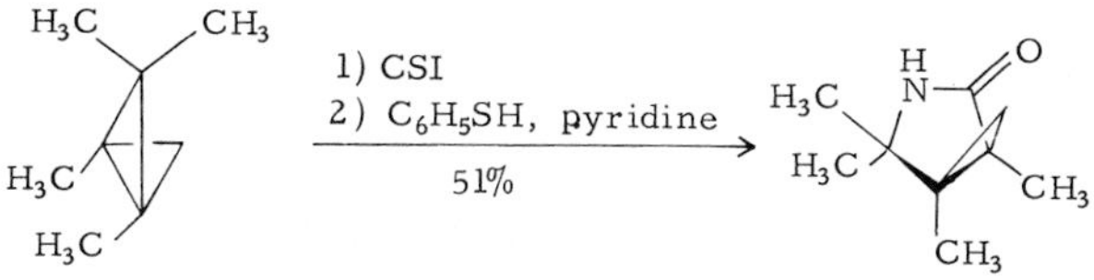

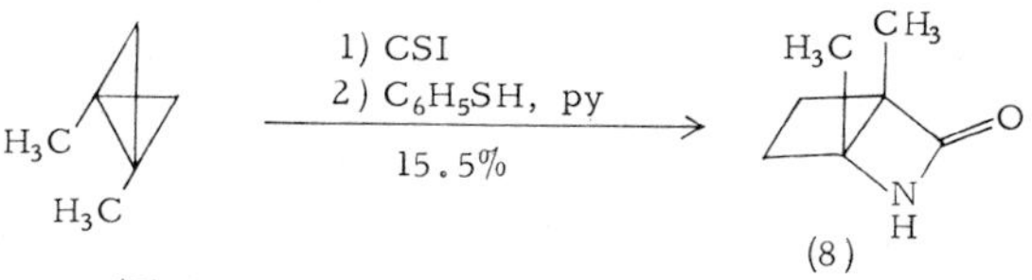

[1] H. Bestian, H. Biener, K. Clauss, and H. Heyn, *Ann.*, **718**, 94 (1968).
[2] J. R. Malpass and N. J. Tweddle, *J.C.S. Chem. Commun.*, 1244 (1972).
[3] T. Sasaki, S. Eguchi, and H. Yamada, *J. Org.*, **38**, 679 (1973); *see also* J. R. Malpass, *Tetrahedron Letters*, 4951 (1972).

[4]L. A. Paquette, G. R. Allen, Jr., and M. J. Broadhurst, *Am. Soc.*, **93**, 4503 (1971), and references cited therein.
[5]L. A. Paquette, private communication.

**Chlorotrifluoroethylene** (2), 1, 1072.

The addition of chlorotrifluoroethylene (2) to *p*-nitrophenylacetylene or *p*-methoxyphenylaeetylene yields the cyclobutenes (3a) and (3b), which on treatment with sulfuric acid afford the corresponding cyclobutenediones (4a,b).[1]

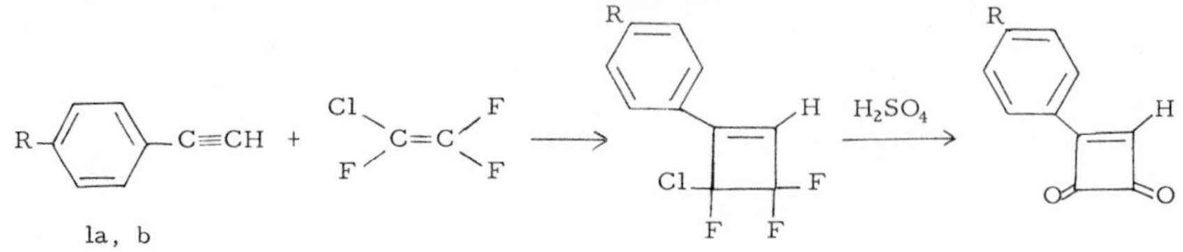

la, b

a = $NO_2$          (2)                    (3)                    (4)

b = $OCH_3$

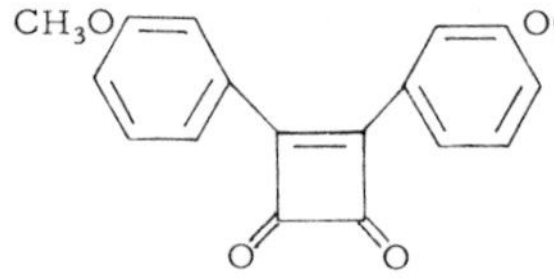

(5)

The dione (5) was prepared but not tested for quinonelike properties.

[1]W. Ried, A. H. Schmidt, and W. Kuhn, *Ber.*, **104**, 2622 (1971).

**1-Chloro-N,N,2-trimethylpropenylamine**, $(CH_3)_2C{=}C\overset{\displaystyle Cl}{\underset{\displaystyle N(CH_3)_2}{\big\backslash}}$          Mol. wt.   133.63,

b.p. 40°/25 mm.

*Preparation.* This α-chloro enamine (1) is prepared in about 80% yield by the reaction of phosgene with N,N-dimethylisobutyramide followed by elimination of hydrogen chloride with triethylamine.[1]

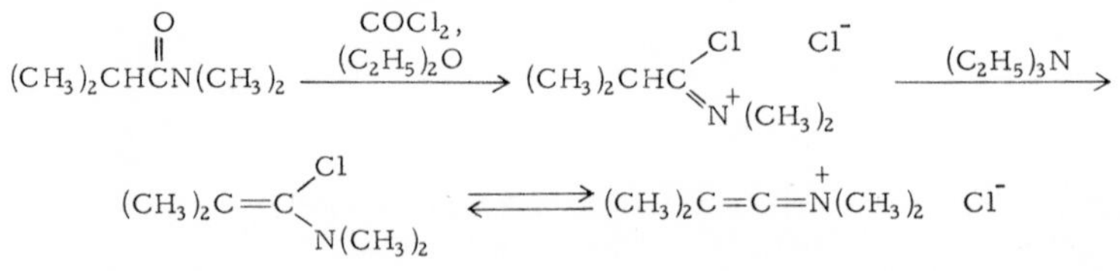

(1)                              (la)

*Aminoalkenylation.*[2] The reagent, in the polar form (1a), reacts readily with electron-rich aromatics to give products of aminoalkenylation:

$$\text{ArH} + (1a) \xrightarrow{(C_2H_5)_3N} (CH_3)_2C=C\overset{Ar}{\underset{N(CH_3)_2}{}} \xrightarrow{H_3O^+} (CH_3)_2CH\overset{O}{\overset{\|}{C}}Ar$$

Thus the reagent reacts readily with furane, pyrrole, and N,N-dimethylaniline to give substitution products in yields of 85–95%.

*Alkylated cyclobutanones.*[3] This chloro enamine on reaction with silver fluoroborate in methylene chloride at −60° is converted into tetramethylketenimmonium

$$(CH_3)_2C=C\overset{Cl}{\underset{N(CH_3)_2}{}} \xrightarrow[-\text{AgCl}]{\text{AgBF}_4} (CH_3)_2C=C=\overset{+}{N}(CH_3)_2 \ BF_4^-$$

(1)                         (2)

fluoroborate (2). This salt readily undergoes [2+2] cycloaddition to olefins to give

$$(2) + \ >C=C< \ \longrightarrow \ (3) \ BF_4^- \xrightarrow{OH^-} (4)$$

(3)                    (4)

iminium salts (3), which are easily hydrolyzed to alkylated cyclobutanones (4). Yields of cycloadducts are high (80–95%).

[1]L. Ghosez, B. Haveaux, and H. G. Viehe, *Angew. Chem., internat. Ed.*, **6**, 454 (1969).
[2]J. Marchand-Brynaert and L. Ghosez, *Am. Soc.*, **94**, 2869 (1972).
[3]*Idem, ibid.*, **94**, 2871 (1972).

**Chromic acid, 1,** 142–144; **2**, 70–72; **3**, 54.

*Jones reagent in acetone* (**1**, 142–143; **2**, 70–71; **3**, 54). Benzoins are oxidized to benzils in about 90–95% yield by oxidation with Jones reagent in acetone.[1] The two-

$$C_6H_5CH\underset{OH}{\overset{}{C}}OC_6H_5 \xrightarrow[95\%]{\overset{H_2CrO_4 \ in}{acetone}} C_6H_5COCOC_6H_5$$

phase method of Brown and Garg (**1**, 143–144; **2**, 71) can also be used, but the yields are somewhat lower (75–90%).

*Two-phase oxidation* (**1**, 143; **2**, 71). Brown *et al.*[2] have published the definitive paper on the two-phase oxidation of secondary alcohols in diethyl ether with aqueous chromic acid. A number of water-immiscible solvents were examined, but diethyl ether is clearly the most satisfactory. Two standard procedures were developed. In one, the stoichiometric amount of sodium dichromate and sulfuric acid in water is added to the ethereal solution of the alcohol at 25–30° and the reaction is allowed to proceed for 2 hr. The following equation is applicable:

$$3 \ R_2CHOH \ + \ Na_2Cr_2O_7 \ + \ 4 \ H_2SO_4 \ \longrightarrow \ 3 \ R_2CO \ + \ Na_2SO_4 \ + \ Cr_2(SO_4)_3 \ + \ 7 \ H_2O$$

Yields of ketones are, in general, 85–97%. Yields, however, are significantly lower in the case of strained bicyclic alcohols. In this case, a modified procedure is recommended. In this procedure a 100% excess of the oxidant is used and the reaction is carried out at 0° for 15 min.

[1]T.-L. Ho, *Chem. Ind.*, 807 (1972).
[2]H. C. Brown, C. P. Garg, and K.-T. Liu, *J. Org.*, **36**, 387 (1971).

**Chromic anhydride, 1**, 144–147; **2**, 72–75; **3**, 54–57.

*CrO₃–pyridine complex* (**1**, 145–146; **2**, 74–75; **3**, 55), Ratcliffe[1] (**3**, 56, ref. 4) has described complete details for the preparation of the chromium trioxide–pyridine complex *in situ* and the use for oxidation of 1-decanol (1) to 1-decanal (2). A solution

$$CH_3(CH_2)_8CH_2OH \ \xrightarrow[\substack{CH_2Cl_2 \\ 83\%}]{CrO_3 \cdot 2 \ Py} \ CH_3(CH_2)_8CHO$$

$$(1) \qquad\qquad\qquad\qquad\qquad (2)$$

of pyridine (1.2 mole) in methylene chloride is cooled to 5°, and chromium trioxide (0.6 mole, J. T. Baker) is added. The deep-burgundy solution is stirred at 5° for five more minutes and then allowed to warm slowly to 20°. The alcohol (1, 0.1 mole) in methylene chloride is added rapidly. Workup affords the aldehyde (2) in 83% yield. The actual yield is higher, but decanal is very volatile.

The reagent is comparable to the dipyridine–chromium(VI) oxide of Collins (this volume); however, pure Collins reagent is very hygroscopic.

Conjugated acetylenic ketones can be prepared in about 40% yield by oxidation of alkynes with chromium trioxide–pyridine complex:

$$RC{\equiv}CCH_2R \ \xrightarrow{CrO_3 \cdot Py_2} \ RC{\equiv}CCR$$
$$\overset{\|}{O}$$

For example, oxidation of 4-octyne (0.015 mole) with $CrO_3 \cdot Py_2$ (0.225 mole) in methylene chloride at room temperature for 24 hr. (N₂) gives 4-octyne-3-one in 42% yield. Anhydrous sodium chromate ($Na_2CrO_4$) can also be used, but yields are lower. Considerable amounts of starting alkyne are recovered.[2]

*Chromic anhydride in graphite.*[3] Supplier: Alfa Inorganics ("Seloxcette"). Croft[4] showed that chromic acid can be intercalated in graphite by diffusion of vapors of $CrO_3$ in the lattice of graphite at reduced pressure and high temperature. The method has been improved by Lalancette et al.[3] to give a product containing 55–60% by weight of $CrO_3$; the material resembles graphite in appearance.

The new reagent is a very specific oxidizing reagent for conversion of primary alcohols to the corresponding aldehydes. The alcohol is treated with the reagent in refluxing toluene; after 24 hr., the solvent is evaporated. The aldehyde is obtained in high yield, 70–100%. Residual chromium salts are retained in the lattice of graphite. Secondary and tertiary alcohols are not oxidized. (Commercial material is not so selective.) 1,2-Diols are oxidized with rupture of the carbon chain to give aldehydes.

$$\text{Benzyl alcohol} \longrightarrow \text{benzaldehyde} \quad (98\%)$$
$$\text{Cinnamic alcohol} \longrightarrow \text{cinnamaldehyde} \quad (100\%)$$
$$\text{Phenylethanediol} \longrightarrow \text{benzaldehyde} \quad (80\%)$$
$$\text{Cyclohexanol} \longrightarrow \text{cyclohexanone} \quad (2\%)$$

[1] R. W. Ratcliffe, *Org. Syn.*, submitted 1973.
[2] J. E. Shaw and J. J. Sherry, *Tetrahedron Letters*, 4379 (1971).
[3] J.-M. Lalancette, G. Rollin, and P. Cumas, *Canad. J. Chem.*, **50**, 3058 (1972).
[4] R. C. Croft, *Australian J. Chem.*, **9**, 201 (1956).

**Chromium(II)–Amine complex,** 3, 57–59.

*Hydrogenolysis of carbon–halogen bonds.*[1]

A mixture of dimethylformamide and ethylenediamine is stirred magnetically under nitrogen and an aqueous solution of chromium(II) perchlorate is added with a hypodermic syringe to form a purple solution complex. A solution of 1.66 g. of 1-bromonaphthalene in oxygen-free dimethylformamide is added, and the mixture is stirred until the colour changes from purple to deep red, and then poured into aqueous hydrochloric acid; extraction with ether and workup affords 0.96–1.00 g. of naphthalene.

[1] R. S. Wade and C. E. Castro, *Org. Syn.*, **52**, 62 (1972).

**Chromous acetate,** 1, 147–149; 2, 75–76; 3, 59–60. Additional supplier: Lancaster.

*Reduction of α,β-oxidoketones.* In a study of potential routes to the A/B ring system of cardiac-active steroids (periplogenin, strophanthidin), Robinson and Henderson[1] studied the reduction of 4β,5β-oxidocholestane-3-one (1) with chromous acetate. The best results were obtained by reduction of (1) with a large excess of freshly

(1)                          (2)                          (3)

prepared chromous acetate in aqueous acetone. In this case the desired product (2) could be obtained in about 50% yield. The minor product (3) can be reconverted into (1); hence (2) can be obtained in satisfactory yield by recycling.

Reduction of the 3-keto group of (2) to the desired 3$\beta$-axial alcohol is best accomplished with W-2 Raney nickel in refluxing ethanol.

[1]C. H. Robinson and R. Henderson, *J. Org.*, **37**, 565 (1972).

**Chromyl chloride**, **1**, 151; **2**, 79; **3**, 62.

*Oxidation of terminal olefins to aldehydes.* In a detailed procedure[1] a mixture of 1.0 mole of 2,4,4-trimethyl-1-pentene (Eastman or MCB) and 1 l. of methylene chloride is stirred mechanically in a 5-l. three-necked flask fitted with a thermometer and a

dropping funnel with a calcium chloride drying tube and immersed in an ice-salt bath. The stirred solution is brought to 0–5° and a solution of 158 g. (1.02 mole) of freshly distilled chromyl chloride (Alfa Inorganics) in 200 ml. of methylene chloride is added dropwise with stirring from the dropping funnel at 0–5° (in about 60 min.). The reaction mixture is stirred for 15 min., and 184 g. of 90–95% technical zinc dust is added. This approximate fivefold excess is necessary to reduce the chromium higher valence salts and thereby eliminate overoxidation and double-bond cleavage. The mixture is stirred for 5 min., 1 l. of ice water and 400 g. of ice are added as rapidly as possible (temperature rise to 8–10°), and the mixture is stirred for an additional 15 min. The ice-salt bath is replaced by a heating mantle, and the flask is fitted for steam distillation. After distillation of the methylene chloride the residue is steam distilled until the distillate gives a negative test with the 2,4-dinitrophenylhydrazine reagent. The combined organic material (undried) is distilled through a 56-cm. vacuum-jacketed Vigreux column to remove the solvent. The product is transferred to a 250-ml. round-bottomed flask, a small amount of methylene chloride is removed, and the fraction boiling at 45–52° (15 mm.) is collected to give 90–100 g. (70–78%) of 2,4,4-trimethylpentanal.

*α-Chloroketones.*[2] Di- and trisubstituted olefins react with chromyl chloride in acetone solution at low temperatures to give α-chloroketones in good yield (40–90%):

$$\underset{R}{\overset{H}{\diagdown}}C=C\underset{R^2}{\overset{R^1}{\diagup}} \xrightarrow{CrO_2Cl_2} R-\overset{\overset{\displaystyle O}{\|}}{C}-\overset{\overset{\displaystyle Cl}{|}}{\underset{\diagdown R^2}{C}}\diagup^{R^1}$$

The success of the method appears to be the use of acetone as solvent. Addition of zinc dust to the crude reaction mixture results in reduction to the corresponding ketone. Yields of chloroketones are highest if the reaction is carried out at $-70°$, but are still reasonable at reaction temperatures of $-5$ to $3°$. Thus oxidation of cyclododecene at $-75°$ with chromyl chloride gave $\alpha$-chlorocyclododecanone in 79 % yield; the yield of the chloroketone was 69 % from a reaction conducted at $-5°$.

*Oxidation of cyclohexene to the oxide.* Decomposition of the chromyl chloride–cyclohexene complex, suspended in $CH_2Cl_2$, with cold 5 % aqueous sodium bicarbonate allowed isolation of cyclohexene oxide in 1.5 % yield.[3]

[1]F. Freeman, R. H. DuBois, and T. G. McLaughlin, *Org. Syn.*, **51**, 4 (1971).
[2]K. B. Sharpless and A. Y. Teranishi, *J. Org.*, **38**, 185 (1973).
[3]F. W. Bachelor and U. O. Cheriyan, *J.C.S. Chem. Comm.*, 195 (1973).

**Cobalt(II) acetate**, $Co(OAc)_2 \cdot 4H_2O$. Mol. wt. 249.08. Suppliers: Alfa, Baker, Eastman, Fisher, Howe and French, MCB, Sargent, Fluka, E. Merck, Riedel de Haën.

*Oxidation of alkyltoluenes.*[1] A Shell Oil group has studied the oxidation of alkyl-toluenes by oxygen in the presence of cobalt(II) acetate as catalyst. The reaction involves the continuous conversion of Co(II) ions to Co(III) ions promoted by methyl ethyl ketone. The methyl group is preferentially oxidized. Thus *p*-cymene affords *p*-isopropyl-benzoic acid (90 % yield) and *p*-acetobenzoic acid (10 % yield). The relative ease of

(1)        (2, 90%)        (3, 10%)

oxidation of alkyl groups is methyl > ethyl > isopropyl ~ *sec*-butyl. An electron-transfer mechanism involving radical cations is proposed.

*Oxidation of n-butane.*[2] In the presence of oxygen, Co(II) is converted into Co(III), the actual catalyst for oxidation of alkanes by oxygen; thus oxidation of *n*-butane by Co(III) ion at 100° at a pressure of 17–24 atm. gives acetic acid (83.5 % yield) together with traces of *n*-butyric acid, propionic acid, and methyl ethyl ketone. Oxidation of *n*-pentane under similar conditions gives acetic acid (48 % yield) and propionic acid (27 % yield). Isobutane is relatively inactive. The reaction involves electron transfer in which cobalt ions function as chain carriers.

[1]A. Onopchenko, J. G. D. Schulz, and R. Seekircher, *J. Org.*, **37**, 1414 (1972).
[2]A. Onopchenko and J. G. D. Schulz, *ibid.*, **38**, 909 (1973).

**Copper(I) acetylacetonate,** $CuC_5H_7O_2$. Rose red. Mol. wt. 162.65.

*Preparation.*[1] The reagent is prepared by the reaction of acetylacetone with an ammoniacal solution of cuprous hydroxide.

*1,4-Diketones.*[2] Copper(I) acetylacetonate is a very effective copper catalyst for the insertion of an α-ketocarbene into C–C double bonds. Thus (3) can be obtained in 55% yield from the reaction of 1-diazo-2-butanone (1) with isopropenyl acetate (2). When (3) is refluxed with 4% methanolic sodium hydroxide, 2,3-dimethylcyclopentene-2-one (5) is obtained in 85% yield.

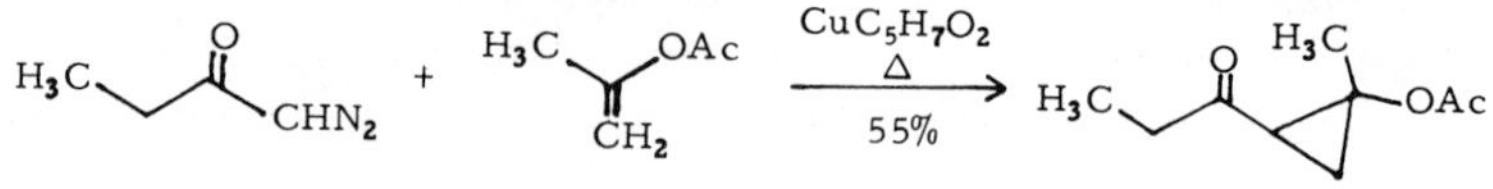

This new synthesis of 1,4-diketones was used in a synthesis of *cis*-jasmone (6).

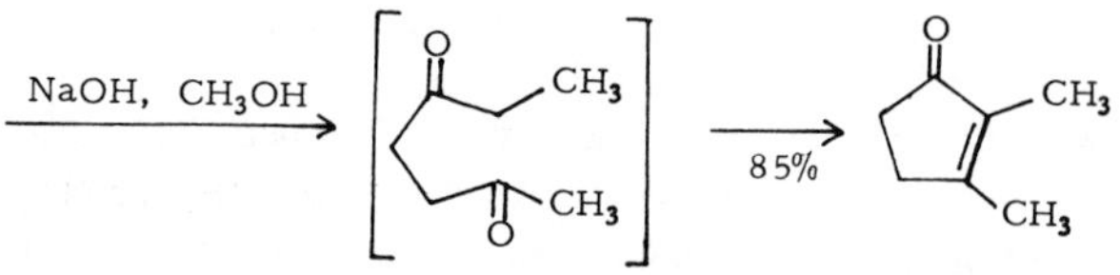

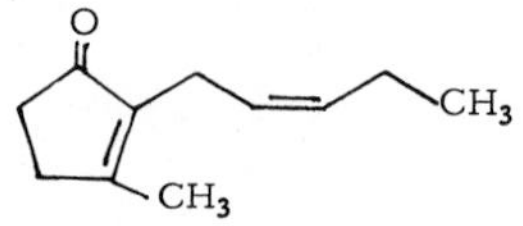

[1] B. Emmert, W. Gsottschneider, and H. Stanger, *Ber.*, **69B**, 1319 (1936).
[2] J. E. McMurry and T. E. Glass, *Tetrahedron Letters*, 2575 (1971).

**Copper bronze.**

Copper bronze is available at paint stores; it is activated by the method of Vogel.[1]

*Ullmann coupling.* The Ullmann reaction can be carried out in a sealed tube at 90–115° (12 hr.) without a solvent. 2,2′-Dipyrimidine has been prepared in this way in about 45% yield.[2]

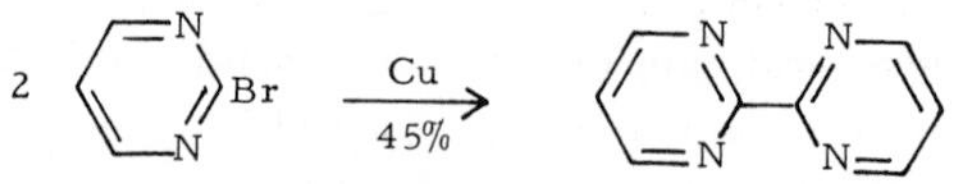

[1] A. L. Vogel, *A Textbook of Practical Organic Chemistry*, 3rd Ed., Wiley, New York, 192 (1957).
[2] T. R. Musgrave and P. A. Westcott, *Org. Syn.*, submitted 1972.

**Copper carbonate, basic,** $CuCO_3Cu(OH)_2$. Mol. wt. 205.11. Suppliers: E. Merck, Sargent.

*Oxidative decarboxylation.*[1] Heating fluorene-9-carboxylic acid (1) with 2.0 molar eq. of basic copper carbonate affords fluorenone (2, 56% yield). This reaction is the first report of a direct oxidative decarboxylation with copper salt catalysis.

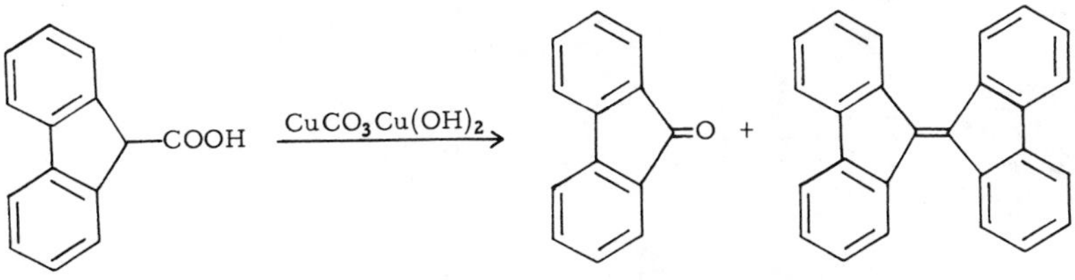

(1)                              (2)                        (3)

[1]B. M. Trost and P. L. Kinson, *J. Org.*, **37**, 1273 (1972).

**Copper(I) oxide–$t$-Butyl isonitrile,** $Cu_2O–(CH_3)_3CNC$.

$t$-Butyl isonitrile can be prepared by the procedure of Ugi and Meyr.[1]

*Dimerization of α,β-unsaturated carbonyl and nitrile compounds.*[2] This copper–isocyanide system effects dimerization of α,β-unsaturated carbonyl and nitrile compounds:

$$2\ RR'CHCH=CHX \xrightarrow{Cu_2O - (CH_3)_3CNC} \begin{array}{c} RR'CHCH=CX \\ | \\ RR'CHCHCH_2X \end{array}$$

X = CN, COOR'', COR''

*Synthesis of cyclopropanes.*[3] Cyclopropanes can be synthesized in fair yield by the reaction of an olefin with an α-chloro compound catalyzed by copper(I) oxide–$t$-butyl isonitrile.

$$ClCH_2Y + \rangle C=C \langle \xrightarrow{Cu_2O - (CH_3)_3CNC} $$

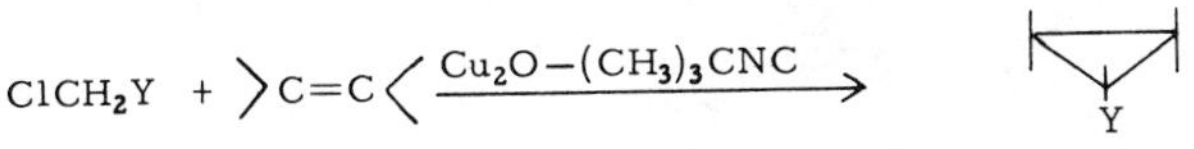

Y = COOR, COR, CN

[1]I. Ugi and R. Meyr, *Ber.*, **93**, 239 (1960).
[2]T. Saegusa, Y. Ito, S. Tomita, and H. Kinoshita, *J. Org.*, **35**, 670 (1970).
[3]T. Saegusa, Y. Ito, K. Yonezawa, Y. Inubushi, and S. Tomita, *Am. Soc.*, **93**, 4049 (1971).

**Copper(I) phenylacetylide,** $C_6H_5–C\equiv C–Cu$. Mol. wt. 164.66.

*Preparation.*[1] A solution of copper(II) sulfate pentahydrate in concentrated aqueous ammonia is stirred under nitrogen, cooled, and treated with hydroxylamine hydrochloride to effect reduction. Then a solution of phenylacetylene in ethanol is added to the pale-blue solution. The copper(I) phenylacetylide separates as a copious yellow precipitate, which is collected and dried in a rotary evaporator.

$$Cu^{II}(NH_3)_4^{+2} \xrightarrow{\ HONH_3^+Cl^-\ } Cu^I(NH_3)_2^+$$

$$Cu^I(NH_3)_2^+ + C_6H_5C\equiv CH \longrightarrow C_6H_5C\equiv C-Cu + NH_4^+ + NH_3$$

*Synthesis of 2-phenylfuro[3,2-b]pyridine.*[1]

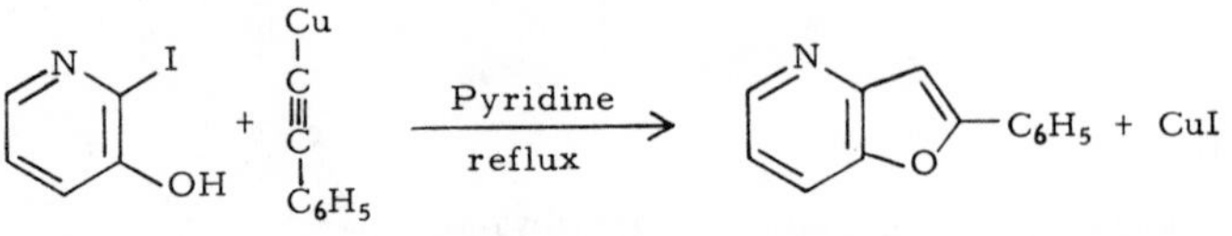

(1)                                                              (2)

A charge of copper(I) phenylacetylide is placed in a reaction flask, the system is purged with nitrogen, and pyridine is added, followed by 2-iodo-3-pyridinol. The mixture, which changes in color from yellow to dark green, is warmed in an oil bath to 110–120° for 9 hr., with continuous stirring under nitrogen. Workup and purification affords 75–82% of 2-phenylfuro[3,2-*b*]pyridine (2).

[1]D. C. Owsley and C. E. Castro, *Org. Syn.*, **52**, 128 (1972).

**Copper powder, 1**, 157–158; **2**, 82–84; **3**, 63–65.

*Ullmann reaction.* Cohen and Poeth[1] have presented evidence that the Ullmann reaction[2] proceeds through organocopper intermediates rather than radicals. Thus coupling of iodofumarate (1) at 100° with activated[3] copper powder gives pure *trans,-trans*-1,2,3,4-tetracarboethoxy-1,3-butadiene (2) in 96% yield. The coupling of iodomaleate (3) yields mainly *cis,cis*-1,2,3,4-tetracarboethoxy-1,3-butadiene (4) and a trace of (2).

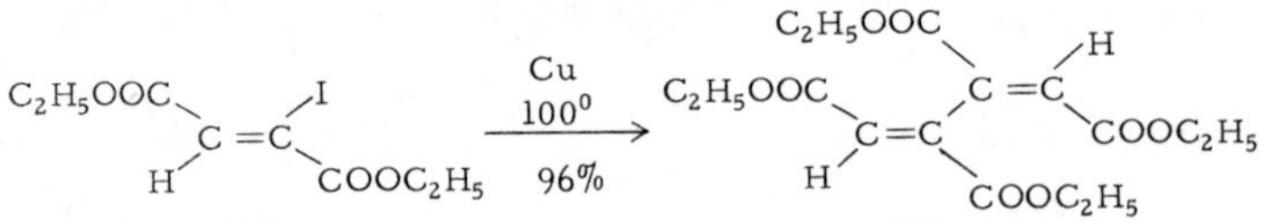

(1)                                                              (2)

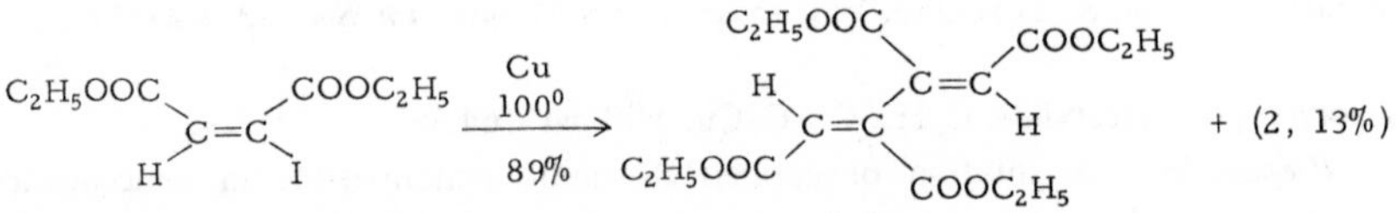

(3)                                                (4, 87%)

When two parts of (3) and one part of (1) are heated with copper at 75°, the product is the pure *trans,trans*-ester (2). When (1) or (3) is heated separately with copper and benzoic acid, the major product is diethyl fumarate or diethyl maleate, respectively. These diesters are probably formed by protonolysis of an organocopper intermediate, with complete retention of configuration.

*α-Diazoketones* (**2**, 82–84; **3**, 63–65). The key step in a total synthesis of the sesquiterpene (±)-*β*-cubebene (4) involved the intramolecular cyclization of the α-diazoketone (1).[4] This reaction was carried out by refluxing (1) in cyclohexane in the presence of

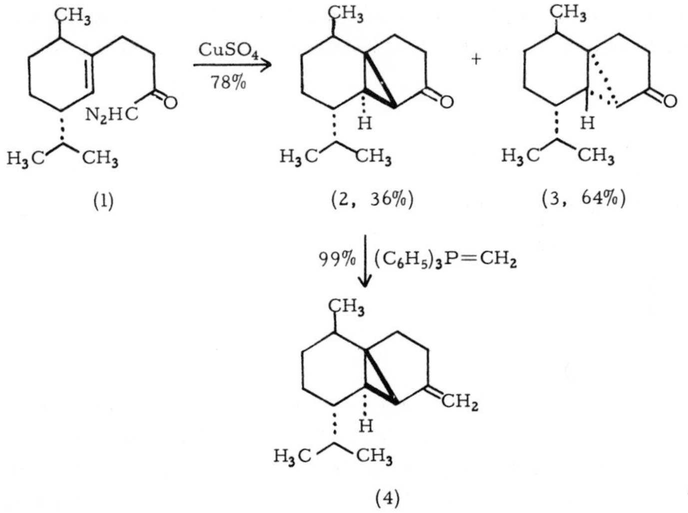

cupric sulfate for 1.5 hr.; two isomeric ketones, (2) and (3), were obtained in the approximate ratio of 3:5. The minor isomer, (±)-*β*-cubebene norketone, was transformed into the natural sesquiterpene (4) by a Wittig reaction.

[1]T. Cohen and T. Poeth, *Am. Soc.*, **94**, 4363 (1972).
[2]P. E. Fanta, *Chem. Rev.*, **64**, 613 (1964).
[3]A. H. Lewin, M. J. Zovko, W. H. Rosewater, and T. Cohen, *Chem. Commun.*, 80 (1967).
[4]E. Piers, R. W. Britton, and W. de Waal, *Canad. J. Chem.*, **49**, 12 (1971).

**Corey's reducing reagent** (1),

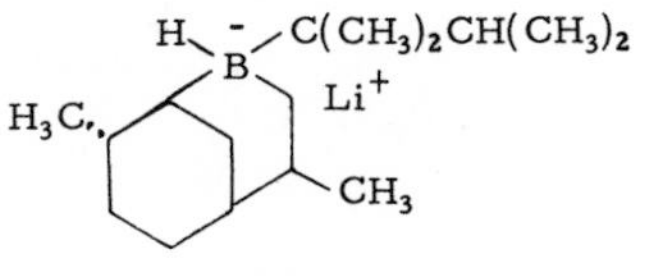

Mol. wt. 239.15.

*Preparation.*[1] The reagent is prepared by the reaction of (+)-limonene with thexylborane (**2**, 148; this volume) to give the trialkylborane. This is converted into the borohydride ion (1) by treatment with *t*-butyllithium in THF.

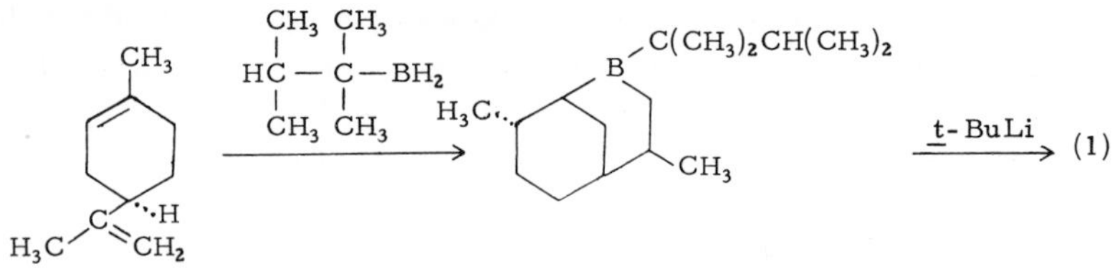

$$\text{(1)}$$

**Stereoselective reduction of ketones.**[2] Corey's synthesis of prostaglandins utilizes as a key intermediate the enone (I). One major synthetic problem is the stereoselective reduction of the carbonyl group to the desired 15S alcohol (IIa). Reduction with various borohydride reagents or various trialkylborohydrides ($R_1R_2R_3BH^-Li^+$) affords about

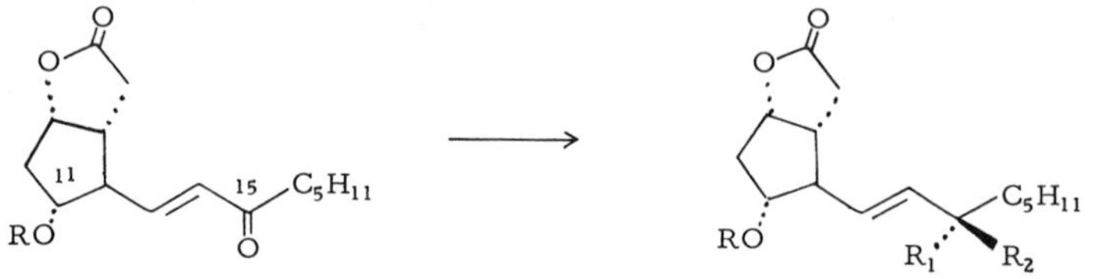

I

IIa (15 S) $R_1$ = OH; $R_2$ = H
IIb (15 R) $R_1$ = H; $R_2$ = OH

equal amounts of (IIa) and (IIb). Corey et al.[2] were able to effect highly stereoselective reduction of (I) (R = $p$-$C_6H_5C_6H_4$NHCO) to (IIa) and (IIb) in the ratio 92:8 by use of reagent (1) in THF; the reaction was conducted first for 4 hr. at $-130°$ and then for 2 hr. at $-115°$. The total yield of (II) was quantitative. Tri-$sec$-butylborohydride (this volume) is almost as selective as reagent (1). With this reagent the urethane (1) (R = $p$-$C_6H_5C_6H_4$NHCO) is reduced to (IIa) and (IIb) in the ratio 89:11.

Although the choice of the reducing reagent is important for control of reduction of the carbonyl group of (I), the choice of the R group is equally important. One reason for the difficulty in achieving selective reduction of (I) is that the enone unit can assume both s-$cis$ and s-$trans$ conformations. The desired rear attack of hydride (axis $a$, III) demands that (I) assume the s-$cis$ conformation as shown. Examination of molecular

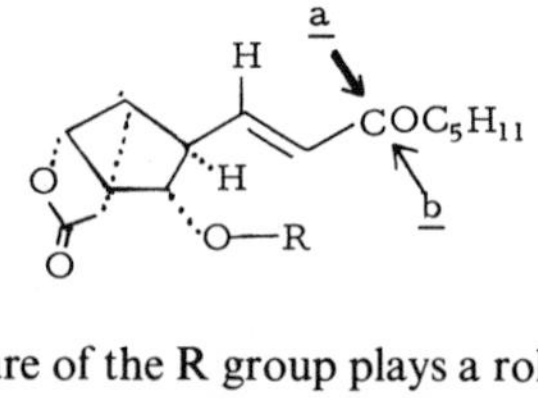

models suggested that the nature of the R group plays a role in the conformation of the enone. The R group must also have sufficient steric bulk to block approach along the undesired axis $b$. Of a number of R groups examined the $p$-phenylphenyl-carbamoyl group proved to provide the most effective stereochemical control of the carbonyl group. The preparation of (I) (R = $p$-C$_6$H$_5$C$_6$H$_4$NHCO) was carried out by reaction of the alcohol (I) (R = H) with $p$-phenylphenyl isocyanate ($p$-C$_6$H$_5$C$_6$H$_4$N= C=O, m.p. 58–58.5°[3]) in THF and triethylamine (1.2 eq.) at 25° for 3 hr. (yield > 90%). Hydrolysis of the protective group is accomplished in > 90% yield by treatment with 1 $M$ aqueous lithium hydroxide at 120° for 72 hr., extraction of the basic reaction mixture at 0° with ether–ethyl acetate to remove neutral and basic components, and relactonization by addition of ethyl chloroformate (2 eq.) to the aqueous phase which had been neutralized with carbon dioxide.

[1] E. J. Corey, S. M. Albonico, U. Koelliker, T. K. Schaaf, and R. K. Varma, *Am. Soc.*, **93**, 1491 (1971).
[2] E. J. Corey, K. B. Becker, and R. K. Varma, *ibid.*, **94**, 8616 (1972).
[3] M. J. van Gelderen, *Rec. trav.*, **52**, 969 (1933).

**Cupric acetate monohydrate, 1, 159–160; 2, 84; 3, 65.**

   *Oxidative coupling of 1,5-hexadiyne to cyclooctadeca-1,3,7,9,13,15-hexayne.*[1] A 6-l. three-necked, round-bottomed flask fitted with a stirrer having two 7.5-cm. paddles 9 cm. apart, a reflux condenser, and a 500-ml. dropping funnel (with a pressure equalizing arm) is charged with 600 g. of cupric acetate monohydrate and 3.8 l. of

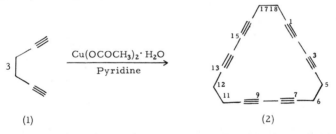

(1)                                    (2)

pyridine. The flask is immersed in a water bath preheated to 55° and the mixture is slowly stirred at this temperature for 1 hr. A solution of 50 g. of 1,5-hexadiyne (b.p. 84–85°) in 400 ml. of dried (KOH) pyridine is then added during 30 min. to the green suspension at 55° with vigorous stirring, and the mixture is stirred vigorously at 55° for a further 2 hr.. Workup affords about 35 g. of the crude cyclooctadecahexayne (2) as a dark-brown residue.

[1] K. Stöckel and F. Sondheimer, *Org. Syn.*, submitted 1971.

**Cupric chloride, 1, 163; 2, 84–85; 3, 66.**

   *Aryl iodides.*[1] Aryl iodides can be prepared in fair to good yields from aromatic hydrocarbons by the action of iodine and a copper salt. The observed reactivity order of various copper salts is CuCl$_2$ > CuF$_2$ > CuCl > Cu(OOCCH$_3$)$_2$. The reaction is considered to involve a two-step sequence:

$$ArH + I_2 \xrightarrow{CuCl_2} ArI + HI$$

$$ArH + HI + 2\ CuCl_2 \longrightarrow ArI + 2\ HCl + 2\ CuCl$$

The overall stoichiometry is shown in equation I. In some cases (benzene, alkylbenzenes) a more reactive iodine donor is required such as aluminum(III) iodide, iron(II) iodide, or copper(I) iodide.

(I)    $2 \text{ ArH} + \text{I}_2 + 2 \text{ CuCl}_2 \longrightarrow 2 \text{ ArI} + 2 \text{ HCl} + 2 \text{ CuCl}$

*Polychlorinated cyclohexanediones.*[2] Cyclohexanone and its methyl derivatives react with a large excess of $\text{CuCl}_2 \cdot 2\text{H}_2\text{O}$ in refluxing 50% acetic acid or 50% dioxane (2 hr.) to give dichloro or trichloro derivatives of cyclohexane-1,2-dione in 60–70% yield.

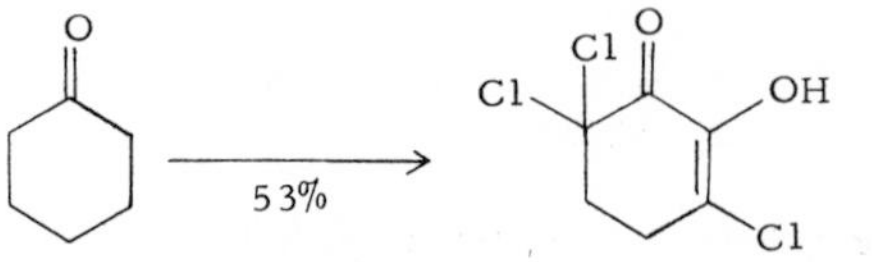

*Oxidative ring contraction.*[3] 2-Hydroxy-3,6-di-*t*-butyl-1,4-benzoquinone (1) on treatment with $\text{CuCl}_2 \cdot 6\text{H}_2\text{O}$ in hot acetic acid yields 2-chloro-2,4-di-*t*-butyl-cyclo-pentene-1,3-dione (2).

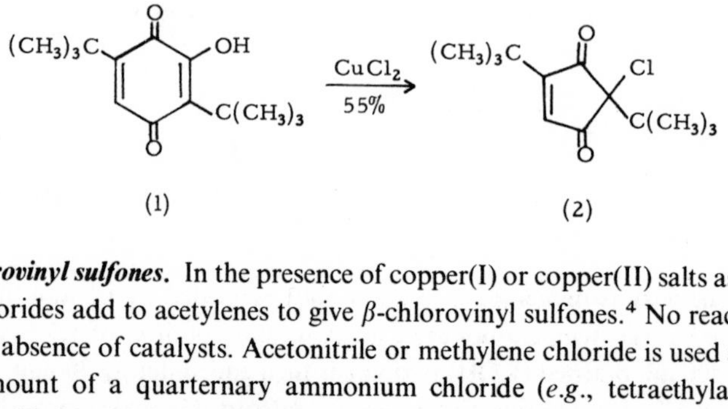

*β-Chlorovinyl sulfones.* In the presence of copper(I) or copper(II) salts as catalysts, sulfonyl chlorides add to acetylenes to give β-chlorovinyl sulfones.[4] No reaction takes place in the absence of catalysts. Acetonitrile or methylene chloride is used as solvent. A small amount of a quarternary ammonium chloride (*e.g.*, tetraethylammonium chloride) is added to solubilize the copper salts. The reaction is carried out at reflux or, preferably, in a sealed tube. The reaction is stereoselective; thus the reaction of benzene-sulfonyl chloride (2) with phenylacetylene (1) leads to two isomeric 2-benzenesulfonyl-1-chlorostyrenes (3) and (4). The main product (3) results from *trans* addition, the

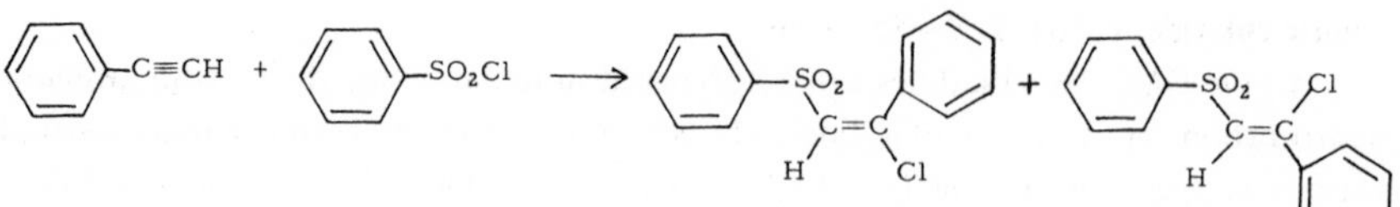

minor product (4) from *cis* addition. The *cis* addition product can assume a coplanar structure, whereas the *trans* addition product cannot attain coplanarity owing to steric hindrance. The addition takes place by a free-radical chain reaction in which the copper catalyst functions as a chlorine atom transfer agent:

$$RSO_2Cl \longrightarrow RSO_2\cdot$$
$$RSO_2\cdot + HC\equiv CR' \longrightarrow RSO_2CH=\dot{C}R'$$
$$RSO_2CH=\dot{C}R' + CuCl_2 \longrightarrow RSO_2CH=CClR' + CuCl$$
$$CuCl + RSO_2Cl \longrightarrow CuCl_2 + RSO_2\cdot$$

The *cis–trans* distribution can be varied by change in the experimental conditions. Excess chloride ions or highly polar solvents favor *trans* addition; solvents of low polarity (*e.g.*, carbon disulfide) favor *cis* addition.[5]

[1] W. C. Baird, Jr. and J. H. Surridge, *J. Org.*, **35**, 3436 (1970).
[2] J. Y. Satoh and K. Nishizawa, *J.C.S. Chem. Comm.*, 83 (1973).
[3] H. W. Moore and R. J. Wikholm, *ibid.*, 1073 (1971).
[4] Y. Amiel, *J. Org.*, **36**, 3691 (1971).
[5] *Idem, ibid.*, **36**, 3697 (1971).

**Cupric dimethoxide**, $Cu(OCH_3)_2$. Mol. wt. 125.61.
Cupric dimethoxide is prepared from $CuCl_2$ and $LiOCH_3$.[1]

*Reaction with carbon monoxide.*[2] Cupric dimethoxide reacts with carbon monoxide in pyridine solution at 35–70° to produce dimethyl carbonate in yields as high as 84%.

$$Cu(OCH_3)_2 + CO \longrightarrow (CH_3O)_2CO$$

An unstable carbomethoxycupric species is suggested. Cu(II) is reduced to Cu(I).
If the reaction is carried out in the presence of a secondary amine, carbamates are formed in good yield:

$$Cu(OCH_3)_2 + R_2NH \xrightarrow{\ CO\ } R_2N\underset{\underset{O}{\|}}{C}OCH_3$$

Other cupric alkoxides undergo carbonylation, for example, $Cu(OCH_3)Cl$, $Cu(OC_3H_5)_2$, and $Cu(acac)(OCH_3)$.

[1] C. H. Brubaker, Jr. and M. Wicholas, *J. Inorg. Nucl. Chem.*, **27**, 59 (1965).
[2] T. Saegusa, T. Tsuda, and K. Isayama, *J. Org.*, **35**, 2976 (1970).

**Cupric nitrate–Pyridine complex**, $Cu(NO_3)_2\cdot(C_5H_5N)_4$, 1, 164.
*Oxidation.*[1] Oxidation of methoxycyclopropanol (1) with cupric nitrate[2] proceeds *via* the methyl β-propionate radical (2) to give methyl acrylate (3) and dimethyl adipate (4). Ferric nitrate, $Fe(NO_3)_3$, can also be used.

Reaction of (1) with methyl vinyl ketone (5) in the presence of cupric nitrate–pyridine complex gives methyl 6-ketoheptanoate (6) in 45% yield. [Use of ferric nitrate gives (6) in 30% yield.]

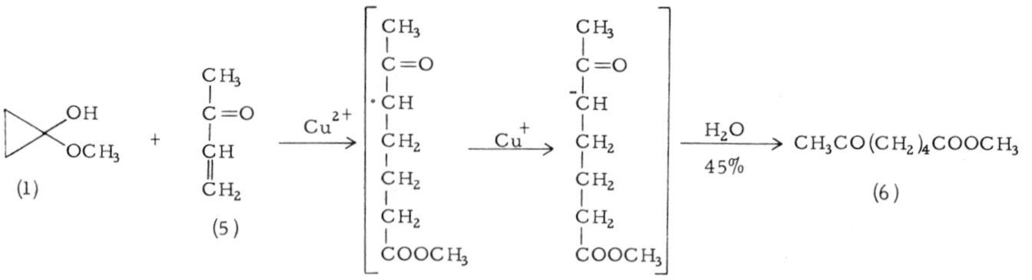

Addition of the radical (2) to butadiene (in $CH_3OH$) gives the allylic radical (7), which dimerizes to a mixture of unsaturated $C_{14}$ diesters.

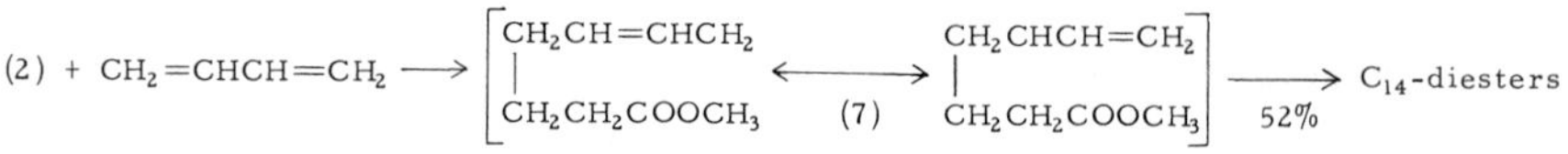

[1]S. E. Schaafsma, R. Jorritsma, H. Steinberg, and Th. J. de Boer, *Tetrahedron Letters*, 827 (1973).
[2]Supplier: ROC/RIC.

**Cupric oxide, CuO.**

*Vapor-phase oxidation of alcohols.* Vapor-phase oxidation of alcohols over a variety of metals and metal oxides has been used mainly in the industry. Sheikh and Eadon[1] now report a simple laboratory procedure for oxidation of primary and secondary alcohols to the corresponding aldehydes and ketones. The alcohol vapor and an inert carrier gas (usually helium) are passed through a 6-ft. column loosely packed with cupric oxide wire worms.[2] A vapor-phase chromatograph provides temperature and flow rate control. Yields are generally high, but allylic and homoallylic alcohols give mixtures of products.

[1]M. Y. Sheikh and G. Eadon, *Tetrahedron Letters*, 257 (1972).
[2]Supplier: Allied Chemical Co., General Chemical Division, New York City.

**Cuprous bromide, 1,** 165–166; **2,** 90–91; **3,** 67.

*Conjugate additions.* Conjugate addition of phenyllithium to 3-methylcyclopent-2-enone (1) was effected in the presence of cuprous bromide. Presumably the reactive species is diphenylcopperlithium.

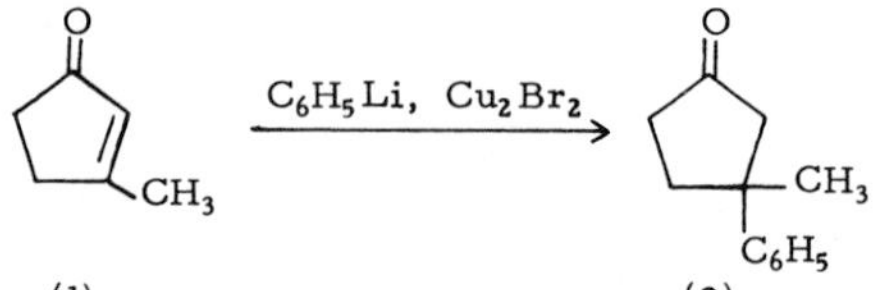

[1]S. Wolff and W. C. Agosta, *J.C.S. Chem. Comm.*, 226 (1972).

**Cuprous *tert*-butoxide**, $(CH_3)_3COCu$. Mol. wt. 136.65, sublimes about 170° (1 mm.).

*Preparation.*[1] The reagent is prepared by the reaction of anhydrous $Cu_2Cl_2$ with *tert*-BuOLi in THF under $N_2$. It is purified by sublimation.

*Metalation.*[1] Cuprous *tert*-butoxide is a useful metalation reagent because the basic alkoxide group effectively abstracts an active hydrogen to form an alcohol. There is thus no need for a basic or buffered reaction solution. The reaction of the reagent with a large excess of phenylacetylene gives the bright-yellow cuprous acetylide in 97% yield.

$$C_6H_5C{\equiv}CH \; + \; (CH_3)_3COCu \longrightarrow C_6H_5C{\equiv}CCu \; + \; (CH_3)_3COH$$

excess                                         97%              97%

[1] T. Tsuda, T. Hashimoto, and T. Saeguse, *Am. Soc.*, **94**, 658 (1972).

**Cuprous chloride**, **1**, 166–169; **2**, 91–92; **3**, 67–69.

*trans*-**Cyclooctene**.[1] *cis*-Cyclooctene (1) is partially isomerized to *trans*-cyclooctene (2) by irradiation in the presence of cuprous chloride. The two isomers are separated by

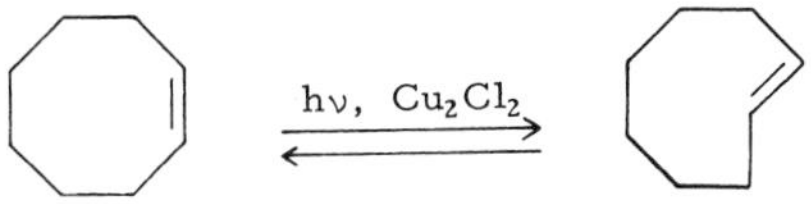

(1)                                        (2)

use of aqueous silver nitrate[2] and (2) can be obtained in 19% yield in > 99% purity. The isomerization proceeds through the Cu(I)–olefin complex. 1,5-*cis,cis*-Cyclo-octadiene can be converted in the same way into 1,5-*cis,trans*-cyclooctadiene in 30–40% yield. Cuprous bromide is much less effective than cuprous chloride.

*Oxidative coupling of terminal acetylenes* (**1**, 168–169).[3] Yamamoto and Sond-heimer have reported the synthesis of a tetraalkylated tetradehydro[18]annulenedione (5), of interest as a possible quinone, from (1). The acetylene (1) is oxidatively coupled

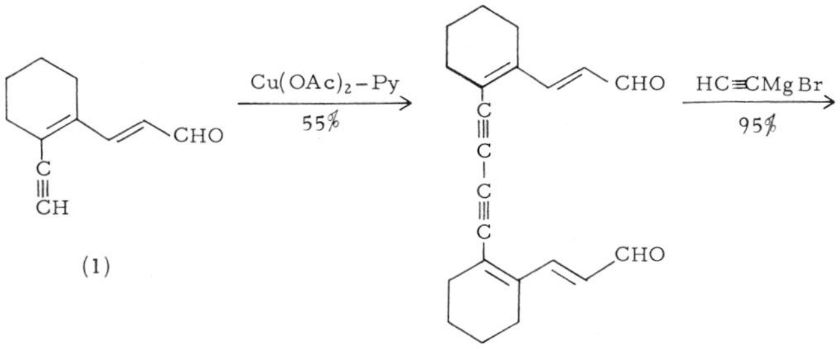

(1)

(2)

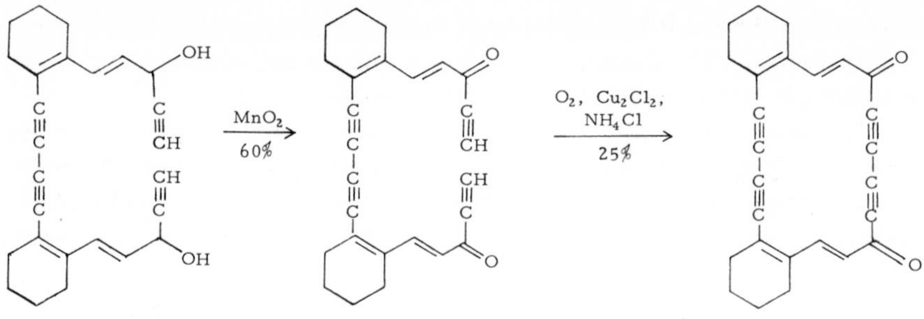

(3)                    (4)                    (5)

with cupric acetate in pyridine (1, 159) to the "dimer" (2).[4] Treatment of (2) with ethynyl-magnesium bromide (1, 389) gives the diol (3) in high yield. This is oxidized with active manganese dioxide to the diketone (4). This diketone cannot be oxidatively coupled with cupric acetate in pyridine, since it decomposes in pyridine. The desired reaction, however, was carried out with oxygen, cuprous chloride, and ammonium chloride in aqueous ethanol and benzene; (5) is obtained in 25% yield, together with 50% of (4).

[1] J. A. Deyrup and M. Betkouski, *J. Org.*, **37**, 3561 (1972).
[2] A. C. Cope and R. D. Bach, *Org. Syn.*, **49**, 39 (1969).
[3] K. Yamamoto and F. Sondheimer, *Angew. Chem., internat. Ed.*, **12**, 68 (1973).
[4] R. H. McGirk and F. Sondheimer, *ibid.*, **11**, 834 (1972).

**Cyanogen bromide**, 1, 174–176; 2, 93.

*trans*-**Olefins.** Zweifel *et al.*[1] have reported a convenient synthesis of *trans*-di- and *trans*-trisubstituted olefins from alkynes. First, the alkyne is hydroborated with a dialkylborane in THF to give a dialkylvinylborane. After removal of the THF, methylene chloride and cyanogen bromide (iodide) are added at 0°. After stirring for 2 hr. sodium hydroxide is added and the olefin extracted into pentane and purified by distillation. Yields of *trans*-olefins are in the range 60–75%.

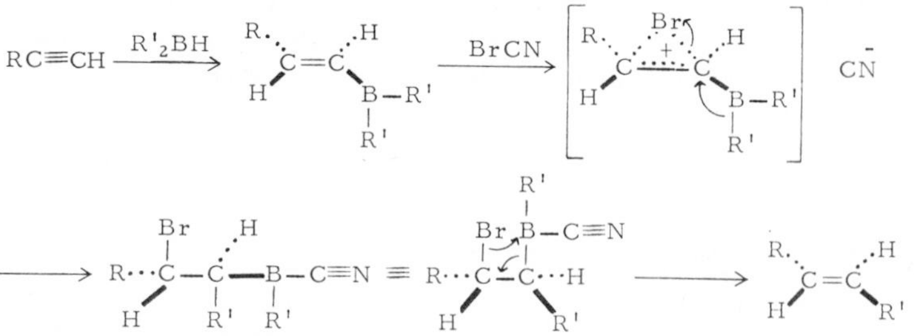

[1] G. Zweifel, R. P. Fisher, J. T. Snow, and C. C. Whitney, *Am. Soc.*, **94**, 6560 (1972).

**Cyanogen chloride**, 1, 176–177.

*Aliphatic isocyanates.*[1] Alkylation of cyanogen chloride with an alkyl chloride and ferric chloride gives an alkyl isocyanide dichloride–FeCl$_3$ complex. For example, cyanogen chloride reacts with isopropyl chloride in the presence of ferric chloride to give the hygroscopic complex (1) in high yield. The complex reacts with ethanol or sodium salt of benzyl alcohol to form the carbamates (3) and (4), respectively.

$$3 \; ClCN \; + \; 3 \; (CH_3)_2CHCl \; + \; 2 \; FeCl_3 \; \longrightarrow \; [\,(CH_3)_2CH-N=CCl_2\,]_3\,(FeCl_3)_2$$
$$(1)$$

$$(1) \xrightarrow{\quad H_2O \quad} (CH_3)_2CH-NH_2$$
$$(2)$$

$$(1) \xrightarrow{\quad C_2H_5OH \quad} (CH_3)_2CHNHCOOC_2H_5$$
$$(3)$$

$$(1) \xrightarrow{\quad C_6H_5CH_2ONa \quad} (CH_3)_2CHNHCOOCH_2C_6H_5$$
$$(4)$$

$$(1) \xrightarrow{\quad ZnO \quad} 3 \; (CH_3)_2CHN=C=O \; + \; 3 \; ZnCl_2 \; + \; 2 \; FeCl_3$$
$$(5, \; 85\%)$$

It reacts with zinc oxide (**1**, 1294; additional suppliers: Alfa Inorganics, ROC/RIC) to give isopropyl isocyanate (5). The yield is high if the complex (1) is not isolated owing to its high sensitivity to moisture. The reaction of (1) with silver oxide was not successful owing to the violence of reaction. Reaction of (1) with sodium acetate gave (5) in 24% yield.

[1]R. Fuks and M. Hartemink, *Tetrahedron*, **29**, 297 (1973).

**Cyanomethylcopper**, $CuCH_2CN$. Mol. wt. 103.59.

The reagent is prepared[1] in THF by the reaction of cyanomethyllithium[2] with 1 eq. of cuprous iodide at $-25°$ for 10 min. under nitrogen or argon.

*γ,δ-Unsaturated nitriles.*[1] The reagent reacts with allylic halides to give γ,δ-unsaturated nitriles in high yield:

$$\underset{/}{\overset{\backslash}{>}}C=C-\underset{|}{\overset{|}{C}}Br \; + \; CuCH_2CN \; \longrightarrow \; \underset{/}{\overset{\backslash}{>}}C=C-\underset{|}{\overset{|}{C}}-CH_2CN \; + \; CuBr$$

For example, the reagent reacts with *trans*-geranyl bromide (1) to give *trans*-homogeranyl cyanide (2) in 92% yield. The reaction is carried out in the following way. A cooled solution of *n*-butyllithium (2.0 ml., 1.3 *M*) in pentane is added to a solution of

(1)                                            (2)

acetonitrile (3.34 mmole) in 5 ml. of THF at $-78°$ under argon. After 40 min. at $-78°$ the reaction is allowed to warm to $-25°$ and cuprous iodide (3.37 mmole) is added. After stirring for 15 min. at $-25°$ the brick-colored solution of cyanomethylcopper is treated with *trans*-geranyl bromide in THF (3 ml.). After 1 hr. at $-25°$, aqueous ammonium chloride is added and the product extracted with ether and then isolated by chromatography or distillation under reduced pressure.

The reagent does not react with benzyl bromide or unactivated alkyl bromides. The reaction with unsaturated bromo esters of the type $BrCH_2CH=CHCOOCH_3$ appears to be too complex to be useful.

[1]E. J. Corey and I. Kuwajima, *Tetrahedron Letters*, 487 (1972).
[2]Prepared from acetonitrile and *n*-butyllithium at $-78°$: E. M. Kaiser and C. R. Hauser, *J. Org.*, **33**, 3402 (1968).

**Cyanomethylidenebis(triphenylphosphonium) dibromide,** $(C_6H_5)_3\overset{+}{P}CHP(C_6H_5)_3 2Br^-$.
$\qquad\qquad\qquad\qquad\qquad\qquad\qquad\qquad\qquad\qquad\qquad\qquad\qquad\qquad\ \ \ \ \ |$
Mol. wt. 723.42, m.p. 268–269°. $\qquad\qquad\qquad\qquad\qquad\qquad\qquad\qquad\quad\ CN$

The reagent (1) is prepared[1] in 93.4% yield by the reaction of triphenylphosphine with dibromoacetonitrile.[2]

$$Br_2CHCN + P(C_6H_5)_3 \xrightarrow[45^0]{C_6H_6} (C_6H_5)_3\overset{+}{P}CHP(C_6H_5)_3\ 2\ Br^-$$
$$|$$
$$CN$$

(1)

*Wittig reaction.*[1] Aldehydes when refluxed with the bisphosphonium salt (1) in a benzene–water solvent containing sodium hydroxide are converted into α,β-unsaturated

$$Ar(R)CHO + (1) \xrightarrow[40-85\%]{\overset{OH^-,\ C_6H_6}{H_2O}} Ar(R)CH=CHCN + 2(C_6H_5)_3P=O$$

(2)

nitriles (2). Ketones do not react with (1). In the case of aliphatic aldehydes the product was partly or wholly isomerized to the β,γ-unsaturated nitrile.

[1]J. W. Wilt and A. J. Ho, *J. Org.*, **36**, 2026 (1971).
[2]J. W. Wilt and J. L. Diebold, *Org. Syn.*, **38**, 16 (1958).

**Cyclohexyl metaborate trimer,** $(C_6H_{11}O_2B)_3$, 3, 72.
The correct formula is shown in (1).

$$\begin{array}{c} OC_6H_{11} \\ | \\ O^{\diagup}{}^B{}^{\diagdown}O \\ |\qquad\quad | \\ H_{11}C_6O^{\diagup}{}^B{}^{\diagdown}{}_O{}^{\diagup}{}^B{}^{\diagdown}OC_6H_{11} \end{array}$$

(1)

**Cyclopentadienyl tribenzyltitanium,** $C_5H_5Ti(CH_2C_6H_5)_3$. Mol. wt. 386.37.
Prepared from $C_5H_5TiCl_3$ and $C_6H_5CH_2MgCl$.

***Cyclocodimerization of ethylene and 1,3-butadiene.***[1] This homogeneous titanium catalyst promotes the cyclocodimerization of ethylene and 1,3-butadiene to vinylcyclobutane. The usual product of codimerization is 1,4-hexadiene.

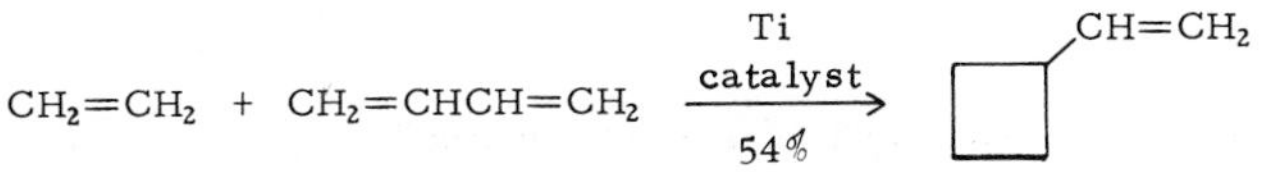

$CH_2=CH_2$ + $CH_2=CHCH=CH_2$
Ti
catalyst
54%
$CH=CH_2$

Using this catalyst norbornene (1) and butadiene codimerize to (2) in yields of about 90%. Less strained olefins give lower yields of vinylcyclobutane products.

+ $CH_2=CHCH=CH_2$
Ti
catalyst
~90%
$CH=CH_2$

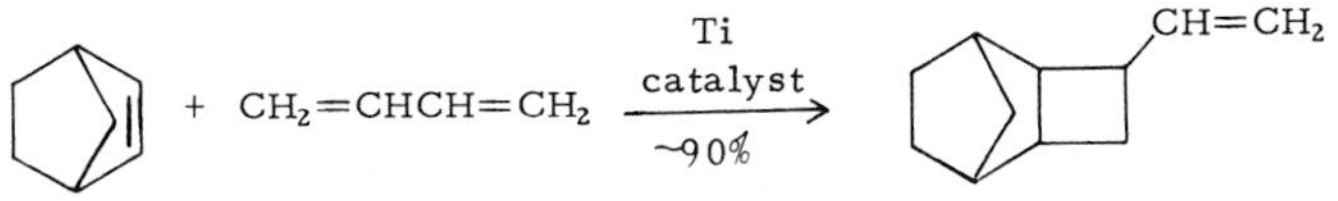

(1) (2)

[1]L. G. Cannell, *Am. Soc.*, **94**, 6867 (1972).

**Cyclopropene,** . Mol. wt. 40.06; unstable.

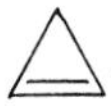

***Preparation***, **1**, 1040. Note that formula (3) is incorrect.

***Spectra.***[1]

***Diels–Alder reactions***. The Diels–Alder reaction of cyclopropene with cyclopentadiene has already been mentioned (**1**, 1040). In fact, cyclopropene is one of the most reactive dienophiles. It reacts with tropone (1) in methylene chloride at 0° to give the 1,4-addition product (2). Similarly, reaction with tropolone (3) affords the adduct (4).

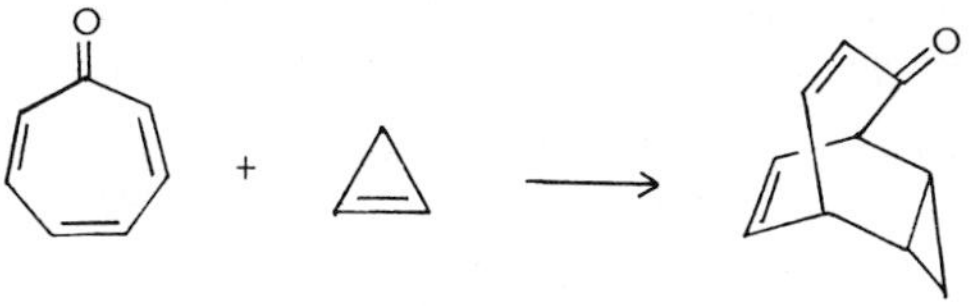

O
+
O

(1) (2)

O
OH
+
O
OH

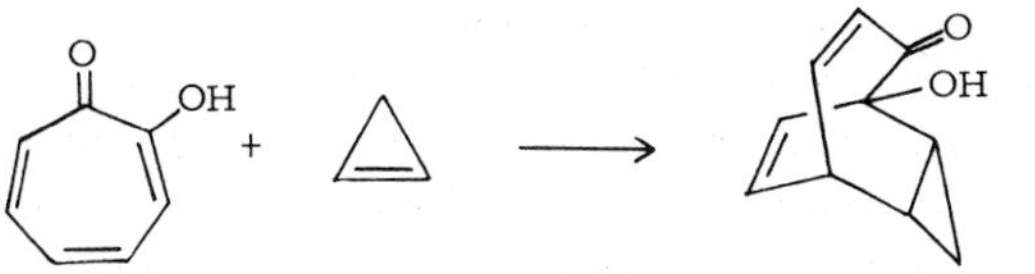

(3) (4)

This reaction is the first example of a Diels–Alder reaction with tropolone. The *endo* selectivity seems to be a general property of cyclopropene.[2]

[1]K. B. Wiberg and B. J. Nist, *Am. Soc.*, **83**, 1226 (1961).
[2]T. Uyehara, N. Sako, and Y. Kitahara, *Chem. Ind.*, 41 (1973).

# D

**Dehydro-N,N′,N″-tricyclohexylguanidinohexacarbonyldiiron(0)** (1). Mol. wt. 583.23.
This complex (1) is obtained as red diamagnetic crystals by refluxing dicyclohexyl-carbodiimide (DCC) with $Fe_2(CO)_6$ in heptane for three days.

**Dichlorocarbene.**[1] When the complex (1) is treated with chloroform containing 0.75% of ethanol, all six CO groups are evolved with formation of the salt (2) and

$$
\begin{array}{c}
N-C_6H_{11} \\
\parallel \\
C_6H_{11}-N \overset{C}{\diagdown} N-C_6H_{11} \\
\parallel \qquad \parallel \\
(CO)_3Fe \longrightarrow Fe(CO)_3
\end{array}
\quad \xrightarrow{\ CHCl_3\ }\quad :CCl_2 \;+\; [C(NHC_6H_{11})_3][FeCl_4]
$$

$$(1) \qquad\qquad\qquad\qquad\qquad\qquad\qquad\qquad (2)$$

dichlorocarbene. Thus in the presence of cyclohexene, dichloronorcarane is formed along with isomeric products formed by insertion into the C–H bonds of $C_2$ and $C_3$. The behavior of the dichlorocarbene generated in this way differs therefore from that of conventionally generated dichlorocarbene, where insertion reactions are generally rare.

[1]N. J. Bremer, A. B. Cutcliffe, and M. F. Farona, *Chem. Commun.*, 932 (1970).

**Dialkoxycarbonium fluoroborates**, $(RO)_2\overset{+}{C}HBF_4{}^-$.
Borch[1] prepared these reagents in about 85% yield by the reaction of orthoformate esters with boron trifluoride etherate in methylene chloride at −30° (3, 303). They are low-melting crystalline solids. Borch reported that these reagents are even more potent alkylating reagents than trialkyloxonium fluoroborates.

*Stevens rearrangement–elimination.* Boekelheide and Anderson[2] have reported a synthesis of [2.2]metaparacyclophane-1,9-diene (5) in which the key step is replacement of a sulfide linkage by a carbon–carbon double bond via the Stevens rearrangement followed by an elimination step. Thus reaction of 2,11-dithia[3.3]metaparacyclophane (1) with dimethoxycarbonium fluoroborate gave an essentially quantitative yield of the disulfonium salt (2). Stevens rearrangement[3] of (2) with sodium hydride in DMSO gave (3) as a mixture of stereoisomers in 67% yield. This product was again alkylated to give a mixture of isomers (4) in 87% yield. The final step involved elimination of dimethyl sulfide, accomplished with *n*-butyllithium in 50% yield.

Essentially the same procedure was used for conversion of (6) into [2.2]metacyclophane-1,9-diene (7). In this case, a mixture of (7) and the valence tautomer (8, 15,16-dimethyldihydropyrene) is formed in the final step. The mixture is converted entirely

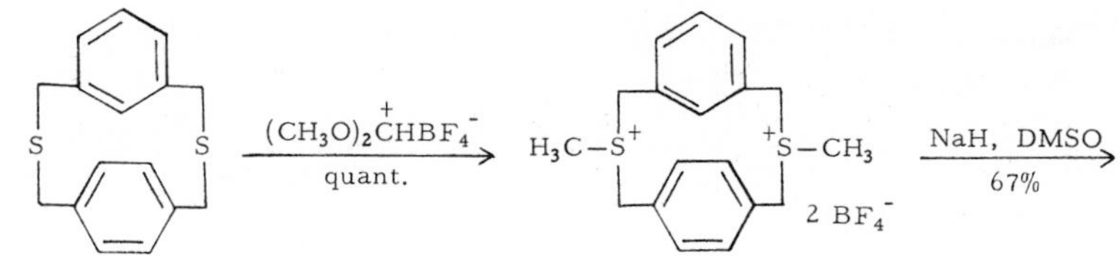

(1)    (2)

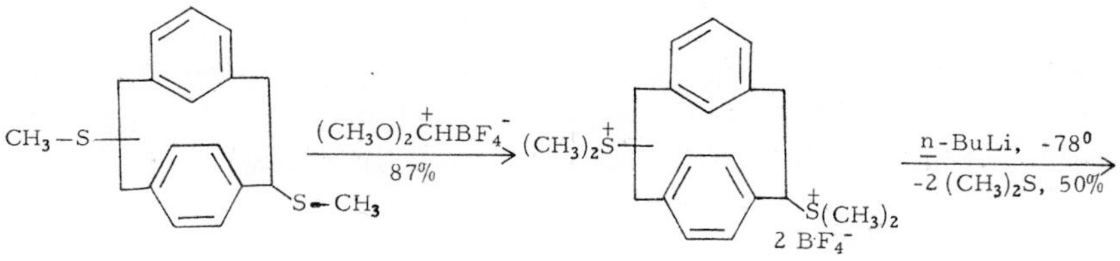

(3)    (4)

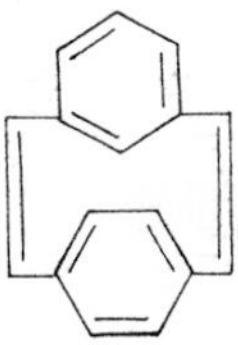

(5)

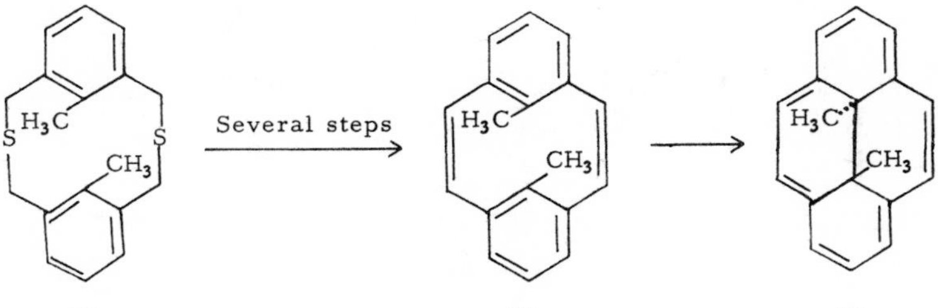

(6)    (7)    (8)

into (8) either on heating or on chromatography over silica gel.[4] For an alternate synthesis of (8) see **2**, 74.

[1] R. F. Borch, *J. Org.*, **34**, 627 (1969).
[2] V. Boekelheide and P. H. Anderson, *Tetrahedron Letters*, 1207 (1970).
[3] C. K. Ingold, *Structure and Mechanism in Organic Chemistry*, Cornell University Press, Ithaca, New York, 524 (1953).
[4] R. H. Mitchell and V. Boekelheide, *Tetrahedron Letters*, 1197 (1970).

**1,5-Diazabicyclo[4.3.0]nonene-5 (DBN)**, **1**, 189–190; **2**, 98–99; **4**, 16–18.

*Dehydrohalogenation.* Alkyl-1,3,5-hexatrienes can be prepared conveniently in high purity in 30–40% yield by dehydrohalogenation of dienyl halides with 1,5-diaza-bicyclo[4.3.0]nonene-5 in DMSO (25–80°). Use of potassium *t*-butoxide–DMSO yields only polymeric material.[1]

$$R_1CH=CHCHCH_2C=CHR_2 \xrightarrow{\text{DBN}} R_1CH=CHCH=CHC=CHR_2$$

with Br and $R_3$ substituents on the left structure and $R_3$ on the right structure.

One step in a synthesis of *dl*-muscone (4) from exaltone (1) involved dehydrobromination of (2). This was effected with 1,5-diazabicyclo[4.3.0]nonene-5 in 70% yield. Treatment with potassium *t*-butoxide in *t*-butanol or DMSO afforded only the dioxine (5).[2]

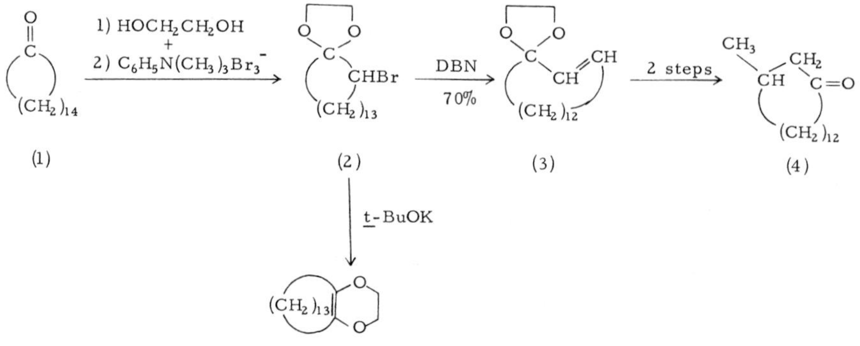

Dehydrohalogenation of the bromide (6) with the reagent gave a mixture of isomers (7) and (8), separated by gas chromatography.[3]

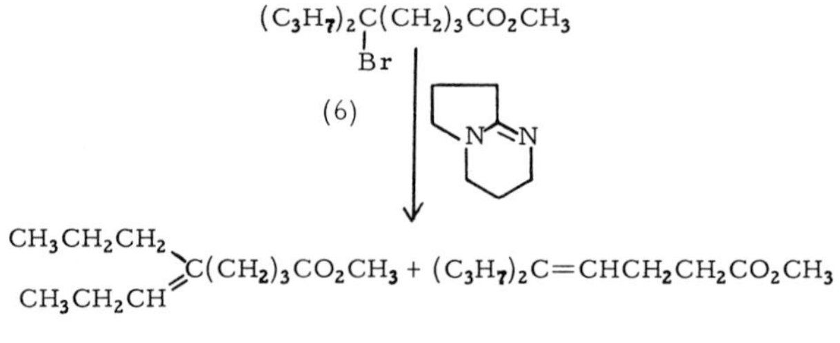

$$(C_3H_7)_2C(CH_2)_3CO_2CH_3$$
with Br substituent

(6)

$$CH_3CH_2CH_2 \diagdown C(CH_2)_3CO_2CH_3 + (C_3H_7)_2C=CHCH_2CH_2CO_2CH_3$$
$$CH_3CH_2CH \diagup$$

(7)                                    (8)

*Triphenylphosphirene oxide (2).*[4] This potentially aromatic heterocycle has been obtained by treatment of bis-(α-bromobenzyl)phenylphosphine oxide (1)[5] in benzene solution with DBN. Stille *et al.*[4] suggest that the conversion of (1) to (2) proceeds by a

1,3-elimination of hydrogen bromide to form a phosphirane oxide intermediate, which then undergoes 1,2-elimination of hydrogen bromide. The phosphacyclopropene (2) is

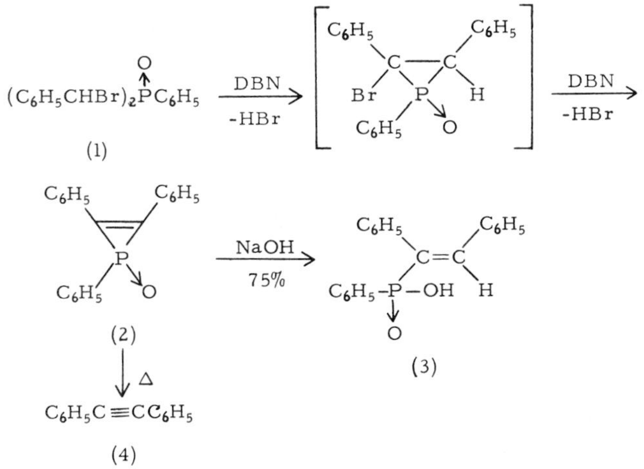

readily hydrolyzed by a base to the 1,2-diphenylvinylphosphinic acid (3). Pyrolysis of (2) at 120° at $10^{-5}$–$10^{-6}$ torr gives diphenylacetylene (4).

*One-step dehydrobromination–decarbomethoxylation; O-alkyl cleavage of methyl esters.* Miles and Parish[6] treated the bromoketone (1, derived from the diterpene podocarpic acid) with DBN with the hope of obtaining the lactone (2). This reaction had been observed previously using collidine as the base, but the yield was low. Surprisingly, use of DBN in refluxing *o*-xylene (165°, 6 hr.) gave the $\alpha,\beta$-unsaturated ketone

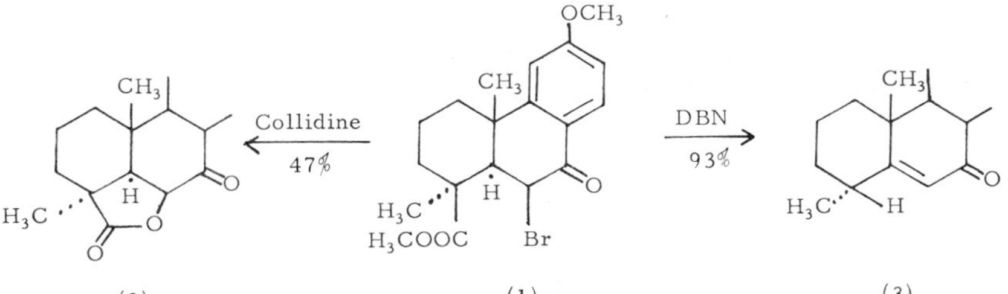

(3) in 93% yield. DBN is known to effect dehydrohalogenation, but this is the first instance of decarbomethoxylation with the reagent. This reaction involves two steps: O-methyl cleavage of the methyl ester, and subsequent decarboxylation. Actually, DBN is a useful reagent for cleavage of hindered methyl esters. Thus methyl mesitoate is cleaved to mesitoic acid in about 90% yield when refluxed with DBN in *o*-xylene (165°) for 6 hr. Methyl triisopropylacetate can be hydrolyzed to the acid under the same conditions in 94% yield. It also effects selective ester cleavage of methyl 3$\beta$-acetoxy-$\Delta^5$-

etienate (4). Treatment of (4) with 4.0 eq. of DBN in refluxing *o*-xylene gives 27% of

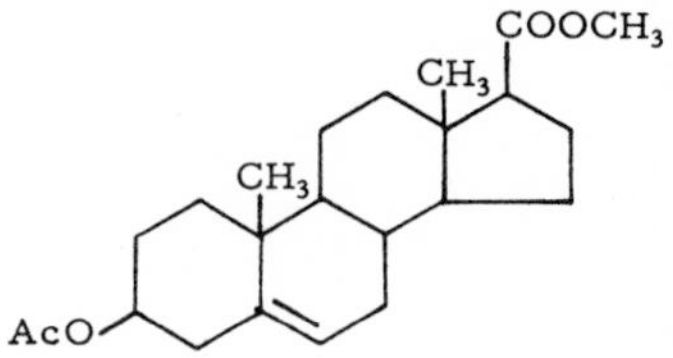

(4)

starting material, 56% of the desired acetoxy acid, 7% of the hydroxy acid, and 5% of the diene resulting from elimination of the acetate group. DBN is thus somewhat less selective than lithium *n*-propyl mercaptide (**3**, 188), but more selective than lithium iodide in refluxing lutidine (**1**, 615–616).

**3,6-Dehydrooxepin** (**3**). Treatment of *erythro*-3-hydroxy-4-*p*-toluenesulfonyloxy-hexa-1,5-diyne (1) with DBN in ether leads to a mixture of *cis*- and *trans*-1,2-diethynyl-oxiranes (2a) and (2b) in 53% yield.[7] Pyrolysis of (2a) or (2b) in a nitrogen flow system

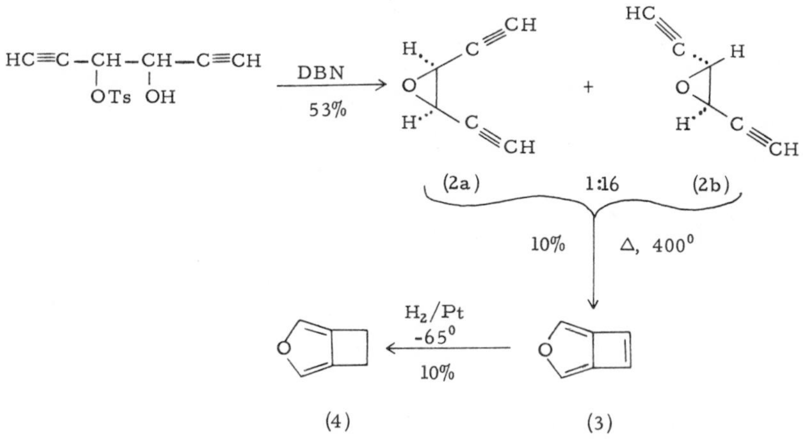

(400°, contact time ~ 20 sec.) gives 3,6-dehydrooxepin (3), a cyclic 8π-electron molecule. The oxepin (3) is extremely sensitive to oxygen.

The oxepin (3) can be partially hydrogenated to 3-oxabicyclo[3.2.0]hepta-1,4-diene (4) with reduced platinum oxide at −65°. Compound (4) is more stable than the precursor (3).[8]

**Stibabenzene.** The reagent was used for the conversion of 1,4-dihydro-1-chloro-stibabenzene (1) into stibabenzene (2).[9]

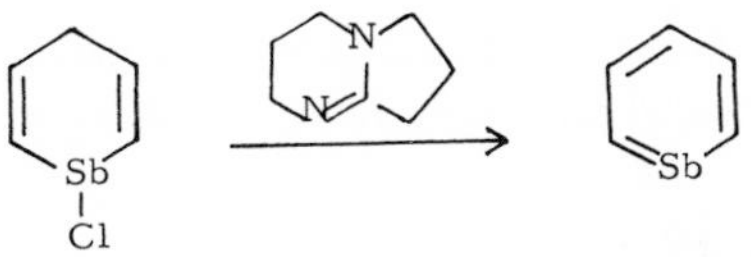

(1)                              (2)

[1]C. W. Spangler, R. Eichen, K. Silver, and B. Butzlaff, *J. Org.*, **36**, 1695 (1971).
[2]B. D. Mookherjee, R. R. Patel, and W. O. Ledig, *ibid.*, **36**, 4124 (1971).
[3]K. Eiter, E. Truscheit, and M. Boness, *Ann.*, **709**, 29 (1967).
[4]E. W. Koos, J. P. Vander Koot, E. E. Green, and J. K. Stille, *Chem. Commun.*, 1085 (1972).
[5]Prepared by bromination of dibenzylphenylphosphine oxide.
[6]D. H. Miles and E. J. Parish, *Tetrahedron Letters*, 3987 (1972).
[7]K. P. C. Vollhardt and R. G. Bergman, *Am. Soc.*, **94**, 8950 (1972).
[8]R. G. Bergman and K. P. C. Vollhardt, *J.C.S. Chem. Comm.*, 214 (1973).
[9]A. J. Ashe, III, *Am. Soc.*, **93**, 6690 (1971).

**1,4-Diazabicyclo[2.2.2]octane (DABCO), 2, 99–101.**

*Dealkylation of quaternary ammonium salts.*[1] Treatment of a quaternary ammonium salt (1 mmole) with DABCO (2 mmole) in refluxing ethanol or DMF effects dealkylation to the tertiary amine in yields of 50–90%.

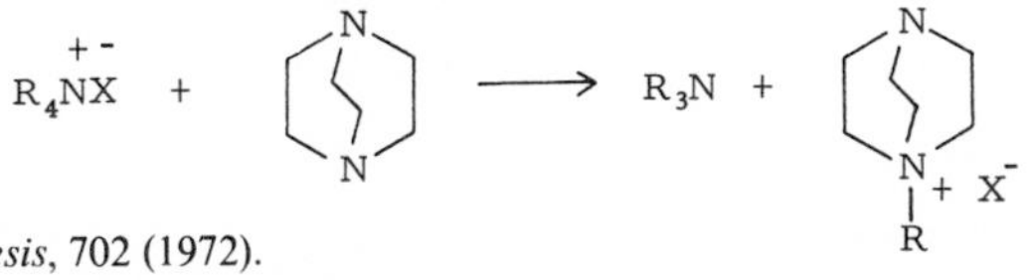

[1]T.-L. Ho, *Synthesis*, 702 (1972).

**1,5-Diazabicyclo[5.4.0]undecene-5 (DBU), 2, 101; 4, 16–18.**

*O-Alkyl cleavage of methyl esters.*[1] Treatment of the bromoketone (1) with 2 eq. of DBU in *o*-xylene at 165° for 5 hr. gives (2) in 92% yield. Treatment of the ester (3,

(1)                              (2)

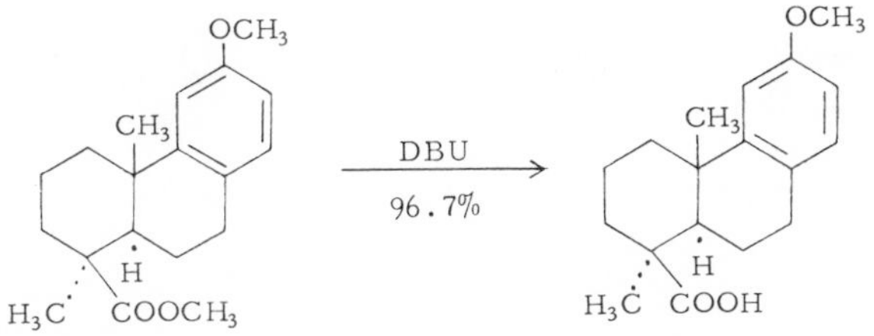

(3)                              (4)

methyl O-methylpodocarpate) under the same conditions gives (4) in high yield. The highly hindered ester methyl mesitoate can be hydrolyzed in this way to mesitoic acid (94.6% yield).

[1]E. J. Parish and D. H. Miles, *J. Org.*, **38**, 1223 (1973).

**Diazoacetaldehyde, 2**, 101–102; **3**, 73.

*Preparation.* For a recent preparation, *see* **Acetic-formic anhydride**, this volume.

**Diazoacetic acid azide**, $\overline{N}=\overset{+}{N}=CH\overset{O}{\overset{\|}{C}}N=\overset{+}{N}=\overline{N}$ (1). Mol. wt. 111.07, b.p. (danger of explosion) 20–21°/0.2 mm., m.p. 7–8°.

*Preparation:* by diazotization of glycine hydrazide.[1]

*Addition of dimethyl acetylenedicarboxylate* to give pyrazole-3,4,5-tricarboxylic acid dimethyl ester monoazide:

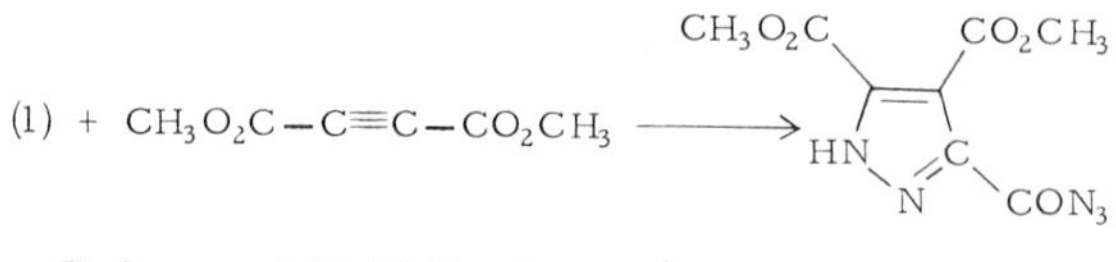

(1) + CH3O2C—C≡C—CO2CH3
CH3O2C
CO2CH3
HN
N
C
CON3


[1]H. Neunhoeffer, G. Cuny, and W. K. Franke, *Ann.*, **713**, 96 (1968).

**Diazomethane, 1**, 191–195; **2**, 102–104; **3**, 74.

*Addition to C–N multiple linkages.*[1]

*Reaction with conjugated diynes and enynes*[2]:

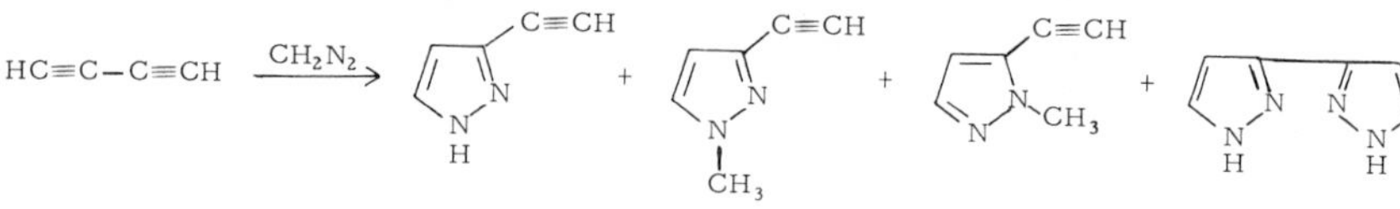

HC≡C—C≡CH
CH2N2
C≡CH
N
H
CH3


(1) (2) (3) (4)

R—C≡C—C=C—CO2CH3
CH2N2
RC≡C
CO2CH3
N
H
R
C=C—CO2CH3


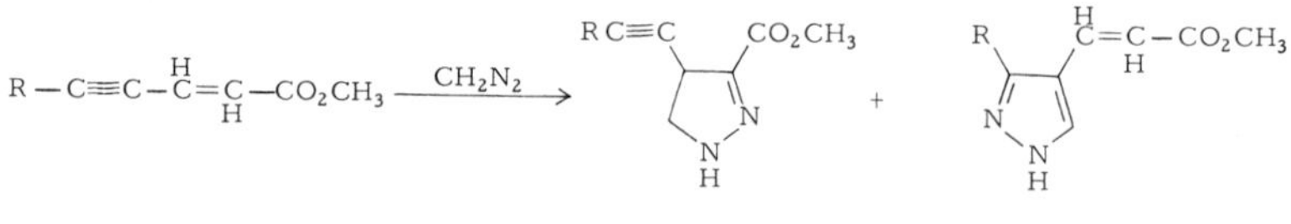

(5) (6)

*Synthesis of* p*-disubstituted tetraphenylbicyclo[3.1.0]hexenones*[3]

(p)R—C6H4
C6H4—R(p)
C6H5
O
CH2N2
R—C6H4
C6H4—R
CH2
N
-N2

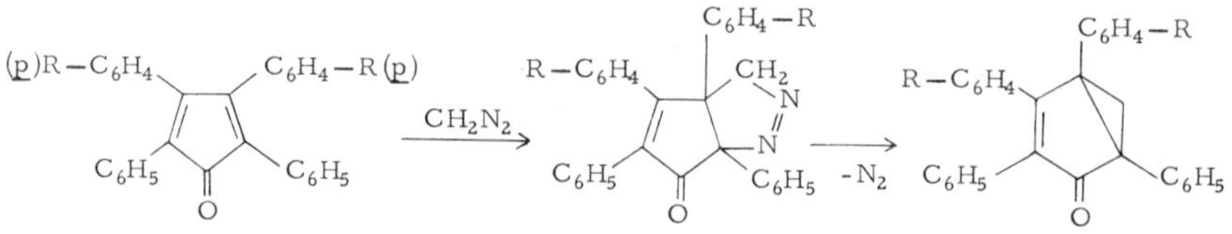

*Addition to di-t-butylquinones[4] :*

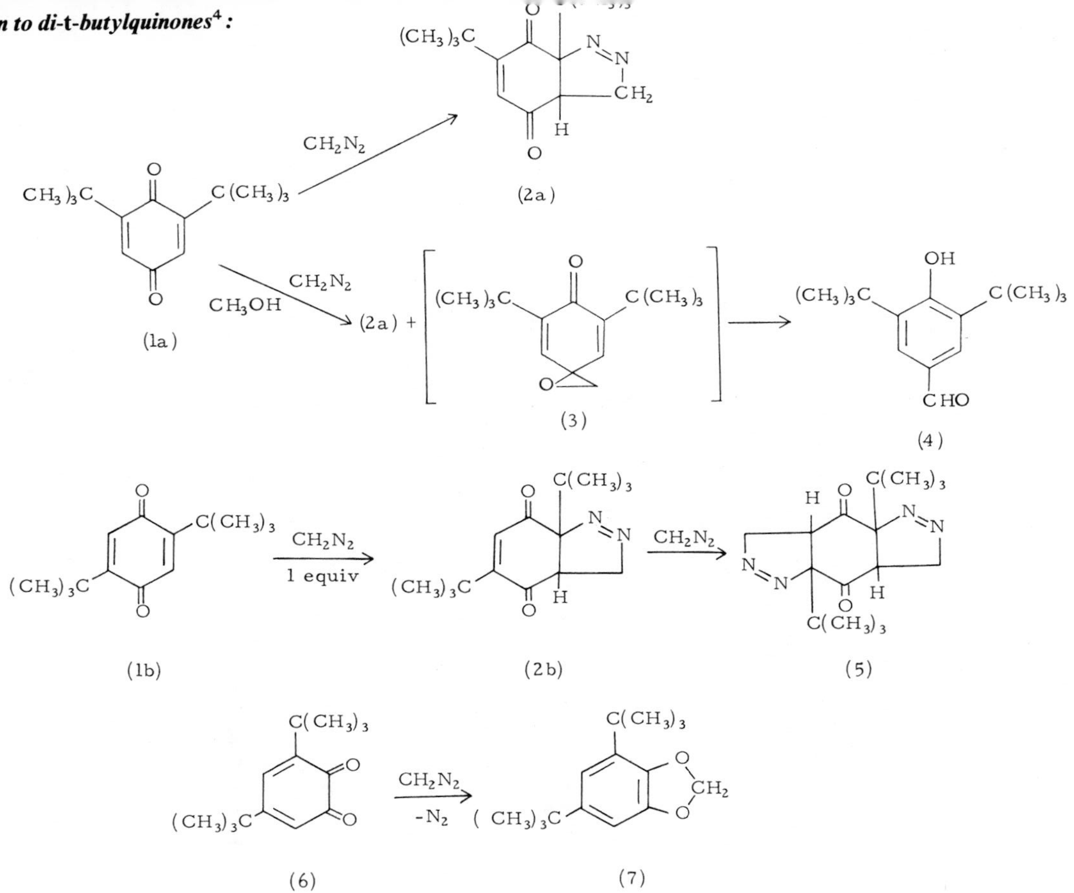

*α-Chloro sulfoxides.*[5] Diazomethane reacts readily with sulfinyl chlorides[6] to form chloromethyl sulfoxides in high yield:

$$H_2CN_2 \ + \ RS(O)Cl \ \longrightarrow H_2CClS(O)R$$

Other diazoalkanes can be used also in this reaction, but yields are somewhat lower. *Synthesis of diazoacetophenone.*[7] When an acid chloride is added slowly to 2 eq. of diazomethane, the hydrogen chloride liberated (equation 1) is then consumed

$$C_6H_5COCl \ + \ CH_2N_2 \ \longrightarrow C_6H_5COCHN_2 \ + \ HCl \quad (1)$$

$$CH_2N_2 \ + \ HCl \longrightarrow CH_3Cl \ + \ N_2 \quad (2)$$

$$C_6H_5COCHN_2 \ + \ HCl \longrightarrow C_6H_5COCH_2Cl \ + \ N_2 \quad (3)$$

according to equation 2. When the order of addition is reversed and only 1 mole of diazomethane is employed, the diazoketone reacts with hydrogen chloride (equation 3) to form the α-chloroketone. The method described here, discovered by Newman and Beal,[8] employs triethylamine to react with the hydrogen chloride; thus only 1 eq. of diazomethane is necessary. This modification is restricted to the preparation of

$$C_6H_5COCl \ + \ CH_2N_2 \ + \ (C_2H_5)_3N \longrightarrow C_6H_5COCHN_2 \ + \ (C_2H_5)_3\overset{+}{N}HCl^-$$

aromatic α-diazoketones; an aliphatic acid chloride gives a mixture of products.

[1]H. Hoberg, *Ann.*, **707**, 147 (1967).
[2]H Reimlinger, J. J. M. Vandewalle, and A. van Overstraeten, *ibid.*, **720**, 124 (1968).
[3]H. Dürr and P. Heitkämper, *ibid.*, **716**, 212 (1968).
[4]W. Rundel and P. Kästner, *ibid.*, **737**, 87 (1970).
[5]C. G. Venier, H.-H. Hsieh, and H. J. Barager, III, *J. Org.*, **38**, 17 (1973).
[6]Preparation: I. B. Douglass and R. V. Norton, *ibid.*, **33**, 2104 (1968).
[7]J. N. Bridson and J. Hooz, *Org. Syn.*, submitted 1971.
[8]M. S. Newman and P. Beal, III, *Am. Soc.*, **71**, 1506 (1949).

**Dibenzoyl peroxide, 1,** 196–198.

*Free-radical cyclization.*[1] δ-Ethylenic compounds can be cyclized to alicyclic compounds by treatment with peroxides to generate carbon radicals. Thus ethyl 1-cyano-2-methylcyclohexanecarboxylate (3) can be prepared from the tosylate of *trans*-4-hexene-1-ol (1) by the procedure formulated. The tosylate is condensed with

ethyl cyanoacetate (sodium hydride, DMF) to give ethyl *trans*-2-cyanooctene-6-oate
(2). Treatment of (2) with dibenzoyl peroxide as a radical initiator gives 1-cyano-2-
methylcyclohexanecarboxylate (3) in 88% yield. Use of di-*t*-butyl peroxide gives (3)
in 68% yield.[2]

  *α-Methylene-γ-butyrolactones.* Ourisson et al.[3] have described a method for conversion of *trans*-fused α-methyl-γ-butyrolactones into α-methylene-γ-butyrolactones.
For example, the lactone (1), obtained in several steps from α-santonin, is treated with
triphenylmethyllithium in dimethoxyethane (DME) and then with dibenzoyl peroxide
in DME. The benzoxylactone (2) is obtained in about 50% yield. Pyrolysis of (2) at
about 450° gives the α-methylene-γ-butyrolactone (3, (+)-arbusculin-B) in 35% yield.

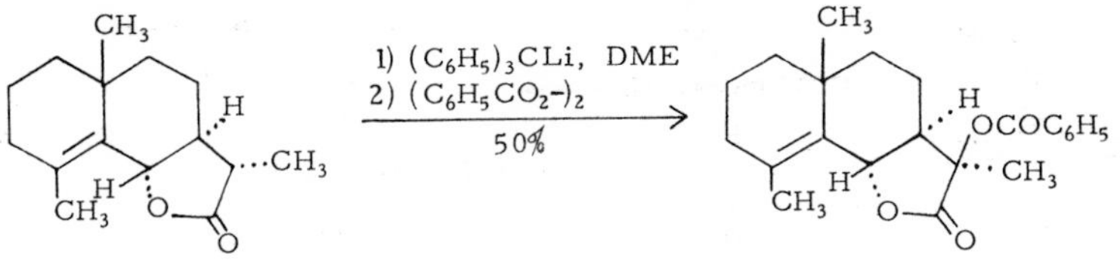

$$1)\ (C_6H_5)_3CLi,\ DME$$
$$2)\ (C_6H_5CO_2-)_2$$
$$50\%$$

(1)          (2)

$$\sim 450°$$
$$35\%$$

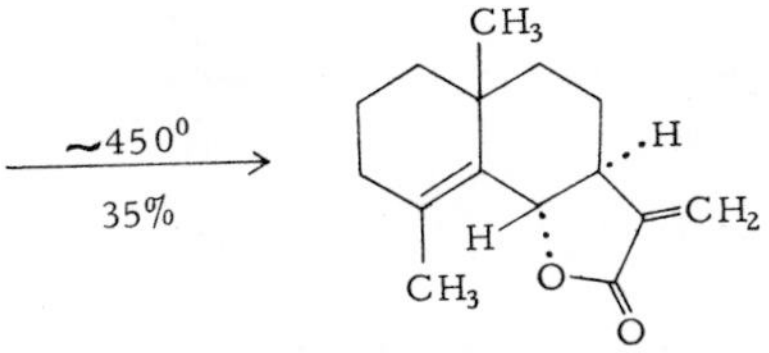

(3)

  For the conversion of *cis*-fused lactones into α-methylene-γ-butyrolactones, *see*
**1,2-Dibromoethane**, this volume.

[1]M. Julia, J. M. Surzur, and L. Katz, *Bull soc.*, 1109 (1964); M. Julia and M. Maumy, *ibid.*,
2415, 2427 (1969); M. Julia, *Accts. Chem. Res.*, **4**, 386 (1971).
[2]M. Julia and M. Maumy, *Org. Syn.*, submitted 1973.
[3]A. E. Greene, J. C. Muller, and G. Ourisson, *Tetrahedron Letters*, 3375 (1972).

**Dibenzyl ketone, 1, 198.**
  *Condensation with phenylcyclobutenedione[1]:*

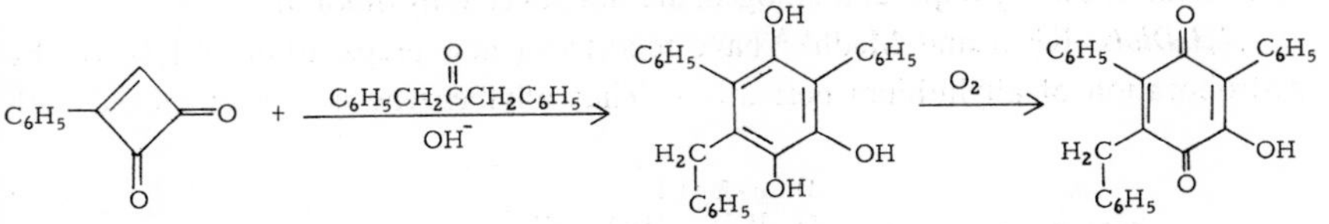

[1]W. Ried and W. Kunkl, *Ann.*, **717**, 54 (1968).

**Diborane,** 1, 199–207; 2, 106–108; 3, 76–77.

Two stable diborane complexes, borane–tetrahydrofurane and borane–dimethyl sulfide, are available from Aldrich. $BH_3 \cdot THF$ should be stored at $0°$; $BH_3 \cdot (CH_3)_2S$ can be stored at room temperature.

*Optically active amines.*[1] Esterification of an oxime, for example 2-hydroximino-1-phenylpropane (1), with a chiral carboxylic acid chloride, S-(+)-2-phenylbutyric acid chloride (2), gives an oximoester (3), which on reduction with diborane gives a mixture of optically active amines (4a, 4b) in which the S-enantiomer (4a) predominates slightly. (The optical yield is about 6%.)

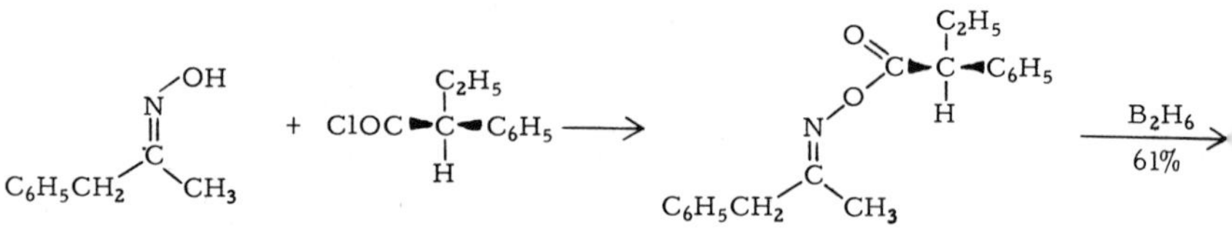

**(1)**          **(2)**              **(3)**

**(4a)**                **(4b)**

*Reduction of amides to amines.*[2] Primary, secondary, and especially tertiary amides are reduced by diborane to the corresponding amines rapidly and almost quantitatively. The reaction is carried out at $25°$ in the case of tertiary amines, and at reflux for primary and secondary amines. The method is not suitable for unsaturated

$$(Ar)RCNR'_2 \xrightarrow[83-97\%]{\begin{array}{c}BH_3\\THF\end{array}} (Ar)RCH_2NR'_2$$

$$R' = H, \text{ or alkyl}$$

amides, since diborane adds rapidly to double bonds. However, other substituents, such as nitro, ester groups, and halogen, are not affected by diborane.

*1,3-Diols.* Klein and Medlik[3] have reported a new preparation of 1,3-diols by hydroboration of allyllithium derivatives. Thus allylbenzene is first metalated with

$$C_6H_5CH_2CH=CH_2 \xrightarrow[65\%]{\begin{array}{c}1)\ \underline{n}\text{-BuLi}\\2)\ B_2H_6;\ H_2O_2,\ OH^-\end{array}} C_6H_5CHCH_2CH_2OH$$
$$\qquad\qquad\qquad\qquad\qquad\qquad\qquad\qquad\qquad | \\ \qquad\qquad\qquad\qquad\qquad\qquad\qquad\qquad\quad OH$$

*n*-butyllithium in ether; after standing overnight the solution is added to a cooled solution of diborane in THF. Subsequent oxidation with alkaline hydrogen peroxide in water gives 1-phenyl-1,3-propanediol, isolated as the diacetate.

**trans-1,2-Diols.**[4] Hydroboration–oxidation of sodium or lithium enolates gives *trans*-1,2-diols. For example, the sodium enolate of cyclohexanone (prepared by heating the ketone with sodium hydride in DME) was treated with diborane (3 moles) in THF

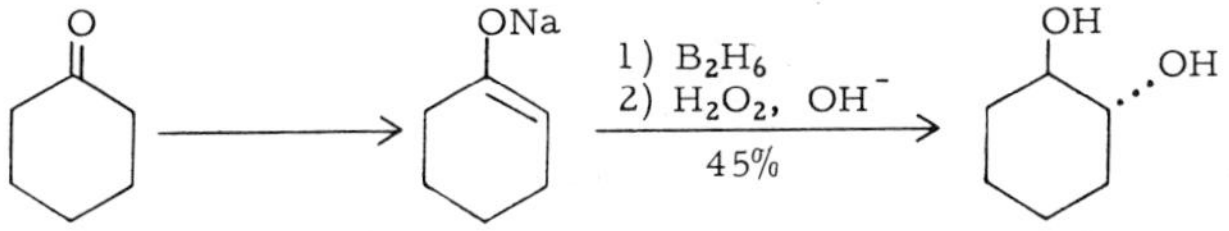

(1)

in an ice bath. Oxidation with alkaline hydrogen peroxide gives *trans*-1,2-cyclohexane-diol (1) in 45% yield. The reaction with the sodium enolate of 4-methylcyclohexanone gave a mixture of *trans*-diols (2) and (3) in the ratio of 3:2 (total yield 55%).

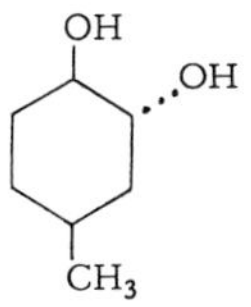

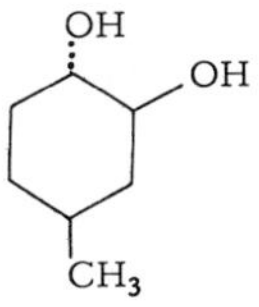

Treatment of 2-methylcyclohexanone with triphenylmethyllithium gives the enolate (4) formed by kinetic control. Hydroboration–oxidation of (4) gives a mixture of *trans*-diols (5) and (6) in a ratio of 42:58 (total yield 70%). The thermodynamically stable enolate (7) gives the *trans*-diol (8) in 60% yield.

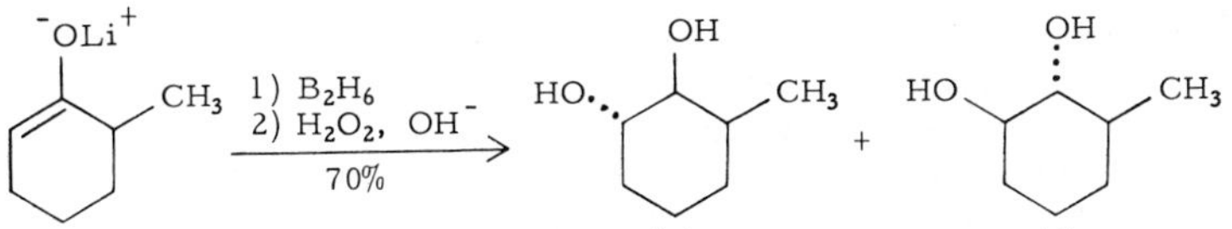

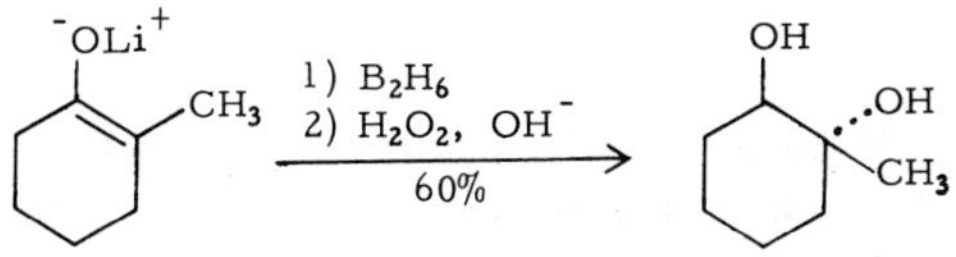

In the same way, trimethylsilyl enol ethers are also converted into *trans*-1,2-diols.

**Phenols.**[5] Organomercury halides of the type ArHgX react with diborane in THF under normal hydroboration procedures to give intermediate organoboranes which are not isolated but oxidized with alkaline hydrogen peroxide to give phenols in good yield.

*Reaction with Grignard reagents.*[6] Diborane reacts with Grignard reagents to give organoboranes, which, without isolation, are oxidized to the corresponding alcohols by alkaline hydrogen peroxide. Yields are highest when the Grignard reagent

$$>BH \ + \ RMgBr \longrightarrow HMgBr \ + \ >BR$$
$$\downarrow OH^-/H_2O_2$$
$$ROH$$

is formed in the presence of $BH_3$. The standard reaction conditions involve reaction of 10 mmole of halide with 10 milliatoms of magnesium and 15 mmole of $BH_3$ in refluxing THF. The formation of the organoborane is essentially complete after 45 min. The scope of the reaction appears to be limited by the efficiency of formation of the Grignard reagent, which, once formed, is converted quantitatively into the organoborane. The reaction offers a simple, convenient method for replacement of halogen by hydroxyl in both aliphatic and aromatic compounds.

Examples:

Bromobenzene $\longrightarrow$ phenol (67%)

m-Chloroiodobenzene $\longrightarrow$ m-chlorophenol (77%)

1-Bromobutane $\longrightarrow$ n-butanol (73%)

Benzyl bromide $\longrightarrow$ benzyl alcohol (99%)

Cyclohexyl bromide $\longrightarrow$ cyclohexanol (67%)

1-Bromoadamantane $\longrightarrow$ 1-hydroxyadamantane (48%)

[1]U. Busser and R. Haller, *Tetrahedron Letters*, 231 (1973).
[2]H. C. Brown and P. Heim, *J. Org.*, **38**, 912 (1973).
[3]J. Klein and A. Medlik, *Am. Soc.*, **93**, 6313 (1971).
[4]J. Klein, R. Levene, and E. Dunkelblum, *Tetrahedron Letters*, 2845 (1972).
[5]S. W. Breuer, M. J. Leatham, and F. G. Thorpe, *Chem. Commun.*, 1475 (1971).
[6]S. W. Breuer and F. A. Broster, *J. Organometal. Chem.*, **35**, C5 (1972).

**Dibromo(bistriphenylphosphine)cobalt(II)**, $CoBr_2 \cdot 2P(C_6H_5)_3$, **3**, 89–90.

*Norbornadiene dimerization to "Bisnor-S"* (**2**, 123–124; **3**, 89–90.)[1] The dimerization of norbornadiene to "Bisnor-S" with this catalyst in conjunction with boron trifluoride etherate can be effected in 80% yield if the initial temperature is controlled carefully (exothermic reaction).

[1]T. Courtney, D. E. Johnston, M. A. McKervey, and J. J. Rooney, *J.C.S. Perkin I*, 2691 (1972).

**1,2-Dibromoethane (Ethylene dibromide)**, $BrCH_2CH_2Br$. Mol. wt. 187.87, m.p. 9–10°, b.p. 132°. Suppliers: Aldrich, J. T. Baker, Columbia, Eastman, Fisher, Howe and French, MCB, E. Merck, P and B, Riedel de Haën, and Sargent.

*α-Methylene-γ-butyrolactones.* Ourisson *et al.*[1] have described a new approach to the synthesis of α-methylene-γ-butyrolactones from α-methyl-γ-butyrolactones. The method was used for the conversion of (1, obtained in several steps from 6-episantonin) into ( − )-frullanolide (3), a natural sesquiterpene. The lactone (1) was treated with

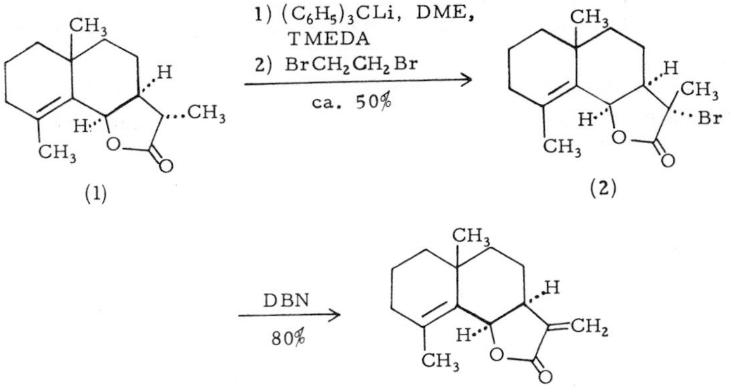

excess triphenylmethyllithium in dimethoxyethane (DME) in the presence of tetra-methylethylenediamine (TMEDA) to give the corresponding enolate, which, on quenching with 1,2-dibromoethane at 5°, afforded the α-bromolactone (2) in about 50% yield. Dehydrobromination with excess 1,5-diazabicyclo[4.3.0]nonene-5 (DBN, **1**, 189–190; **2**, 98–99) in refluxing toluene afforded ( − )-frullanolide (3) in about 80% yield. This procedure is apparently the first example of the use of a *vic*-dihalide for formation of an α-haloketone.

Note that this method is limited to *cis*-fused lactones. *Trans*-fused bromolactones undergo *trans* elimination of HBr to afford exclusively endocyclic olefins (4 → 5).

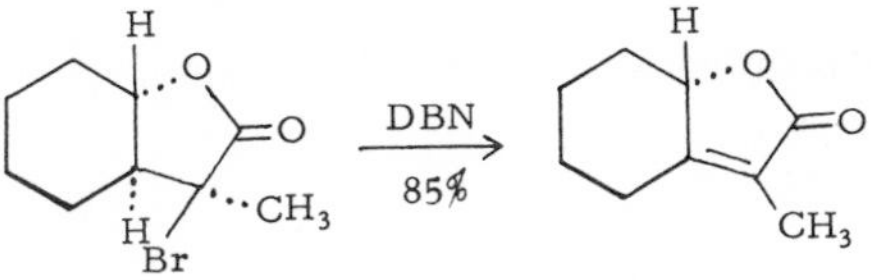

Ourisson *et al.* have developed a method for conversion of *trans*-fused lactones into α-methylene-γ-butyrolactones (*see* **Dibenzoyl peroxide**, this volume).

[1] A. E. Greene, J.-C. Muller, and G. Ourisson, *Tetrahedron Letters*, 2489 (1972).

**Di-*n*-butylcopperlithium, 2**, 151; **3**, 79.
    *Ketone synthesis*[1]

$$2 \, \underline{n}\text{-}C_4H_9Li + CuI \longrightarrow (\underline{n}\text{-}C_4H_9)_2CuLi \xrightarrow[\text{Ether}]{CH_3O_2CCH_2CH_2COCl} CH_3O_2CCH_2CH_2COC_4H_9\text{-}\underline{n}$$

*Ketones.* The diversity of reagent structural types, the simplicity and mildness of experimental conditions, and the functional group selectivity and stereoselectivity of organocopper reagents suggest that these are generally the reagents of choice for conversion of acid chlorides into ketones.[2]

[1]G. H. Posner and C. E. Whitten, *Org. Syn.*, submitted 1971.
[2]H. Reinheckel, K. Haage, and D. Jahnke, *Organometal. Chem. Rev. (A)*, **4**, 55 (1969).

**Di-*t*-butyl dicarbonate**, $(CH_3)_3COCO_2CO_2C(CH_3)_3$. Mol. wt. 209.17, m.p. 21–22°.
*Preparation.*[1] The reagent is prepared as shown from di-*t*-butyl tricarbonate.

$$(CH_3)_3COK + CO_2 \longrightarrow (CH_3)_3COCO_2K \xrightarrow[64\%]{COCl_2}$$

$$(CH_3)_3COCO_2CO_2CO_2C(CH_3)_3 \xrightarrow[84-89\%]{\substack{DABCO \\ -CO_2}} (CH_3)_3COCO_2CO_2C(CH_3)_3 \quad (1)$$

*Protection of amino groups.*[2] The reagent (1) reacts with amino acid esters to give N-*t*-butoxycarbonyl (*t*-BOC) derivatives (2) in good yield. In a typical procedure glycine ethyl ester hydrochloride is suspended in chloroform and $NaHCO_3$ in water

$$(1) + \underset{\substack{| \\ NH_2}}{R'CHCOOR''} \longrightarrow \underset{(2)}{(CH_3)_3CO\overset{\overset{O}{\|}}{C}-NH\underset{\underset{\substack{| \\ R'}}{}}{CH}COOR''}$$

is added, followed by addition of NaCl. The reagent (1, in $CHCl_3$) is added, and the mixture refluxed for 90 min. *t*-BOC-Glycine ethyl ester is obtained in 89% yield.

[1]B. M. Pope, Y. Yamamoto, and D. S. Tarbell, *Org. Syn.*, submitted 1973.
[2]*Idem, Proc. Nat. Acad. Sci. U.S.A.*, **69**, 730 (1972).

**2,6-Di-*tert*-butyl-α-dimethylamino-*p*-cresol** ("Ethyl" antioxidant 703). Mol. wt. 263.4, m.p. 94°, b.p. 179°/40 mm.
Solubility (wt.% at 18°):

Toluene.........22    Water .............. < 0.0007
Ethanol.........28    10% NaOH .......... < 0.002

Suppliers: Ethyl Corporation and Pfatz and Bauer.

Oxidation inhibitor in natural and synthetic rubbers, polyolefin plastics, resins, adhesives, petroleum oils, and waxes.

**Di-*t*-butyl nitroxide, 1, 211; 2, 81.**
The mechanism of triplet quenching by di-*t*-butyl nitroxide has been discussed by Schwerzel and Caldwell.[1]

[1]R. E. Schwerzel and R. A. Caldwell, *Am. Soc.*, **95**, 1382 (1973).

**Dichlorobis(benzonitrile)palladium [Bis(benzonitrile)palladium(II) chloride],**
$(C_6H_5CN)_2PdCl_2$. Mol. wt. 357.85. Supplier: ROC/RIC.
   *2-Substituted benzofuranes.*[1] Treatment of the sodium salt of 2-allylphenol (1)[2] with 1 eq. of the organometallic reagent in refluxing benzene (3 hr.) gives 2-methyl-

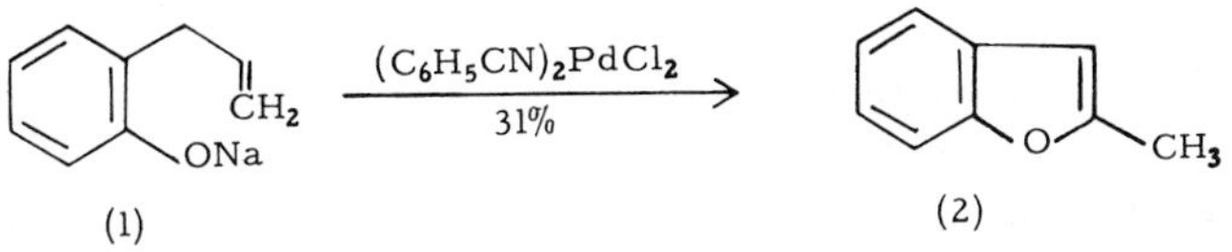

(1)    →    (2)
$(C_6H_5CN)_2PdCl_2$
31%

benzofurane (2) in 31 % yield. 2-Benzylbenzofurane was obtained in the same way in 53 % yield by use of 2-cinnamylphenol.
   2-Methyl[2,1-*b*]naphthofurane (4) was obtained by this method from 1-allyl-2-naphthol (3).

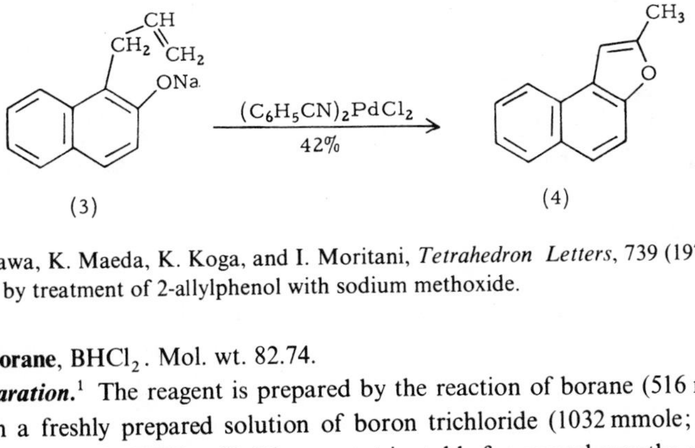

(3)    →    (4)
$(C_6H_5CN)_2PdCl_2$
42%

[1]T. Hosokawa, K. Maeda, K. Koga, and I. Moritani, *Tetrahedron Letters*, 739 (1973).
[2]Prepared by treatment of 2-allylphenol with sodium methoxide.

**Dichloroborane, $BHCl_2$.** Mol. wt. 82.74.
   *Preparation.*[1] The reagent is prepared by the reaction of borane (516 mmole) in THF with a freshly prepared solution of boron trichloride (1032 mmole; **1**, 67–68; **2**, 34–35; **3**, 31–32) in THF at 0°. The reagent is stable for several months at 0°.

$$2\ BCl_3\ +\ BH_3\ \longrightarrow\ 3\ HBCl_2$$

   *Deoxygenation of sulfoxides to sulfides.*[1] Dichloroborane reduces aliphatic sulfoxides to sulfides almost quantitatively in a few minutes at 0° in THF. Reduction of aryl sulfoxides is much slower (24 hr., 25°, excess $BHCl_2$) and yields are lower. Ester,

$$\underset{R}{\overset{R}{\diagdown}}S{=}O \ + \ BHCl_2 \ \longrightarrow \ \left[ \underset{R}{\overset{R}{\diagdown}}\overset{+}{S}{-}O{-}\overset{-}{B}HCl_2 \right] \longrightarrow \ \underset{R}{\overset{R}{\diagdown}}S \ + \ HOBCl_2$$

acid chloride, nitrile, nitro, and even carbonyl groups are not reduced under the conditions used for aliphatic sulfoxides.

[1]H. C. Brown and N. Ravindran, *Synthesis*, 42 (1973).

**Dichloroborane diethyl etherate**, $BHCl_2 \cdot O(C_2H_5)_2$. Mol. wt. 156.86.

*Preparation.*[1,2] The reagent is prepared by the reaction of lithium borohydride with boron trichloride (5–10% excess) in diethyl ether at 0°. The reagent can be stored for several weeks at 0–5°.

$$LiBH_4 \ + \ 3 \ BCl_3 \ + \ 4 \ (C_2H_5)_2O \ \longrightarrow 4 \ BHCl_2 \cdot O(C_2H_5)_2 \ + \ LiCl$$

*Synthesis of alkyldichloroboranes.* Dichloroborane diethyl etherate reacts with olefins very slowly in diethyl ether or THF. However, addition of boron trichloride (Lewis acid) results in a rapid reaction at 0° to give alkyldichloroboranes in 80–90% yield:

$$BHCl_2 \cdot O(C_2H_5)_2 \ + \ CH_3(CH_2)_5CH{=}CH_2 \ + \ BCl_3 \ \longrightarrow \ CH_3(CH_2)_6CH_2BCl_2 \ + \ BCl_3 \cdot O(C_2H_5)_2$$

Alkylboronic acid esters, $RB(OR')_2$, can be prepared by addition of an excess of an alcohol to the reaction above.[2]

[1]H. C. Brown and P. A. Tierney, *J. Inorg. Nucl. Chem.*, **9**, 51 (1959).
[2]H. C. Brown and N. Ravindran, *Am. Soc.*, **95**, 2396 (1973).

**Dichlorocarbene.**

*6-Chlorofulvene.*[1] The reaction of cyclopentadiene (1) with dichlorocarbene (generated from chloroform and potassium *t*-butoxide) gives 6-chlorofulvene (2) as the major product (10% yield); chlorobenzene (3) is also formed (1–2% yield).

(1)                    (2)              (3)

[1]M. B. D'Amore and R. G. Bergman, *Chem. Commun.*, 461 (1971).

**2,3-Dichloro-5,6-dicyano-1,4-benzoquinone** (DDQ), **1**, 215–219; **2**, 112–117; **3**, 83–86. Additional supplier: E. Merck, G.M.B.H., 61 Darmstadt, West Germany.

*Dehydrogenation.* Vogel *et al.*[1] present procedures for the synthesis of tricyclo-[4.4.1.0$^{1,6}$]undeca-3,8-diene (C) by reduction of naphthalene to isotetralin (A), addition of dichlorocarbene (B), and dechlorination (C). Dehydrogenation with DDQ in dioxane–acetic acid (D) gives the highly unsaturated bicyclo product, 1,6-methano[10]-annulene, melting at 27–28° (yield 85–87 %).

A. 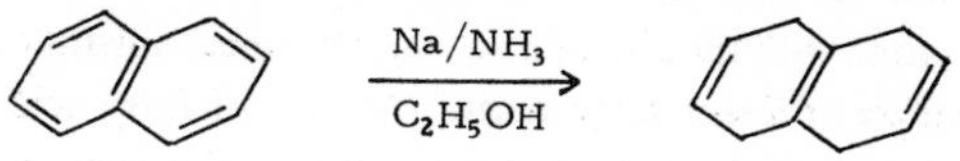 $\xrightarrow[\text{C}_2\text{H}_5\text{OH}]{\text{Na/NH}_3}$

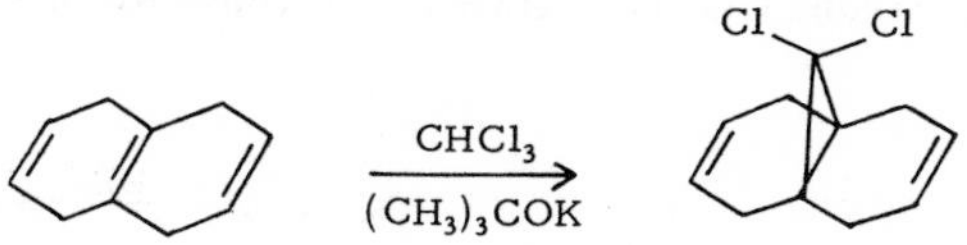

B. $\xrightarrow[(\text{CH}_3)_3\text{COK}]{\text{CHCl}_3}$

C. $\xrightarrow{\text{Na/NH}_3}$

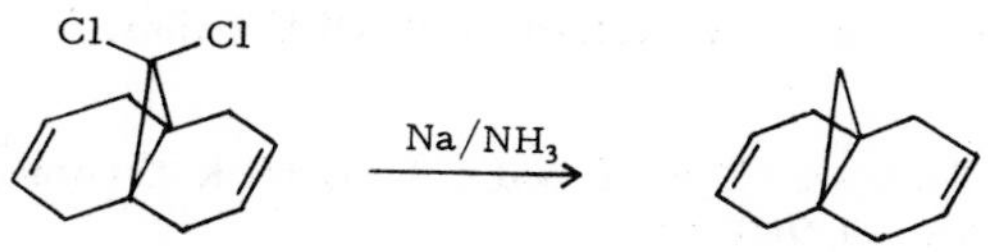

D. 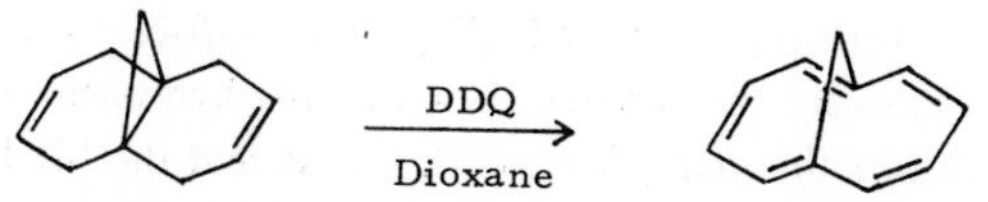 $\xrightarrow[\text{Dioxane}]{\text{DDQ}}$

***Dehydrogenation.*** Dihydrodictamnine (1) is dehydrogenated to the furoquinoline alkaloid dictamnine (2) in nearly quantitative yield by DDQ in refluxing dioxane

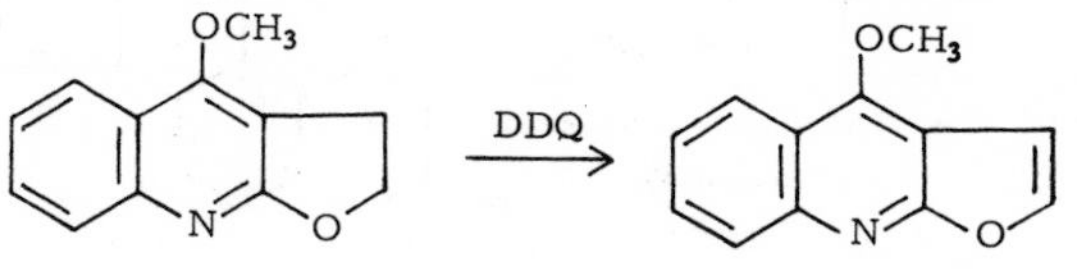

$\xrightarrow{\text{DDQ}}$

(1)                    (2)

solution (36 hr.). No dehydrogenation occurs in benzene, acetonitrile, or methanol solution.[2]

The new nonalternating aromatic hydrocarbon (4, benzo[4.5]cyclohepta[de-1,2,3]-naphthalene) was prepared by dehydrogenation of (3) in benzene solution at 80° by

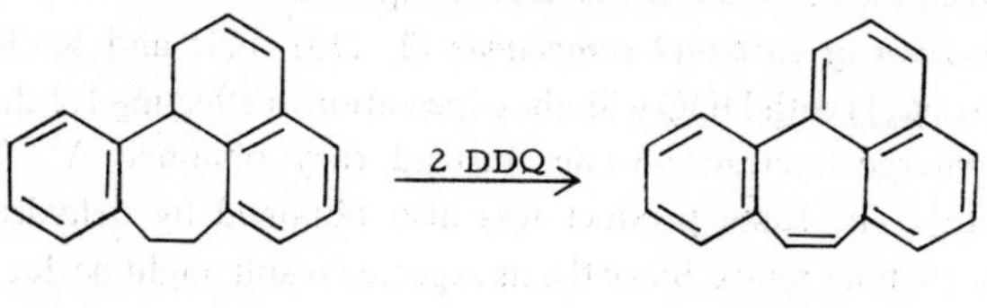

$\xrightarrow{\text{2 DDQ}}$

(3, 232 mg.)              (4, 150 mg.)

use of DDQ or chloranil. Use of palladium-on-charcoal at elevated temperatures led to a mixture of fluorescent resins.[3]

Dehydrogenation of estrogens and 3-acetaminoestratriene-1,3,5(10)-ol-17$\beta$ with DDQ was investigated with the aim of disclosing the mechanisms of the main reactions.[4]

*Dehydrogenation of linear fatty acids.*[5] Anions of fatty acids, prepared by treatment of the sodium salt of the fatty acid with lithium diisopropylamide (**1**, 611; **2**, 249; **3**, 184–185; this volume) in THF–HMPT, are converted into the corresponding

$$R\,CH_2CH_2COONa \longrightarrow \left[ R\,CH_2CHCO_2 \right]^{=} Na^{+}Li^{+} \xrightarrow{\ DDQ\ } R\,CH{=}CHCOOH$$

(E)[a]-$\alpha,\beta$-unsaturated acids by treatment with DDQ (reflux, 3 hr.). Yields are about 30%.

[a]For E, Z nomenclature, see J. L. Blackwood, C. L. Gladys, K. L. Loening, A. E. Petrarca, and J. E. Rush, *Am. Soc.*, **90**, 509 (1968).

*Dehydrogenation of neoergosterol.*[6] Neoergosterol (1) is dehydrogenated in 80% yield by DDQ to 19-norergosta-5,7,9,14,22-pentaene-3$\beta$-ol (2). Thus hydride abstraction involves exclusively the tertiary, benzylic 14$\alpha$-hydrogen. Brown and Turner[6] suggest several factors that may play a part. One is that the 14$\alpha$-hydrogen is virtually

orthogonal to the plane of the aromatic B ring, thereby allowing efficient $\sigma$–$\pi$ overlap in the transition state. Also formation of a tertiary carbonium ion at $C_{14}$ is attended by relief of strain in the C/D-*trans*-ring junction. A further factor may involve $\pi$-complex formation between the aromatic B ring and the quinone.

*Dehydrogenation of carbonyl compounds* (**2**, 113). Pelc and Kodicek[7] dehydrogenated ergosterone (1) with DDQ with the expectation of effecting 1,2-dehydrogenation to give $\Delta^{1,4,7,22}$-ergostatetraene-3-one. Instead they obtained $\Delta^{4,6,8(14),22}$-ergostatetraene-3-one (2). The same product was also obtained by dehydrogenation with chloranil in xylene under reflux. Since the unexpected result might be due to the presence of the 7,8-double bond in (1), they then examined dehydrogenation of isoergosterone (3) with DDQ catalyzed by *p*-toluenesulfonic acid and obtained the expected $\Delta^{1,4,6,22}$-ergostatetraene-3-one (4).

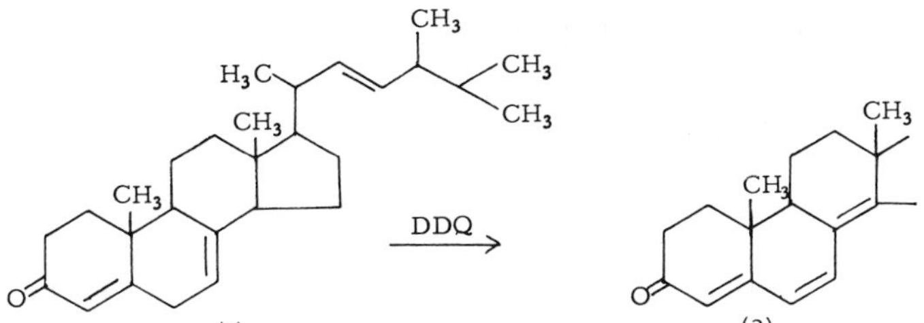

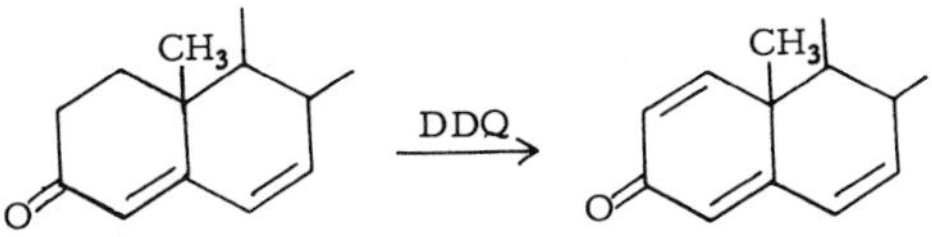

***Ring B oxidation in the podocarpic acid series.***[8] Phenols of the podocarpic acid series undergo exclusive oxidation in ring B with 2,3-dichloro-5,6-dicyanobenzoquinone in alcoholic solvents at room temperature. Thus oxidation of methyl podocarpate (1) with 2.5 eq. of DDQ in methanol gives a mixture of two products, (2) and (3), the latter being the major product (77% yield). Use of 2 eq. of reagent results mainly in

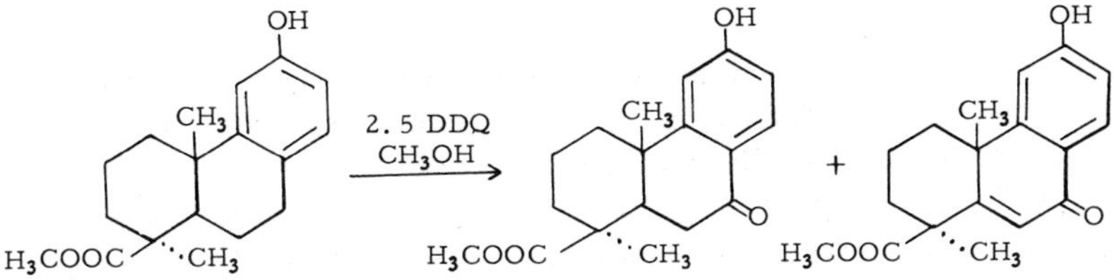

attack at the benzylic methylene group to give the 7-ketone (2); whereas use of 3 eq. leads to the 5,6-dehydro-7-ketone (3) as the predominant product. The saturated 7-ketone is assumed to be formed by way of quinone methide intermediates (4), since the acetate of (1) is not oxidized to a 7-ketone.

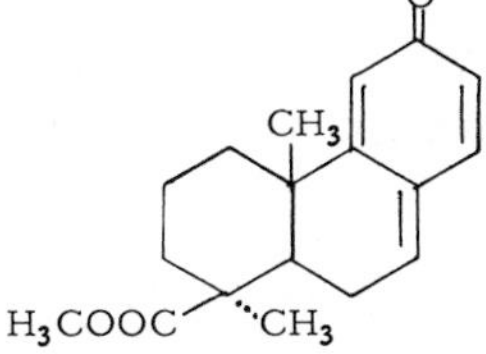

[1]E. Vogel, W. Klug, and A. Breuer, *Org. Syn.*, submitted 1972.
[2]F. Piozzi, P. Venturella, and A. Bellino, *Org. Prep. Proc. Int.*, **3**, 223 (1971).

[3] J. F. Muller, D. Cagniant, and P. Cagniant, *Bull. soc.*, 4364 (1972).
[4] A. Bodenberger and H. Dannenberg, *Ber.*, **104**, 2389 (1971).
[5] G. Cainelli, G. Cardillo, and A. U. Ronchi, *J.C.S. Chem. Commun.*, 94 (1973).
[6] W. Brown and A. B. Turner, *J. Chem. Soc. (C)*, 2057 (1971).
[7] B. Pelc and E. Kodicek, *ibid.*, 859 (1971).
[8] J. W. A. Findlay and A. B. Turner, *J. Chem. Soc. (C)*, 547 (1971).

**Dichloroketene, 1**, 221–222; **2**, 118; **3**, 87–88.

*Dichloromethylenecycloalkanes.*[1] Zinc dechlorination of trichloroacetyl chloride in the presence of cyclic ketones results in formation of dichloroketene and *in situ* cycloaddition to give 3,3-dichloro-2-oxetanones (1). The 2-oxetanones are easily

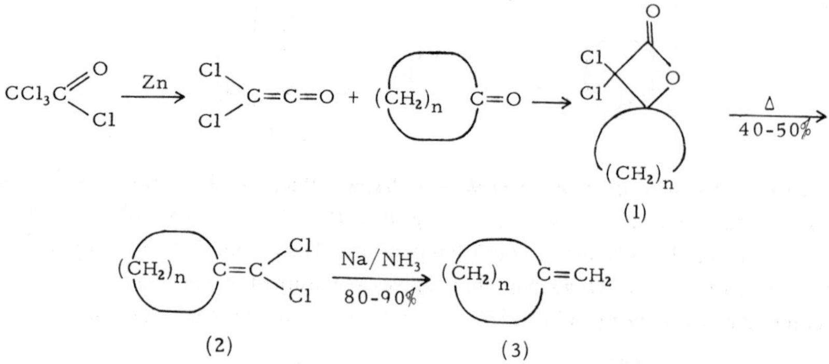

(1)

(2)                    (3)

decarboxylated when vacuum distilled to give dichloromethylenecycloalkanes (2) in 40–50% overall yield. Dechlorination of (2) to methylenecycloalkanes (3) can be effected in high yield with sodium in liquid ammonia at $-78°$.[2]

*1,2-Cycloaddition* (**1**, 221–222; **2**, 118; **3**, 87–88). Two reviews have been published recently on the 1,2-cycloaddition of dichloroketene to olefins and dienes.[3,4] In general, the best yields of cyclobutanones are obtained when the dehydrochlorination of dichloroacetyl chloride with triethylamine is conducted between 30 and 50° in a hydrocarbon solvent containing an excess of the ketenophile. The following reactivity sequence has been observed in the behavior of various ketenes with cyclopentene:

$$Cl_2C=C=O > (C_6H_5)_2C=C=O > (CH_3)_2C=C=O > H_2C=C=O$$

The cycloaddition of dichloroketene to trisubstituted olefins has been reported by Jeffs and Molina.[5] Thus dichloroketene reacts with 1-methylcyclohexene (1) to give

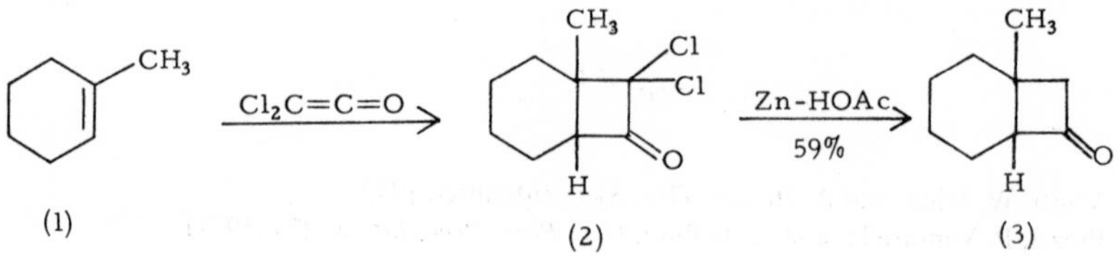

(1)                    (2)                    (3)

a dichlorocyclobutanone (2), which on reduction affords (3) in 59% yield. Similar cyclobutanones were obtained from 1-phenylcyclohexene and from 1-methylcyclopentene.

[1] W. T. Brady and A. D. Patel, *Synthesis*, 565 (1972).
[2] M. C. Hoff, K. W. Greenlee, and C. E. Boord, *Am. Soc.*, **73**, 3329 (1951).
[3] W. T. Brady, *Synthesis*, 415 (1971).
[4] L. Ghosez, R. Montaigne, A. Roussel, H. Vanlierde, and P. Mollet, *Tetrahedron*, **27**, 615 (1971).
[5] P. W. Jeffs and G. Molina, *J.C.S. Chem. Comm.*, 3 (1973).

**Dichloromaleic anhydride,** [structure] . Mol. wt. 166.95. Supplier: Aldrich.

*Isomerization of 1-methoxycyclohexa-1,4-dienes into 1-methoxycyclohexa-1,3-dienes.* 1-Methoxycyclohexa-1,4-dienes, available by Birch reduction of anisoles, have generally been isomerized to the more stable cyclohexa-1,3-dienes with potassium amide in ammonia.[1] The observation that the 1,4-dienes react with dienophiles, usually at elevated temperatures, to give Diels–Alder adducts of the corresponding 1,3-dienes led Birch and Dastur[2] to investigate possible dienophiles as catalysts for the isomerization without undergoing Diels–Alder condensation. Dichloromaleic anhydride was found to be suitable for this purpose. Thus, when the nonconjugated diene (1) was refluxed gently with dichloromaleic anhydride (0.1%) for 2–3 hr. a mixture of (1) and the desired 1,3-diene (2) was obtained in the ratio 17:38 (total recovery 75–85% with some polymer).

(1)                    (2)

Dichloromaleic anhydride can also be used as a catalyst to effect Diels–Alder reactions between the 1,4-dienes and some dienophiles at lower temperatures and shorter reaction periods than required in its absence.

[1] A. J. Birch and G. Subba Rao, *Advan. Org. Chem.*, **8**, 1 (1972).
[2] A. J. Birch and K. P. Dastur, *Tetrahedron Letters*, 4195 (1972).

**Dichloromethylenedimethylammonium chloride,**   $C=\overset{+}{N}(CH_3)_2Cl^-$   (1). Mol. wt. 162.45, m.p. about 180° dec. Supplier: Aldrich ("Phosgene Immonium Chloride").

The reagent is obtained[1] in about 90% yield by chlorination of tetramethylthiuram disulfide[2] in $CH_2Cl_2$.

The reagent reacts readily with protic compounds. For example, it reacts with phenol (2 molar eq.) to give the ammonium salt (2), which on hydrolysis yields diphenyl carbonate (3).

$$2\ C_6H_5OH\ +\ (1)\ \xrightarrow[90\%]{CH_2Cl_2,\ 40^0}\ \begin{array}{c} C_6H_5-O \\ \phantom{xx} \\ C_6H_5-O \end{array}\!\!\!C=\overset{+}{N}(CH_3)_2\ \overset{Cl^-}{\phantom{x}}\ \xrightarrow{H^+\ or\ OH^-}\ \begin{array}{c} C_6H_5-O \\ \phantom{xx} \\ C_6H_5-O \end{array}\!\!\!C=O$$

$$(2)\qquad\qquad\qquad\qquad (3)$$

Reaction of (1) with 2 moles of dimethylamine yields tetramethylurea dichloride (4), which forms tetramethylthiourea (5) on treatment with sodium sulfide.

$$2\ (CH_3)_2NH\ +\ (1)\longrightarrow\ \underset{\underset{Cl\ \ \ Cl}{}}{(CH_3)_2NCN(CH_3)_2}\ \xrightarrow{Na_2S}\ \underset{\underset{S}{\parallel}}{(CH_3)_2NCN(CH_3)_2}$$

$$(4)\qquad\qquad\qquad (5)$$

The immonium salt (1, 2 moles) reacts with N,N-dimethylcarboxamides (e.g., 6, 1 mole) in refluxing methylene chloride or chloroform (1–3 hr.) to form 1,3-dichloro-malonylcyanines (7) in high yield; these are hydrolyzed by sodium bicarbonate to malonylamides (e.g., 8).[3]

$$\underset{(6)}{C_6H_5CH_2\overset{\overset{O}{\parallel}}{C}-N(CH_3)_2}\ \xrightarrow{(1)}\ \left[\underset{Cl^-}{C_6H_5CH_2\overset{\overset{Cl}{|}}{C}=\overset{+}{N}(CH_3)_2}\right]\ \xrightarrow[90\%]{(1)}$$

$$\underset{\underset{Cl^-}{(7)}}{\underset{\underset{Cl\ \ \ Cl}{}}{(H_3C)_2N\cdots\overset{\overset{\overset{C_6H_5}{|}}{C}}{C}\overset{+}{\cdots}C\cdots N(CH_3)_2}}\ \xrightarrow[H_2O]{NaHCO_3}\ \underset{(8)}{\underset{H}{\overset{C_6H_5}{\phantom{x}}}\!\!C\!\!\overset{CON(CH_3)_2}{\underset{CON(CH_3)_2}{\phantom{x}}}}$$

A further example of the reactivity of (1) with carbonyl-activated methylene compounds is the reaction of (1) with cycloalkanones.[4] Thus (1) reacts with cycloalkanones (9) in refluxing chloroform to give immonium chlorides (10), which on hydrolysis give amides (11) in overall yields of about 70–85%. The reagent (1) reacts with enamines

$$\underset{(9,\,n=2,\,3,\,4)}{\overset{O}{\phantom{x}}}\ \xrightarrow{(1)}\ \underset{(10)}{\underset{Cl^-}{\overset{Cl}{\phantom{x}}\ \overset{\overset{Cl}{|}}{C}=\overset{+}{N}(CH_3)_2}}\ \xrightarrow{H_2O}\ \underset{(11)}{\overset{Cl}{\phantom{x}}\ \overset{\overset{O}{\parallel}}{C}-N(CH_3)_2}$$

containing a $\beta$-hydrogen atom (12) to give amide chlorides (13) in high yield.[5] These have been transformed into aminopyrimidines (14) and aminopyrazoles (15) as shown.

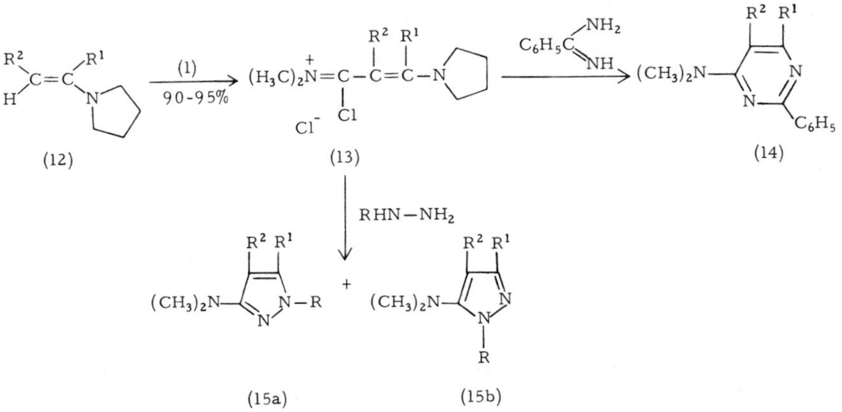

The reagent (1) reacts with secondary amides (16) in refluxing chloroform to give chloroformamidinium salts (17) in high yield (85–95 %).[6] These salts can be hydrolyzed first to N-(1-chloroalkenyl)ureas (18) and then to N-acylureas (19). Dehydrochlorination of (18) with potassium $t$-butoxide gives the hitherto unknown ethynylureas (20).

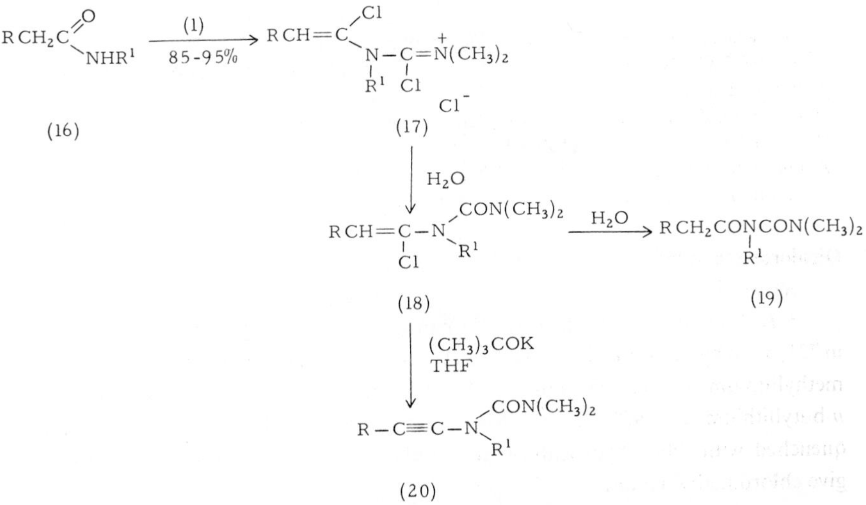

*"Biuret trichlorides".*[7] Dimethylcyanamide (1) reacts with the reagent in methylene chloride to give almost quantitative yields of N,N,N',N'-tetramethyl-1,3-dichloro-2-azatrimethinecyanine (2), a "biuret trichloride." This is hydrolyzed under mild conditions to $N^1,N^1$-dimethyl-$N^2$-(dimethylcarbamoyl)chloroformamidine (3), hydrolyzable under stronger conditions to the biuret derivative (4).

$$(CH_3)_2N-C\!\!\equiv\!\!N \;+\; (CH_3)_2\overset{+}{N}=CCl_2Cl^- \xrightarrow{\;CH_2Cl_2\;}$$
$$\qquad\qquad (1) \qquad\qquad\qquad\qquad\qquad 93\%$$

$$\left[ \begin{array}{c} (CH_3)_2N \underset{Cl}{\overset{N}{\underset{C}{\diagup}}} \underset{Cl}{\overset{N}{\underset{C}{\diagup}}} N(CH_3)_2 \end{array} \right]^+ Cl^- \xrightarrow[71\%]{\begin{array}{c}H_2O,\; NaHCO_3\\ 0^0\end{array}} (CH_3)_2N-\underset{Cl}{\overset{}{C}}=N-\underset{O}{\overset{}{C}}-N(CH_3)_2$$

$$\qquad\qquad\qquad (2) \qquad\qquad\qquad\qquad\qquad\qquad\qquad\qquad (3)$$

$$\Big\downarrow H_2O$$

$$(CH_3)_2N-\underset{O}{\overset{}{C}}-NH-\underset{O}{\overset{}{C}}-N(CH_3)_2$$

$$(4)$$

The azacyanine (2) is useful for synthesis of heterocycles. Thus reaction of (2) with phenylhydrazine gives the 1,2,4-triazole (5) in 93% yield.

$$(2) \;+\; C_6H_5NHNH_2 \xrightarrow[93\%]{\begin{array}{c}1)\;(C_2H_5)_3N\\ 2)\;KOH/H_2O\end{array}} (CH_3)_2N-\!\!\!\underset{\underset{C_6H_5}{N-N}}{\overset{N}{\diagup\diagup}}\!\!\!-N(CH_3)_2$$

$$(5)$$

[1]H. G. Viehe and Z. Janousek, *Angew. Chem., internat. Ed.*, **10**, 573 (1971).
[2](CH₃)₂NCS₂CS₂N(CH₃)₂, mol. wt. 240.43, m.p. 146–148 ; suppliers: Aldrich, Fluka.
[3]Z. Janousek and H. G. Viehe, *Angew. Chem., internat. Ed.*, **10**, 574 (1971).
[4]H. G. Viehe, Z. Janousek, and M.-A. Defrenne, *ibid.*, **10**, 575 (1971).
[5]H. G. Viehe, T. van Vyve, and Z. Janousek, *ibid.*, **11**, 916 (1972).
[6]Z. Janousek, J. Collard, and H. G. Viehe, *ibid.*, **11**, 917 (1972).
[7]Z. Janousek and H. G. Viehe, *ibid.*, **12**, 74 (1973).

**Dichloromethyllithium,** 1, 223–224; 2, 119; 3, 89.

   *Review.*[1]

   *α-Haloketones.*[2] Benzaldehyde (1) can be converted into phenacyl chloride (2) in 72% yield by the following procedure: The aldehyde is added to a solution of dichloromethyllithium (1.20 eq.) in THF at −95°; the mixture is then treated with an excess of *n*-butyllithium at −95°. The resulting solution is allowed to warm to 0° and then quenched with dilute hydrochloric acid. Other aldehydes react in the same way to give chloromethyl ketones.

$$C_6H_5CHO \xrightarrow{Cl_2CHLi} \left[ C_6H_5\overset{H}{\underset{\underset{Li^+}{O^-}}{C}}CHCl_2 \xrightarrow{n-BuLi} C_6H_5\overset{H}{\underset{\underset{2\,Li^+}{O^-}}{C}}-\overset{-}{C}Cl_2 \xrightarrow{\;-LiCl\;} \right.$$

$$(1)$$

$$\begin{bmatrix} C_6H_5\underset{\underset{\text{Li}^+}{O^-}}{C}=CHCl \end{bmatrix} \xrightarrow[72\%]{H^+} C_6H_5COCH_2Cl$$

(2)

The conversion is also applicable to ketones. Thus cyclopentanone can be converted into 2-chlorocyclohexanone in 64% yield and cyclododecanone into 2-chlorocyclo-tridecanone in 46% yield.

[1]G. Köbrich, *Angew. Chem., internat. Ed.*, **11**, 473 (1972).
[2]H. Taguchi, H. Yamamoto, and H. Nozaki, *Tetrahedron Letters*, 4661 (1972); see also J. Villieras, C. Bacquet, and J. F. Normant, *J. Organometal. Chem.*, **40**, C1 (1972).

**Dichlorosilane**, $H_2SiCl_2$. Mol. wt. 101.02. Supplier: Union Carbide Corporation, Sistersville, W. Va.

*Internally substituted alkylsilanes.*[1] Dichlorosilane adds directly to internal olefins when catalyzed by chloroplatinic acid to give internally substituted alkylsilanes in high yield. The addition is carried out in a bomb heated in an oil bath at 130–150° for 8–24 hr.

$$CH_3(CH_2)_3CH=CHCH_3 \xrightarrow[81\%]{\overset{H_2SiCl_2}{H_2PtCl_6}} \underset{\underset{36\%}{SiHCl_2}}{CH_3(CH_2)_3\overset{|}{C}HCH_2CH_3} + \underset{\underset{64\%}{SiHCl_2}}{CH_3(CH_2)_3CH_2\overset{|}{C}HCH_3}$$

[1]R. A. Benkeser and W. C. Muench, *Am. Soc.*, **95**, 285 (1973).

**N,N-Dichlorourethane (DCU)**, **2**, 121–122.

*Review.*[1] Neale has reviewed the addition of N,N-dichlorourethane and of N-chlorourethane, $ClNHCOOC_2H_5$ (MCU), to olefins.

[1]R. S. Neale, *Synthesis*, 1 (1971).

**Dicobalt octacarbonyl**, **1**, 224–228; **3**, 89.

*Reaction with gem-dihalides.*[1] Dicobalt octacarbonyl reacts at 50° with activated *gem*-dihalides in THF or benzene first to give a coupling product, which is then dehalogenated to the "dimer" olefin:

$$R_2CX_2 \xrightarrow{Co_2(CO)_8} R_2-\underset{X}{\overset{|}{C}}-\underset{X}{\overset{|}{C}}-R_2 \xrightarrow{Co_2(CO)_8} R_2C=CR_2$$

Dichlorodiphenylmethane and 9,9-dibromofluorene react in this way to give the "dimer" olefin in high yield.

[1]D. Seyferth and M. D. Millar, *J. Organometal. Chem.*, **38**, 373 (1972).

**Di(cobalttetracarbonyl)zinc**, **2**, 123–124; **3**, 89–90.

*Diamantane.* Use of the reagent to catalyze a dimerization preparatory to the synthesis of diamantane is described by Schleyer *et al.*[1] The resulting Bisnor-S is hydrogenated and the tetrahydride is isomerized by aluminum bromide, with boron trifluoride as cocatalyst.

$$2 \; \text{[Norbornadiene]} + \frac{\text{Zn[Co(CO)}_4]_2}{80-85\%} \longrightarrow \text{[Bisnor-S]} \; \frac{\text{H}_2, \; \text{PtO}_2/\text{AcOH-HCl}}{90-97\%}$$

(1) Norbornadiene        (2) Bisnor-S

$$C_{14}H_{20} \quad \frac{\text{AlBr}_3(\text{BF}_3) - \text{CS}_2}{62-71\%} \longrightarrow \text{[Diamantane]}$$

(3) Tetrahydro-Bisnor-S

(4) Diamantane

[1]T. M. Gund, W. Thielecke,and P. v. R. Schleyer, *Org. Syn.*, submitted 1971.

**Dicyanoacetylene, 1,** 230–231; **2,** 124–125.

*Diels–Alder reaction with thiophene.* Helder and Wynberg[1] have reported the first successful Diels–Alder reaction of thiophenes using dicyanoacetylene as the dienophile. Thus 2,5-dimethylthiophene (1) when heated for 12 hr. in a closed tube at 100° with an excess of freshly prepared dicyanoacetylene yields 3,6-dimethylphthalonitrile (2)

$$\text{(1)} + \text{NC}-\text{C}{\equiv}\text{C}-\text{CN} \longrightarrow \left[ \text{intermediate} \right] \xrightarrow{49\%} \text{(2)} + \text{S}$$

(1)                                                                                (2)

in 49% yield. Alkyl substituents activate the thiophene ring in this reaction; thus the reaction with thiophene itself proceeds in only 8% yield.

More recently, Wynberg and Helder[2] have reported that the reaction of 2,3,4,5-tetramethylthiophene (3) with dicyanoacetylene in methylene chloride at room temperature is dramatically influenced by catalysis with aluminum chloride (**1,** 31–32). The product, obtained in 57% yield, is a bright-yellow solid, shown to be the thiepin (4).

$$\text{(3)} \; \frac{\text{NC}-\text{C}{\equiv}\text{C}-\text{CN}}{\underset{57\%}{\text{AlCl}_3}} \longrightarrow \text{(4)} \; \frac{300°}{(\text{C}_6\text{H}_5)_3\text{P}} \; \text{(5)}$$

(3)                        (4)                        (5)

The thiepin when heated at 300° for 7 min. with triphenylphosphine gives tetramethylphthalonitrile (5) in quantitative yield.

[1]R. Helder and H. Wynberg, *Tetrahedron Letters*, 605 (1972).
[2]H. Wynberg and R. Helder, *ibid.*, 3647 (1972).

**Dicyclohexylborane, 3,** 90–91.

***Conjugated* cis,cis-*dienes.***[1] Dihydroboration of a conjugated diyne, for example dodeca-5,7-diyne (1), with dicyclohexylborane followed by protonolysis of the organoborane intermediate with acetic acid leads to a conjugated *cis,cis*-diene (*cis,cis*-dodeca-

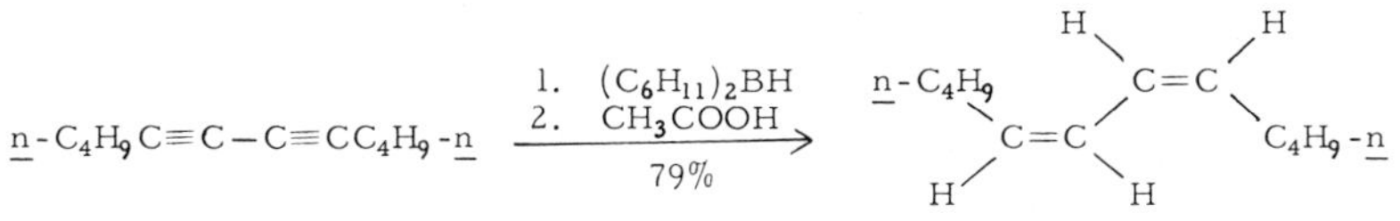

n-C4H9C≡C−C≡CC4H9-n
1. (C6H11)2BH
2. CH3COOH
79%
n-C4H9
H
C=C
H
H
C=C
H
C4H9-n

(1) (2)

5,7-diene). In the case of hindered diynes such as 2,2,7,7-tetramethylocta-3,5-diyne, reduction is carried out in two steps. The diyne is first converted into the enyne by the reaction with 1 eq. of dicyclohexylborane followed by protonolysis. Further reaction with disiamylborane followed by protonolysis affords the *cis,cis*-diene.

[1]G. Zweifel and N. L. Polston, *Am. Soc.*, **92**, 4068 (1970).

**Dicyclohexylcarbodiimide (DCC), 1,** 231–236; **2,** 126; **3,** 91.

***α,β-Unsaturated and α-cyclopropyl ketones.***[1] β-Ketols when treated with DCC and a trace of cuprous chloride undergo intramolecular dehydration to give α,β-unsaturated ketones. The reaction provides a useful route to methylenecycloalkanones:

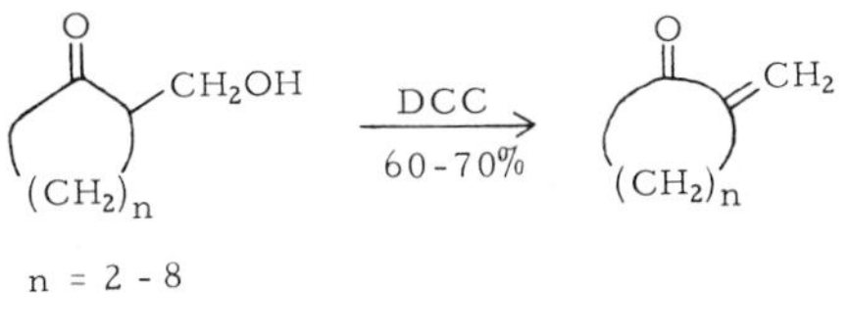

O
CH2OH
(CH2)n
DCC
60-70%
O
CH2
(CH2)n
n = 2 - 8

γ-Ketols under the same conditions are transformed into α-cyclopropyl ketones:

O
OH
DCC
O
C6H11
N
H
NHC6H11
C
O
Δ
O
60%
+ O=C
NHC6H11
NHC6H11

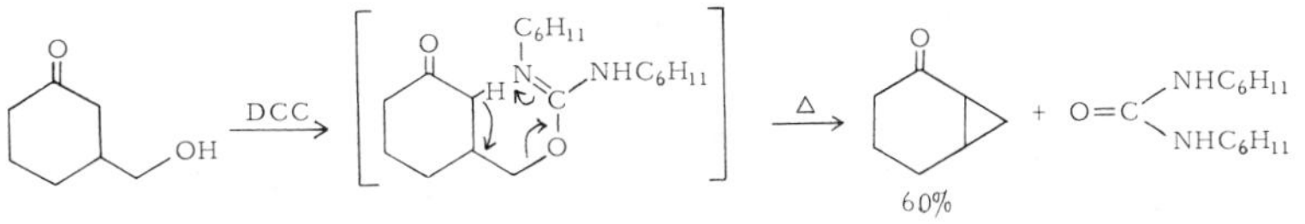

[1]C. Alexandre and F. Rouessac, *Bull. soc.*, 1837 (1971).

**Dicyclohexyl-18-crown-6** (2). *Caution:* Toxic. Supplier: Aldrich.

Du Pont chemists have been interested for several years in macrocyclic polyethers.[1] More than 60 of these polyethers have been prepared, but the two which have been examined the most are (1) and (2). The systematic names of these polyethers are

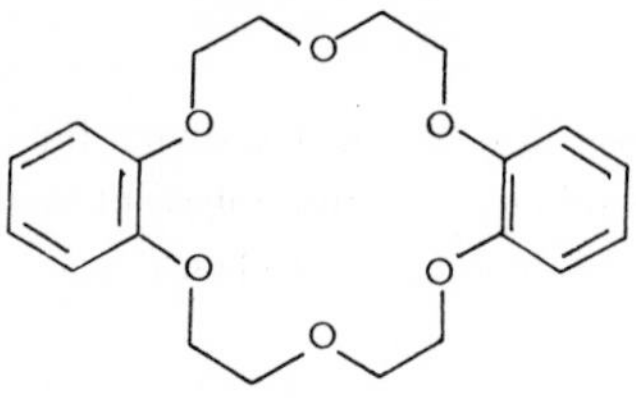

(1) 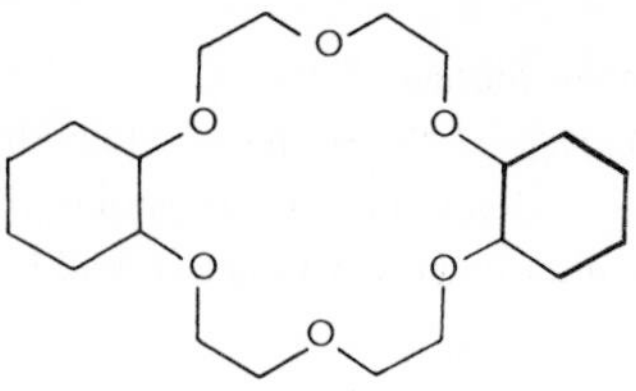                    (2)

exceedingly cumbersome; hence they are referred to by their "crown" names, derived from the appearance of the molecular models. Thus (1) is known as dibenzo-18-crown-6 since it contains 18 atoms in the polyether ring and 6 oxygen atoms in the polyether ring. The common name for (2) is dicyclohexyl-18-crown-6. Dibenzo-18-crown-6 (1) is obtained as white crystals, m.p. 164°, in 80% yield by the reaction of bis[2-(*o*-hydroxy-phenoxy)ethyl] ether with bis(2-chloroethyl) ether in the presence of sodium hydroxide. Dicyclohexyl-18-crown-6 (2) is obtained by hydrogenation of (1) with a ruthenium catalyst. Two isomers are formed, one melting at 61–62.5°, the other at 69–70°. The mixture of isomers can be used for most purposes.[2]

An improved procedure has been described more recently.[3] Note that bis(2-chloro-ethyl) ether is available from Eastman.

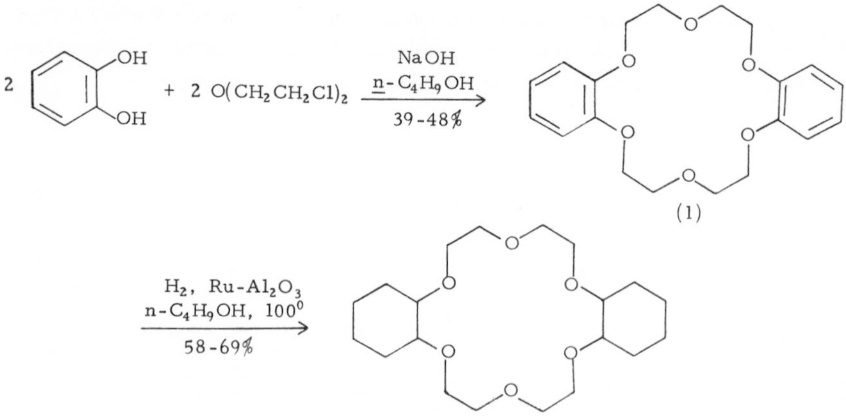

(1)

(2)

The most remarkable property of cyclic polyethers is their ability to form complexes with metal salts, which are soluble in organic solvents. These complexes show enhanced reactivity. Thus sterically hindered esters of 2,4,6-trimethylbenzoic acid which resist saponification by potassium hydroxide in hydroxylic solvents can be hydrolyzed by the

complex of potassium hydroxide with dicyclohexyl-18-crown-6 (2) in aromatic hydrocarbons. It is probable that the activity of these solutions of the complex is due to the
presence of unsolvated hydroxyl ions that are more reactive than ordinary solvated
hydroxyl ions. An even more spectacular example of the use of such complexes is
described below.

**Potassium permanganate oxidations in benzene.** Du Pont chemists[4] have reported
that potassium permanganate complexes with dicyclohexyl-18-crown-6 (2) to form an

(3)

efficient oxidant (3), which is soluble in benzene. The complex is prepared by stirring
equimolar amounts of (2) and potassium permanganate in benzene at 25° to give a clear,
purple solution with concentrations of potassium permanganate as high as 0.06 $M$. The
complex can be isolated as a purple solid, but is more stable in benzene solution. The
complex oxidizes olefins, alcohols, aldehydes, and alkylbenzenes in high yield. Thus

(4)    (5)

trans-stilbene is oxidized to benzoic acid (100% yield). An even more striking example
is the oxidation of α-pinene (4) to cis-pinonoic acid (5) in 90% yield. The yield in this
oxidation with aqueous potassium permanganate is 40–60%. Toluene is oxidized to
benzoic acid in quantitative yield. Oxidations are carried out at 25° since at higher
temperatures the reagent decomposes to adipic acid and other products. The mechanism of oxidations with (3) is presumably similar to that postulated for aqueous
systems. The advantage is that the permanganate anion when complexed is unusually
reactive.

[1] For a recent review see C. J. Pedersen and H. K. Frensdorff, *Angew. Chem., internat. Ed.,* **11,**
16 (1972).
[2] C. J. Pedersen, *Am. Soc.,* **89,** 7017 (1967).
[3] *Idem, Org. Syn.,* **52,** 66 (1972).
[4] D. J. Sam and H. E. Simmons, *Am. Soc.,* **94,** 4024 (1972).

**Diethoxycarbonium fluoroborate,** $HC(OCH_2CH_3)_2{}^+BF_4{}^-$, see **3**, 303. Mol. wt. 189.96. The reagent is prepared by the reaction of triethyl orthoformate with boron trifluoride etherate.[1]

$$HC(OCH_2CH_3)_3 + BF_3 \cdot C_2H_5OC_2H_5 \longrightarrow HC(OCH_2CH_3)_2{}^+BF_4{}^-$$

*Alkylation.* The reagent has been used to prepare N-ethylbenzylamine (2) from benzonitrile (1).

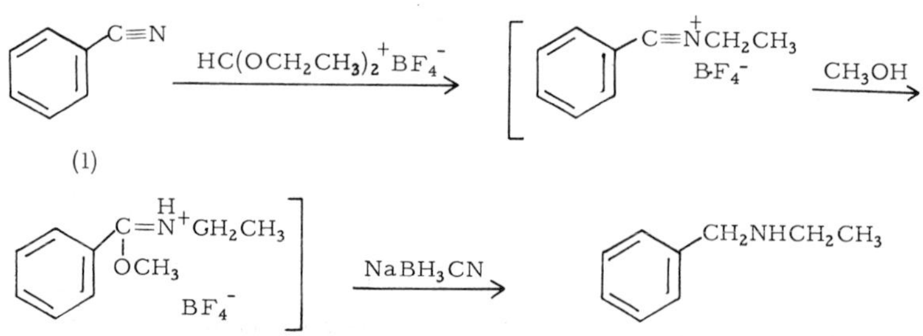

(1)

(2)

[1]R. F. Borch, *Org. Syn.*, submitted 1972.

**Diethylaluminum chloride,** $ClAl(C_2H_5)_2$. Mol. wt. 120.56. Suppliers: ROC/RIC, Texas Alkyls, Inc.

*Alkynyldiethyl alanes.* Alkynyldiethyl alanes, also known as 1-diethylaluminum alkynes (3), are readily prepared[1] by conversion of a terminal alkyne (1) in toluene solution into the lithium derivative (2) with n-butyllithium in hexane. Addition of a 20% solution of diethylaluminum chloride in toluene leads to formation of the reagent (3) in solution with precipitation of LiCl.

$$RC{\equiv}CH \xrightarrow{\text{n-}C_4H_9Li} RC{\equiv}CLi \xrightarrow[-LiCl]{(C_2H_5)_2AlCl} RC{\equiv}CAl(C_2H_5)_2$$

(1)                    (2)                         (3)

*Alkynylation of alicyclic epoxides.*[1] Alkynyldiethyl alanes (2 eq.) react with alicyclic epoxides (1 eq.) to form, after hydrolysis, β-hydroxyacetylenes in fair to excellent yields:

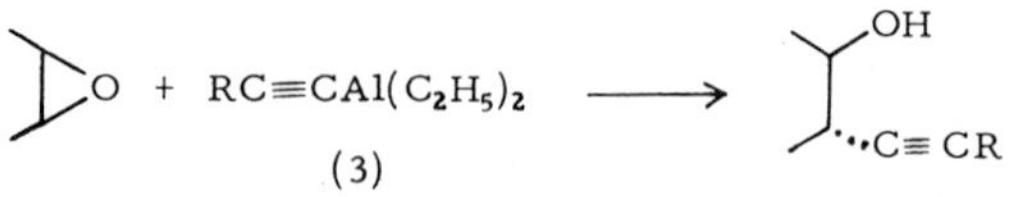

(3)

Examples:

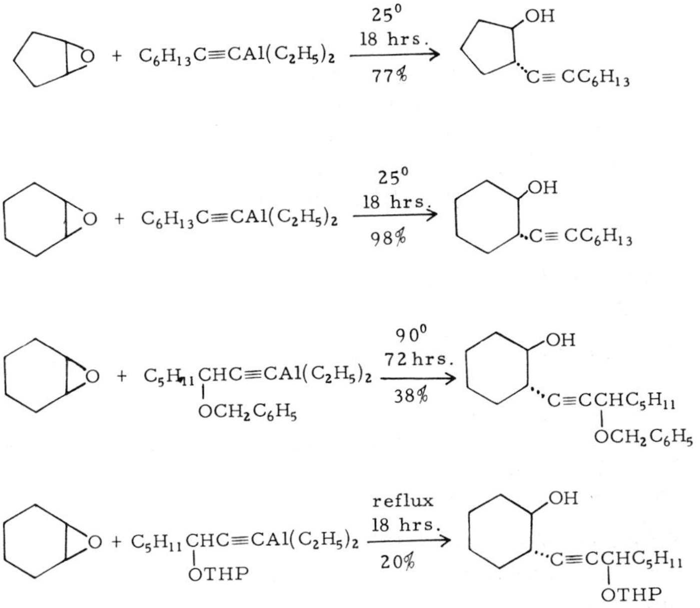

$$\text{(cyclopentene oxide)} + C_6H_{13}C\equiv CAl(C_2H_5)_2 \xrightarrow[77\%]{\substack{25^0 \\ 18 \text{ hrs.}}} \text{product} \quad OH, \ C\equiv CC_6H_{13}$$

$$\text{(cyclohexene oxide)} + C_6H_{13}C\equiv CAl(C_2H_5)_2 \xrightarrow[98\%]{\substack{25^0 \\ 18 \text{ hrs.}}} \text{product} \quad OH, \ C\equiv CC_6H_{13}$$

$$\text{(epoxide)} + C_5H_{11}\underset{OCH_2C_6H_5}{CHC}\equiv CAl(C_2H_5)_2 \xrightarrow[38\%]{\substack{90^0 \\ 72 \text{ hrs.}}} \text{product} \quad OH, \ C\equiv C\underset{OCH_2C_6H_5}{C}HC_5H_{11}$$

$$\text{(epoxide)} + C_5H_{11}\underset{OTHP}{CHC}\equiv CAl(C_2H_5)_2 \xrightarrow[20\%]{\substack{reflux \\ 18 \text{ hrs.}}} \text{product} \quad OH, \ C\equiv C\underset{OTHP}{C}HC_5H_{11}$$

This reaction of an epoxide with an organoalane was used in one step of a synthesis of ($\pm$)-7-oxaprostaglandin $F_{1\alpha}$ (7).[2] Thus reaction of the epoxide (4) with the organoalane (5) gave (6) in "respectable" yield. This was transformed in several steps into (7).

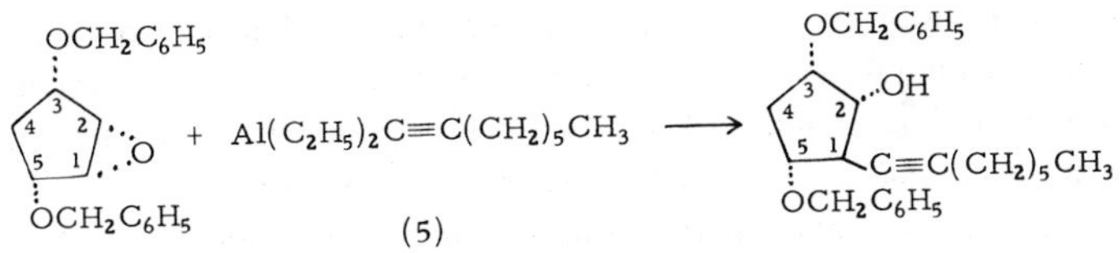

$$(4) + Al(C_2H_5)_2C\equiv C(CH_2)_5CH_3 \longrightarrow (6)$$

(4)        (5)        (6)

$$\xrightarrow{\text{several steps}} (7)$$

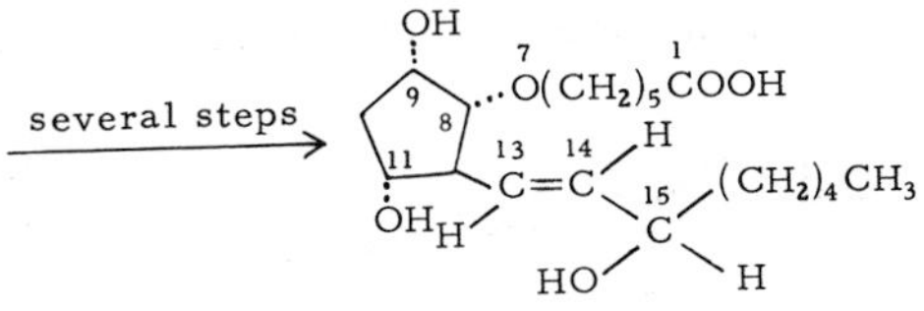

(7)

*γ,δ-Acetylenic ketones.*[3] 1-Diethylaluminum alkynes (3) react with α,β-unsaturated ketones (8) to give, after acid hydrolysis, fair to excellent yields of 1,4-addition products

$$(3) \quad + \quad -\overset{|}{\underset{}{C}}=\overset{|}{\underset{}{C}}\overset{|}{\underset{}{C}}=O \quad \longrightarrow \quad RC\equiv C\overset{|}{\underset{|}{C}}\overset{|}{\underset{}{C}}HC=O$$

$$(8) \qquad\qquad\qquad (9)$$

(9). The success of the reaction depends critically on experimental conditions. Ether–ligroin mixtures usually are suitable as the solvent. The reaction is carried out at temperatures of $-15$ to $25°$. The reaction is restricted to ketones that can adopt a cisoid configuration. Ketones which must have a transoid configuration (e.g., 3-cyclo-hexenone) react to give products of 1,2-addition of the acetylenic unit.

Examples:

$$\underline{n}\text{-}C_4H_9C\equiv CAl(C_2H_5)_2 \;+\; CH_2=CH\overset{O}{\overset{\|}{C}}CH_3 \xrightarrow[48\%]{} \underline{n}\text{-}C_4H_9C\equiv C-CH_2CH_2\overset{O}{\overset{\|}{C}}CH_3$$

$$C_6H_5C\equiv CAl(C_2H_5)_2 \;+\; C_6H_5CH=CH\overset{O}{\overset{\|}{C}}CH_3 \xrightarrow[95\%]{} C_6H_5C\equiv C-\overset{C_6H_5}{\overset{|}{C}}HCH_2\overset{O}{\overset{\|}{C}}CH_3$$

[1]J. Fried, C.-H. Lin, and S. H. Ford, *Tetrahedron Letters*, 1379 (1969).
[2]J. Fried, S. Heim, S. J. Etheredge, P. Sunder-Plassman, T. S. Santhanakrishnan, J. Himizu, and C. H. Lin, *Chem. Commun.*, 634 (1971).
[3]J. Hooz and R. B. Layton, *Am. Soc.*, **93**, 7320 (1971).

**Diethylaluminum cyanide**, 1, 244; 2, 127–128.

The preparation of the reagent has been published.[1] In contrast to alkylaluminums it is not pyrophoric.

$$(C_2H_5)_3Al \;+\; HCN \xrightarrow[80\%]{} (C_2H_5)_2AlCN \;+\; C_2H_6$$

*Hydrocyanation.* Nagata et al.[1-4] have published in detail their results on hydrocyanation with diethylaluminum cyanide (method B) and with a trialkylaluminum and hydrogen cyanide ($R_3Al$–HCN,[5] method A). The main difference between the two methods is that method A hydrocyanation is irreversible and thus controlled kinetically, whereas method B is reversible and therefore can be controlled both kinetically and thermodynamically; that is, the product is kinetically controlled in the early stages.

The original papers contain many examples and comparisons of both types of hydrocyanation.

This reaction provided a key step in a stereocontrolled total synthesis of *dl*-gibberellin $A_{15}$ (3).[6] Thus reaction of the enone (1) with excess diethylaluminum cyanide in methylene chloride at room temperature gave the C/D *cis*-cyano ketone (2) in 87% yield. Note

$$(1) \qquad\qquad\qquad\qquad (2)$$

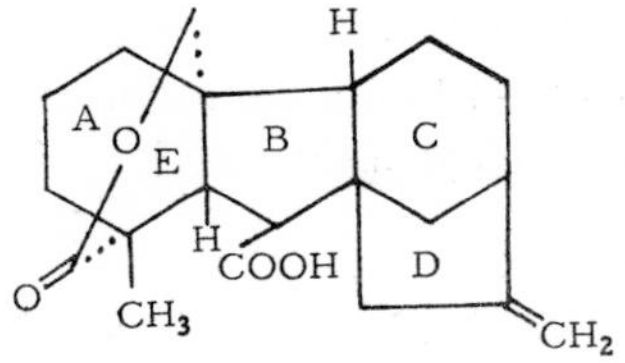

(3)

that the enone (4) resisted hydrocyanation presumably owing to a retarding effect of the neighboring oxygen group on the 9a center.

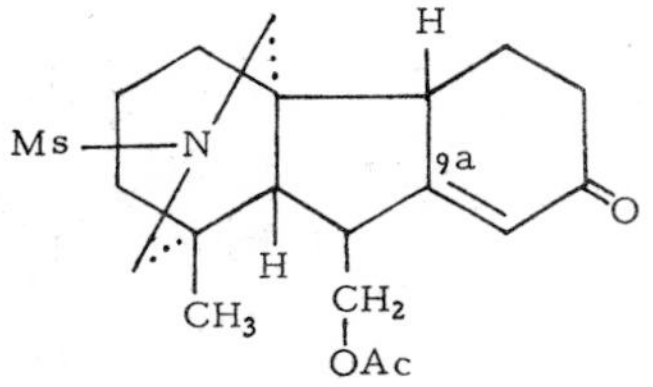

(4)

*Conversion of a carbonyl compound of low reactivity into a cyanohydrin.*[7] The starting material does not react with hydrogen cyanide.

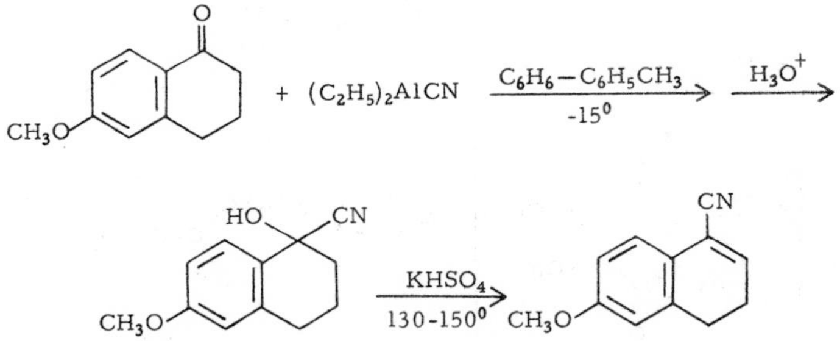

[1]W. Nagata, M. Yoshioka, and S. Hirai, *Am. Soc.*, **94**, 4635 (1972); W. Nagata and M. Yoshioka, *Org. Syn.*, **52**, 90 (1972).

[2]W. Nagata, M. Yoshioka, and M. Murakami, *Am. Soc.*, **94**, 4644 (1972).

[3]*Idem, ibid.*, **94**, 4654 (1972).

[4]W. Nagata, M. Yoshioka, and T. Terasawa, *ibid.*, **94**, 4672 (1972).

[5]Triethylaluminum usually used. Alkylaluminums are available from Ethyl Corp. They are very pyrophoric and must be handled with great care in an inert atmosphere.

[6]W. Nagata, T. Wakabayashi, M. Narisada, Y. Hayase, and S. Kamata, *Am. Soc.*, **93**, 5740 (1971).

[7]W. Nagata, M. Yoshioka, and M. Murakami, *Org. Syn.*, **52**, 96 (1972).

**Diethylamine, 1,** 292; 388–389, 397, 1127.

The basic reagent is used for dehydrochlorination of a quinone dichloride in ether in two steps of a synthesis of *t*-butylcyanoketene.[1]

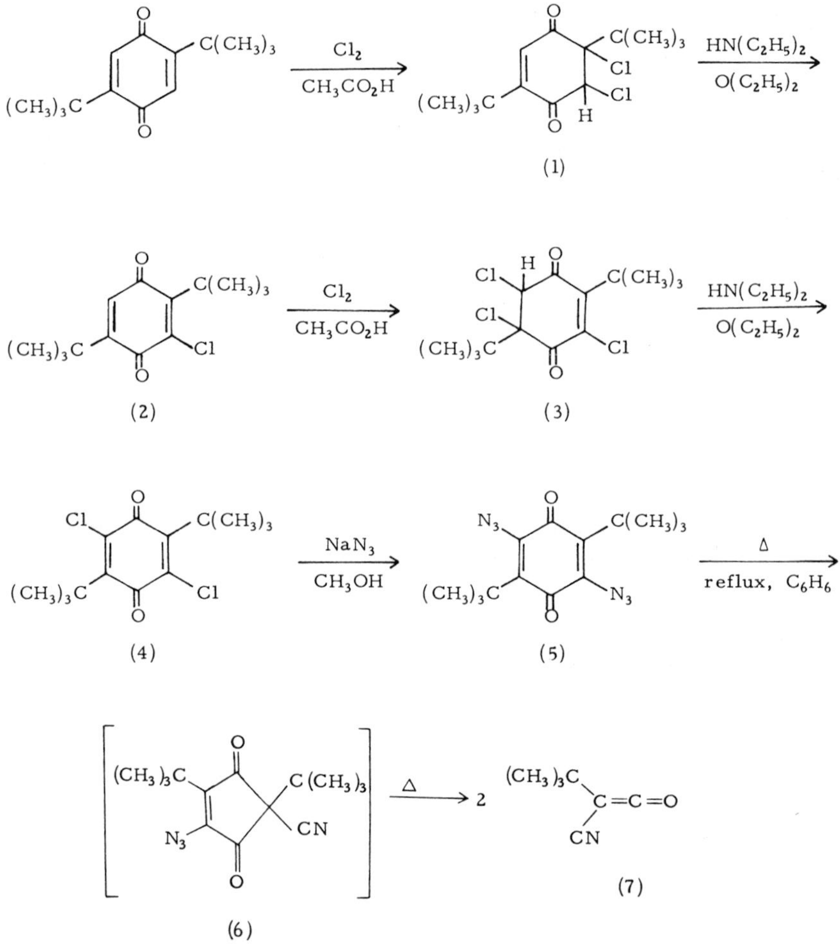

(1)

(2)    (3)

(4)    (5)

(6)    (7)

[1]W. Weyler, Jr., W. G. Duncan, M. B. Liewen, and H. W. Moore, *Org. Syn.*, submitted 1972.

**Diethyl azodicarboxylate, 1,** 245–247; **2,** 128–129.

*Carbodiimides.*[1] Diethyl azodicarboxylate (1) reacts with N,N'-disubstituted thioureas (2) in THF at room temperature overnight to form a 1:1 adduct (3) [an N$^1$,N$^2$-disubstituted-S-(N$^3$,N$^4$-biscarboethoxy)hydrazinoisothiourea], which on treatment with triphenylphosphine at room temperature affords the corresponding carbo-

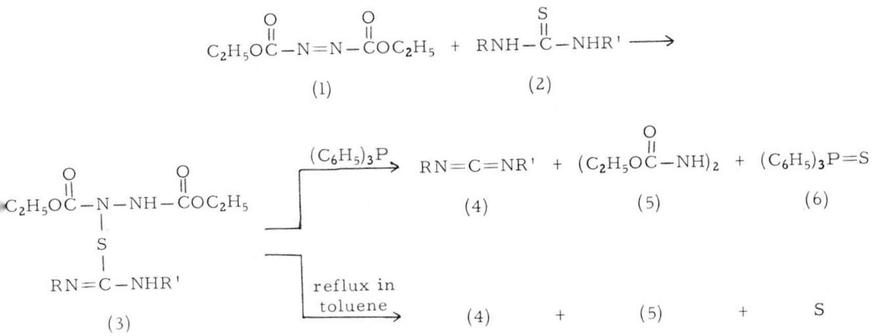

$$C_2H_5O\overset{\overset{O}{\|}}{C}-N=N-\overset{\overset{O}{\|}}{C}OC_2H_5 \ + \ RNH-\overset{\overset{S}{\|}}{C}-NHR' \longrightarrow$$

$$(1) \qquad\qquad\qquad (2)$$

$$C_2H_5O\overset{\overset{O}{\|}}{C}-N-NH-\overset{\overset{O}{\|}}{C}OC_2H_5$$
$$\underset{S}{|}$$
$$RN=C-NHR'$$
$$(3)$$

$$\xrightarrow{(C_6H_5)_3P} RN=C=NR' \ + \ (C_2H_5O\overset{\overset{O}{\|}}{C}-NH)_2 \ + \ (C_6H_5)_3P=S$$
$$(4) \qquad\qquad (5) \qquad\qquad (6)$$

$$\xrightarrow[\text{toluene}]{\text{reflux in}} \quad (4) \quad + \quad (5) \quad + \quad S$$

diimide (4). Alternatively, (3) can be converted into (4) by refluxing in benzene or toluene for 8 hr. Yields of carbodiimides for the most part are 80%.[1] The method has the added merit that the diethyl hydrazodicarboxylate (5) can be oxidized to (1) by hypochlorous acid in 81–83% yield.[2]

**Dehydroabietic acid.** Methyl dehydroabietate (2) can be prepared conveniently and in 85% yield by dehydrogenation of methyl levopimarate (1) with diethyl azodicarboxylate in refluxing benzene.[3]

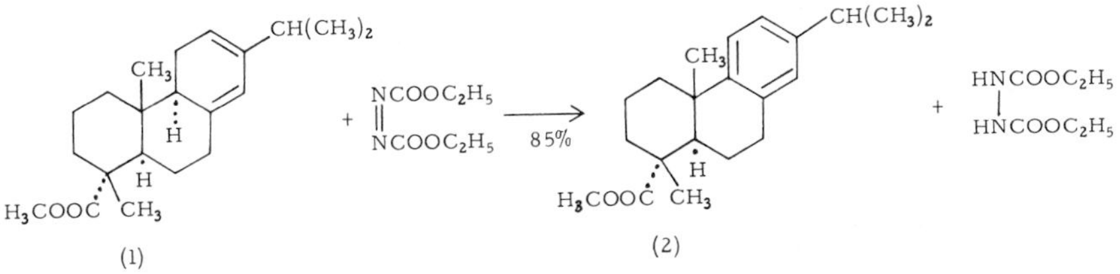

[1]O. Mitsunobu, K. Kato, and F. Kakese, *Tetrahedron Letters*, 2473 (1969); O. Mitsunobu, K. Kato, and M. Tomari, *Tetrahedron*, **26**, 5731 (1970).
[2]N. Rabjohn, *Org. Syn., Coll. Vol.*, **3**, 375 (1955).
[3]G. Mehta and S. K. Kapoor, *Org. Prep. Proc. Int.*, **4**, 257 (1972).

**Diethyl(2-chloro-1,1,2-trifluoroethyl)amine** (2-Chloro-1,1,2-trifluorotriethylamine), 1, 249; 2, 130; 3, 95–96.

**Allenic acids.**[1] Reaction of 17α-difluorocyclopropenyl-5α-androstane-3β,17β-diol 3-acetate (1) with the reagent in dry methylene chloride affords a mixture of three isomeric products (2–4).

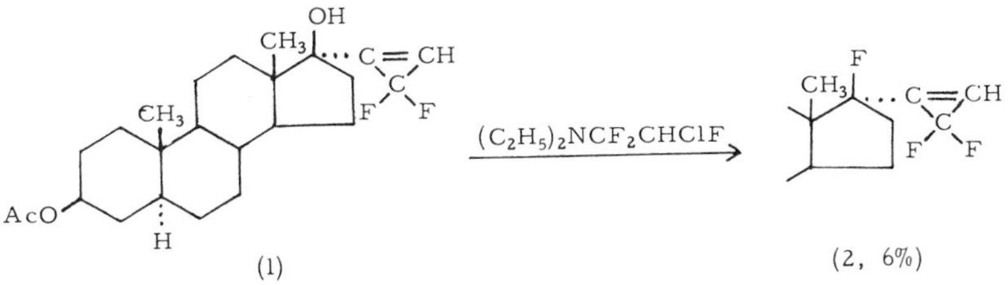

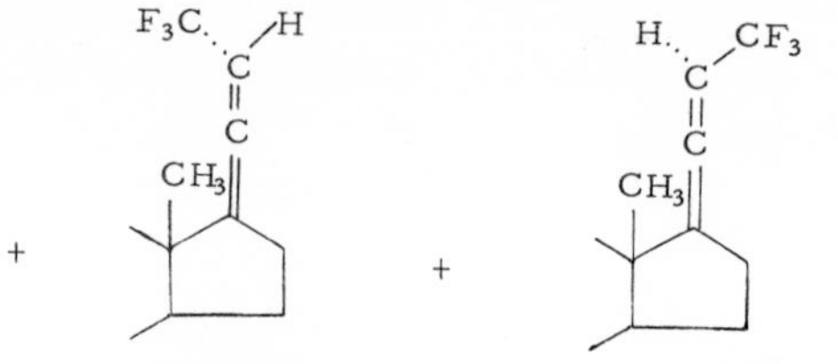

(3, 25%)         (4, 1%)

Treatment of the cyclopropenonecarbinol (5) with the fluoramine gives the allenic acid fluoride (6) in 80% yield. Treatment of (6) with sodium methoxide in methanol

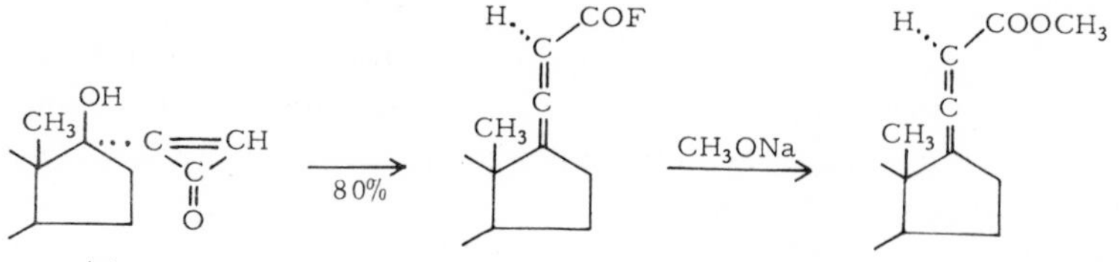

(5)                    (6)                         (7)

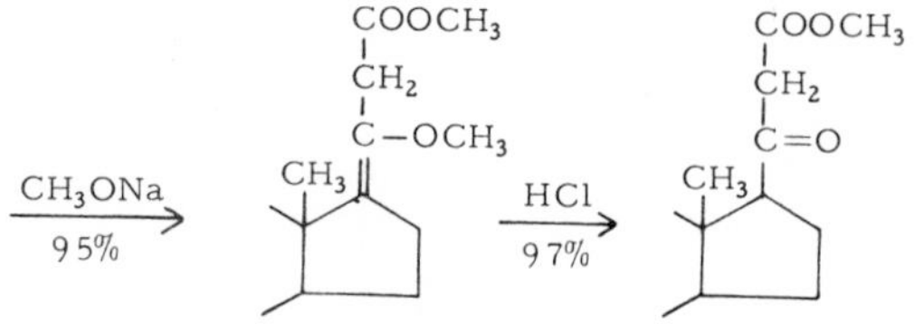

(8)                (9)

for 16 hr. affords (8) via (7). Finally, hydrochloric acid hydrolysis of the 17,20-enol ether (8) affords the $\beta$-keto ester (9) in high yield.

The reactions thus afford a method for synthesis of allenic acids and $\beta$-keto esters.

[1]P. Crabbé, H. Carpio, and E. Velarde, *Chem. Commun.*, 1028 (1971).

**Diethyl cyanomethylphosphonate, 1**, 250; **2**, 130–131; **3**, 96.

An important step in a synthesis of 3$\beta$-acetoxy-5$\beta$,14$\alpha$-bufa-20,22-dienolide from dehydroepiandrostene includes the condensation of the carbanion derived from diethyl cyanomethylphosphonate with the ketone (1).[1]

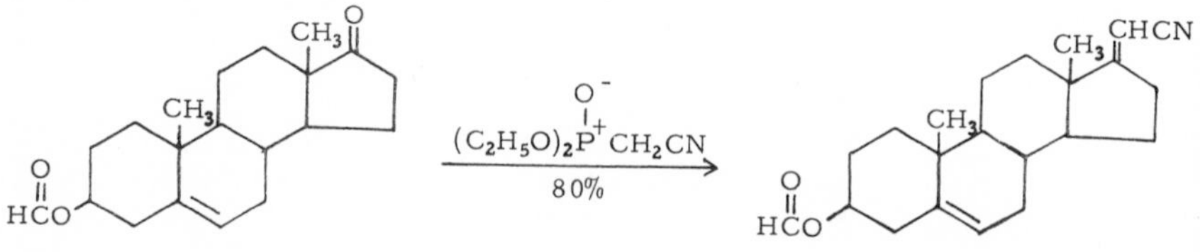

(1)                               (2)

[1]G. R. Pettit and J. R. Dias, *J. Org.*, **36**, 3207 (1971).

$$O$$

**Diethyl 1-(methylthio)ethylphosphonate**, $CH_3SCHP(OC_2H_5)_2$. Mol. wt. 202.26, b.p.
$$CH_3$$

68–70°/0.20 mm.

The reagent is prepared[1] from diethyl methylthiomethylphosphonate (3, 97) by treatment with *n*-butyllithium under argon at −78° followed by addition of methyl iodide (yield 77%).

The reagent was used to transform the ketodiene (1) into the sesquiterpene occidentalol (5).[2] Thus a Wadsworth–Emmons reaction of (1) with diethyl 1-(methylthio)ethyl-phosphonate (3 eq.) in HMPT–dimethoxyethane (62°, 12 hr.) gave the vinyl thioether (2). This was hydrolyzed with mercuric chloride in aqueous acetonitrile to a 2.9 : 1 mixture of the 7α- and 7β-acetyl dienes (3) and (4), respectively, in 70% yield. The epimers were separated by preparative thick-layer chromatography. (±)-Occidentalol (5) was obtained by addition of methyllithium to the 7α-isomer (3) (72% yield).

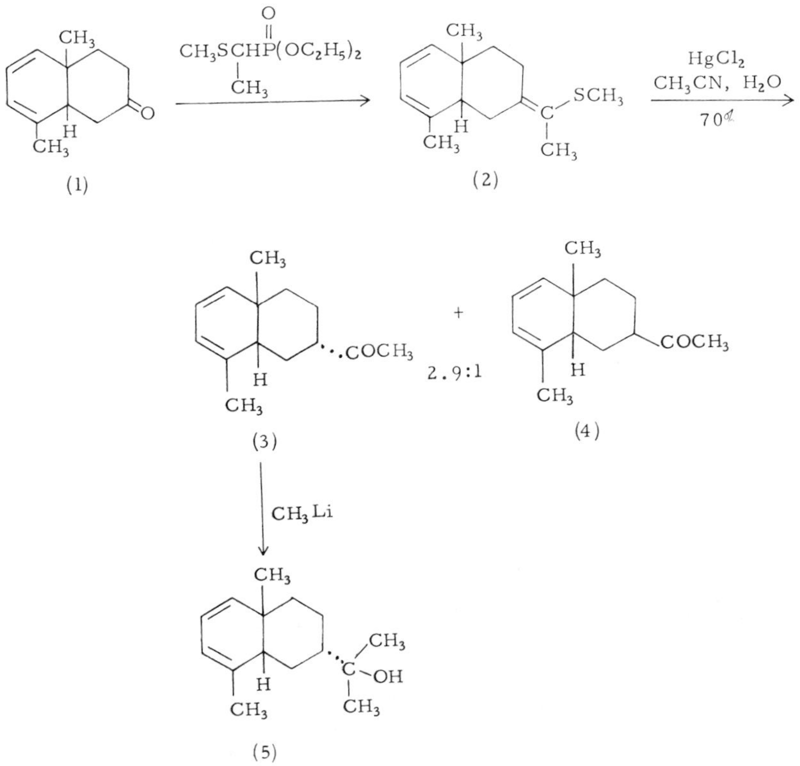

[1]E. J. Corey and J. I. Shulman, *J. Org.*, **35**, 777 (1970).
[2]D. S. Watt and E. J. Corey, *Tetrahedron Letters*, 4651 (1972).

**Diethyl peroxide,** $C_2H_5OOC_2H_5$. Mol. wt. 90.12, b.p. 62–63°, $n_D$ 1.3724.

*Preparation.*[1] The reagent is prepared in 50% yield by the reaction of ethyl methanesulfonate[2] with alkaline 30% hydrogen peroxide with use of stearic acid as dispersing agent.

*Fragmentation of phosphoranes.*[3] 3-Methyl-1-phenyl-3-phospholene (1) reacts with diethyl peroxide to give the phosphorane (2), which fragments spontaneously at room temperature to give isoprene (3) and diethyl phenylphosphonite (4).

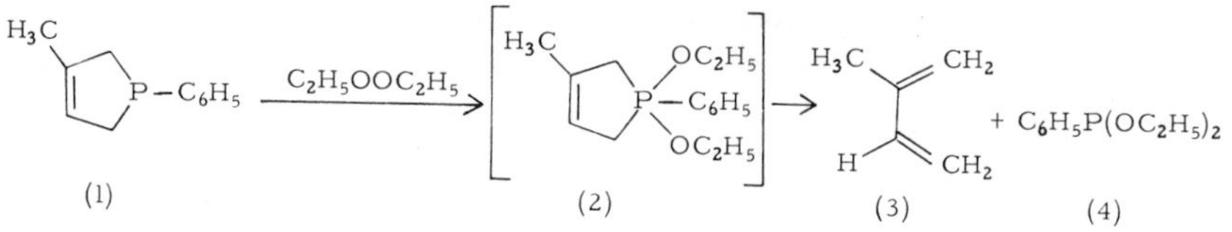

Application of the fragmentation reaction to a mixture of the isomeric 1,2,5-trimethyl-3-phospholenes (5a) and (5b) gives *trans,trans*-hexadiene-2,4 (6) in quantitative yield together with diethyl methylphosphonite (7). Since the diene (6) is the most stable isomer, it is not possible to comment on the stereospecificity of the reaction.

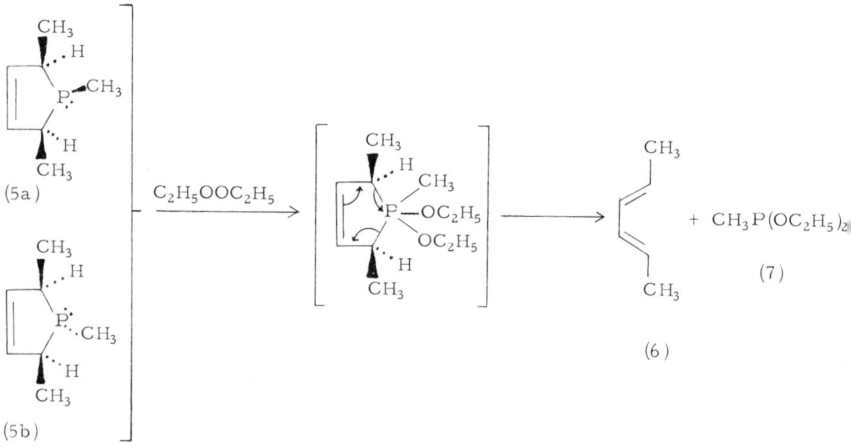

[1]W. A. Pryor and D. M. Huston, *J. Org.*, **29**, 512 (1964).
[2]H. R. Williams and H. S. Mosher, *Am. Soc.*, **76**, 2987 (1954).
[3]C. D. Hall, J. D. Bramblett, and F. F. S. Lin, *ibid.*, **94**, 9264 (1972).

**Diethyl phosphorochloridate, 1,** 248; **2,** 98.

*Aminodehydroxylation.* Rossi and Bunnett[1] have described a method for conversion of phenols into anilines. Two steps are involved. First the phenol is converted

$$ArOH + (C_2H_5O)_2POCl + NaOH \xrightarrow[80-90\%]{} ArOPO(OC_2H_5)_2$$

into the corresponding aryl diethyl phosphate ester by reaction with diethyl phosphoro-chloridate and sodium hydroxide. The second step involves reaction of the ester with potassium amide and potassium metal in liquid ammonia. In the three cases examined the yields in this step were 56–78 %.

$$ArOPO(OC_2H_5)_2 \; + \; KNH_2 \; + \; K \; \xrightarrow[56-78\%]{} ArNH_2$$

The present method of aminodehydroxylation is in some respects complementary to the procedure of Scherrer and Beatty (*see* **4-Chloro-2-phenylquinazoline**, this volume). This latter method involves high temperatures and an alkaline hydrolysis step and hence is not suitable in the case of thermally labile or alkali-sensitive compounds. However, it is applicable to halo- and nitro-substituted phenols. Such substituents are destroyed in the procedure of Rossi and Bunnett.

[1] R. A. Rossi and J. F. Bunnett, *J. Org.*, **37**, 3570 (1972).

### Diethylzinc–Iodoform.

*Ring expansion of arenes.*[1] The iodocarbenoid reagent formed from diethylzinc and iodoform converts benzene into 7-ethylcyclohepta-1,3,5-triene (1). Mixtures of

<table>
<tr><td>(a)</td><td>(b)</td><td>(1)</td></tr>
</table>

alkyl-substituted 7-ethylcyclohepta-1,3,5-trienes are formed by this method from alkylbenzenes, as shown in the example.

|   57%   |   22%   |   21%   |

[1] S. Miyano and H. Hashimoto, *J.C.S. Chem. Comm.*, 216 (1973).

### Diethylzinc–Methylene iodide, 1, 253; 2, 134.

*Cyclopropanation of olefins* (1, 253; 2, 134). Cyclopropanation of olefins with the diethylzinc–methylene iodide reagent is markedly accelerated by oxygen. In addition yields are improved. Thus the reaction of cyclohexene with the carbenoid in the presence of air is complete within 30 min.; norcarane is obtained in 90 % yield. The yield is 53 % when the reaction is carried out under nitrogen.[1]

[1] S. Miyano and H. Hashimoto, *Chem. Commun.*, 1418 (1971).

**1,2-Dihydroxy-3-bromopropane (α-Monobromohydrin)**, $HOCH_2CHCH_2Br$ (1). Mol.
$$\underset{OH}{}$$
wt. 155.01, b.p. 106–110°/4 mm.

The reagent is prepared[1] from epibromohydrin.

$$H_2C\overset{O}{-\!\!\!-}CHCH_2Br \xrightarrow[64.5\%]{\underline{p}-TsOH;\ H_2O} (1)$$

*Protection of carbonyl groups.*[2] The reagent (1) reacts with aldehydes or ketones in refluxing benzene (*p*-TsOH catalysis) to give bromomethylethylene ketals (2) in

$$\underset{R^2}{\overset{R^1}{}}C=O\ +\ (1)\xrightarrow{H^+} \underset{R^1\ \ \ R^2}{\overset{CH_2Br}{\underset{O\ \ \ O}{}}}\xrightarrow{\underset{CH_3OH}{Zn,}} R^1COR^2\ +\ CH_2=CHCH_2OH$$
$$(2)$$

excellent yield. The carbonyl group can be regenerated in high yield by treatment with activated zinc dust[3] in refluxing methanol (12 hr., argon).

Note that removal of ketal protective groups usually necessitates acid catalysis.

[1]S. Winstein and L. Goodman, *Am. Soc.*, **76**, 4368 (1954).
[2]E. J. Corey and R. A. Ruden, *J. Org.*, **38**, 834 (1973).
[3]The zinc dust is activated by brief treatment with acetic acid followed by methanol wash.

**Diimide, 1**, 257–258; **2**, 139; **3**, 99–101.

*Isolation.*[1] Diimide has been isolated by low-temperature ($-196°$, liquid $N_2$) condensation of the pyrolysis products of the lithium derivative of *p*-toluenesulfonyl hydrazine (2). This is prepared by the reaction of tosyl hydrazine with lithium bis-(trimethylsilyl) amide[2] in benzene at room temperature. Pyrolysis of (2) is carried out under high vacuum ($< 10^{-4}$ torr).

$$TsNHNH_2 \xrightarrow[-HN[Si(CH_3)_3]_2]{LiN[Si(CH_3)_3]_2} \underset{Li}{\overset{Ts}{}}N-N\underset{H}{\overset{H}{}} \xrightarrow[-LiTs]{\Delta} HN=NH$$

$$(1)\qquad\qquad\qquad\qquad (2)\qquad\qquad (3)$$

*Reduction of double and triple bonds.*[3] The final step in a highly stereoselective synthesis of 4-methylhex-3-*cis*-ene-1-ol(2) involved reduction of the dienol (1). This

$$\underset{H_3C}{\overset{H_2C}{}}\diagup\!\!\!\diagdown\!\!\!\diagup OH \xrightarrow[74\%]{HN=NH} \underset{H_3C}{\overset{H_3C}{}}\diagup\!\!\!\diagdown\!\!\!\diagup OH$$

$$(1)\qquad\qquad\qquad\qquad\qquad (2)$$

reaction was carried out in 74 % yield with a large excess of hydrazine hydrate and $H_2O_2$ in ethanol. Some of the fully reduced alcohol was also obtained. In this case catalysis by $Cu^{2+}$ increased the amount of polymeric by-product. The enol (2) was also obtained in good yield by diimide reduction of the acetylene (3) under the same conditions. This

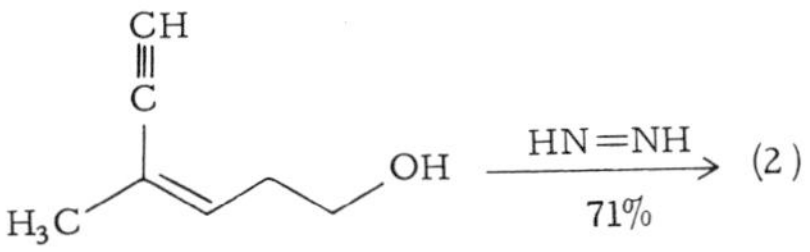

(3)

procedure is more convenient for preparation of (2) than hydrogenation of (3) with Lindlar's catalyst followed by diimide reduction.

[1]N. Wiberg, H. Bachhuber, and G. Fischer, *Angew. Chem., internat. Ed.*, **11**, 829 (1972).
[2]U. Wannagut and H. Niederprüm, *Ber.*, **94**, 1540 (1961).
[3]K. Mori, M. Ohki, A. Sato, and M. Matsui, *Tetrahedron*, **28**, 3739 (1972).

**Diiminosuccinonitrile (DISN)**, Mol. wt. 106.09, m.p. 165–166° dec.,

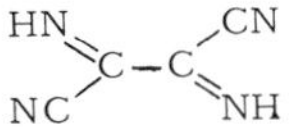

sublimes at 100° (1 mm.). Supplier: PCR.

*Caution:* DISN produces hydrogen cyanide when wet or in contact with a hydroxylic solvent; it causes mild skin and nose irritation.

*Preparation.*[1,2] DISN is formed in 96% yield by base-catalyzed addition of hydrogen cyanide to cyanogen at −40°. It can also be prepared by passing chlorine into a toluene solution of HCN and trimethylamine at −15° (65% yield).

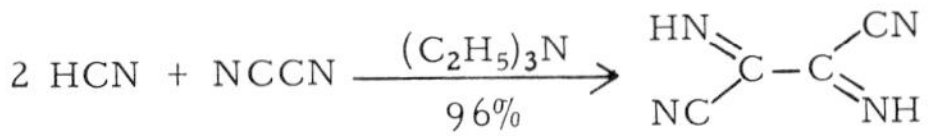

(1)

DISN is readily reduced quantitatively by hydrogen (Pd/C) to diaminomaleonitrile (DAMN), the tetramer of HCN.[3]

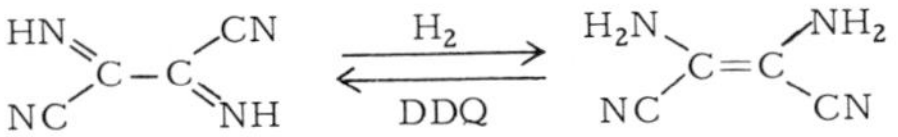

(1)                (2, DAMN)

*Reactions.* DISN (1) when heated with water is hydrolyzed to oxalic acid; controlled hydrolysis with *p*-TsOH in THF gives oxalyl cyanide. This substance is highly reactive

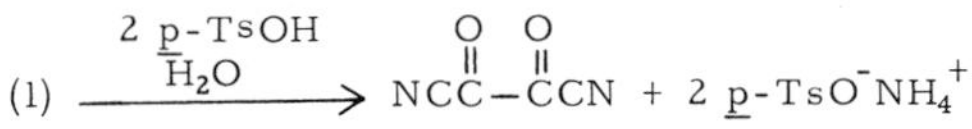

(2)

and is used in subsequent reactions without isolation. DISN reacts with methanol under either acid or basic catalysis to give dimethyl oxaldiimidate (3) in 79.5% yield. DISN reacts with amines, for example aniline, to give diphenyloxamidine (4).

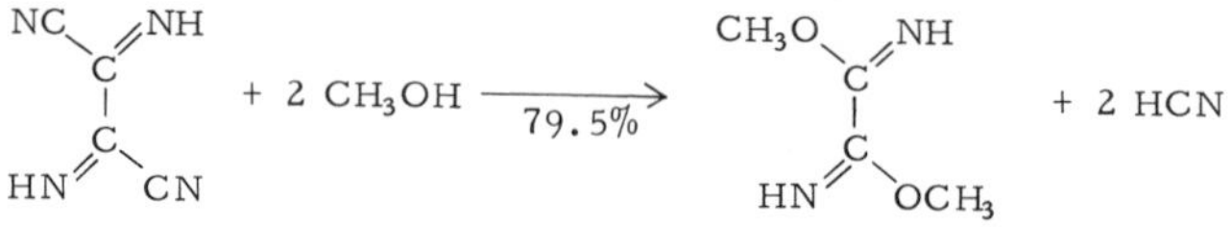

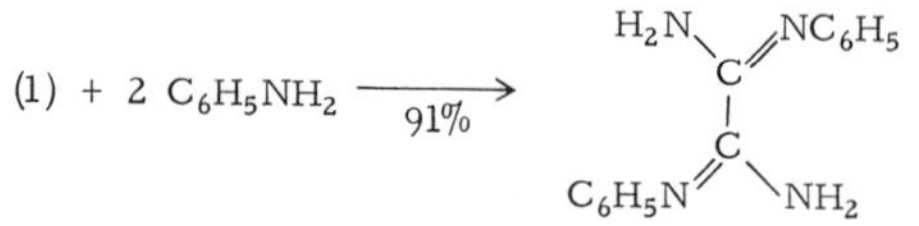

DISN reacts with acetyl chloride to give N,N'-diacetyldiiminosuccinonitrile (5) in very low yield. Use of acetic anhydride gives (5) in 11% yield.

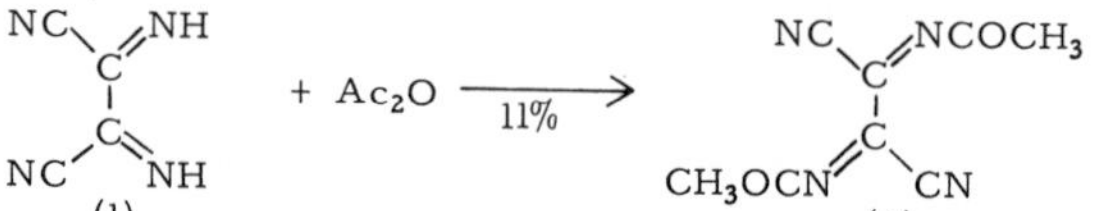

Addition of excess chlorine to (1) at −40 to −20° in acetonitrile gives N,N'-dichloro-diiminosuccinonitrile (6, $Cl_2$DISN) in quantitative yield.

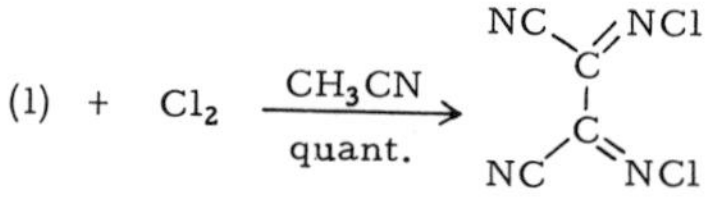

**Nitrogen heterocycles.**[1,4] DISN (1) is extremely useful for synthesis of nitrogen heterocycles. For example, DISN undergoes [2 + 4] cycloaddition reactions with electron-rich olefins. Thus reaction of (1) with cis-1,2-dimethoxyethylene (2) in acetonitrile results in formation of 2,3-dimethoxy-5,6-dicyano-1,2,3,4-tetrahydropyrazine (3) with retention of configuration in 76% yield.

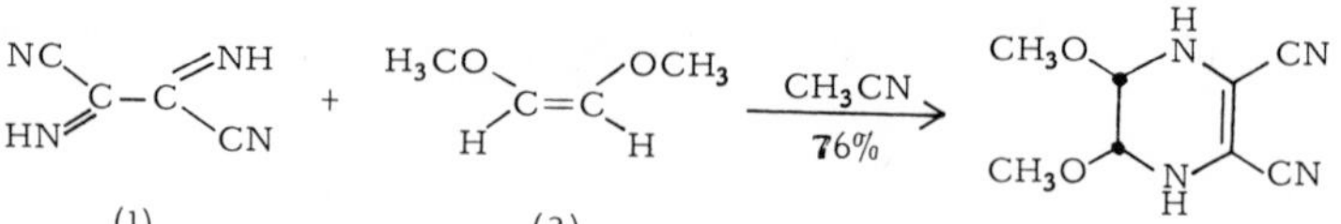

DISN reacts with phosgene at $-20°$ in THF to give 4,5-dichloro-4,5-dicyano-2-imidazolidone (4) in high yield. This 1:1 adduct is unstable; it reacts readily with alcohols to give 4,5-dicyano-4,5-dialkoxy-2-imidazolidones (5) in good yield. Reaction with amines gives an imidazolone (6).

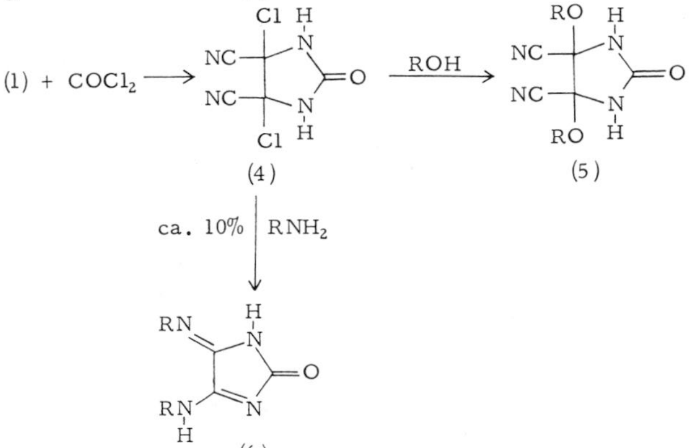

(1) + COCl$_2$ ⟶  (4)   ROH ⟶  (5)

ca. 10% | RNH$_2$

(6)

DISN reacts with sulfur dichloride to give 3,4-dicyano-1,2,5-thiadiazole (7), which is readily hydrolyzed to the diacid (8).

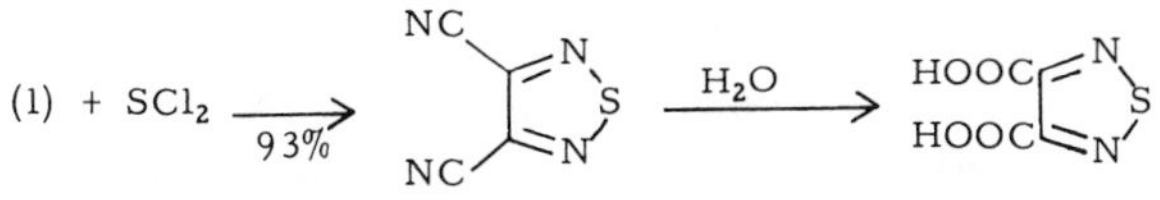

(1) + SCl$_2$ $\xrightarrow{93\%}$ (7) $\xrightarrow{H_2O}$ (8)

DISN adds to 2,2-dimethoxypropane in the presence of sulfuric acid as catalyst to give 2,2-dimethyl-4,5-dicyanoisoimidazole (9) in 80% yield. Reaction of (1) under oxalic acid catalysis also gives (9) in 80% yield. Reaction with acetone itself gives (9) in low yield.

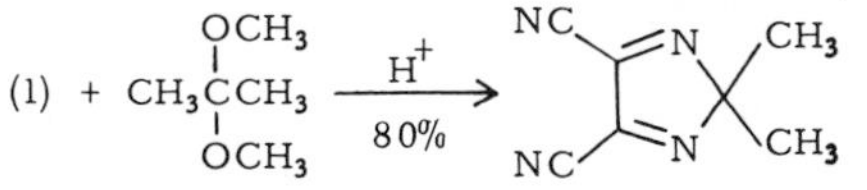

(1) + CH$_3$CCH$_3$ (with OCH$_3$ groups) $\xrightarrow[80\%]{H^+}$ (9)

[1] R. W. Begland, A. Cairncross, D. S. Donald, D. R. Hartter, W. A. Sheppard, and O. W. Webster, *Am. Soc.*, **93**, 4953 (1971).
[2] O. W. Webster, D. R. Hartter, R. W. Begland, W. A. Sheppard, and A. Cairncross, *J. Org.*, **37**, 4133 (1972).
[3] H. Bredereck, G. Schmötzer, and E. Oehler, *Ann.*, **600**, 81 (1956).
[4] R. W. Begland and D. R. Hartter, *J. Org.*, **37**, 4136 (1972).

**Diiron nonacarbonyl, 1**, 259–260; **2**, 139–140; **3**, 101.

*Cycloheptenones.*[1] Dehalogenation of $\alpha,\alpha'$-dibromoketones with diiron nona-carbonyl in the presence of a 1,3-diene provides a direct route to seven-membered cyclic ketones. Thus the reaction of 2,4-dibromo-2,4-dimethylpentane-3-one (1), diiron

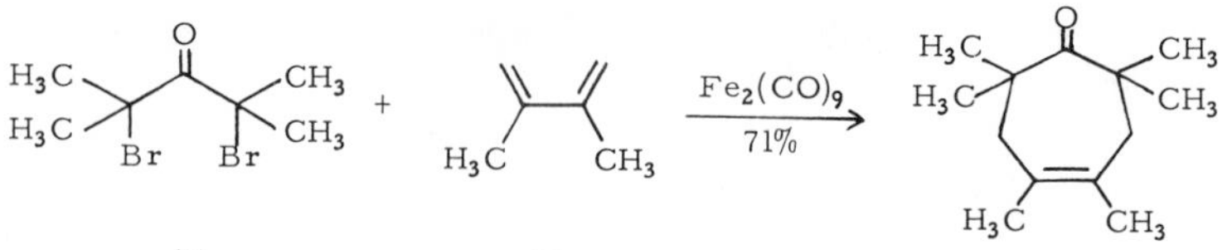

(1)                    (2)                              (3)

nonacarbonyl, and 2,3-dimethylbutadiene (2) in the mole ratio 1.0:1.2:9.0 at 60° for 40 hr. ($N_2$) gives 2,2,4,5,7,7-hexamethyl-4-cycloheptenone (3) in 71% yield. Products obtained from secondary bromides and open-chain 1,3-dienes can be converted into tropones by bromination followed by dehydrobromination. Thus treatment of (4) with 4 eq. of pyrrolidone hydrotribromide (3, 240–241) in THF followed by dehydro-bromination with lithium chloride (1, 609; 2, 246) in DMF gives 2,7-dimethyltropone (5) in 64% yield.

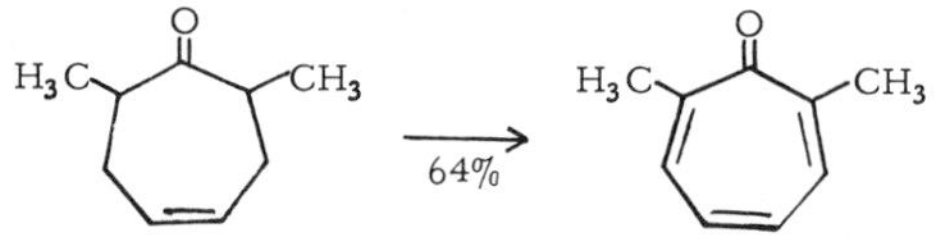

(4)                              (5)

*Cyclopentenone synthesis.*[2] The reaction of α,α'-dibromoketones with enamines in the presence of diiron nonacarbonyl gives cyclopentenone derivatives in 50–100% yield. Thus the reaction of (1) with α-morpholinostyrene (2) with diiron nonacarbonyl as reducing agent gives the cyclopentenone (3) in 94% yield.

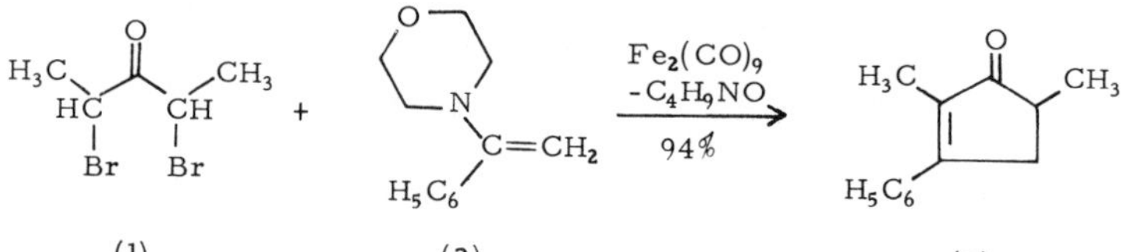

(1)                    (2)                                        (3)

[1] R. Noyori, S. Makino, and H. Takaya, *Am. Soc.*, **93**, 1272 (1971); R. Noyori, Y. Hayakawa, M. Funakura, H. Takaya, S. Murai, R. Kobayashi, and S. Tsutsumi, *ibid.*, **94**, 7202 (1972).
[2] R. Noyori, K. Yokoyama, S. Makino, and Y. Hayakawa, *ibid.*, **94**, 1772 (1972).

**Diisobutylaluminum hydride (DIBAH), 1**, 260–262; **2**, 140–142; **3**, 101–102.

*trans,trans-1,3-Dienes.*[1] Hydroalumination of terminal acetylenes with diisobutyl-aluminum hydride to give terminal vinylalanes followed by treatment with cuprous chloride in THF solution gives isomerically pure *trans,trans*-1,3-dienes in about 70% yield. The vinylalanes probably react with cuprous chloride to give vinylcopper(I) compounds which then decompose to the diene products.

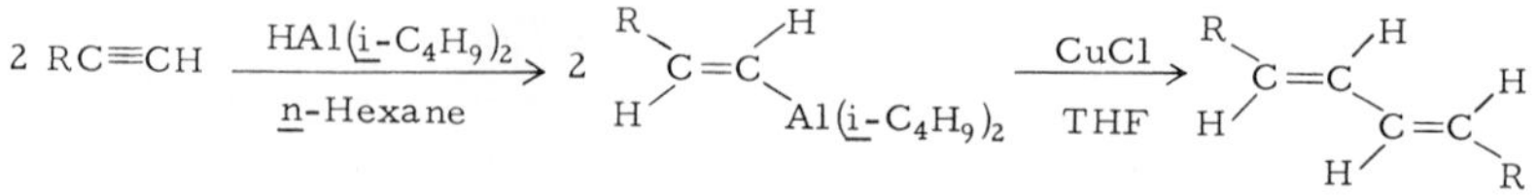

The procedure is also applicable to vinylalanes derived from disubstituted alkynes. Thus 3-hexyne was converted into 4,5-diethyl-*trans,trans*-3,5-octadiene in 71 % yield.

$$C_2H_5C\equiv CC_2H_5 \xrightarrow{\text{HAl}(\underline{i}\text{-}C_4H_9)_2} \overset{C_2H_5}{\underset{H}{>}}C=C\overset{C_2H_5}{\underset{\text{Al}(\underline{i}\text{-}C_4H_9)_2}{<}} \xrightarrow[71\%]{\text{CuCl}} \overset{C_2H_5}{\underset{H}{>}}C=C\overset{C_2H_5}{\underset{H_5C_2}{<}}\overset{H}{\underset{C_2H_5}{>}}C=C\overset{H}{\underset{C_2H_5}{<}}$$

**Cyclopropanes.** 1-Alkynes (1) can be converted into alkylcyclopropanes (4) by the following sequence.[2] Reaction with diisobutylaluminum hydride gives a 1-alkenylalane

$$C_4H_9C\equiv CH \xrightarrow{(C_4H_9)_2AlH} \overset{C_4H_9}{\underset{H}{>}}C=C\overset{H}{\underset{Al(C_4H_9)_2}{<}} \xrightarrow[\text{Zn-Cu}]{CH_2Br_2}$$

(1)                  (2)

$$\overset{C_4H_9}{\underset{H}{>}}C\overset{\overset{H_2}{C}}{-}C\overset{H}{\underset{Al(C_4H_9)_2}{<}} \xrightarrow[\substack{62\% \\ \text{overall}}]{H_3O^+} \overset{C_4H_9}{\underset{H}{>}}C\overset{\overset{H_2}{C}}{-}C\overset{H}{\underset{H}{<}}$$

(3)                  (4)

(2) by *cis* addition. Reaction with methylene bromide in the presence of a zinc–copper couple (**1**, 1020) produces a cyclopropylalane (3) which can be hydrolyzed to an alkyl-cyclopropane (4). The intermediate (3) can also be converted into a *trans*-1-halo-2-alkylcyclopropane (5) by reaction with bromine or iodine.

$$(3) \xrightarrow[58\%]{3\ Br_2} \overset{C_4H_9}{\underset{H}{>}}C\overset{\overset{H_2}{C}}{-}C\overset{H}{\underset{Br}{<}}$$

(5)

**trans-*Allylic alcohols*.**[3] The *trans*-vinylalanes (1), obtained by *cis* addition of diisobutylaluminum hydride to 1-alkynes (**2**, 141), react with aldehydes or ketones in ether or benzene to give, after acidification, the corresponding allylic alcohols (2) in

$$RC\equiv CH \xrightarrow{\text{AlH}(\underline{i}\text{-}C_4H_9)_2} \overset{R}{\underset{H}{>}}C=C\overset{H}{\underset{Al(\underline{i}\text{-}C_4H_9)_2}{<}} \xrightarrow[\substack{2)\ H_3O^+}]{\substack{1)\ \overset{R^1}{\underset{R^2}{>}}C=O}}$$

(1)

$$\overset{R}{\underset{H}{>}}C=C\overset{H}{\underset{\overset{\displaystyle C}{\underset{HO}{\underset{R^2}{<}}}}{<}}\!\!\!\overset{R^1}{}$$

(2)

30–50% yield. The new method of alkenylation of aldehydes and ketones is convenient, since vinylalanes are more readily prepared than vinyllithium or vinyl Grignard reagents.

*Conjugate reduction of α,β-unsaturated epoxides.*[4] Reduction of 2-methyl-1,2-oxido-3-butene (1)[5] with DIBAH in refluxing hexane results in highly stereoselective conjugate addition to give the Z alcohol (2); the E isomer (3) is a minor product.

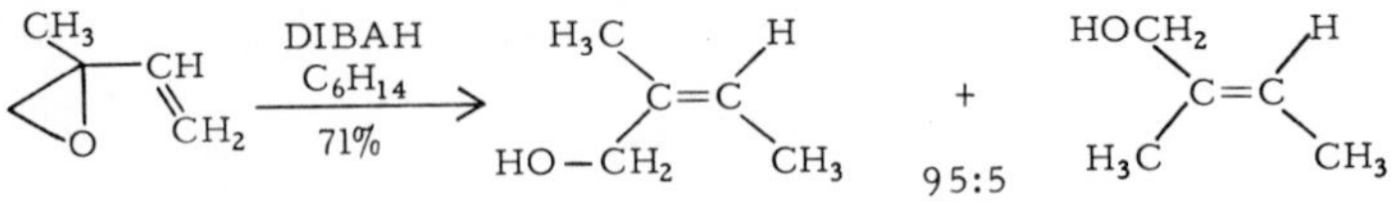

(1)    (2)    (3)

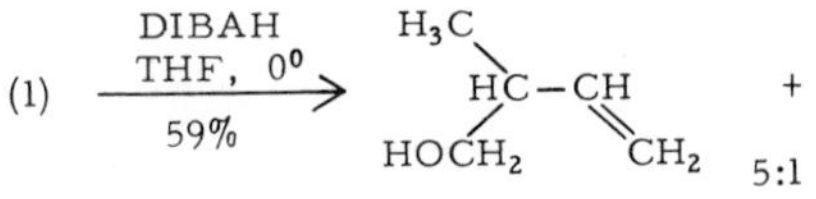

(4)

Reduction of (1) with DIBAH in THF leads mainly to the homoallylic alcohol (4) formed by direct, Markownikoff addition of the hydride to the epoxide. Reduction of (1) with lithium aluminum hydride gives exclusively the tertiary allylic alcohol (5).

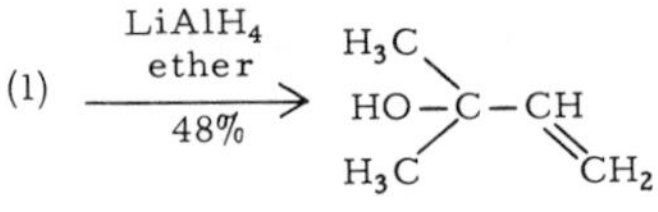

(5)

In contrast reduction of (1) by calcium in liquid ammonia gives, almost exclusively, the E isomer (3). Stereoselectivity is decreased using either lithium or sodium as the metal.

Extension of these studies to other α,β-unsaturated epoxides indicates that the course of reduction, conjugate versus direct, depends on the hindrance of the two functional groups.

*Selective reduction.*[6] The nitrile (1) was reduced to the corresponding aldehyde (2) by diisobutylaluminum hydride in 75–85% yield (*cf.* **1**, 262).

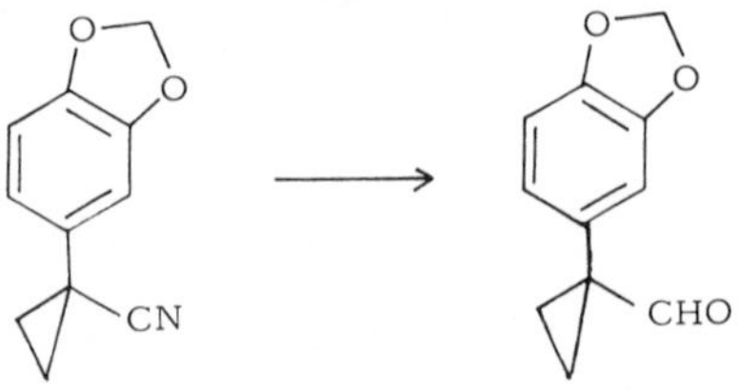

(1)    (2)

[1]G. Zweifel and R. L. Miller, *Am. Soc.*, **92**, 6678 (1970).

[2]G. Zweifel, G. M. Clark, and C. C. Whitney, *ibid.*, **93**, 1305 (1971).

[3]H. Newman, *Tetrahedron Letters*, 4571 (1971).

[4]R. S. Lenox and J. A. Katzenellenbogen, *Am. Soc.*, **95**, 957 (1973).

[5]Prepared from isoprene according to the procedure of E. J. Reist, I. G. Junga, and B. R. Baker, *J. Org.*, **25**, 1673 (1960).

[6]R. V. Stevens, L. E. DuPree, Jr., and P. L. Loewenstein, *ibid.*, **37**, 977 (1972).

## Diisopinocampheylborane (Tetra-3-pinanyldiborane), 1, 262–263.

The reagent actually exists as the diborane derivative (1) rather than as the monomer.[1] The *Chemical Abstracts* name for this reagent is tetra-3-pinanyldiborane. It has been referred to as $Pn_4B_2H_2$.

(1)

*Reactions with allenes.* Caserio et al.[2] have reported that partial hydroboration of racemic 1,3-dimethylallene and 1,3-diphenylallene with $(+)$-$Pn_4B_2H_2$ [from $(-)$-α-pinene] gave recovered $(-)$-allene of moderate activity. The method has been studied in greater detail by Moore et al.[3] and they report that in every case examined the recovered allene is enriched in the R enantiomer. Brown's reagent is thus useful for preparation of optically active allenes; however, only moderate activities result and, of course, some of the allene is irrevocably lost in the process.

*Asymmetric hydroboration.* A key step in an asymmetric synthesis[4] of loganin (1), a key intermediate in the biosynthesis of indole and monoterpene alkaloids, involved

(1)

asymmetric hydroboration of 5-methylcyclopentadiene (2). This was accomplished by use of $(+)$-[5] or $(-)$-[6]-diisopinocampheylborane. Use of the $(+)$-borane gave the alcohol (3a, $\alpha D - 169°$), whereas use of the $(-)$-borane gave the alcohol (3b, $\alpha D + 170°$),

(3a)                    (2)                    (3b)

both obtained in over 30% yield. No *cis*-alcohols were detected in either case. Both alcohols were shown to be at least 98–99% optically pure.

[1]H. C. Brown and G. J. Klender, *Inorg. Chem.*, **1**, 204 (1962).
[2]W. L. Waters and M. C. Caserio, *Tetrahedron Letters*, 5233 (1968); W. L. Waters, W. S. Linn, and M. C. Caserio, *Am. Soc.*, **90**, 6741 (1968).
[3]W. R. Moore, H. W. Anderson, and S. D. Clark, *ibid.*, **95**, 835 (1973).
[4]J. J. Partridge, N. K. Chadha, and M. R. Uskoković, *ibid.*, **95**, 532 (1973).
[5]Prepared from ( – )-α-pinene (Chemical Samples Co.).
[6]Prepared from ( + )-α-pinene (Aldrich).

## N,N'-Diisopropylhydrazine, $(CH_3)_2CHNHNHCH(CH_3)_2$. Mol. wt. 116.20, b.p. 124–124.5°.

This hydrazine is available by catalytic reduction of acetone azine.[1]

*Protection of carboxylic acids.*[2] The reagent reacts with carboxylic acid derivatives (the acyl chloride or mixed anhydride) to give a monoacylhydrazide, $RCON(CHMe_2)\cdot NH(CHMe_2)$. The derivatives are stable to both acids and bases. They are reconverted into carboxylic acids by selective oxidation, preferably with lead tetraacetate. The new method of protection has been used for penicillins.

[1]H. R. Lochte, J. R. Bailey, and W. A. Noyes, *Am. Soc.*, **43**, 2597 (1921).
[2]D. H. R. Barton, M. Girijavallabhan, and P. G. Sammes, *J.C.S. Perkin I*, 929 (1972).

## 2,2'-Dilithiumdiphenyl, (1).

### Preparation.[1]

*Formation of the atropisomeric o-hexaphenylenes.*[2] When the chromium ate complex (2), prepared from 2,2'-dilithiumdiphenyl and chromic chloride, is treated with a transition metal halide ($CoCl_2$, $FeCl_3$), the atropisomers of *o*-hexaphenylene (3a) and (3b) are formed:

(1)                    (2)

(3a)                              (3b)
Centrosymmetric form              Screw form
Atropisomeric forms of *o*-hexaphenylene.

[1]H. Gilman, J. E. Kirby, and C. R. Kinney, *Am. Soc.*, **51**, 2252 (1929); J. Collette, D. McCreer, R. Crawford, F. Chubb, and R. B. Sandin, *ibid.*, **78**, 3819 (1956); G. Wittig and W. Herwig, *Ber.*, **87**, 1511 (1954).
[2]G. Wittig and K.-D. Rümpler, *Ann.*, **751**, 1 (1971).

**Dilithium tetrachlorocuprate**, $Li_2CuCl_4$. Mol. wt. 219.25.

*Preparation.*[1] The reagent is prepared in THF solution by reaction of lithium chloride (0.2 mole) and copper(II) chloride (0.1 mole).

*Cross-coupling of Grignard reagents and alkyl bromides.*[1] Copper(I) catalyzes cross-coupling between Grignard reagents and alkyl bromides when carried out in THF solution at 0° or below. The yield of homodimers, $R^1$–$R^1$ and $R^2$–$R^2$, under these

$$R^1 - MgX + R^2X \xrightarrow{Li_2CuCl_4} R^1 - R^2 + MgX_2$$

conditions is negligible. The reaction involves nucleophilic displacement of the halide ion from $R^2X$ by the Grignard reagent. Yields are highest with primary alkyl halides; iodides > bromides ≫ chlorides. The structure of the Grignard reagent is not as important.

Examples:

$$\underline{n}\text{-}C_6H_{13}MgBr + \underline{n}\text{-}C_4H_9Br \xrightarrow[73\%]{\overset{Li_2CuCl_4}{0°, \ 3\,hrs.}} \underline{n}\text{-}C_{10}H_{22}$$

$$H_2C=CH(CH_2)_4MgBr + C_2H_5Br \xrightarrow[45\%]{\overset{Li_2CuCl_4}{0°, \ 3\,hrs.}} H_2C=CHC_6H_{13}$$

This new coupling procedure was used in a recent stereoselective synthesis of a sex attractant of the codling moth, *trans,trans*-8,10-dodecadiene-1-ol(4).[2] Thus coupling of (1) with (2) catalyzed by dilithium tetrachlorocuprate gave (3) in 85 % yield. This product consisted of 78 % of the desired *trans,trans*-isomer. Hydrolysis of the protective group (*p*-TsOH in aqueous methanol) gave the free alcohol (4) containing about 80 % of the desired *trans,trans*-alcohol. The pure alcohol (4) was obtained by crystallization from pentane at −5°.

Silver is also an effective catalyst for the coupling of Grignard reagents and alkyl halides, but is useful only when both alkyl groups are the same:

$$\underline{n}\text{-}C_4H_9MgBr \; + \; \underline{n}\text{-}C_4H_9Br \; \xrightarrow[79\%]{\substack{Ag, \; THF \\ 25^0, \; 5\,hrs.}} \; \underline{n}\text{-}C_8H_{18}$$

When the alkyl groups are different a mixture of three coupled products is obtained. The soluble silver catalyst is prepared by the reaction of silver nitrate (5.0 mmole) with a 1.0 $M$ solution (30 ml.) of ethylmagnesium bromide at 0° for 3 hr. The supernatant is 7.5 $M$ in soluble silver and is stable for some time in the refrigerator.[1]

[1]M. Tamura and J. Kochi, Synthesis, 303 (1971).
[2]C. Descoins and C. A. Henrick, Tetrahedron Letters, 2999 (1972).

**2,4-Dimethoxybenzylamine,** $(CH_3O)_2C_6H_3CH_2NH_2$. Mol. wt. 167.20. Pierce supplies the hydrochloride, mol. wt. 203.45, m.p. 185–186°.

*Preparation.* The amine is prepared in 86% yield by reduction of 2,4-dimethoxybenzaldoxime with sodium bis-(2-methoxyethoxy)aluminum hydride (**3**, 260–261).

*Protection of glutamine and asparagine.*[1] The 2,4-dimethoxybenzyl group (Dmb) has been recommended for protection of the amide groups of asparagine and glutamine residues in peptide synthesis. The derivatives are nicely crystalline, and the Dmb group can be removed by trifluoroacetic acid or anhydrous hydrogen fluoride.

[1]P. G. Pietta, P. Cavallo, and G. R. Marshall, J. Org., **36**, 3966 (1971).

**2-(2,2-Dimethoxyethyl)-1,3-dithiane,**

(1)    Mol. wt. 192.34.

The reagent is prepared[1] by the reaction of 1,3-propanedithiol (1 eq.) with malondialdehyde bisdimethylacetal (Aldrich), in a manner similar to that used for synthesis of 1,3-dithiane.[2]

$$HS(CH_2)_3SH \; + \; CH_2[CH(OCH_3)_2]_2 \longrightarrow$$

$+ \; 2 \; CH_3OH$

(1)

*Dihydro-γ-pyrones and dihydro-3-furanones.*[1] A new synthesis of dihydro-γ-pyrones (5) and dihydro-3-furanones (8) involves conversion of (1) into 2-lithio-2-(2′,2′-dimethoxyethyl)-1,3-dithiane (2) by treatment with *n*-butyllithium in THF at −30°. Treatment of (2) with an epoxide gives (3). On treatment with *p*-toluenesulfonic

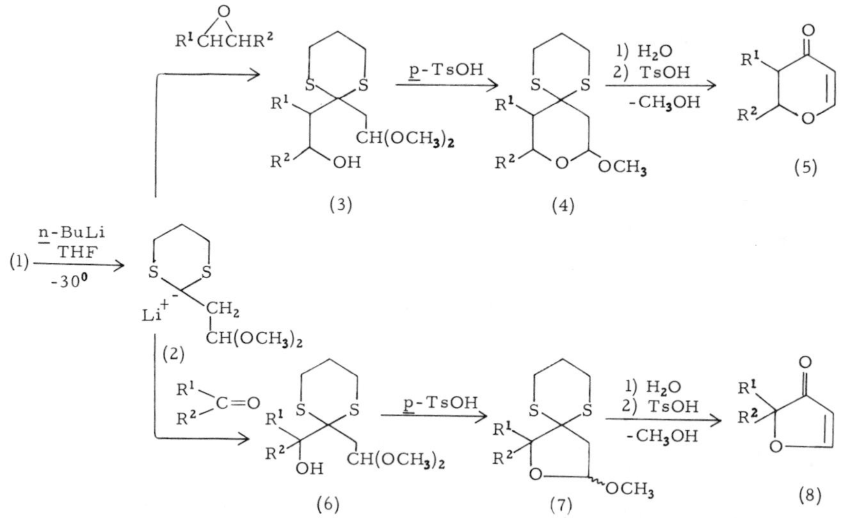

acid in refluxing THF or benzene, (3) is cyclized to the acetal (4). The dihydro-γ-pyrone (5) is then generated by removal of the thioacetal group followed by elimination of methanol (*p*-TsOH in refluxing aqueous THF). Dihydro-3-furanones (8) are obtained in the same way from (2) by substitution of a ketone for an epoxide. Overall yields are only moderate, mainly owing to difficulty in removal of the thioacetal function. Established methods lead to poor yields; in some cases use of NBS (see this volume) improved the yield.

[1] F. Sher, J. L. Isidor, H. R. Taneja, and R. M. Carlson, *Tetrahedron Letters*, 577 (1973).
[2] E. J. Corey and D. Seebach, *Org. Syn.*, **50**, 72 (1970).

### Dimethylacetamide, 1, 270–271; 2, 144–145.

*Vinylation of carbazole* (solvent effect).[1] DMA, DMSO, and N-methyl-2-pyrrolidone were found to be the most satisfactory solvents for reaction of acetylene with alkali metal carbazoles (1) to give vinylcarbazole (2).

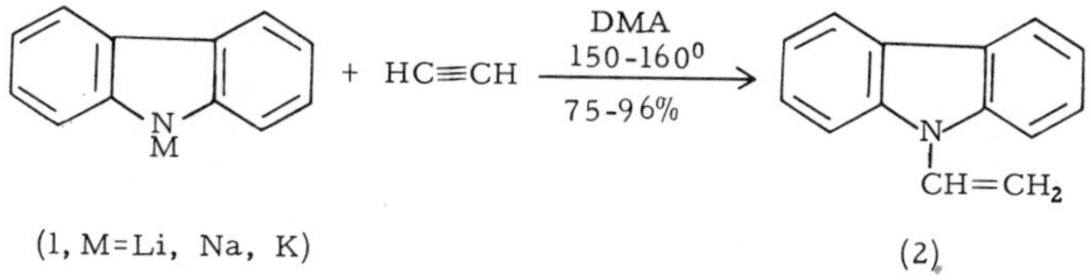

[1] S. R. Sandler, *Chem. Ind.*, 134 (1973).

**N,N-Dimethylacetamide dimethyl acetal,** $\underset{(CH_3)_2NC(OCH_3)_2}{\overset{CH_3}{|}}$ . Mol. wt. 133.19.
Supplier: Fluka.

Meerwein[1] describes the preparation of N,N-dimethylacetamide diethyl acetal.

*Reaction with vicinal* **cis-diols.**[2] N,N-Dimethylacetamide dimethyl acetal (2) reacts with vicinal *cis*-diols such as methyl β-D-ribofuranoside (1) in 1,1,1-trichloroethane to give the acetal (3) in over 80% yield. This derivative is stable under anhydrous

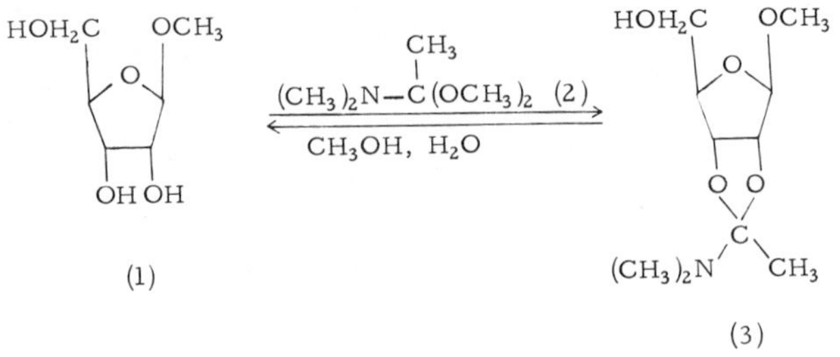

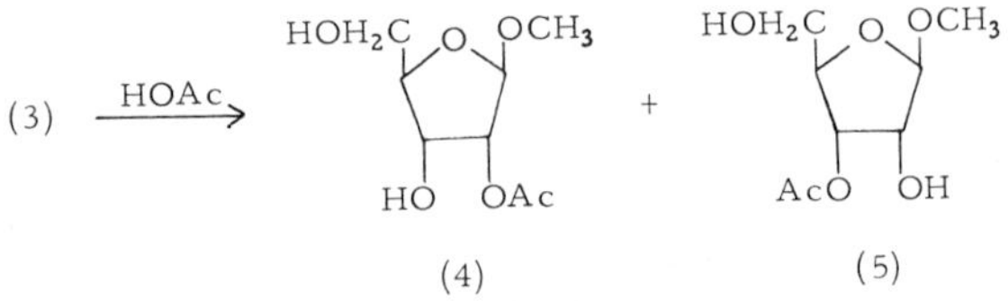

conditions but is hydrolyzed quantitatively when refluxed in 3% aqueous methanol (the reaction can also be carried out at room temperature). The acetal (2) is thus useful for protection of *vic*-diols. At pH 4–5 in 95% aqueous acetic acid (3) is cleaved to give a mixture of 2- and 3-monoacetates (4) and (5) in approximately equal amounts.

N,N-Dimethylbenzamide dimethyl acetal can be used in the same way.

*Amide–Claisen rearrangement.*[3] Satisfactory Δ⁴-cholestene-3β-ol, m.p. 130.5–131° (from ethanol), can be prepared by the procedure of **Burgstahler and Nordin**[4]; a lower melting point indicates contamination with some of the 5α-hydroxy isomer. A 50-ml.

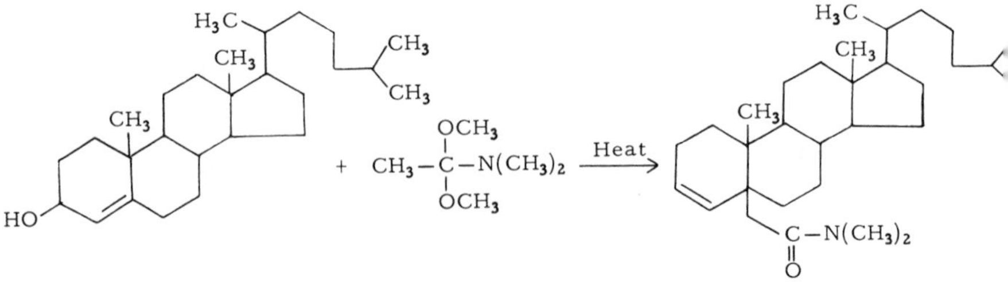

round-bottomed flask equipped with a Teflon-covered magnetic stirring bar and a reflux condenser connected to a gas inlet tube is charged with 0.97 g. (2.5 mmole) of $\Delta^4$-cholestene-3$\beta$-ol and 30 ml. of *o*-xylene and the mixture is stirred to effect solution. To this solution is added 1.67 g. (12.5 mmole) of N,N-dimethylacetamide dimethyl acetal. The flask is flushed with argon and then heated with a sand bath in an electric heating mantle at reflux under a positive pressure of argon with vigorous stirring for 65 hr. After cooling, the volatile materials are removed by rotary evaporation followed by vacuum (0.1 mm.) drying for 1 hr. to give 1.2 g. of a yellow oil that is chromatographed on 60 g. of silica gel with ether. Elution of the column with 200 ml. of ether gives a mixture of cholestadiene which is discarded; further elution with 500 ml. of ether produces 0.74 g. of 5$\beta$-N,N-dimethylcarboxamidomethyl-3-cholestene as a clear colorless oil which on trituration with acetone affords 0.74 g. (65%) of the amide as white plates, m.p. 128–129.5°.

*Reaction with allylic alcohols* (1, 271–272).[5] The reaction of *trans*-3-penten-2-ol (1) with N,N-dimethylacetamide dimethyl acetal and/or its synthetic equivalent 1-methoxy-1-dimethylaminoethylene[6] in refluxing xylene for 17 hr. gives the N,N-dimethylamide of 3-methyl-4-*trans*-hexenoic acid (2) in 80% yield. When the reaction was carried out

with (S)-(−)-*trans*-3-pentene-2-ol, (2) was formed in greater than 90% optical purity and with inversion of configuration. Correspondingly high asymmetric induction was observed in the reaction of optically active (1) with triethyl orthoacetate (3, 300–302) to give the ethyl ester (3).

[1]H. Meerwein, W. Florian, N. Schön, and G. Stopp, *Ann.*, **641**, 1 (1961).
[2]S. Hanessian and E. Moralioglu, *Tetrahedron Letters*, 813 (1971); *Canad. J. Chem.*, **50**, 233 (1972).
[3]R. E. Ireland and D. J. Dawson, *Org. Syn.*, submitted 1971.
[4]A. W. Burgstahler and I. C. Nordin, *Am. Soc.*, **83**, 198 (1961).
[5]R. K. Hill, R. Soman, and S. Sawada, *J. Org.*, **37**, 3737 (1972).
[6]H. Bredereck, F. Effenberger, and G. Simchen, *Ber.*, **96**, 1350 (1963).

## Dimethyl acetylenedicarboxylate (DMAD), 1, 272–273; 2, 145–146; 3, 103–104.

### Reaction with $\Delta^3$-oxazolinone[1]:

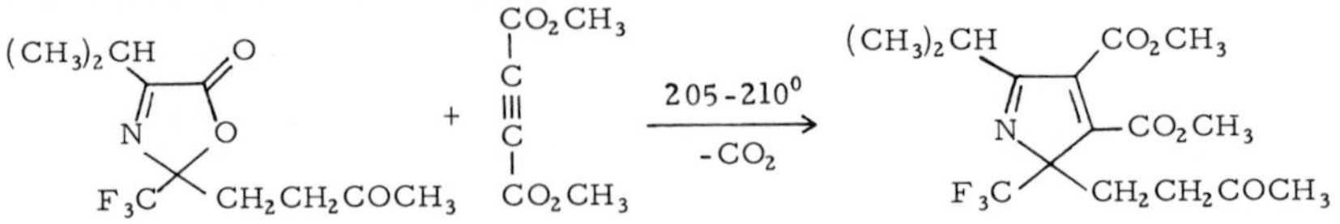

$$
(CH_3)_2CH \quad \overset{O}{\underset{N \quad O}{\bigvee}} \quad + \quad \overset{CO_2CH_3}{\underset{CO_2CH_3}{\overset{|}{\underset{|}{\overset{C}{\underset{C}{\parallel\parallel\parallel}}}}}} \quad \xrightarrow[-CO_2]{205-210^0} \quad (CH_3)_2CH \quad \overset{CO_2CH_3}{\underset{CO_2CH_3}{\bigvee}}
$$

F$_3$C  CH$_2$CH$_2$COCH$_3$      F$_3$C  CH$_2$CH$_2$COCH$_3$

**Reaction with α-diazoketones.**[2] DMAD reacts with 2-diazo-3-butanone (1) to give the N-acetylpyrazole (2). The reagent reacts with 2-diazocyclopentanone (3) to form (4), dimethyl-4,5-dihydro-7(6H)-oxopyrazolo[1,5-*a*]pyridine-2,3-dicarboxylate.

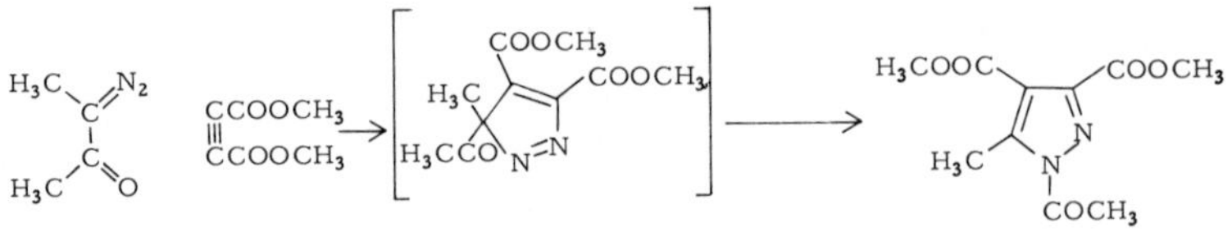

(1)  (2)

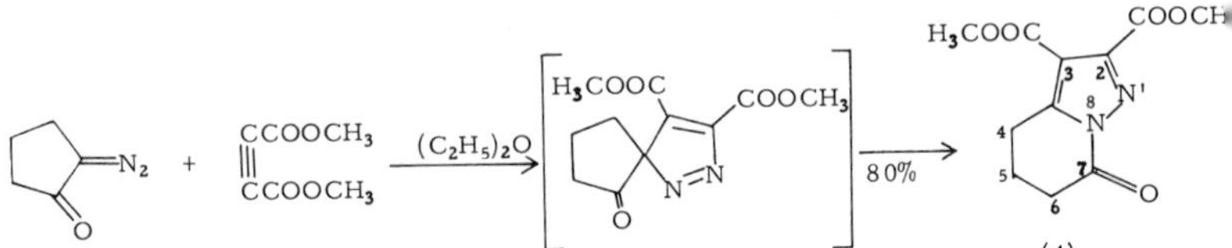

(3)  (4)

**Reaction with 3-amino-1,2,4-triazole.**[3] The ester (1) reacts with 3-amino-1,2,4 triazole (2) in boiling ethanol to give the products (3), (4), and (5).

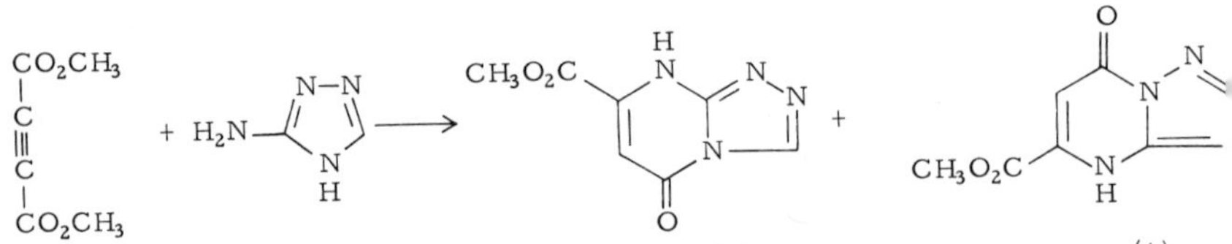

(1)  (2)  (3)  (4)

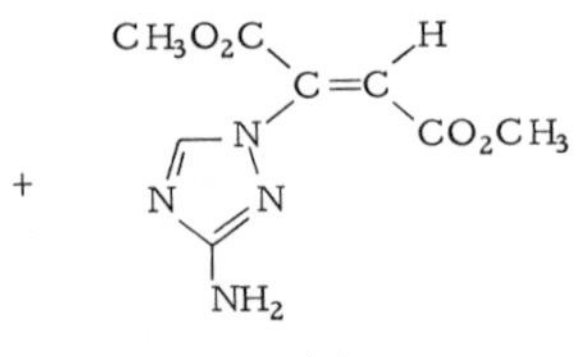

(5)

*Aromatic annelation.*[4] The reaction of dibenzoyl peroxide (1 mole) with a large excess of dimethyl acetylenedicarboxylate (10 mole) at 80° gives tetramethyl naphthalene-1,2,3,4-tetracarboxylate (0.5 mole/mole peroxide). The reaction proceeds through a phenyl radical, since other sources of phenyl radicals (N-nitrosoacetanilide or acetanilide–pentyl nitrite) can replace the dibenzoyl peroxide. The reaction is believed to proceed by addition of the phenyl radical to the acetylene followed by further addition

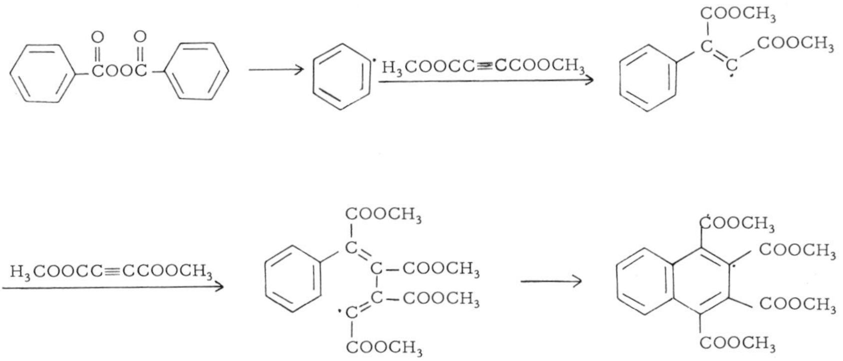

of the substituted styryl radical thus formed to another molecule of the acetylene with subsequent cyclization as formulated.

*Benzcyclobutadiene as an intermediate.*[5] *trans*-3,4-Di(phenylethynyl)-1,2,3,4-tetramethylcyclobutene-1 (1) reacts with dimethyl acetylenedicarboxylate in benzene under nitrogen to give dimethyl 1,2-diphenyl-5,6,7,8-tetramethylnaphthalene-3,4-dicarboxylate (2) in 85% yield. This unusual reaction is believed to proceed through the intermediates (a–c) shown in the formulation.

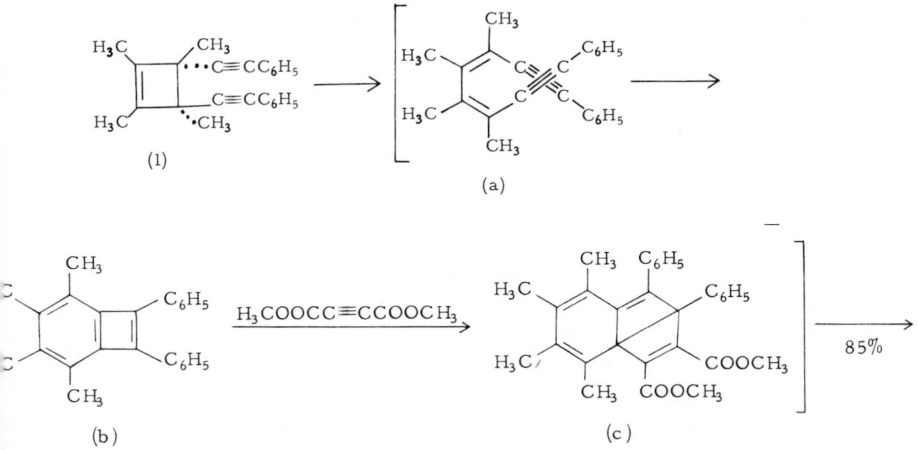

(1)

(a)

(b)

(c)

85%

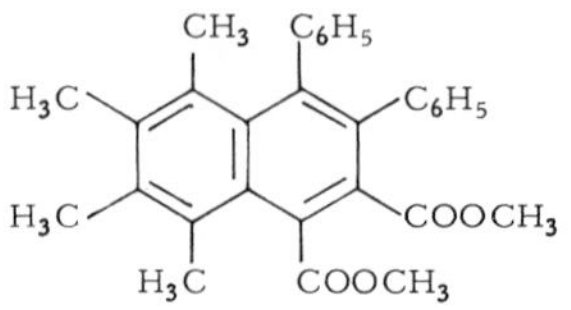

(2)

*Reaction with enamines and related substances.* Brannock *et al.*[6] found that t
enamines derived from acyclic aldehydes and ketones react with dimethyl acetyler
dicarboxylate to give products derived by rearrangement of cyclobutenes initia
formed by 1,2-cycloaddition. Thus, N,N-dimethylisobutenylamine (1) reacts wi
dimethyl acetylenedicarboxylate in refluxing ether to give dimethyl 2-dimethylamin
methylene-3-isopropylidenesuccinate (2) in 49 % yield.

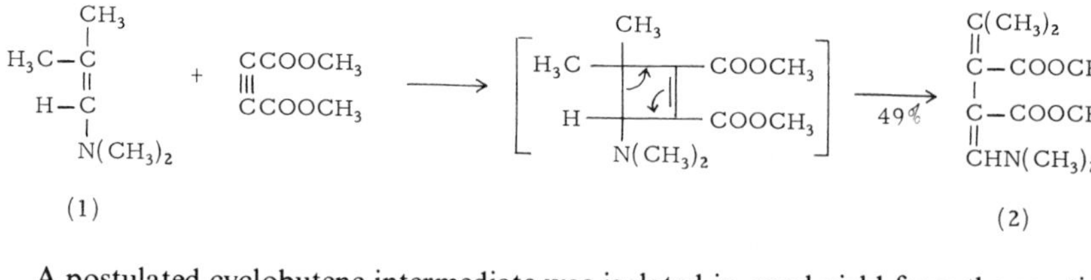

A postulated cyclobutene intermediate was isolated in good yield from the reactic
of the pyrrolidine enamine of cyclohexanone (3) with dimethyl acetylenedicarboxyla
in ether at 25–35°. The cyclobutene (4) is heat sensitive and when warmed on the stea:

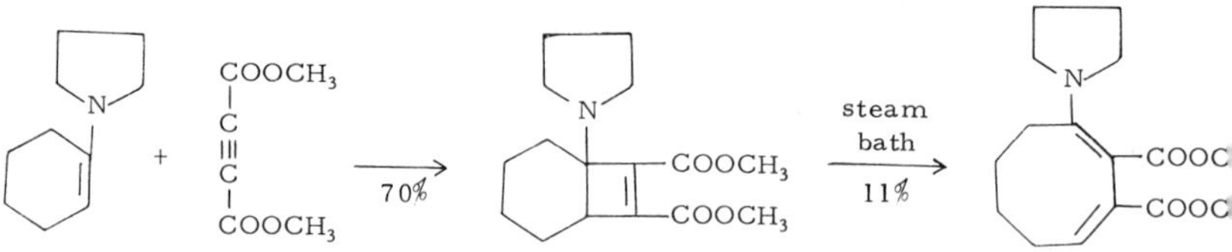

bath (11 hr.) rearranges to the ring enlargement product (5) in low yield. Ring enlarg(
ment products were obtained directly in good yield from enamines derived from cycl(
pentanone, cycloheptanone, and cyclooctanone.

The reaction of 1-methylindole (6) with dimethyl acetylenedicarboxylate in acet(
nitrile under reflux gives, as the major product, dimethyl 1-methylbenz[*f*]azepine-3,
dicarboxylate (8) as the major product.[7] Evidently the heterocyclic nitrogen atom an
the 2,3-double bond function as an enamine system, and the reaction involves a cycl(
butene (7) as an intermediate.

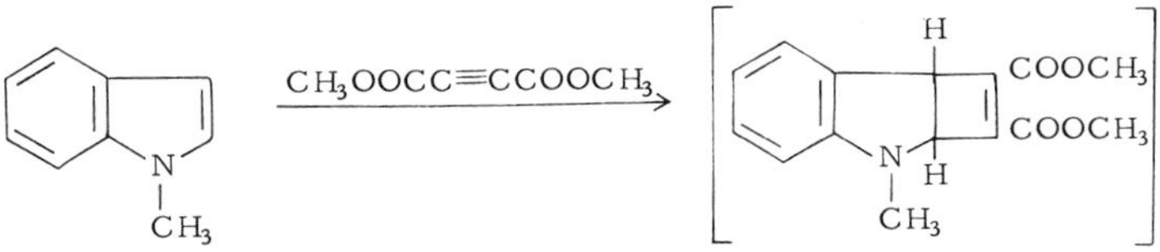

$$CH_3OOCC \equiv CCOOCH_3$$

(6)    (7)

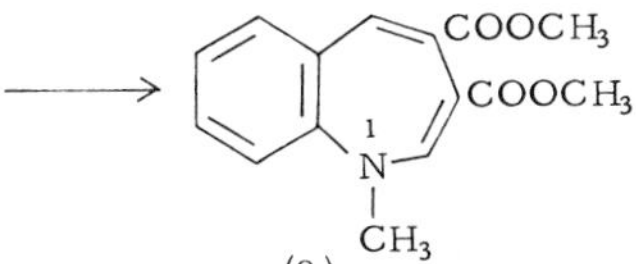

(8)

Plieninger and Wild[8] obtained a 2-ethoxybenzazepine (11) from the reaction of 2-ethoxy-1-methylindole (9), and also suggested a cyclobutene intermediate (10).

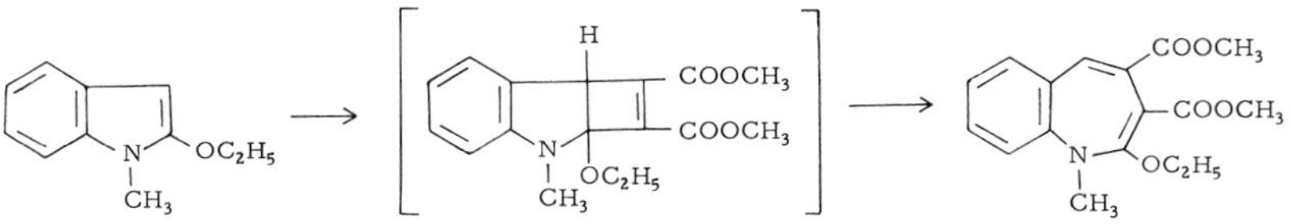

(9)    (10)    (11)

Lehman[9] has recently reported the isolation of a cyclobutene adduct (13) from the reaction of 1-methyl-1,4-dihydroquinoline (12) with dimethyl acetylenedicarboxylate in acetonitrile under $N_2$. The adduct (13) rearranges when heated under reflux in dry benzene for 8 hr. into 1-methyl-3,4-dicarbomethoxy-1,6-dihydrobenz[g]azocine (14) in 77.9% yield.

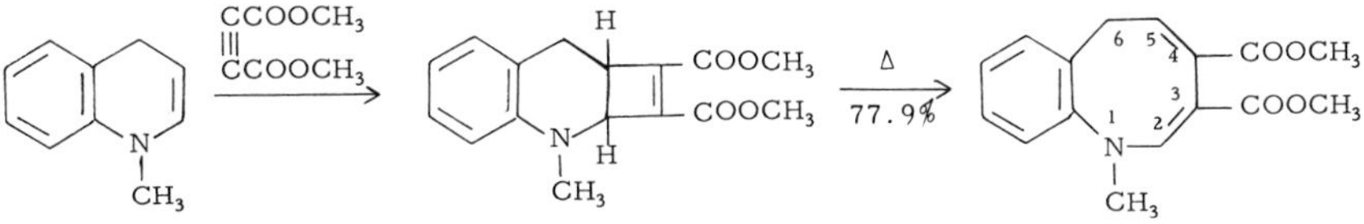

(12)    (13)    (14)

*Azulene synthesis.*[10]   The reaction between 6-(2-dimethylaminovinyl)fulvene (1)[11] and dimethyl acetylenedicarboxylate (2) affords dimethyl 4,5-azulenedicarboxylate (3, 11%) and dimethyl dimethylaminomaleate (4, *ca* 20%).

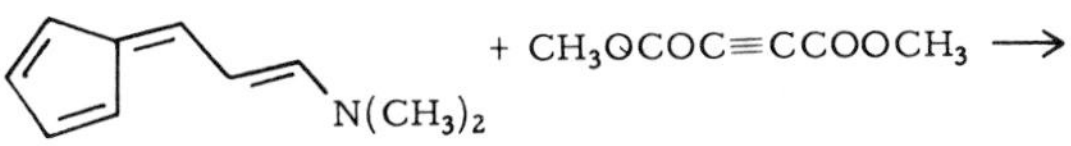

$$+ CH_3OCOC \equiv CCOOCH_3 \longrightarrow$$

(1)    (2)

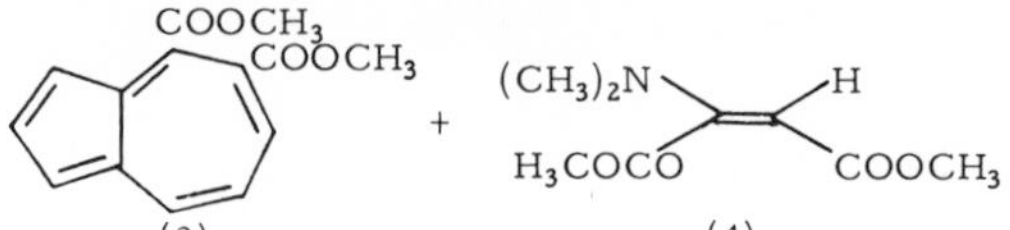

(3)                    (4)

**[2.2]-p-Cyclophanes.** Diels–Alder reaction of 1,2,4,5-hexatetraene (1, biallenyl) with dimethyl acetylenedicarboxylate gives the [2.2]-p-cyclophane (3) via dimerization of the initially formed quinone methide (2).[12]

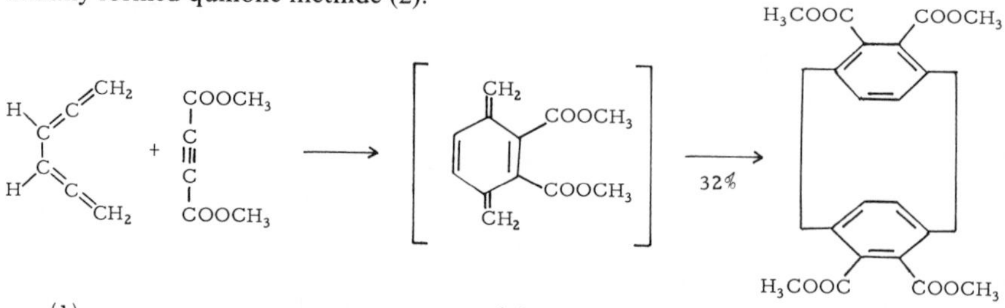

(1)                    (2)                    (3)

*Reaction with phenylhydrazones of aldehydes and ketones.*[13] Benzaldehyde phenylhydrazone (1) reacts with dimethyl acetylenedicarboxylate to give (2), (3), and (4). Reaction of acetophenone phenylhydrazone (5) gives the pyrazole (3) in 57% yield.

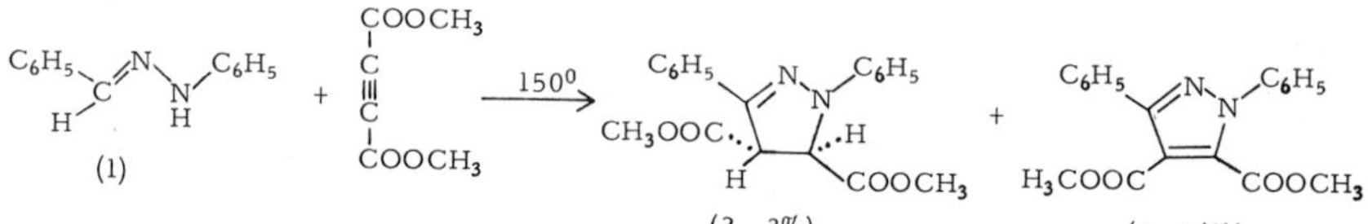

(1)

(2, 3%)            (3, 16%)

+ (4, 5%)

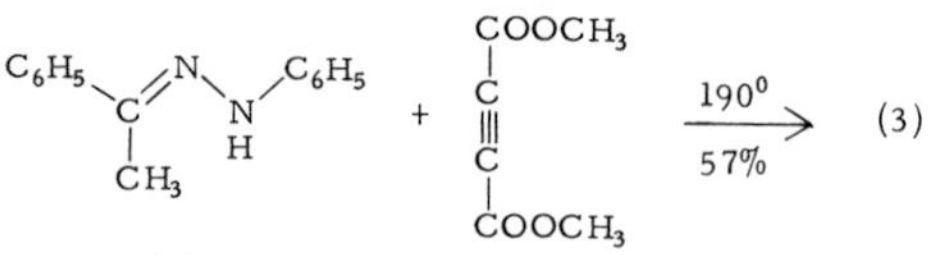

(5)                    (3)

[1]W. Steglich, P. Gruber, H. U. Heininger, and F. Kneidl, *Ber.*, **104**, 3816 (1971).
[2]A. S. Katner, *J. Org.*, **38**, 825 (1973).
[3]H. Reimlinger, R. Jacquier, and J. Daunis, *Ber.*, **104**, 2702 (1971).
[4]B. D. Baigrie, J. I. G. Cadogan, J. Cook, and J. T. Sharp, *J.C.S. Chem. Comm.*, 1318 (1972).
[5]E. Müller and A. Huth, *Tetrahedron Letters*, 1035 (1972); 4359 (1972).
[6]K. C. Brannock, R. D. Burpitt, V. W. Goodlett, and J. G. Thweatt, *J. Org.*, **28**, 1464 (1963).
[7]R. M. Acheson, J. N. Bridson, and T. S. Cameron, *J.C.S. Perkin I*, 968 (1972).
[8]H. Plieninger and D. Wild, *Ber.*, **99**, 3070 (1966).
[9]P. G. Lehman, *Tetrahedron Letters*, 4863 (1972).
[10]R. W. Alder and G. Whittaker, *Chem. Commun.*, 776 (1971).

[11]K. Hafner, *et al.*, *Angew. Chem.*, **75**, 35 (1963); H. Bredereck, F. Effenberger, and D. Zeyfang, *ibid.*, **77**, 219 (1965).
[12]H. Hopf, *Angew. Chem., internat. Ed.*, **11**, 419 (1972).
[13]M. K. Saxena, M. N. Gudi, and M. V. George, *Tetrahedron*, **29**, 101 (1973).

**4-Dimethylamino-1-*t*-butyloxycarbonylpyridinium chloride** (1). Mol. wt. 258.75, m.p. 50° dec., water soluble.

The reagent is prepared by addition of 4-dimethylaminopyridine[1] to an excess of *t*-butyloxycarbonyl chloride[2] in dry ether at 0°.

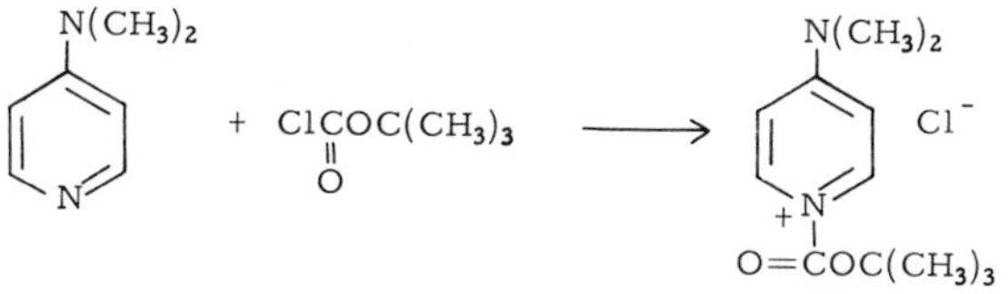

The reagent reacts with the sodium salts of amino acids in aqueous solution at 25° to give good yields of *t*-butyloxycarbonylamino acids (BOC-amino acids).[3]

[1]Aldrich, mol. wt. 122.17, m.p. 108–110°.
[2]R. B. Woodward, *et al.*, *Am. Soc.*, **88**, 852 (1966).
[3]E. Guibé-Jampel and M. Wakselman, *Chem. Commun.*, 267 (1971).

**N,N-Dimethylaminocyclopropylphenyloxosulfonium fluoroborate**, Mol. wt. 309.15, m.p. 121–122°.

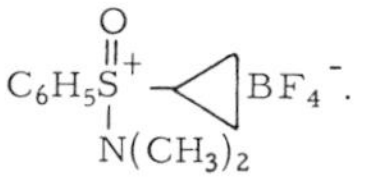

*Preparation.* The reagent can be prepared from phenyl 3-chloropropyl sulfide[1] or from cyclopropyl phenyl sulfide.[2]

*Spiro compounds.*[3] Reaction of (1) with sodium hydride in DMSO affords the cyclopropylide (2), which reacts with ketones to afford spiro compounds. Thus reaction of (2) with mesityl oxide (3) affords a 1-acetyl-2,2-dimethylspiropentane (4), which rearranges at 70° to 1-acetonyl-1-isopropenylcyclopropane (5).

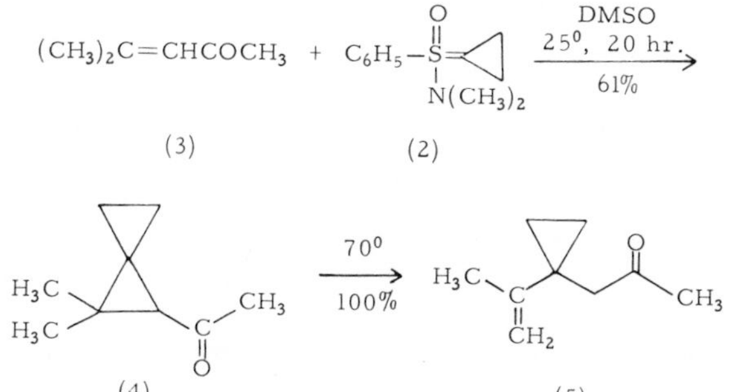

Reaction of the ylide (2) with cyclohexanone produced the unstable dispiro epoxide (6), which rearranged on preparative glc to (7).

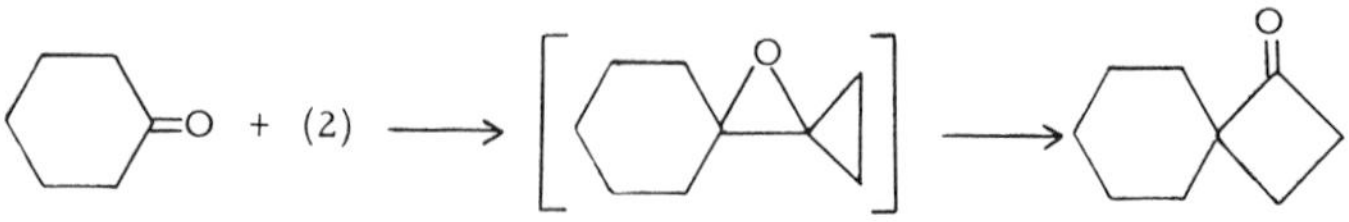

(6)                              (7)

[1]H. E. Zimmerman and B. S. Thyagarajan, *Am. Soc.*, **82**, 2505 (1960).
[2]W. E. Truce, K. R. Hollister, L. B. Lindy, and J. E. Parr, *J. Org.*, **33**, 43 (1968).
[3]C. R. Johnson, G. F. Katekar, R. F. Huxol, and E. R. Janiga, *Am. Soc.*, **93**, 3771 (1971).

**(Dimethylamino)-phenyl-(2-phenylvinyl)-oxosulfonium fluoroborate,**

Mol. wt. 345.19, m.p. 130–131.5°.

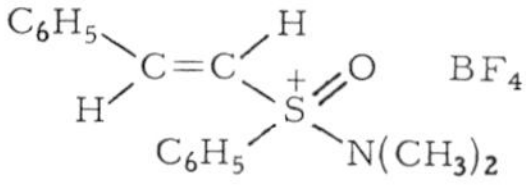

(1)

*Preparation.*[1] The reaction of lithium N-methylbenzenesulfonimidoylmethide with benzaldehyde gives (2).[2] This is converted by dehydration (TsOH) and methylation with trimethyloxonium fluoroborate into the reagent (1).

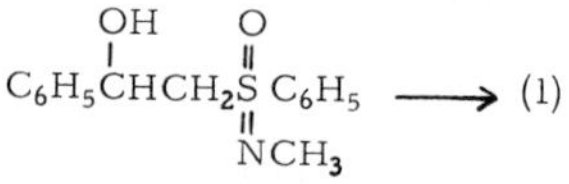

(2)

*Ethylene transfer to dibasic nucleophiles.* The reagent reacts with primary amines, enamines, and active methylene compounds:

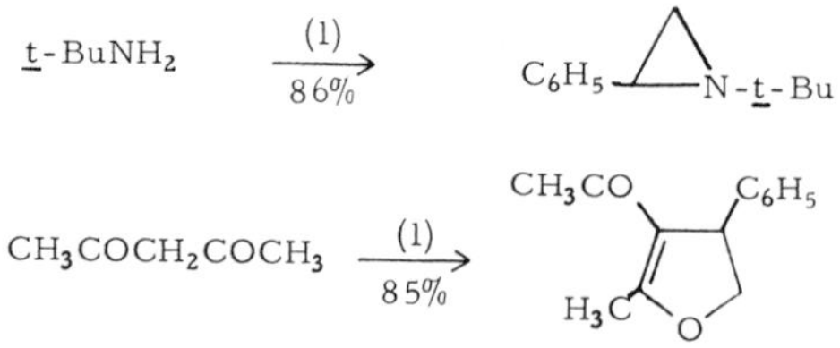

[1]C. R. Johnson and J. P. Lockard, *Tetrahedron Letters*, 4589 (1971).
[2]E. W. Yankee and D. J. Cram, *Am. Soc.*, **92**, 6329 (1970).

**Dimethylbromosulfonium bromide (DMBS),** $(CH_3)_2SBr_2$. Mol. wt. 221.97, orange crystals, m.p. 81–82° dec.

*Preparation.* The reagent is prepared by the reaction of dimethyl sulfide with bromine. It can be stored for several weeks at room temperature in the absence of light and moisture.

*Alkyl bromides.*[1] DMBS reacts with alcohols (used in excess) at *ca.* 80° for 4–5 hr. to give alkyl bromides in yields of 40–80%. The reaction proceeds with inversion.

$$(CH_3)_2SBr_2 + ROH \longrightarrow [(CH_3)_2\overset{+}{S}OR]Br^- \longrightarrow RBr + (CH_3)_2SO$$

Thus the reaction of (+)-*sec*-octyl alcohol with DMBS gives (−)-*sec*-octyl bromide (70% yield, optical yield 91%). See also 191–192, this volume.

[1] H. Furukawa, T. Inoue, T. Aida, and S. Oae, *J.C.S. Chem. Comm.*, 212 (1973).

**2,3-Dimethyl-2-butylborane (Thexylborane),** $ThBH_2$, **1**, 276; **2**, 148–149.

*Selective reductions.* Brown *et al.*[1] have reported a detailed study of the reduction of a large number of organic compounds with thexylborane, and compared results with this reagent with those obtained with diborane ($BH_3$) and disiamylborane ($Sia_2BH$, **1**, 57–59; **2**, 29; **3**, 22–23). The three reagents show similar reducing characteristics with some minor differences. Thus $ThBH_2$ and $Sia_2BH$ show less stereospecificity in reduction of ketones than $BH_3$. Toward esters, the order of reactivity is $BH_3 > ThBH_2 > Sia_2BH$.

*Multicarbon homologation of olefins.*[2] In a new general ketone synthesis, thexylborane is allowed to react with an equimolar quantity of an olefin to give a thexylmonoalkylborane (1). Addition of an ω-alkenyl acetate yields a thexyldialkylborane (2).

(1)

(2)    (3)

Carbonylation followed by oxidation gives the homologated ketone (3). Yields are in the range of 60–75%. An overall process can be presented as:

$$CH_3CH_2CH_2CH=CH_2 + CH_2=CH(CH_2)_9OAc + CO \longrightarrow CH_3(CH_2)_4\overset{O}{\overset{\|}{C}}(CH_2)_{11}OAc$$

*Synthesis of* **trans**-*disubstituted olefins.* Brown *et al.*[3] have reported a new stereoselective synthesis of *trans*-disubstituted olefins (4) utilizing thexylmonoalkylboranes (1) as intermediates. These are prepared by treating an olefin with thexylborane under $N_2$ at −25° for 1 hr. A 1-bromo-1-alkyne is then added, and after 1 hr. the resulting

$$\underset{\substack{| \ | \\ CH_3 \ CH_3}}{\overset{\substack{CH_3 \ CH_3 \\ | \ |}}{H-C-C-BH_2}} \xrightarrow{>C=C<, \ -25^0} \underset{\substack{| \ | \\ CH_3 \ CH_3}}{\overset{\substack{CH_3 \ CH_3 \\ | \ |}}{H-C-C-B}} \overset{\overset{\backslash | \quad |/}{C-CH}}{\underset{H}{\big\langle}} \equiv \underset{\substack{| \ | \\ CH_3 \ CH_3}}{\overset{\substack{CH_3 \ CH_3 \\ | \ |}}{H-C-C-B}}\overset{R^1}{\underset{H}{\big\langle}}$$

(1)

$$\xrightarrow{Br-C\equiv CR^2, \ -25^0} \underset{\substack{| \ | \\ CH_3 \ CH_3}}{\overset{\substack{CH_3 \ CH_3 \\ | \ |}}{H-C-C-B}}\overset{R^1}{\underset{\underset{\underset{Br}{C=C}}{\big\backslash}}{\big\langle}}H \xrightarrow{NaOCH_3, \ 25^0}$$

(2)

$$\underset{\substack{| \ | \\ CH_3 \ CH_3}}{\overset{\substack{CH_3 \ CH_3 \\ | \ |}}{H-C-C-B}}\overset{\overset{R^1}{\big\backslash} \quad H}{\underset{OCH_3}{\overset{C=C}{\big\langle}}R^2} \xrightarrow[reflux]{i-C_3H_7COOH} \overset{R^1}{\underset{H}{\big\backslash}}C=C\overset{H}{\underset{R^2}{\big/}}$$

(3)     (4)

thexylalkyl-1-bromo-1-alkenylborane (2) is treated with sodium methoxide at 25° f 1 hr. The resulting intermediate (3) is then hydrolyzed by treatment with isobutyric ac under reflux, 1 hr. Yields are high (80–95%). For example, *trans*-4-ethyl-5-decene ( was obtained in this way from *trans*-3-hexene and 1-bromo-1-hexyne in 78% yie (isolated).

$$\underset{\substack{| \\ H}}{\overset{\substack{CH_3CH_2CH_2 \\ | \\ CH_3CH_2CH}}{}}\overset{\big\backslash \quad H}{\underset{(CH_2)_3CH_3}{\overset{C=C}{\big/}}}$$

(5)

[1]H. C. Brown, P. Heim, and N. M. Yoon, *J. Org.*, **37**, 2942 (1972).
[2]E. Negishi and H. C. Brown, *Synthesis*, 196 (1972).
[3]E. Negishi, J.-J. Katz, and H. C. Brown, *ibid.*, 555 (1972).

**Dimethyl-*t*-butylchlorosilane,** $\underset{(CH_3)_2SiCl}{\overset{C(CH_3)_3}{|}}$ . Mol. wt. 151.73, m.p. 92.5°.

*Preparation.* The reagent can be prepared in 70% yield by dropwise addition commercial *t*-butyllithium in pentane to a 1 *M* solution of dimethyldichlorosila (1.15 eq.) in pentane at 0° under $N_2$ with stirring; the temperature is maintained at for 1.5 hr. and then at 25° for 48 hr. Distillation at atmospheric pressure (b.p. 125 affords the pure reagent.[1] See also the preparation by Sommer and Tyler.[2]

*Protection of hydroxyl groups.*[1] Trimethylsilyl ethers are too susceptible solvolysis in protic media to be widely useful in synthesis. However, the dimethyl-*t*-buty silyloxy group is about $10^4$ times more stable than the trimethylsilyloxy grou Dimethyl-*t*-butylchlorosilane reacts only slowly with alcohols in THF in the presen of excess pyridine. However, if imidazole (**1**, 492–494; **2**, 220) is used as catalyst an DMF as solvent, dimethyl-*t*-butylsilyl ethers are obtained in high yield under mil conditions. N-Dimethyl-*t*-butylsilylimidazole is probably the actual silylating reagen

The ethers are stable to aqueous or alcoholic base, to hydrogenolysis ($H_2$–Pd), and to mild chemical reduction (Zn–$CH_3OH$). The dimethyl-$t$-butylsilyl group is an alternative to the tetrahydropyranyl group but has the advantage that it does not have a chiral center.

Dimethyl-$t$-butylsilyl ethers are cleaved by 2:1 acetic acid–water at 25° or by treatment with tetra-$n$-butylammonium fluoride (this volume) in THF at 25°.

The new ethers were developed in the course of researches on the synthesis of prostaglandins and are especially valuable in this area.

[1]E. J. Corey and A. Venkateswarlu, *Am. Soc.*, **94**, 6190 (1972).
[2]L. H. Sommer and L. J. Tyler, *ibid.*, **76**, 1030 (1954).

**3,5-Dimethyl-4-chloromethylisoxazole, 1**, 276–277; **2**, 150–151.

*Pyridine synthesis.* The definitive paper on the synthesis of pyridines via 4-(3-oxo-alkyl)isoxazoles (**2**, 150) has now been published.

[1]G. Stork, M. Ohashi, H. Kamachi, and H. Kakisawa, *J. Org.*, **36**, 2784 (1971).

**Dimethylcopperlithium (Lithium dimethyl cuprate), 2**, 151–153; **3**, 106–113.

*Conjugate addition.* Conjugate addition reactions of organocopper reagents have been reviewed by Posner.[1]

Conjugate addition of dimethylcopperlithium to 3,4,5,6,7,8-hexahydronaphthalene-1(2H)-one-7-carboxylic acid (1) affords a mixture of three stereoisomeric 4$a$-methyl-3,4,4$a$,5,6,7,8,8$a$-octahydronaphthalene-1(2H)-one-7-carboxylic acids, (2), (3), and (4),

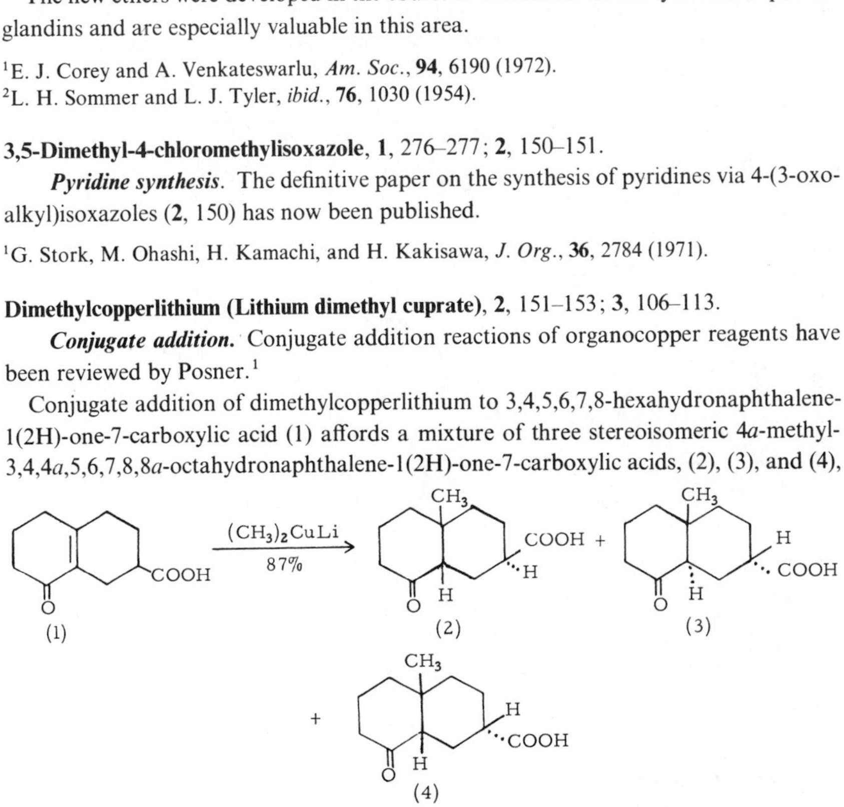

in the ratio 4:3:2. The reaction was used in a nonannelation synthesis of the sesquiterpene $\beta$-eudesmol.[2]

The reagent has been used[3] in a highly stereoselective total synthesis of the racemic form of the tricyclic sesquiterpene seychellene (8), a minor component of commercial

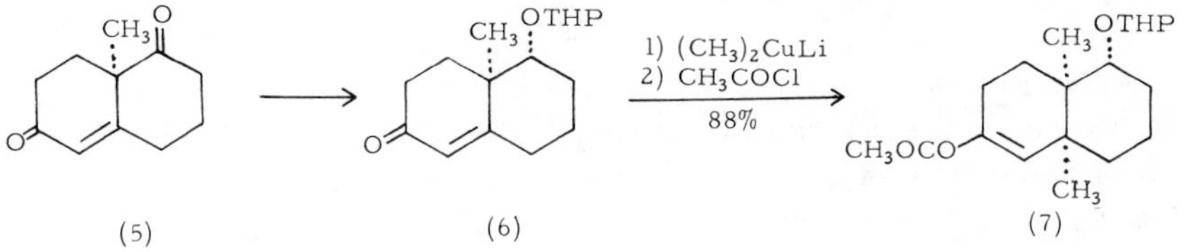

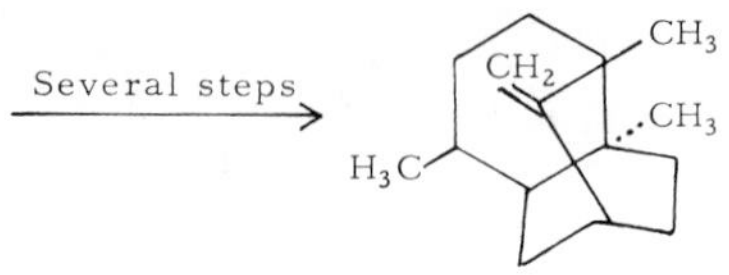

(8)

patchouli oil.[4] The starting material was the Wieland–Miescher ketone (5). This w
converted into (6) by selective reduction by sodium borohydride followed by conversi
to the tetrahydropyranyl ether. Addition of dimethylcopperlithium was effected at $-2$
followed by acetylation with acetyl chloride. The product (7) was converted in
seychellene in about 10 steps.

The reaction of dimethylcopperlithium with the enone (9) in ether at 0° affor
principally the expected 1,4-adduct (10, 55%); two cyclopropane cleavage produc

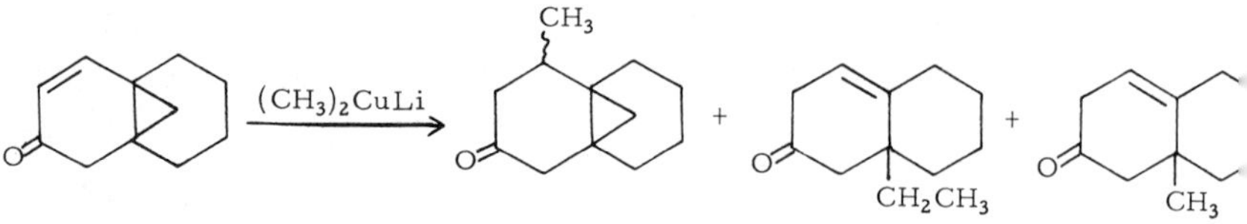

(9)                    (10,  55%)          (11,  39%)          (12,  6%)

(11, 39%, and 12, 6%) are also formed. The 1,4-adduct is largely the *trans*-adduct
The conjugated double bond appears to be essential for the cyclopropane cleava
reaction. Thus bicyclo[4.1.0]heptane-2-one (13) does not react with the reagent.[6]

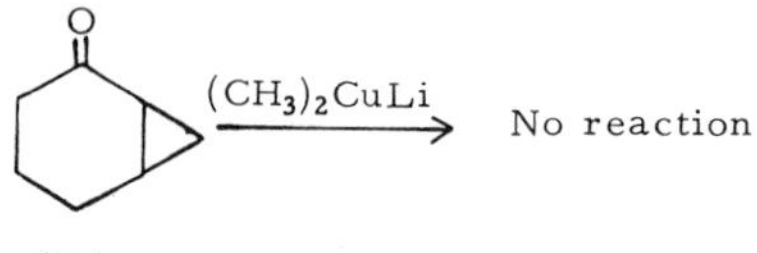

(13)

The conjugate addition of dimethylcopperlithium to the 2-methyl-6-keto-octahydr
isoquinoline (14) leads uniquely to the 2,10-dimethyl-6-keto-decahydroisoquinoline (1
with *cis* ring junction.[7]

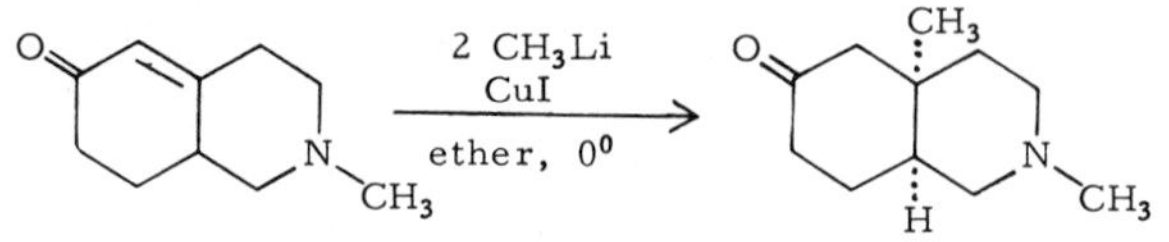

(14)                                      (15)

**Conjugate addition to conjugated dienones.** Marshall and co-workers[8] have studie
the reaction of the reagent with cyclohexadienones and conclude that steric factor
account for the ratio of 1,4- to 1,6-addition. The last two examples illustrate the virtu
absence of angular methylation.

Examples:

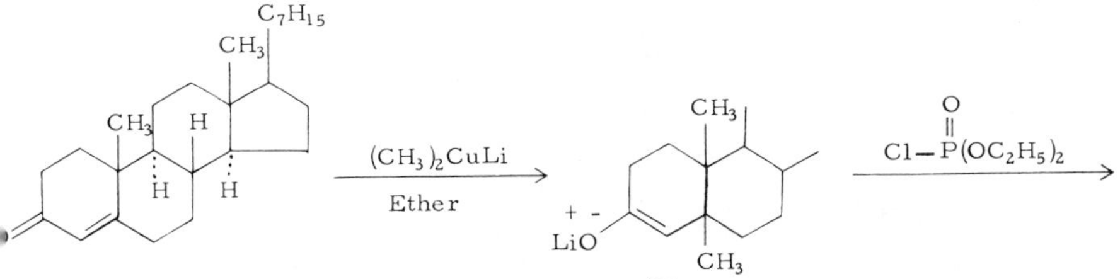

*Cyclization.*[9] The reaction of the iodo ether (1) with dimethylcopperlithium gives in addition to the expected methylation product (2, 30% yield), three products identified as (3), (4), and (5), in the ratio 50:45:5, respectively. The formation of (3) is particularly interesting since formally it involves an intramolecular vinyl–allyl coupling.

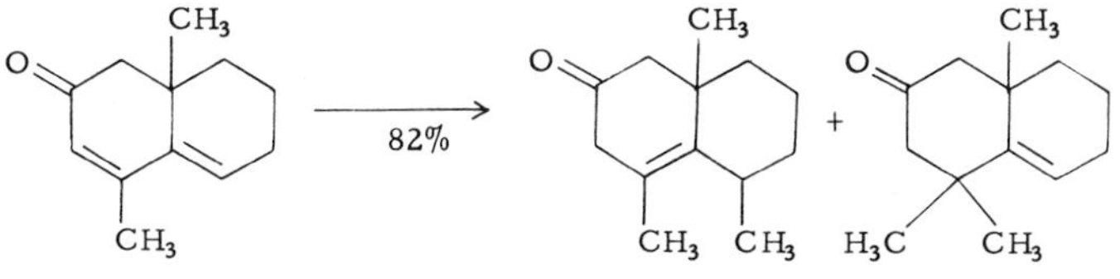

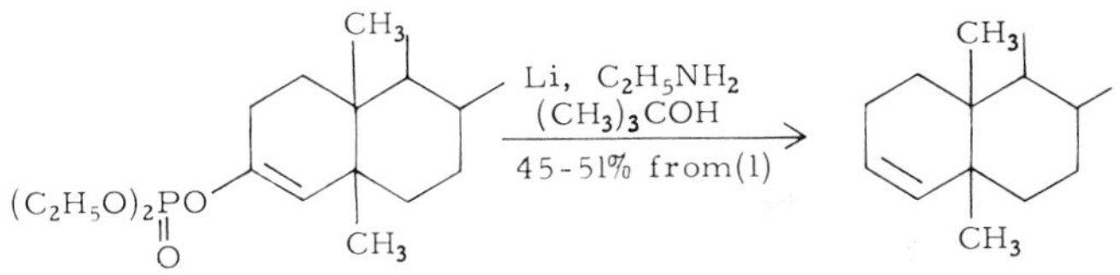

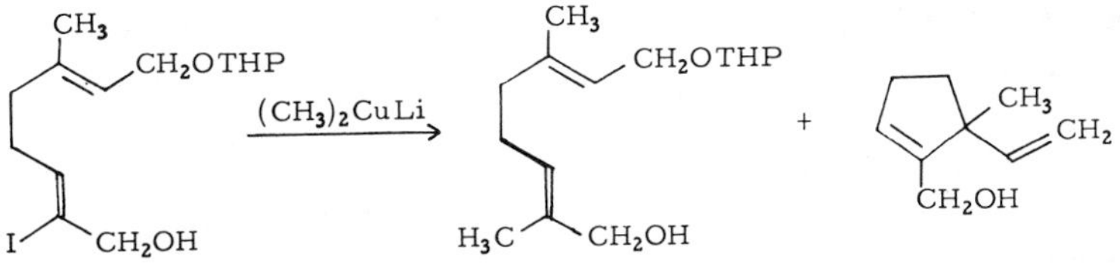

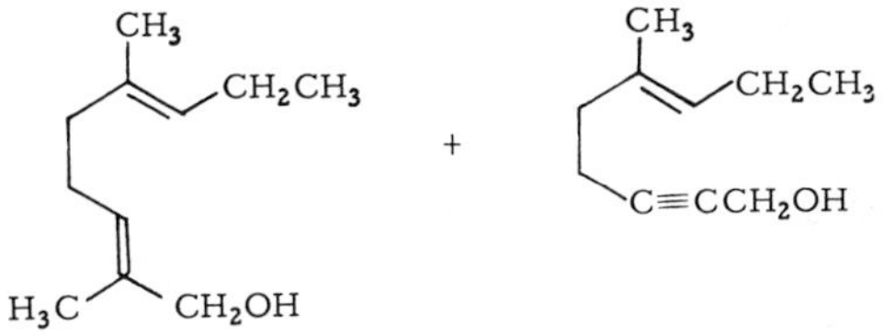

*Eudalene sesquiterpenes.* Syntheses of eudalene-type sesquiterpenes commoᵢ utilize the Robinson annelation reaction to construct the bicyclic ring system. Howevᵢ this annelation reaction often proceeds in low yields and with stereochemical difficultiᵢ Huffman and Mole[10] have recently reported a new stereoselective synthesis. 8-Methoᵢ tetralin-2-carboxylic acid (1) is reduced under Birch conditions followed by aᵢ

(1)                          (2)

(3) 3 isomers                          (4)

treatment to give a mixture of four compounds from which the unsaturated ketoacid can be obtained by chromatography in 49 % yield. The angular methyl group is intr duced by reaction with dimethylcopperlithium to give approximately equal amounts three isomers (3) in approximately quantitative yield. The next step, a Wittig reactiᵢ with methylenetriphenylphosphorane, is carried out on the mixture of (3), since tᵢ Wittig reaction of both *cis*- and *trans*-1-decalones leads predominantly to products wᵢ a *trans* ring fusion. In this case, the reaction gives a mixture of three acids in the approᵢ imate ratio of 1:3:6. Esterification, equilibration with methanolic sodium methoxiᵢ and hydrolysis gives the desired 2β-acid (4) in 75 % yield from (3, mixture). The acid ᵢ has been converted into β-eudesmol, which in turn has been converted into crypt meridiol and neomeridiol.

*Hindered ketones.* Dimethylcopperlithium (and related reagents) undergo crosᵢ coupling with α-bromoketones such as (1) to give sterically hindered ketones such as (ᵢ Even tertiary lithium cuprates can be employed.[11]

(1)                          (2)

*Preparation of enol phosphates.*[12] The reagent is used in ether solution to convᵢ Δ⁴-cholestene-3-one (1) into the enol form for reaction with diethyl phosphoᵢ chloridate and triethylamine to give the phosphate ester (3). Reaction with lithium aᵢ ethylamine, and treatment with *t*-butyl alcohol then affords 5-methyl-Δ³-coprostene (ᵢ

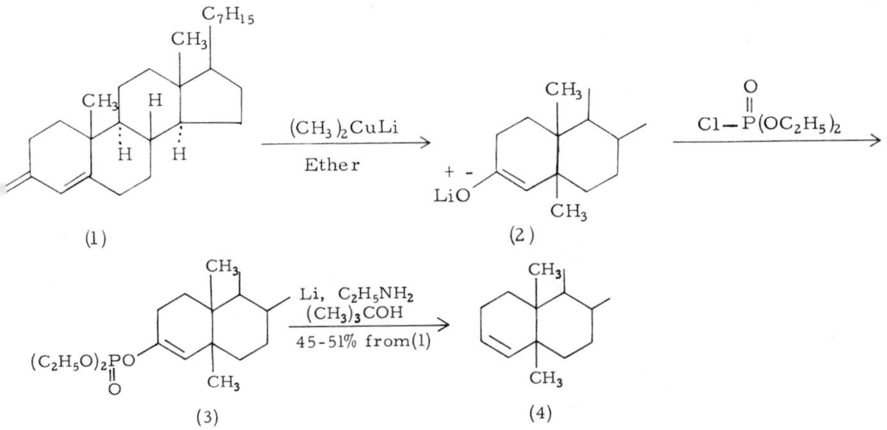

**Reaction with α,β-ethylenic sulfones.**[13]  Aldehydes and ketones can be transformed into *gem*-dialkylalkanes (3) by conversion first into the α,β-ethylenic sulfone (1, *see* *n*-**Butyllithium**, this volume). This is then treated with a dialkylcopperlithium reagent

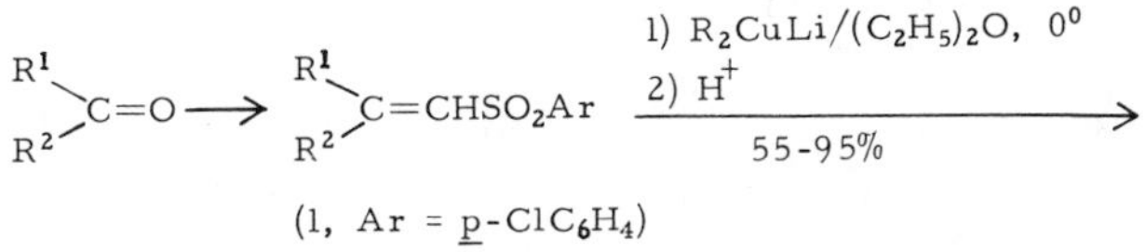

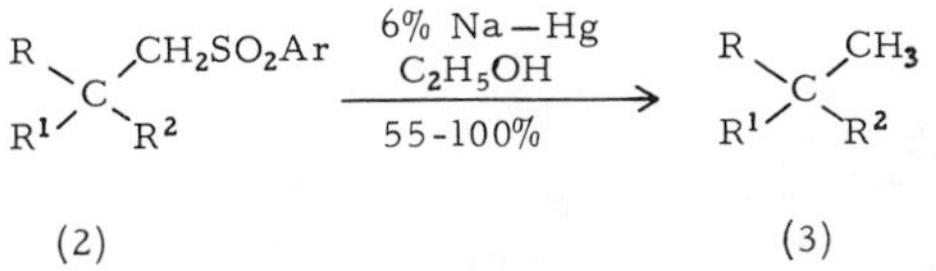

to give an alkyl aryl sulfone (2) in good yield. Hydrogenolysis of (2) is carried out with 6% sodium amalgam in boiling ethanol.[14]

*Methyl ketones from carboxylic acid chlorides* (3, 112). Posner *et al.*[15] report that the synthesis of ketones from acid chlorides and organocopper reagents is applicable to acid chlorides containing iodo, cyano, acyl, and carboalkoxy groups. The reaction, however,

$$R'COCl + R_2CuLi \xrightarrow[-78°]{(C_2H_5)_2O} R'COR$$

has some limitations. Aldehydo groups are reactive to dialkylcopperlithium even at −90°. Three equivalents of $R_2CuLi$ are required for optimal yields. Another disadvantage is that organolithium reagents are generally less readily available than organomagnesium reagents.

*Reaction with benzylic chlorides.* Posner and Brunelle[16] report the following examples of the reaction of the reagent with benzylic chlorides in diethyl ether.

$$C_6H_5CHCl_2 \xrightarrow[0^0]{(CH_3)_2CuLi} C_6H_5CH(CH_3)_2 + C_6H_5\overset{|}{C}H-\overset{|}{C}HC_6H_5$$

4 0%                     $CH_3$  $CH_3$

                                          40%

$$C_6H_5\overset{|}{\underset{CH_3}{C}}HCl \xrightarrow[23^0]{(CH_3)_2CuLi} C_6H_5CH(CH_3)_2 + C_6H_5\overset{|}{C}H-\overset{|}{C}HC_6H_5$$

4 0%                     $CH_3$  $CH_3$

                                          4 0%

$$C_6H_5CH_2Cl \xrightarrow[0^0]{(CH_3)_2CuLi} C_6H_5CH_2CH_3$$

8 0%

$$(C_6H_5)_2CCl_2 \xrightarrow[0^0]{(CH_3)_2CuLi} (C_6H_5)_2C=C(C_6H_5)_2$$

81%

**Reaction with 1,3- and 1,4-cyclohexadiene monoepoxides.** The reagent reacts wi
1,3-cyclohexadiene monoepoxide (1) to give the product of direct opening (2) and t
conjugate addition product (3) with *trans* stereospecificity.[17]

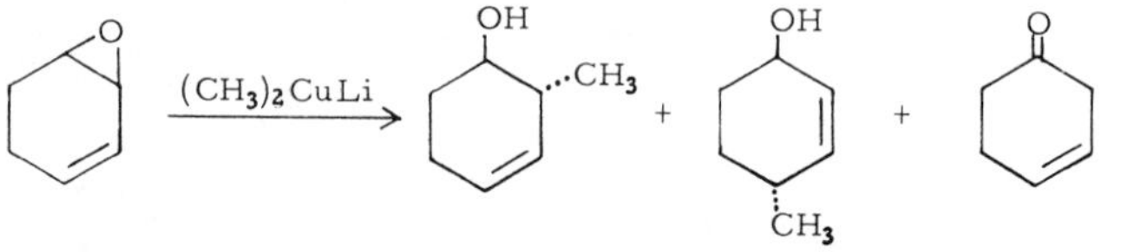

(1)                    (2, 35%)        (3, 42%)        (4, 23%)

Similar results have been reported by Wieland and Johnson.[18]

On the other hand, the reagent reacts with 1,4-cyclohexadiene monoepoxide (5)
give the *trans* opening product (6) exclusively.[17]

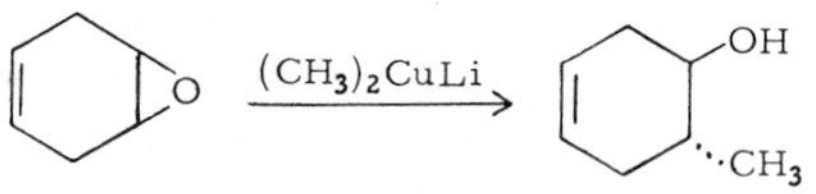

(5)                              (6)

**Reaction with vinylic bromides.** The reaction of a vinylic bromine atom wit
dimethylcopperlithium proceeds with retention of configuration.[19] Thus *trans-*
bromocinnamic acid (1) reacts with the reagent to give, after treatment with acidifie
water, *trans*-α-methylcinnamic acid (2). This behavior contrasts with the reaction

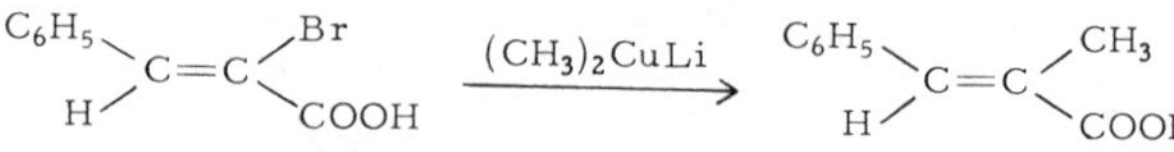

(1)                              (2)

saturated alkyl halides with dialkylcopperlithium reagents where inversion of configuration is observed.[20]

[1]G. H. Posner, *Org. Reac.*, **19**, 1 (1972).

[2]J. W. Huffman and M. L. Mole, *J. Org.*, **37**, 13 (1972).

[3]E. Piers, W. de Waal, and R. W. Britton, *Am. Soc.*, **93**, 5113 (1971).

[4]G. Wolff and G. Ourisson, *Tetrahedron*, **25**, 4903 (1969).

[5]J. A. Marshall and R. A. Ruden, *J. Org.*, **37**, 659 (1972).

[6]J. A. Marshall and R. A. Ruden, *Tetrahedron Letters*, 2875 (1971).

[7]S. Sicsic and N.-T. Luong-Thi, *ibid.*, 169 (1973).

[8]J. A. Marshall, R. A. Ruden, L. K. Hirsch, and M. Phillippe, *ibid.*, 3795 (1971).

[9]J. A. Katzenellenbogen and E. J. Corey, *J. Org.*, **37**, 1441 (1972).

[10]J. W. Huffman and M. L. Mole, *Tetrahedron Letters*, 501 (1971).

[11]J.-E. Dubois, C. Lion, and C. Moulineau, *ibid.*, 177 (1971).

[12]D. C. Muchmore, *Org. Syn.*, **52**, 109 (1972).

[13]G. H. Posner and D. J. Brunelle, *Tetrahedron Letters*, 935 (1973).

[14]R. E. Dabby, J. Kenyon, and R. F. Mason, *J. Chem. Soc.*, 4881 (1952).

[15]G. H. Posner, C. E. Whitten, and P. E. McFarland, *Am. Soc.*, **94**, 5106 (1972).

[16]G. H. Posner and D. J. Brunelle, *Tetrahedron Letters*, 293 (1972).

[17]J. Staroscik and B. Rickborn, *Am. Soc.*, **93**, 3046 (1971).

[18]D. M. Wieland and C. R. Johnson, *ibid.*, **93**, 3047 (1971).

[19]J. Klein and R. Levene, *ibid.*, **94**, 2520 (1972).

[20]G. M. Whitesides, J. San Fillippo, Jr., C. P. Casey, and E. J. Panek, *ibid.*, **89**, 5302 (1967); G. M. Whitesides, W. F. Fischer, Jr., J. San Fillippo, Jr., R. W. Bashe, and H. O. House, *ibid.*, **91**, 4871 (1969).

**Dimethyldichlorosilane**, $(CH_3)_2SiCl_2$. Mol. wt. 129.07, b.p. 70.3°. Supplier: Pierce Chemical Company.

*Pinacol cyclization.*[1] A key step in a new method for carbocyclic synthesis applicable to gibberellic acids involves the pinacol cyclization of (1) to (2). This was accomplished, in 75% yield, by addition of (1) to a mixture of magnesium amalgam (7.5 eq.) and dimethyldichlorosilane (2 eq.) in dry THF at 25° followed by subsequent alkaline

desilylation. The product (2) is obtained as a mixture of *cis-* and *trans-*diols in a ratio of 80:20. The use of the chlorosilane is crucial to the success of the cyclization, since in its absence a complex mixture of products is formed. Compare the use of trimethylchlorosilane in related reactions (**1**, 1232; **2**, 435–438; **3**, 310–312; this volume).

[1]E. J. Corey and R. L. Carney, *Am. Soc.*, **93**, 7318 (1971).

**N,N-Dimethylformaldimmonium trifluoroacetate, 3,** 114–115.
The definitive paper of the French investigators has now been published.[1]

[1]A. Ahond, A. Cavé, C. Kan-Fan, and P. Potier, *Bull. soc.*, 2707 (1970).

**N,N-Dimethylformamide, 1,** 278–281; 1110; **2,** 153–154; **3,** 115.
*Aminoanthraquinones.*[1] The reaction of 1-chloroanthraquinone (1) with DMF (reflux, 24 hr.) gives 1-methylaminoanthraquinone (2) as the major product, accompanied by 1-dimethylaminoanthraquinone. 2-Chloroanthraquinone under the same conditions gives 2-dimethylaminoanthraquinone as the only product.

The reaction is not limited to DMF, but is applicable to compounds of the general formula $R^1CONR^2R^3$, in which $R^1$ = H or alkyl, $R^2$ and $R^3$ = H, alkyl, or aryl, but not both = H.[2] Particularly high yields of substituted aminoanthraquinones are obtained from 1-chloroanthraquinone using N,N-diethylformamide, N-methylformamide, and N,N-dimethylacetamide.

*Demethylation of quaternized nitrogen heterocycles.*[3] N-Methylpyridinium, -quinolinium, and -isoquinolinium iodide are demethylated in high yield when heated in DMF for 6–95 hr.

[1]W. M. Lord and A. T. Peters, *J. Chem. Soc.* (*C*), 783 (1968).
[2]*Idem, Chem. Ind.*, 227 (1973).
[3]D. Aumann and L. W. Deady, *J.C.S. Chem. Comm.*, 32 (1973).

**Dimethylformamide dialkyl acetals,** $\begin{array}{c} H_3C \\ \phantom{} \\ H_3C \end{array} N\overset{OR}{\underset{OR}{\overset{|}{C}H}}$ . Supplier: Pierce Chemical Company

supplies DMF dimethyl, diethyl, dipropyl, n-dibutyl, and t-dibutyl acetals.

**N,N-Dimethylformamide dimethyl acetal.** (*See* **N,N-Dimethylformamide diethyl acetal,** **1,** 281–282; **2,** 154; **3,** 115–116.) Suppliers: Pierce, Fluka.
*Olefin synthesis.* Eastwood *et al.*[1] have reported a new method for conversion of *vic*-diols into alkenes. For example, *rac*-1,2-diphenylethane-1,2-diol (1) is heated with N,N-dimethylformamide dimethyl acetal to give 2-dimethylamino-*trans*-4,5-diphenyl-1,3-dioxolane (2). When this dioxolane is heated with acetic anhydride at 165–180° *trans*-stilbene (3) is formed in 80% yield. Similarly, *meso*-1,2-diphenylethane-1,2-diol is converted into *cis*-stilbene (75%) and a trace of *trans*-stilbene if the elimination

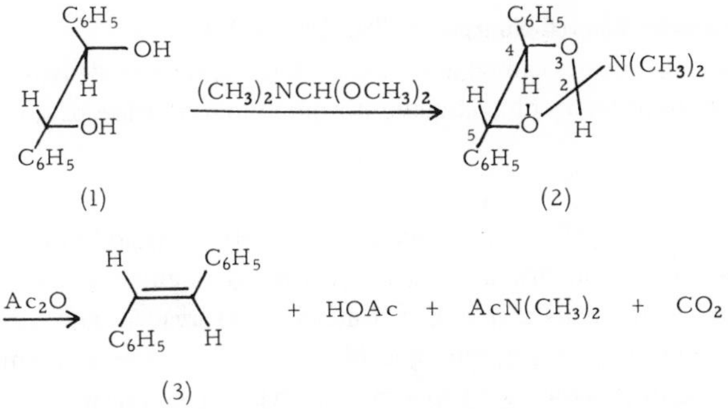

$$(1) \xrightarrow{(CH_3)_2NCH(OCH_3)_2} (2)$$

$$\xrightarrow{Ac_2O} (3) + HOAc + AcN(CH_3)_2 + CO_2$$

reaction is conducted at 90°. The latter alkene becomes the principal product if the temperature is increased to 150°. The reaction appears to be stereospecific, but the resulting alkene may undergo acid-catalyzed isomerization.

*Alkylation of thiols.*[2] Dimethylformamide dimethyl acetal reacts with heterocyclic mercapto derivatives in boiling benzene or acetonitrile to give the corresponding alkylthio derivatives in excellent yields:

$$RSH + (CH_3)_2N-CH(OCH_3)_2 \longrightarrow RSCH_3 + CH_3OH + (CH_3)_2NCHO$$

For example, when 2-mercaptopyrimidine and dimethylformamide dimethyl acetal are refluxed for 1 hr. in benzene, 2-methylthiopyrimidine is obtained in 86% yield.

*Chromone.* A convenient procedure for the preparation of chromone (4)[3] involves the reaction of 2-hydroxyacetophenone (1) with dimethylformamide dimethyl acetal (2) in refluxing xylene with continuous distillation of the methanol formed. The enaminoketone (3) formed in this way is cyclized to chromone by treatment with aqueous sulfuric acid at 100°.

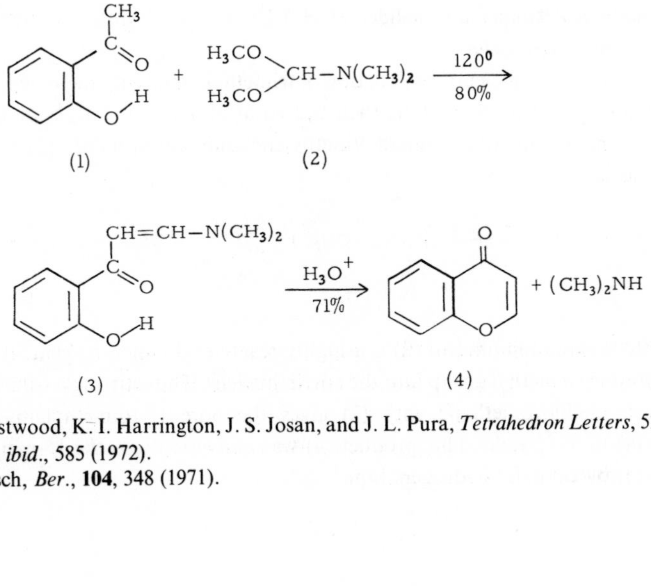

[1]F. W. Eastwood, K. I. Harrington, J. S. Josan, and J. L. Pura, *Tetrahedron Letters*, 5223 (1970).
[2]A. Holy, *ibid.*, 585 (1972).
[3]B. Föhlisch, *Ber.*, **104**, 348 (1971).

**Dimethylformamide–Thionyl chloride**, 1, 286–289; 3, 116.

*Isonitriles.*[1] Isonitriles can be prepared in 70–95 % yield by dehydration of form\
mides in a DMF solution of chlorodimethylformiminium chloride, prepared *in situ* fr(

$$RNHCHO \xrightarrow{-H_2O} R\overset{..}{N}=C:$$

DMF and thionyl chloride. Low temperatures ($-50°$) and addition of sodium c\
bonate (to react with the hydrochloric acid formed) are essential for satisfactory yiel\
*2-Dimethylaminobenzo-1,3-dioxole.*[2] Catechol (1) reacts with dimethylchlor\
formiminium chloride in methylene chloride (40°) to give the iminium chloride (\
This is cyclized to the dioxole (3) by treatment with triethylamine in methylene chlor\

(1)                    (2)

(3)

(40°). The transformation of (1) to (3) can be carried out in one step in somewhat low\
yield (69 %) by treatment of (1) with the reagent and triethylamine in methylene chlori\

[1]H. M. Walborsky and G. E. Niznik, *J. Org.*, **37**, 187 (1972).\
[2]G. Ege and H. O. Frey, *Tetrahedron Letters*, 4217 (1972).

**Dimethyl(methylene)ammonium iodide**, $(CH_3)_2\overset{+}{N}=CH_2I^-$. Mol. wt. 185.01, d(\
*ca.* 240°. Supplier: Lancaster.

The salt is prepared efficiently by reaction of trimethylamine with methylene iodide\
dioxane–ethanol. The product (1) is then heated at about 150° for about 12 min. \
tetrahydrothiophene dioxide; dimethyl(methylene)ammonium iodide (2) is obtain\
in 81 % yield.[1]

$$(CH_3)_3N + CH_2I_2 \xrightarrow[89\%]{} (CH_3)_3\overset{+}{N}CH_2I \ I^- \xrightarrow[81\%]{150°, \ -CH_3I} (CH_3)_2\overset{+}{N}=CH_2I^-$$

(1)                    (2)

The methyleneammonium salt (2) is a highly reactive Mannich reagent. It has be\
used to introduce a methyl group into the corrin nucleus. Thus stirring a solution of t\
corrin (3) in dichloromethane with (2) gives the pure 15-(dimethylamino)meth\
derivative (4) in 85 % yield. This product (4) was converted into the 15-methylcorr\
derivative (5) by catalytic hydrogenation.[1]

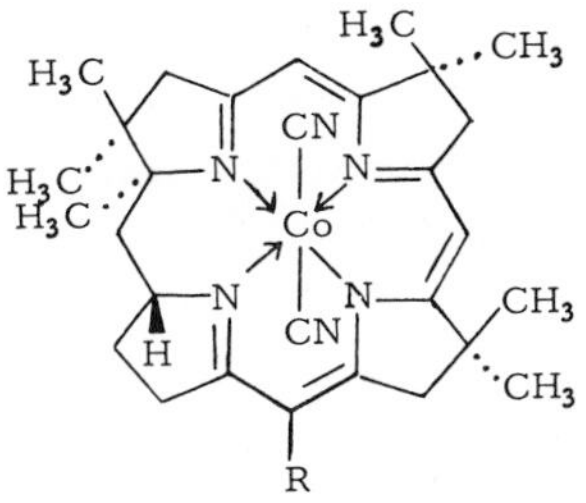

(3), R = H

(4), R = CH$_2$-N(CH$_3$)$_2$

(5), R = CH$_3$

*Regiospecific synthesis of Mannich bases.* Hooz and Bridson[2] have described a new regiospecific[3] synthesis of certain Mannich bases which involves first reaction of a trialkylborane with an α-diazoketone in THF to form an enol borinate (1)[4]; after evolution of nitrogen ceases, dimethyl(methylene)ammonium iodide (2) in DMSO is added. The Mannich base (3) is obtained in 85–100% yield after hydrolytic workup. Cosolvent DMSO is crucial for high yields.

$$R_3B + R'COCHN_2 \xrightarrow[-N_2]{} R_2BO\overset{R'}{\underset{}{C}}=CHR + (CH_3)_2\overset{+}{N}=CH_2\overset{-}{I} \longrightarrow R'\underset{OCH_2N(CH_3)_2}{\overset{O}{\underset{}{C}}CHR}$$

$$(1) \qquad\qquad (2) \qquad\qquad (3)$$

For example, sequential treatment of triethylborane (Alfa Inorganics) with diazomethyl *n*-propyl ketone (THF) followed by addition of (2, 1.15 eq. in DMSO) affords 3-dimethylaminomethyl-4-heptanone (4) in 91% yield.

$$(C_2H_5)_3B + CH_3(CH_2)_2COCHN_2 \longrightarrow \xrightarrow[91\%]{(2)} CH_3(CH_2)_2CO\underset{CH_2N(CH_3)_2}{CHCH_2CH_3}$$

$$(4)$$

Enol borinates are also intermediates in the reaction of organoboranes with α,β-unsaturated ketones and aldehydes. Thus the reaction of triethylborane and 3-methyl-3-butene-2-one (5) followed by addition of (2, in DMSO) affords the Mannich base (6) in 87% yield.

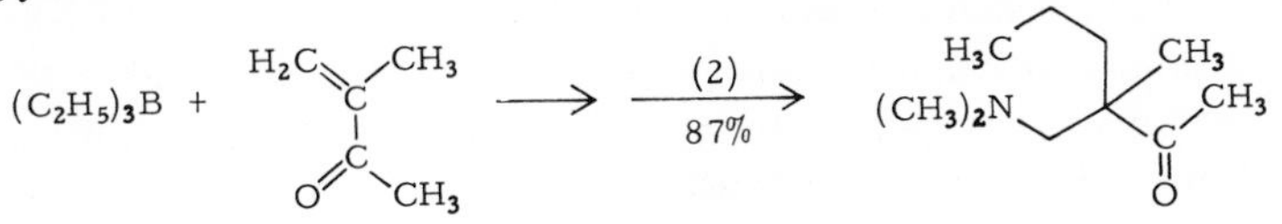

$$(5) \qquad\qquad\qquad\qquad (6)$$

[1] J. Schreiber, H. Maag, N. Hashimoto, and A. Eschenmoser, *Angew. Chem., internat. Ed.*, **10**, 330 (1971).

[2] J. Hooz and J. N. Bridson, *Am. Soc.*, **95**, 602 (1973).

[3] For a definition of the term regio, see A. Hassner, *J. Org.*, **33**, 2684 (1968).

[4] J. Hooz and J. N. Bridson, *Canad. J. Chem.*, **50**, 2387 (1972), and references cited therein.

**Dimethyl phosphite, 1**, 293.

*Etherification.*[1] When cholesterol (1) is heated for several hours in dimet[
phosphite, cholesteryl methyl phosphite (2) is obtained as the main product. Howev
if *p*-TsOH is present, 3β-methoxy-Δ⁵-cholestene (3) is obtained as the main produc[
about 60% yield. Cholestanol is converted by this method into 3β-methoxycholesta[
in 60–70% yield.

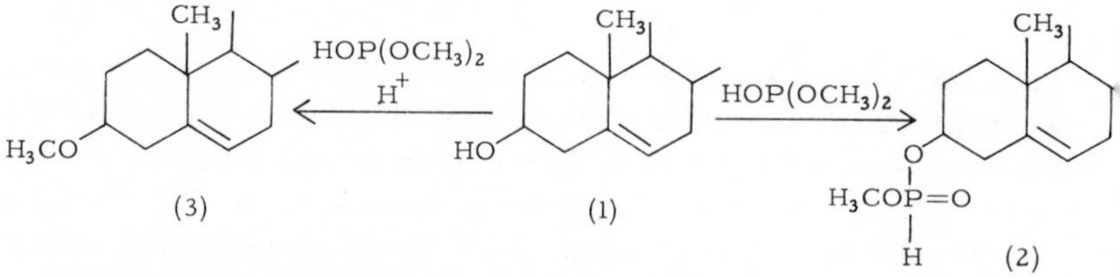

(3)                                    (1)                                    (2)

Use of diphenyl phosphite for etherification is limited to Δ⁵-3-hydroxysteroids
homoallylic cation is probably involved in this case). Thus cholesterol is converted in
3β-phenoxy-Δ⁵-cholestene in 50% yield by treatment with diphenyl phosphite in t
presence of an acid catalyst.

*Reaction with N,N'-bisbenzenesulfonyl-p-quinonediimine*[2]:

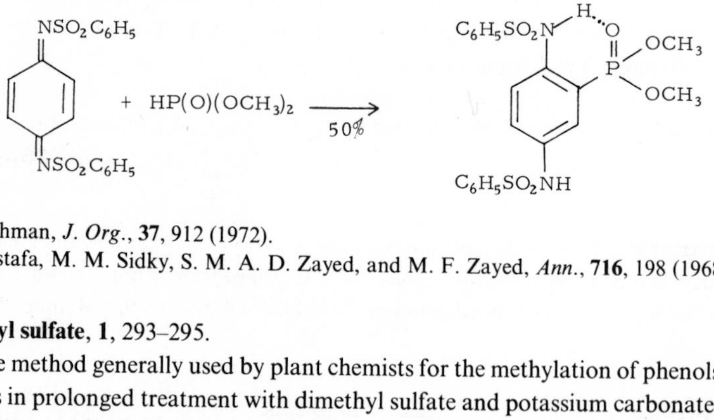

[1]Y. Kashman, *J. Org.*, **37**, 912 (1972).
[2]A. Mustafa, M. M. Sidky, S. M. A. D. Zayed, and M. F. Zayed, *Ann.*, **716**, 198 (1968).

**Dimethyl sulfate, 1**, 293–295.

In the method generally used by plant chemists for the methylation of phenols, whi[
consists in prolonged treatment with dimethyl sulfate and potassium carbonate in bo[
ing acetone, an alcoholic group in the phenol can lead to a carbonate by the action [
dimethyl carbonate formed by interaction of dimethyl sulfate with potassium ca[
bonate:

$$(CH_3)_2SO_4 \;+\; K_2CO_3 \longrightarrow (CH_3)_2CO_3$$

$$(CH_3)_2CO_3 \;+\; ROH \longrightarrow R-O-COOCH_3 \;+\; CH_3OH$$

Thus one of the products of the methylation of laccaic acid B (I) proved to be (II[
This was confirmed by heating β-phenylethanol with dimethyl sulfate and potassiu[

CH₂CH₂OH ... structures (I and II) ...

I                                    II

carbonate in boiling acetone for 36 hr., when $\beta$-phenylethylmethylcarbonate was obtained in 45 % yield; $\beta$-phenylethanol with dimethyl carbonate at 100° for 48 hr. gave the same product in 90 % yield.

*Esterification.* Dimethyl sulfate has been used widely for O-methylation of phenols and alcohols, but has been used only a few times for esterification of carboxylic acids[2]; the method has not attracted much attention. Actually dimethyl sulfate is a useful reagent for this purpose, especially for esterification of hindered acids such as (1).[3] This acid has been converted into the methyl ester in over 95 % yield by either of two methods. Concentrated aqueous NaOH (1.1 moles) is added at room temperature to a well-stirred mixture of (1, 1 mole) and dimethyl sulfate in dioxane followed by a

(1)

terminal reflux of 0.5 hr. Alternatively, an acetone solution of (1) and dimethyl sulfate is refluxed for 3 hr. in the presence of $K_2CO_3$ (10 % excess).

The method has been used successfully for esterification of 2,6-bistrifluoromethylbenzoic acid and of 2,6-dinitrobenzoic acid. Esterification can be achieved under the same conditions with diethyl sulfate.[4]

[1] E. D. Pandhare, A. V. Rama Rao, I. N. Shaikh, and K. Venkataraman, *Tetrahedron Letters*, 2437 (1967).
[2] R. Stoermer, *Ber.*, **44**, 637 (1911); C. N. Riiber, *ibid.*, **48**, 823 (1915); G. E. Ullyott, H. W. Taylor, and N. Dawson, *Am. Soc.*, **70**, 542 (1948).
[3] J. Grundy, B. G. James, and G. Pattenden, *Tetrahedron Letters*, 757 (1972).
[4] G. Hallas and J. D. Hepworth, *Chem. Ind.*, 691 (1972).

**Dimethyl sulfate–Potassium carbonate, 1, 295–296.**

Carboxylic acids are converted into their methyl esters in high yield by this combination of reagents in refluxing acetone.[1] The method is useful for esterification of hindered acids.[2] Tropolone is converted into 2-methoxytropone in 96 % yield under the same conditions.[3]

[1] M. Pailer and P. Bergthaller, *Monatshefte*, **99**, 103 (1968).
[2] J. Grundy, B. G. James, and G. Pattenden, *Tetrahedron Letters*, 757 (1972).
[3] D. H. Evans and R. B. Greenwald, *Org. Prep. Proc. Int.*, **4**, 75 (1972).

**Dimethyl sulfide, 2,** 156–157.

*Ortho-Alkylation of aromatic amines.* Gassman and Gruetzmacher[1] have described a simple, high-yield procedure for specific *ortho*-methylation of anilines. The aniline (1) is treated with 1 eq. of *t*-butyl hypochlorite (**1,** 90; **2,** 50; **3,** 38) or of N-chlorosuccinimide (**1,** 139; **2,** 69–70; **3,** 36, 142) in methylene chloride at −65°. Then dimethyl sulfide (3 eq.) is added at the same temperature. After 40 min., sodium methoxide (1.2 eq.) in

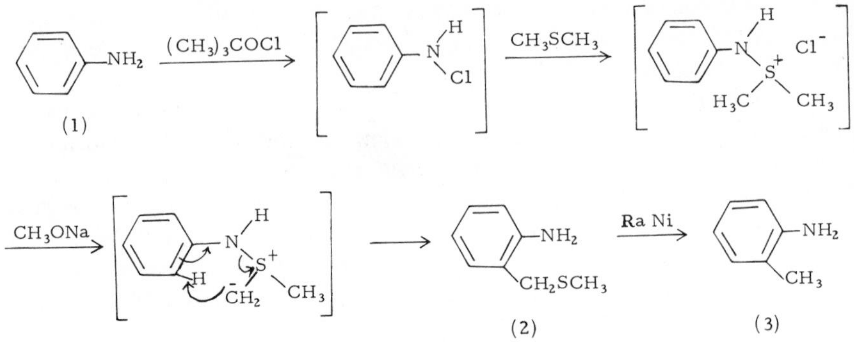

methanol is added. 2-Thiomethoxymethylanilines (2) are obtained in good yield (about 50–90%). Raney nickel desulfurization gives the *o*-toluidine derivatives (3) in 60–88% yield.

The process is not limited to methylation. Thus reaction of aniline (1) with tetramethylene sulfide in the general procedure followed by Raney nickel desulfurization gives *ortho-n*-butylaniline (5) in satisfactory overall yield.

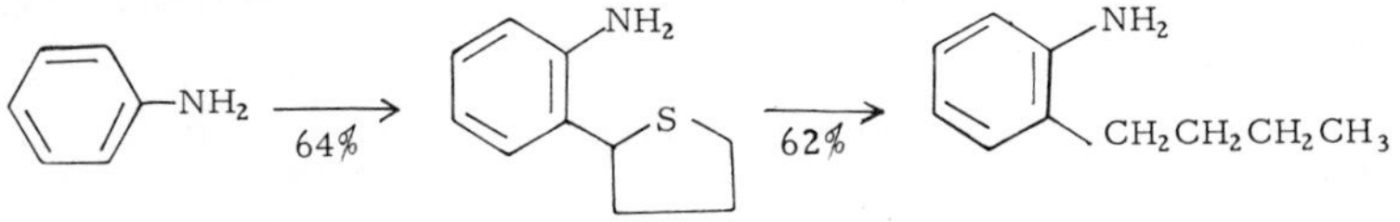

*Indole synthesis.* Gassman and van Bergen[2] have used a modification of the above-mentioned procedure for synthesis of 2-substituted indoles from anilines. The aniline is treated as above with a chlorinating reagent at −65° and then an equivalent of methylthio-2-propanone[3] at the same temperature. An equivalent of a base (usually triethylamine) is added. Workup affords the indole derivative (2) in 60–70% yield. The thiomethyl group is removed with Raney nickel (> 70% yield). The keto sulfide can be varied; thus use of methyl phenacyl sulfide [$CH_3SCH_2C(=O)C_6H_5$][4] in the synthesis leads to 2-phenylindoles.

Indole itself (or indoles carrying substituents in the aromatic ring) can be synthesized by the route just described[5] if the keto sulfide is replaced by methylthioacetaldehyde

see    P 78

(1)                        (2)                              (3)

[CH$_3$SCH$_2$(C=O)H] or the corresponding dimethyl ketal.[6] Thus treatment of aniline
(1) with the reagents indicated in the formulations gave (4) in 57% yield. Brief treatment of (4) with hydrochloric acid gave the thiomethyl indole (5) in 97% yield.

(1)                           (4)                         (5)

[1]P. G. Gassman and G. Gruetzmacher, *Am. Soc.*, **95**, 588 (1973).
[2]P. G. Gassman and T. J. van Bergen, *ibid.*, **95**, 590 (1973).
[3]C. K. Bradsher, F. C. Brown, and R. J. Grantham, *ibid.*, **76**, 114 (1954).
[4]V. Prelog, V. Hahn, H. Brauchli, and H. C. Beyerman, *Helv.*, **27**, 1209 (1944).
[5]P. G. Gassman and T. J. van Bergen, *Am. Soc.*, **95**, 591 (1973).
[6]E. H. Wick, T. Yamanishi, H. C. Wertheimer, Y. E. Hoff, B. E. Proctor, and S. A. Goldblith,
*J. Agr. Food. Chem.*, **9**, 289 (1961).

**Dimethyl sulfide–Borane,** $(CH_3)_2S \cdot BH_3$. Mol. wt. 75.98, m.p. $-40$ to $-38°$. Stable
liquid at room temperature under $N_2$. Supplier: Aldrich.
    The reagent is prepared by condensing equal gas volumes of $(CH_3)_2S$ and $B_2H_6$ in
a tensimeter at room temperature.[1]
    Diborane is not stable at room temperature and is usually generated *in situ*. It is
available from Alfa Inorganics but only in a very dilute solution in THF. Adams *et al.*[2]
report that dimethyl sulfide–borane provides a useful substitute for diborane. It is less
reactive than diborane, but can be used for most hydroboration and reduction applications.

[1]A. B. Burg and R. I. Wagner, *Am. Soc.*, **76**, 3307 (1954).
[2]L. M. Braun, R. A. Braun, H. R. Crissman, M. Opperman, and R. M. Adams, *J. Org.*, **36**,
2388 (1971).

**Dimethyl sulfide–Chlorine,** $(CH_3)_2S^+ClCl^-$. Mol. wt. 133.05.
    *Oxidation of primary and secondary alcohols to carbonyl compounds.*[1] Treatment
of dimethyl sulfide in carbon tetrachloride at 0° with 1 eq. of chlorine in the same solvent
results in rapid formation of the partially insoluble complex (1). This complex has been
used by Corey and Kim for oxidation of primary and secondary alcohols. In a typical

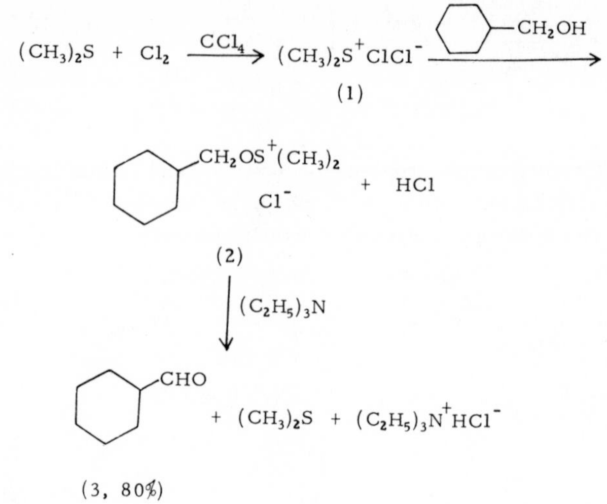

$$(CH_3)_2S + Cl_2 \xrightarrow{CCl_4} (CH_3)_2S^+ClCl^- \underset{(1)}{\underbrace{\hspace{1cm}}} \text{[cyclohexyl]}-CH_2OH \longrightarrow$$

[cyclohexyl]$-CH_2OS^+(CH_3)_2$  $Cl^-$  $+$  HCl

(2)

$\downarrow (C_2H_5)_3N$

[cyclohexyl]$-CHO$

$+ (CH_3)_2S + (C_2H_5)_3N^+HCl^-$

(3, 80%)

procedure the complex is cooled to $-25°$ (Dry Ice–$CCl_4$ bath) and treated with cyclo-hexylcarbinol for 2 hr. with stirring at this temperature. The sulfoxonium complex (2) is formed; this on treatment with 2 eq. of triethylamine is converted into cyclohexyl-carboxaldehyde (3, 80% yield) and dimethyl sulfide.

This oxidation can be effected in higher yield (93%) using the complex of N-chloro-succinimide and dimethyl sulfide (this volume).

See also **Dimethyl sulfoxide–Chlorine**, this volume.

[1]E. J. Corey and C. U. Kim, *Am. Soc.*, **94**, 7586 (1972).

**Dimethyl sulfoxide, 1,** 296–310; **2,** 157–166; **3,** 119–123.

*Solvent Effect*

**Dehydrobromination.** Grimshaw *et al.*[1] report considerable difficulty in dehydro-bromination of the $\beta$-bromonorpinone (3) to apoverbenone (4, 6,6-dimethylnorpin-3-ene-2-one) by the usual basic reagents. Under these conditions the $\beta$-isomer (3) is converted into a mixture of the $\alpha$-(2) and $\beta$-(3) compounds. The elimination reaction is difficult for two reasons. The basic medium is not favorable to a *trans*-diaxial arrange-ment of the departing groups. In addition the *gem*-dimethyl system hinders approach of the base. Thus reaction at 140–150 in DMF or DMA containing lithium carbonate and

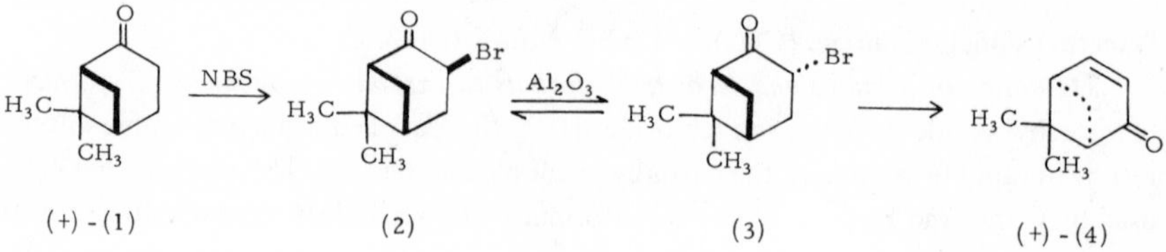

$$\underset{(+)-(1)}{\text{[structure]}} \xrightarrow{NBS} \underset{(2)}{\text{[structure]}} \underset{Al_2O_3}{\rightleftharpoons} \underset{(3)}{\text{[structure]}} \longrightarrow \underset{(+)-(4)}{\text{[structure]}}$$

lithium bromide (**2**, 245–246) gave mainly a mixture of (2) and (4), which are difficult to separate. The desired dehydrobromination to (4) was achieved in 77 % yield by use of DMSO containing the lithium salts. Dimethyl sulfoxide is effective because of preferential solvation of cations, thus allowing the bromide ion to function as a stronger Lewis base.

### Reactions

*Dehydration* (**1**, 301–303). Selective dehydration of 2,4-di(2-hydroxy-2-propyl)-cyclohexene-1 (1) to the diene (2) can be achieved by heating the diol in DMSO at 130°.[2] The reaction is apparently acid catalyzed because it is completely inhibited by addition

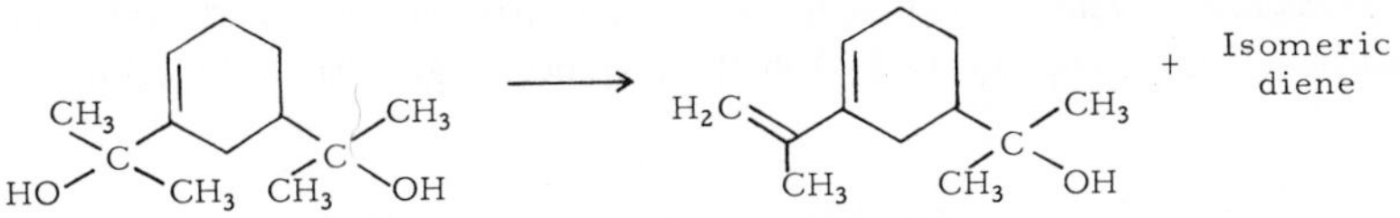

of 1,5-diazabicyclo[4.3.0]nonene-5 (**1**, 189–190). In this case dehydration by heating with pyridine-treated alumina[3] was found to be difficult to reproduce.

*Reaction with epoxides.*[4] Dimethyl sulfoxide and a strong acid (2,4,6-trinitro-benzenesulfonic acid, TNBSA, was used) reacts with epoxides (1) to form alkoxysulfonium salts (2) in good yield. Since hydrolysis of alkoxysulfonium salts proceeds exclusively by attack at sulfur, the reaction provides a stereospecific synthesis of 1,2-glycols. Thus the reaction of *cis*- and *trans*-9,10-epoxystearic acids with DMSO–

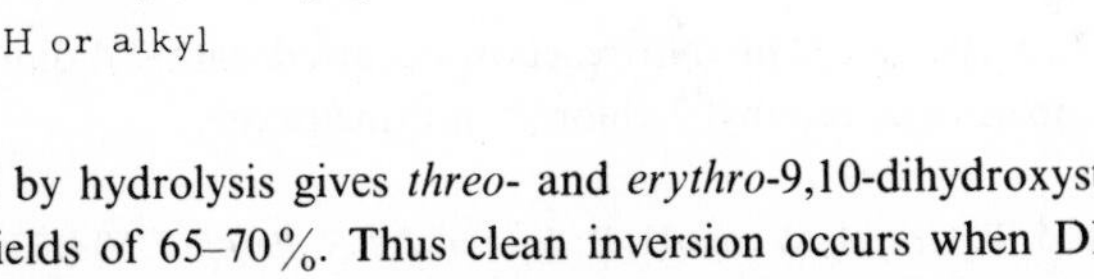

R = H, alkyl, or $C_6H_5$

R' = H or alkyl

TNBSA followed by hydrolysis gives *threo*- and *erythro*-9,10-dihydroxystearic acids, respectively, in yields of 65–70 %. Thus clean inversion occurs when DMSO reacts with the oxide ring.

*Oxidation of alkyl 4-nitrobenzenesulfonates.*[5] The oxidation of primary alcohols to aldehydes by dimethyl sulfoxide requires temperatures around 100° or introduction of acids or heavy metal ions. In a new procedure the alcohol is converted into the 4-nitrobenzenesulfonate ester by reaction with 4-nitrobenzenesulfonyl chloride and

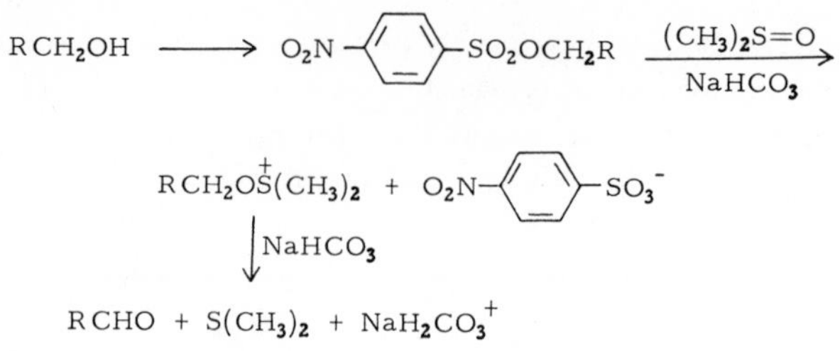

$$RCH_2OH \longrightarrow O_2N-\langle\bigcirc\rangle-SO_2OCH_2R \xrightarrow[NaHCO_3]{(CH_3)_2S=O}$$

$$RCH_2O\overset{+}{S}(CH_3)_2 + O_2N-\langle\bigcirc\rangle-SO_3^-$$

$$\Big\downarrow NaHCO_3$$

$$RCHO + S(CH_3)_2 + NaH_2CO_3^+$$

pyridine and then oxidized by dimethyl sulfoxide in the presence of sodium hydroger carbonate. The reaction is carried out at room temperature and requires about 100- 200 hr. of stirring.

*Oxidation of thiocarbonyl compounds.*[6] Thiocarbonyl compounds are oxidized to carbonyl compounds by DMSO in the presence of acid catalysts (sulfuric acid

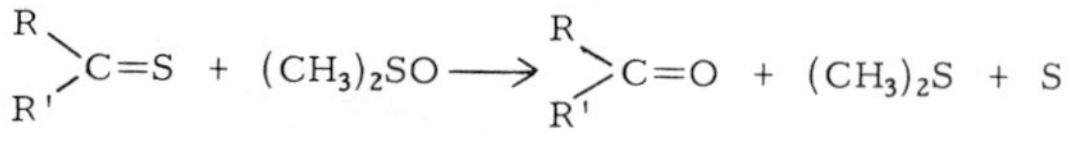

$$\underset{R'}{\overset{R}{>}}C=S + (CH_3)_2SO \longrightarrow \underset{R'}{\overset{R}{>}}C=O + (CH_3)_2S + S$$

*p*-TsOH, trifluoroacetic acid) or of boron trifluoride. Yields are in the range 65–90%. The oxidation is also applicable to selenocarbonyl compounds.

*Reaction with 6,7-disubstituted 4-chloroquinoline-3-carboxylates.*[7] Quinolines of this type (1) are converted into the corresponding 4-hydroxyquinolines (2) when heated

$$\text{DMSO, } 100^\circ, \text{ 2 hrs.}$$
$$70\text{-}90\%$$

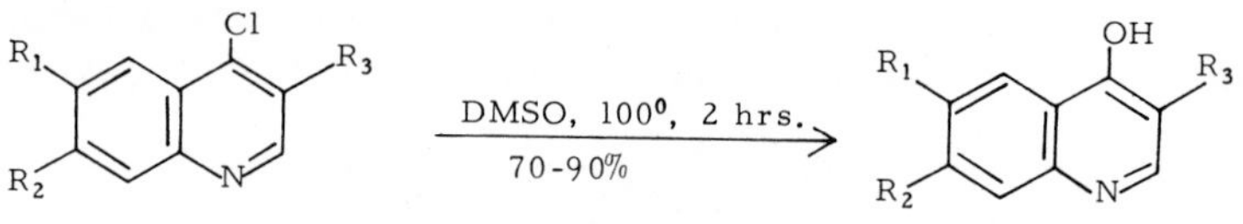

(1)                    (2)

$R_1$ = alkoxy, alkyl
$R_2$ = alkoxy
$R_3$ = H, COOR

with DMSO at 100° for 2 hr. No reaction occurred with 2,4-dinitrochlorobenzene, ethyl 2-chlorobenzoate, or ethyl 2-chloro-5-nitrobenzoate.

[1]J. Grimshaw, J. T. Grimshaw, and H. R. Juneja, *J.C.S. Perkin I*, 50 (1972).
[2]P. S. Wharton, C. E. Sundin, D. W. Johnson, and H. C. Kluender, *J. Org.*, **37**, 34 (1972).
[3]E. von Rudloff, *Canad. J. Chem.*, **39**, 1860 (1961).
[4]M. A. Khuddus and D. Swern, *Tetrahedron Letters*, 411 (1971).
[5]C. H. Snyder, P. L. Gendler, and H.-H. Chang, *Synthesis*, 655 (1971).
[6]M. Mikolajczyk and J. Luczak, *Chem. Ind.*, 76 (1972).
[7]N. D. Harris, *Synthesis*, 625 (1972).

**Dimethyl sulfoxide-derived reagent (a). Sodium methylsulfinylmethylide (Dimsylsodium),** 1, 310–313; 2, 166–169; 3, 123–124.

*Octamethylnaphthalene.*[1] The definitive paper on the preparation of octamethyl-naphthalene (2, 168, ref. 11b) has now been published. The conversion of (3) to (4) with dimsylsodium is essentially quantitative. Isobutyric acid has been identified as a second product. Use of other bases (sodium methoxide in methanol, sodium hydride) was unsuccessful.

*1-Methoxy-4,5;10,11-bis(tetramethylene)-6,8-didehydro[13]annulenyl  anion* (2). Le Goff and Sondheimer[2] used dimsylsodium in DMSO to abstract hydrogen from (1)

to form the 13-membered carbocyclic system (2). The NMR spectrum of (2) demonstrates the aromaticity of this system (it contains $4n + 2$ π-electrons).

*Preparation of ethers.*[3] Treatment of secondary and tertiary alcohols with dimsyl-sodium in DMSO followed by alkylation with a dialkyl sulfate gives ethers in 60–90% yield. The presence of ester and tertiary amino groups and of double bonds does not interfere with the reaction. However, alkylation of α- or β-hydroxyketones was not successful by this procedure.

*Synthesis of olefins.*[4] In a new synthesis of olefins, a nonenolizable aldehyde or ketone, for example benzophenone, is treated with dimsyllithium in THF; o-phenylene phosphorochloridite (2, 321–322) is added to the solution at $-80°$. The reaction is

allowed to warm to room temperature and finally is refluxed for 1 hr. 1,1-Diphenyl-ethylene is obtained in 91% yield. Yields are low with enolizable carbonyl compounds.

*Tschugaeff reaction* (3, 123). Meurling et al.[5] have prepared xanthates in good yield by treating alcohols with an equivalent amount of dimsylsodium in DMSO followed by carbon disulfide and an alkylating agent. The decomposition temperature of the xanthates is lowered when pyrolysis is carried out in DMSO.

*Cyclization.* In a synthesis of ($\pm$)-$\beta$-copaene (3) Corey and Watt[6] effected cyclization of (1) to the tricyclic ketone (2) with this reagent. Use of potassium *t*-butoxide or 1,5-diazabicyclo[4.3.0]nonene-5 proved unsuccessful.

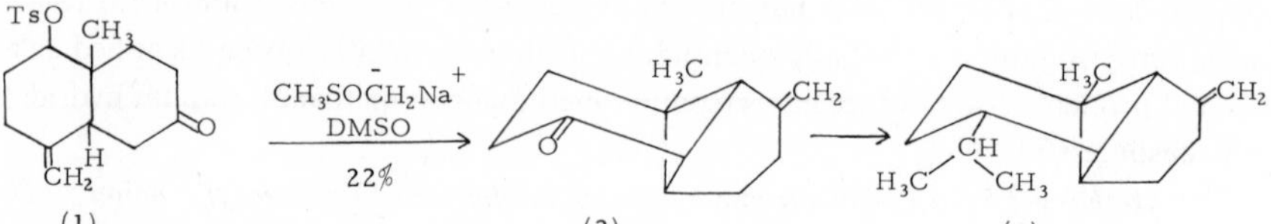

(1)                    (2)                    (3)

[1]H. Hart and A. Oku, *J. Org.*, **37**, 4269 (1972).
[2]E. Le Goff and F. Sondheimer, *Angew. Chem., internat. Ed.*, **11**, 926 (1972).
[3]B. Sjöberg and K. Sjöberg, *Acta Chem. Scand.*, **26**, 275 (1972).
[4]I. Kuwajima and M. Uchida, *Tetrahedron Letters*, 649 (1972).
[5]P. Meurling, K. Sjöberg, and B. Sjöberg, *Acta Chem. Scand.*, **26**, 279 (1972).
[6]E. J. Corey and D. S. Watt, *Am. Soc.*, **95**, 2303 (1973).

**Dimethyl sulfoxide-derived reagent (b). Dimethylsulfonium methylide, 1,** 314–315; **2,** 169–171; **3,** 124–125.

*Synthesis of furanes.*[1] 3- or 3,4-Substituted furanes can be prepared in good yield by the reaction of the *n*-butylthiomethylene derivatives[2] of ketones with dimethylsulfonium methylide[3] (equation I).

I.

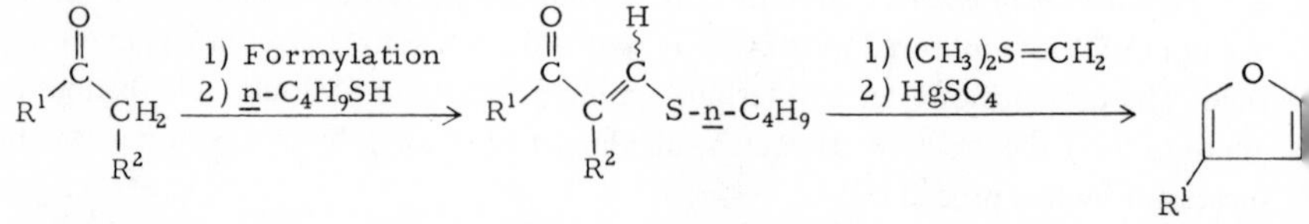

The reaction with dimethylsulfonium methylide probably proceeds through the intermediates (a) and (b). Intermediates of type (b) have been detected spectrally.

(a)                    (b)

Examples:

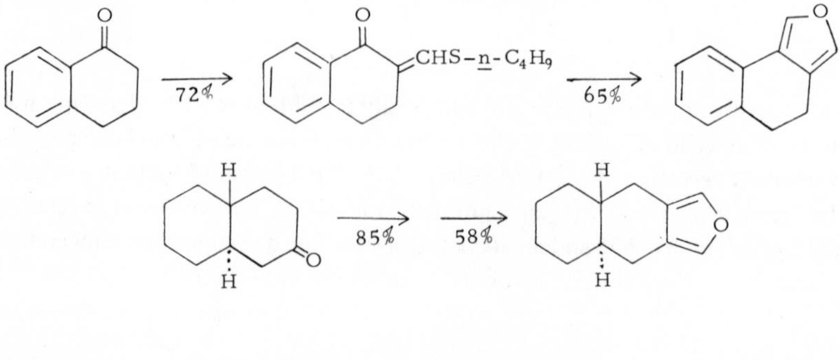

In most cases the product furane is accompanied by unreacted $n$-butylthiomethylene derivative (*ca.* 20–30% yield).

The reaction was used to synthesize the naturally occurring furane perillene (2) from the ketone (1).

(1)                    (2)

*N-Methylation of indoles.*[4] Indoles can be selectively N-methylated by dimethyl-sulfonium methylide in THF at room temperature.

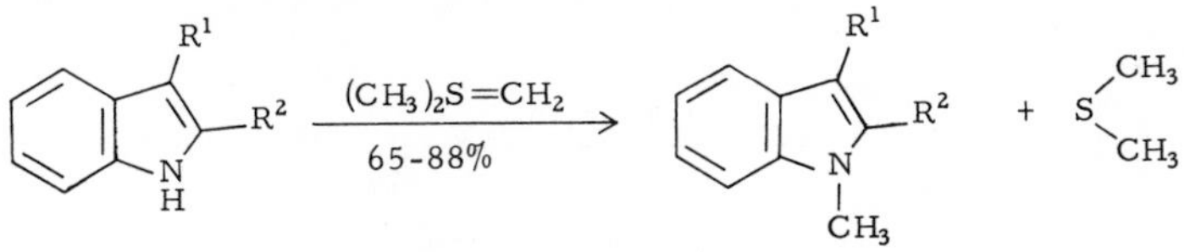

[1]M. A. Garst and T. A. Spencer, *Am. Soc.*, **95**, 250 (1973).
[2]See *n*-**Butyl mercaptan**, **2**, 53–54; this volume.
[3]Prepared by the reaction of *n*-butyllithium in hexane with trimethylsulfonium fluoroborate [H. Teichmann and G. Hilgetag, *Ber.*, **96**, 1454 (1963)] in dimethoxyethane (DME). In this case use of trimethylsulfonium iodide is unsuitable.
[4]P. Bravo, G. Gaudiano, A. Umani-Ronchi, *Gazz. Chim. Ital.*, **100**, 652 (1970).

**Dimethyl sulfoxide-derived reagent (c). Dimethyloxosulfonium methylide, 1,** 315–318; **2,** 171–173; **3,** 125–127.

*Correction* (**1**, 317). The structure of (5) should be

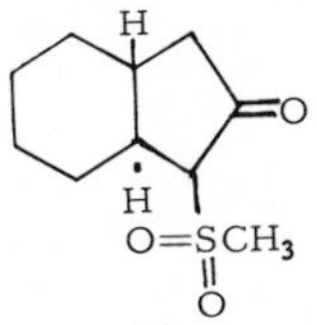

(5)

*Adamantane-2-carboxylic acid.*[1] In a new, convenient synthesis of adamantane-2-carboxylic acid (4), adamantanone (1) is treated with dimethyloxosulfonium methylide to give 2-methyleneadamantane epoxide (2). This is rearranged by treatment with boron trifluoride etherate[2] to the aldehyde (3, unstable). Oxidation with Jones reagent gives (4) in an overall yield of about 70% from (1).

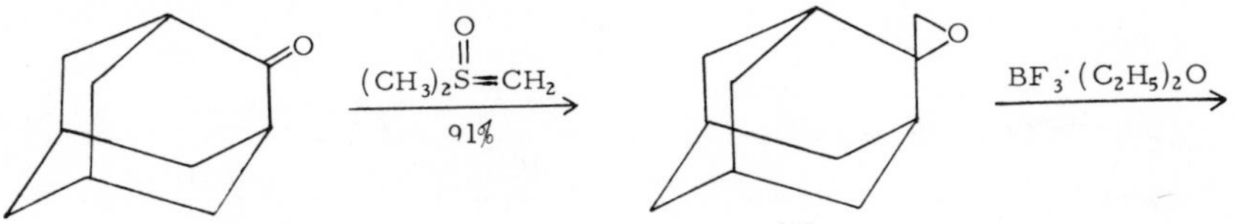

(1)                    (2)

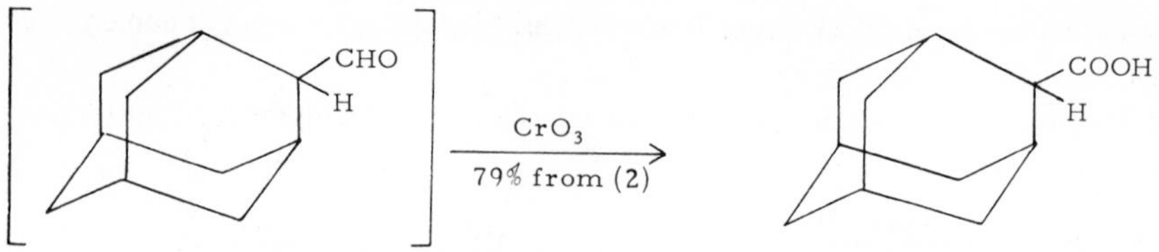

$$\xrightarrow[\text{79\% from (2)}]{\text{CrO}_3}$$

(3)                                    (4)

***4,5-Benzohomotropone.***[3] 4,5-Benzohomotropone (2) has been prepared in unspecified yield by the reaction of 4,5-benzotropone (1) with this reagent in THF. Further homologation of (2) was not successful.

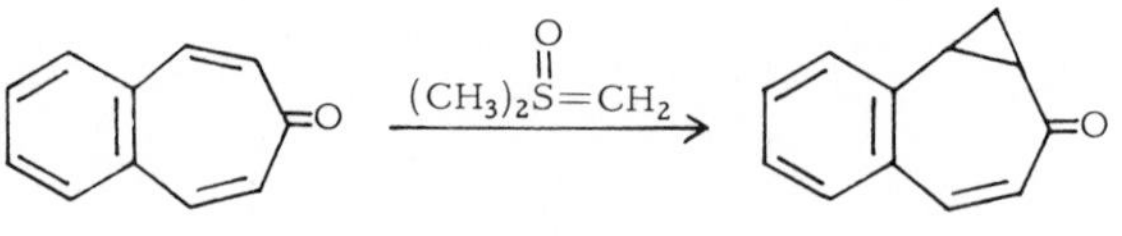

(1)                              (2)

***Epoxyolefins.***[4] Epoxyolefins can be obtained in fair yield by addition of aldehydes in DMSO dropwise to the reagent. Polymers are also formed by aldol-type condensation. A mixture of unseparable diastereoisomeric epoxides is formed. The reaction of citronellal (1) to give 8,9-epoxy-2,6-dimethyl-2-nonene (2a, 2b) is formulated.

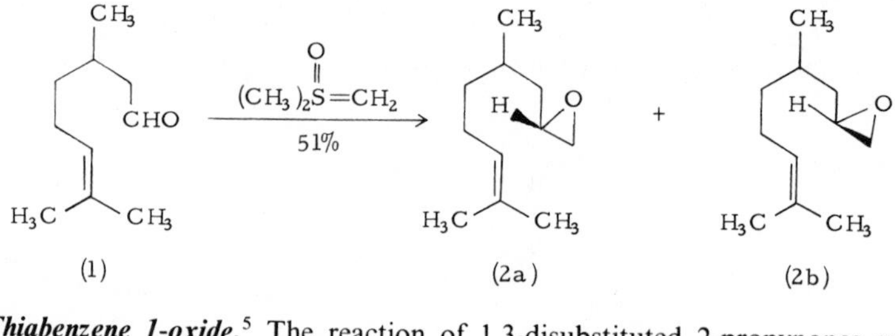

(1)                    (2a)            +            (2b)

***Thiabenzene 1-oxide.***[5] The reaction of 1,3-disubstituted 2-propynones such as 1,3-diphenyl-2-propyn-1-one (1) with dimethyloxosulfonium methylide in DMSO at 16.5° gives 1-methyl-3,5-disubstituted thiabenzene 1-oxides; in the case of (1) the product is 1-methyl-3,5-diphenylthiabenzene 1-oxide (3), obtained in 76% yield. If the reaction is carried out in THF–DMSO at −8°, the intermediate (2) can be isolated; it probably arises by Michael addition of the reagent to (1).

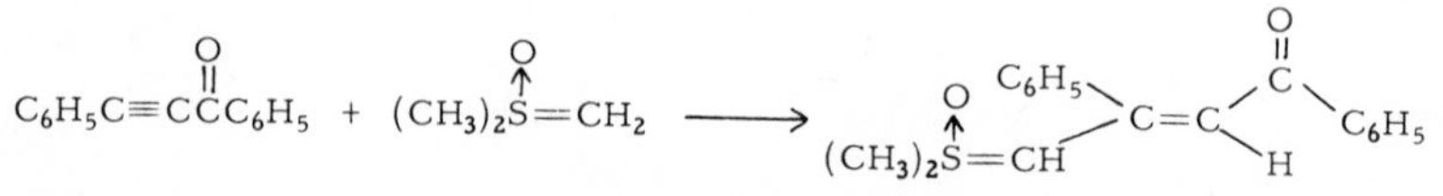

(1)                                    (2) isolated in 63% yield

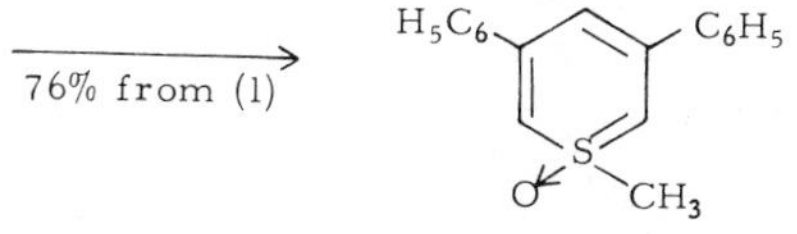

$$\xrightarrow[\text{76\% from (1)}]{}$$

(3)

[1]D. Fărcaşiu, *Synthesis*, 615 (1972).
[2]H. O. House, *Am. Soc.*, **77**, 3075 (1955).
[3]Y. Sugimura, N. Soma, and Y. Kishida, *Tetrahedron Letters*, 91 (1971).
[4]B. C. Clark, Jr., and D. J. Goldsmith, *Org. Prep. Proc. Int.*, **4**, 113 (1972).
[5]A. G. Hortmann and R. L. Harris, *Am. Soc.*, **93**, 2471 (1971).

**Dimethyl sulfoxide–Acetic anhydride**, **1**, 305; **2**, 163–165; **3**, 121–122.
Polyporic acid (1) is converted in 95% yield into pulvinic acid dilactone (2) on

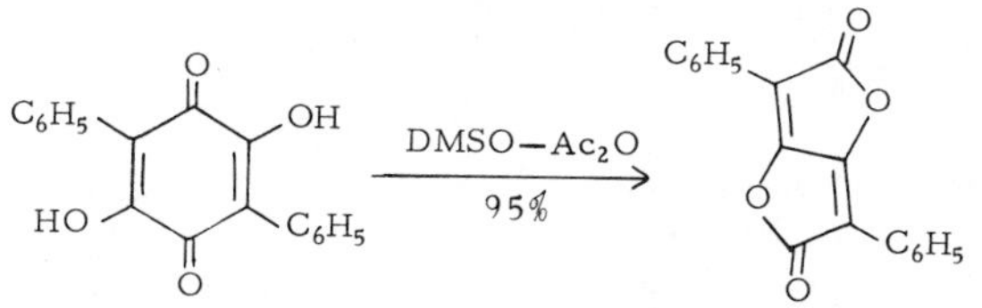

(1)                              (2)

treatment with DMSO–acetic anhydride (2:1) at 60° for 15 min.[1] This reaction has
been demonstrated *in vivo*.[2] The original paper should be consulted for a plausible
mechanism of this oxidation and for a discussion of possible biosynthetic implications.

[1]R. J. Wikholm and H. W. Moore, *Am. Soc.*, **94**, 6152 (1972).
[2]W. S. G. Maass and A. C. Neish, *Canad. J. Bot.*, **45**, 59 (1967).

**Dimethyl sulfoxide–N-Bromosuccinimide.**
*Methylene acetals.*[1] Methylene acetals are obtained in good yield by the reaction
of alcohols and diols with DMSO and N-bromosuccinimide or N-chlorosuccinimide.
For example, cyclohexanol (1) in dry DMSO containing 1–2 molar eq. of NBS is heated
at 50° with stirring overnight. After workup, dicyclohexyloxymethane (2) is obtained
in 86% yield.

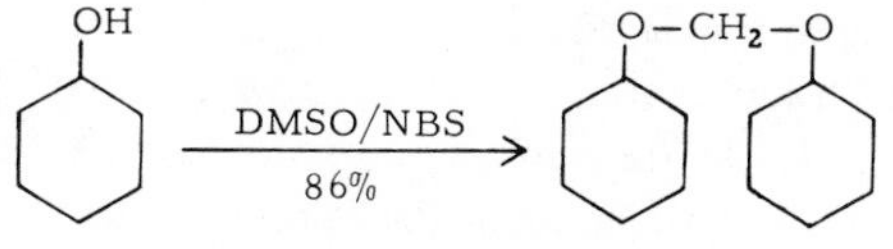

(1)                              (2)

Other examples:

$$CH_3OH \longrightarrow CH_3OCH_2OCH_3 (quant.)$$

(cyclohexane-1,2-diol with two OH groups) $\xrightarrow{62\%}$ (cyclohexane fused dioxolane, O–CH$_2$–O)

(cyclohexane-1,2-diol, OH and ··OH) $\xrightarrow{70\%}$ (cyclohexane fused O–CH$_2$, O–CH$_2$ dioxane)

Both hydrogen atoms of the methylene group originate from DMSO; if the reaction is carried out with DMSO-$d_6$, the resulting acetals have a CD$_2$ unit.

[1] S. Hanessian, G. Yang-Chung, P. Lavallee, and A. G. Pernet, *Am. Soc.*, **94**, 8929 (1972).

**Dimethyl sulfoxide–Chlorine, $(CH_3)_2SO$–$Cl_2$.**

*Oxidation of primary and secondary alcohols.*[1] Addition of dimethyl sulfoxide to a solution of chlorine in methylene chloride at $-45°$ produces an unstable complex

$$(CH_3)_2SO + Cl_2 \xrightarrow{-45°} \underset{CH_3}{\overset{CH_3}{\phantom{.}}} \hspace{-0.3em} S^+ \hspace{-0.3em} \begin{matrix} O \\ Cl \end{matrix} \quad Cl^-$$

(1)

with the probable formula (1). This complex, like the complex of dimethyl sulfide–chlorine (this volume) oxidizes primary and secondary alcohols to carbonyl compounds. The reaction is carried out with about 100% excess of the complex at $-45°$ (2.4 hr.) followed by addition of excess triethylamine. Yields are in the range 94–98%.

$$R^1R^2CHOH \xrightarrow[-45°]{(CH_3)_2SO-Cl_2} R^1R^2CH-O\overset{O}{\overset{||}{S}}{}^+(CH_3)_2Cl^-$$

$$\Big\downarrow \begin{matrix} N(C_2H_5)_3 \\ -45 \text{ to } -10° \end{matrix}$$

$$R^1\overset{O}{\overset{||}{C}}R^2 + (C_2H_5)_3NHCl + (CH_3)_2SO$$

The reagent has one drawback: It reacts rapidly with olefins to form *vic*-dichlorides. Thus cyclooctene is converted by the reagent into 1,2-dichlorocyclooctane. Therefore the reagent is not suitable for oxidation of unsaturated alcohols.

[1] E. J. Corey and C. U. Kim, *Tetrahedron Letters*, 919 (1973).

**Dimethyl sulfoxide–Sulfur trioxide, 2**, 165–166.

The reagent of Parikh and Doering was used successfully for oxidation of tricarbonyl-(7-norbornadienol)iron (1) to tricarbonyl(7-norbornadienone)iron (2).[1] In this case use of chromium trioxide–pyridine or of the Pfitzner-Moffatt procedure resulted in fragmentation to benzaldehyde and benzene.

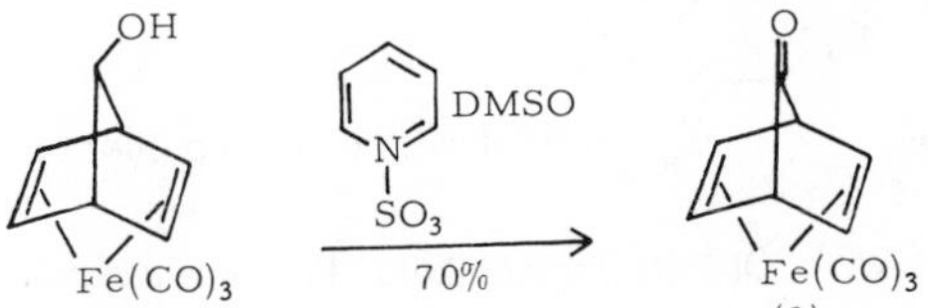

Goh and Harvey[2] effected the oxidation of the *cis*-diol (3) to the *o*-quinone (4) efficiently by use of the sulfur trioxide–pyridine complex in DMSO–triethylamine. Use of DMSO–acetic anhydride (**1**, 305; **2**, 163–165; **3**, 121–122) gave fair, but erratic yields of the quinones. Other reagents were completely unsuccessful. Note that oxidation of aromatic 1,2-diols of this type to 1,2-diones is complicated by the relative facility of both dehydration and oxidative cleavage of the carbon–carbon bond.

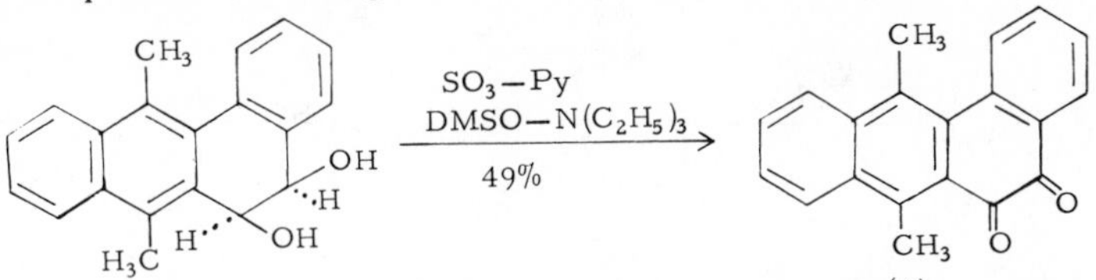

*Oxidation of carbohydrate derivatives.* Oxidation of partially acetylated carbohydrates with the reagent of Parikh and Doering leads to both oxidation and elimination of the elements of acetic acid, and thus provides a high-yield route to novel unsaturated carbohydrates.[3] Apparently the oxidation step is followed by spontaneous elimination of a β-acetoxy group. Hydroxy groups protected by benzyl or trimethylsilyl groups are not eliminated.

Examples:

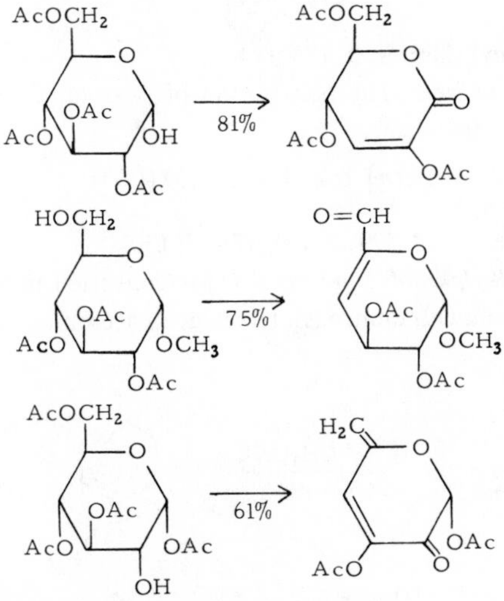

[1]J. M. Landesberg and J. Sieczkowski, *Am. Soc.*, **93**, 972 (1971).
[2]S. H. Goh and R. G. Harvey, *ibid.*, **95**, 242 (1973).
[3]G. M. Cree, D. W. Mackie, and A. S. Perlin, *Canad. J. Chem.*, **47**, 511 (1969).

**2,4-Dimethylthiazole,**    H$_3$C structure    . Mol. wt. 113.18, b.p. 144–145°/719 mm. Supplier: Aldrich.

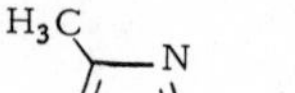

*Aldehyde synthesis.*[1] 2,4-Dimethylthiazole (1) when treated with *n*-butyllithium in dry ThF and then with benzyl chloride yields 2-(2-phenylethyl)-4-methylthiazole (2) in 94% yield. This product is then alkylated with trimethyloxonium fluoroborate and reduced with sodium borohydride to give 2-(2-phenylethyl)-3,4-dimethylthiazolidine (3) in high yield. The final step is hydrolysis using mercuric chloride as catalyst; 3-phenyl-propanal (4) is obtained in 60% yield.

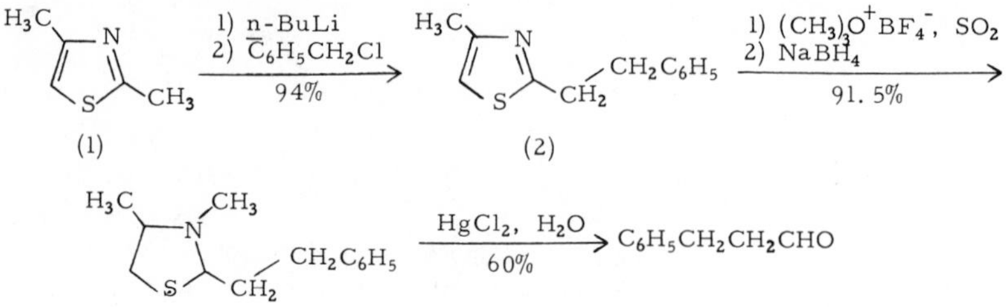

The sequence provides an alternative to the aldehyde synthesis of Meyers using 2,4,4,6-tetramethyl-5,6-dihydro-1,3-(4H)-oxazine (**3**, 280–282; this volume) but has the advantage that strongly acidic conditions are not necessary.

[1] L. J. Altman and S. L. Richheimer, *Tetrahedron Letters*, 4709 (1971).

**Dimethylthiocarbamyl chloride, 2,** 173–174.

*Phenols → thiophenols.* The preparation of 2-naphthalenethiol from β-naphthol has been checked and published.[1]

[1] M. S. Newman and F. W. Hetzel, *Org. Syn.,* **51,** 139 (1971).

**Dinitrogen tetroxide, 1,** 324–329; **2,** 175–176; **3,** 130.

*Nitration of alkylphenols.*[1] *p*-Cresol (1) is converted into 2,6-dinitro-*p*-cresol in 66% yield by reaction with dinitrogen tetroxide in acetonitrile (25–30°).

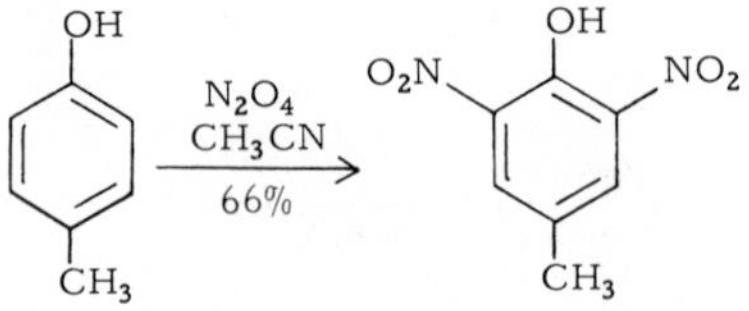

*Alkyl nitrates.*[2] Primary amines are deaminated by dinitrogen tetroxide to give alkyl nitrates. The reaction is strongly dependent on the solvent. THF and other ethers which as Lewis bases can form complexes with $N_2O_4$ are preferred. Deamination occurs with a high degree of retention of configuration and thus involves an ionic mechanism.

$$RNH_2 \ + \ N_2O_4 \ \xrightarrow{\text{THF, } -60^0} \ RONO_2 \ + \ N_2 \ + \ H_2O$$

[1]E. V. P. Tao and C. F. Christie, Jr., *Org. Prep. Proc. Int.*, **4**, 300 (1972).
[2]F. Wudl and T. B. K. Lee, *Am. Soc.*, **93**, 271 (1971).

**Diperoxo-oxohexamethylphosphoramidomolybdenum(VI)**,  $MoO_5 \cdot HMPT$.  Mol.  wt.
371.14.

*Preparation.* Mimoun et al.[1] prepared this complex of $MoO_5$ and HMPT by dissolving molybdenum trioxide ($MoO_3$, ROC/RIC, 50 g.) at 40° in 30 % $H_2O_2$ (250 cc.). The yellow solution so obtained is cooled to 10° and HMPT (62.3 g.) is added with efficient stirring. A yellow precipitate is formed immediately and is collected on a filter and washed several times with ether. The dried material is crystallized twice from methanol; it is obtained as yellow crystals in 84 % yield. The French chemists showed that the complex has the structure shown in formula (1).

(1)

The reagent should be stored at low temperatures; an explosion has been observed when material was stored at room temperature for a month.[2]

The water-free complex, $MoO_5 \cdot HMPT$, can be obtained by dehydration of (1) with phosphorous pentoxide *in vacuo*.[3]

*Epoxidation.* Mimoun et al.[3] report that the $MoO_5 \cdot HMPT$ complex reacts with olefins to form epoxides in high yield. Alkyl substituents on the double bond increase the rate of epoxidation. Aprotic solvents also enhance the rate; highest rates are observed in methylene chloride. The reaction is very slow in DMF or THF. The epoxidation is stereospecific with retention of configuration of the olefin. Thus *cis*-butene-2 is converted into *cis*-2,3-epoxybutane and *trans*-butene-2 into *trans*-2,3-epoxybutane. The French chemists proposed the mechanism shown in scheme I.

Scheme I:

Sharpless *et al.*[4] have confirmed this mechanism in part by labeling experimen which demonstrated that the epoxide oxygen is derived exclusively from the perox ligands of the complex and not from the oxo oxygen. However, the reactivity of th molybdenum complex toward olefins closely parallels that of peracids, for which three-membered cyclic transition state is favored.[5]

*Hydroxamic acids.*[6] Oxidation of N-trimethylsilylamides (1, prepared by th reaction of a secondary amide with hexamethyldisilazane) with the $MoO_5 \cdot HMP$ complex in methylene chloride at room temperature for several hours up to sever days affords dioxomolybdenum complexes (2) in moderate yields (15–40%). The fre hydroxamic acids (3) are liberated by treatment of (2) with ethylenediaminetetraaceti acid (EDTA).

$$R^1CON\begin{subarray}{l}R^2\\ Si(CH_3)_3\end{subarray} \xrightarrow{MoO_5 \cdot HMPT} \left(\begin{array}{c} R^1C\diagdown^O \\ | \diagdown \\ R^2N \diagdown_O \diagup \end{array}\right)_2 MoO_2 \xrightarrow{EDTA} R^1CON\begin{subarray}{l}R^2\\ OH\end{subarray}$$

(1)                                    (2)                                    (3)

[1]H. Mimoun, I. S. de Roch, and L. Sajus, *Bull soc.*, 1481 (1969).
[2]P. L. Fuchs, personal communication.
[3]H. Mimoun, I. S. de Roch, and L. Sajus, *Tetrahedron*, **26**, 37 (1970).
[4]K. B. Sharpless, J. M. Townsend, and D. R. Williams, *Am. Soc.*, **94**, 295 (1972).
[5]K. D. Bingham, G. D. Meakins, and G. H. Whitham, *Chem. Commun.*, 445 (1966).
[6]S. A. Matlin and P. G. Sammes, *J.C.S. Chem. Comm.*, 1222 (1972).

**N,N-Diphenylcarbamoyl pyridinium chloride,** $(C_6H_5)_2NCON^+$ ⬡ $Cl^-$ . Mol. wt. 310.7 m.p. 105–110°.

*Preparation.*[1]

*Mixed anhydrides.*[2] Mixed carbamic anhydrides can be prepared in aqueous solu tion by treatment of a carboxylic acid with the reagent and triethylamine:

$$RCOOH + (C_6H_5)_2NCON^+ \bigcirc Cl^- \xrightarrow[60-90\%]{(C_2H_5)_3N} RCO_2CON(C_6H_5)_2$$

The anhydrides are stable, crystalline products.

[1]J. Herzog, *Ber.*, **40**, 1831 (1907).
[2]K. L. Shepard, *Chem. Commun.*, 928 (1971).

**Diphenyldiazomethane, 1**, 338–339.

*Addition to cyclooctyne*[1]:

$$\text{cyclooctyne} + \begin{subarray}{l}C_6H_5\\ C_6H_5\end{subarray}CN_2 \longrightarrow \text{product}$$

[1]G. Wittig and J. J. Hutchison, *Ann.*, **741**, 79 (1970).

**Diphenyldi(1,1,1,3,3,3-hexafluoro-2-phenyl-2-propoxy)sulfurane** (1). Mol. wt. 672.53, m.p. 107–109°.

*Preparation.*[1,2] The sulfurane (1) is prepared most conveniently in nearly quantitative yield by treatment of an ethereal solution of the potassium salt of hexafluoro-2-phenyl-2-propanol ($R_FOH$, prepared by the reaction of the alcohol with potassium metal) and diphenyl sulfide with chlorine at −78°. Removal of the potassium chloride by filtration and of the ether by evaporation under vacuum leaves the white, crystalline sulfurane (1), which can be crystallized from pentane. Moisture must be avoided

$$(C_6H_5)_2S \ + \ 2\ C_6H_5\underset{\underset{CF_3}{|}}{\overset{\overset{CF_3}{|}}{C}}OK \quad \xrightarrow[-2\,KCl]{Cl_2} \quad$$

(R$_F$OK)

(1)

throughout since the sulfurane is readily hydrolyzed to diphenyl sulfoxide and $R_FOH$. Unlike most sulfuranes, (1) is stable indefinitely at room temperature if protected from moisture. It decomposes only slowly in solution at room temperature.

*Dehydration of alcohols.*[3] A preliminary observation that the alkoxy ligands of (1) rapidly exchange with added alcohols suggested that (1) might be a dehydrating reagent. Indeed *t*-butanol is dehydrated by sulfurane (1) in chloroform at −50° within seconds to give isobutylene in quantitative yield. The dehydration probably proceeds by ligand exchange to give the sulfurane (2) followed by abstraction of a β-proton.

$$(CH_3)_3COH \ + \ (1) \ \rightleftharpoons \ (C_6H_5)_2\underset{\underset{OC(CF_3)_2C_6H_5}{|}}{\overset{\overset{OC(CH_3)_3}{|}}{S}} \quad + \quad C_6H_5\underset{\underset{CF_3}{|}}{\overset{\overset{CF_3}{|}}{C}}OH$$

(2)

(R$_F$OH)

$$(C_6H_5)_2SO \ + \ (CH_3)_2C{=}CH_2 \ + \ R_FOH$$

Sulfurane (1) even dehydrates tricyclopropylcarbinol (3) to the olefin (4); in this case use of sulfuric acid or phosphorus pentoxide fails to yield any of the olefin. The method for isolation of (4) involved removal of the acidic $R_FOH$ with sodium hydroxide;

removal of the solvent chloroform by distillation; and removal of diphenyl sulfoxide by flash distillation at $10^{-2}$ mm.

Tertiary alcohols are dehydrated instantaneously by sulfurane (1) at room temperature. Most secondary alcohols are also dehydrated rapidly at room temperature. There is evidence for a preferred *trans*-diaxial disposition of leaving groups. Thus *cis*-4-*t*-butylcyclohexanol is converted into 4-*t*-butylcyclohexene at least 150 times more rapidly than the *trans*-isomer. Primary alcohols (ROH) are generally not dehydrated but react rapidly and quantitatively with (1) at $-50°$ to give ethers, $R_F OR$; dehydration occurs only if the $\beta$-proton is sufficiently acidic.

*Cleavage of secondary amides.*[4] The reaction of sulfurane (1) with a suitably substituted secondary amide results in a facile single-step cleavage of the amide at room temperature or below. Thus treatment of benzanilide with 0.5 mole of (1) in DMF at 41° for 3 min. gives the sulfilimine (2, isolated in 72% yield) and the benzoate

$$\underset{\text{(1)}}{\begin{array}{c} C_6H_5 \\ | \\ CF_3-C-CF_3 \\ | \\ O \\ | \\ C_6H_5-S-C_6H_5 \\ | \\ O \\ | \\ CF_3-C-CF_3 \\ | \\ C_6H_5 \end{array}} + C_6H_5CONHC_6H_5 \longrightarrow (C_6H_5)_2S=NC_6H_5 + C_6H_5CO_2C(CF_3)_2C_6H_5 + C_6H_4$$

$$\hspace{6cm} (2) \hspace{3cm} (3, \ C_6H_5COOR_F) \hspace{1cm} (R_F$$

ester (3). The reaction is general, but the rate is sensitive to steric bulk of either the N-alkyl group or the acyl group.

*Epoxides.* Martin et al.[5] at an A.C.S. meeting mentioned a single-step conversion of glycols to epoxides using sulfurane (1). The reaction proceeds in high yield in the case of diols in which the geometry permits a *trans* coplanar positioning of the hydroxyl groups.

[1] J. C. Martin and R. J. Arhart, *Am. Soc.*, **93**, 2341 (1971).

[2] An improved procedure for the preparation has been submitted to *Org. Syn.* by J. C. Martin, R. J. Arhart, J. A. Franz, E. F. Perozzi, and L. J. Kaplan (1972). Note that hexafluoro-2-phenyl-2-propanol is available from PCR, Inc., or can be prepared by the method of B. S. Farah, E. E. Gilbert, and J. P. Sibilia, *J. Org.*, **30**, 998 (1965). In the new procedure bromine is used rather than chlorine. The overall yield of the sulfurane is 65–75%.

[3] J. C. Martin and R. J. Arhart, *Am. Soc.*, **93**, 4327 (1971).
[4] J. A. Franz and J. C. Martin, *ibid.*, **95**, 2017 (1973).
[5] R. J. Arhart, J. A. Franz, and J. C. Martin, A.C.S. National Meeting, Washington, Sept. 1971, Abstract ORGN 175.

**Diphenyliodonium-2-carboxylate monohydrate** (4), m.p. 220–222° (dec.), first described by Le Goff,[1] can be prepared [2,3] in yields of 72–79% by oxidation of *o*-iodobenzoic acid with potassium persulfate in sulfuric acid to the iodonium salt (2) and Friedel–

Crafts-like reaction of this substance with benzene to form the diphenyliodonium salt (3); neutralization with ammonia then liberates the inner salt (4) which crystallizes from water as the monohydrate.

*Generation of benzyne for Diels–Alder reaction.* **(a)** *Org. Syn. procedure.*[2] A 100-ml. round-bottomed flask equipped with an 11-cm. water-cooled Ace Glass bearing No. 8244 to serve as a condenser is charged with 60 ml. of Eastman's *mp*-diethylbenzene; the flask is heated with the free flame of a microburner until the liquid rises well into the condenser. Moisture is removed with an applicator stick wrapped with absorbent cotton. The same technique is used later for removal of water of hydration that appears in the early stages of the reaction and causes hissing and eruption if allowed to drop back into the flask. The flame is removed and 10 g. of tetraphenylcyclopentadienone and 11.8 g. of diphenyliodonium carboxylate monohydrate are added to the flask. Material adhering to the flask and spatula is dissolved in boiling methylene chloride and the solution transferred to a tared 500-ml. flask and evaporated to dryness. The solid product is added to the flask and the combined solid is brought to constant weight at steam-bath temperature and water-aspirator pressure. The yield of crude product is 23.3–24.4 g.

Boiling water (275 ml.) is poured into the flask, the product brought into solution at the boiling point, and 0.4 g. of Norit is added carefully to the slightly cooled solution. The solution is again heated to boiling, filtered, and allowed to stand for crystallization overnight, eventually at 0°. The colorless prisms of diphenyliodonium-2-carboxylate monohydrate are collected and air-dried to constant weight at room temperature. The yield of product, m.p. 220–222° (dec.), is 20–22 g. (72–79%).

(b) *1,2,3,4-Tetraphenylnaphthalene.* To a 100-ml. round-bottomed flask equipped with an 11-cm. water-cooled condenser there is added 60 ml. of diethylbenzene (*meta* and *para* mixture) and the flask is heated in a fume hood with the free flame of a micro-burner until refluxing liquid rises well into the condenser. If a cloudy zone of condensate appears at the top of the condenser, the moisture is removed with an applicator stick wrapped with absorbent cotton. The same technique is used later for removal of water of hydration which appears in the early stages of the reaction and causes hissing and eruption if allowed to drop back into the flask. The flame is removed and 10 g. (0.026 mole) of tetraphenylcyclopentadienone and 11.8 g. (0.035 mole) of diphenyl-iodonium carboxylate monohydrate are added to the flask. The mixture is heated over a microburner at a rate such as to maintain vigorous gas evolution and gentle refluxing. The water of hydration is eliminated in 8–10 min. The flask is then fitted with a normal reflux condenser and the heating is continued. After 30 min. considerable undissolved diphenyliodonium carboxylate can still be seen, under illumination, at the bottom of the flask. In another 5 min. the color changes to transparent red, and in a minute or two longer the solution becomes pale amber. Refluxing is continued until no solid remains (10–15 min). The flask is then fitted for distillation, and 55 ml. of liquid (diethylbenzene and iodobenzene, b.p. 188°) is removed by distillation. The residue is cooled and dissolved in 25 ml. of dioxane. The solution is rinsed into a 125-ml. Erlenmeyer flask and diluted with 25 ml. of 95% ethanol. The solution is heated to boiling, and water (6–7 ml.) is added gradually until a few shiny prisms remain undissolved on boiling. Crystallization is allowed to proceed at room temperature and then for several hours at 0°. The precipitate is removed by filtration, and the mother liquor upon further standing deposits a small second crop of crystals (0.3 g., m.p. 1° low). The main product melts initially in the range 196–199°, solidifies on cooling, and remelts sharply at 203–204°. The total yield is 9.2–10.2 g. (82–90%).

(c) *Procedure for students.*[3] Place 1.0 g. of diphenyliodonium-2-carboxylate monohydrate and 1.0 g. of tetraphenylcyclopentadienone in a 25 × 150-cm. test tube and add 6 ml. of triethyleneglycol dimethyl ether (b.p. 222°), using the solvent to rinse down the walls. Support the test tube vertically, insert a thermometer, and heat with a microburner. When the temperature reaches 200° remove the burner and note the time. Then keep the mixture at 200–205° by intermittent heating until the purple color is discharged, the evolution of gas ($CO_2$ + CO) subsides, and a pale-yellow solution is obtained. In case a purple or red color persists after 3 min. at 200–205°, add additional small amounts of the benzyne precursor and continue to heat until all the solid is dissolved. Let the yellow solution cool to 90° while heating 6 ml. of 95% ethanol to the boiling point. Pour the yellow solution into a 25-ml. Erlenmeyer flask and use portions of the hot ethanol, drawn into a capillary dropping tube, to rinse the test tube. Add the remainder of the ethanol and heat the mixture at the boiling point. If shiny prisms do not separate at once, add a little water by microdrops at the boiling point until prisms begin to separate. Let crystallization proceed at room temperature and then at 0°. Collect the product and wash it with a little methanol; yield 0.8–0.9 g. The pure hydrocarbon melts first in the range 196–199°, solidifies, and remelts at 203–204°.

[1]E. Le Goff, *Am. Soc.*, **84**, 3786 (1962).
[2]L. F. Fieser and M. J. Haddadin, *Org. Syn.*, **46**, 107 (1966).
[3]L. F. Fieser, *Org. Expts.*, 2nd Ed., D. C. Heath and Co., Boston, 303–306 (1968).

**Diphenyl isobenzofurane, 1,** 342–343, **2,** 178–179.

*Preparation.* Potts and Elliott[1] have reported an improved procedure for conversion of *o*-dibenzoylbenzene (1) into diphenylisobenzofurane (4). Reduction of (1) with excess sodium borohydride, followed by treatment with hot acetic anhydride, provides (4) in yields around 70%. Isolation of the intermediate alcohol (2) is not necessary.

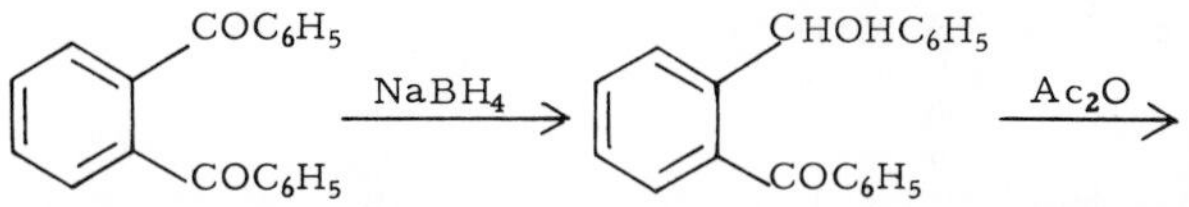

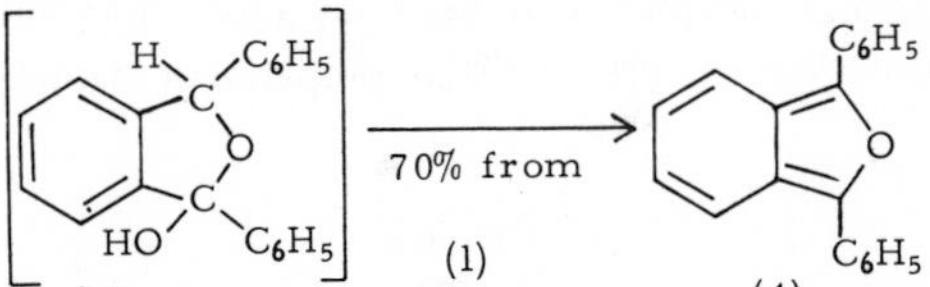

Diphenylisobenzothiophene can be obtained in 74% yield by treatment of (1) with $P_2S_5/Py$ (reflux 3 hr.).

*Cyclopentyne.* Use as trapping agent to demonstrate the formation of cyclopentyne.[2]

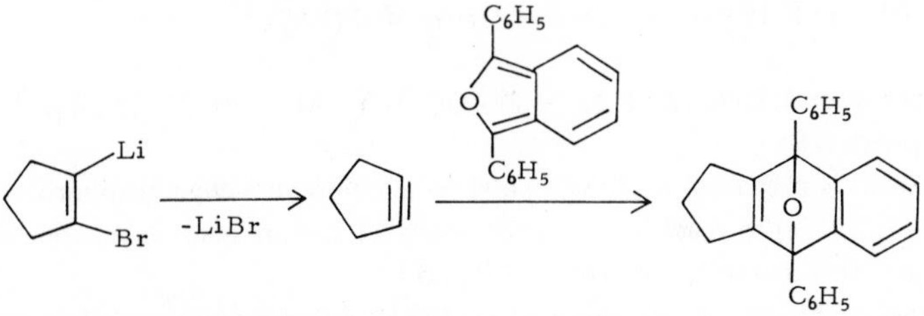

[1]K. T. Potts and A. J. Elliott, *Org. Prep. Proc. Int.*, **4**, 269 (1972).
[2]G. Wittig and J. Heyn, *Ann.*, **726**, 59 (1969).

**Diphenylketene, 1**, 343–345; **3**, 131–132.

*Reaction with thiocarbonyl ylides.* Pyrolysis of the thiocarbonyl ylide precursors (1a, 1b) in hydrocarbon solvents in the presence of diphenylketene leads to the cyclo-adducts (3a, 3b) in 85–94% yield. The reaction, therefore, involves 1,3-addition of the

(1a, $R^1 = R^4 = \underline{t}\text{-}Bu$;
$\quad R^2 = R^3 = H$)
(1b, $R^1 = R^4 = C_2H_5$;
$\quad R^2 = R^3 = H$)

(2a, 2b)

(3a, 3b)

*trans*-thiocarbonyl ylide to the carbonyl group of the ketene with retention of configuration of the alkenic component (suprafacial cycloaddition).

[1]R. M. Kellogg, *J. Org.*, **38**, 844 (1973).

**Diphenyl phosphite**, $\overset{O}{\overset{\|}{H}}P(OC_6H_5)_2$. Mol. wt. 234.19, m.p. 12°. Suppliers: Aldrich, Eastman, Fisher, P and B, Sargent.

*Peptide synthesis.* Japanese chemists[1] have reported a convenient synthesis of peptides by means of diphenyl phosphite and a tertiary amine (pyridine was used) as summarized in the equation. Various peptides were prepared in this way in high yield

$$\underline{t}\text{-Boc-NHR}^1\text{CHCOOH} + \text{NH}_2\text{R}^2\text{CHCOOX} \xrightarrow[\text{Pyridine}]{\overset{O}{\overset{\|}{H}}P(OC_6H_5)_2} \underline{t}\text{-Boc-NHR}^1\text{CHCONHR}^2\text{CHCOO}$$

$$\overset{O}{\overset{\|}{H}}P\overset{OH}{\underset{OC_6H_5}{\diagdown}} + C_6H_5OH$$

and with high optical purity. The method can be applied also to synthesis of active esters of amino acids.

[1]N. Yamazaki and F. Higashi, *Tetrahedron Letters*, 5047 (1972).

**Diphenylphosphoryl azide (DPPA)**, $N_3PO(OC_6H_5)_2$. Mol. wt. 275.20, b.p. 157°/0.17 mm. Supplier: WBL.

The reagent is prepared in > 90% yield by allowing diphenylphosphorochloridate (**1**, 345–346; **2**, 180) to react with a slight excess of sodium azide in acetone at room temperature. It is a stable, nonexplosive liquid.[1]

*Urethanes.* Urethanes are obtained readily by refluxing an equimolar mixture of a carboxylic acid, triethylamine, an alcohol, and the reagent for 5–25 hr. This modified Curtius reaction is less laborious than the classical procedure.

$$RCOOH + R'OH \xrightarrow[\text{(C}_2\text{H}_5)_3\text{N}]{\text{N}_3\text{PO(OC}_6\text{H}_5)_2} RNHCOOR'$$

*Peptide synthesis.* In the above-mentioned synthesis of urethanes the carboxylic acid azide may be the intermediate, and this possibility prompted the Japanese chemists to investigate the usefulness of diphenylphosphoryl azide in peptide synthesis. Indeed the reagent allows coupling of acylamino acids or peptides with amino acid esters or peptide esters in the presence of a base in high yield and with practically no racemization. The new method is compatible with various functional groups.

[1]T. Shioiri, K. Ninomiya, and S. Yamada, *Am. Soc.*, **94**, 6203 (1972).

**Diphenylsulfonium cyclopropylide**, $(C_6H_5)_2\overset{+}{S}\!\!-\!\!\overset{-}{\triangleleft}$ . Mol. wt. 227.34.
   *Preparation*[1]:

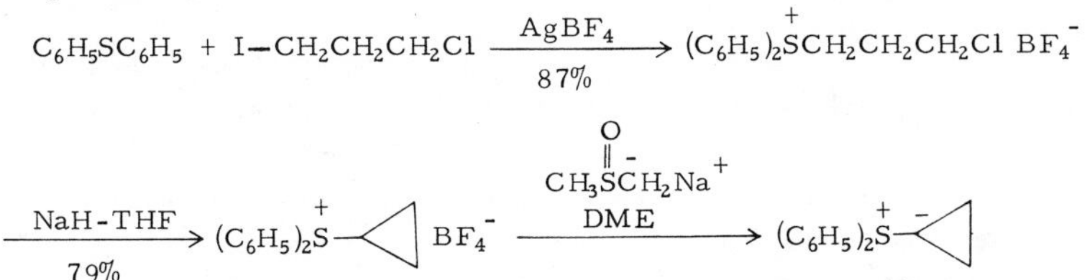

$$C_6H_5SC_6H_5 + I-CH_2CH_2CH_2Cl \xrightarrow[87\%]{AgBF_4} (C_6H_5)_2\overset{+}{S}CH_2CH_2CH_2Cl\ BF_4^-$$

$$\xrightarrow[79\%]{NaH-THF} (C_6H_5)_2\overset{+}{S}\!\!-\!\!\triangleleft\ BF_4^- \xrightarrow[DME]{\overset{O}{\overset{\|}{CH_3\overset{-}{S}CH_2Na}}\overset{+}{}} (C_6H_5)_2\overset{+}{S}\!\!-\!\!\overset{-}{\triangleright}$$

(1)

*Spiroalkylation.*[1] The reagent (1) reacts with an $\alpha,\beta$-unsaturated ketone (2) to form a spiropentane (3):

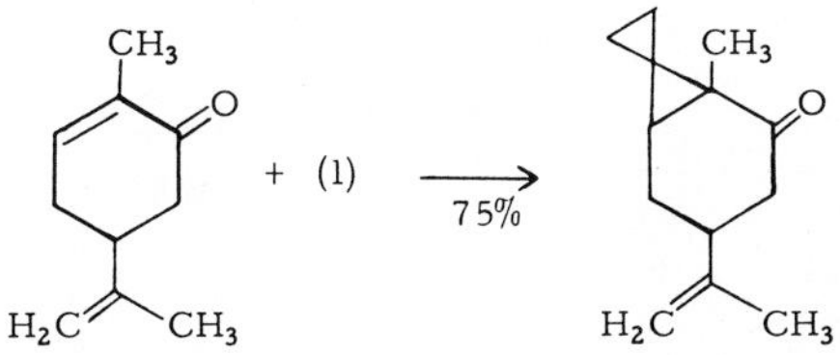

Spirobutanones are formed from the reaction of (1) with aldehydes and ketones:

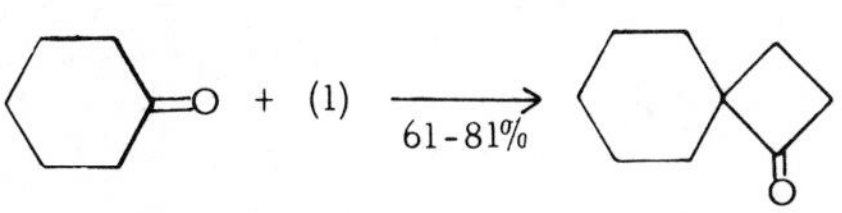

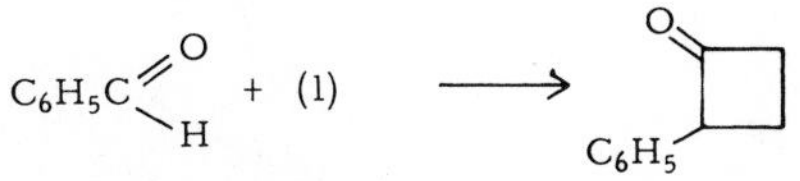

The reaction can proceed with a high degree of stereoselectivity.[2] Thus treatment of 4-*t*-butylcyclohexanone with this sulfur ylide followed by rearrangement of the intermediate oxaspiropentane with Eu(fod)$_3$ gives the spirobutanone (4) in 80% yield

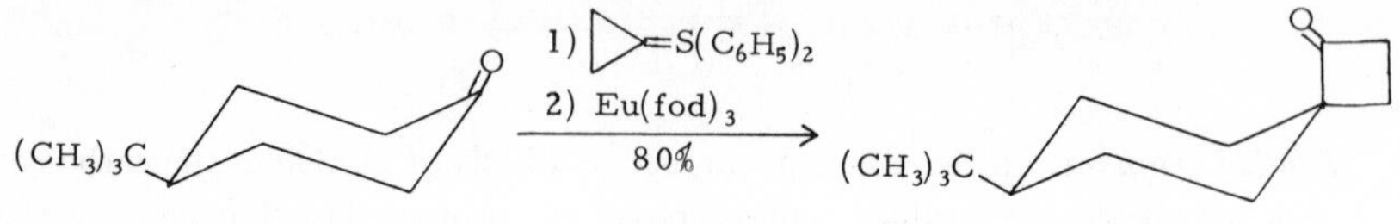

$(CH_3)_3C$

1) ▷=$S(C_6H_5)_2$
2) $Eu(fod)_3$
   80%

$(CH_3)_3C$

(4)

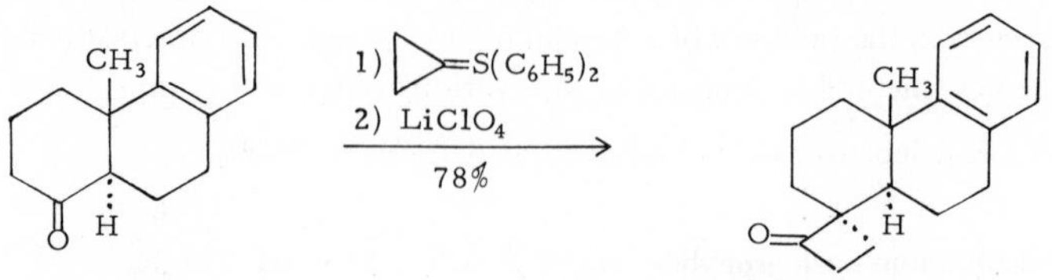

(5)

1) ▷=$S(C_6H_5)_2$
2) $LiClO_4$
   78%

(6)

uncontaminated with its isomer. Similarly, the spirobutanone (6) was obtained exclusively from the tricyclic ketone (5) with use of lithium perchlorate as the rearranging agent.

**Geminal alkylation.**[2] Trost and Bogdanowicz have reported an interesting method for effecting geminal alkylation via spirobutanones, prepared as in the preceding section. Thus treatment of the ketone (1) with the reagent gives (2); this is brominated ($Br_2$ or, preferably, pyridinium hydrobromide perbromide, **1**, 967–970) to give the

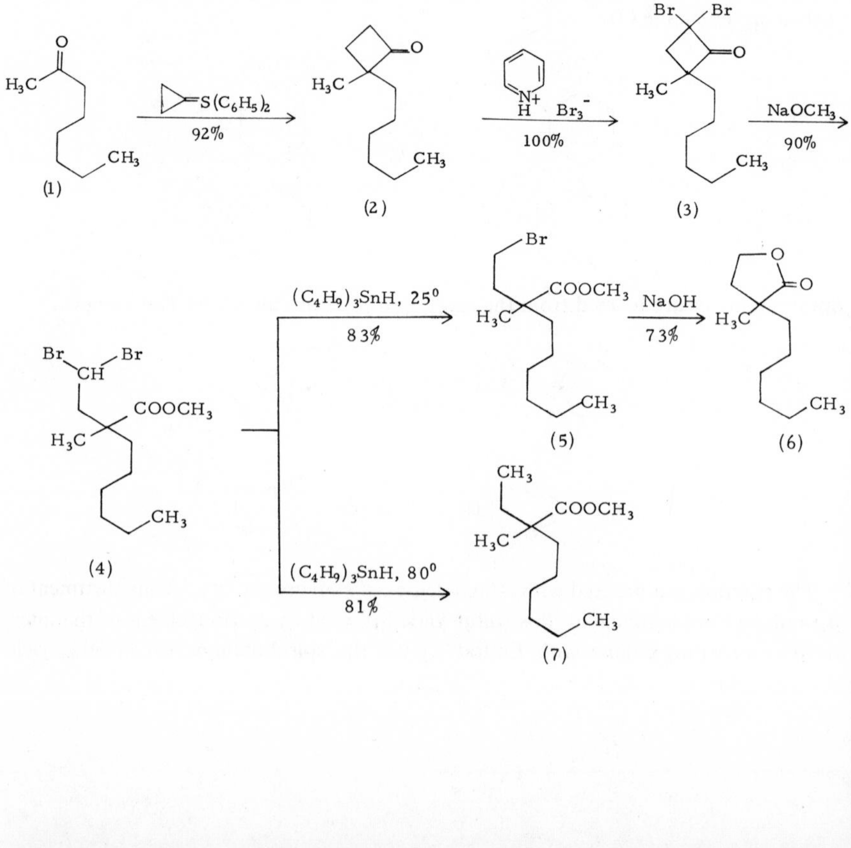

$H_3C$

$CH_3$

(1)

▷=$S(C_6H_5)_2$
   92%

$H_3C$

$CH_3$

(2)

$H_3C$
$Br_3^-$
   100%

$H_3C$

$CH_3$

(3)

$NaOCH_3$
   90%

$Br$  $Br$
$CH$

$COOCH_3$

$H_3C$

$CH_3$

(4)

$(C_4H_9)_3SnH, 25°$
   83%

$H_3C$

$COOCH_3$

$CH_3$

(5)

$NaOH$
   73%

$H_3C$

(6)

$(C_4H_9)_3SnH, 80°$
   81%

$CH_3$

$H_3C$
$COOCH_3$

$CH_3$

(7)

α,α-dibromoketone (3). Treatment of (3) with methanolic sodium methoxide cleaves the ring to give (4). Reduction of (4) with tri-*n*-butyltin hydride (**1**, 1192–1193; **2**, 424; **3**, 294; this volume) at room temperature effects monodebromination to give (5), which on treatment with base yields the lactone (6). Treatment of (4) with the tin hydride at 80° yields (7).

Alternatively, the *gem*-dibromo group represents a masked carbonyl group. Thus treatment of (8) with sodium methoxide gives (9); when this is allowed to react with methanolic silver nitrate, the bromine atoms are replaced by methoxyl groups to generate the succinaldehyde ester (10). Hydrolysis of (10) leads to the aldehyde (11). Reduction with tris(triphenylphosphine)chlororhodium (**1**, 1252; **2**, 448–453; **3**, 325–329; this volume) effects decarbonylation to give (12).

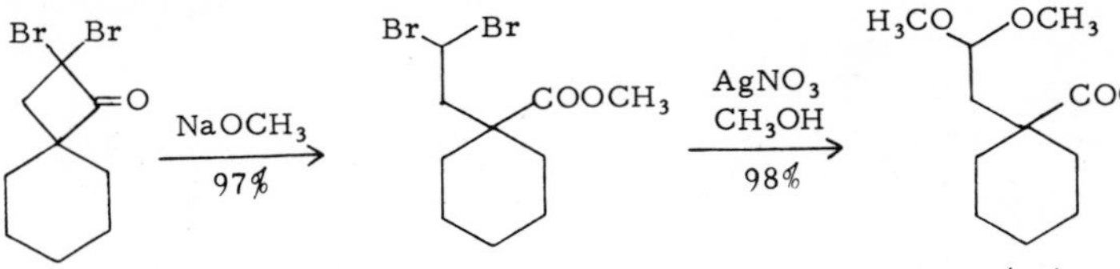

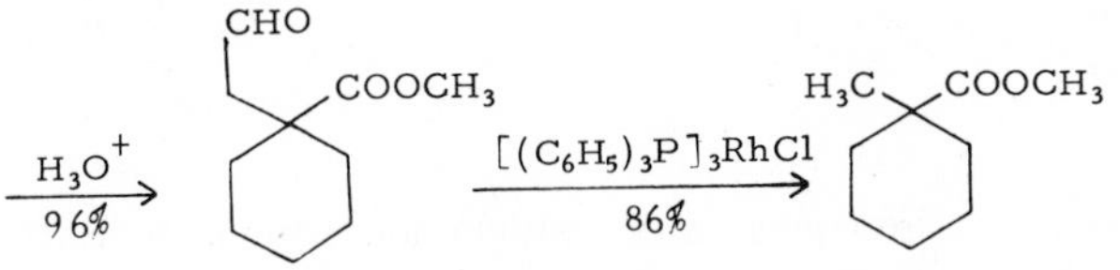

***Oxaspiropentanes.***[3] Oxaspiropentanes (2) were suggested as intermediates in the synthesis of cyclobutanones formulated above. These have now been isolated in about 60–80 % yield by treatment of a carbonyl compound with cyclopropyldiphenylsulfonium fluoroborate (1)[4] and solid potassium hydroxide at 25° in DMSO followed by flash distillation of the crude extract under vacuum.

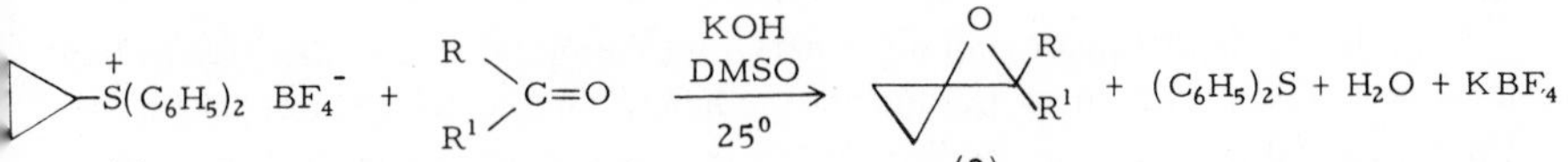

Examples:

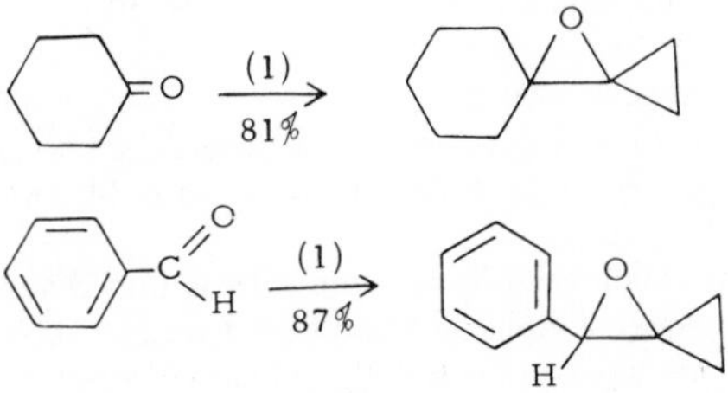

Only rearranged cyclobutanones were obtained in the case of benzophenone and cyclopropyl methyl ketone.

**Cyclopentane annelation.**[5] Trost and Bogdanowicz have reported a useful cyclopentanone synthesis from oxaspiropentanes, prepared as above. An example is the synthesis of bicyclo[3.3.0]octane-2-one (5) from cyclopentanone (1). The ketone is treated with cyclopropyldiphenylsulfonium fluoroborate (3 hr.) to give the oxaspiropentane (2). Treatment of the reaction mixture with trimethylchlorosilane in 1,2-dimethoxyethane (DME) gives the silyl ether (3) in 94 % yield. Passage of a hexane solution of (3) through a tube packed with glass helices[6] at 330° with a contact time of 4 sec. leads to quantitative rearrangement to the enol silyl ether (4). Hydrolysis of (4) gives bicyclo[3.3.0]octane-2-one (5).

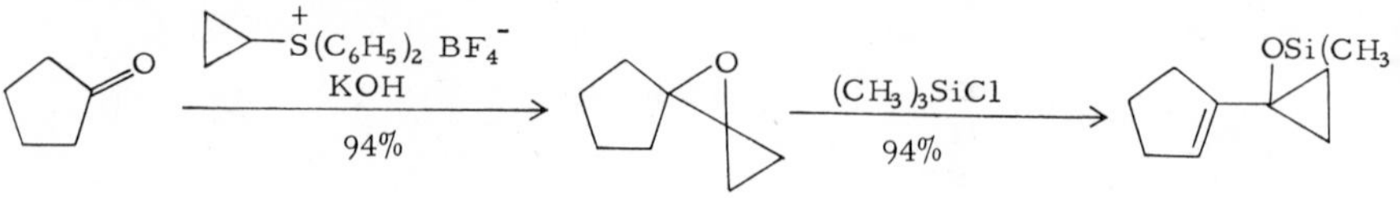

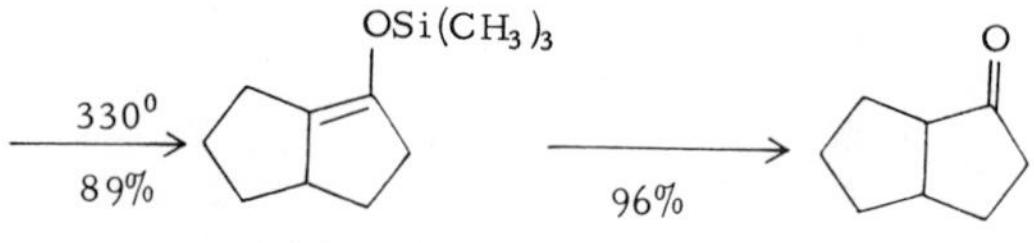

A similar sequence of reactions starting with cycloheptanone (6) leads to the perhydroazulenone derivative (7).

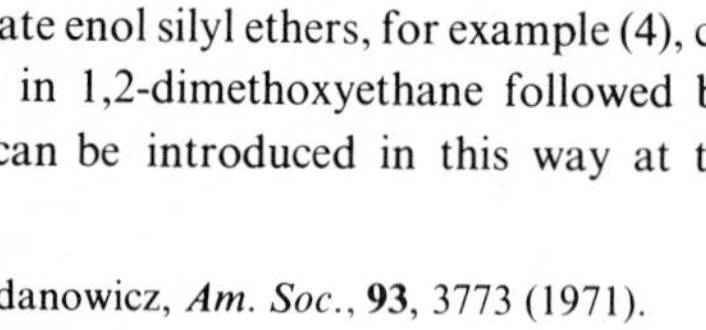

Note that the intermediate enol silyl ethers, for example (4), can be alkylated by treatment with methyllithium in 1,2-dimethoxyethane followed by addition of an alkyl halide. An alkyl group can be introduced in this way at the bridgehead position exclusively.

[1] B. M. Trost and M. J. Bogdanowicz, *Am. Soc.*, **93**, 3773 (1971).
[2] *Idem, ibid.*, **95**, 2038 (1973).
[3] *Idem, Tetrahedron Letters*, 887 (1972).
[4] A detailed procedure for the preparation of this reagent (mol. wt. 314.16, m.p. 137–139°) has been submitted to *Org. Syn.* by M. J. Bogdanowicz and B. M. Trost (1973).
[5] *Idem, Am. Soc.*, **95**, 289 (1973).
[6] The hot column was conditioned by washing with saturated aqueous sodium bicarbonate solution, followed by water, acetone, and hexane. Then either O,N-bistrimethylsilylacetamide or trimethylchlorosilane is passed through the column, followed by diethylamine.

**Dipotassium cyanodithioimidocarbonate,** $NCN=C\begin{smallmatrix}SK\\SK\end{smallmatrix}$  (1). Mol. wt. 194.37.

The reagent is prepared[1] by the reaction of cyanamide with carbon disulfide followed by addition of potassium hydroxide (86% yield).

$$NCNH_2 + CS_2 + 2\ KOH \xrightarrow[86\%]{C_2H_5OH} NCN=C\begin{smallmatrix}SK\\SK\end{smallmatrix}$$

*3-Halo-1,2,4-thiadiazoles.* On treatment with sulfuryl chloride, (1) undergoes oxidative cyclization to the 1,2,4-thiadiazole (2) in 80–100% yield.[2]

$$(1) \xrightarrow[80-100\%]{SO_2Cl_2} (2)$$

The reagent (1) can be monoalkylated to (3) prior to oxidative cyclization.[1,3]

$$(1) \xrightarrow{RX} NCN=C\begin{smallmatrix}SK\\SR\end{smallmatrix}\ (3) \xrightarrow{SO_2Cl_2} (4)$$

[1] L. S. Wittenbrook, G. L. Smith, and R. J. Timmons, *J. Org.*, **38**, 465 (1973).
[2] W. A. Thaler and J. R. McDivitt, *ibid.*, **36**, 14 (1971).
[3] R. J. Timmons and L. S. Wittenbrook, *ibid.*, **32**, 1566 (1967).

**Dipotassium platinum tetrachloride, 2,** 182.

*Deformylation.* In a study of the homogeneous deuteration of α-hydroxy acids with this catalyst, Calf and Garnatt[1] found that benzilic acid undergoes deformyla-

$$(C_6H_5)_2C\begin{smallmatrix}OH\\COOH\end{smallmatrix} \longrightarrow (C_6H_5)_2C=O$$

tion in high yield to give benzophenone. Under these conditions benzhydrol is not converted into benzophenone.

[1] G. E. Calf and J. L. Garnatt, *Tetrahedron Letters*, 511 (1973).

**Dipyridine chromium(VI) oxide (Collins reagent),** $CrO_3(C_5H_5N)_2$, **2,** 74–75; **3,** 55–56. Mol. wt. 258.21, red crystals. Supplier: Eastman.

*Preparation.*[1] Collins and Hess have published a detailed procedure for preparation of the reagent from chromium(VI) oxide (dried over phosphorus pentoxide) and pyridine (reagent grade). *Caution*: The reaction is extremely exothermic. The chromium(VI) oxide should be added to pyridine at such a rate that the temperature does not exceed 20° and in such a way that the oxide mixes rapidly with pyridine. As the chromic oxide is added a yellow flocculate precipitates. When the addition is complete, the mixture is allowed to warm slowly to room temperature with stirring. Within an hour the initial yellow product changes to deep red crystals. The supernatant pyridine is decanted, and the crystals washed by decantation with anhydrous petroleum ether. The product is collected and washed with more solvent. The complex is dried at 10 mm. (yield 85–91%); it is extremely hygroscopic. It should be stored at 0° in a brown bottle.

*Oxidation of alcohols.*[1]   The oxidation of alcohols with the reagent is conducted in methylene chloride with a sixfold molar excess of oxidant. The oxidation usually proceeds to completion in 5–15 min. at 25°. The reagent is particularly recommended for oxidation of primary alcohols to aldehydes; in this case use of the Sarett reagent (**1**, 145–146; **2**, 74–75) usually gives low yields.[2] Thus 1-heptanol can be oxidized by the Collins reagent in methylene chloride to 1-heptanal in 70–84% yield.

The oxidation of *exo*-7-hydroxybicyclo[4.3.1]decatriene-2,4,8 (1) to bicyclo[4.3.1]-decatriene-2,4,8-one-7 (2) was effected with Collins reagent in 64% yield.[3]

$$CrO_3(C_5H_5N)_2$$
$$\overrightarrow{64\%}$$

(1)    (2)

[1]J. C. Collins and W. W. Hess, *Org. Syn.*, **52**, 5 (1972).
[2]J. R. Holum, *J. Org.*, **26**, 4814 (1961).
[3]G. Schröder, U. Prange, B. Putze, J. Thio, and J. F. M. Oth, *Ber.*, **104**, 3406 (1971).

**1,3-Dithiane, 2**, 187; **3**, 135–136. Note 9 (**2**, 187): D. Seebach and A. K. Beck, *Org. Syn.*, **51**, 76 (1971). Additional supplier: Eastman.

*Oxidative hydrolysis of 2-acyl-1,3-dithiane derivatives.*[1] A study of the hydrolysis of 2-acyl-1,3-dithiane derivatives by mercuric chloride ($HgCl_2$) has been published. However, hydrolysis of 2-acyl-1,3-dithianes is slow. In this case use of N-halosuccinimide reagents is recommended. Either N-bromosuccinimide or N-chlorosuccinimide–silver nitrate is suitable for oxidative hydrolysis of 1,3-dithiane derivatives. The N-halosuccinimide reagents are useful for hydrolysis of 2-acyl-1,3-dithianes to 1,2-dicarbonyl compounds.

*Oxidation of aldehydes to esters and acids.*[2] Treatment of 2-lithio-2-alkyl-1,3-dithianes (1) with methyl disulfide in THF gives the corresponding orthothioformates (2) in about 90% yield. These are converted in high yield into esters (3) when refluxed in an aqueous alcohol in the presence of mercuric chloride and mercuric oxide as catalysts. Conversion to acids is accomplished by refluxing (2) in aqueous acetone for 24 hr. with catalysis by mercuric chloride and mercuric oxide. Yields in this case are in the range 40–65%.

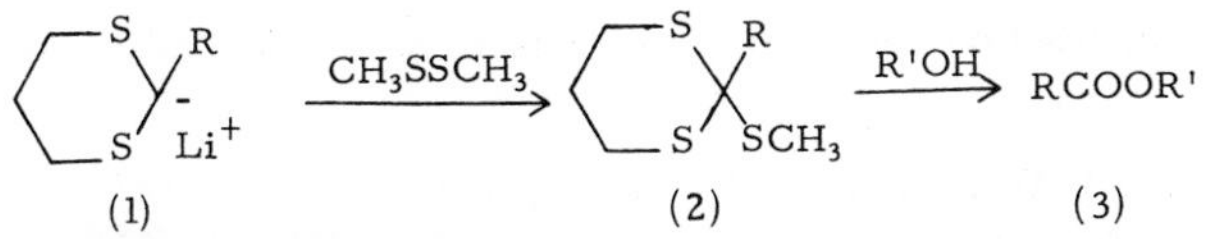

$$(1) \xrightarrow{CH_3SSCH_3} (2) \xrightarrow{R'OH} RCOOR' \quad (3)$$

The present procedure may be useful in the case of aldehydes that are sensitive to conventional oxidizing reagents.

*Synthesis of β,γ-unsaturated aldehydes.*[3] Treatment of the allylic bromide (1) with 1,3-dithiane gives the sulfonium bromide (2) in high yield. Treatment of (2) at −78° in THF with *n*-butyllithium gives the ylide (3), which rearranges when warmed to 20°

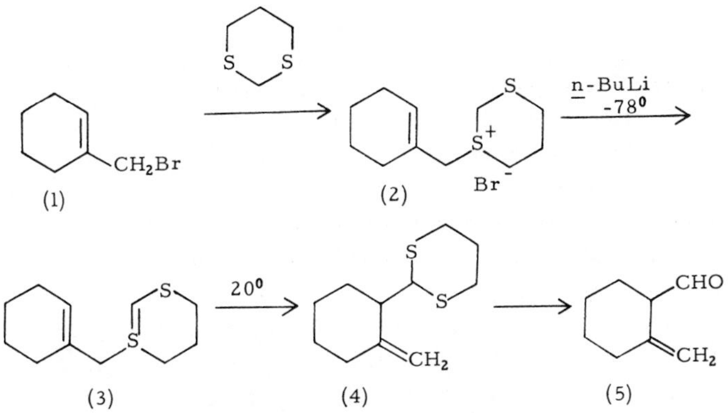

to the 2-substituted 1,3-dithiane (4). Hydrolysis of (4) by usual methods gives the β,γ-unsaturated aldehyde (5).

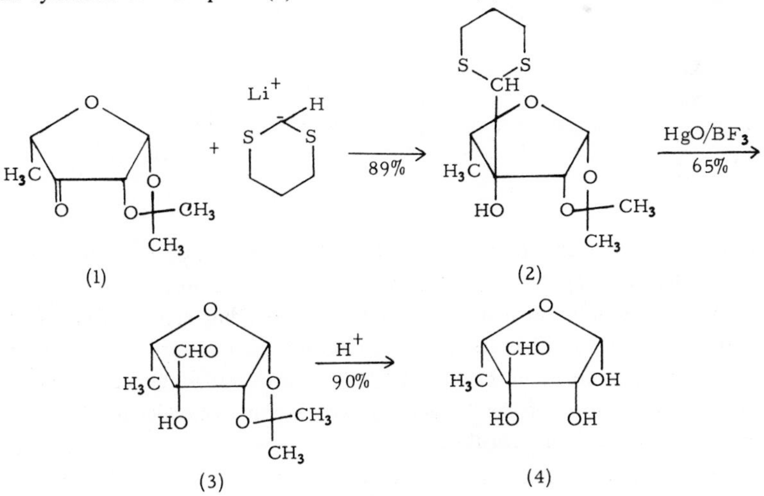

L-*Streptose.*[4] 2-Lithio-1,3-dithiane has been used in the synthesis of branched sugars with aldehyde or keto groups in the side chain. The reaction sequence is illustrated for the synthesis of L-streptose (4) from the 3-ulose (1). The most critical step is the desulfurization reaction [(2) → (3)], which must be carried out under carefully controlled conditions.

[1] E. J. Corey and B. W. Erickson, *J. Org.*, **36**, 3553 (1971).
[2] R. A. Ellison, W. D. Woessner, and C. C. Williams, *ibid.*, **37**, 2757 (1972).
[3] E. Hunt and B. Lythgoe, *Chem. Commun.*, 757 (1972).
[4] H. Paulsen, V. Sinnwell, and P. Stadler, *Angew. Chem., internat. Ed.*, **11**, 149 (1972).

**1,3-Dithienium fluoroborate** (1),    . Mol. wt. 206.05, m.p. 188–189° dec.

(1)

*Preparation.*[1] The reagent (1) is obtained in 92 % yield as a yellow, crystalline solid by the reaction of 1,3-dithiane (**2**, 182–187; **3**, 135–136; this volume) with trityl fluoroborate (**1**, 1256–1258; this volume) in dry methylene chloride (heating at reflux for 30 min., evaporation of solvent under reduced pressure, trituration with cold ether, and drying under vacuum). The reagent can be stored with exclusion of moisture at −20° for several months without appreciable decomposition.

$\Delta^3$-*Cyclopentenones.*[1] The reagent (1) undergoes a Diels–Alder type of reaction with 1,3-dienes (2) at 0–25° to afford adducts (3) in high yield. The reaction can be carried out in either dry methylene chloride or, in the case of dienes that are not very soluble in methylene chloride, a dry mixture of methylene chloride–nitromethane (2:1). On treatment with 1 eq. of *n*-butyllithium first at −78° and then at −78° to 25°, (3)

rearranges in high yield to the vinylcyclopropane derivatives (4). These products rearrange when heated in benzene solution (200°) into the cyclopentenone thioketals (5). The thioketals are smoothly hydrolyzed to the corresponding ketones (6) by treatment in acetone–water (8:1) with excess mercuric chloride–calcium carbonate at 25°. Yields for the several steps are all in the range 90–95 %. The overall process in a formal sense provides the equivalent of the 1,4-addition of carbon monoxide to a 1,3-diene, a process that is probably not feasible.

[1] E. J. Corey and S. W. Walinsky, *Am. Soc.*, **94**, 8932 (1972).

**Divinylcopperlithium**, $(CH_2=CH)_2CuLi$. Mol. wt. 124.57.

*Preparation.*[1] This organometallic reagent is prepared from a solution of cuprous iodide in isopropyl sulfide (3.1 mole eq.) by reaction at $-25$ to $-15°$ with a 2.2 $M$ solution of vinyllithium (1.92 eq.) in THF. The solution is then cooled to $-72°$ to allow complete reaction.

*Stereoselective introduction of the angular vinyl grouping.*[1] Divinylcopperlithium is useful for the introduction of a $\beta$- and even an angular $\beta$-vinyl substituent. Thus addition of the enone (1) to a solution of 1.3 mole eq. of divinylcopperlithium in THF at $-72°$ under argon leads to an exothermic reaction with formation of a single $\beta$-vinylated

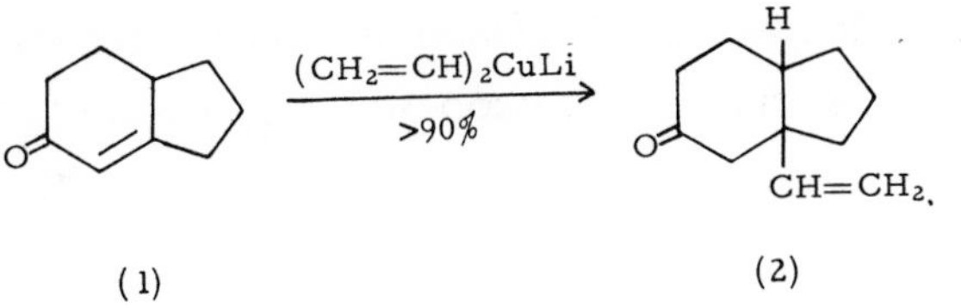

(1)                                        (2)

ketone (2) in $>90\%$ yield. The product was shown to be useful for carbocyclic synthesis applicable to the gibberellic acids (*see* **Dimethyldichlorosilane**, this volume).

*Homoconjugate addition to cyclopropanes.* Corey and Fuchs[2] have investigated the reaction of cyclopropanes with organocopper reagents as a possible synthetic route to prostanoids. For example, the tricyclic lactone ester (1) reacts with divinylcopperlithium (2.0 eq.) in ether at $-12°$ (19 hr.) to give the vinylcyclopentane lactone ester (2). This product was treated directly with lithium iodide (5 eq.) in pyridine (**1**, 615–616) at reflux for 3 hr. to give the lactone (3) in about 37% yield.

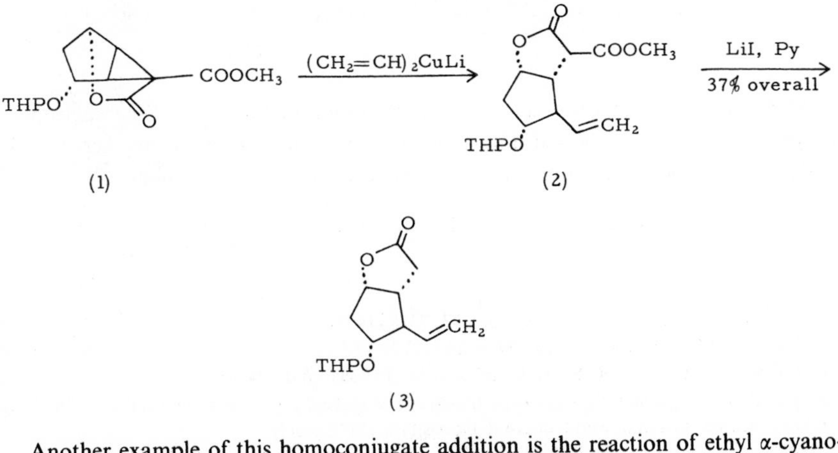

(1)                                        (2)

(3)

Another example of this homoconjugate addition is the reaction of ethyl α-cyano-cyclopropanecarboxylate (4) with 2 eq. of divinylcopperlithium to give ethyl 2-cyano-5-hexenoate (5) in 70% yield.

$$\underset{(4)}{\overset{\displaystyle \text{CN}}{\underset{\displaystyle \text{COOC}_2\text{H}_5}{\bigtriangleup}}} \quad \xrightarrow[70\%]{(\text{CH}_2=\text{CH})_2\text{CuLi}} \quad \underset{(5)}{\text{H}_2\text{C}=\text{HC}\diagdown\diagup\underset{\displaystyle \text{COOC}_2\text{H}_5}{\overset{\displaystyle \text{CN}}{\diagdown}}}$$

The addition reaction was also carried out successfully with dimethylcopperlithium. **Stereospecific synthesis of 1,3-dienes.** Corey et al.[3] have described a new synthesis of 1,3-dienes based on the highly stereospecific *cis* addition of alkylcopper reagents to α,β-acetylenic carbonyl compounds (3, 108). Thus the reaction of methyl 4-trimethyl-siloxy-2-nonynoate (1)[4] in THF with divinylcopperlithium (1.25 eq.) at −90° and then at −78° affords the pure *cis* adduct (2) in > 90% yield. Treatment of (2) with methanolic hydrochloric acid effects cleavage of the trimethylsilyl ether and lactonization to give (3).

$$\underset{(1)}{\text{n-C}_5\text{H}_{11}\text{CHO} + \text{HC}\equiv\text{CCOOCH}_3 \xrightarrow[71\%]{} \underset{(\text{CH}_3)_3\text{SiO}}{\text{n-C}_5\text{H}_{11}\text{CHC}\equiv\text{CCOOCH}_3}}$$

$$\xrightarrow[>90\%]{(\text{CH}_2=\text{CH})_2\text{CuLi}} \quad \underset{(2)}{\overset{\displaystyle \text{OSi}(\text{CH}_3)_3}{\underset{\displaystyle \text{H}_2\text{C}=\text{CH}}{\text{n-C}_5\text{H}_{11}\overset{|}{\text{C}}\text{H}}} \diagdown \underset{\displaystyle \text{H}}{\overset{\displaystyle \text{COOCH}_3}{\text{C}=\text{C}}}} \quad \xrightarrow{\text{H}^+} \quad \underset{(3)}{\text{H}_2\text{C}=\text{CH} \diagdown \text{C}_5\text{H}_{11}\text{-n}}$$

Addition of methyl 2-butynoate (4) to vinylcopper[5] affords methyl 3-methyl-*trans*-2,4-pentadienoate (5) in 74% yield.

$$\underset{(4)}{\text{CH}_3\text{C}\equiv\text{CCOOCH}_3} \xrightarrow[74\%]{\text{CH}_2=\text{CHCu}} \underset{(5)}{\text{H}_2\text{C}\diagup^{\text{CH}}\diagdown\underset{\displaystyle \text{CH}_3}{\text{C}}\diagup^{\text{CH}}\diagdown\text{COOCH}_3}$$

An analogous process for stereospecific generation of 1,4-dienes using allylcopper reagents has also been reported. Thus reaction of 2 eq. of allylcopper[6] in ether at −78° under $N_2$ with methyl propynoate (6) gives methyl *trans*-2,5-hexadienoate (7).

$$\underset{(6)}{\text{HC}\equiv\text{CCOOCH}_3} \xrightarrow{\text{CH}_2=\text{CHCH}_2\text{Cu}} \underset{(7)}{\text{H}_2\text{C}\diagup^{\text{CH}_2}\diagdown_{\text{CH}}\diagup^{\text{CH}}\diagdown\text{COOCH}_3}$$

[1] E. J. Corey and R. L. Carney, *Am. Soc.*, **93**, 7318 (1971).
[2] E. J. Corey and P. L. Fuchs, *ibid.*, **94**, 4014 (1972).
[3] E. J. Corey, C. U. Kim, R. H. K. Chen, and M. Takeda, *ibid.*, **94**, 4395 (1973).
[4] Prepared in a single step by sequential treatment of methyl propynoate in THF at −78° with n-butyllithium, hexanal, and trimethylchlorosilane (71% yield).
[5] Prepared from 2 eq. of vinyllithium and 2 eq. of cuprous iodide under $N_2$ at −78°.
[6] Prepared from 1 eq. of allylmagnesium chloride and 1 eq. of cuprous iodide at −30 to −40° for 2 hr. (deep red color).

**Divinylcopperlithium–Tri-*n*-butylphosphine complex,** $(CH_2{=}CH)_2CuLiP(n\text{-}C_4H_9)_3$.[1]
Mol. wt. 302.86.

*Preparation.*[2] A solution of 2 eq. of vinyllithium (Alfa Inorganics) in THF is added slowly to a THF solution of 1 eq. of tetrakis[iodo-(tri-*n*-butylphosphine)copper(I)] (**2**, 400; **3**, 278) at $-78°$ under $N_2$ with stirring. The complex is blue-black.

*γ,δ-Unsaturated ketones.* The complex undergoes conjugate addition to α,β-unsaturated ketones to give γ,δ-unsaturated ketones in yields of 65–90%. The last example, the reaction with isophorone to give 3-vinyl-3,5,5-trimethylcyclohexanone, is noteworthy. Only a yield of 8% is obtained of the 1,4-addition product when isophorone is treated with vinylmagnesium chloride under cuprous iodide catalysis.[2]

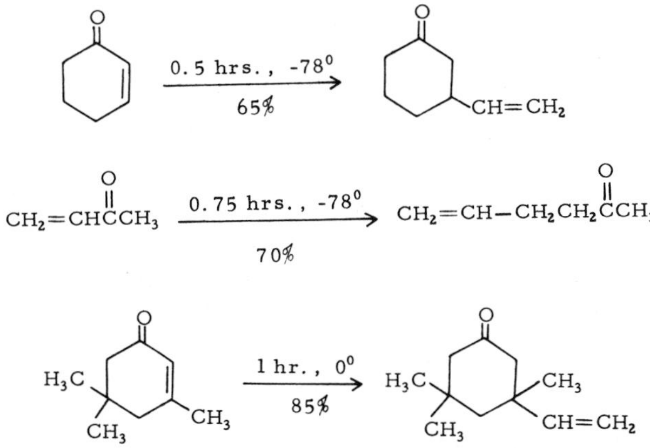

The phosphine-free reagent, divinylcopperlithium, $(CH_2{=}CH)_2CuLi$, can be prepared by treatment of vinyllithium (2 eq.) with cuprous iodide (1 eq.), but the rate of formation of this reagent and its rate of reaction with enones are considerably slower than those of the phosphine-complexed species.

A similar complex has been used in a recent synthesis of ($\pm$)-15-desoxyprostaglandin $E_1$.[3] Thus reaction of (**1**) with 2 molar eq. of 1-lithio-*trans*-octene-1 in the presence of 1 eq. of tetrakis[iodo-(tri-*n*-butylphosphine)copper(I)] in ether at 0° gives, after hydrolysis of the tetrahydropyranyl protecting group, ($\pm$)-15-desoxyprostaglandin $E_1$ ethyl ester (**2**) in 60% yield.

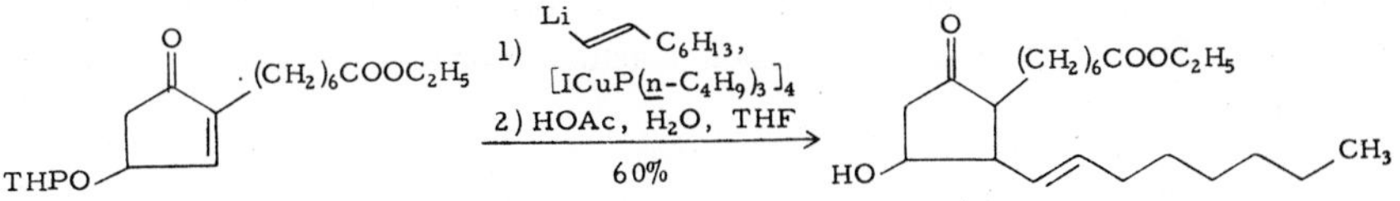

---

[1] This formulation is used for simplicity with no structural implication.
[2] J. Hooz and R. B. Layton, *Canad. J. Chem.*, **48**, 1626 (1970).
[3] C. J. Sih, R. G. Salomon, P. Price, G. Peruzzoti, and R. Sood, *Chem. Commun.*, 240 (1972).

# E

**Epichlorohydrin, 1**, 355.

*Scavenger for hydrogen bromide.* Oxidation of acyclic α-haloketones with DMSO yields α-diketones (**1**, 303–304). The oxidation proceeds abnormally, however, in the case of cyclic α-bromoketones. Thus DMSO oxidation of 5-bromo-2-carboethoxy-2-methylcyclopentanone (1) gives 3-bromo-5-carboethoxy-2-hydroxy-5-methylcyclopent-

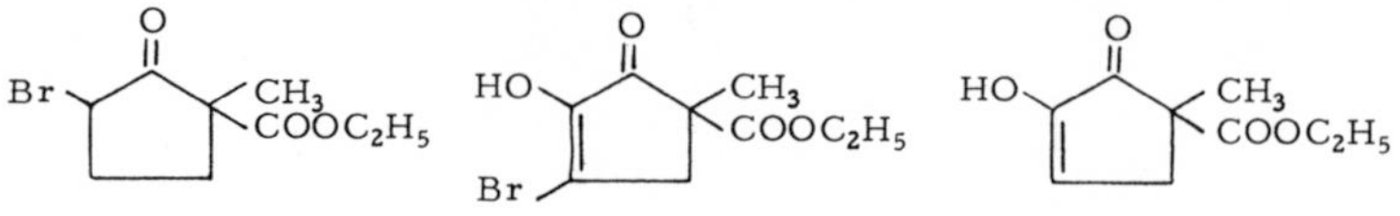

(1)           (2)           (3)

2-ene-1-one (2) in 70 % yield.[1] This abnormal result is probably due to oxidation of the liberated hydrogen bromide to bromine by DMSO. Indeed, the normal product (3) can be obtained in 58.5 % yield if the oxidation of (1) is carried out in the presence of epichlorohydrin, which serves as scavenger for the hydrogen bromide.[2]

[1] K. Sato, S. Suzuki, and Y. Kojima, *J. Org.*, **32**, 339 (1967).
[2] K. Sato, Y. Kojima, and H. Sato, *ibid.*, **35**, 2374 (1970).

**Ethanolamine, 1**, 357; **2**, 189–190.

*Selective halogenation of methyl ketones.*[1] Pregnenolone (1) reacts with ethanolamine in toluene containing some Dowex-20 ion-exchange resin (acid form) to give the ketimine (2) in 94 % yield. Treatment of (2) with either NCS or NBS in ether, followed by

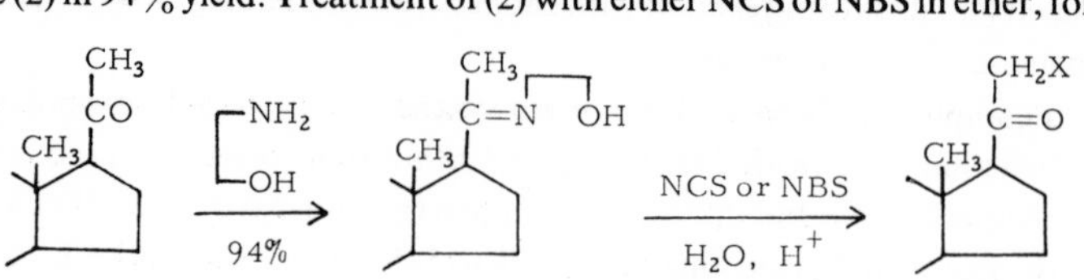

(1)           (2)           (3)

hydrolysis with dilute hydrochloric acid, gives the corresponding 21-halo derivative in high yield (90–100 %). The halogenation reaction probably proceeds through the tautomeric enamine form of (2):

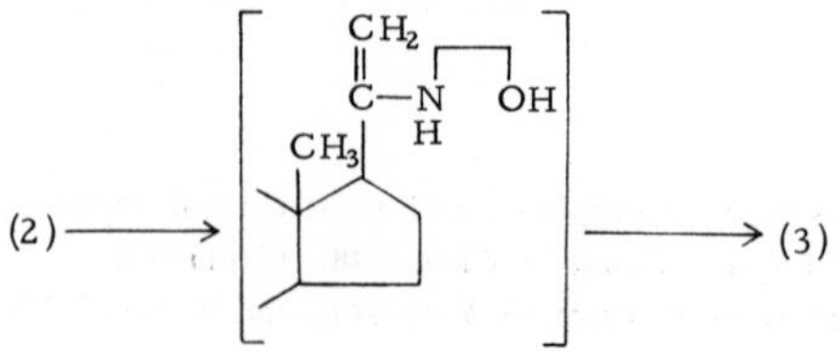

*222*

It is noteworthy that the 5,6-double bond need not be protected.

The procedure was also applied to 2-pentanone; the major products resulted from halogenation of the methyl group.

[1]J. F. W. Keana and R. R. Schumaker, *Tetrahedron*, **26**, 5191 (1970).

## N-Ethoxycarbonyl-2-ethoxy-1,2-dihydroquinoline (EEDQ), 2, 191; 3, 137.

*Merrifield peptide synthesis.*[1] The reagent is preferred to DCC for the coupling of Boc-amino acids and peptide resins in the Merrifield solid-phase peptide synthesis.

*Solid-phase coupling reagent for peptide synthesis.*[2] EEDQ has been incorporated into an insoluble polymer derived from styrene and divinylbenzene. The polymeric form is somewhat less effective for peptide synthesis than the monomeric form, but is comparable to the widely used dicyclohexylcarbodiimide–N-hydroxysuccinimide combination.

*Steroidal amides.* Herz and Mantecón[3] have prepared amides of lithocholic acid 3-formate in satisfactory yields by the EEDQ method.

[1]F. Sipos and D. W. Gaston, *Synthesis*, 321 (1971).
[2]J. Brown and R. E. Williams, *Canad. J. Chem.*, **49**, 3765 (1971).
[3]J. E. Herz and R. E. Mantecón, *Org. Prep. Proc. Int.*, **4**, 129 (1972).

## Ethoxycarbonyl isothiocyanate, $C_2H_5OOCN=C=S$. Mol. wt. 131.15, b.p. 44–46°/ 10 mm.

*Preparation.*[1] The reagent is prepared by the reaction of potassium thiocyanate and ethyl chloroformate in acetonitrile (steam bath).

*Synthesis of heterocycles.* The reagent reacts with tertiary enamines such as (1) to form adducts (2), which when treated with primary amines (or ammonia) undergo amine exchange and cyclization to form 4-thiouracils (3).[1]

The reagent reacts with pyrrole (4) at temperatures below 40° to give N'-ethoxy-carbonylpyrrole-2-thiocarboxamide (5) in 93% yield.[2] When (5) is heated with quinoline it is converted into 2-thiopyrrole-1,2-dicarboximide (6). Reaction of pyrrlylpotassium (7, prepared by reaction of pyrrole with potassium in THF) with the reagent leads to N'-ethoxycarbonylpyrrole-1-thiocarboxamide (8) in 45% yield. Brief boiling of (8) with quinoline yields 1-thiopyrrole-1,2-dicarboximide (9) in 58% yield.

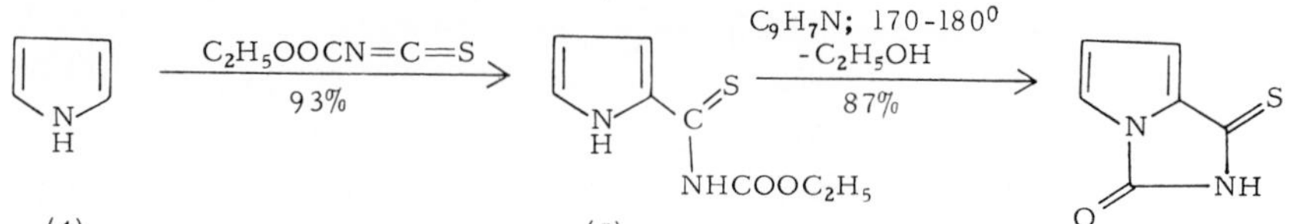

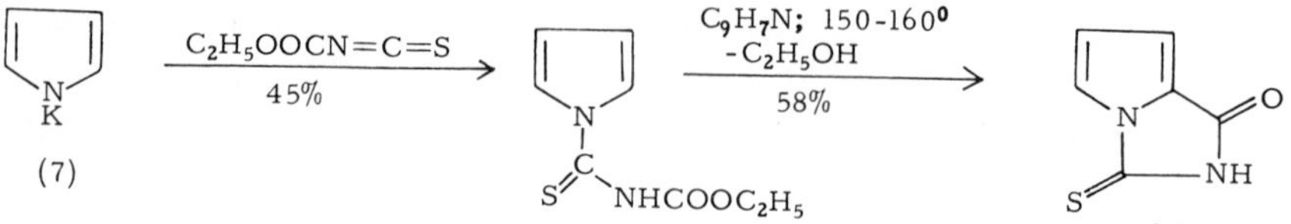

[1]R. W. Lamon, *J. Heterocyclic Chem.*, **5**, 837 (1968).
[2]E. P. Papadopoulos, *J. Org.*, **38**, 667 (1973).

**Ethoxycarbonylmethylcopper**, $CuCH_2COOC_2H_5$. Mol. wt. 150.64.

*Preparation.*[1] This organocopper compound is prepared by addition of an equimolar amount of lithium diisopropylamide to ethyl acetate in the presence of cuprous iodide at $-110°$ under $N_2$; the mixture is then allowed to warm gradually to $-30°$ when a light-brown homogeneous solution is obtained. The temperature is critical; the yield of reagent is much lower if the initial reaction is carried out at $-78°$.

$$CH_3COOC_2H_5 \xrightarrow[73\%]{LiN[CH(CH_3)_2]_2,CuI} CuCH_2COOC_2H_5$$

*γ,δ-Unsaturated esters.*[1] The reagent reacts with allylic halides in THF to give coupling products, that is, γ,δ-unsaturated esters.
Examples:

$$CH_2=CBrCH_2Br \xrightarrow{83\%} CH_2=CBrCH_2CH_2COOC_2H_5$$

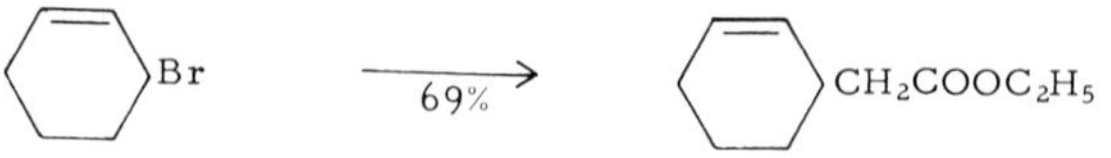

$$CH_3OOCCH=CHCH_2Br \xrightarrow{89\%} CH_3OOCCH=CHCH_2CH_2COOC_2H_5$$

$$C_6H_5CH_2Br \xrightarrow{62\%} C_6H_5CH_2CH_2COOC_2H_5$$

[1]L. Kuwajima and Y. Doi, *Tetrahedron Letters*, 1163 (1972).

**Ethyl azidoformate, 1**, 363–364; **2**, 191–192; **3**, 138.

*Reaction with ethoxyacetylene.*[1] The cycloaddition of ethyl azidoformate with ethoxyacetylene (room temperature, 35 days) is nonregiospecific and leads to the formation of about equal amounts of 1-carboethoxy-5-ethoxy-1,2,3-triazole (1) and 1-carboethoxy-4-ethoxy-1,2,3-triazole (2). In addition, a trace of the N-2 triazole (3)

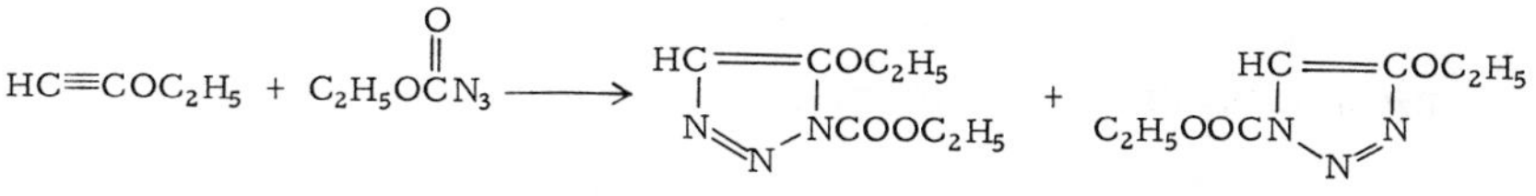

$$HC{\equiv}COC_2H_5 \; + \; C_2H_5O\overset{\overset{O}{\|}}{C}N_3 \; \longrightarrow$$

(1)                                    (2)

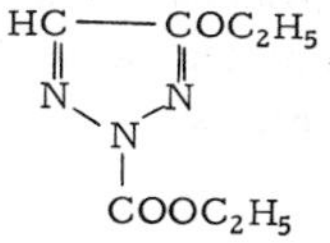

(3)

is formed by isomerization (probably thermal) of triazoles (1) and (2). Both (1) and (2) are readily isomerized to (3) by 1,4-diazabicyclo[2.2.2]octane (DABCO, **2**, 99–101).

The reaction of excess ethyl azidoformate with the ynamine N,N-diethylamino-propyne, however, is regiospecific; if it is carried out in $CCl_4$ at room temperature for 20 min., only the triazole (4) is observed by NMR spectral measurements. If the reaction

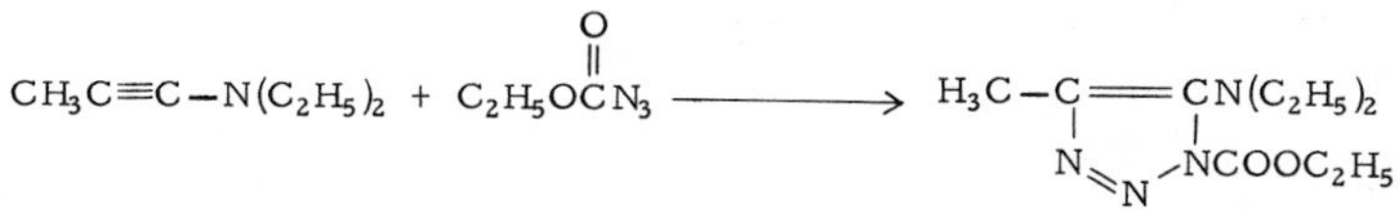

$$CH_3C{\equiv}C-N(C_2H_5)_2 \; + \; C_2H_5O\overset{\overset{O}{\|}}{C}N_3 \; \longrightarrow$$

(4)

is carried out with equimolar amounts of the two reagents in methylene dichloride at room temperature, a mixture of (4) and (5) is obtained in the ratio of 95:5. Addition of the ynamine to this mixture results in further conversion of (4) to (5). Thus ynamines, like DABCO, catalyze isomerization of N-1 acyl-1,2,3-triazoles to N-2 acyl-1,2,3-triazoles.

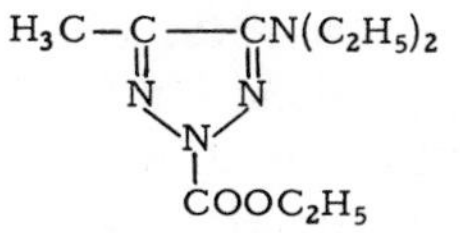

(5)

[1]P. Ykman, G. L'abbé, and G. Smets, *Chem. Ind.*, 886 (1972).

**Ethyl 3-bromopropyl acetaldehyde acetal (1-Ethoxyethyl 3-bromopropyl ether),**

$BrCH_2CH_2CH_2OCHCH_3$ .   Mol. wt. 212.12, b.p. 49–51° (1 mm.).
$\qquad\qquad\quad |$
$\qquad\qquad OCH_2CH_3$

The reagent (1) is prepared in 92% yield by the addition of 3-bromopropanol (Eastman) to ethyl vinyl ether catalyzed by dichloroacetic acid.[1]

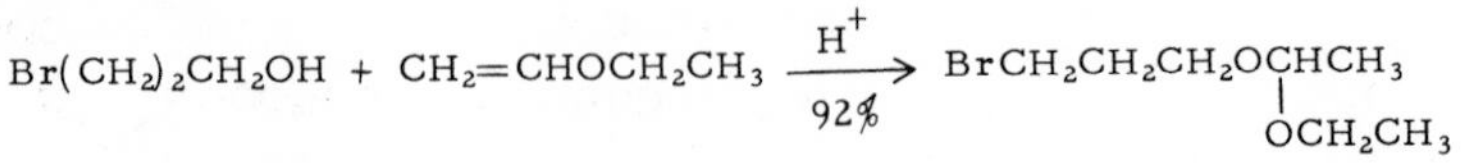

$$Br(CH_2)_2CH_2OH + CH_2{=}CHOCH_2CH_3 \xrightarrow[92\%]{H^+} BrCH_2CH_2CH_2OCHCH_3$$
$$\qquad\qquad\qquad\qquad\qquad\qquad\qquad\qquad\qquad\qquad\qquad\qquad |$$
$$\qquad\qquad\qquad\qquad\qquad\qquad\qquad\qquad\qquad\qquad\qquad OCH_2CH_3$$

(1)

*Hydroxypropylation.*[1]   Reaction of (1) in ether with lithium wire (1% sodium) at 0° gives the organometallic reagent (2), which is a satisfactory carrier of the hydroxypropyl group. Thus (2) reacts with cyclohexanone in ether to give the adduct (3) in 90% yield.

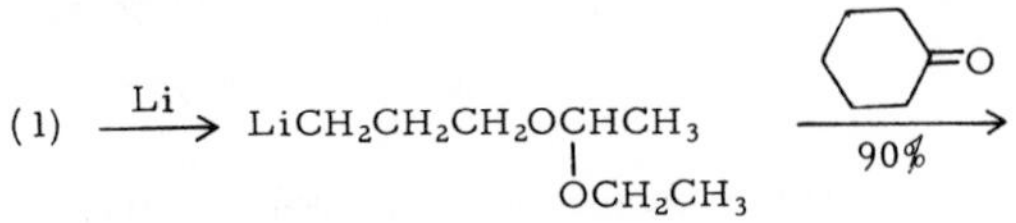

$$(1) \xrightarrow{\;Li\;} LiCH_2CH_2CH_2OCHCH_3 \xrightarrow{\;90\%\;}$$

(2)

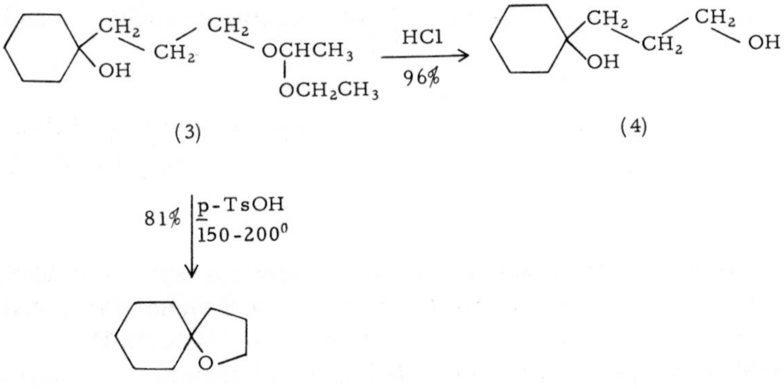

(3)                                        (4)

$$81\% \left|\begin{array}{c} p\text{-TsOH} \\ 150\text{-}200° \end{array}\right.$$

(5)

Hydrolysis of (3) with hydrochloric acid in aqueous ethanol gives 1-(3-hydroxypropyl)-cyclohexanol (4) in 96% yield. Treatment of the adduct (3) with *p*-toluenesulfonic acid under reduced pressure at 150–200° gives 1-oxaspiro[4.5]decane (5).

Reaction at −60° of the lithium reagent (2) with 0.5 eq. of cuprous iodide gives the lithium organocuprate reagent (6). This reagent undergoes conjugate addition to cyclopentenone to give the adduct (7), which on hydrolysis with dichloroacetic acid gives 3-(3-hydroxypropyl)cyclopentanone (8).

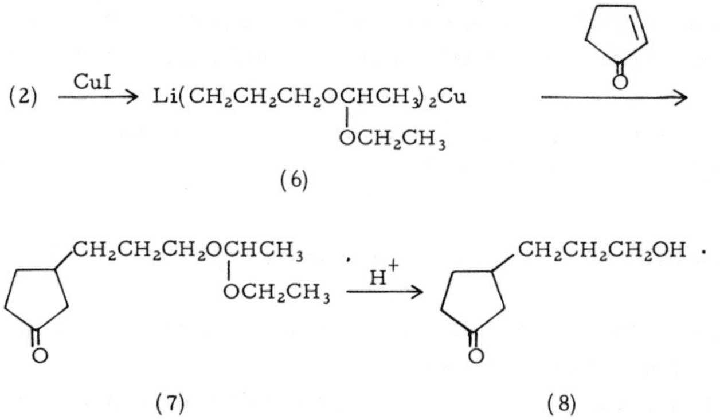
(2)
CuI
Li(CH2CH2CH2OCHCH3)2Cu
OCH2CH3
CH2CH2CH2OCHCH3
OCH2CH3
H+
CH2CH2CH2OH

(6) (7) (8)

[1]P. E. Eaton, G. F. Cooper, R. C. Johnson, and R. H. Mueller, *J. Org.*, **37**, 1947 (1972).

**Ethyl(carboxysulfamoyl)triethylammonium hydroxide inner salt,** $C_2H_5O_2CN^-SO_2N^+$-$(C_2H_5)_3$. Mol. wt. 252.34, m.p. 66–69°.

***Preparation.***[1] The reagent is prepared in 81% yield by the reaction of carboethoxysulfamoyl chloride with 2 eq. of triethylamine in benzene solution at 30°.

$$C_2H_5O_2CNHSO_2Cl \xrightarrow[81\%]{2\,(C_2H_5)_3N} C_2H_5O_2C\bar{N}SO_2\overset{+}{N}(C_2H_5)_3 + (C_2H_5)_3N\cdot HCl$$

***Dehydration.*** Burgess *et al.*[2] reported that the reagent is useful for dehydration of simple alcohols. The reaction is a stereospecific *cis* elimination and follows Saytzeff's rule. Crabbé and León[3] have used this procedure with various steroidal secondary and tertiary alcohols. They conclude that the nature of the alcohol group, the configuration, and the environment are the primary factors governing the course of dehydration. The reactions are carried out at room temperature in anhydrous benzene for 2 hr. followed

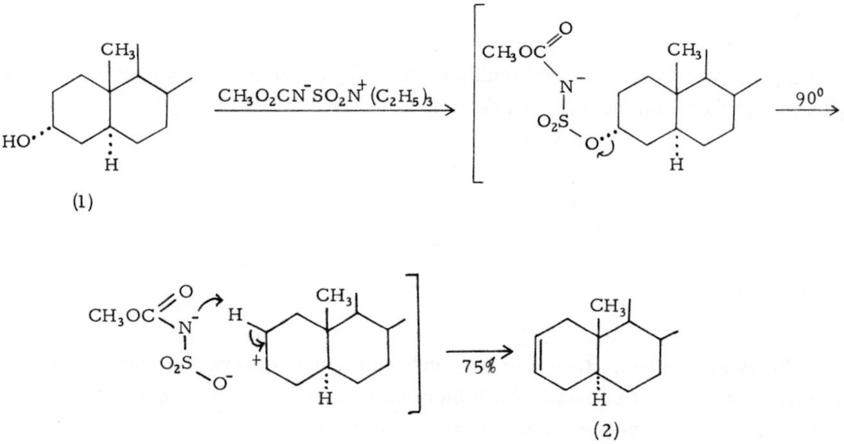
CH3
HO
H
CH3O2CN⁻SO2N⁺(C2H5)3
CH3OC
O
N⁻
O2S
O
90°
CH3OC
N⁻
H
O2S
O⁻
+
75%

(1) (2)

by decomposition at 90° under vacuum. Yields are often satisfactory. Thus 3α-hydroxy-5α-androstene-17-one (1) is converted into $\Delta^2$-5α-androstene-17-one (2) in 75% yield.

[1]G. M. Atkins, Jr. and E. M. Burgess, *Am. Soc.*, **90**, 4744 (1968).
[2]E. M. Burgess, E. A. Taylor, and H. P. Penton, Jr., 159th National Meeting, A.C.S. Houston, Texas, Feb. 1970, Abstracts ORGN-105.
[3]P. Crabbé and C. León, *J. Org.*, **35**, 2594 (1970).

**Ethyl chloroformate** (also known as **Ethyl chlorocarbonate**), $ClCO_2C_2H_5$, **1**, 364–367; **2**, 193.

The reagent is used in the first step of the mixed carboxylic–carbonic anhydride synthesis of 1-phenylcyclopentylamine.[1]

*Alkylation of imides.*[2] Ethyl chloroformate in DMF reacts with lithium phthalimide at temperatures of 60–110° to give N-ethylphthalimide (2) in 80% yield.

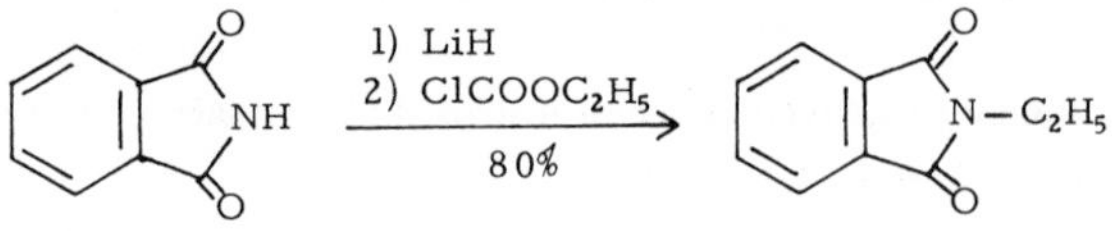

(1)                    (2)

[1]C. Kaiser and J. Weinstock, *Org. Syn.*, **51**, 48 (1972).
[2]J. A. Vida, *Tetrahedron Letters*, 3921 (1972).

**Ethyl diazoacetate**, **1**, 367–370; **2**, 193–194; **3**, 138–139.

Addition of ethyl diazoacetate catalyzed by cupric acetate to Cu(II) complexes of porphyrin derivatives allows identification of their vinyl groups on a milligram scale.[1]

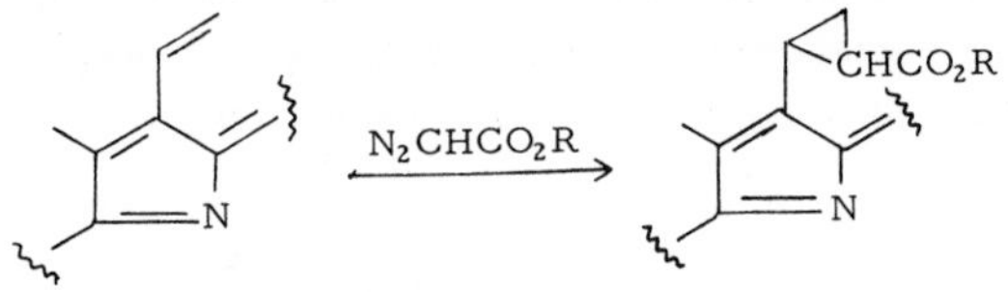

Ethyl diazoacetate reacts with undiluted N-methylisatin at room temperature in the presence of diethylamine in 2–3 weeks to give the crystalline adduct (2)[2]:

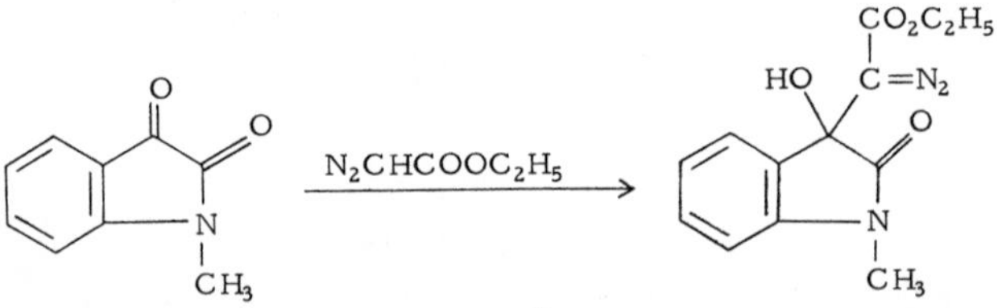

(1)                    (2)

*Propargylic esters.* Hooz and Layton[3] report that trialkynylboranes (2) can be prepared readily by treatment of a lithium acetylide (1, from a terminal alkyne and *n*-butyllithium) with boron trifluoride etherate in THF at −20°. Subsequent treatment of (2) with ethyl diazoacetate gives, after hydrolysis, propargylic esters (3) in yields of about 80–90% (isolated).

$$3 \ RC{\equiv}CLi \ + \ BF_3{\cdot}O(C_2H_5)_2 \ \xrightarrow[-20^0]{THF} \ (RC{\equiv}C)_3B$$

(1)                                                    (2)

$$(2) \ + \ N_2CHCOOC_2H_5 \ \xrightarrow[\text{2) } H_2O]{\text{1) THF, } -20^0} \ RC{\equiv}CCH_2COOC_2H_5$$

(3)

This reaction also provides an attractive route to $\gamma$-keto esters, since mercuric ion-promoted hydration of $\beta,\gamma$-acetylenic esters proceeds with regiospecificity to give $\gamma$-keto esters (4).

$$(3) \ \xrightarrow{Hg^{2+}, \ H_2O} \ RCOCH_2CH_2COOC_2H_5$$

(4)

**Reaction with dialkylchloroboranes.**[4] The reagent reacts with dialkylchloroboranes[5] at $-78°$ to give ethyl alkylacetates by two-carbon homologation. Yields are in the range 90–95%. Brown et al. suggest that the reaction involves the following steps: coordination of the diazo compound and the borane (equation 1), followed by

$$(1) \ R_2BCl \ + \ N_2CHCOOC_2H_5 \ \longrightarrow \ R-\underset{\underset{\overset{|}{N_2^+}}{\overset{|}{Cl}}}{\overset{\overset{\displaystyle R}{|}}{B^-}}-CHCOOC_2H_5$$

I

$$(2) \ R-\underset{\underset{\overset{|}{N_2^+}}{\overset{|}{Cl}}}{\overset{\overset{\displaystyle R}{|}}{B^-}}-CHCOOC_2H_5 \ \longrightarrow \ R\underset{\overset{|}{Cl}}{\overset{\overset{\displaystyle R}{|}}{B}}CHCOOC_2H_5 \ + \ N_2$$

II

$$(3) \ R-\underset{\underset{\overset{|}{N_2^+}}{\overset{|}{Cl}}}{\overset{\overset{\displaystyle R}{|}}{B^-}}-CHCOOC_2H_5 \ \longrightarrow \ R\underset{\overset{|}{Cl}}{\overset{\overset{\displaystyle R}{|}}{B}}CHCOOC_2H_5 \ + \ N_2$$

III

$$(4) \ II \ or \ III \ \xrightarrow{CH_3OH} \ R\underset{\overset{|}{OCH_3}}{\overset{\overset{\displaystyle R}{|}}{B}}CHCOOC_2H_5 \ + \ HCl$$

$$\Big\downarrow CH_3OH$$

$$RB(OCH_3)_2 \ + \ RCH_2COOC_2H_5$$

loss of nitrogen and migration of an alkyl group (equation 2) or of chlorine (equation 3). The final step (equation 4) involves methanolysis of either intermediate.

[1] H. Budzikiewicz and K. Taraz, *Ann.*, **737**, 128 (1970).
[2] B. Eistert and G. Borggrefe, *ibid.*, **718**, 142 (1968).
[3] J. Hooz and R. B. Layton, *Canad. J. Chem.*, 1105 (1972).
[4] H. C. Brown, M. M. Midland, and A. B. Levy, *Am. Soc.*, **94**, 3662 (1972).
[5] *Idem, ibid.*, **94**, 2114 (1972).

**Ethyldiisopropylamine**, 1, 371.

*Pyrrolo[1,2-d]-as-triazine.*[1] This $10\pi$-heteroaromatic system (2) has been synthesized by the base-catalyzed cyclodehydration of pyrrole-2-carboxaldehyde formylhydrazone (1) in refluxing xylene in 56% yield.

$$[(CH_3)_2CH]_2NC_2H_5$$
$$56\%$$

(1)                                                              (2)

[1] J. P. Cress and D. M. Forkey, *J.C.S. Chem. Comm.*, 35 (1973).

**Ethyl 1,3-dithiane-2-carboxylate** (1). Mol. wt. 192.30, b.p. 75–77°/0.2 mm.

*Preparation.*[1] The reagent (1) is prepared in 70% yield by the reaction of ethyl diethoxyacetate[2] with 1,3-propanedithiol in the presence of boron trifluoride etherate.

$$(C_2H_5O)_2CHCOOC_2H_5 + HS(CH_2)_3SH \xrightarrow[70\%]{\substack{BF_3 \cdot (C_2H_5)_2O \\ CHCl_3}}$$

(1)

*α-Keto esters.*[1] Ethyl 1,3-dithiane-2-carboxylate (1) can be converted into the sodium salt by reaction with sodium hydride in DMF–benzene. Reaction of this with a primary or secondary halide gives an ethyl 2-alkyl-1,3-dithiane-2-carboxylate (2) in 75–95% yield. Desulfurization of (2) with Raney nickel affords an alkylated acetic ester (3). Of more significance, treatment of (2) with N-bromosuccinimide (*see* **1,3-Dithiane**, this volume) affords an α-keto ester (4) in 60–85% yield. This new α-keto ester synthesis has the advantage that the precursor (1) is readily available and that alkyllithium reagents are not used.

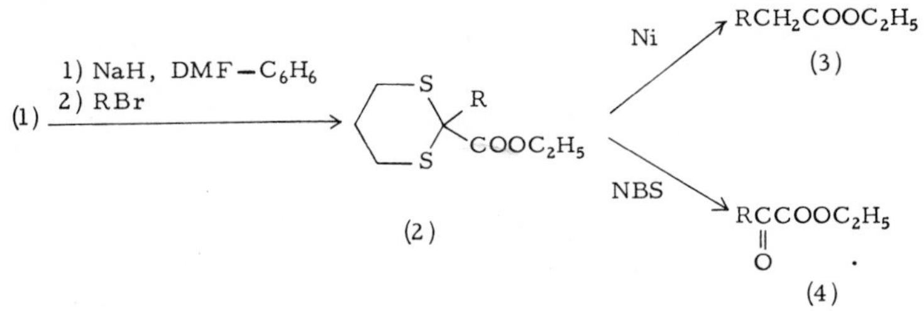

$$(1) \xrightarrow{\substack{1)\ \text{NaH, DMF} - \text{C}_6\text{H}_6 \\ 2)\ \text{RBr}}} \underset{(2)}{\text{(ring with S, S, R, COOC}_2\text{H}_5)}$$

$$\xrightarrow{\text{Ni}} \underset{(3)}{\text{RCH}_2\text{COOC}_2\text{H}_5}$$

$$\xrightarrow{\text{NBS}} \underset{(4)}{\underset{\text{O}}{\overset{\text{RCCOOC}_2\text{H}_5}{\underset{\|}{}}}}$$

[1]E. L. Eliel and A. A. Hartmann, *J. Org.*, **37**, 505 (1972).
[2]R. B. Moffett, *Org. Syn., Coll. Vol.*, **4**, 427 (1963).

**Ethylenediamine (EDA)**, 1, 372–373.

*Propargylic rearrangement.*[1] Strong bases (sodium amide, *n*-butyllithium) do not rearrange 3-hexyne within 72 hr. at room temperature. However, the combination of such bases and EDA rearranges 3-hexyne rapidly. With the molar ratio of 3-hexyne to sodium amide to EDA being 18:1:18, the products are 2-hexyne (80%), 3-hexyne (12%), 1-hexyne (4%), and 2,3-hexadiene (4%). The actual rearrangement reagent is NaEDA. A concerted mechanism involving a nine-membered ring transition state (a) is suggested for the rearrangement.

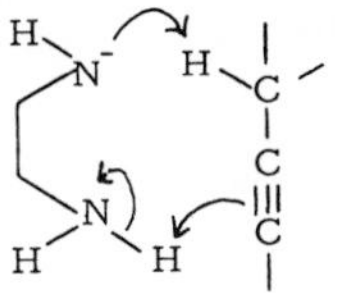

(a)

[1]J. H. Wotiz, P. M. Barelski, and D. F. Koster, *J. Org.*, **38**, 489 (1973).

**Ethylene oxide**, 1, 377–378; 2, 196–197; 3, 140.

*Generation of dihalocarbenes* (2, 196–197). Buddrus[1] has reviewed the use of ethylene oxide for generation of dihalocarbenes according to the formulation:

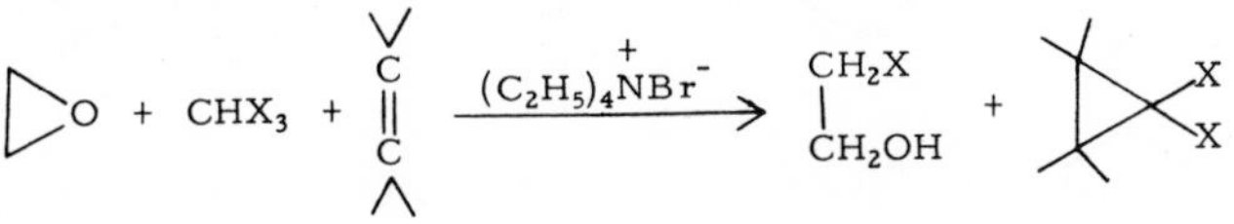

*2,3,3a,8a-Tetrahydrofuro[2,3-b]indoles.* A small number of alkaloids from Calabar beans contain this ring system. One of these, physovenine (3), has been synthesized recently.[2] The key step is the reaction of 5-methoxyskatolylmagnesium iodide (1, prepared from 5-methoxyskatole and methylmagnesium iodide) with anhydrous ethylene oxide in ether. The product (2) is obtained in 13% yield.

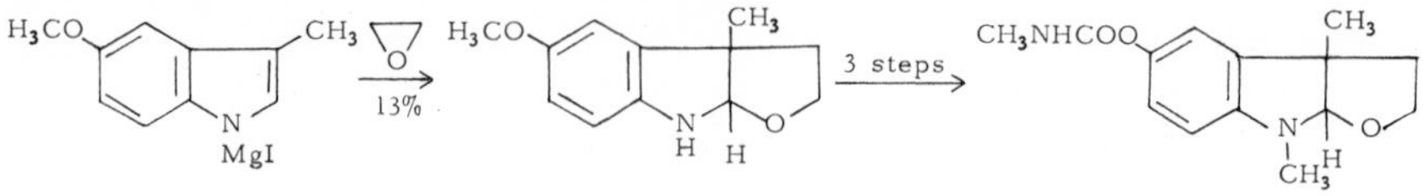

(1)                    (2)                    (3)

[1]J. Buddrus, *Angew. Chem., internat. Ed.*, **11**, 1041 (1972).
[2]T. Onaka, *Tetrahedron Letters*, 4391 (1971).

**Ethyl 3-ethyl-5-methyl-4-isoxazolecarboxylate.** (Compare **1**, 276–277; **2**, 150–151.)
   *Synthesis[1]:*

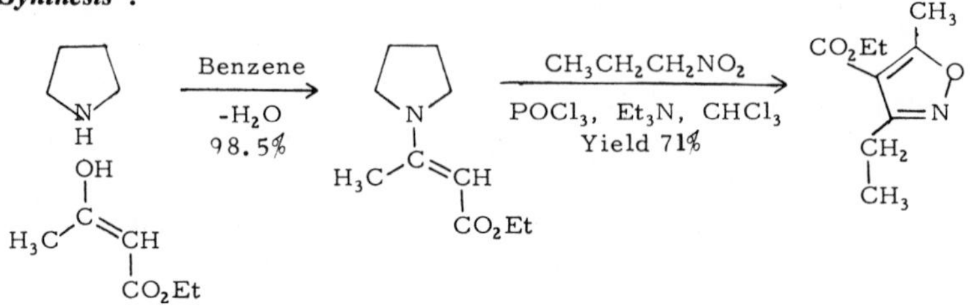

*Isoxazole annelation reaction[2]:*

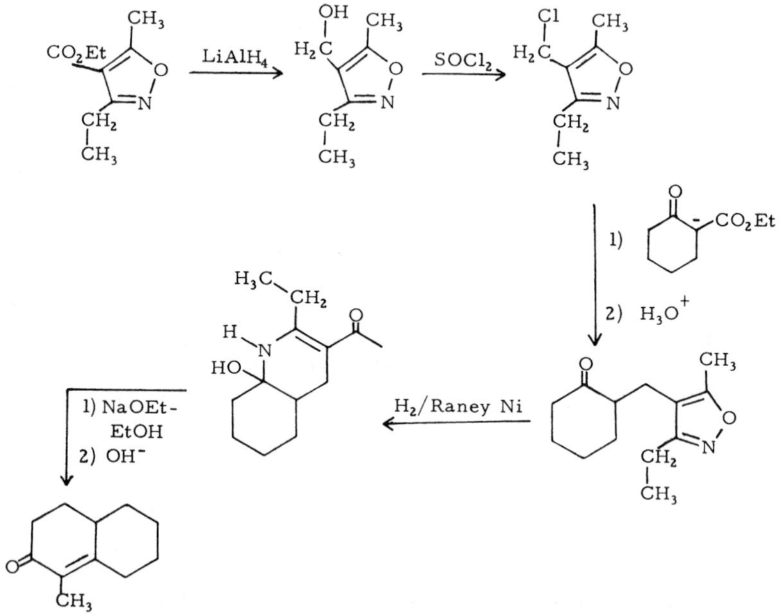

1-Methyl-4,4a,5,6,7,8-hexahydro-
   naphthalene-2(3H)-one

[1]J. E. McMurry, *Org. Syn.*, submitted 1972.
[2]*Idem, ibid.*, submitted 1972.

**Ethyl formate, 1,** 380–383; **2,** 197.

*Hydroxymethylation of ketones.*[1] A ketone can be converted into the α-hydroxy-methyl derivative in two steps: acylation with ethyl formate, followed by aluminum hydride reduction. The sequence is illustrated for the conversion of 4-*tert*-butylcyclo-hexanone (1) into 2-hydroxymethyl-4-*tert*-butylcyclohexanone (3).

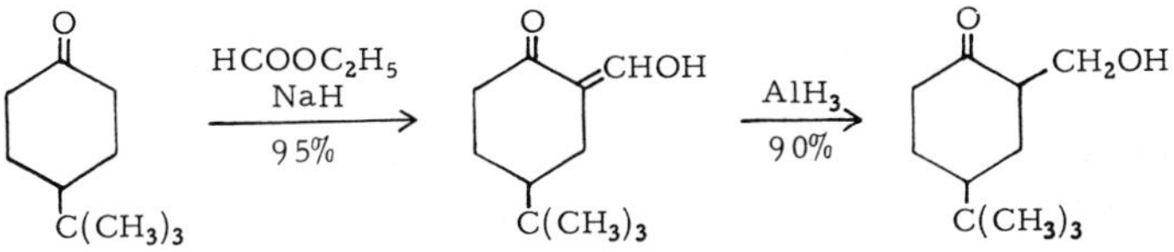

(1)                    (2)                    (3)

[1] E. J. Corey and D. E. Cane, *J. Org.*, **36,** 3070 (1971).

**Ethyl trichloroacetate, 1,** 386; **3,** 143. Additional suppliers: Aldrich, K and K, Koch-Light, Pfaltz and Bauer, Sargent, Schuchardt.

*Dichlorocarbene* (**3,** 143). This reaction is employed in the first of a three-step synthesis of bicyclo[3.2.1]octane-3-one from norbornene[1] (yield of initial adduct 74–88%). A solution of 52.5 g. (0.56 mole) of norbornene in 400 ml. of petroleum ether with 112 g. of NaOCH$_3$ in a 1-1. four-necked, round-bottomed flask is stirred

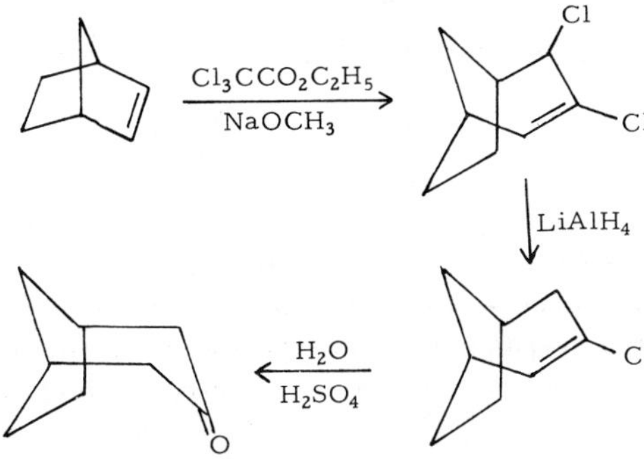

mechanically in a salt-ice bath and 112 g. (2.06 moles) of ethyl trichloroacetate is placed in an addition funnel and allowed to drip slowly into the stirred mixture at a rate such that the temperature does not rise above 0°. The addition requires about 4 hr. and the initially white mixture becomes increasingly yellow in color. The mixture is allowed to come to room temperature overnight and poured onto crushed ice and water. The organic layer is separated and the aqueous layer is extracted with four portions of ether, neutralized with hydrochloric acid, and extracted further with ether. Finally the ethereal extracts are combined, washed with saturated sodium chloride solution, dried over magnesium sulfate, concentrated, and distilled to give 72.5–87.0 g. (74–88%) of *exo*-3,4-dichlorobicyclo[3.2.1]oct-2-ene, b.p. 72–73° (0.9 mm.).

The next step of selective replacement of chlorine on saturated $C_4$ by hydrogen without disturbance of the chlorine on unsaturated $C_3$ is accomplished efficiently by lithium aluminum hydride in ether to give 3-chlorobicyclo[3.2.1]oct-2-ene (yield 74–75%). The final step, involving stirring with concentrated sulfuric acid and addition of ice, undoubtedly proceeds through the enol:

$$RCH=\underset{|}{C}-Cl \longrightarrow RCH=\underset{|}{C}-OH \longrightarrow RCH_2\underset{|}{C}=O$$

***Reformatsky reagent.***[2] Ethyl trichloroacetate reacts with zinc in THF at $-15°$ to form a stable chlorozinc enolate (1). This reagent can be condensed with a number of

$$Cl_3C-C\overset{O}{\underset{OC_2H_5}{\diagdown}} \xrightarrow{\text{Zn, THF}} Cl_2C=C\overset{OZnCl}{\underset{OC_2H_5}{\diagdown}}$$

(1)

electrophiles such as carbonyl compounds, acid anhydrides, and alkyl halides.

(1)  +  $ClCH_2OC_2H_5$  $\xrightarrow[56\%]{}$  $C_2H_5OCH_2CCl_2COOC_2H_5$

(1)  +  $CH_3CH_2CHO$  $\xrightarrow[59\%]{}$  $CH_3CH_2CHOHCCl_2COOC_2H_5$

(1)  +  $(CH_3CO)_2O$  $\xrightarrow[65\%]{}$  $CH_3COCCl_2COOC_2H_5$

[1]C. W. Jefford, J. Gunsher, D. T. Hill, P. Brun, J. Le Gras, and B. Waegell, *Org. Syn.*, **51**, 60 (1971).
[2]B. Castro, J. Villieras, and N. Ferracutti, *Bull. soc.*, 3521 (1969).

**Ethyl vinyl ether, 1,** 386–388; **2,** 198.

***γ,δ-Unsaturated aldehydes.***[1] Vinyl ethers react with tertiary vinylcarbinols to give γ,δ-unsaturated aldehydes in good yield. Phosphoric acid is used as catalyst. The reaction is illustrated for the preparation of 5-methyl-4-hexenal (5) from 2-methyl-3-butene-2-ol (1) and ethyl vinyl ether (2). The reaction is believed to proceed via the mixed acetal

(1)          (2)                    (3)                              (4)

+  $CH_3CH(OC_2H_5)_2$

(5)                              (6)

(3), which when heated in the presence of an acid catalyst loses ethanol which reacts with excess ethyl vinyl ether to form the acetal (6). The resulting allyl vinyl ether (4) then undergoes Claisen rearrangement to afford the product (5) in 52–60% yield. Compare the synthesis of $\gamma,\delta$-unsaturated ketones from tertiary vinylcarbinols and isopropenyl methyl ether (**2**, 231).

*Introduction of functionalized angular methyl groups.*[2] The reaction of ethyl vinyl ether with 10-methyl-$\Delta^{1(9)}$-2-octalol (1) catalyzed by mercuric acetate in a sealed Carius tube for 12 hr. at 200° gives 9-formylmethyl-10-methyl-$\Delta^1$-octalin (2, 85% yield) together with some dienes (3, 15% yield). Use of phosphoric acid leads only to the dienes (3). The one-step procedure involves formation of the vinyl ether followed by Claisen

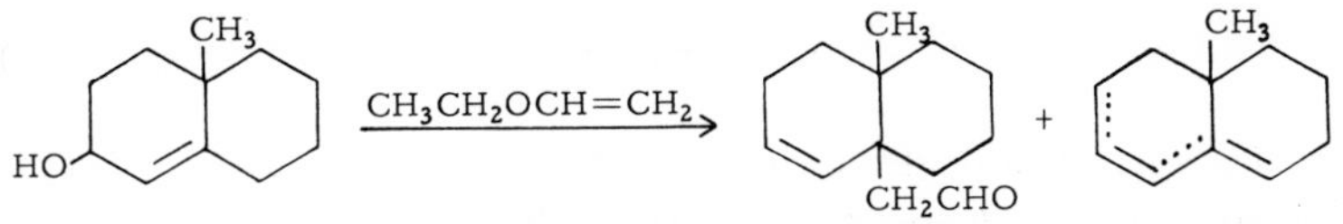

|   (1)   |   (2, 85%)   |   (3, 15%)   |

rearrangement. The method fails with the hydrindenyl ring system; in this case the elimination leading to dienes is the only reaction. The difficulty can be overcome by a two-step procedure: formation of the vinyl ether, followed by thermolysis.

[1]R. Marbet and G. Saucy, *Helv.*, **50**, 2095 (1967); *Org. Syn.*, submitted 1971.
[2]W. G. Dauben and T. J. Dietsche, *J. Org.*, **37**, 1212 (1972).

# F

**Ferric chloride, 1,** 390–392; **2,** 199; **3,** 145.

*Alkenylation of Grignard reagents.*[1] Facile vinylation of a Grignard reagent can be achieved with vinyl bromide with ferric chloride as catalyst and THF as solvent. For example, the reaction of 5-hexenylmagnesium bromide with vinyl bromide in THF at 25° catalyzed by ferric chloride gives 1,7-octadiene in 64% yield:

$$H_2C=CH(CH_2)_4MgBr \ + \ BrCH=CH_2 \ \xrightarrow[64\%]{FeCl_3} \ H_2C=CH(CH_2)_4CH=CH_2 \ + \ MgBr_2$$

The reaction appears to be stereoselective because the reaction of methylmagnesium bromide with *cis*- and *trans*-1-bromopropene affords only *cis*- and *trans*-butene-2, respectively.

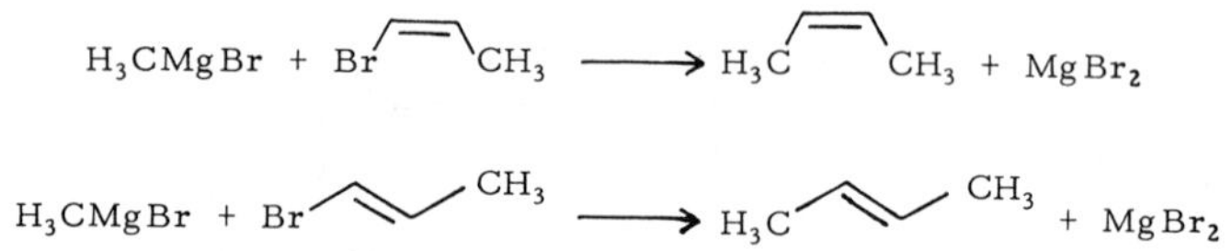

$$H_3CMgBr \ + \ Br\overset{}{\diagup}\diagdown CH_3 \ \longrightarrow \ H_3C\overset{}{\diagup}\diagdown CH_3 \ + \ MgBr_2$$

$$H_3CMgBr \ + \ Br\diagup\diagdown\diagup CH_3 \ \longrightarrow \ H_3C\diagup\diagdown\diagup CH_3 \ + \ MgBr_2$$

[1]M. Tamura and J. Kochi, *Synthesis,* 303 (1971).

**Ferric chloride–Dimethylformamide.**

*Oxidative coupling.* In an investigation of oxidative coupling of phenols, Japanese chemists[1] obtained promising results with ferric chloride in DMF. On further investigation they isolated a new crystalline complex of ferric chloride and DMF with the formula $[Fe(DMF)_3Cl_2][FeCl_4]$, mol. wt. 543.74, m.p. 220°. The complex is obtained in 95.2% yield by addition of ferric chloride (1 mole) in dry ether to DMF (1.5 moles). This complex is very effective for both intramolecular and intermolecular oxidative coupling of phenols. A typical reaction is carried out by addition of the complex (10 mmole) in water to a solution of the phenol (1 mmole) in ether; the mixture is refluxed with stirring for 1 hr. In this way the phenol (1) was oxidized to the dienone (2) in 67% yield.

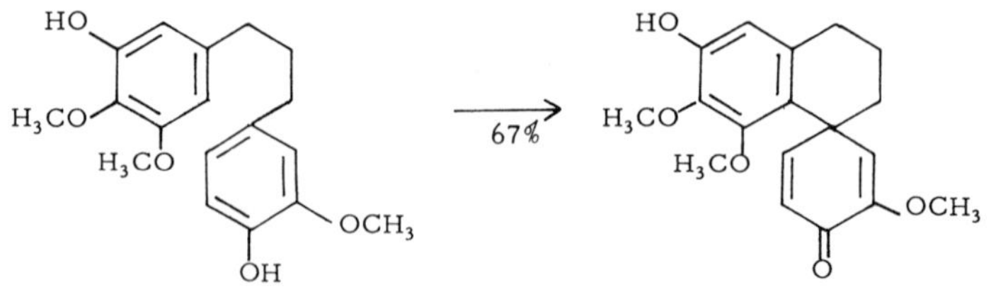

(1)  (2)

236

An example of intermolecular oxidative coupling with the new complex is the oxidation of $p$-cresol (3) to Pummerer's ketone (4) in 28% yield.

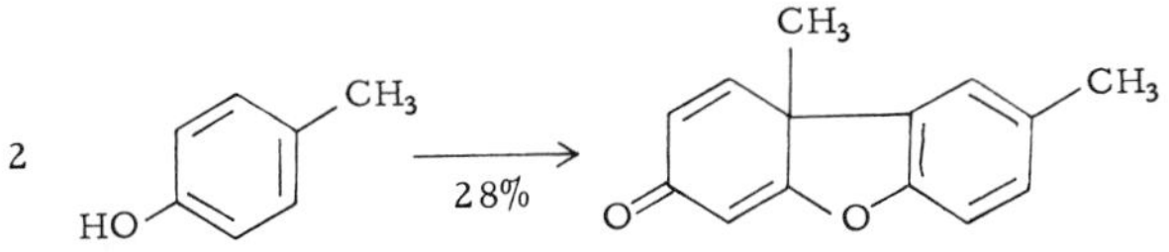

(3)                                    (4)

[1]S. Tobinaga and E. Kotani, *Am. Soc.*, **94**, 309 (1972).

**Ferrous sulfate, 1**, 393. Additional supplier: ROC/RIC.

*Carboxylation of heteroaromatic bases.*[1] Heterocyclic aromatic bases can be carboxylated by the redox decomposition of oxyhydroperoxides of α-keto esters. The

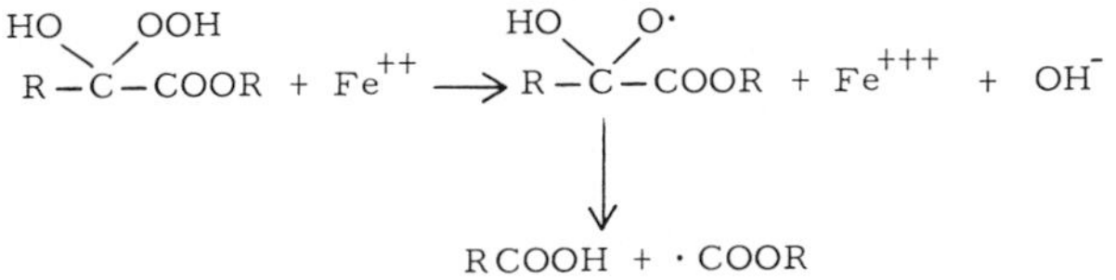

reaction is easily carried out, and the hydroperoxide need not be isolated. For example, ethyl pyruvate, $CH_3COCOOC_2H_5$, is treated, with cooling ($-10°$), with aqueous hydrogen peroxide. This mixture and a solution of $FeSO_4 \cdot 7H_2O$ in water is then added to a stirred solution of the heterocyclic base in water containing $H_2SO_4$ (0–5°). In this way,

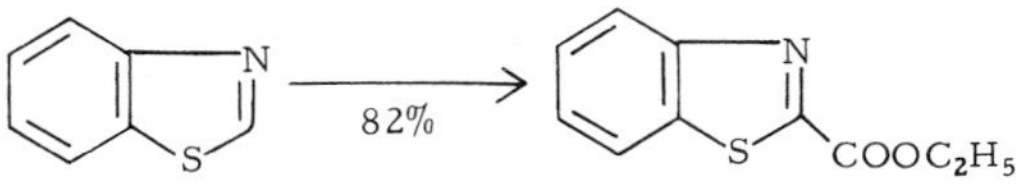

(1)                                    (2)

benzothiazole (1) was converted into 2-carbethoxybenzothiazole (2) in 82% yield. Pyridine is converted by this method to 2-carbethoxypyridine (58%) and 4-carbethoxypyridine (34%).

[1]R. Bernardi, T. Caronna, R. Galli, F. Minisci, and M. Perchinunno, *Tetrahedron Letters*, 645 (1973).

**9-Fluorenylmethyl chloroformate, 3**, 145–146.

The definitive paper on the 9-fluorenylmethyloxycarbonyl group (FMOC) for protection of an amino group has been published.[1]

[1]L. A. Carpino and G. Y. Han, *J. Org.*, **37**, 3404 (1972).

**Fluoroxytrifluoromethane, 2**, 200; **3**, 146–147.

*Fluorinated steroids.* The action of fluoroxytrifluoromethane on the enol acetates of 3-keto-$\Delta^4$- and 3-keto-$\Delta^{4,6}$-steroids yields derivatives fluorinated at the 6- and 2-positions, respectively.[1]

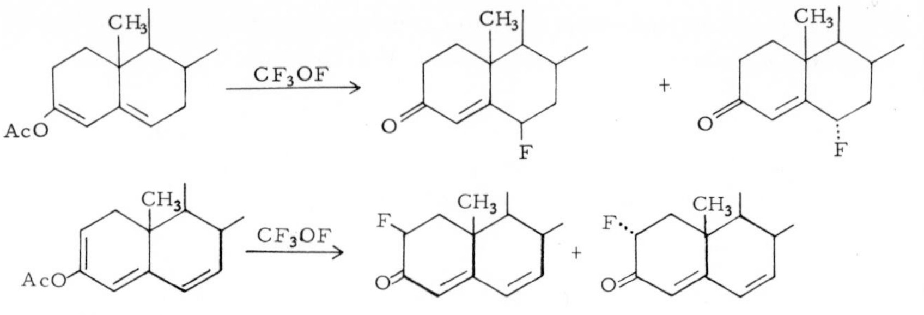

[1] C. Chavis and M. Mousseron-Canet, *Bull. soc.*, 632 (1971).

**Formaldehyde, 1,** 397–402; **2,** 200–201.

*Methylation of amines.* The classical method for methylation of primary or secondary amines is the Eschweiler[1]–Clarke[2] reaction. The reaction involves treatment of an amine with formaldehyde and formic acid:

$$R^1_{\phantom{1}}NH + H_2C{=}O + HCOOH \longrightarrow R^1_{\phantom{1}}NCH_3 + CO_2 + H_2O$$

(with $R^2$ below $R^1$ on both sides)

In some cases, complex mixtures are obtained.[3]

Recently, two new methods have been introduced for this transformation. In one,[4] formic acid is replaced by sodium borohydride. Thus treatment of a steroidal amine of type (1) with a large excess of formaldehyde in methanol followed by reduction with sodium borohydride gives the methylated amine (2) in about 85% yield.

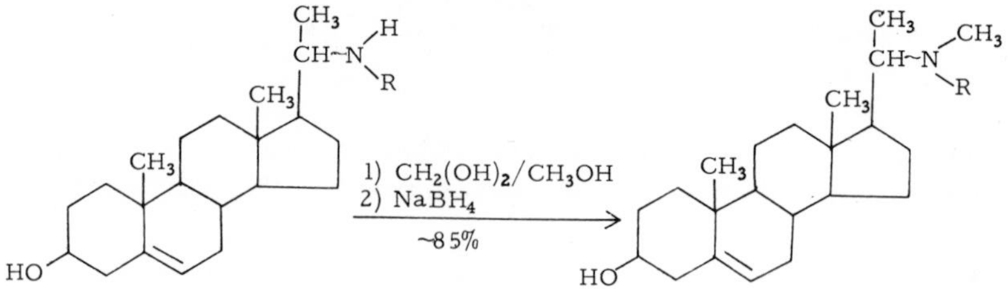

1) $CH_2(OH)_2/CH_3OH$
2) $NaBH_4$
~85%

(1)                                   (2)

In a second method,[5] the amine is treated with aqueous formaldehyde in acetonitrile and the resulting imine reduced with sodium cyanoborohydride (this volume). Yields for the most part are high, 80–90%. Even the very weak base *p*-nitroaniline ($pK_a$ 1.00) is converted into a mixture of mono- and dimethylated products. Hindered amines can be used (*e.g.*, N-isopropylcyclohexylamine is methylated in 87% yield).

*Chloromethylation.* The reagent[6] can be used for the preparation of benzyl chloromethyl ether.[7]

$$C_6H_5CH_2OH + HCl + HCHO \xrightarrow[88-97\%]{10^0} C_6H_5CH_2OCH_2Cl$$

[1] W. Eschweiler, *Ber.*, **38**, 880 (1905).
[2] H. T. Clarke, H. B. Gillespie, and S. Z. Weisshaus, *Am. Soc.*, **55**, 4571 (1933).
[3] S. H. Pine and B. L. Sanchez, *J. Org.*, **36**, 829 (1971).
[4] B. L. Sondengam, J. H. Hémo, and G. Charles, *Tetrahedron Letters*, 261 (1973).
[5] R. F. Borch and A. I. Hassid, *J. Org.*, **37**, 1673 (1972).
[6] *s*-Trioxane (0.493 mole) may be substituted for aqueous 37% formaldehyde solution with no change in the procedure.
[7] D. S. Connor, G. W. Klein, and G. N. Taylor, *Org. Syn.*, **52**, 16 (1972).

**Formic acid, 1,** 404–407; **2,** 202–203.

*Carboxylation* (**1,** 406–407). cis-Bicyclo[3.3.0]octane-1-carboxylic acid (2) can be prepared[1] conveniently in about 25% yield by treatment of cis-bicyclo[3.3.0]octene-2

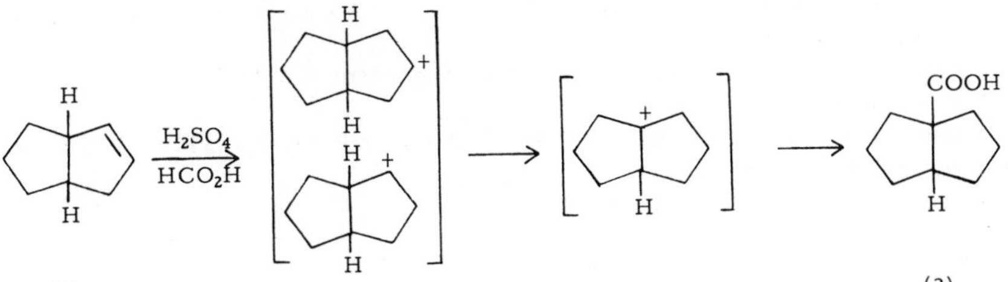

(1)                                                                                      (2)

(1)[2] with 98–100% formic acid and sulfuric acid at room temperature. The reaction involves hydride transfers to the more stable tertiary cation.

*Bicyclic ethers.*[3] Treatment of 3-aryl- or 3-alkyl-3-hydroxyhepta-1,6-dienes (1) with a mixture of formic acid and 70% sulfuric acid at room temperature gives 1-aryl- or 1-alkyl-8-oxabicyclo[3.2.1]octanes (2) in 60–70% yield.

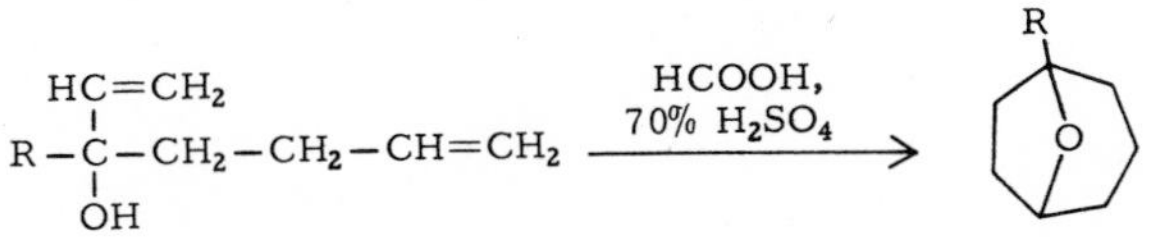

(1)                                                                                      (2)

*Ring C aromatization of steroids.*[4] Treatment of 17$\alpha$-methyl-$\Delta^{6,9(11)}$-testosterone (1) with 90% formic acid under reflux for 15 hr. results in formation of the rearranged 17,17–dimethylandrosta-4,8,11,13-tetraene-3-one (2) in 45% yield.

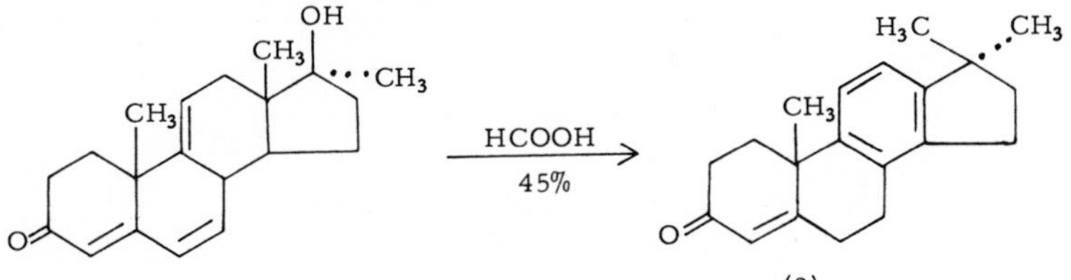

(1)                                                                                      (2)

*Rearrangement of an arylprotoadamantanol* (1) in boiling formic acid affords the corresponding adamantane formate ester (2).[5]

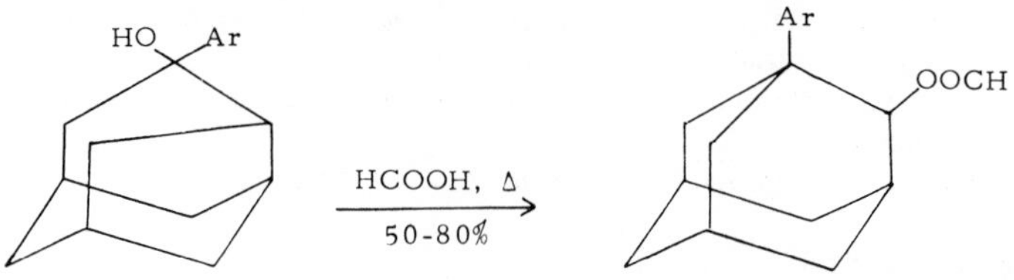

HO   Ar

HCOOH, Δ
──────────→
50-80%

(1)                                    (2)

Ar

OOCH

[1]M. A. McKervey, H. A. Quinn, and J. J. Rooney, *J. Chem. Soc.* (*C*), 2430 (1971).
[2]Prepared in 55% yield by isomerization of cyclooctadiene-1,3 with potassium [P. R. Stapp and R. F. Kleinschmidt, *J. Org.*, **30**, 3006 (1965)].
[3]S. Watanabe, K. Suga, T. Fujita, and Y. Takahashi, *Synthesis*, 422 (1972).
[4]A. B. Turner, *Chem. Ind.*, 932 (1972).
[5]D. Lenoir, *Ber.*, **106**, 78 (1973). See also D. Lenoir, R. Glaser, P. Mison, and P. von R. Schleyer, *J. Org.*, **36**, 1821 (1971).

# G

**Gallium oxide, $Ga_2O_3$.** Mol. wt. 187.44. Suppliers: Alfa Inorganics, ROC/RIC.

Gallium oxide catalyzes the exchange of vinylic and allylic hydrogens by deuterium with high selectivity. It also interconverts *cis*- and *trans*-isomers of butenes and pentenes at relatively low temperatures (110°).

[1]F. B. Charlton, H. A. Quinn, and J. J. Rooney, *J.C.S. Chem. Comm.*, 231 (1973).

**Gases, inert, 1,** 409–410.

*Argon.* An *Organic Syntheses* procedure for the dialkylation of 1,3-dithiane with 1-bromo-3-chloropropane, as a first step in the synthesis of cyclobutanone, specified that argon, if available, be used as the inert gas in preference to nitrogen because of its greater density.[1]

[1]D. Seebach and A. K. Beck, *Org. Syn.*, **51**, 76 (1971).

**Glutaronitrile,[1] Succinonitrile.[2]**

*Nitrile synthesis.* Conversion of acids into nitriles by reaction with acetonitrile at high temperatures (150–300°) has been reported in the patent literature.[3] Klein[4] now reports that the use of succinonitrile, glutaronitrile, or α-methylglutaronitrile leads to improved yields and avoids use of pressure equipment. The reaction can be catalyzed by sulfonic, sulfuric, or phosphoric acids. He has used this method for synthesis of aliphatic dinitriles. Thus, 1,12-dodecanedinitrile can be prepared in high yield by refluxing 1,12-dodecanedioic acid with 2 eq. of succinonitrile, glutaronitrile, or α-methylglutaronitrile for 18 hr. in the presence of an acid catalyst.

$$HOOC(CH_2)_{10}COOH + 2\ NC(CH_2)_nCN \xrightarrow{H^+} NC(CH_2)_{10}CN + 2$$

n = 2, 3

n = 2, 3

[1]$NC(CH_2)_3CN$. Mol. wt. 94.12, b.p. 122–124°/5 mm., b.p. 137–140°/10 mm. Suppliers: Aldrich, Eastman.
[2]$NC(CH_2)_2CN$. Mol. wt. 80.09, b.p. 265–267°, m.p. 53–60°. Suppliers: Aldrich, Eastman.
[3]D. J. Loder, U.S. Patent 2,377,795 (1945); French Patent 1,525,498 (1968).
[4]D. A. Klein, *J. Org.*, **36**, 3050 (1972).

# H

**Hexaethylphosphorous triamide** [**Tris(diethylamino)phosphine**], $[(C_2H_5)_2N]_3P$, **1**, 425; **2**, 207; **3**, 148–149.

*Partial desulfurization of disulfides* (**2**, 207). Harpp and Gleason[1] have reported that alkyl, aralkyl, and alicyclic disulfides undergo ready desulfurization to the corresponding sulfides on reaction with the reagent [tris(dimethylamino)phosphine is equally effective]. The process is stereospecific in that inversion of configuration occurs at one of the carbon atoms α to the disulfide group. Thus desulfurization of (1), *cis*-3,6-dicarbomethoxy-1,2-dithiane, gives a quantitative yield of *trans*-2,5-dicarbomethoxythiolane (2). Similarly (3) is converted quantitatively into (4).

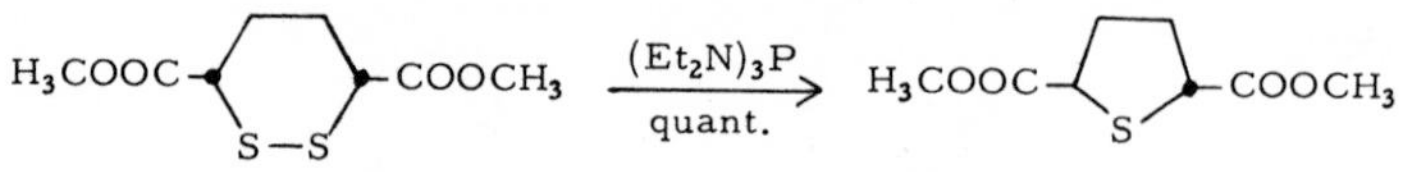

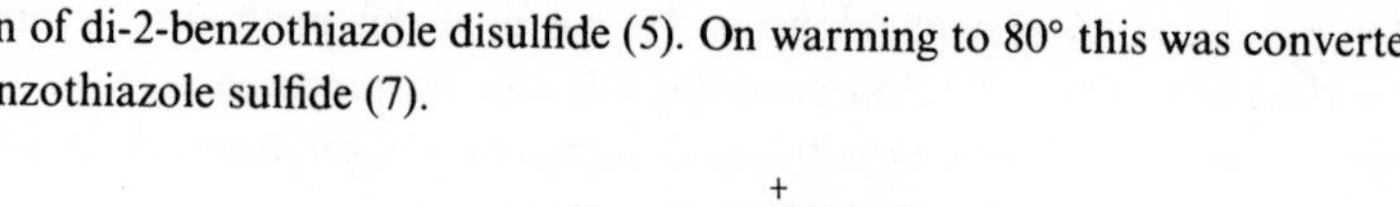

A thermally labile phosphonium salt intermediate (6) was obtained as an oil in the reaction of di-2-benzothiazole disulfide (5). On warming to 80° this was converted into di-2-benzothiazole sulfide (7).

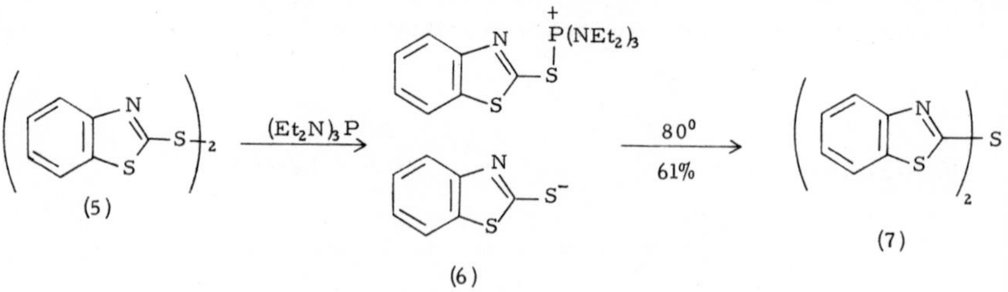

Mechanistic studies suggest that nucleophilic scission of the disulfide bond is the rate-controlling step.

[1] D. N. Harpp and J. G. Gleason, *Am. Soc.*, **93**, 2437 (1971).

**Hexamethylenetetramine, 1,** 427–428; **2,** 208.

*Duff reaction* **(1,** 430). Review.[1] In the classical procedure highly activated aromatic compounds are converted into their formyl derivatives by treatment with hexamethylenetetramine and glyceroboric acid; yields are generally low. Smith[2] finds that a variety of aromatic compounds, including simple hydrocarbons, when treated with hexamethylenetetramine in conjunction with trifluoroacetic acid at reflux temperature (82–90°) are converted into imine products which yield aldehydes on hydrolysis:

$$ArH \; + \; C_6H_{12}N_4 \; \xrightarrow{\begin{array}{c}1)\ CF_3COOH\\ 2)\ H_2O\end{array}} \; ArCHO$$

Even benzene can be converted into benzaldehyde in 32 % yield (sealed tube 125–150°). Yields are high in the case of activated aromatics; thus 2,6-xylenol (1) is converted into 3,5-dimethyl-4-hydroxybenzaldehyde (2) in 95 % yield. The reaction shows *para*

(1)                    (2)

regioselectivity.[3] Thus toluene is converted into *p*-tolualdehyde (50 % yield) and *o*-tolualdehyde (11 % yield).

*Sommelet reaction.* Newman and Hung[3] report an improved synthesis of 3-methyl-2-naphthoic acid (4). 2,3-Dimethylnaphthalene (1) was treated with NBS in carbon

(1)                    (2)

(3)                    (4)

tetrachloride (benzoyl peroxide) to give a mixture rich in 2-bromomethyl-3-methyl-naphthalene (2). The solvent was removed and the mixture treated with hexamethylene-tetramine in acetic acid–water. The aldehyde (3) was obtained in 78 % yield. The final step involved oxidation with silver nitrate under alkaline conditions. The overall yield of (4) from (1) was 69 %.

[1]L. N. Ferguson, *Chem. Rev.*, **38,** 230 (1946).
[2]W. E. Smith, *J. Org.*, **37,** 3972 (1972).
[3]M. S. Newman and W. M. Hung, *Org. Prep. Proc. Int.*, **4,** 227 (1972).

## Hexamethylphosphoric triamide (HMPT), 1, 430–431; 2, 208–210; 3, 149–153.

*Dehydration of alcohols.*[1] Primary and secondary alcohols undergo dehydration when heated in HMPT at temperatures of 220–240°. Primary alcohols give modest yields of 1-alkenes and 1-dimethylaminoalkanes. Secondary alcohols are converted into olefins (no dimethylaminoalkane formation).

Sterically hindered olefins can be obtained by dehydration of tertiary alcohols with HMPT at elevated temperatures. Thus (2) can be obtained from (1) in this way in 77%

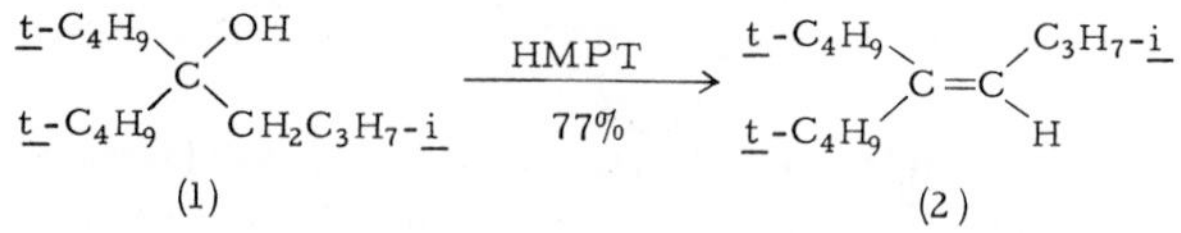

yield. This method, an alternative to ester pyrolysis, is useful when esters are difficult to prepare.[2]

*Nitriles.*[3] Aliphatic and aromatic amides (carboxamides) are dehydrated by HMPT at 220–240° to give nitriles in good yield. A phosphorodiamidate derivative (a) was suggested as an intermediate.

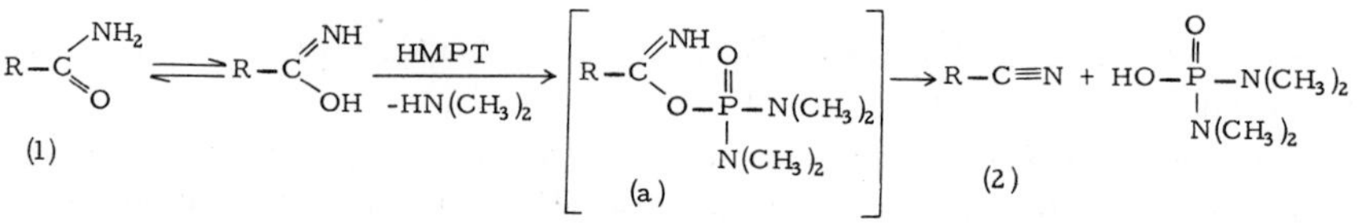

*Amidines.*[4] Secondary amides (carboxamides) (1) when refluxed in HMPT give N,N-dimethylamidines (2) in fair yield.

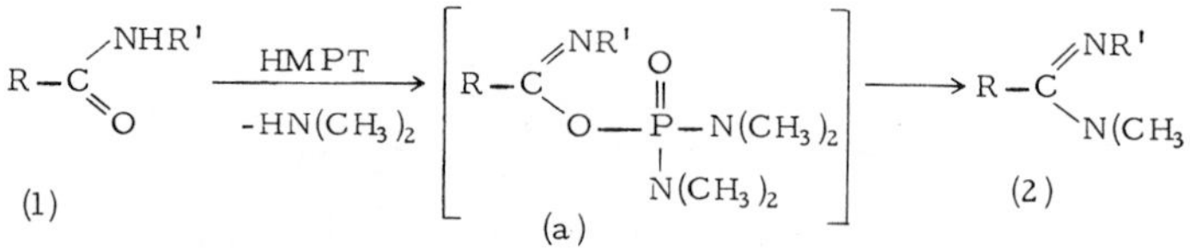

*1,3,2,4-Diazadiphosphetidines.* When aniline is refluxed in HMPT for 3–7 hr. the 1,3,2,4-diazadiphosphetidine (1) is obtained in 76% yield.[5]

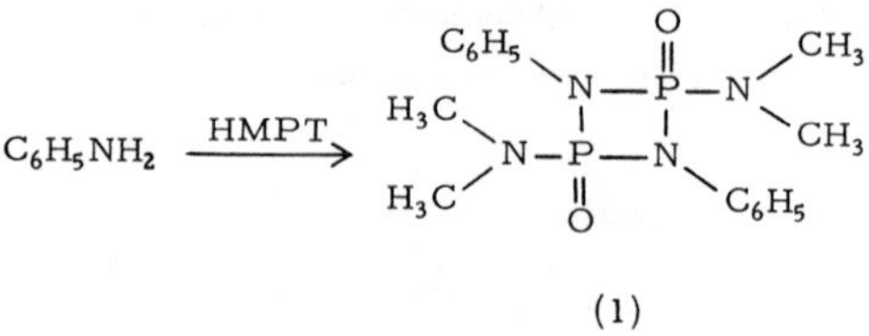

*Dehydrohalogenation.*[6] Normal alkyl halides undergo ready dehydrohalogenation when heated in hexamethylphosphoric triamide at 180–210° and afford unrearranged 1-alkenes in yields of 60–65%.

*Decyanation.*[7] Aliphatic nitriles undergo reductive decyanation in good yield when treated with potassium in HMPT and ether. If the nitrile is tertiary, a protic cosolvent (alcohol) is not necessary. Yields are in the range 60–90%.

$$\text{RCN} \xrightarrow{\text{K, HMPT}} \text{RH} + \text{KCN}$$

*5′-Halogenated ribonucleosides.*[8] Ribonucleosides can be converted conveniently into 5′-chlorinated and 5′-brominated ribonucleosides by reaction with either thionyl chloride or thionyl bromide in combination with HMPT. For example, 5′-chloro-5′-desoxycytidine (1) and 5′-bromo-5′-desoxyadenosine (2) have been prepared in satisfactory yields in this way.

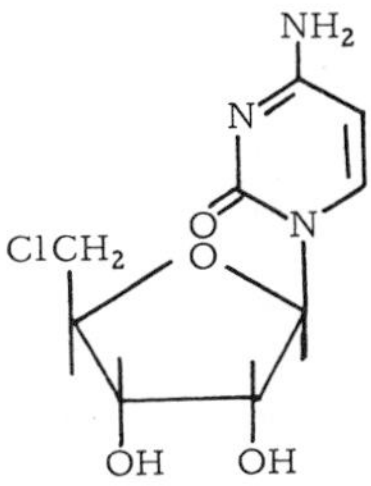

(1) 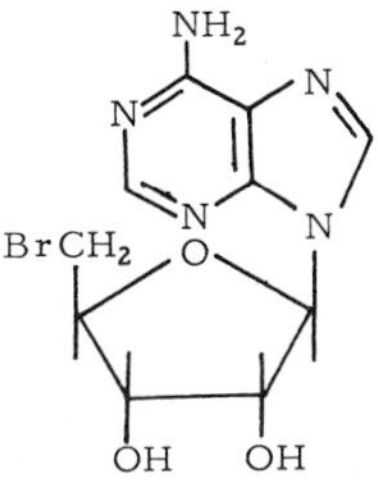    (2)

*Reaction with cycloalkanones.*[9] If cyclohexanone is refluxed in HMPT (215–220°) for 40 min., the enamine 1-dimethylaminocyclohexene is formed to a substantial extent. If the mixture is then cooled to room temperature, the enamine can be alkylated in the same pot:

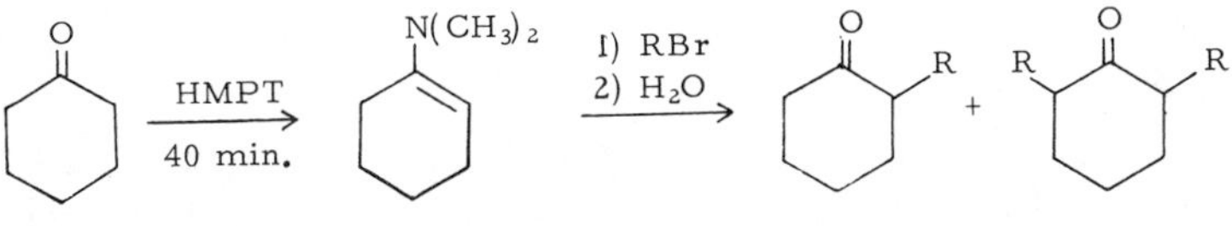

If the reflux period is extended, the enamine is destroyed and cyclohexyldimethylamine (1) is obtained as the major product (45.2% yield, 4-hr. reflux). In addition, 1,2,3,4,5,6,7,8-octahydroacridine (2) is obtained in 4.8% yield. The unexpected formation of octahydroacridines from dimethylamine enamines is being investigated.

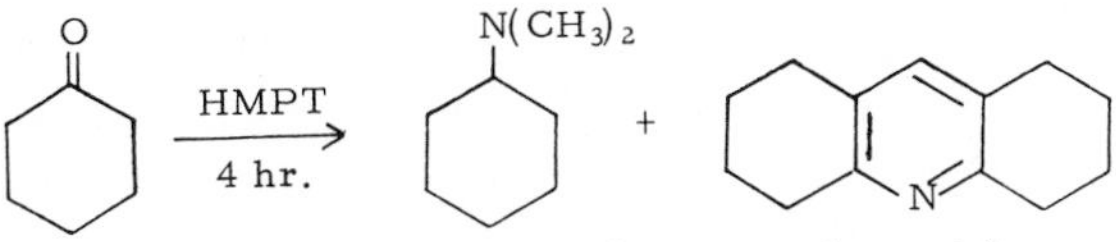

(1, 45.2%)    (2, 4.8%)

*Reduction of carbonyl compounds.*[10] Aldehydes and ketones are reduced to the corresponding alcohols by alkali metals (Li, Na, K) in HMPT in the presence of a protic cosolvent such as *t*-butanol.

*Reduction of aromatic hydrocarbons.* Anthracene is reduced quantitatively to 9,10-dihydroanthracene by sodium in HMPT with THF as cosolvent.[11] The reaction has been extended to benzanthracene, tetracene, and 9-alkyl- and 9,10-dialkylanthracenes to give the corresponding *meso*-dihydro derivatives. Water or methanol diluted in THF is used as the proton source.[12]

*Ynyl tetrahydropyranyl ethers.* Nearly all sex attractants of lepidopterous insect species are long-chain unsaturated alcohols or their acetates. They are usually synthesized via ynyl tetrahydropyranyl ethers (2). Schwarz and Waters[13] report that these ethers can be prepared in nearly quantitative yield by treatment of an acetylene compound of type (1) dissolved in THF with 1 eq. of *n*-butyllithium in hexane at $-65°$. After addition is complete a solution of the alkyl halide in HMPT is added with control

$$THP-O-(CH_2)_n-C\equiv CH \quad \xrightarrow[\text{2)}\ \underline{X}-(CH_2)_m-CH_3]{\text{1)}\ \underline{n}-C_4H_9Li} \quad THP-O-(CH_2)_n-C\equiv C-(CH_2)_m-CH_3$$

(1)                                    (2)

of the temperature below 25°. In the classical procedure (2) is prepared from (1) using sodamide in liquid ammonia followed by alkylation. However, some acetylenides are not very soluble in liquid ammonia.

The insect attractants are readily prepared from (2) by reduction of the triple bond to a double bond followed by hydrolytic cleavage of the THP–O bond.

*Alkylation of indole sodium salt.*[14] Excellent yields of N-alkylindoles can be obtained by formation of indole sodium salt by treatment with sodium hydride in

HMPT, followed by addition of an alkyl halide. Use of THF as solvent gives 3-alkyl- and 1,3-dialkylindoles in approximately 10% yield as by-products. Addition of benzene to HMPT also leads to these by-products.

*Esterification of hindered carboxylic acids.*[15] Hindered carboxylic acids can be esterified in high yield by alkylation of their potassium salts in ethanol–HMPT. In a typical procedure an acid such as (1) is dissolved in 50% ethanol–HMPT; powdered potassium hydroxide is added and the mixture is heated at 50° until solution is complete.

$$CH_3(CH_2)_2\overset{\overset{\displaystyle CH_3}{\displaystyle |}}{\underset{\underset{\displaystyle (CH_2)_2CH_3}{\displaystyle |}}{C}}-COOH$$

(1)

The alkyl halide is added and the mixture stirred at 50° for 30 min. After acidification the ester is extracted with hexane. The method appears to be most satisfactory for hindered acids, since slower rates were observed in the case of pelargonic acid, $CH_3(CH_2)_7COOH$. Omission of HMPT as cosolvent results in significantly lower conversions. Sodium carboxylates tend to react more slowly than potassium carboxylates.

*Quantitative esterification of carboxylic acids.*[16] To a solution of 10 mmole of mesitoic acid in 25 ml. of HMPT in a separatory funnel was added 20 mmole of sodium hydroxide in a 25% aqueous solution. After shaking for 5 min., 40 mmole of methyl

$$R-\underset{\underset{O}{\|}}{C}-OH \xrightarrow[\text{HMPT}]{\text{NaOH}} R-\underset{\underset{O}{\|}}{C}-ONa \xrightarrow[95\text{-}100\%]{\text{R'-X}}{\text{HMPT}} R-\underset{\underset{O}{\|}}{C}-OR'$$

iodide was added and shaking was continued for another 5 min. Then the mixture was acidified (HCl) and worked up by ether extraction. Infrared and glpc analyses revealed the presence of methyl mesitoate in yield of 99%.

[1]R. S. Monson, *Tetrahedron Letters*, 567 (1971).
[2]J. S. Lomas, D. S. Sagatys, and J.-E. Dubois, *ibid.*, 165 (1972).
[3]R. S. Monson and D. N. Priest, *Canad. J. Chem.*, **49**, 2897 (1971).
[4]E. B. Pedersen, N. O. Vesterager, and S.-O. Lawesson, *Synthesis*, 547 (1972).
[5]N. O. Vesterager, R. Dyrnesli, E. B. Pedersen, and S.-O. Lawesson, *ibid.*, 548 (1972).
[6]R. S. Monson, *Chem. Commun.*, 113 (1971).
[7]T. Cuvigny, M. Larchevêque, and H. Normant, *Compt. rend.*, **274**, 797 (1972).
[8]K. Kikugawa and M. Ichino, *Tetrahedron Letters*, 87 (1971).
[9]R. S. Monson, D. N. Priest, and J. C. Ullrey, *ibid.*, 929 (1972).
[10]M. Larchevêque and T. Cuvigny, *Compt. rend.*, **272**, 794 (1971).
[11]P. Labandibar, R. Lapouyade, and H. Bouas-Laurent, *ibid.*, **269** (C), 701 (1969).
[12]*Idem, ibid.*, **272** (C), 1257 (1971).
[13]M. Schwarz and R. M. Waters, *Synthesis*, 567 (1972).
[14]G. M. Rubottom and J. C. Chabala, *ibid.*, 566 (1972).
[15]P. E. Pfeffer, T. A. Foglia, P. A. Barr, I. Schmeltz, and L. S. Silbert, *Tetrahedron Letters*, 4063 (1972).
[16]J. E. Shaw, D. C. Kunerth, and J. J. Sherry, *ibid.*, 689 (1973).

**Hexamethylphosphorous triamide, 1**, 431–432.

*Epoxides.* The unstable epoxide (2) was prepared in 75% yield from the dialdehyde (1) with Mark's reagent using appropriate care.[1] The method had previously been reported to fail in this case.[2]

*Carboxylic acid dimethylamides.*[3] Anhydrides of carboxylic acids when heated with excess hexamethylphosphorous triamide at 150–170° are converted into carboxylic acid dimethylamides (40–80% yield).

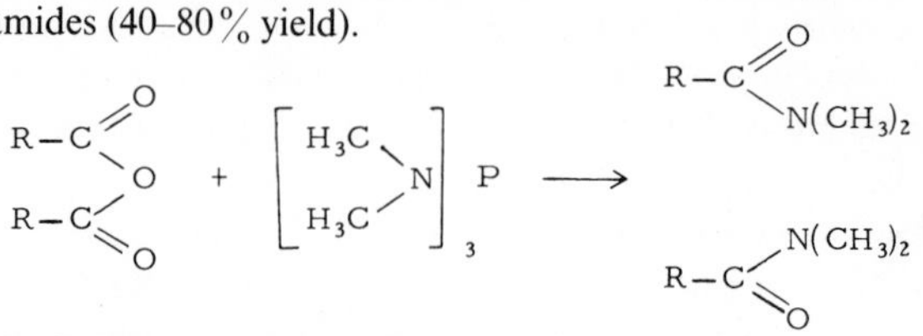

[1]S. H. Goh and R. G. Harvey, *Am. Soc.*, **95**, 242 (1973).
[2]M. S. Newman and S. Blum, *ibid.*, **86**, 5598 (1964).
[3]H. Schindlbauer and S. Fischer, *Synthesis*, 634 (1972).

**Hydrazine, 1,** 434–445; **2,** 211; **3,** 153.

*Huang-Minlon–Wolff-Kishner reduction* (**1,** 1197)[1]. Reduction of 3,4,5-trimethoxy-benzaldehyde:

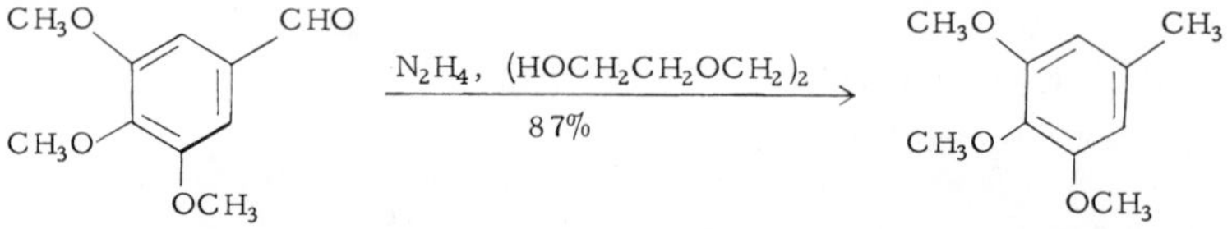

In the synthesis of 3,4-benzpyrene from pyrene by the steps (1 → 7), Schlude[2] carried out the Wolff-Kishner–Huang-Minlon reaction of (2a → 3) in 80% yield in lots of 100 g. by using *n*-butanol as solvent. Succinoylation of pyrene as previously described

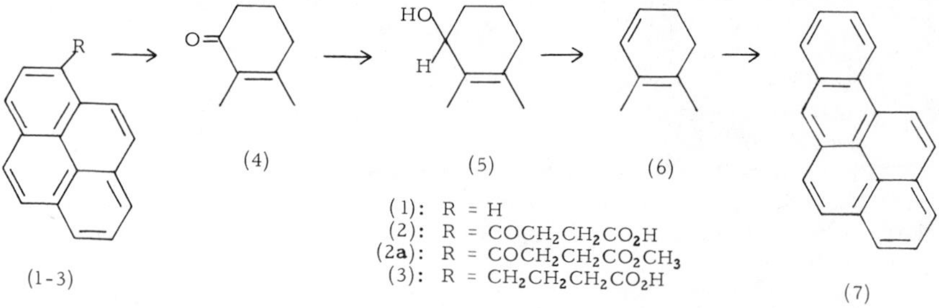

(4)    (5)    (6)

(1): R = H
(2): R = COCH$_2$CH$_2$CO$_2$H
(2a): R = COCH$_2$CH$_2$CO$_2$CH$_3$
(3): R = CH$_2$CH$_2$CH$_2$CO$_2$H

(1-3)

(7)

by others gives the keto acid (2) in yields up to 95%, and Fischer esterification gives the methyl ester (2a) in 95% yield.

*Reaction with 1,5-diphenylpentadiynone*[3]:

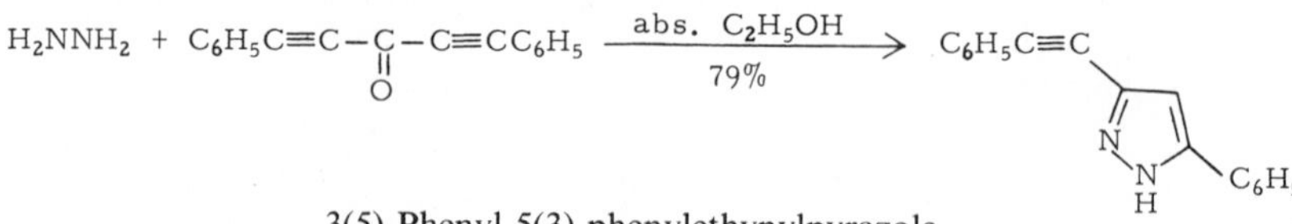

3(5)-Phenyl-5(3)-phenylethynylpyrazole

[1]H. Aquila, *Ann.*, **721**, 220 (1969).
[2]H. Schlude, *Ber.*, **104**, 3995 (1971).
[3]H. Reimlinger and J. J. M. Vandewalle, *Ann.*, **720**, 117 (1968).

**Hydridotris(triphenylphosphine)carbonylrhodium(I)**,  HRh(CO)[(C$_6$H$_5$)$_3$P]$_3$  (1).  Mol. wt. 918.77. Suppliers: Alfa Inorganics, ROC/RIC, Strem.

*Trisubstituted olefins.*[1] The reagent adds stereospecifically to dimethyl acetylene-dicarboxylate (DADC) to give the *cis*-vinylrhodium(I) compound (2). This reacts with neat methyl iodide at 25° (30 min.) to give dimethyl[iodocarbonylbis(triphenyl-

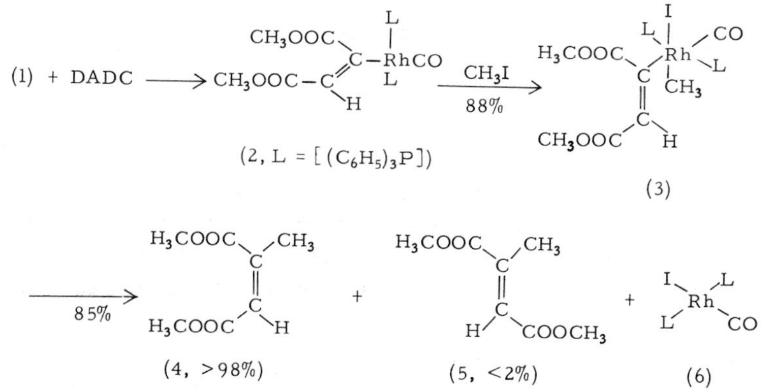

(1) + DADC  $\longrightarrow$

(2, L = [(C$_6$H$_5$)$_3$P])

(3)

(4, >98%)          (5, <2%)          (6)

phosphine)methylrhodium(III)]maleate (3). To effect C–C coupling with C=C *cis* stereochemistry (3) is placed in a tube evacuated to 80 $\mu$, and heated to 115°. Dimethyl citraconate (4) and dimethyl mesaconate (5) are formed in the yields indicated. If (3) is allowed to warm slowly to 115° in an evacuated tube, (5) is obtained in >75% yield together with (4, <25% yield). The transition metal is recovered as IRh(CO)[(C$_6$H$_5$)$_3$P]$_2$.

The organometallic reagent (1) reacts with ethyl tetrolate in the sequence above to give ethyl $\beta$,$\beta$-dimethylacrylate (7).

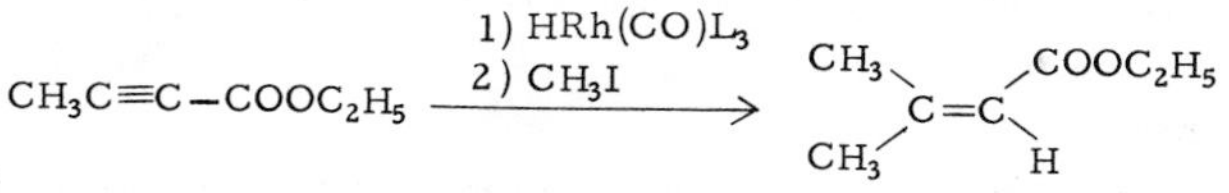

CH$_3$C≡C—COOC$_2$H$_5$  $\xrightarrow{\begin{array}{c}1)\ HRh(CO)L_3\\2)\ CH_3I\end{array}}$

(7)

[1]J. Schwartz, D. W. Hart, and J. L. Holden, *Am. Soc.*, **94**, 9269 (1972).

**Hydrobromic acid,** 1, 450–452; 2, 214–215; 3, 154.

*Trisubstituted olefins.*[1] The oxirane (1)[2] is converted into (2a), (E)-5-bromo-2-methyl-2-pentene-1-ol, in 81% yield (pure) when treated with 48% hydrobromic acid at 0° for 1 hr. Alternatively, (1) is transformed into (2b) by treatment with sodium

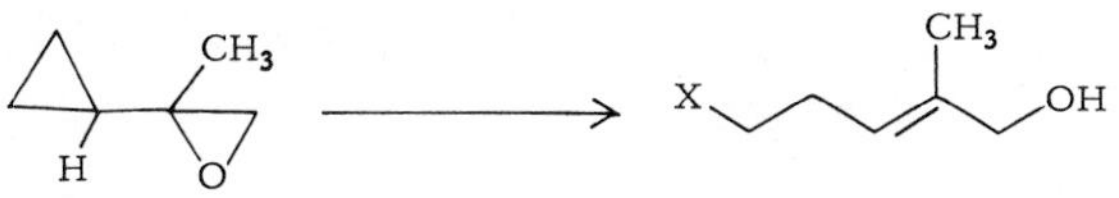

(1)          (2a, X = Br)

(2b, X = I)

iodide in acetic acid–propionic acid–sodium acetate (**1**, 881, 1116, 1284) at $-18°$ for 30 min.

Similar treatment of the isomeric oxirane (**3**) fails; however, treatment with anhydrous zinc bromide (**2**, 463–464) in ether at 0° for 3 hr. gives (**4**), (E)-5-bromo-3-methyl-2-pentene-1-ol, in 73% yield (pure).

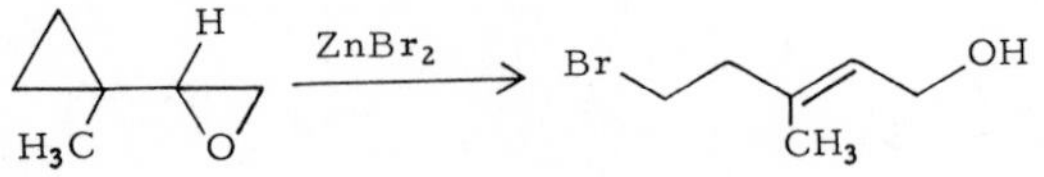

Note that 2- or 3-methyl-2-butene-1-ols are the chain terminal units of isoprenoids.

*Demethylation.*[3] Demethylation of 4,5-dimethoxybenzocyclobutene-1,2-dione (1) was effected by treatment with 48% hydrobromic acid. 4,5-Dimethoxybenzocyclobutene-1,2-dione is a benzolog of squaric acid. It gives a positive ferric chloride test.

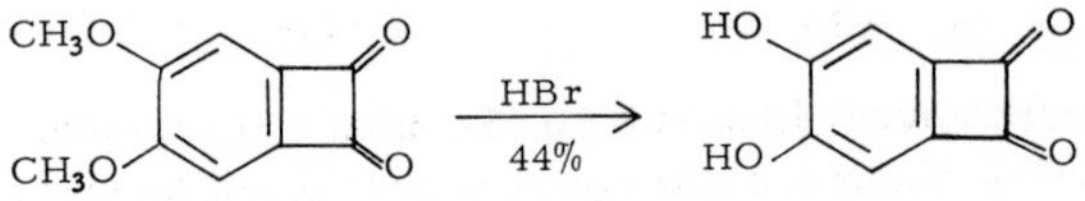

[1]H. Nakamura, H. Yamamoto, and H. Nozaki, *Tetrahedron Letters*, 111 (1973).
[2]Prepared from cyclopropyl methyl ketone and dimethylsulfonium methylide (**1**, 314–315; **2**, 169–171; **3**, 124–125; this volume) in DMSO at $-5°$ for 20 min. and 25° for 1 hr.
[3]J. F. W. McOmie and D. H. Perry, *J.C.S. Chem. Comm.*, 248 (1973).

**Hydrochloric acid.**

*Selective O-demethylation.* Selective O-demethylation of the hydrochloride of $(\pm)$-O-methylanhalonide (1) to the hydrochloride of $(\pm)$-7,8-dihydroxy-6-methoxy-1-methyl-1,2,3,4-tetrahydroisoquinoline (2) was effected by refluxing in 20% hydrochloric acid. The reaction was a key step in the synthesis of two tetrahydroisoquinoline cactus alkaloids, anhalonine and lophophorine.[1]

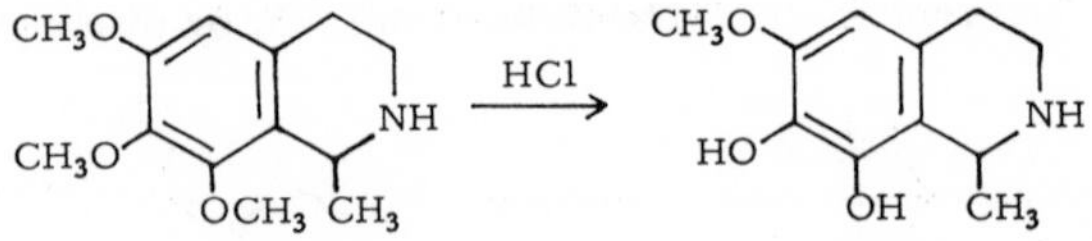

[1]A. Brossi, J. F. Blount, J. O'Brien, and S. Teitel, *Am. Soc.*, **93**, 6248 (1971).

**Hydrogen bromide, 1**, 450–452.

*Ring expansion of 1-vinylcyclopropanols to cyclobutanones.*[1] 1-Vinylcyclopropanol, obtained by the reaction of (1) with vinylmagnesium bromide in refluxing THF, rearranges to 2-methylcyclobutanone (3) when treated with excess HBr in methylene chloride. The rearrangement probably involves an intermediate cyclopropylcarbinyl cation. The ring expansion can also be accomplished by heating (*ca.* 100°).[2]

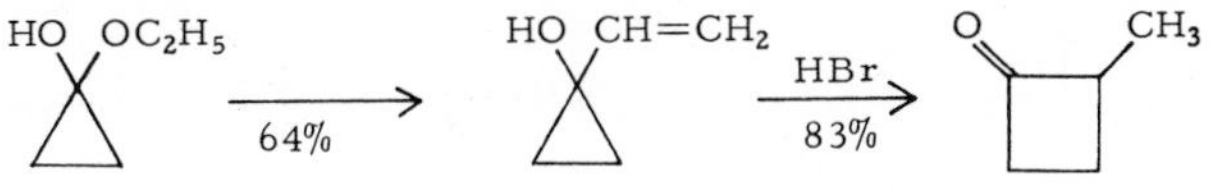

(1)                    (2)                    (3)

The cyclopropanol (2) also undergoes ring expansion when treated with other electrophilic reagents. Thus (2) is converted into 2-hydroxymethylcyclobutanone when treated with perbenzoic acid in ether.

[1] H. H. Wasserman, R. E. Cochoy, and M. S. Baird, *Am. Soc.*, **91**, 2375 (1969).
[2] J. R. Salaün and J. M. Conia, *Tetrahedron Letters*, 2849 (1972).

**Hydrogen bromide, anhydrous, 1**, 453.

*Rearrangement.* 3-Methoxyestratriene-1,3,5(10)-one-17 (1) on treatment with anhydrous hydrogen bromide in dimethylformamide at 130° for 6 hr. affords 3-hydroxy-17β-methyl-14β-gonapentaene-1,3,5(10),6,8 (2)[1] in good yield.

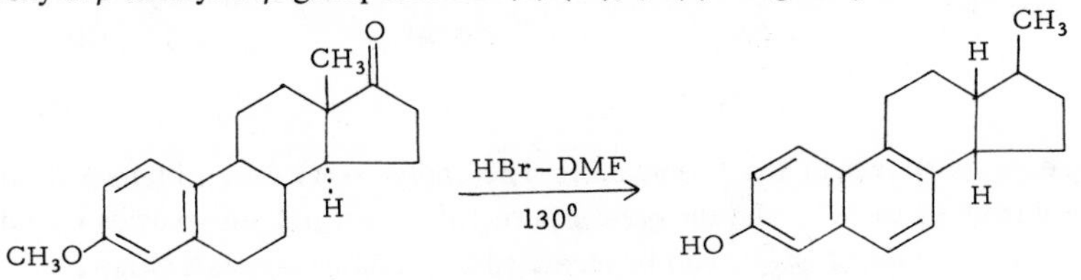

(1)                                        (2)

*Addition (1,4- and 1,6-) to 4-methyl-o-benzoquinone*[2] :

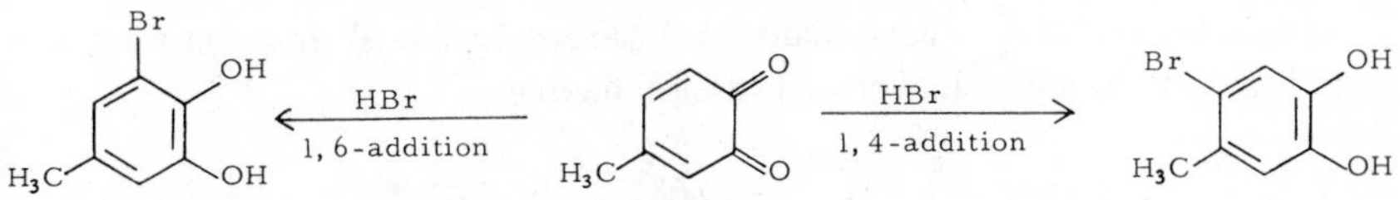

Hydrogen chloride adds in the same way.

[1] J.-C. Hilscher, *Ber.*, **104**, 2341 (1971).
[2] L. Horner and T. Burger, *Ann.*, **708**, 105 (1967).

**Hydrogen bromide (1**, 453), **Trifluoroacetic acid (2**, 305).

The velocity of cleavage of the principal peptide N- and C-protective groups by acids of varying concentrations in acetic acid and trifluoroacetic acid is reported.[1]

[1] G. Losse, D. Zeidler, and T. Grieshaber, *Ann.*, **715**, 196 (1968).

**Hydrogen chloride, 2, 215.**

*Steroid aromatic rearrangement.*[1] When (1), derived from cholic acid, is dissolved in methanol, and hydrogen chloride is passed in for 4 hr., the aromatic steroid (2) is obtained in high yield.

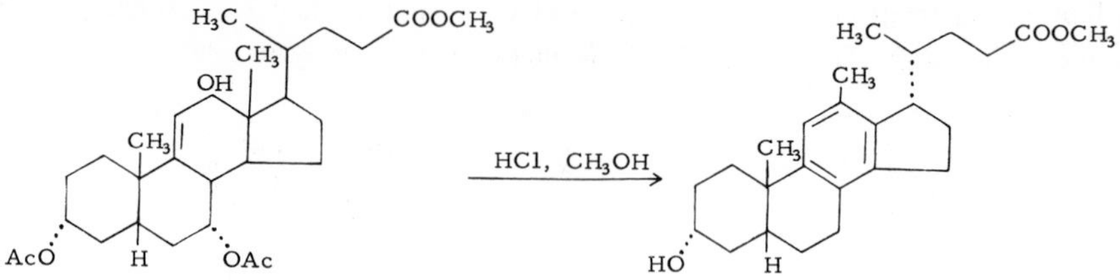

(1, 12α- and 12β-isomers)                    (2, 1.85 g.)
(1.935 g.)

[1]J. Meney, Y.-H. Kim, R. Stevenson, and T. N. Margulis, *Tetrahedron*, **29**, 21 (1973).

**Hydrogen cyanide, 1, 454–455.**

*Asymmetric synthesis of amino acids.*[1] Optically active amino acids can be prepared by addition of hydrogen cyanide to Schiff bases prepared from aliphatic aldehydes (representing the R group) and optically active benzylic amines (representing the

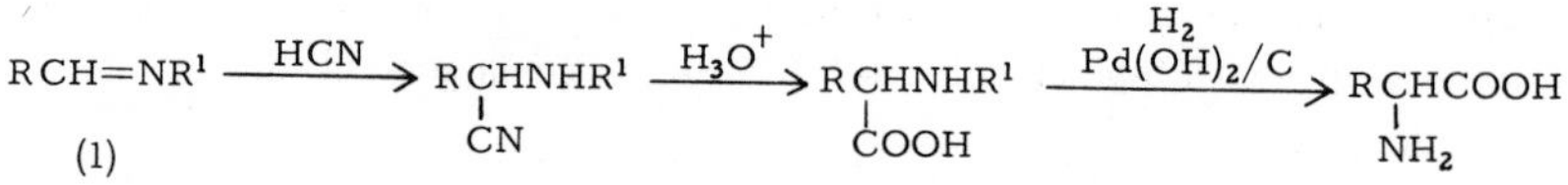

$R^1$ group) as shown in the formulation. The amino acids were obtained in yields ranging from 19 to 58%, and the optical purity of the natural amino acids was in the range 22–58%. Optical purity can be increased by fractional crystallization.

[1]K. Harada and T. Okawara, *J. Org.*, **38**, 707 (1973).

**Hydrogen fluoride, anhydrous, 1, 455–456; 2, 215–216.**

*Oxazoles.* N-Aroyl-α-amino ketones (1) can be cyclized to substituted oxazoles (2) in high yield by use of anhydrous hydrogen fluoride.[1]

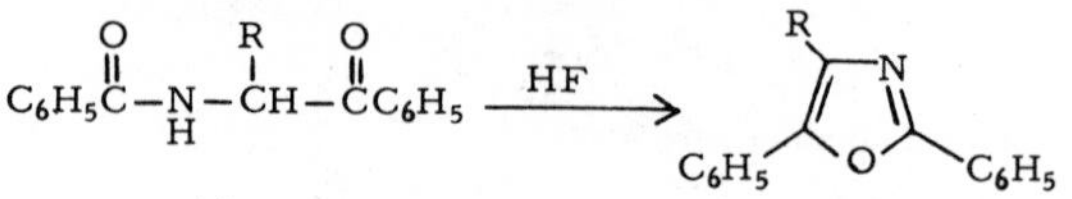

[1]G. H. Daub, M. E. Ackerman, and F N. Hayes, *J. Org.*, **38**, 828 (1973).

**Hydrogen fluoride–Boron trifluoride.**

*Isomerization of acetylenes and allenes.*[1] HF and $BF_3$ presumably form the acid $HBF_4$ while HF and $PF_5$ form $HPF_6$. Table II shows the approximate times for 20%

isomerization of the five straight-chain $C_6$ isomers hex-1-yne, hex-2-yne, hex-3-yne, and hexa-1,2- and -2,3-diene when treated with each of the four catalysts in dry sulfolane.

**Table II**    Approximate times for 20% isomerization at 25°

| Catalyst | Hex-1-yne | Hexa-1,2-diene | Hex-2-yne | Hexa-2,3-diene | Hex-3-yne |
|---|---|---|---|---|---|
| 0.1 $M$ HBF$_4$ | 30 min. | 25 sec. | 40 min. | 35 sec. | 25 min. |
| 0.1 $M$ HPF$_6$ | 25 min. | < 5 sec. | 2 hr. | 6 hr. | 30 min. |
| 0.96 $M$ H$_2$SO$_4$ | 24 hr. | 2 hr. | > 100 hr. | 20 min. | 40 hr. |
| 2.07 $M$ H$_2$SO$_4$ | — | — | — | — | 2 hr. |

The results support the sequential reaction shown in the following scheme:

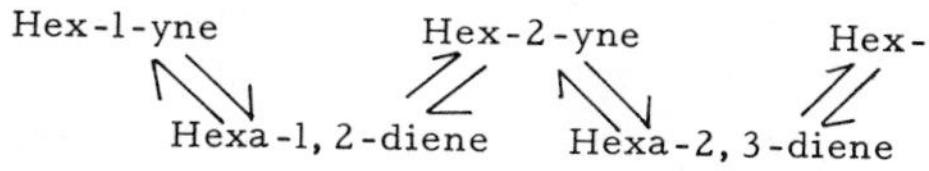

[1] B. J. Barry, W. J. Beale, M. D. Carr, S.-K. Hei, and I. Reid, *J.C.S. Chem. Comm.*, 177 (1973).

**Hydrogen peroxide, H$_2$O$_2$.**
Concentrated hydrogen peroxide (about 90%) is not readily available. Hart *et al.*[1] have reported a method for increasing the oxidizing effectiveness of 30% hydrogen peroxide (generally available). For example, 30% hydrogen peroxide is added in the cold to excess acetic anhydride and concentrated sulfuric acid. The addition is exothermic but easily controlled. The resulting solution readily oxidized hexamethylbenzene (1) to hexamethyl-2,4-cyclohexadienone (2) in a yield (86%) that is comparable with that obtained with 90% hydrogen peroxide or with pertrifluoroacetic acid/BF$_3$.

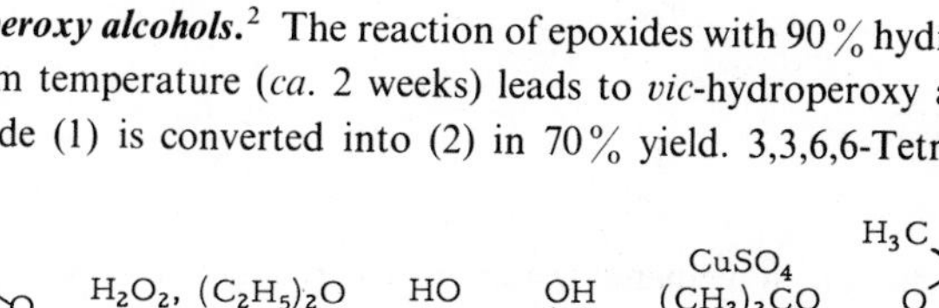

**vic-*Hydroperoxy alcohols*.**[2] The reaction of epoxides with 90% hydrogen peroxide in ether at room temperature (*ca.* 2 weeks) leads to *vic*-hydroperoxy alcohols. Thus isobutylene oxide (1) is converted into (2) in 70% yield. 3,3,6,6-Tetramethyl-1,2,4-

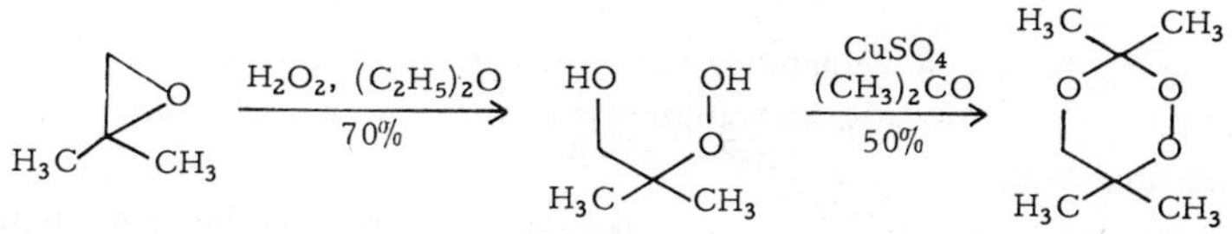

trioxane (3) can be prepared from (2) by treatment with anhydrous cupric sulfate and acetone.

[1]J. Blum, Y. Pickholtz, and H. Hart, *Synthesis*, 195 (1972).
[2]W. Adam and A. Rios, *Chem. Commun.*, 822 (1971).

**Hydrogen peroxide, acidic, 1,** 457–465; **2,** 216; **3,** 154.

*Isochromane-3-one.* Isochromane-3-one (2) can be prepared in 68% yield (pure) by Baeyer–Villiger oxidation of indane-2-one (1) with 90% hydrogen peroxide, sulfuric

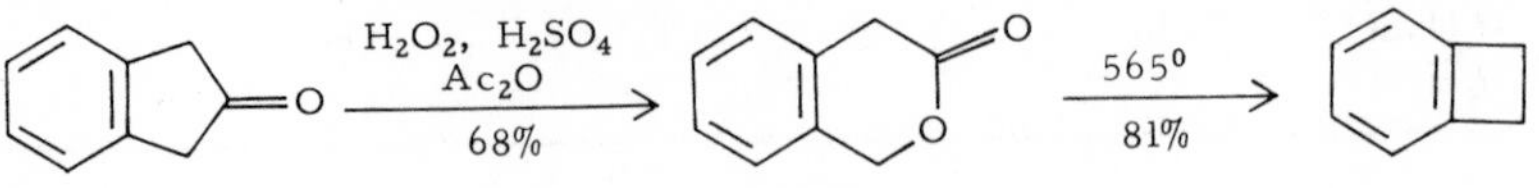

(1)                        (2)                        (3)

acid, and acetic anhydride.[1,2] Pyrolysis of (2) at 565° (N$_2$) gives benzocyclobutene (3) in high yield. This two-step procedure is the most convenient route to (3). For example, a recent synthesis of (3) reported by Markgraf *et al.*[3] by the route formulated gives (3) in about 20% overall yield.

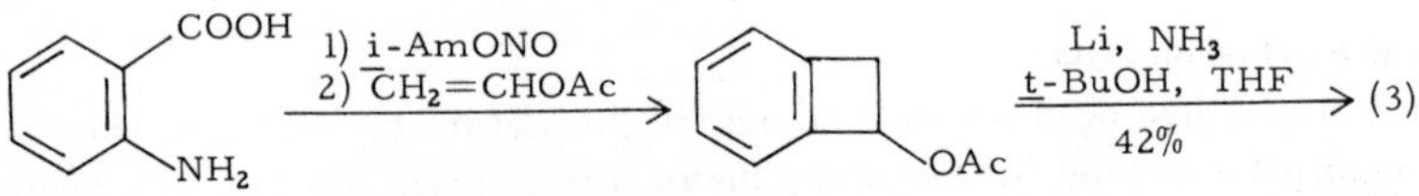

*Oxidation of cyclobutanones to γ-lactones* (**3,** 155). For another example, see E. J. Corey and T. Ravindranathan, *Tetrahedron Letters*, 4753 (1971).

[1]J. H. Markgraf and S. J. Basta, *Syn. Commun.*, **2,** 139 (1972).
[2]R. J. Spangler and J. H. Kim, *Synthesis*, 107 (1973).
[3]J. H. Markgraf, S. J. Basta, and P. M. Wege, *J. Org.*, **37,** 2361 (1972).

**Hydrogen peroxide, basic, 1,** 466–471; **2,** 216–217; **3,** 155.

*γ-Butyrolactones.*[1] Cyclobutanones, now readily available from carbonyl compounds by use of diphenylsulfonium cyclopropylide (this volume), undergo Baeyer–Villiger oxidation to give γ-butyrolactones by treatment with 30% hydrogen peroxide

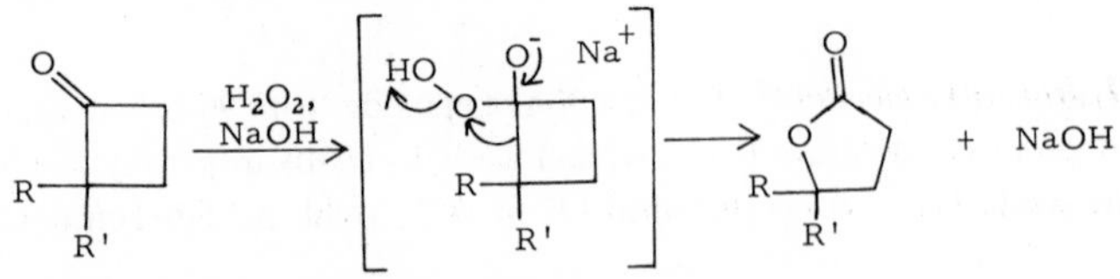

and sodium hydroxide in methanol–water. Yields are high (80–100%). The rearrangement can also be conducted with sodium hypobromite generated *in situ* from sodium hydroxide and bromine.

The migratory preference is tertiary > secondary > primary; the migration occurs with retention of configuration at the migrating carbon atom.

*Hydration of allenic nitriles.*[2] 4,4-Dialkyl-substituted allenic nitriles (1) are converted into allenic amides (2) in 60–70% yield by reaction with hydrogen peroxide in alcoholic sodium hydroxide.

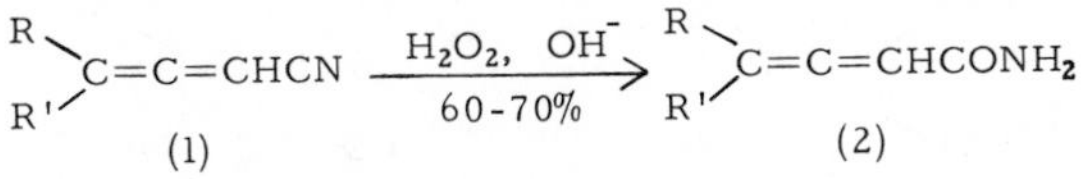

$$\underset{(1)}{\overset{R}{\underset{R'}{\diagup}}C=C=CHCN} \xrightarrow[60-70\%]{H_2O_2, \ OH^-} \underset{(2)}{\overset{R}{\underset{R'}{\diagup}}C=C=CHCONH_2}$$

Hydration can also be accomplished by the Ritter reaction[3] (treatment with *t*-butanol in sulfuric acid–acetic acid).

[1]M. Bogdanowicz, T. Ambelang, and B. M. Trost, *Tetrahedron Letters*, 923 (1973).
[2]P. M. Greaves, P. D. Landor, S. R. Landor, and O. Odyck, *ibid.*, 209 (1973).
[3]J. J. Ritter and J. Kalish, *Am. Soc.*, **70**, 4045, 4048 (1948).

## Hydrogen peroxide–Aluminum chloride.

*Aromatic hydroxylation.*[1] Addition of hydrogen peroxide (0.05 mole of 90%, Du Pont) to a stirred solution or slurry of aluminum chloride (0.075–0.10 mole) in

$$ArH + H_2O_2 + AlCl_3 \longrightarrow ArOH + HCl + AlCl_2OH$$

excess aromatic compounds at 0–5° gives phenols after hydrolysis on ice. The more available 30% hydrogen peroxide can also be used if a 2:1 molar ratio of $AlCl_3$ to $H_2O_2$ is used. Nuclear chlorination, presumably via hypochlorous acid, is a minor competing reaction. The yield of phenol from benzene is only 10%; nitrobenzene fails to react.

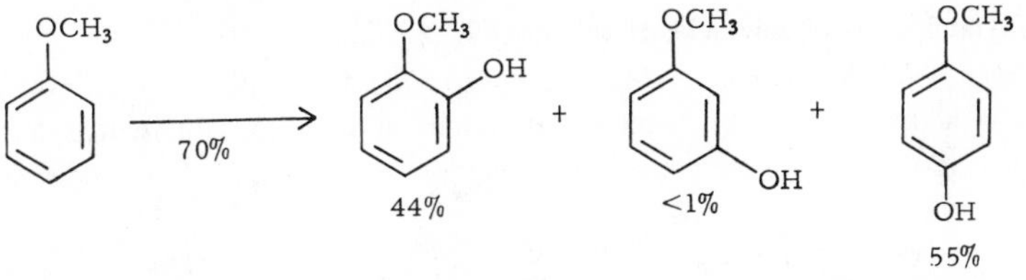

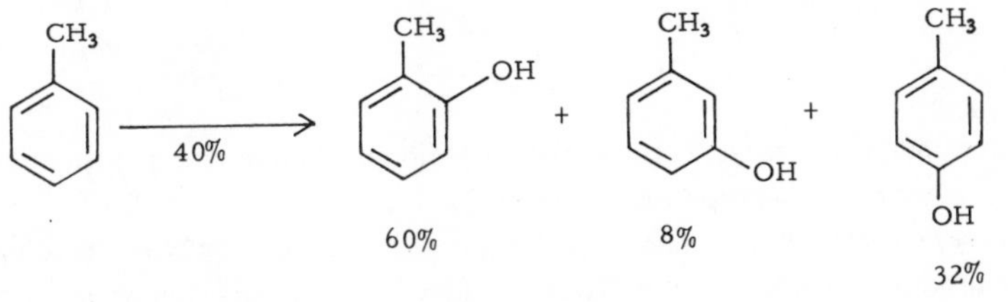

[1]M. E. Kurz and G. J. Johnson, *J. Org.*, **36**, 3184 (1971).

**Hydrogen sulfide,** $H_2S$. Mol. wt. 34.08. Suppliers: Fisher, Sargent, Schuchardt.
*Nucleophilic substitutions of chloro-cyano benzenes*[1]:

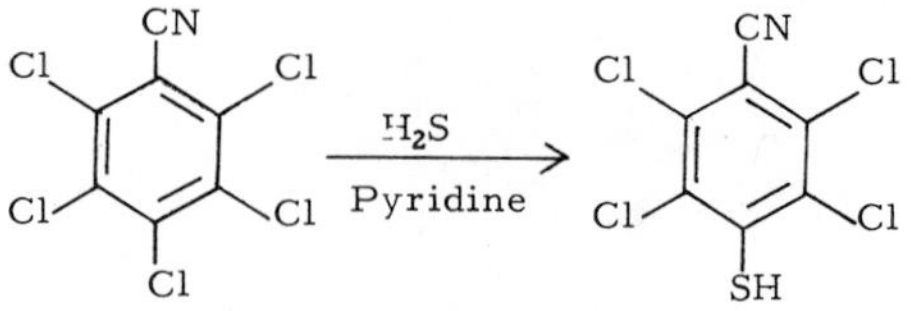

[1]G. Beck, E. Degener, and H. Heitzer, *Ann.*, **716**, 47 (1968).

**2-Hydroperoxy-2-nitropropane,** $(CH_3)_2C{<}^{OOH}_{NO_2}$. Mol. wt. 121.10.

*Preparation.* 2-Hydroperoxy-2-nitropropane is prepared *in situ* by oxidation of 2-nitropropane (suppliers: Aldrich, J. T. Baker, Eastman, Fisher, Howe and French, MCB, P and B, Sargent, Koch-Light, Schuchardt) with oxygen catalyzed by copper(I) chloride.

*N-Nitrosoamines.* The reagent (2) reacts with tertiary amines (1) in pyridine at 50° to give dialkylnitrosoamines (3) and aldehydes (4).[1]

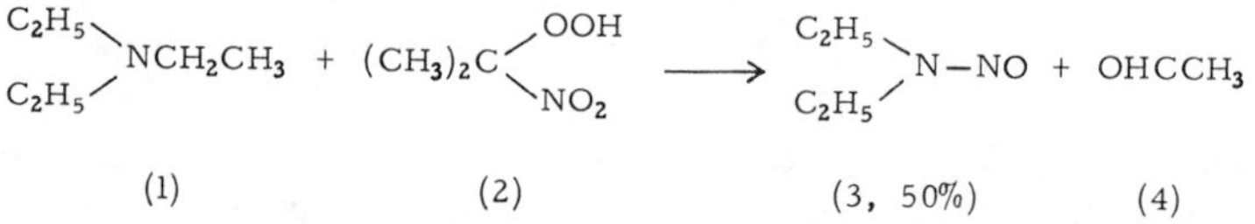

[1]B. Franck, J. Conard, and P. Misbach, *Angew. Chem., internat. Ed.*, **9**, 892 (1970).

**Hydroxylamine-O-sulfonic acid,** 1, 481–484; 2, 217–219; 3, 156–157.

*Dibenzo-1,4-diazepines.*[1] 1-Methyldibenzo[b,f]-1,4-diazepine (2) is obtained in 71.5% yield by reaction of N-methylacridinium iodide (1) with hydroxylamine-O-sulfonic acid in absolute methanol containing 30% ammonia (3–4 hr., room tempera-

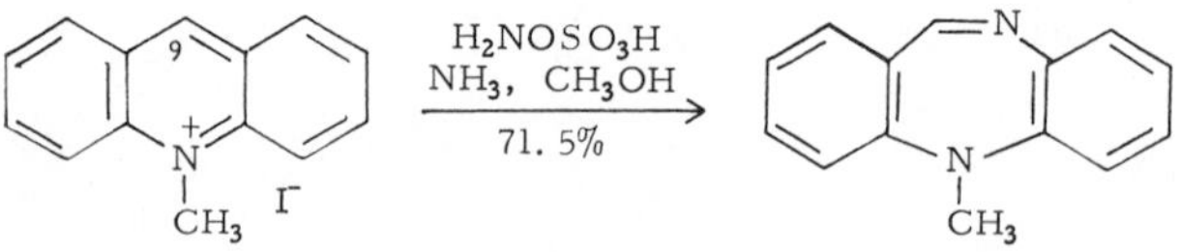

ture). Presumably the reagent attacks the 9-position of (1) followed by formation of an aziridine ring; ring expansion then leads to (2).

*Knorr pyrrole synthesis.* The reagent reacts with some active methylene compounds to give products of C-amination. Thus 2,4-dimethyl-3,5-dicarbethoxypyrrole (2) can be prepared in one step by the reaction of hydroxylamine-O-sulfonic acid with ethyl acetoacetate (1).[2]

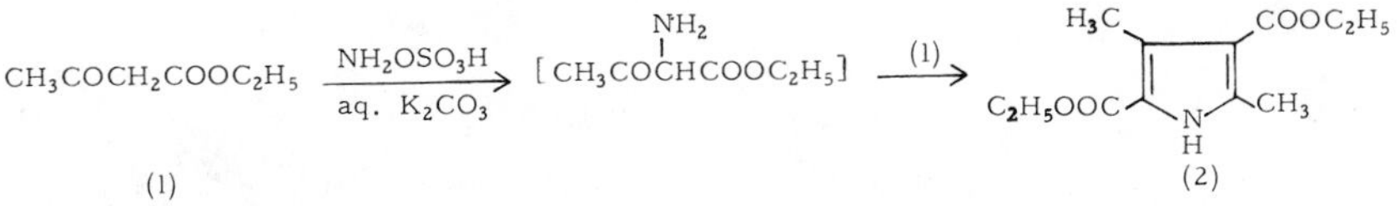

CH$_3$COCH$_2$COOC$_2$H$_5$ $\xrightarrow[\text{aq. K}_2\text{CO}_3]{\text{NH}_2\text{OSO}_3\text{H}}$ [ CH$_3$COCHCOOC$_2$H$_5$ ] $\xrightarrow{(1)}$

$\overset{\text{NH}_2}{|}$

(over CH of middle structure)

H$_3$C, COOC$_2$H$_5$ / C$_2$H$_5$OOC, CH$_3$ ring with N—H

(1)                                                    (2)

[1]M. Hirobe and T. Ozawa, *Tetrahedron Letters*, 4493 (1971).
[2]Y. Tamura, S. Kato, and M. Ikeda, *Chem. Ind.*, 767 (1971).

**4-Hydroxymethyl-2,6-di-*t*-butylphenol ("Ethyl" antioxidant 754).** Mol. wt. 218.33, m.p. 141°, b.p. 162°/2.6 mm. Solubility (wt. % at 20°):

| | | | |
|---|---|---|---|
| Methanol | 39.2 | Benzene | 7.2 |
| Acetone | 37.1 | Isopentane | 0.5 |
| Ethanol | 23.4 | Water | < 0.0035 |
| Isopropyl alcohol | 18.7 | | |

Supplier: Ethyl Corporation.

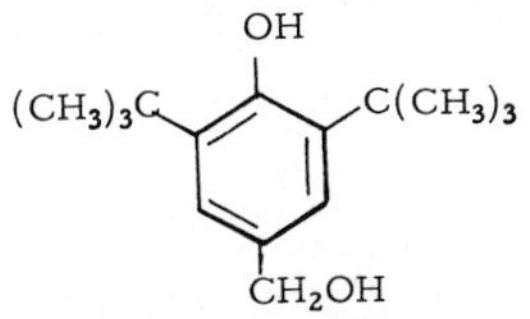

OH

(CH$_3$)$_3$C        C(CH$_3$)$_3$

CH$_2$OH

An effective antioxidant for plastics, rubber, and waxes. It is useful for stabilizing fats and oils in food applications—especially where low volatility is needed and where a tasteless and odorless antioxidant is desired.

# I

"**Indanocyclone**" (1). Mol. wt. 286.31, m.p. 205–206°. Prepared from ninhydrin and dibenzyl ketone and ethanolic potassium hydroxide.[1]

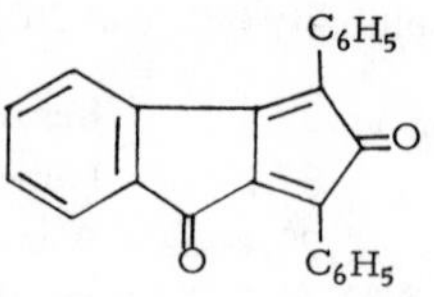

(1)

***Diels–Alder syntheses*** with alkenes at high temperatures and under pressure[2]:

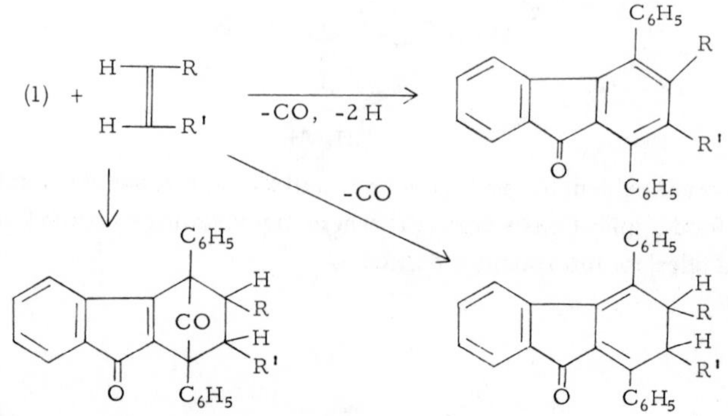

[1]W. Ried and D. Freitag, *Ber.*, **99**, 2675 (1966).
[2]W. Ried and R. Wagner, *Ann.*, **716**, 186 (1968).

**Iodine, 1**, 495–500; **2**, 220–222; **3**, 159–160.

*Oxidation* (**1**, 497–498; **2**, 220–221). Iodine in methanolic sodium hydroxide oxidizes 1,3-dibenzoylpropane to *trans*-1,2-dibenzoylcyclopropane, probably by the mechanism formulated:

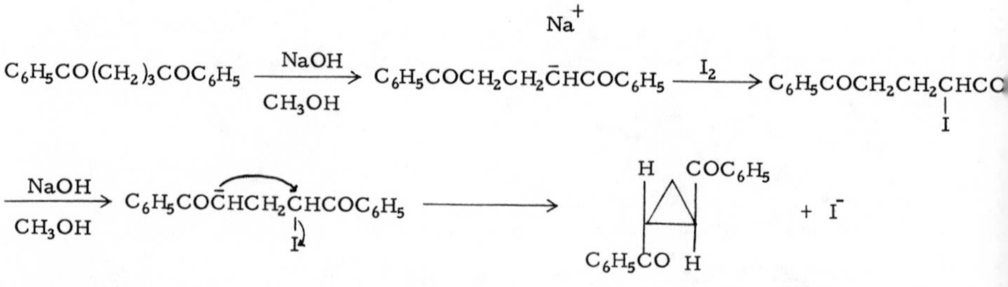

258

Pure, crystalline dehydro-D($-$)-ascorbic acid (2) is obtained in pure form as prisms, m.p. 191–193° dec., in 25 % yield by oxidation of D($-$)-ascorbic acid (1) with iodine in acetic acid–methanol.[2]

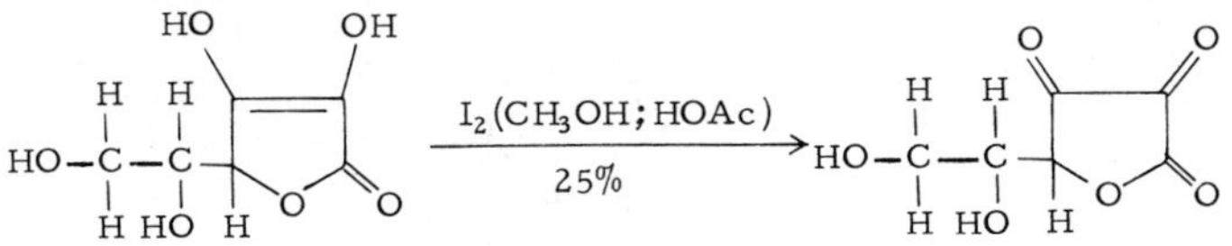

$$(1) \qquad\qquad\qquad\qquad (2)$$

*Aromatic iodination* (**2**, 220). In a procedure for the preparation of iododurene,[3] a mixture of 13.4 g. (0.1 mole) of durene, 4.56 g. (0.02 mole) of periodic acid dihydrate, and 10.2 g. (0.04 mole) of iodine is stirred magnetically under reflux and treated with a

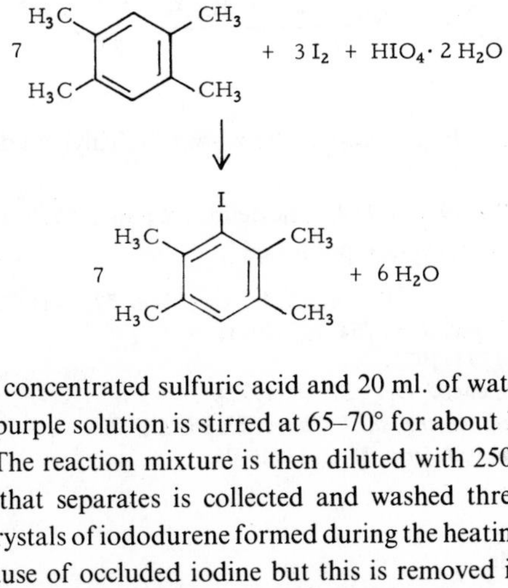

solution of 3 ml. of concentrated sulfuric acid and 20 ml. of water in 100 ml. of acetic acid. The resulting purple solution is stirred at 65–70° for about 1 hr. until the color of iodine disappears. The reaction mixture is then diluted with 250 ml. of water and the white-yellow solid that separates is collected and washed three times with 100-ml. portions of water (crystals of iododurene formed during the heating process may take on a purple color because of occluded iodine but this is removed in the crystallization). The product is dissolved in the minimum amount of boiling acetone (about 125 ml.) and the solution let stand, eventually at 0°. The yield of iododurene in the form of colorless fine needles, m.p. 78–80°, is 20.8–22.6 g. (80–87 %).

*Dehydration.*[4] An approximately 1:1 mixture of the *dl* and *meso* forms of 3,4-dimethyl-3,4-hexanediol (1) is dehydrated when heated with iodine in propionic

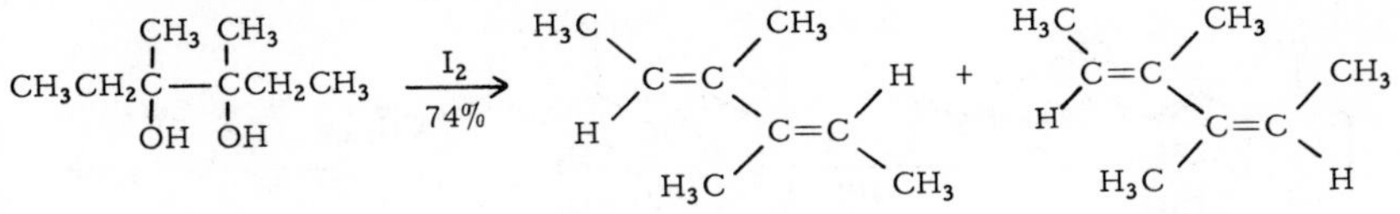

$$(1) \qquad\qquad\qquad (2,\ 35\%) \qquad\qquad (3,\ 38\%)$$

anhydride to give mainly *cis,cis*-3,4-dimethyl-2,4-hexadiene (2) and *cis,trans*-3,4-dimethyl-2,4-hexadiene (3) by a Saytzeff pathway. Use of phenyl isocyanate, $C_6H_5NCO$, as the reagent leads predominantly to *cis*-2-ethyl-3-methyl-1,3-pentadiene (4, Hofmann pathway).

$$
\begin{array}{c}
\underset{H}{\overset{H}{\diagdown}} C=C \overset{CH_2CH_3}{\diagup} \\
\diagup \qquad \diagdown \\
H \qquad \qquad \underset{H_3C}{C}=C \overset{H}{\diagup} \\
\qquad \qquad \diagdown CH_3
\end{array}
$$

(4)

**gem-*Diiodoalkanes*.**[5] The reaction of hydrazones (1) with iodine in ether in the presence of triethylamine results in formation of *gem*-diiodoalkanes (2) in 40–60% yield.

$$
\underset{R^2}{\overset{R^1}{\diagdown}} C=NNH_2 \xrightarrow{I_2;\ (C_2H_5)_3N} \underset{R^2}{\overset{R^1}{\diagdown}} C \overset{I}{\underset{I}{\diagup}}
$$

(1)                    (2)

If $R^1$ or $R^2$ possesses a hydrogen atom in the α-position, vinyl iodides may be formed as by-products.

*Alkylmagnesium fluorides* (3, 159). The definitive paper on the preparation of alkylmagnesium fluorides has now been published.[6]

[1] I. Colon, G. W. Griffin, and E. J. O'Connell, Jr., *Org. Syn.*, **52**, 33 (1972).
[2] W. Weiss and H. Staudinger, *Ann.*, **754**, 152 (1971).
[3] H. Suzuki, *Org. Syn.*, **51**, 94 (1971).
[4] W. Reeve and D. M. Reiche, *J. Org.*, **37**, 68 (1972).
[5] A. Pross and S. Sternhell, *Australian. J. Chem.*, **23**, 989 (1970).
[6] S. W. Yu and E. C. Ashby, *J. Org.*, **36**, 2123 (1971).

**Iodine–Potassium iodide.**[1]

*Isoxazoles*.[2] Oximes of certain α,β-unsaturated ketones are oxidized by iodine–potassium iodide in aqueous THF containing sodium bicarbonate to isoxazoles. Thus treatment of the oxime of either (E)- or (Z)-β-ionone (1) with this combination of reagents gives the isoxazole (2) in 91% yield. The expensive potassium iodide is needed only in catalytic amount. Reduction of (2) with sodium and 3 eq. of *t*-butanol in liquid

(1)                    (2)

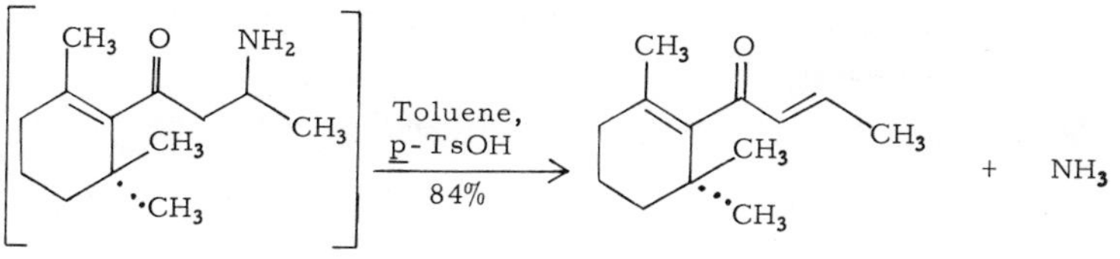

(3)                              (4)

ammonia generates the enolate, which is converted into the β-amino ketone (3) on aqueous workup. Treatment of (3) in refluxing toluene containing a trace of p-toluene-sulfonic acid gives (4, β-damascone) in 84% yield from (2).

The method thus interconverts α,β-unsaturated ketones. The method is also suitable for conversion of a ketone to an aldehyde, but an aldoxime cannot be employed since the resulting isoxazole is very unstable. Isopropenyl ethyl ketone (5) was converted by this procedure into 2-methylpentene-2-al (9).

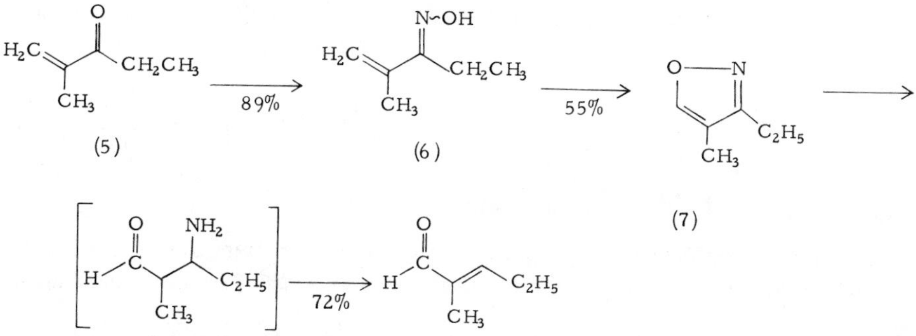

(5)                    (6)                           (7)

(8)                (9)

[1]Potassium iodide is supplied by Alfa, Baker, Fisher, Howe and French, K and K, MCB, E. Merck, P and B, Riedel-de Haën, Sargent, Schuchardt.
[2]G. Büchi and J. C. Vederas, *Am. Soc.*, **94**, 9128 (1972).

## Iodine–Silver acetate.

*Addition to the 22(23)-double bond of ergosterol.*[1] Iodine–silver acetate in moist acetic acid, a typical electrophilic reagent, reacts with 3α,5α-cycloergosta-7,22-diene-

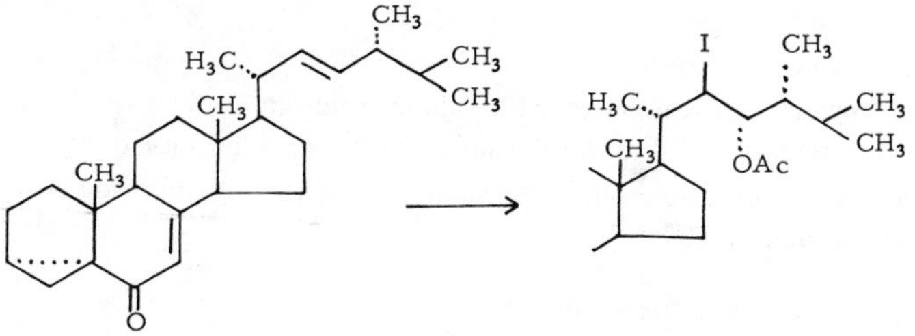

(1)                              (2)

6-one (1) to give only one iodoacetate, shown to have structure (2). The addition is thus stereospecific and regiospecific.[2]

[1]D. H. R. Barton, J. P. Poyser, P. G. Sammes, M. B. Hursthouse, and S. Neidle, *Chem. Commun.*, 715 (1971).
[2]A. Hassner, *J. Org.*, **33**, 2684 (1968).

**Iodine azide, 1,** 500–501; **2,** 222–223; **3,** 160–161.

*Reaction with 2-cyclopentenone and 2-cyclohexenone.*[1] The reaction of 2-cyclo-pentenone (1) gives 2-iodo-2-cyclopentenone (2) as the only identified product (30% yield). The reaction of 2-cyclohexenone (3) also affords 2-iodo-2-cyclohexenone (4),

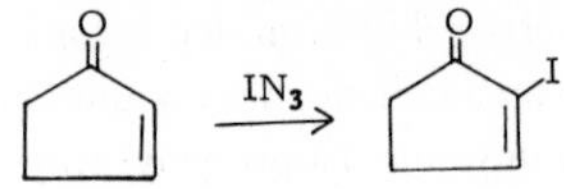

but in this case the expected addition products, (5) and (6), are formed as well in 20% yield.

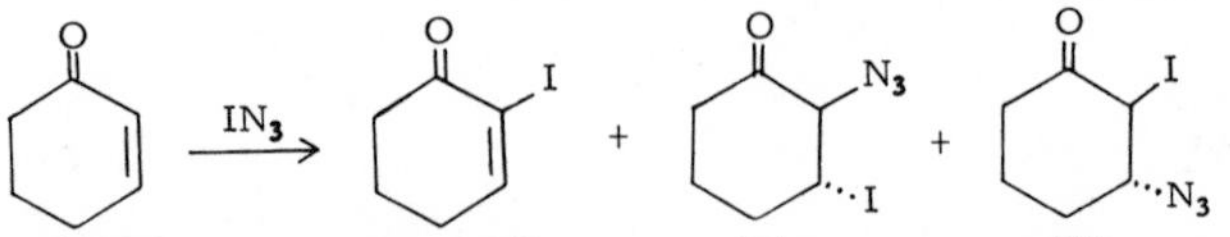

*α-Azidovinyl ketones.*[2] Treatment of the *trans*-α,β-unsaturated ketone (1) with iodine azide in acetonitrile solution gives the *erythro*-iodoazide (2) in nearly quantitative

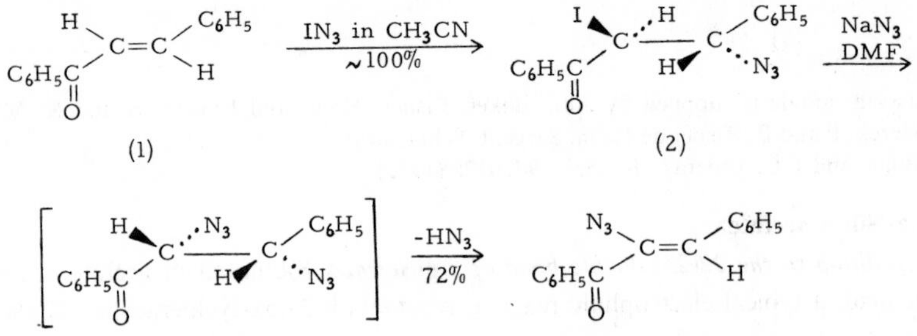

yield. This product on treatment with sodium azide in DMF is converted into the azidovinyl ketone (4) in 72% yield. Presumably the reaction proceeds by an $S_N2$ attack of azide ion on the iodine-bearing carbon atom of (2) to give a bisazide (3), which then eliminates hydrazoic acid.

[1]J. M. McIntosh, *Canad. J. Chem.*, **49**, 3045 (1971).
[2]G. L'abbé and A. Hassner, *J. Org.*, **36**, 258 (1971).

**Iodine thiocyanate**, ISCN. Mol. wt. 185.00.

*Preparation.*[1] Ethereal solutions of iodine thiocyanate are prepared essentially by the method of Raby.[2] Bromine (0.025 mole) is added all at once to a slurry of lead thiocyanate (0.025 mole, Eastman). The mixture is stirred at room temperature until the bromine fades (several minutes). A solution of iodine (0.025 mole) in dry ether is then added to the solution of thiocyanogen, and the mixture is stirred for 15 min. at room temperature with protection from light.

*Episulfides.*[3] Hinshaw reasoned that addition of an alkene to such solutions should give β-iodothiocyanates in analogy with the reaction of iodine isocyanate (**1**, 501; **2**, 223–224; **3**, 161–163) with alkenes. Indeed treatment of cyclohexene with iodine thiocyanate gives a β-iodothiocyanate (2), which on treatment with methanolic

(1)                    (2)                                                      (3)

potassium hydroxide is converted into cyclohexene episulfide (3) in 57% yield. The intermediate iodothiocyanates need not be isolated.

For reaction with Δ²-cholestene, iodine thiocyanate was prepared in methylene chloride; 2β,3β-epithiocholestane was obtained in 46% yield.

Unfortunately the reaction of iodine thiocyanate with acyclic olefins does not give useful yields of episulfides.

[1]J. C. Hinshaw, personal communication.
[2]C. Raby, *Ann. chim.*, **6**, 481 (1961).
[3]J. C. Hinshaw, *Tetrahedron Letters*, 3567 (1972).

**Iodine tris(trifluoroacetate)**, $I(OCOCF_3)_3$. Mol. wt. 465.9, m.p.120° dec.

*Preparation.*[1] The reagent is prepared by the reaction of iodine, fuming nitric acid, and trifluoroacetic anhydride at $-30°$.

$$I_2 + 6\ HNO_3 + 6\ (CF_3CO)_2O \xrightarrow[99.5\%]{} 2\ I(OCOCF_3)_3 + 6\ CF_3COOH + 6\ NO_2$$

*Stereoselective oxidation of alkenes.*[2] The reagent oxidizes alkenes at room temperature to give esters of *vic*-diols in 50–70% yield. Thus isobutene (1) is converted into 1,1-dimethylethylene bis(trifluoroacetate) (2).

The oxidation is highly stereoselective. Thus *cis*-2-butene gives mainly the *meso*-compound (3), and *trans*-2-butene gives mainly the *dl*-compound (4). Cyclohexene is converted mainly into the *cis*-compound (5).

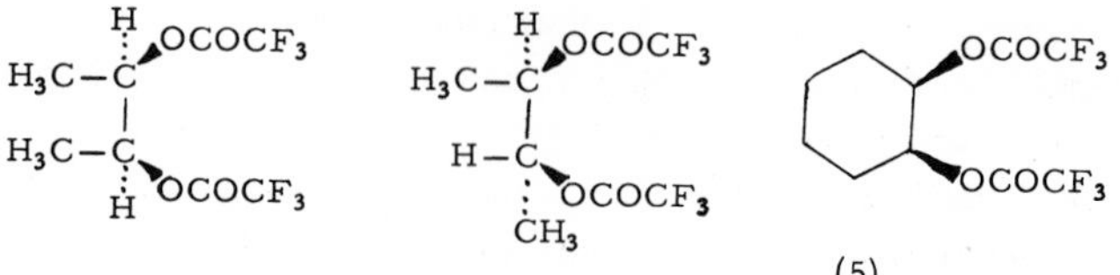

(3)                    (4)                    (5)

A two-step mechanism is proposed:

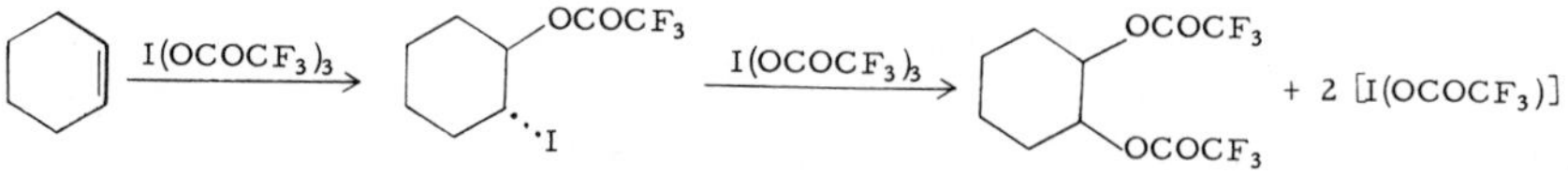

[1]M. Schmeisser, K. Dahmen, and P. Satori, *Ber.*, **100**, 1633 (1967).
[2]J. Buddrus, *Angew. Chem., internat. Ed.*, **12**, 163 (1973).

**Iodobenzene dichloride**, 1, 505–506; 2, 225–226; 3, 164–165.

*Selective free-radical chlorination.*[1] Irradiation of the steroid (1) and an excess of iodobenzene dichloride in benzene with a 275-W sunlamp for 10 min., followed by treatment with silver perchlorate in aqueous acetone, gives an approximately 1:1

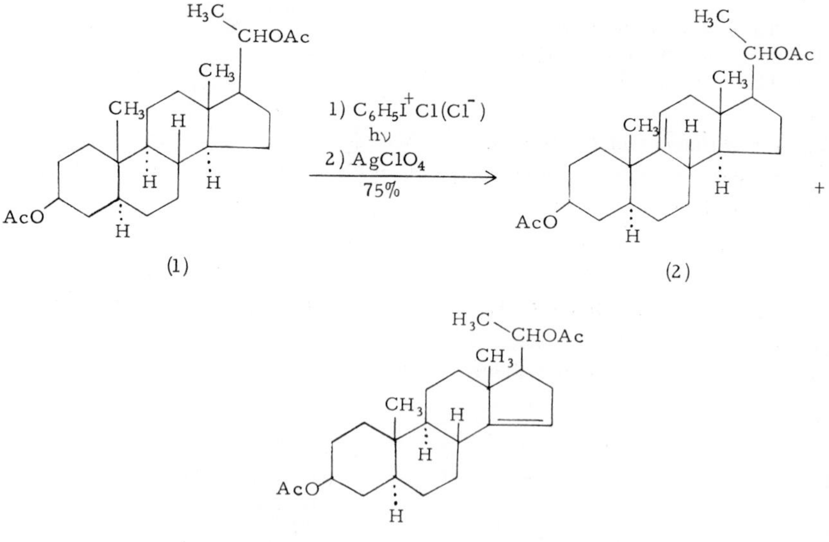

(1)                                        (2)

(3)

mixture of the olefins (2) and (3) in 75% yield. Similarly, $3\beta$-cholestanyl acetate is converted into a 1:1 mixture of the $\Delta^{9(11)}$- and $\Delta^4$-cholestenol acetates. The reaction involves $C_6H_5ICl\cdot$ radicals[2]; note that the attack is highly selective for tertiary, axial hydrogens.

This method for introduction of a $\Delta^{9(11)}$-double bond was used by Djerassi *et al.*[3] for synthesis of the diacetate (5) of the natural starfish genin, $\Delta^{9(11)}$-$5\alpha$-pregnene-$3\beta,6\alpha$-diol-20-one,[4] from (4).

[1] R. Breslow, J. A. Dale, P. Kalicky, S. Y. Liu, and W. N. Washburn, *Am. Soc.*, **94**, 3276 (1972).
[2] Iodobenzene dichloride can be replaced by $BrCCl_3$ as the radical source.
[3] J. E. Gurst, Y. M. Sheikh, and C. Djerassi, *Am. Soc.*, **95**, 628 (1973).
[4] Y. M. Sheikh, B. M. Tursch, and C. Djerassi, *ibid.*, **94**, 3278 (1972).

**Iodoethynyl(trimethyl)silane,** $IC\equiv CSi(CH_3)_3$. Mol. wt. 132.67, b.p. 55°/20 mm., $n_D^{20}$ 1.5110.

*Preparation.*[1] The reagent is prepared in 95% yield by the reaction of iodine chloride and bis(trimethylsilyl)acetylene[2] in methylene chloride at room temperature.

$$(CH_3)_3SiC\equiv CSi(CH_3)_3 + ICl \xrightarrow{95\%} (CH_3)_3SiC\equiv CI + (CH_3)_3SiCl$$

*Arylacetylenes.*[3] Arylacetylenes can be prepared simply and in fairiy good yields (30–80%) by reaction of an arylcopper reagent with iodoethynyl(trimethyl)silane at 0° (3 hr.) and then at 20° (6 hr.). The protective trimethylsilyl group can be removed quantitatively by treatment with alkali.[4]

$$ArCu + IC\equiv CSi(CH_3)_3 \longrightarrow ArC\equiv CSi(CH_3)_3 \xrightarrow{OH^-} ArC\equiv CH$$

The arylcopper reagents are best prepared by addition of an ethereal or THF solution of a Grignard or an aryllithium reagent (1 mole) to a vigorously stirred suspension of freshly prepared cuprous bromide (1.1 mole) in ether. The reagent (0.8 mole) in THF is then added.

[1] D. R. M. Walton and M. J. Webb, *J. Organometal. Chem.*, **37**, 41 (1972).
[2] Prepared in 86% yield from dilithiumacetylide and trimethylchlorosilane in ether/THF [D. R. M. Walton and F. Waugh, *ibid.*, **37**, 45 (1972)].
[3] R. Oliver and D. R. M. Walton, *Tetrahedron Letters*, 5209 (1972).
[4] C. Eaborn and D. R. M. Walton, *J. Organometal. Chem.*, **4**, 217 (1966).

**Iodoform, 3,** 165–166.

*Photochemical reaction with 2,3-dimethyl-2-butene* to give 2,2,3,3-tetramethyliodo-cyclopropane[1]:

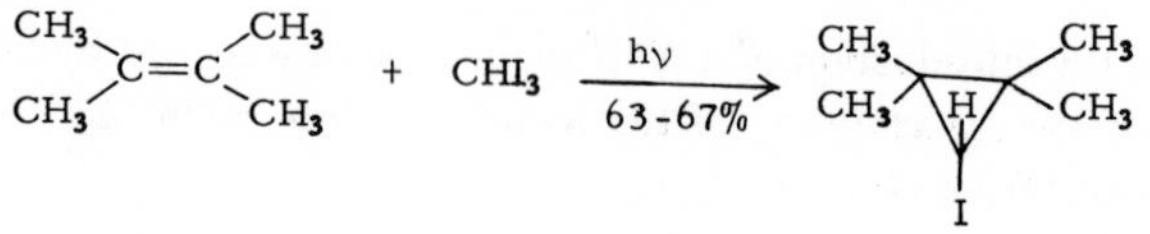

[1]T. A. Marolewski and N. C. Yang, *Org. Syn.*, **52**, 132 (1972).

**Iodosobenzene diacetate, 1,** 508–509; **3,** 166.

*o-Quinones.* Pyrocatechols (1) react with iodosobenzene diacetate at room temperature to give *o*-quinones (3) and iodobenzene. 2-Phenylbenzo-1,3,2-dioxaiodoles (2) are intermediates.[1]

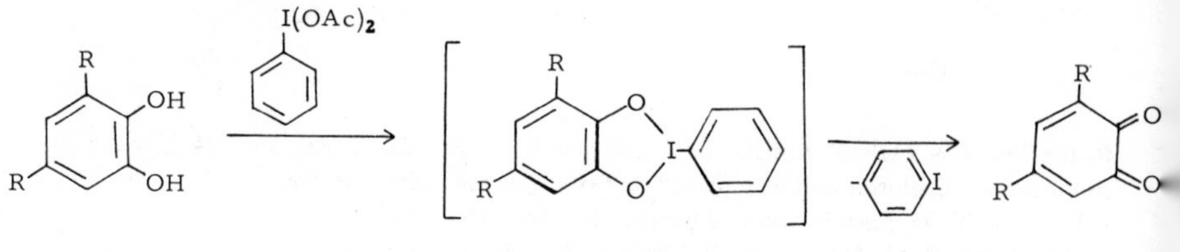

(1)                              (2)                              (3)

[1]A. T. Balaban, *Rev. Roumaine Chim.*, **14**, 1281 (1969).

**Ion-exchange resins, 1,** 511–519; **2,** 227–228.

*Cyclodehydration.*[1] The sulfonic acid resin Amberlite-15 was found to be more satisfactory than Bradsher's reagent (**2,** 214), sulfuric acid, or formic acid for cyclodehydration of 2-methyl-4-phenyl-2-butanol (1) to 1,1-dimethylindane (2).[1]

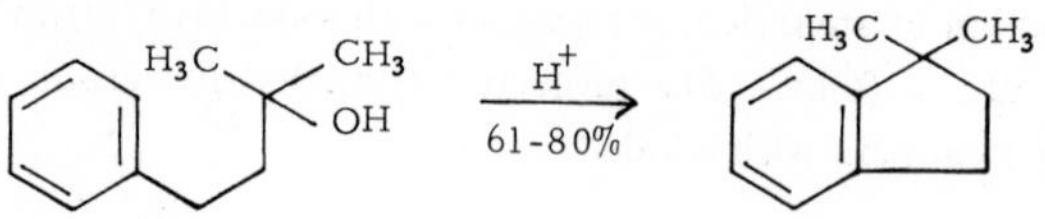

(1)                              (2)

*Esterification.*[2] Acids are readily esterified by use of a combination of an acid ion-exchange resin[3] and a drying agent such as calcium sulfate. The resin is dried in an oven at 100° before use and the calcium sulfate is dried at 180°. Esterification proceeds at room temperature, usually in good yield. Workup is simple because both components of the catalytic dehydration are insoluble in the reaction medium.

*Acetal and ketal synthesis.* The combination of an acidic ion-exchange resin and a drying agent is also effective for synthesis of acetals and ketals.[4]

*Raney nickel catalysts* (**1**, 723). The usual preparation of Raney nickel catalysts W-2[5] and W-6[6] requires a tedious washing process with distilled water to remove the last traces of salt. Wynberg *et al.*[7] report that demineralization can be accomplished readily with an ion-exchange resin (Amberlite IR 120, hydrogen form). The resulting catalysts are somewhat more active than those prepared by the *Organic Syntheses* procedures.

[1]W. M. Harms and E. J. Eisenbraun, *Org. Prep. Proc. Int.*, **3**, 239 (1971).
[2]G. F. Vesley and V. I. Stenberg, *J. Org.*, **36**, 2548 (1971).
[3]Rexyn 101(H) R-231 and Rexyn 101(H) R-204 (Fisher Scientific Co., Fairlawn, N.J.) were used.
[4]V. I. Stenberg, G. F. Vesley, and D. Kubik, *J. Org.*, **36**, 2550 (1971).
[5]R. Mozingo, *Org. Syn., Coll. Vol.*, **3**, 181 (1955).
[6]H. R. Billica and H. Adkins, *ibid.*, **3**, 176 (1955).
[7]H. van Driel, J. W. van Reijendam, J. Buter, and H. Wynberg, *Syn. Commun.*, **1**, 25 (1971).

**Ion-exchange resins, basic, 1**, 514–517.

*Cyclopropanes via abnormal Hofmann elimination.* This procedure[1] provides ethyl 1-benzylcyclopropanecarboxylate in significantly improved yield and is generally applicable to the synthesis of 1-substituted cyclopropanecarboxylic esters.

$$C_6H_5CH_2CH(CO_2C_2H_5)_2 \xrightarrow[Nah]{ClCH_2CH_2N(CH_3)_2} C_6H_5CH_2C(CO_2C_2H_5)_2 \xrightarrow{CH_3I} $$
$$CH_2CH_2N(CH_3)_2$$

$$C_6H_5CH_2C(CO_2C_2H_5)_2 \xrightarrow[\substack{\text{Ion exchange}\\\text{resin, EtOH}}]{OH^-} C_6H_5CH_2C(CO_2C_2H_5)_2$$
$$CH_2CH_2\overset{+}{N}(CH_3)_3I^- \qquad CH_2CH_2\overset{+}{N}(CH_3)_3, (\overset{-}{O}C_2H_5)$$

$$\xrightarrow{\Delta} \quad C_6H_5CH_2C\overset{CO_2C_2H_5}{\underset{CH_2}{\diagdown}}CH_2$$

In a previously flame-dried 1-l. three-necked, round-bottomed flask equipped with a mechanical stirrer, dropping tube, and reflux condenser fitted with a calcium chloride tube are placed 20.5 g. (0.5 mole) of a 58.5% dispersion of sodium hydride in mineral oil (Metal Chemicals Division, Ventron Corporation) and 200 ml. of dimethyl sulfoxide. The mixture is stirred at 20–25°, with occasional ice-bath cooling, and a solution of 125.1 g. (0.5 mole) of diethyl benzylmalonate (Aldrich) is added dropwise. Stirring is continued at 20–25° for about 30 min. (until the evolution of hydrogen is complete), then a solution of 53.8 g. (0.5 mole) of 2-dimethylaminoethyl chloride[2] is added dropwise. After the addition is completed the stirred mixture is gradually heated to 100°. After 30 min. the mixture is cooled and poured into 1 l. of ice water. The mixture is extracted with three 150-ml. portions of ether and the combined ethereal solution is dried over magnesium sulfate and transferred to a 1-l. three-necked, round-bottomed flask equipped with a mechanical stirrer, dropping funnel, and air condenser, and then

78.1 g. (0.6 mole) of methyl iodide is added dropwise. The mixture is stirred at room temperature for 30 min., then the water is decanted from the syrupy methiodide; this is washed with 100 ml. of ether, and the washings decanted. To the syrupy methiodide in the round-bottomed flask are added 100 ml. of ethanol and then an approximately twofold excess (400 g.) of anion-exchange resin (OH⁻ form). Fisher's Rexyn 201 and Mallinckrodt's Amberlite IRA-400 are equally satisfactory. A recovered anion-exchange resin can be converted to the hydroxide form by passing 10% aqueous sodium hydroxide through a column of the recovered resin until free of halide ion (AgNO$_3$–HNO$_3$ test) and then washing with water until neutral. The mixture of exchange resin and ethanol is filtered and the filter cake is washed with three 100-ml. portions of ethanol. The combined filtrates are transferred to a 1-l. one-necked, round-bottomed flask fitted with a 6-in. Vigreux distillation head and the nearly colorless solution is concentrated under reduced pressure (*caution:* foaming). The residual liquid is gradually heated to 150–200° under a water aspirator vacuum (10–25 min.). After thermal decomposition as evidenced by gas evolution is complete, the residue is dissolved in 150 ml. of ether and the solution is dried over magnesium sulfate, filtered, and the filtrate is concentrated. Distillation of the residue gives 66.5–75.5 g. (65–75% overall) of ethyl 1-benzylcyclopropane-carboxylate, as a colorless liquid, b.p. 109–120°/2 mm.

[1]C. Kaiser and J. Weinstock, *Org. Syn.*, submitted 1971.

[2]Liberated from its hydrochloride (Aldrich) a short time before it is needed, by dissolving 86.4 g. (0.6 mole) of the hydrochloride in 100 ml. of water and slowly adding excess solid potassium carbonate. The base is extracted from the thick mixture with ether and the ethereal solution is dried (MgSO$_4$) and concentrated *in vacuo* at 25° to a colorless liquid. Quaternization of the base is noted on prolonged storage even at 0°.

**Iron(III) acetylacetonate,** Fe(CH$_3$COCHCOCH$_3$)$_3$. Mol. wt. 353.17. Supplier: ROC/RIC.

*Decyanation.*[1] Alkyl nitriles can be reductively decyanated to hydrocarbons by treatment with iron(III) acetylacetonate and sodium sand under argon in dry benzene

$$RCN \longrightarrow RH$$

at room temperature for 50–70 hr. Yields of 58–100% can be obtained with saturated primary, secondary, or tertiary cyanides. Yields are somewhat lower in the case of allyl and aryl cyanides.

[1]E. E. van Tamelen, H. Rudler, and C. Bjorklund, *Am. Soc.*, **93**, 7113 (1971).

**Iron pentacarbonyl, 1,** 519–520; **2,** 229–230; **3,** 167. Additional supplier: Strem Chemicals, Inc.

*Selective hydrogenation of α,β-unsaturated carbonyl compounds.* Japanese chemists[1] have effected selective hydrogenation of α,β-unsaturated carbonyl compounds by a wine-red hydridoiron carbonyl complex generated *in situ* from iron pentacarbonyl and a small amount of base in moist solvents. Three different procedures can be used:

1) Iron pentacarbonyl is treated with a small amount of NaOH in 95% methanol at 0–60° ($N_2$). 2) A mixture of ether and water (4:1 v/v) can be used in place of 95% methanol. 3) 1,4-Diazabicyclo[2.2.2]octane (DABCO, **2**, 99–101; this volume) is used as base in moist DMF or HMPT. All methods give about equally satisfactory results. Reduction is effected by addition of the $\alpha,\beta$-unsaturated carbonyl compound (ketones, aldehydes, esters, and lactones) to solutions of the reagent prepared as above and allowing the mixture to stand at room temperature for about 12 hr. Yields of the corresponding saturated carbonyl compounds are generally in the range 90–98%. Reduction of a ketonic or an ester function is negligible; pinacol reduction is not observed.

Examples:

Benzalacetone $\rightarrow$ benzyl acetone (> 98% yield)

Methyl vinyl ketone $\rightarrow$ methyl ethyl ketone (>98% yield)

$\Delta^4$-Cholestenone $\rightarrow$ coprostanone (32%)

Cinnamaldehyde $\rightarrow$ 3-phenylpropionaldehyde (98% yield)

Dimethyl maleate $\rightarrow$ dimethyl succinate (96% yield)

5-Hydroxy-2-hexenoic acid $\delta$-lactone $\rightarrow$ 5-hydroxyhexanoic acid $\delta$-lactone (90% yield)

*Reductions.* Nelson *et al.*[2] selectively reduced enol acetates, vinyl chlorides, and $\alpha,\beta$-unsaturated aldehydes to the corresponding olefins. The reagent also reduces $\alpha$-acetoxy ketones to the respective ketones. Ketones and esters are not reduced. The

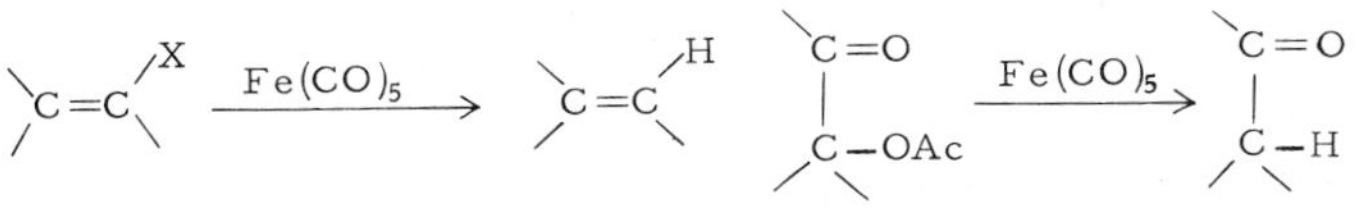

X = OAc, Cl, CHO

best yields are obtained in dibutyl ether or cumene; no olefinic products are observed in DMF or toluene. Yields are in the range 30–75%. In the general procedure, the substrate and iron pentacarbonyl (1:5–10 mole ratio) are refluxed in dibutyl ether under nitrogen. Excess reagent is then destroyed by addition of cupric chloride in acetone or ferric chloride in ethanol.

*Corey–Winter olefin synthesis.* The Corey–Winter olefin synthesis (**1**, 1233–1234) involves treatment of a thionocarbonate with trimethyl phosphite:

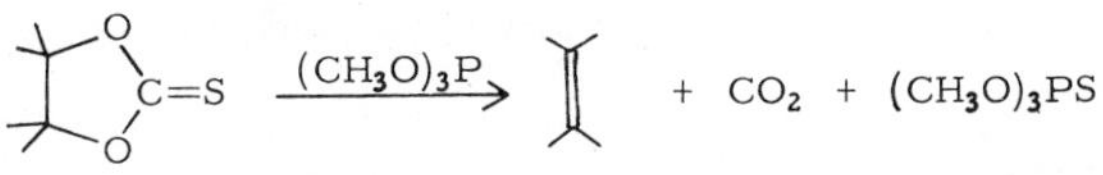

The method gives poor yields in the case of thermally labile olefins. Daub *et al.*[3] have obtained satisfactory yields by use of iron pentacarbonyl in place of trimethyl phosphite:

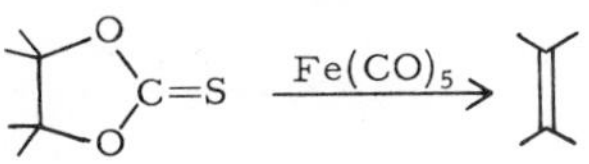

The thionocarbonate and iron pentacarbonyl (molar ratio about 1:1) are heated under nitrogen at a temperature of about 100° for about 2 hr. For example, dibenzobarrelene (1) was obtained by this procedure in 79.1 % yield.

(1)

*Deoxygenation of N-oxides.* Amine oxides are deoxygenated to the corresponding amines by reaction with iron pentacarbonyl in boiling di-*n*-butyl ether.[4]

$$
\begin{array}{c}
R^1 \\
R^2 - N \rightarrow O \\
R^3
\end{array}
\xrightarrow[\substack{45-80\%}]{\substack{Fe(CO)_5 \\ -Fe(CO)_4, \ -CO_2}}
\begin{array}{c}
R^1 \\
R^2 - N \\
R^3
\end{array}
$$

*Amines from N-nitroso compounds.* Aromatic N-nitroso compounds are reduced by iron pentacarbonyl in boiling di-*n*-butyl ether to the corresponding amines. Dialkyl-nitrosoamines afford ureas.[4]

[1] R. Noyori, I. Umeda, and T. Ishigami, *J. Org.*, **37**, 1542 (1972).
[2] S. J. Nelson, G. Detre, and M. Tanabe, *Tetrahedron Letters*, 447 (1973).
[3] J. Daub, V. Trautz, and U. Erhardt, *ibid.*, 4435 (1972).
[4] H. Alper and J. T. Edward, *Canad. J. Chem.*, **48**, 1543 (1970).

**Isoamyl nitrite**, $(CH_3)_2CHCH_2CH_2ONO$. Mol. wt. 117.15, b.p. 99°. Suppliers: Aldrich, Columbia, K and K, E. Merck, P and B, Riedel de Haën.

*α-Substituted α-diazoesters.* α-Substituted α-diazoesters can be obtained simply and in fairly good yields by treatment of an amino acid ester with isoamyl nitrite and a small amount of acid (for example, acetic acid) in refluxing chloroform or benzene.[1]

$$
\begin{array}{c}
R - CHCOOR' \\
| \\
NH_2
\end{array}
\xrightarrow[40-88\%]{}
\begin{array}{c}
R CCOOR' \\
\| \\
N_2
\end{array}
$$

*Diphenoquinones.* 2,6-Disubstituted phenols are oxidized by isoamyl nitrite in methylene chloride to 3,3′,5,5′-tetrasubstituted diphenoquinones.[2]

[1] N. Takamura, T. Mizoguchi, K. Koga, and S. Yamada, *Tetrahedron Letters*, 4495 (1971).
[2] R. A. Jerussi, *J. Org.*, **35**, 2105 (1970).

**Isocyanomethane-phosphonic acid diethyl ester**, $(C_2H_5O)_2P(O)CH_2N{=}C$ (1). Mol. wt. 167.14.

The reagent (1) is prepared as shown in the formulation:

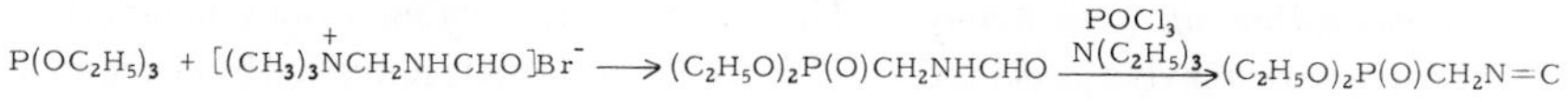

$$P(OC_2H_5)_3 + [(CH_3)_3\overset{+}{N}CH_2NHCHO]Br^- \longrightarrow (C_2H_5O)_2P(O)CH_2NHCHO \xrightarrow[N(C_2H_5)_3]{POCl_3} (C_2H_5O)_2P(O)CH_2N{=}C$$

$$(1)$$

*Vinyl isocyanides.*[1] Reaction of the lithium salt of (1) with carbonyl compounds gives vinyl isocyanides in 70–85% yield:

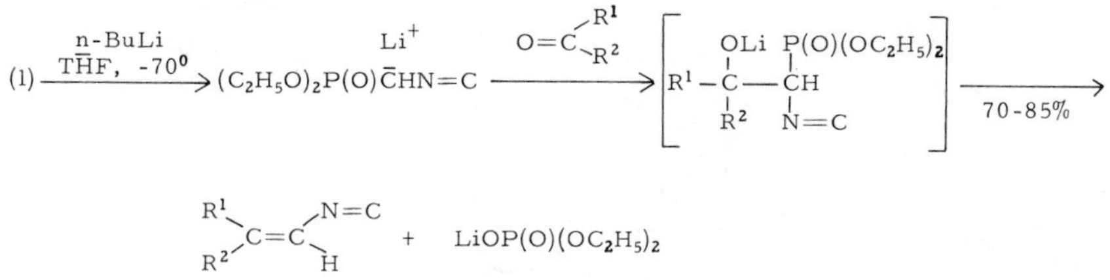

*4-Diethylphosphonyl-2-oxazolines.*[1] Treatment of (1) with about 10% sodium cyanide in ethanol followed by reaction with an aldehyde or ketone gives 4-diethylphosphonyl-2-oxazolines (2) in 70–85% yield.

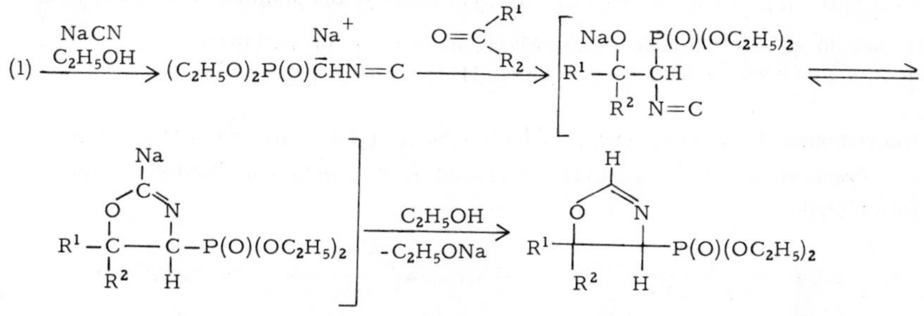

$$(2)$$

*1-Formylamino-1-diethylphosphonylalkenes.*[1] Treatment of (2, above) with potassium *t*-butoxide in THF leads to 1-formylamino-1-diethylphosphonylalkenes in 78–92% yield.

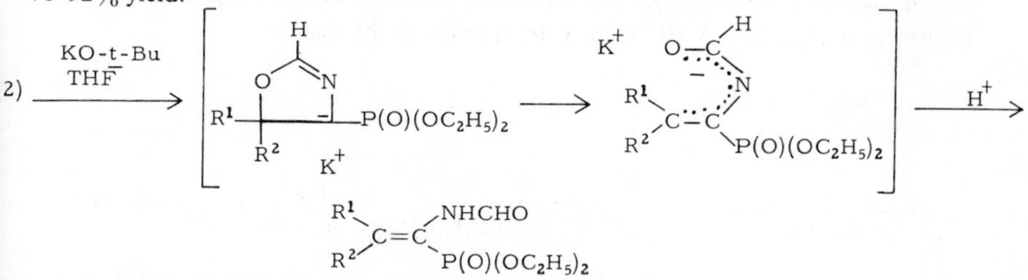

*Cf.* **Tosylmethylisocyanide**, this volume.

[1] U. Schöllkopf and R. Schröder, *Tetrahedron Letters*, 633 (1973).

**Isocyanomethyllithium,** $LiCH_2\ddot{N}{=}\ddot{C}$. Mol. wt. 46.98.

The reagent is prepared by the reaction of methyl isocyanide (0.1 mole) and $n$-butyllithium (0.1 mole) in THF under nitrogen at $-60°$.[1]

*Homologation of cyclic ketones.*[1] The reagent reacts with cyclohexanone (1) to give 1-(isocyanomethyl)-1-cyclohexanol (2) in 77% yield. This is converted by methanolic

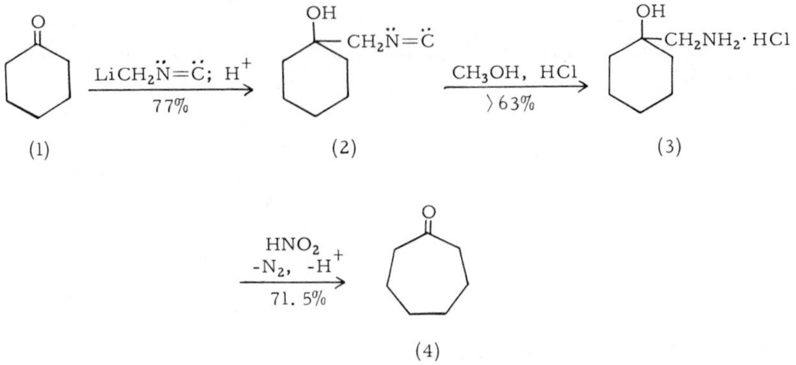

(1)    (2)    (3)

(4)

hydrogen chloride into the amine hydrochloride (3). Application of the Tiffeneau–Demjanov ring expansion reaction[2] to (3) affords cycloheptanone (4) in good yield.

[1] U. Schöllkopf and P. Böhme, *Angew. Chem., internat. Ed.*, **10**, 490 (1971).
[2] P. A. S. Smith and D. R. Baer, *Org. React.*, **11**, 157 (1960).

**Isocyanomethyl *p*-tolyl sulfone,** $p\text{-}CH_3C_6H_4SO_2CH_2NC$. Mol. wt. 195.24, m.p. 110°.

*Preparation.* The reagent (1) is prepared in 66% yield from methyl isocyanide as formulated.

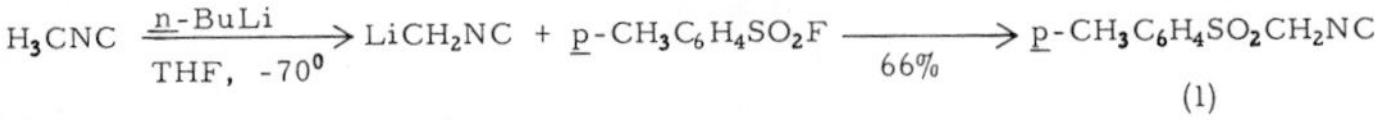

$$H_3CNC \xrightarrow[\text{THF, } -70^0]{n\text{-}BuLi} LiCH_2NC + \underline{p}\text{-}CH_3C_6H_4SO_2F \xrightarrow{66\%} \underline{p}\text{-}CH_3C_6H_4SO_2CH_2NC$$
(1)

*Conversion of a carbonyl compound into the next higher carboxylic acid.*[1] The reagent (1) on treatment with potassium *t*-butoxide in THF at 7–10° is converted into an α-metalated derivative, which reacts with an aldehyde or a ketone to give an adduct (2). Treatment of (2) with 2 *N* HCl (reflux, 10 hr.) affords the acid (3).

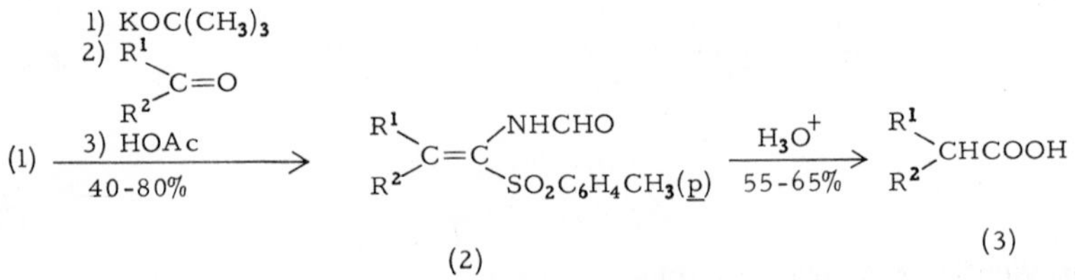

(2)

[1] U. Schöllkopf and R. Schröder, *Angew. Chem., internat. Ed.*, **11**, 311 (1972).

**Isopropenyl acetate, 1,** 524–526; **2,** 191, 230, 309, 325; **3,** 103.

*Condensation with succinic anhydride* (**1,** 526).[1] A detailed procedure has been published.

[1]F. Merényi and M. Nilsson, *Org. Syn.*, **52,** 1 (1972).

**(−)   and   (+)-2,3-O-Isopropylidene-2,3-dihydroxy-1,4-bis(diphenylphosphino)butane (DIOP)** (1). Mol. wt. 498.52, m.p. 91–92°.

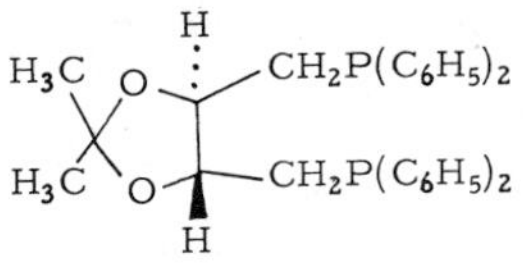

$$(-)-(1),\ \alpha_D\ -12.3^0 \qquad\qquad (+)-(1),\ \alpha_D\ +12.3^0$$

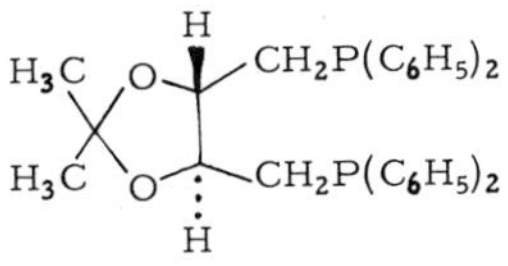

Strem Chemicals supplies both (−)- and (+)-DIOP.

The chiral diphosphine ligand reagents are prepared in several steps from L(+)-tartaric acid and D(−)-tartaric acid, respectively.[1]

*Synthesis of D- and L-amino acids.*[1] Hydrogenation of β-substituted α-acetamido-

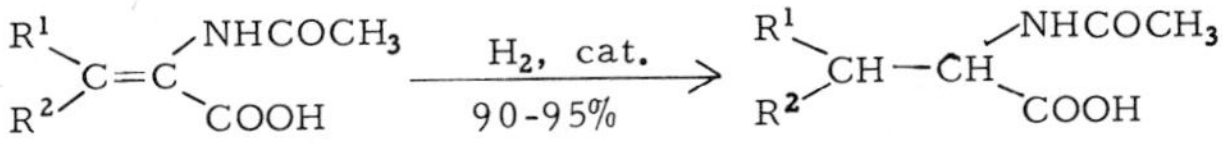

acrylic acids (1) with an optically active Rh(I) complex as catalyst gives optically active N-acetylamino acids (2) in optical yields of 70–80%. The catalyst is prepared by the following displacement reaction:

$$[RhCl(cyclooctene)_2]_2 + 2\ DIOP + 2\ C_6H_6 \longrightarrow 2\ RhCl[DIOP]C_6H_6 + 4\ cyclooctene$$

$[Rh(ethylene)_2Cl]_2$[2] can be used in place of $[RhCl(cyclooctene)_2]_2$. If (+)-DIOP is used as the ligand, L-amino acid derivatives are obtained; if (−)-DIOP is used, D-amino acid derivatives are obtained.

The presence of a carboxyl group in the substrate is not necessary. Thus the enamide (3) is reduced to (4) with an optical yield of 78% by the Rh(I) catalyst containing (−)-DIOP as ligand.

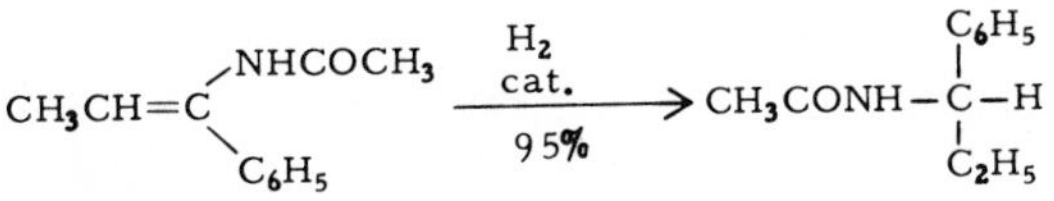

[1]H. B. Kagan and T. P. Dang, *Am. Soc.*, **94,** 6429 (1972).
[2]Available from Strem Chemicals.

**Isovaline (2-Amino-2-methylbutyric acid)**, $\overset{\overset{\displaystyle CH_3}{|}}{\underset{\underset{\displaystyle NH_2}{|}}{CH_3CH_2C}}COOH$ . Mol. wt. 117.15, m.p. > 300°. Supplier: Aldrich.

*Benzylamines.*[1] Benzaldehydes can be reduced to benzylamines in 60–75% yield when refluxed in DMF with *dl*-isovaline followed by acid hydrolysis of the imine

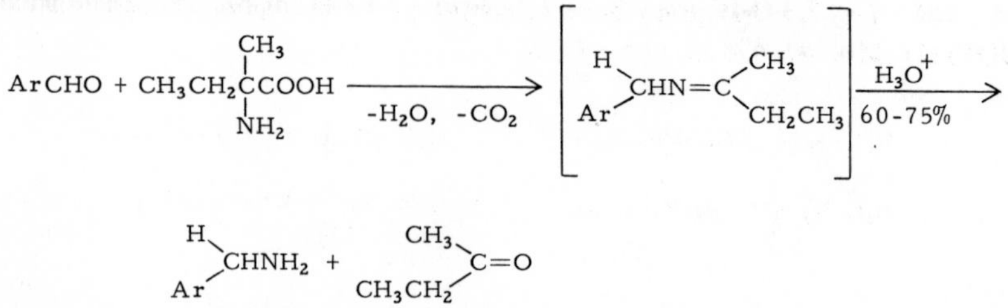

$$ArCHO + \overset{\overset{\displaystyle CH_3}{|}}{\underset{\underset{\displaystyle NH_2}{|}}{CH_3CH_2C}}COOH \xrightarrow[-H_2O, \ -CO_2]{} \left[ \underset{Ar}{\overset{H}{>}}CHN=C\underset{CH_2CH_3}{\overset{CH_3}{<}} \right] \xrightarrow[60-75\%]{H_3O^+}$$

$$\underset{Ar}{\overset{H}{>}}CHNH_2 + \underset{CH_3CH_2}{\overset{CH_3}{>}}C=O$$

intermediate. The reaction proceeds more slowly in diglyme. Yields are lower when α-methylalanine (α-aminoisobutyric acid) is used as the reagent.

[1]G. P. Rizzi, *J. Org.*, **36**, 1710 (1971).

# K

Ketene, **1**, 528–531; **2**, 232.

*Acetylation of CH-acidic compounds.*[1]  CH-acidic compounds can be acetylated by ketene with use of a catalyst (piperidine, sodium methoxide, sodium chloroacetate).
Examples:

$$2\ H_2C{=}C{=}O\ +\ CH_2\underset{\displaystyle COCH_3}{\overset{\displaystyle COOC_2H_5}{\diagdown}}\ \xrightarrow[85\text{-}95\%]{cat.}\ (CH_3CO)_3CCOOC_2H_5$$

$$2\ H_2C{=}C{=}O\ +\ CH_2(COOC_2H_5)_2\ \xrightarrow[85\text{-}95\%]{cat.}\ (CH_3CO)_2C(COOC_2H_5)_2$$

$$H_2C{=}C{=}O\ +\ HC(COOC_2H_5)_3\ \xrightarrow[55\%]{cat.}\ CH_3CO{-}C(COOC_2H_5)_3$$

[1]H. Eck and H. Prigge, *Ann.*, **731**, 12 (1970).

# L

**Lead acetate, Pb(OCOCH$_3$)$_2$.** Mol. wt. 325.30.

The reagent is available as the trihydrate from Fisher Scientific Company.

*Cyclic disulfides.*[1] Reaction of a dithiol (1) with an aqueous solution of lead acetate gives a lead dithiolate (2) in nearly quantitative yield. These react with sulfur in benzene solution at room temperature to give the cyclic disulfide (3) in high yield.

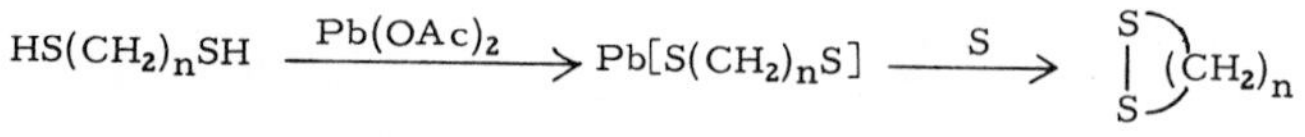

$$HS(CH_2)_nSH \xrightarrow{Pb(OAc)_2} Pb[S(CH_2)_nS] \xrightarrow{S} \overset{S}{\underset{S}{|}} (CH_2)_n$$

(1, n = 3, 4, 5, 6)          (2)          (3)

Sulfur can be replaced by selenium; use of chlorine or iodine results in lower yields.

[1] R. H. Cragg and A. F. Weston, *Tetrahedron Letters*, 655 (1973).

**Lead(IV) acetate azides, Pb(OAc)$_{4-n}$(N$_3$)$_n$.** The reagent decomposes even at $-20°$.

*Review.*[1] The reagent is prepared by mixing lead tetraacetate and trimethylsilyl azide (1, 1236; 3, 316) in aprotic solvents at $-40$ to $-20°$.

The reagent reacts with simple olefins, such as *trans*-stilbene (1), to give principally *vic*-diazido (2) and acetoxyazido compounds (3). Bridged olefins undergo skeletal

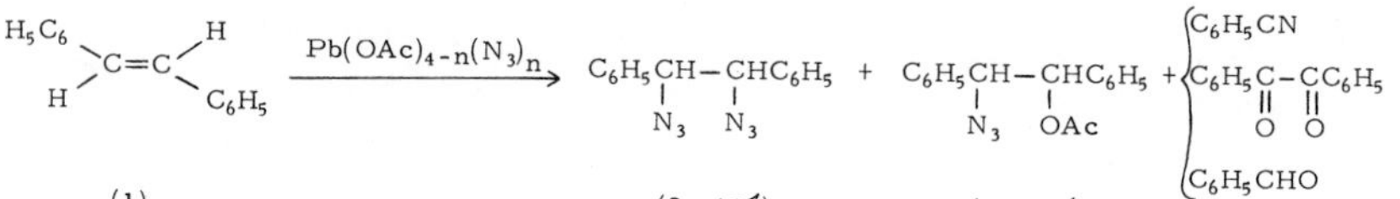

rearrangements. Thus norbornene (4) reacts with lead(IV) acetate azides to give (5) in 75% yield.

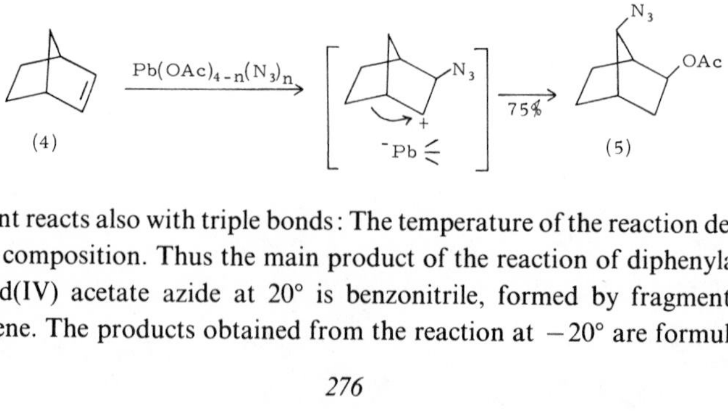

The reagent reacts also with triple bonds: The temperature of the reaction determines the product composition. Thus the main product of the reaction of diphenylacetylene (6) with lead(IV) acetate azide at 20° is benzonitrile, formed by fragmentation of diazidostilbene. The products obtained from the reaction at $-20°$ are formulated.

*276*

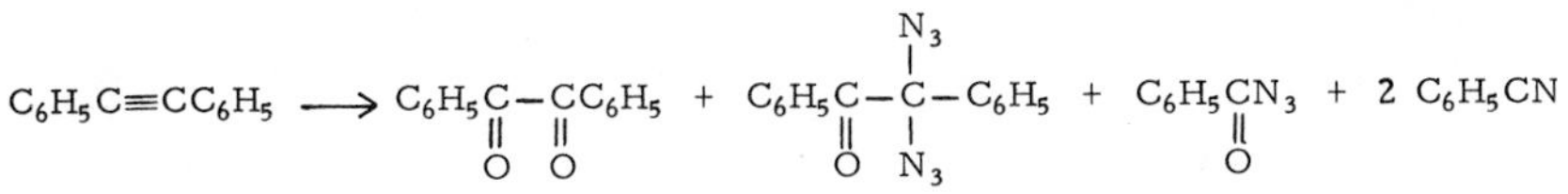

$$C_6H_5C{\equiv}CC_6H_5 \longrightarrow \underset{\substack{\|\ \ \| \\ O\ \ O}}{C_6H_5C-CC_6H_5} \ + \ \underset{\substack{\|\ \ \ | \\ O\ \ N_3}}{C_6H_5C-\overset{\overset{\displaystyle N_3}{|}}{C}-C_6H_5} \ + \ \underset{\substack{\| \\ O}}{C_6H_5CN_3} \ + \ 2\ C_6H_5CN$$

$$\text{(6)} \qquad\qquad 35\% \qquad\qquad\qquad 30\% \qquad\qquad 15\% \qquad\quad 15\%$$

$\Delta^5$-Steroids (7) are cleaved by the reagent at $-15°$ to give 5-keto-6-cyano-5,6-secosteroids (8) in 50–70% yields. The probable intermediates in this fragmentation are shown.

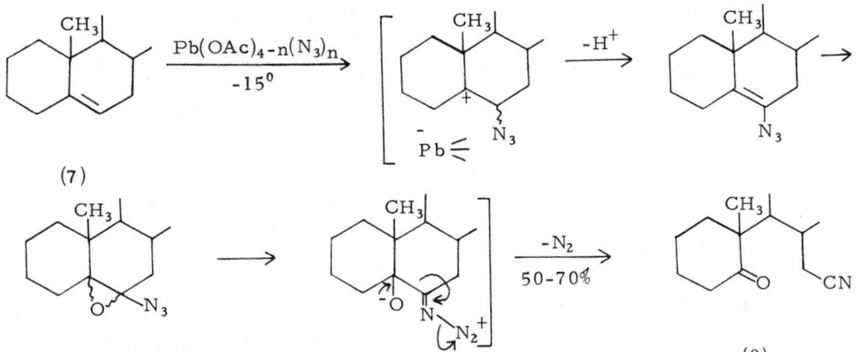

Disubstituted steroid olefins are converted by the reagent into $\alpha$-azido ketones. Thus $\Delta^2$-cholestene (9) reacts with the reagent in dichloromethane at $-20°$ to give 3$\alpha$-azidocholestanone-2 (10) in 50–60% yield. A possible mechanism is formulated. The

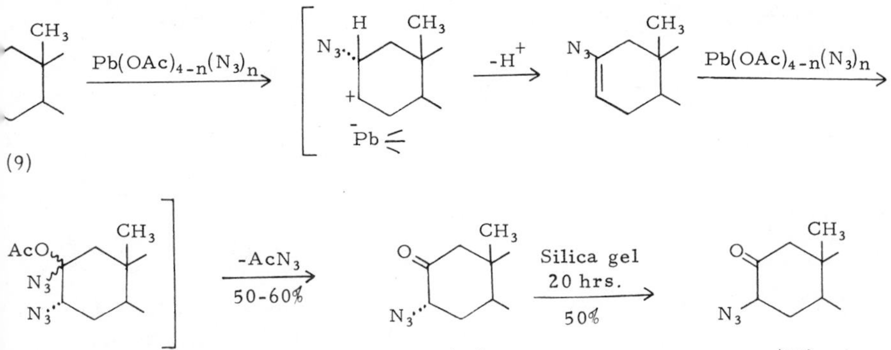

axial azido ketone (10) is rearranged by lengthy treatment with silica gel (acidic) to the more thermodynamically stable 3$\beta$-azidocholestanone-2 (11).

The reaction of $\Delta^5$-steroids (7) with the reagent at room temperature takes a different course than that at $-15°$; the main products are 7$\alpha$-azido-$\Delta^5$-steroids (12) in 35–55% yields. This reaction may involve a radical reaction. $\Delta^6$-Steroids are also converted into 7$\alpha$-azido-$\Delta^5$-steroids.

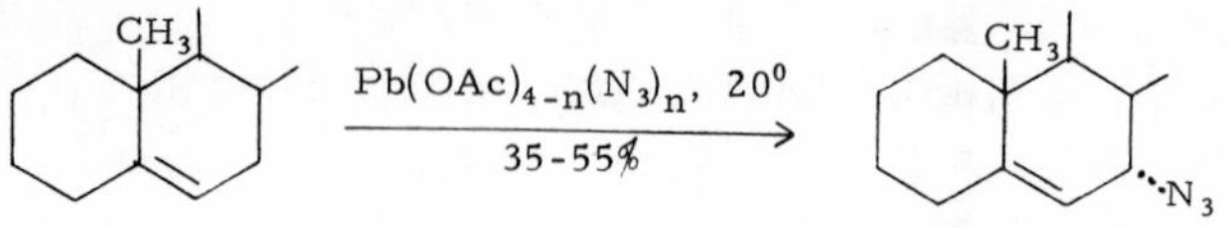

(7)                                    (12)

The iodobenzene diacetate–trimethylsilylazide system, $[C_6H_5I(OAc)_2]-(CH_3)_3SiN_3$, reacts with olefins in a manner similar to that of lead(IV) acetate azides.
[1]E. Zbiral, *Synthesis*, 285 (1972).

**Lead tetraacetate, 1**, 537–563; **2**, 234–238; **3**, 168–171.

*Halodecarboxylation.* Kochi[1] has found that halide salts in the presence of lead tetraacetate convert carboxylic acids into alkyl halides in good yield. The method is particularly convenient for preparation of alkyl chlorides. Unlike the classical Hunsdiecker reaction, the new method is applicable to secondary and tertiary acids as well

eq. 1    $RCOOH + Pb(OAc)_4 + MX \longrightarrow RX + CO_2 + Pb(OAc)_2 + HOAc + MOAc$

as primary acids. The stoichiometry of equation 1 applies. The reaction is carried out in benzene solution and it is desirable to remove oxygen for facile decarboxylation. A free-radical mechanism is proposed.

The method has been used several times since its introduction. Thus Reich and Cram[2] effected the transformation of (1) to (2) in 15% yield.

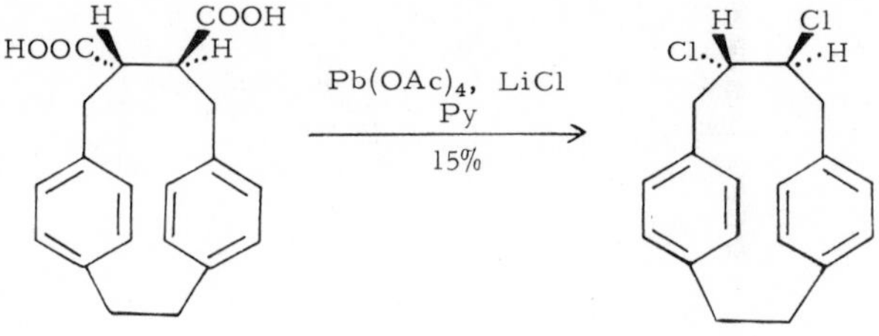

(1)                                    (2)

The reaction was also used by Japanese chemists in a study of the stereochemistry of zizanoic acid.[3] Thus the carboxylic acid (3) was converted into the chloroketone (4) by chlorodecarboxylation.

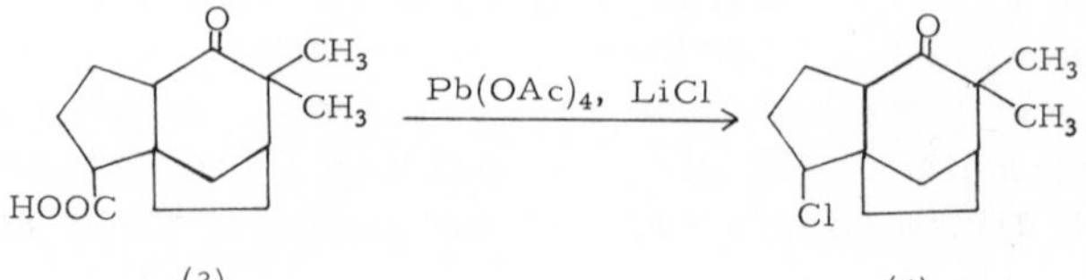

(3)                                    (4)

Australian chemists[4] report an instance where chlorodecarboxylation was effective where the Hunsdiecker reaction (Cristol–Firth modification) failed. Thus treatment of the acid (5) with lead tetraacetate in benzene in the presence of lithium chloride gave a

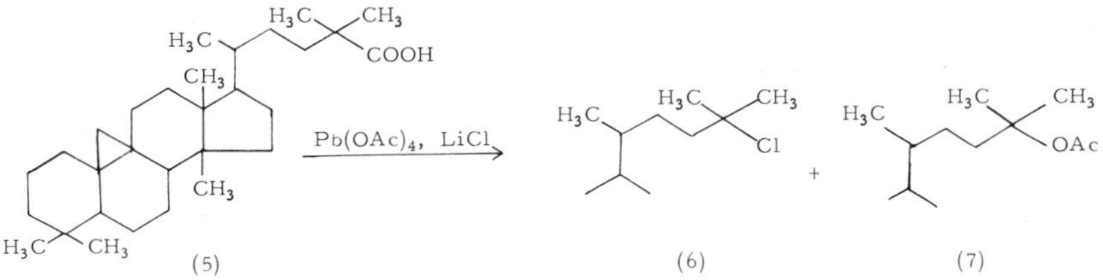

mixture consisting of the desired chloride (6), the corresponding tertiary acetate (7), and an olefin mixture.

Stolow and Giants[5] have examined the stereochemistry of the Kochi reaction. Both *cis*- and *trans*-4-*t*-butylcyclohexanecarboxylic acid (8) and (9) give the same product

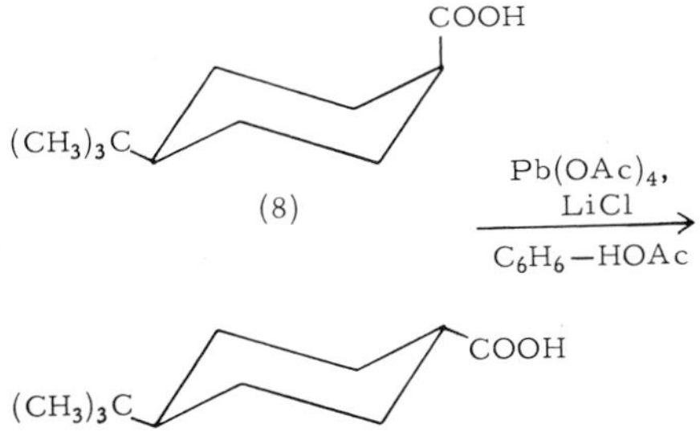

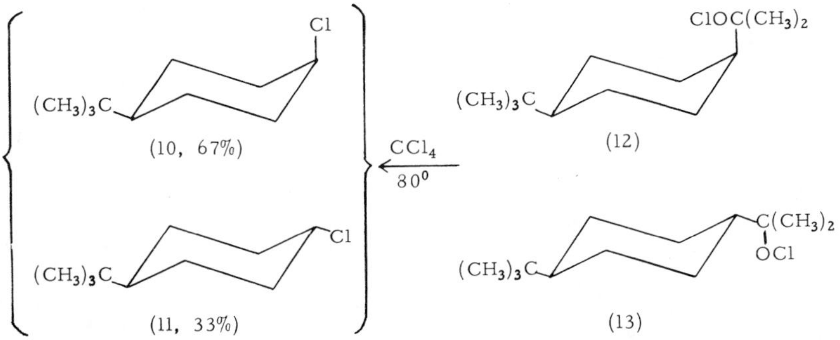

composition: 67% *cis*- and 33% *trans*-4-*t*-butylcyclohexyl chloride (10) and (11). This is the same product distribution found on decomposition in carbon tetrachloride of dimethyl (*cis*- or *trans*-4-*t*-butylcyclohexyl)carbinyl hypochlorite (12) and (13), a reaction known to involve the 4-*t*-butylcyclohexyl radical.[6]

The Kochi reaction has also been used to demonstrate polar effects of remote substituents upon free-radical stereoselectivities in chlorine atom transfer.[7]

*Oxidative decarboxylation of acids* (**1**, 554–557; **2**, 235–237; **3**, 168–169). The oxidative decarboxylation of acids by lead tetraacetate has been reviewed by Sheldon and Kochi.[8]

The oxidative decarboxylation of (1), bicyclo[2.2.0]hex-5-ene-2,3-dicarboxylic anhydride, to (2), bicyclo[2.2.0]hexa-2,5-diene (Dewar benzene) requires rigorously

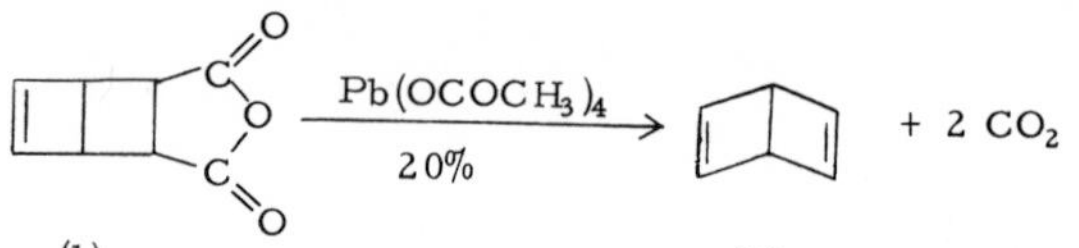

controlled conditions: reaction temperature 43–45°; bath 47–48°; 20-min. reaction time; reduced pressure.[9]

*Oxidation of amides.* Primary amides are oxidized to isocyanates by lead tetraacetate at 50–60° in DMF or benzene in a Hofmann-like reaction.[10]

$$RCONH_2 \xrightarrow{Pb(OAc)_4} RN=C=O$$

Treatment of a β-hydroxy amide in pyridine at room temperature results in a very rapid reaction to give a 2-oxazolidinone (1) in yields of about 95%.[11] Since 2-oxazolidinones are hydrolyzed by base to β-hydroxy amines, the reaction can be used to convert

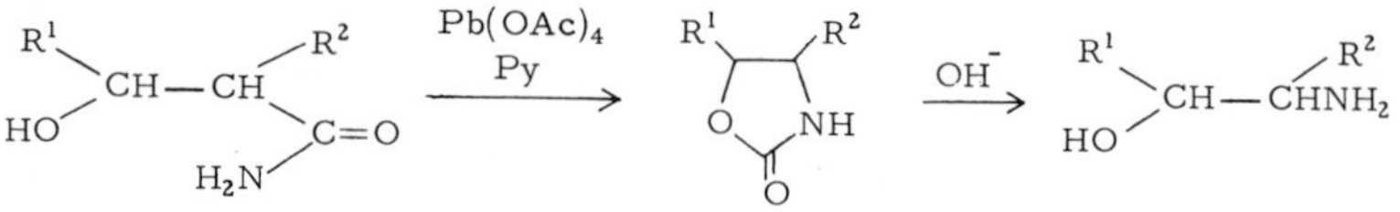

primary amides to amines (2). The reaction proceeds with retention of configuration. Thus (3) is converted into (5) in 72% overall yield.

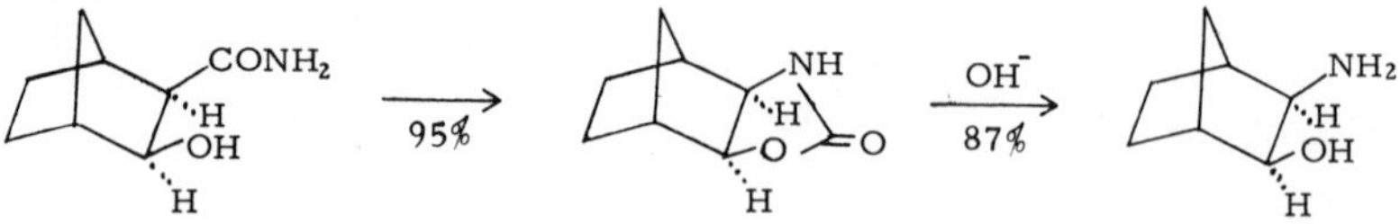

*1,2,3-Benzotriazines.* Oxidation of hydrazones of *o*-aminophenyl alkyl or aryl ketones provides a simple route to 4-alkyl- or 4-aryl-1,2,3-benzotriazines. Thus oxidation of (1) gives two products, (2) and (3).[12]

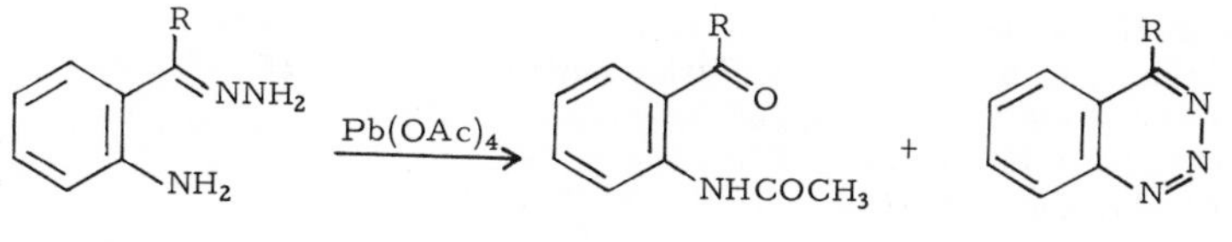

(1, R = CH₃ or C₆H₅)          (2)          (3)

1,2,3-Benzotriazines are also formed rapidly and almost quantitatively by oxidation of 1- and 2-amino-3-substituted indazoles with lead tetraacetate in methylene chloride[13]:

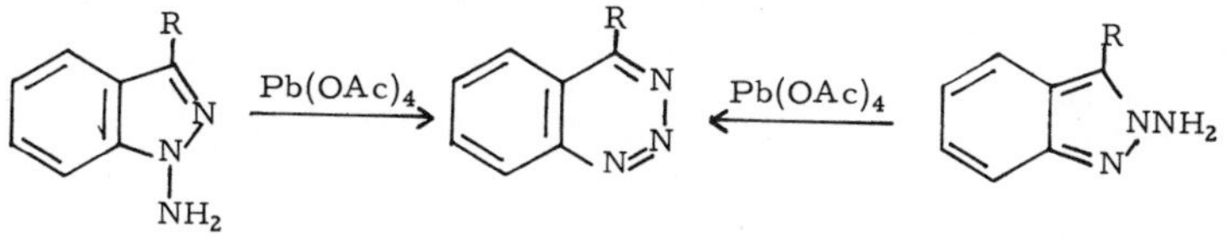

(4)          (5)          (6)

*Reaction with aldo- and ketonitrones.*[14] Lead tetraacetate reacts with aromatic aldonitrones (1) to give N-aroyl-O-acetylhydroxylamines (3).

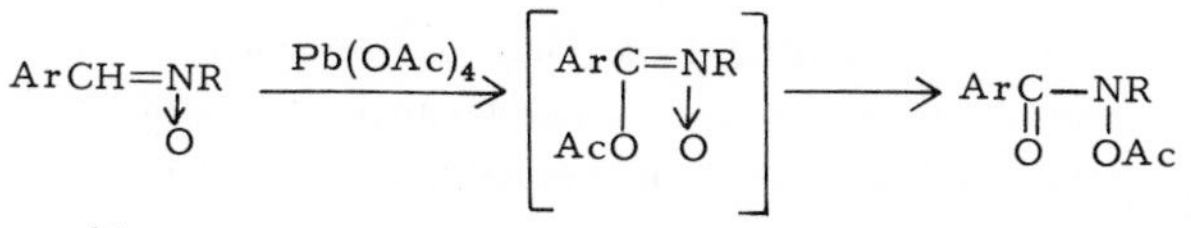

(1)          (2)          (3)

The oxidation takes a different course, cleavage of the C=N bond, in the case of ketonitrones. Thus C,C,N-triphenylnitrone (4) is converted into benzhydrylidene diacetate (5) in about 90% yield.

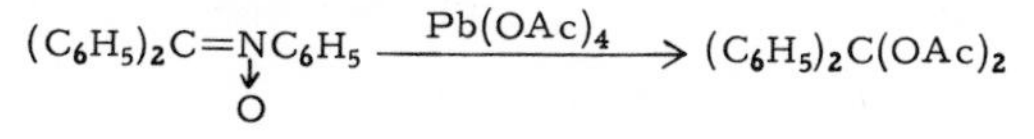

$$(C_6H_5)_2C=NC_6H_5 \xrightarrow{Pb(OAc)_4} (C_6H_5)_2C(OAc)_2$$
$$\qquad\ \downarrow$$
$$\qquad\ O$$

(4)          (5)

*Oxidation of a 4,6-diamino-5-nitrosopyrimidine.*[15] The first step in a new, general synthesis of 2-, 8-, and 9-substituted adenines involves the oxidation of a 4,6-diamino-5-nitrosopyrimidine (1) with a slight excess of lead tetraacetate. The product (2) is a 7-aminofurazano[3,4-d]pyrimidine.

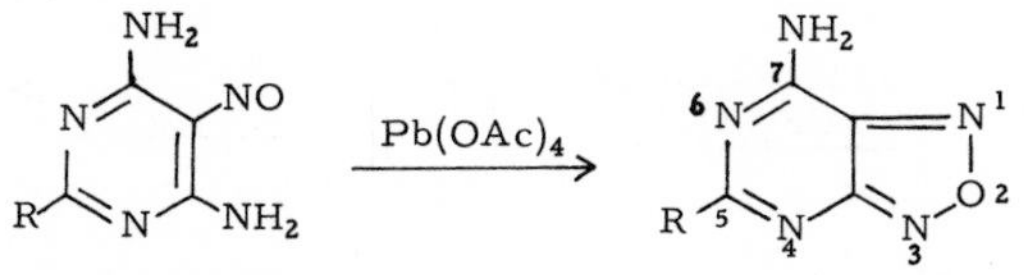

(1)          (2)

[1]J. K. Kochi, *Am. Soc.*, **87**, 2500 (1965); *J. Org.*, **30**, 3265 (1965).
[2]H. J. Reich and D. J. Cram, *Am. Soc.*, **91**, 3517 (1969).

[3]F. Kido, H. Uda, and A. Yoshikoshi, *Tetrahedron Letters*, 1247 (1968).
[4]E. Ritchie, R. G. Senior, and W. C. Taylor, *Australian J. Chem.*, **22**, 2371 (1969).
[5]R. D. Stolow and T. W. Giants, *Tetrahedron Letters*, 695 (1971).
[6]F. D. Greene, C.-C. Chu, and J. Walia, *J. Org.*, **29**, 1285 (1964).
[7]R. D. Stolow and T. W. Giants, *Am. Soc.*, **93**, 3536 (1971).
[8]R. A. Sheldon and J. K. Kochi, *Org. React.*, **19**, 279 (1972).
[9]E. E. van Tamelen, S. P. Pappas, and K. L. Kirk, *Am. Soc.*, **93**, 6092 (1971).
[10]J. B. Aylward, *Quart. Rev. Chem. Soc.*, **25**, 407 (1971).
[11]S. S. Simons, Jr., *J. Org.*, **38**, 414 (1973).
[12]S. Bradbury, M. Keating, C. W. Rees, and R. C. Storr, *Chem. Commun.*, 827 (1971).
[13]D. J. C. Adams, S. Bradbury, D. C. Horwell, M. Keating, C. W. Rees, and R. C. Storr, *ibid.*,
828 (1971).
[14]L. A. Neiman and S. V. Zhukova, *Tetrahedron Letters*, 499 (1973).
[15]E. C. Taylor, G. P. Beardsley, and Y. Maki, *J. Org.*, **36**, 3211 (1971).

**Lead tetra(trifluoroacetate) (LTTFA), 2, 238–239.**

*Trifluoroacetoxylation of benzene derivatives.* Australian chemists[1] studied the reaction of LTTFA in trifluoroacetic acid (TFA) with a number of benzene derivatives and showed that the reaction proceeds by a typical electrophilic substitution mechanism ($S_E$). Thus alkylbenzenes react more rapidly than halobenzenes, whereas nitrobenzene fails to react under moderate conditions. They also obtained evidence (NMR) that, at least in the case of halobenzenes (1), *p*-halophenyllead tristrifluoroacetates (2) are intermediates in the reaction. The *p*-fluoro compound (2, X = F) was actually isolated.

The attack thus involves displacement of a proton by lead (IV). They then explored the possibility of displacing other species by LTTFA and found that metal–metal exchange proceeded particularly smoothly in the case of *p*-substituted phenyltrimethylsilanes (4)[2]; yields of aryl trifluoroacetates (6) were practically quantitative. The reaction thus offers a useful route from an aryl halide to a phenol.

X = CH₃, F, Cl, Br

It is not necessary to prepare the LTTFA (very hygroscopic); it can be generated *in situ* from lead tetraacetate and trifluoroacetic acid.

[1] J. R. Campbell, J. R. Kalman, J. T. Pinhey, and S. Sternhell, *Tetrahedron Letters*, 1763 (1972); J. R. Kalman, J. T. Pinhey, and S. Sternhell, *ibid.*, 5369 (1972).
[2] Preparation: H. Gilman and F. J. Marshall, *Am. Soc.*, **71**, 2066 (1949); V. Chvalovsky and V. Bazant, *Coll. Czech.*, **16**, 580 (1951).

**Lindlar catalyst, 1**, 566–567.

**Reduction.** Hydrogenation of β-damascenone (1) with Lindlar catalyst gives

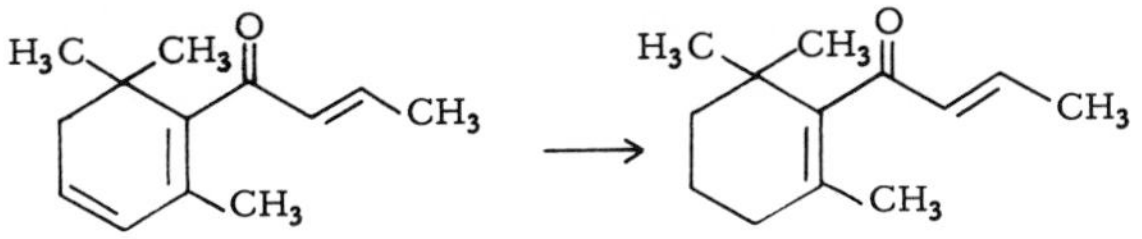

(1) → (2)

β-damascone (2) in nearly quantitative yield. Hydrogenation with palladium-on-carbon gives a mixture of products.[1]

[1] G. Büchi and H. Wüest, *Helv.*, **54**, 1767 (1971).

$$\text{CH}_3$$

**Lithio propylidene-*t*-butylimine,** $\text{Li}\overset{|}{\text{C}}\text{HCH}{=}\text{NC(CH}_3)_3$. Mol. wt. 107.12.

Prepared from propylidene-*t*-butylimine.[1]

**Directed aldol condensation (2, 249).** Büchi and Wüest[2] used Wittig's method of aldol condensation for a highly stereoselective synthesis of the racemic sesquiterpene aldehyde nuciferal (3) by reaction of the aldehyde (1) with lithio propylidene-*t*-butylimine (2). The product was obtained in 83 % yield.

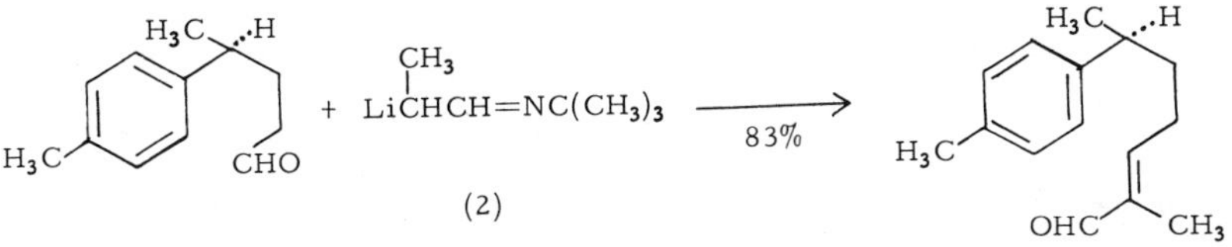

(1)  + $\text{Li}\overset{\text{CH}_3}{\overset{|}{\text{C}}}\text{HCH}{=}\text{NC(CH}_3)_3$ (2)  $\xrightarrow{83\%}$  (3)

The reagent was also used[3] for the synthesis of β-sinensal (5), a flavor constituent of the Chinese orange, from the aldehyde (4).

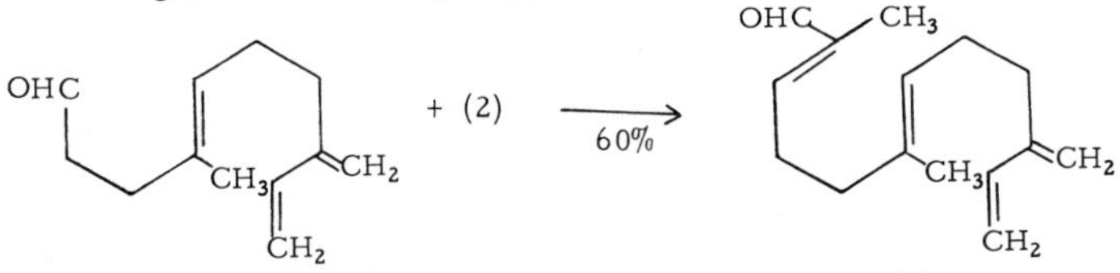

(4)  + (2)  $\xrightarrow{60\%}$  (5)

[1] R. Tiollais, *Bull. soc.*, **14**, 708 (1947).
[2] G. Büchi and H. Wüest, *J. Org.*, **34**, 1122 (1969).
[3] *Idem, Helv.*, **50**, 2440 (1967).

**2-Lithio-2-trimethylsilyl-1,3-dithiane, 2,** 184.    Mol. wt. 198.35.

*Preparation.*[1] The reagent is prepared in essentially quantitative yield by the reaction of 2-lithio-1,3-dithiane (**2,** 182–183) in THF with freshly distilled trimethylchlorosilane at −55°. The mixture is kept at this temperature for 20 min. and then for 5 hr. at 20°. On treatment with water the reagent is converted into 2-trimethylsilyl-1,3-dithiane.

*Ketene thioacetals.* Two laboratories[2,3] have reported a useful, general synthesis of ketene thioacetals (2) by the reaction of (1) in THF with aldehydes or ketones at

$$(1) \ + \ O{=}C\diagdown^{R_1}_{R_2} \xrightarrow{\text{THF}} {}^{R_1}_{R_2}\diagdown C{=}\diagup^{S}_{S} \ + \ (CH_3)_3SiOLi$$

(2)

low temperatures. Yields are generally good (50–90%). Ketene thioacetals are useful synthetic intermediates. Thus on hydrolysis they afford carboxylic acids (3)[4]:

$$(2) \xrightarrow{\text{hydrolysis}} {}^{R_1}_{R_2}\diagdown CHCOOH$$

(3)

Protonation followed by hydride transfer leads to the thioacetal (4), which on hydrolysis with NBS in acetonitrile–water gives the aldehyde (5).[3]

$$(2) \xrightarrow[\text{CH}_2\text{Cl}_2]{\text{CF}_3\text{COOH}} R_1R_2CH{-}\diagup^{S}_{S}{+} \xrightarrow{(C_2H_5)_3SiH} R_1R_2CH{-}\diagup^{H\ S}_{S} \xrightarrow{H_2O} R_1R_2CHCHO$$

(4)                    (5)

*Vinylketene thioacetals.*[5] 2-Lithio-2-trimethylsilyl-1,3-dithiane (1) also reacts with α,β-unsaturated aldehydes and ketones exclusively with the carbonyl group to give vinylketene thioacetals (2). These undergo Diels–Alder cycloaddition, which Carey and

$$(1) \ + \ R^3\diagup^{O}_{}{\overset{\|}{C}}\diagdown^{}_{\underset{R^2}{\overset{}{C}}}{=}{\overset{H}{\underset{}{C}}}\diagdown R^1 \longrightarrow R^3\diagup\diagdown_{R^2}{=}\diagup^{H}\diagdown R^1 \ + \ (CH_3)_3SiOLi$$

(2)

Court have used for a synthesis of a cyclohexenone. Thus the vinylketene thioacetal (3) reacts with maleic anhydride in refluxing xylene (3 hr.) to give the adduct (4) in 60% yield. Hydrolysis of the dithioacetal group was effected with mercuric chloride in 90% aqueous methanol at reflux. The initial product (5), a β-keto acid, undergoes decar-

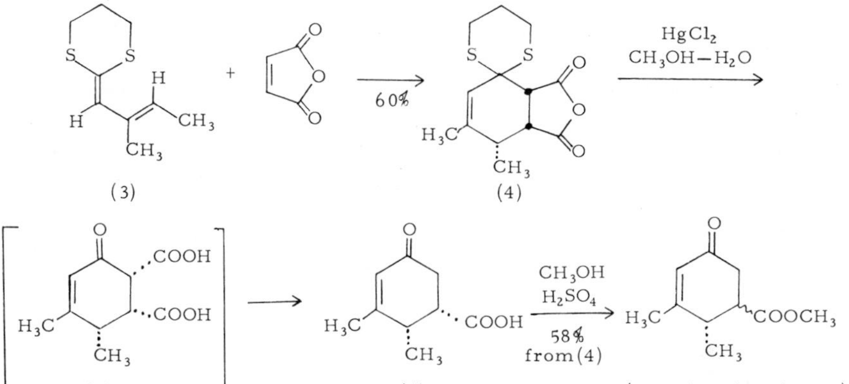

(3)    (4)

(5)    (6)    (7, epimeric mixture)

boxylation under these conditions to (6), obtained as a mixture with the methyl ester (7). The crude product (6) was esterified to give the cyclohexenonecarboxylic ester (7) as a mixture of epimers; the overall yield of (7) from (4) was 58%.

Unfortunately the Diels–Alder reaction of vinylketene thioacetals is limited to reactive dienophiles; thus (3) does not react with diethyl maleate or diphenylacetylene. Also several vinylketene thioacetals were found to be unreactive toward maleic anhydride.

Vinylketene thioacetals react by a Michael-type addition with an alkyllithium in THF at $-80$ to $-20°$. Thus successive addition of n-butyllithium and methyl iodide to (8) gives the thioacetal of the $\alpha,\beta$-unsaturated methyl ketone (9) in almost quantitative

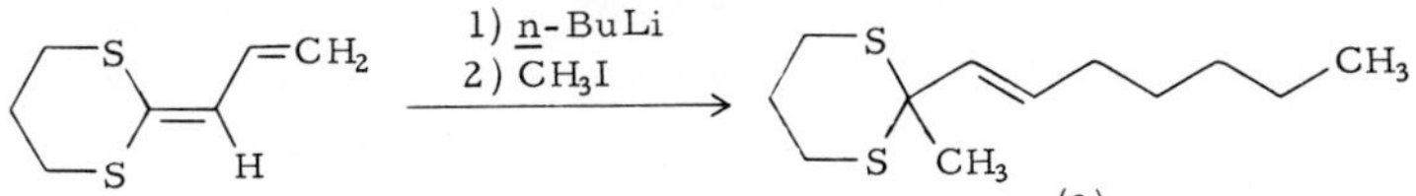

(8)    (9)

yield. Hydrolysis of the thioacetate by mercuric chloride and mercuric oxide in methanol affords the free carbonyl compounds, for example: (10) → (11). The methyl iodide can be replaced by other electrophiles.[6]

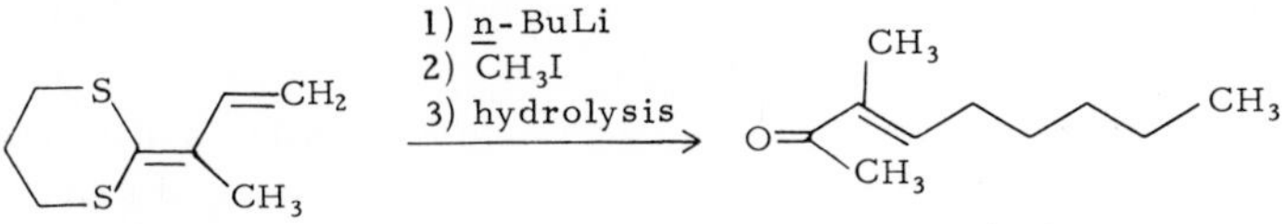

(10)    (11)

[1] D. Seebach, Synthesis, 17 (1969).
[2] D. Seebach, B.-Th. Gröbel, A. K. Beck, M. Braun, and K.-H. Geiss, Angew. Chem. internat. Ed., 11, 443 (1972).
[3] F. A. Carey and A. S. Court, J. Org., 37, 1926 (1972).
[4] J. A. Marshall and J. L. Belletire, Tetrahedron Letters, 871 (1971).
[5] F. A. Carey and A. S. Court, J. Org., 37, 4474 (1972).
[6] D. Seebach, M. Kolb, and B.-T. Gröbel, Angew. Chem., internat Ed., 12, 69 (1973).

**Lithium, 1**, 570–573; **3**, 174–175.

A convenient form of 99.6% pure lithium metal for small-scale reactions can be obtained from Associated Lead Manufacturers Ltd., 14 Gresham Street, London. Preparation for use involves weighing, washing with 40–60° petroleum ether, and cutting the rod by scissors directly into the reaction vessel. Each inch of rod is cut into five pieces, giving an average weight of 0.3 g. per piece (as excess lithium is specified in these reactions, accurate weighing is unnecessary).

*Single-stage equivalent to the Grignard synthesis of 3-butyl-2-methylhept-1-ene-3-ol.*[1]

$$2 \; \underline{n}\text{-}C_4H_9Br \; + \; CH_2{=}\underset{\underset{CH_3}{|}}{C}\text{-}\overset{\overset{O}{\|}}{C}\text{-}OCH_3 \; \xrightarrow{Li} \; CH_2{=}\underset{\underset{CH_3}{|}}{\overset{\overset{C_4H_9\text{-}\underline{n}}{|}}{C}}\text{-}\underset{\underset{C_4H_9\text{-}\underline{n}}{|}}{C}\text{-}OLi \; + \; CH_3OLi$$

$$\downarrow \text{dil. HCl}$$

$$CH_2{=}\underset{\underset{CH_3}{|}}{C}\text{-}\underset{\underset{C_4H_9\text{-}\underline{n}}{|}}{\overset{\overset{C_4H_9\text{-}\underline{n}}{|}}{C}}\text{-}OH \; + \; LiCl$$

A 2-l. four-necked flask equipped with a sealed stirrer, thermometer, gas inlet tube, and dropping funnel is charged with 1200 ml. of dry tetrahydrofurane and 50 g. (7.1 moles) of lithium pieces under an atmosphere of nitrogen. The stirred mixture is cooled to −20° and a mixture of 100 g. (1.0 mole) of methyl methacrylate and 411 g. (3.0 moles) of n-butyl bromide is slowly added dropwise over a period of 3–4 hr. Isothermal conditions required for control of the highly exothermic reaction are maintained by using the Jack-o-matic equipment supplied by Instruments for Research and Industry, Cheltenham, Penn. This consists in a cooling bath seated in a pneumatically operated labjack controlled at −20° by a temperature sensor attached to a thermometer dipping into the reaction vessel, and on completion of the addition the vessel is maintained at this temperature, with stirring, for a further 30 min., during which time the bright, silvery appearance of the lithium metal remaining assumes a duller, mottled aspect. The contents of the flask are then filtered to remove excess lithium and the solvent is removed on a rotary evaporator at reduced pressure. The residual lithium alcoholate is hydrolyzed with dilute hydrochloric acid and the alcohol so liberated is ether extracted and dried over magnesium sulfate. After filtration and removal of the ether on a rotary evaporator, the crude alcohol is distilled through a short Vigreux column to yield 147 g. (80%) of 3-butyl-2-methylhept-1-ene-3-ol, b.p. 80°/1 mm. Purity by glc analysis is > 99%.

The method is generally applicable. The carbonyl compound may be an aldehyde, ketone, or ester; the halide may be chloride, bromide, or iodide, although yields are generally lower with the iodides. Alkyl and aryl halides react with equal facility and the alkyl halide may be primary, secondary, or tertiary.

The table gives the yields observed by glc in reactions conducted with 20% excess halide. The technique is more efficient than the conventional Grignard reaction for three reasons: 1) it is a one-stage process; 2) the yields are generally higher; 3) the final workup is cleaner and more convenient.

Table

| Carbonyl compound | Halide | Product | % yield |
|---|---|---|---|
| Propionaldehyde | Ethyl bromide | 3-Pentanol | 90 |
| Benzaldehyde | Chlorobenzene | Benzhydrol | 100 |
| Di-n-butyl ketone | n-Butyl bromide | Tri-n-butylcarbinol | 91 |
| Ethyl formate | n-Butyl bromide | 5-Nonanol | 91 |
| Acrolein | Ethyl bromide | Pent-1-ene-3-ol | 90 |
| Butyraldehyde | sec-Butyl bromide | 3-Methylheptane-4-ol | 89 |

**cis-Stilbene.**[2] Diphenylacetylene is reduced to cis-stilbene in quantitative yield by treatment with lithium in THF at −78° (1–2 hr.), followed by protonation with methanol also at −78°. The method can be used to prepare $C_6H_5CD{=}CDC_6H_5$.

Reduction with sodium under the same conditions gives a mixture of products, including trans-stilbene, but none of the cis-isomer.

[1]P. J. Pearce, D. H. Richards and N. F. Scilly, Chem. Commun., 1160 (1970); British Patent Application No. 61956 (1969), J.C.S. Perkin I, 1655 (1972); Idem, Org. Syn., **52**, 19 (1972).
[2]G. Levin, J. Jagur-Grodzinski, and M. Szwarc, J. Org., **35**, 1702 (1970).

**Lithium amalgam, 3,** 177–178.

**Dehalogenation.**[1] Dehalogenation of 1,4-dibromo-$\Delta^{2,6}$-cis-hexalin (1) to cis-1,4,9,10-tetrahydronaphthalene (2) is achieved conveniently by treatment with 4 mole

(1)                        (2)

eq. of 0.5% lithium amalgam in ether. The method is superior to that using zinc in refluxing methanol.

[1]E. E. van Tamelen and B. C. T. Pappas, Am. Soc., **93**, 6111 (1971).

**Lithium–Alkylamines, 1,** 574–581; **2,** 241–242; **3,** 175.

**Reduction of carboxylic acids (3,** 175). A detailed procedure for reduction of decanoic acid (1) to decanal (3) and to N-methyldecylamine (4) has been submitted by Bedenbaugh et al. to Organic Syntheses.[1]

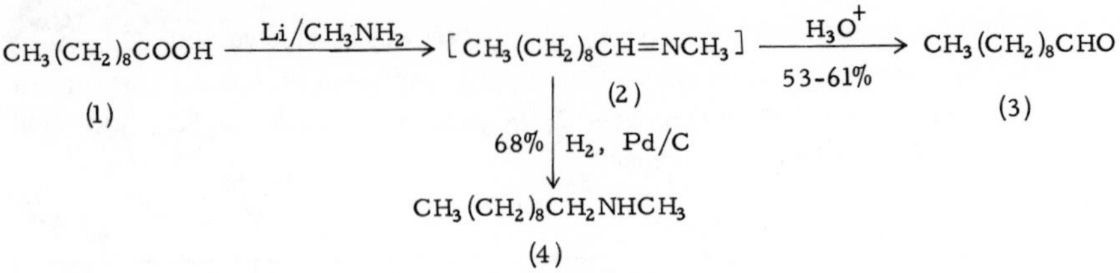

$$CH_3(CH_2)_8COOH \xrightarrow{Li/CH_3NH_2} [CH_3(CH_2)_8CH=NCH_3] \xrightarrow[53-61\%]{H_3O^+} CH_3(CH_2)_8CHO$$

(1)    (2)    (3)

$$68\% \downarrow H_2, \ Pd/C$$

$$CH_3(CH_2)_8CH_2NHCH_3$$

(4)

**3β-Amino steroids.**[2] Reduction of oximes of 3-ketosteroids with lithium in ethylamine is the best procedure for preparation of 3β-amino steroids.

**1-t-Butylcyclohexene.**[3] The most convenient method for preparation of 1-*t*-butylcyclohexene is reduction of *t*-butylbenzene with lithium dissolved in a mixture of ethylenediamine and morpholine to a mixture containing 84 % of the desired product and 16 % of *t*-butylcyclohexane. The products are readily separable by chromatography through a silica gel column. The presence of morpholine ensures a more selective reduction relative to the 3- and 4-*t*-butyl isomers.

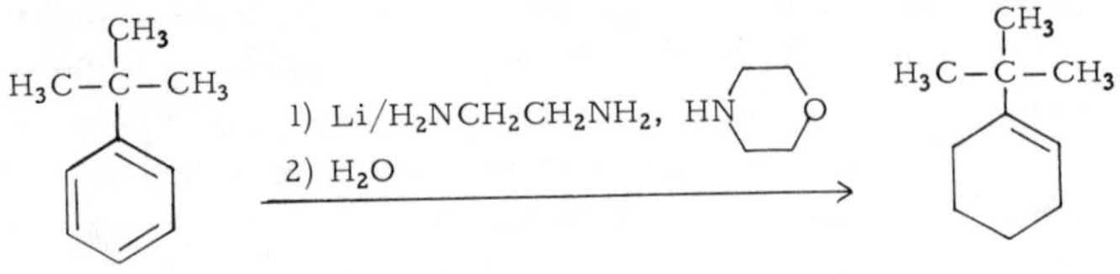

$$\begin{array}{c} 1) \ Li/H_2NCH_2CH_2NH_2, \ HN \ \ O \\ 2) \ H_2O \end{array} \longrightarrow$$

[1] A. O. Bedenbaugh, W. A. Bergin, J. H. Bedenbaugh, and J. D. Adkins, *Org. Syn.*, submitted 1972.
[2] F. Khuong-Huu and M. Tassel, *Bull. soc.*, 4072 (1971).
[3] R. A. Benkeser, R. K. Agnihotri, and T. J. Hoogeboom, *Synthesis*, 147 (1971).

**Lithium–Ammonia, 1,** 601–603; **2,** 205; **3,** 179–182.

*Reductive cleavage of cyclopropane rings.* Cyclopropanes can be reductively cleaved by solutions of an alkali metal in liquid ammonia if a carbonyl group (example I),[1] a carboxylate group (example II),[2] or a phenyl group (example III)[3] is attached to the cyclopropyl ring.

I

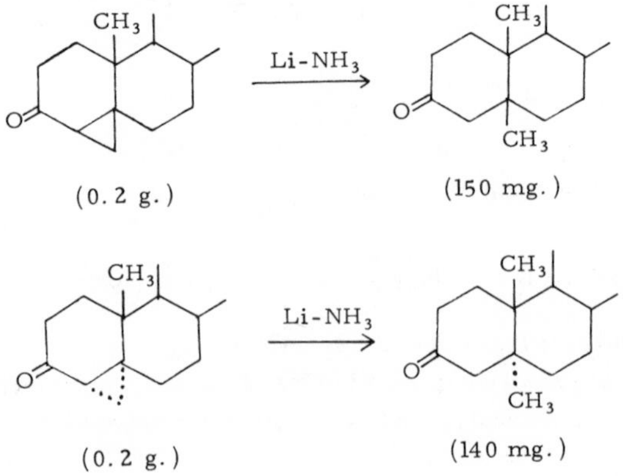

$$\xrightarrow{Li-NH_3}$$

(0.2 g.)    (150 mg.)

$$\xrightarrow{Li-NH_3}$$

(0.2 g.)    (140 mg.)

II

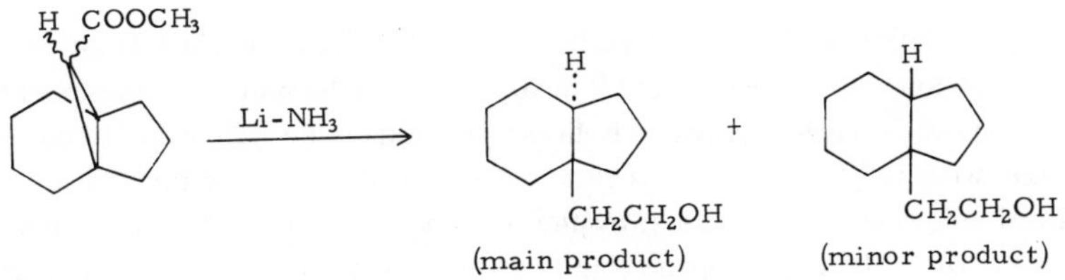

$$\xrightarrow{\text{Li-NH}_3}$$

$CH_2CH_2OH$
(main product)

+

$CH_2CH_2OH$
(minor product)

Present evidence does not establish whether the direction of cleavage of the cyclo-propane ring is determined by stereoelectronic or steric factors.

III

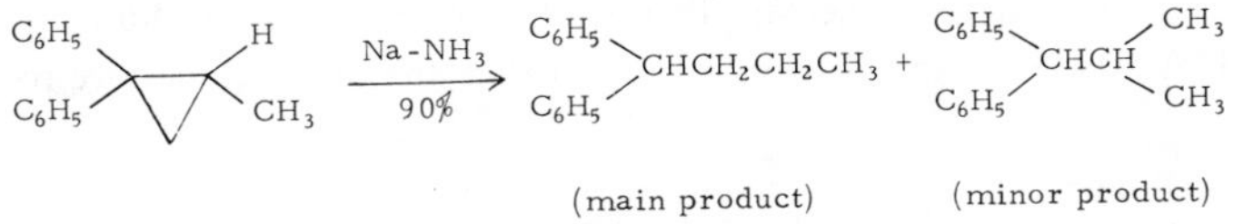

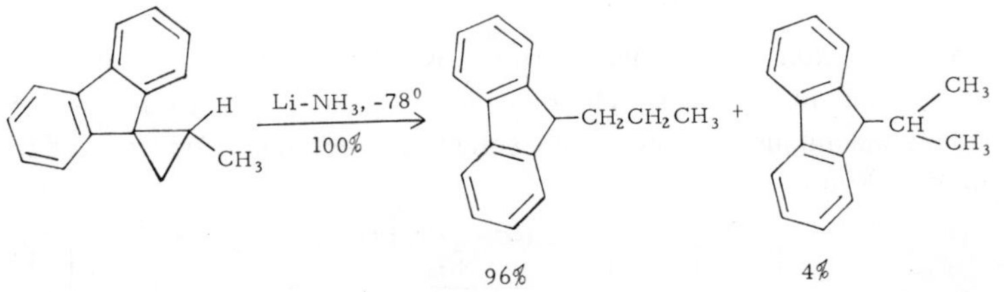

$$\xrightarrow[90\%]{\text{Na-NH}_3}$$

$C_6H_5$ $CHCH_2CH_2CH_3$ $C_6H_5$
(main product)

+

$C_6H_5$ $CHCH$ $CH_3$ $C_6H_5$ $CH_3$
(minor product)

$$\xrightarrow[100\%]{\text{Li-NH}_3, -78^0}$$

$-CH_2CH_2CH_3$

96%

+

$-CH$ $CH_3$ $CH_3$

4%

Walborsky *et al.* [3c] examined a number of reducing systems and obtained highest conversions with lithium–ammonia ($-78°$) and with naphthalene–sodium (**1**, 711–712; **2**, 289; this volume) in glyme (1,2-dimethoxyethane) at 25°. Electrolysis in acetonitrile (25°) was also very effective.

Paquette and Fuhr[4] have used this cleavage reaction for synthesis of strained tri-cyclic systems. Thus treatment of the epoxide (2), obtained from the tricycloheptene

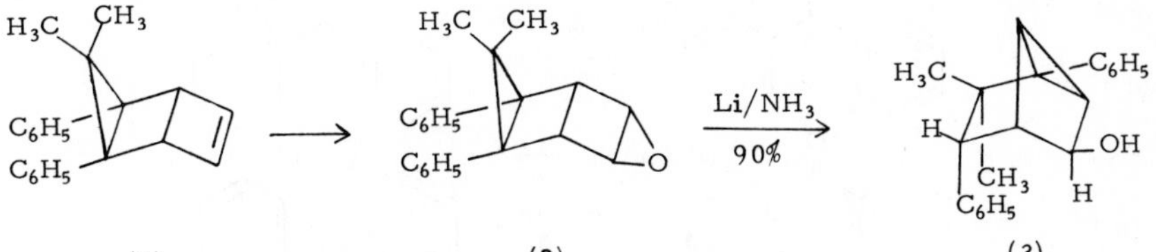

$H_3C$ $CH_3$ $C_6H_5$ $C_6H_5$

(1)

$\longrightarrow$

$H_3C$ $CH_3$ $C_6H_5$ $C_6H_5$ $O$

(2)

$$\xrightarrow[90\%]{\text{Li/NH}_3}$$

$H_3C$ $C_6H_5$ $H$ $OH$ $CH_3$ $C_6H_5$ $H$

(3)

$H_3C$ $CH_3$ $C_6H_5$ $C_6H_5$ $O$

(4)

$$\xrightarrow[95\%]{\text{Li/NH}_3}$$

$H_3C$ $C_6H_5$ $H$ $OH$ $CH_3$ $C_6H_5$ $H$

(5)

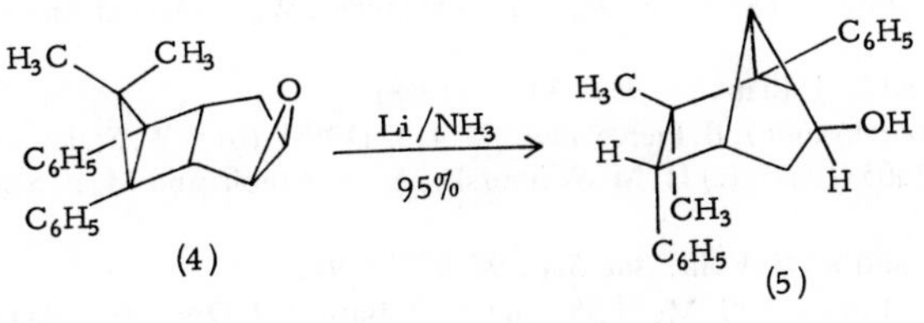

(1) by oxidation with *m*-chloroperbenzoic acid, with 2 equivalents of lithium in liquid ammonia gives the alcohol (3) in 90% yield. In the same way, the epoxide (4) is converted into (5) in 95% yield. In both cases (2) and (4) the cyclopropyl bond common to the two phenyl substituents is ruptured with marked kinetic preference to give a radical anion, which is attacked by the proximate C–O bond to form a new cyclopropane ring. The cyclopropane ring of (3) and of (5) is presumably protected from further reduction by electrostatic factors.

**Reduction of aromatic ketones.**[5] It has been generally assumed that metal–ammonia reduction of aromatic ketones leads to alcohols, mainly because benzophenone (1) is reduced to diphenylmethanol (2, benzhydrol) by sodium in liquid ammonia. However, Hall and co-workers report that aromatic ketones are reduced,

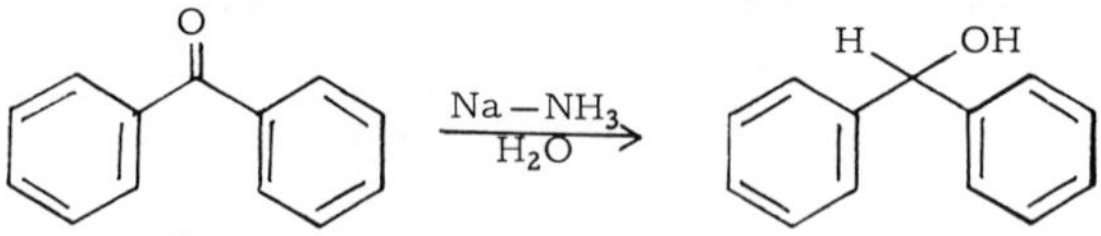

almost quantitatively, to hydrocarbons by lithium–ammonia followed by quenching with ammonium chloride. The reduction is catalyzed by a trace of metals such as cobalt and aluminum. Under these conditions 1-tetralone (3) is reduced to tetralin (4) in 95% yield.

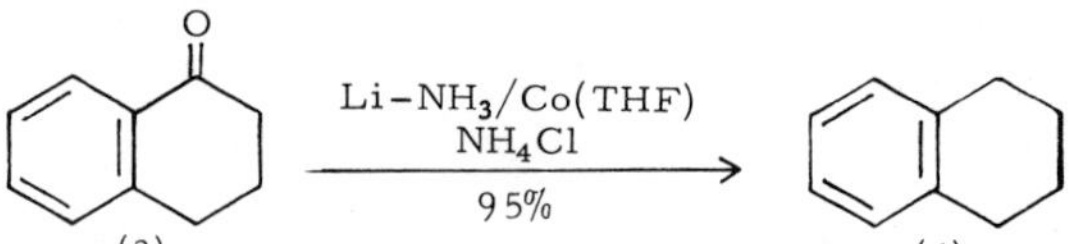

Other examples (isolated yields):

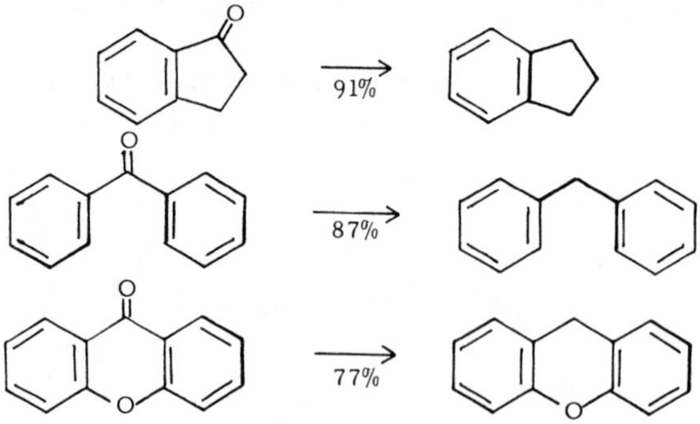

[1]W. G. Dauben and E. I. Deviny, *J. Org.*, **31**, 3794 (1966); W. G. Dauben and R. E. Wolf, *ibid.*, **35**, 374 (1970).
[2]H. O. House and C. J. Blankley, *ibid.*, **33**, 47 (1968).
[3](a) H. M. Walborsky and J. B. Pierce, *ibid.*, **33**, 4102 (1968); (b) S. W. Staley and J. J. Rocchio, *Am. Soc.*, **91**, 1565 (1969); (c) H. M. Walborsky, M. S. Arnoff, and M. F. Schulman, *J. Org.*, **36**, 1036 (1971).
[4]L. A. Paquette and K. H. Fuhr, *Am. Soc.*, **94**, 9221 (1972).
[5]S. S. Hall, S. D. Lipsky, F. J. McEnroe, and A. P. Bartels, *J. Org.*, **36**, 2588 (1971).

**Lithium–*t*-Butanol–Tetrahydrofurane, 1, 604–606.**

*Reductive dechlorination.*[1] When a solution of (1, 0.084 mole), sodium (0.87 g.-atom), and *t*-butanol (20 ml.) is heated in THF (140 ml.), *endo*-2-phenyl-7,7-difluoronorbornene-5 (2) is obtained in 72% yield. In the case of 7,7-difluoronorbornene-2, the yield is low, probably owing to difficulties in the isolation of this very

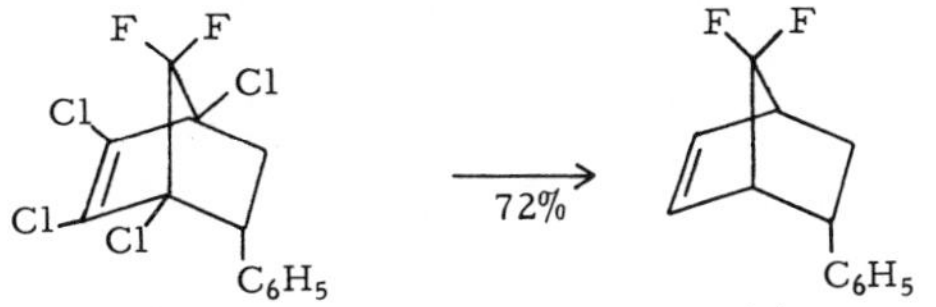

volatile substance. The method is not suitable for preparation of 7,7-difluoronorbornadiene because one double bond is also reduced.

[1]C. W. Jefford and W. Broeckx, *Helv.*, **54**, 1479 (1971).

**Lithium–Ethylenediamine, 1, 614–615; 2, 250; 3, 186.**

*Reductive cleavage of oxetanes.*[1] Oxetanes, like epoxides (3, 186), are readily reduced by lithium in ethylenediamine (EDA). Thus the oxetane (1) is reduced to a mixture of two alcohols, (2) and (3), in the ratio 3:1. The usefulness of this reduction

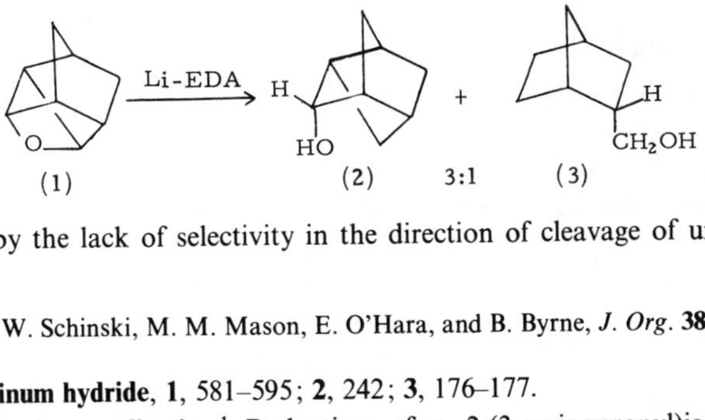

is restricted by the lack of selectivity in the direction of cleavage of unsymmetrical oxetanes.

[1]R. R. Sauers, W. Schinski, M. M. Mason, E. O'Hara, and B. Byrne, *J. Org.* **38**, 642 (1973).

**Lithium aluminum hydride, 1, 581–595; 2, 242; 3, 176–177.**

*Intramolecular cyclization.*[1] Reduction of a 2-(3-aminopropyl)isoquinolinium salt (1, $n = 2$) with lithium aluminum hydride in ether gives the new heterocycle

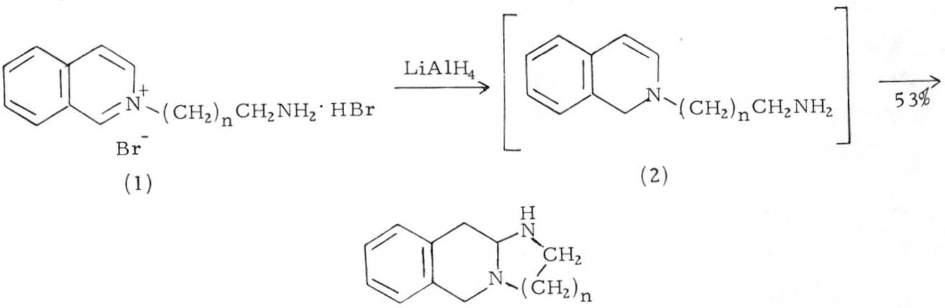

(3, $n = 2$) in 53% yield. A similar reductive cyclization of (1, $n = 1$) gives (3, $n = 1$) in 83% yield. The reactions proceed through the 1,2-dihydroisoquinolines (2, $n = 1$ or 2).

Reduction of N-(3-bromopropyl)isoquinolinium bromide (4) with lithium tri-*t*-butoxyaluminum hydride (**1**, 620–625; **2**, 251–252; **3**, 188) gives the 1,2-dihydroisoquinoline (5), which on treatment with ethanolic methylamine yielded (6), the N-methyl analog of (3, $n = 2$).

(4)                                    (5)

(6)

**Reductive dehalogenation of alkyl halides.**[2] Lithium aluminum hydride has commonly been used only for reductive dehalogenation of reactive substrates; organotin hydrides, for example tri-*n*-butyltin hydride (**1**, 1192–1193; **2**, 424; **3**, 294), have been used for reduction of inert halides. Recently Jefford et al. have reported that supposedly inert halides are reducible by lithium aluminum hydride. Thus the vinyl halide (1) is reduced to (2, *endo*-2-phenylbicyclo[3.2.1]octene-3) by lithium aluminum hydride in refluxing ether (24 hr.). 3-Bromobicyclo[3.2.1]octene-2 is reduced to the parent

(1)                            (2)

hydrocarbon bicyclo[3.2.1]octene-2 in >80% yield when heated with lithium aluminum hydride in dimethoxyethane (DME) for 100 hr.[3] Similar treatment of 2-bromo-2-methylstyrene leads to 2-methylstyrene in 81% yield.

Bridgehead halides can also be reduced by LiAlH$_4$. Thus 1-bromoadamantane is reduced to adamantane in 80% yield when treated with LiAlH$_4$ in boiling ether for 18 hr. Even 1-bromotriptycene is reduced to the parent hydrocarbon when treated with LiAlH$_4$ in DME for 48 hr.

Even cyclopropyl halides are reducible by LiAlH$_4$.[3] Thus treatment of 7,7-dibromo-bicyclo[4.1.0]heptane (3) with LiAlH$_4$ in ether for 18 hr. gives the *anti*- and *syn*-monobromo derivatives (4) and (5) together with the parent hydrocarbon (6) in the

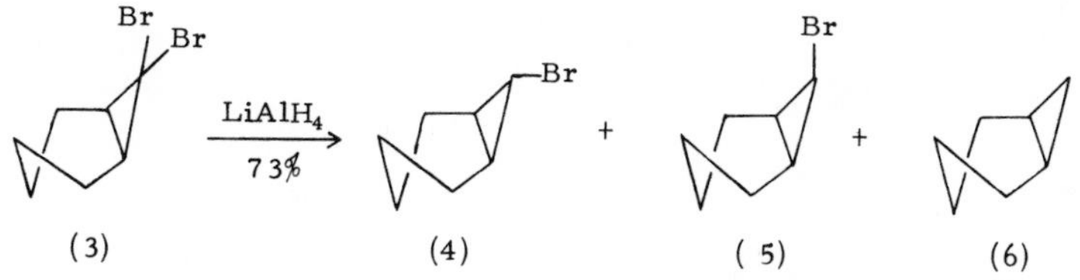

(3)                    (4)              ( 5)              (6)

ratio 23.2:70.0:6.8. In fact, it was shown that geminal halogen is even more reducible than a bridgehead halogen.

**Stereoselective reduction.** Lithium aluminum hydride in ether reduces (1, bicyclo-[4.3.1]decatriene-2,4,8-one-7) to the less hindered alcohol (2, endo-7-hydroxybicyclo-[4.3.1]decatriene-2,4,8).[4]

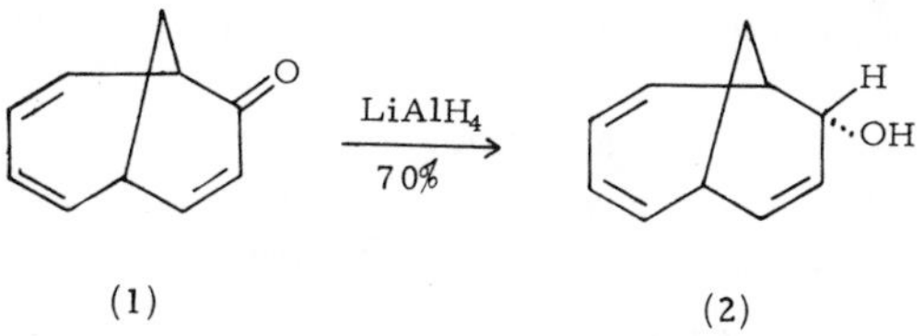

(1)                                    (2)

**3,3-Disubstituted 3H-indoles.**[5] 3,3-Disubstituted 3H-indoles (3) can be prepared readily by lithium aluminum hydride [or sodium bis-(2-methoxyethoxy)aluminum hydride, **3**, 260–261; this volume] reduction of 3,3-disubstituted oxindoles (1), followed by oxidation of the indoline (2) with activated manganese dioxide or potassium permanganate in acetone.

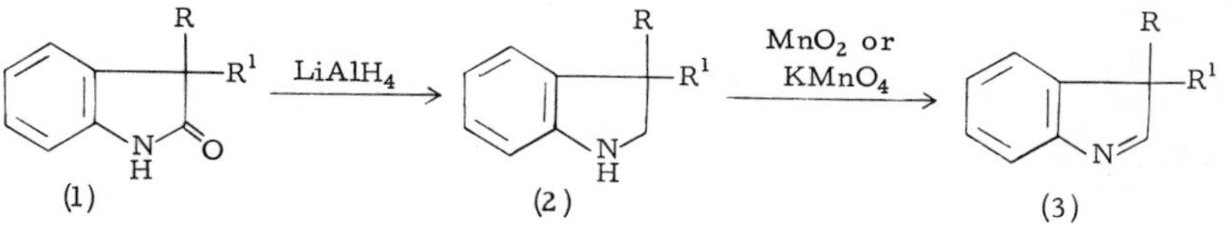

(1)                    (2 )                    (3 )

[1] N. Finch and C. W. Gemenden, *J. Org.*, **38**, 437 (1973).
[2] C. W. Jefford, A. Sweeney, and F. Delay, *Helv.*, **55**, 2214 (1972).
[3] C. W. Jefford, D. Kirkpatrick, and F. Delay, *Am. Soc.*, **94**, 8905 (1972).
[4] G. Schröder, U. Prange, B. Putze, T. Thio, and J. F. M. Oth, *Ber.*, **104**, 3406 (1971).
[5] K. Takayama, M. Isobe, K. Harano, and T. Taguchi, *Tetrahedron Letters*, 365 (1973).

**Lithium aluminum hydride–Aluminum chloride** (Alane, $AlH_3$), **1**, 595–599; **2**, 243; **3**, 176–177.

**Reduction of oxetanes.**[1] Oxetanes, like epoxides (**1**, 598), are reduced by lithium aluminum hydride–aluminum chloride (ratio 3:1) in refluxing ether. Thus oxetane (1) is reduced to (2) and (3) in the ratio 4:1. The alcohol (2) is presumably formed via the aldehyde (a).

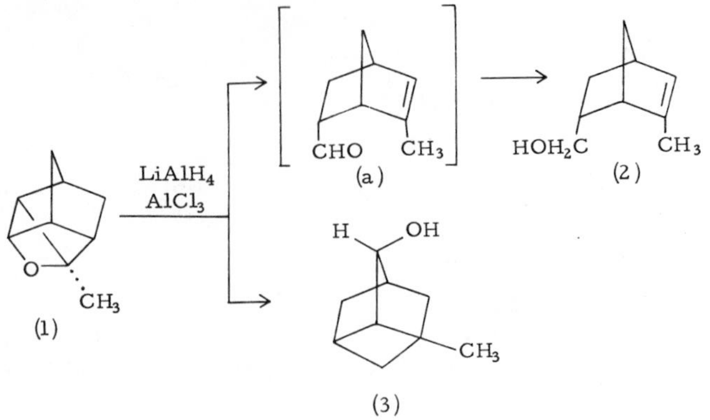

*Cyclobutenes from cyclopropenes.*[2] Treatment of ethyl 1,2-dipropyl-1-cyclopropene-3-carboxylate (1)[3] with lithium aluminum hydride–aluminum chloride (1:1) in ether at −80° affords 1,2-dipropylcyclobutene (2) in 80% yield.

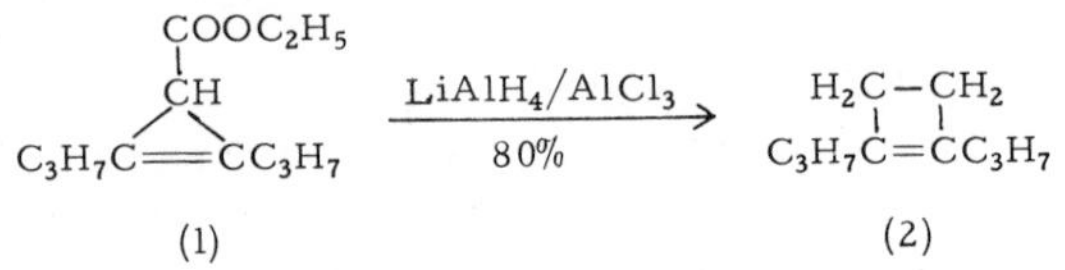

*Reduction of 6β-methoxy-3α,5-cyclo-5α-steroids.*[4] Reduction of 6β-methoxy-3α,5-cyclocholestane (1) with mixed hydride affords as the major product 3α,5-cyclo-5α-cholestane (2).

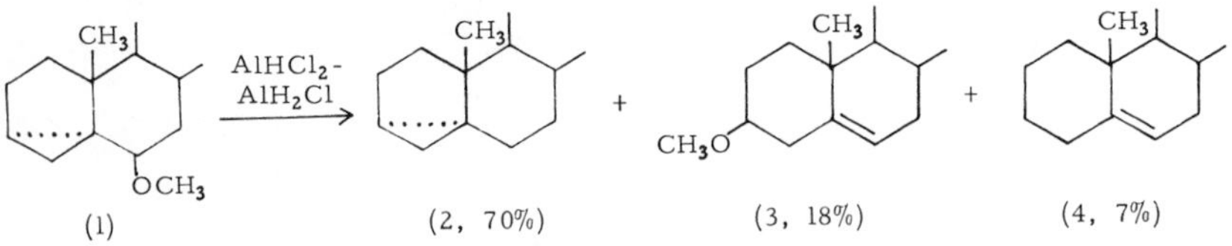

[1]R. A. Sauers, W. Schinski, M. M. Mason, E. O'Hara, and B. Byrne, *J. Org.*, **38**, 642 (1973).
[2]W. J. Gensler, J. J. Langone, and M. B. Floyd, *Am. Soc.*, **93**, 3828 (1971).
[3]Obtained by treatment of 4-octyne with ethyl diazoacetate in ether.
[4]A. Romeo and M. P. Paradisi, *J. Org.*, **37**, 46 (1972).

**Lithium aluminum hydride–Pyridine [Lithium tetrakis(N-dihydropyridyl)aluminate, LDPA], 1, 599–600.**

*3-Substituted pyridines.*[1] When LDPA is heated with an alkyl halide, benzyl chloride, bromine, or other electrophilic reagent, 3-substituted pyridines are obtained in 40–90% yield. The reaction probably involves alkylation of the 1,2-dihydropyridyl moiety of LDPA to give a 2,5-dihydropyridine, which is then oxidized to the final product. The reaction is useful because direct reaction of pyridine with nucleophilic reagents leads to 2- and 4-substituted pyridines.

[1]C. S. Giam and S. D. Abbott, *Am. Soc.*, **93**, 1294 (1971).

## Lithium aluminum hydride–Sodium methoxide.

*Reduction of propargylic alcohols to allylic alcohols.* Corey et al.[1] reported in 1967 that reduction of the propargylic alcohol (1) with 0.1 $M$ lithium aluminum hydride–sodium methoxide (mole ratio 1:2) in THF at reflux (3 hr.) followed by iodination gives specifically the $\gamma$-iodo alcohol (2). Reaction of (2) with dimethylcopperlithium affords *trans,trans*-farnesol.

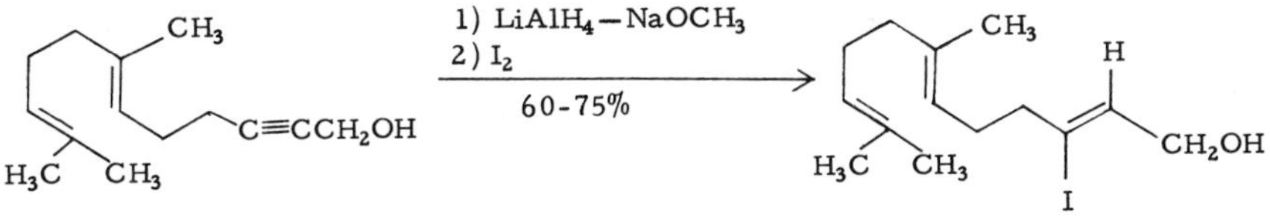

The method was used in a new stereospecific synthesis of 3-ethyl-7-methyl-*trans,cis*-2,6-nonadiene-1-ol (IV), an important intermediate in a synthesis of the $C_{18}$ juvenile hormone of *Cecropia*.[2] Thus reduction of octa-2,6-diyne-1-ol (I) with a large excess of lithium aluminum hydride–sodium methoxide at reflux in dimethoxyethane gave an organoaluminum intermediate (II) which was iodinated to afford 3,7-diiodo-*trans,trans*-2,6-octadiene-1-ol (III). Reaction with 8 eq. of diethylcopperlithium then afforded the desired alcohol (IV).

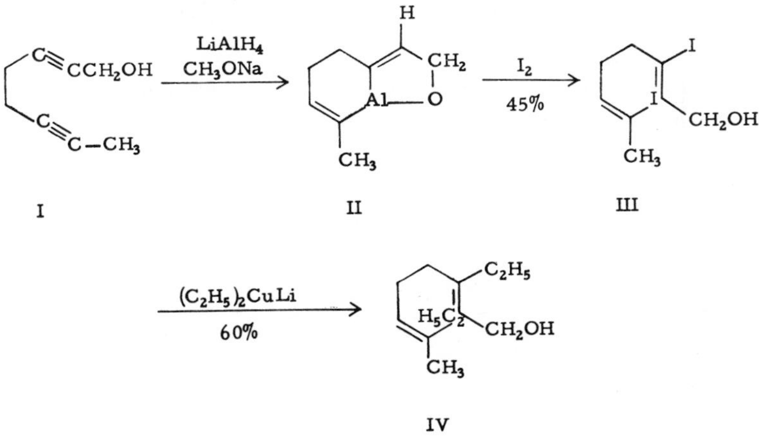

IV

[1]E. J. Corey, J. A. Katzenellenbogen, and G. H. Posner, *Am. Soc.*, **89**, 4245 (1967).
[2]E. J. Corey, J. A. Katzenellenbogen, S. A. Roman, and N. W. Gilman, *Tetrahedron Letters*, 1821 (1971).

**Lithium bis(trimethylsilyl)amide**, $[(CH_3)_3Si]_2NLi$. Mol. wt. 167.33.

*Preparation* (3, 173).

*Synthesis of glycidic esters.* The Darzens reaction[1] fails with compounds such as acetaldehyde which give only self-condensation products. Borch[2] has described a new procedure which is generally applicable. The α-bromo ester anion (2) is smoothly generated by reaction with lithium bis(trimethylsilyl)amide in THF at $-78°$. Addition of the aldehyde or ketone at $-78°$ followed by workup affords the corresponding

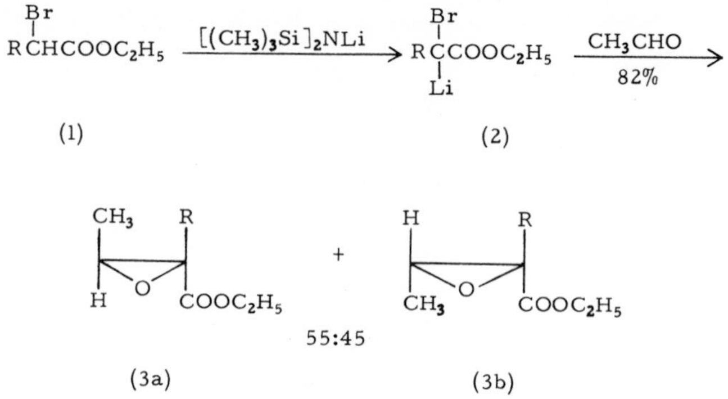

glycidic esters (3) in high yield, the *trans*-isomer (a) being formed generally in greater amount.

[1]M. S. Newman and B. J. Magerlein, *Org. React.*, **5**, 413 (1949).
[2]R. F. Borch, *Tetrahedron Letters*, 3761 (1972).

**Lithium borohydride, 1**, 603.

*5α-Bufalin.*[1] Reduction of the scillarenone (1) with lithium borohydride in pyridine gives the previously unknown 5α-bufalin (2).

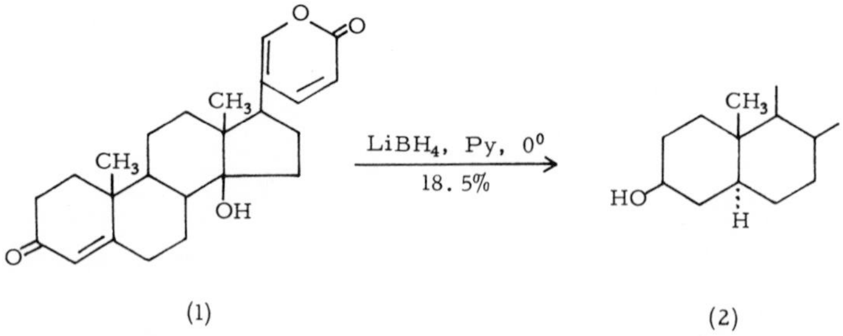

[1]U. Stache, K. Radscheit, W. Fritsch, W. Haede, H. Kohl, and H. Ruschig, *Ann.*, **750**, 149 (1971).

**Lithium bromide**, **1**, 604; **2**, 245; **3**, 75, 95, 306.

Conversion of the tetratosylates (1a) and (2a) into 1,1,6,6-tetrabromomethylspiro-[3.3]heptane (1b) and 1,1,10,10-tetrabromomethyltrispiro[3.1.1.3.1.1]tridecane (2b) is obtained by heating with anhydrous lithium bromide in acetone.[1]

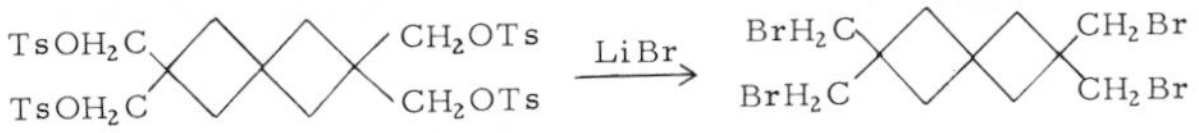

(1a) → LiBr → (1b)

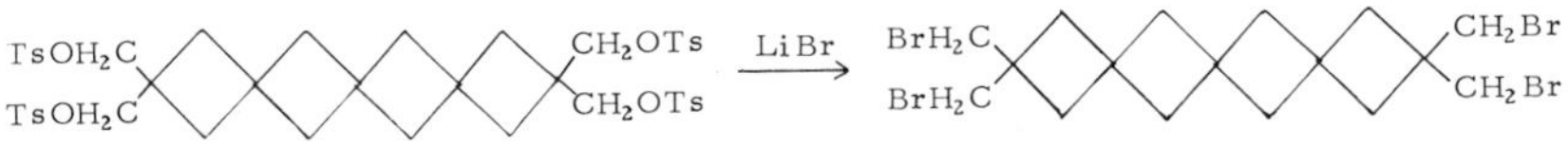

(2a) → LiBr → (2b)

*Epoxide–carbonyl rearrangement.*[2] Lithium bromide effects facile rearrangement of epoxides to aldehydes and/or ketones in benzene solution. The salt is insoluble in benzene but addition of 1 mole of HMPT or tri-*n*-butylphosphine oxide per mole of lithium bromide affords a soluble complex which effects the epoxide rearrangement. Evidence suggests a mechanism involving the salt of the bromohydrin as an intermediate.

Examples:

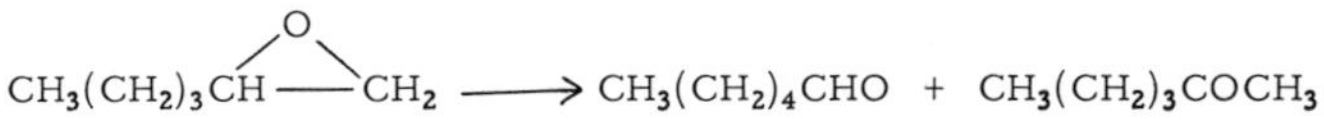

$$CH_3(CH_2)_3CH \overset{O}{-}CH_2 \longrightarrow CH_3(CH_2)_4CHO + CH_3(CH_2)_3COCH_3$$

60%          40%

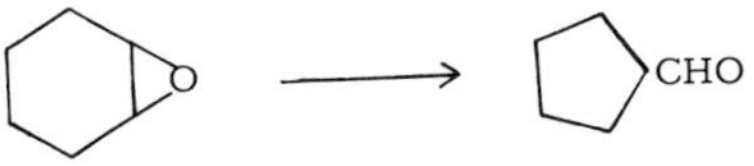

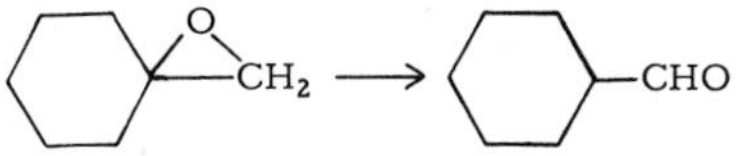

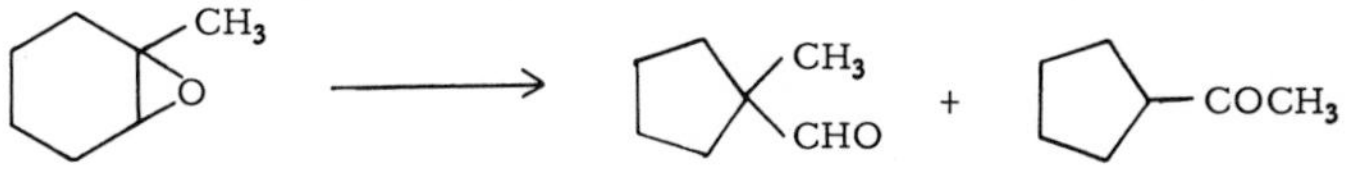

70%          30%

[1] E. Buchta and W. Merk, *Ann.*, **716**, 106 (1968).
[2] B. Rickborn and R. M. Gerkin, *Am. Soc.*, **90**, 4193 (1968); **93**, 1693 (1971).

**Lithium *t*-butoxide**, **3**, 183.

*t-Butyl esters.* The preparation of *t*-butyl *p*-toluate has now been published.[1]

[1] G. P. Crowther, E. M. Kaiser, R. A. Woodruff, and C. R. Hauser, *Org. Syn.*, **51**, 96 (1971).

**Lithium carbonate–Lithium halide**, **1**, 606–608; **2**, 245–246.

*Dehydrochlorination of 2-chloro-5-t-butylcyclohexanone.*[1] Dehydrochlorination of 2-chloro(axial or equatorial)-5-*t*-butylcyclohexanone (1) with lithium carbonate and lithium chloride (method of Joly and Warnant, **1**, 606–608) gives 5-*t*-butylcyclohexene-2-one-1 (2) as essentially the only product. Use of lithium chloride in DMF (method

$$\text{(1)} \xrightarrow[\text{100\%}]{\text{Li}_2\text{CO}_3/\text{LiCl}/\text{DMF}} \text{(2)}$$

of Holysz, **1**, 609) or of collidine gives (2) accompanied by two further products, (3) and (4).

$$\text{(1)} \xrightarrow[\substack{\text{a) LiCl/DMF} \\ \text{b) Collidine}}]{} \underset{67-82\%}{\text{(2)}} + \text{(3) 12-24\%} + \text{(4) 6-9 \%}$$

[1]P. Moreau and E. Casadevall, *Compt. rend.*, **272**, 801 (1971).

**Lithium diethylamide**, **1**, 610–611; **2**, 247–248.

*Reaction with epoxides* (**1**, 610–611; **2**, 247–248). Crandall and Crawley[1] have given a detailed procedure for conversion of α-pinene oxide (1) into pinocarveol (2) by reaction with lithium diethylamide in ether.

$$\text{(1)} \xrightarrow[\text{90-95\%}]{\text{LiN}(\text{CH}_2\text{CH}_3)_2} \text{(2)}$$

[1]J. K. Crandall and L. C. Crawley, *Org. Syn.*, submitted 1972.

**Lithium diisopropylamide**, **1**, 611; **2**, 249; **3**, 184–185.

*α-Hydroxymethylation of γ- and δ-lactones.*[1] Treatment of the *trans*-bicyclic lactone (1) in THF with a slight excess of lithium diisopropylamide in THF at −78° forms the lactone enolate; the mixture is allowed to warm to −20° and gaseous formaldehyde ($N_2$) is passed in. The reaction is quenched by addition of 10% HCl. Workup affords (2) in > 95% yield. This can be transformed into the *trans*-α-methylene-γ-butyrolactone (3) by conversion into the mesylate followed by treatment with refluxing pyridine. The overall yield of (3) from (1) is about 80%.

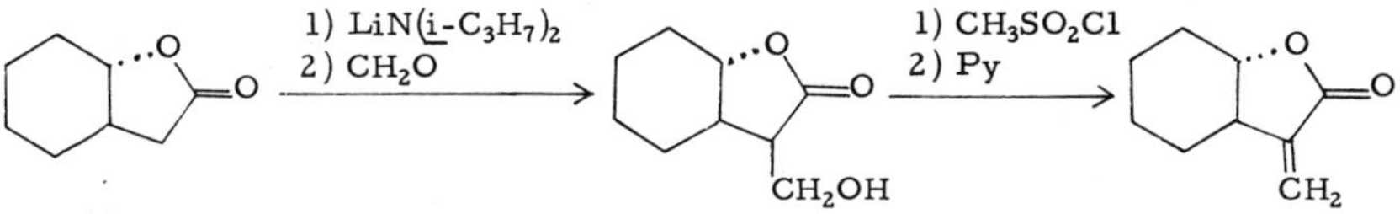

(1)    (2)    (3)

The same procedure applied to δ-valerolactone (4) afforded α-hydroxymethyl-δ-valerolactone (5) in > 95% yield.

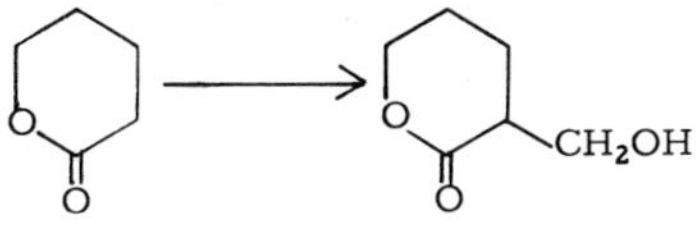

(4)    (5)

*Lithium enolates of unsymmetrical ketones.* House *et al.*[2] find that the less highly substituted lithium enolates of unsymmetrical ketones are best obtained by kinetically controlled deprotonation of the ketone with the hindered base lithium diisopropylamide. Thus treatment of 1-decalone (1) with 1.03 eq. of the base in 1,2-dimethoxyethane for 10 min. gives predominantly the lithium enolate (2); alkylation of the mixture with a

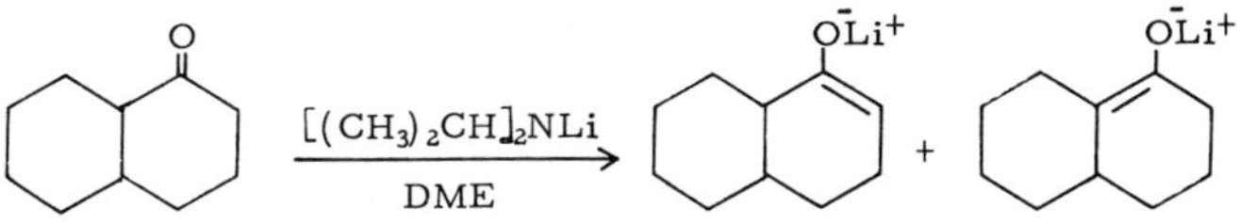

(1, 92% trans,    (2, 98%)    (3, 2%)
8% cis)

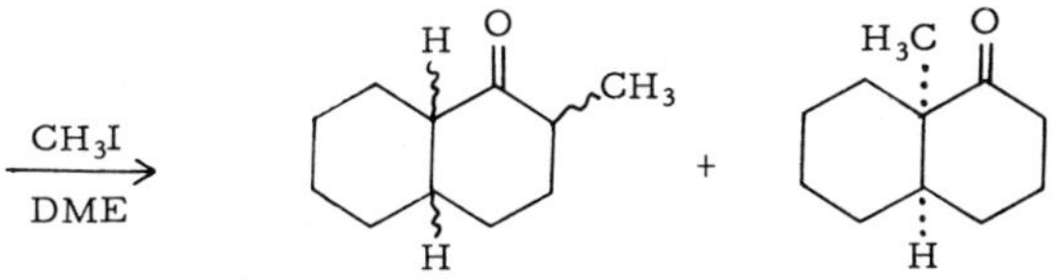

(4, 66%)    (5, 3%)

large excess of methyl iodide gives predominantly the monoalkylated ketone (4). Thus unsymmetrical ketones can be alkylated in this way predominantly at the less hindered position.

*α-Methyl-γ-butyrolactone.*[3] γ-Butyrolactone (1) is converted into the enolate by treatment with lithium diisopropylamide or lithium isopropylcyclohexylamide (this volume) in THF at −78°. Excess methyl iodide is then added at the same temperature and then the reaction mixture is allowed to warm to −30°, at which temperature it is maintained for 2–3 hr. The yield of α-methyl-γ-butyrolactone is 36% if 1 eq. of base is used; it is increased to 56% with use of 1.2 eq. of base. With use of more than 1 eq. of base, however, α,α-dimethylation begins to occur. This report is the first instance of direct conversion of γ-butyrolactone into α-methyl-γ-butyrolactone.

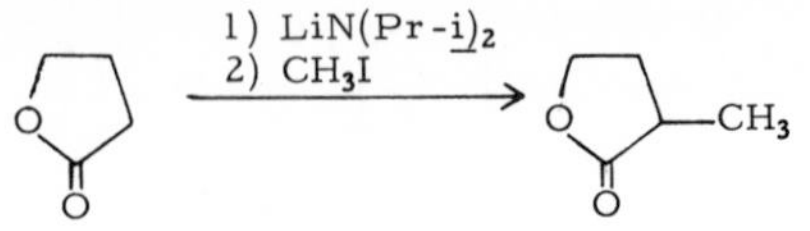

*Monoalkylation of α,β-unsaturated ketones.* Stork and Benaim[4] have reported a method for monoalkylation of α,β-unsaturated ketones. The procedure is illustrated for

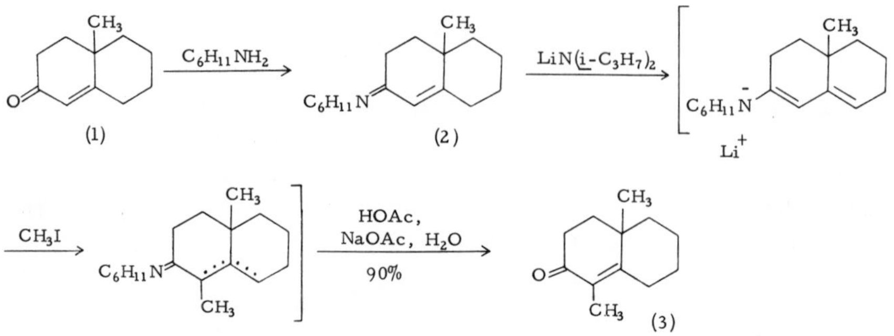

the case of 10-methyl-$\Delta^{1\,(9)}$-2-octalone (1). This ketone is first converted into the N-cyclohexylimine (2) by condensation with cyclohexylamine (p-TsOH catalysis, azeotropic distillation with toluene). The imine is then treated with lithium diisopropylamide[5] in refluxing THF under $N_2$ for 8 hr.; excess methyl iodide is then added and the reaction is allowed to reflux overnight. Hydrolysis then gives 1,10-dimethyl-$\Delta^{1\,(9)}$-2-octalone (3) in about 90% yield; about 10% of the starting enone (1) is recovered.

*Synthesis of ketones.* Stork and Maldonado[6] have disclosed a new synthesis of ketones from aldehydes: $RCHO \rightarrow RCOR'$. The aldehyde is first converted into the cyanohydrin and then the hydroxyl group is protected by reaction with ethyl vinyl ether to give (1). This is then converted into the anion by reaction with lithium diisopropylamide under carefully controlled conditions. The base is generated from butyllithium and diisopropylamine in THF and then (1) in hexamethylphosphoric triamide

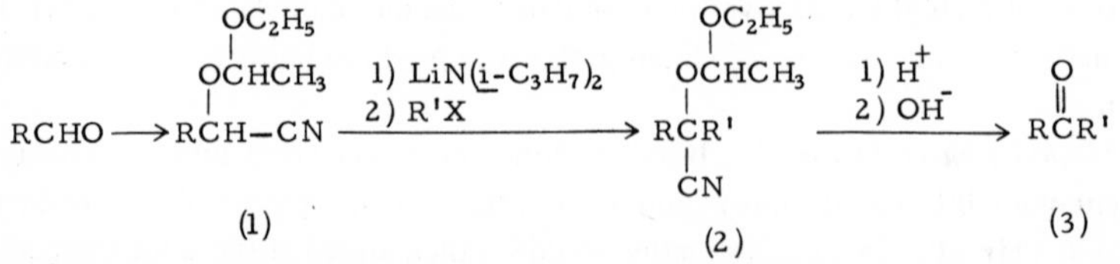

is added dropwise with stirring. The halide R'X is then added dropwise. The product (2) can be purified by distillation or used directly in the next step. The ketone (3) is obtained by hydrolysis with acid and then with base. The sequence can be applied to α,β-unsaturated aldehydes to give α,β-unsaturated ketones.

*C-Alkylation of cyclic α-diketones.*[7] Under normal conditions, alkylation of cyclohexane-1,2-dione (1, which exists predominantly in the enol form as shown) leads only to O-alkylated products. Kende and Eilerman find that, if (1) is treated with 2 eq. of lithium diisopropylamide in THF at −78° under $N_2$ until no starting material remains

(1)                                    (2)

(3)                                    (4)

and then treated with methyl iodide also at −78°, (2) can be obtained in 70% yield. If (1) is treated with the base at −78° followed by addition of methyl iodide (room temperature), (2), (3), and (4) are obtained in the ratio of 6:3:1.

The method is also successful for alkylation of cyclopentane-1,2-dione.

*β-Hydroxy acids.*[8] α-Lithiated lithium salts of aliphatic carboxylic acids can be prepared by the reaction of the carboxylic acid with lithium diisopropylamide (**3,** 184). These react with aldehydes and ketones to give fair to good yields of β-hydroxy acids (the method is an alternative to the classical Reformatsky reaction). Both mono- and

disubstituted acetic acids can be used. However, the reaction is subject to steric hindrance, and a number of acids (*e.g.,* diphenylacetic acid) and ketones (*e.g.,* camphor) failed to react.

*α-Alkylhydracrylic acids and α-alkylacrylic acids.*[9] Treatment of a carboxylic acid of the type $RCH_2COOH$ (1) with 2 eq. of lithium diisopropylamide in THF–HMPT yields the dianion (2).[10] This reacts with formaldehyde (generated from paraformaldehyde, **1,** 397), followed by acid hydrolysis (dilute HCl) to give α-alkylhydracrylic acids

(1)                              (2)                                    (3)

$$\xrightarrow[\substack{90-94\%}]{\substack{H_3O^+ \\ 270^0}} CH_2=\underset{\underset{R}{|}}{C}COOH \quad + \quad H_2O$$

(4)

(3) in high yield. These acids are dehydrated by an acid catalyst (phosphoric acid) at an elevated temperature to give $\alpha$-alkylacrylic acids (4) in high yield.

[1] P. A. Grieco and K. Hiroi, *J.C.S. Chem. Commun.*, 1317, (1972).
[2] H. O. House, M. Gall, and H. D. Olmstead, *J. Org.*, **36**, 2361 (1971).
[3] G. H. Posner and G. L. Loomis, *Chem. Commun.*, 892 (1972).
[4] G. Stork and J. Benaim, *Am. Soc.*, **93**, 5938 (1971).
[5] This base gave better results than *n*-butyllithium, sodium hydride, or the lithium salt of hexamethyldisilazane.
[6] G. Stork and L. Maldonado, *Am. Soc.*, **93**, 5286 (1971).
[7] A. S. Kende and R. G. Eilerman, *Tetrahedron Letters*, 697 (1973).
[8] G. W. Moersch and A. R. Burkett, *J. Org.*, **36**, 1149 (1971).
[9] P. E. Pfeffer, E. Kinsel, and L. S. Silbert, *ibid.*, **37**, 1256 (1972).
[10] P. E. Pfeffer and L. S. Silbert, *ibid.*, **35**, 262 (1970); P. E. Pfeffer, L. S. Silbert, and J. M. Chirinko, *ibid.*, **37**, 451 (1972).

**Lithium dimethylcarbamoylnickel tricarbonylate** (1). Mol. wt. 221.75, air sensitive.

The reagent is prepared by the reaction of dimethylamine with *n*-butyllithium in *n*-hexane at 0°. An ethereal solution of nickel carbonyl is added to an ethereal suspension of lithium dimethylamide below 10° and the mixture is stirred for 1 hr.

$$LiN(CH_3)_2 \quad + \quad Ni(CO)_4 \longrightarrow \underset{(CH_3)_2N}{\overset{LiO}{\underset{}{C}}} C \cdots Ni(CO)_3$$

(1)

The reagent reacts with organic halides or acid chlorides to give acid amides, generally in good yield.[1] Alkyl iodides and *t*-butyl bromide do not undergo dimethylcarbamoylation. However, ArI and alkenyl halides (RCH=CHX) react readily.

$$RX(RCOX) \quad + \quad (1) \longrightarrow RCON(CH_3)_2$$

Benzophenone reacts with (1) in THF to give, after hydrolysis, $\alpha$-phenyl-N,N-dimethylmandelamide:

$$C_6H_5\underset{\underset{O}{\|}}{C}C_6H_5 \quad + \quad (1) \xrightarrow[30\%]{\substack{1)\ THF,\ 67^0 \\ 2)\ H_3O^+}} C_6H_5\underset{\underset{C_6H_5}{|}}{\overset{OH}{\underset{}{C}}}CON(CH_3)_2$$

Benzaldehyde reacts with (1) to give N,N-dimethylbenzamide (84.2% yield).

[1] S. Fukuoka, M. Ryang, and S. Tsutsumi, *J. Org.*, **36**, 2721 (1971).

**Lithium diphenylphosphide (LDP)**, LiP(C$_6$H$_5$)$_2$. Mol. wt. 192.12.

*Preparation.*[1] Chlorodiphenylphosphine (Aldrich) is added dropwise to 4 g.-atoms of lithium wire in dry THF, and the mixture is stirred and cooled as necessary. The reaction is complete after 2 hr. at room temperature. Solutions can be stored for several days under argon.

*Olefin inversion.*[2] Lithium diphenylphosphide in THF opens epoxides stereo-specifically; quaternization of the crude product with methyl iodide leads to betaines, which fragment under mild conditions (25°) to give an olefin and methyldiphenyl-phosphine oxide. The product olefins are formed with inversion of stereochemistry relative to the starting epoxide owing to S$_N$2 epoxide opening followed by *cis* elimina-tion of the phosphine oxide. Thus the oxide of *trans*-stilbene (1) is converted into the betaine (2), which at 25° gives *cis*-stilbene (3) in 95% yield overall. The conversion is stereospecific (> 98%). Similarly *cis*-stilbene can be converted into *trans*-stilbene.

Interconversion of *cis*- and *trans*-2-octenes is effected in 75% yield with high stereo-specificity (> 99.5%).

The phorphorus betaine method is recommended for inversion of the stereo-chemistry of acyclic di- and trisubstituted alkenes. Highly hindered epoxides react very slowly with LDP and alkenes are not obtained in good yield. Keto groups interfere with the sequence owing to enolate formation unless 2 eq. of reagent is used. Epoxy esters cannot be deoxygenated in practical yield.

The new technique is convenient for synthesis of pure *trans*-cyclooctene. The oxide of *cis*-cyclooctene (4) is converted into *trans*-cyclooctene (7) in > 90% yield, > 99.5% isomeric purity.[3] Rigorously aprotic conditions are required for stereoselectivity. In the same way *trans*-1-methylcyclooctene, *cis,trans*-1,4-cyclooctadiene, and *cis,trans*-1,5-cyclooctadiene have been prepared from the corresponding epoxides.

The method fails, however, in the attempted conversion of cycloheptene oxide (8) into *trans*-cycloheptene. In this case, the reaction affords (*trans*-2-hydroxycyclo-heptyl)methyldiphenylphosphonium fluoroborate (9).

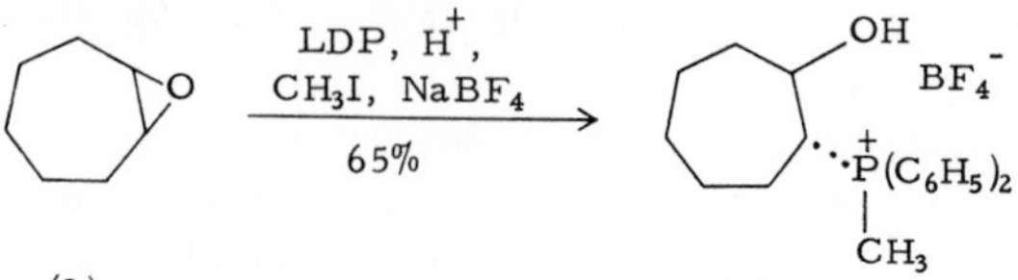

$$\text{(8)} \qquad\qquad \text{(9)}$$

[1]E. Vedejs and P. L. Fuchs, *Am. Soc.*, **94**, 822 (1972).
[2]E. Vedejs and P. L. Fuchs, *ibid.*, **93**, 4070 (1971).
[3]E. Vedejs, K. A. J. Snoble, and P. L. Fuchs, *J. Org.*, **38**, 1178 (1973).

**Lithium hydride**, **2**, 25.

$\Delta^5 \rightarrow \Delta^{5,7}$-***Steroid.*** Dauben and Fullerton[1] chose the following route for conversion of $\Delta^5$-androstene-3$\beta$,17$\beta$-diol diacetate (1) into $\Delta^{5,7}$-androstadiene-3$\beta$,17$\beta$-diol diacetate (4). The starting material was converted into the 7-ketone (2) by oxidation with chromium trioxide–pyridine complex in methylene chloride (**2**, 74–75). The ketone

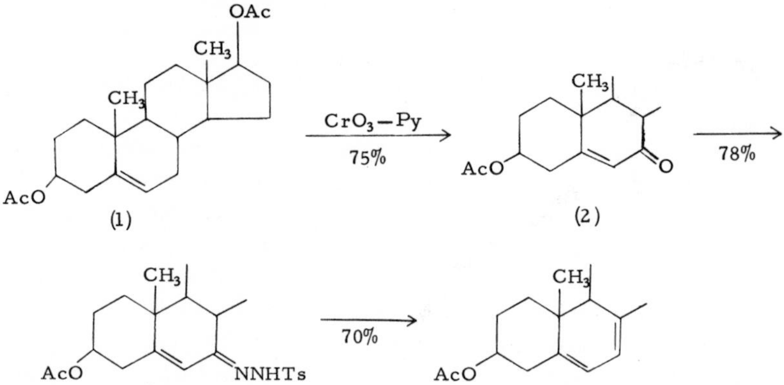

$$\text{(1)} \qquad\qquad\qquad \text{(2)}$$

$$\text{(3)} \qquad\qquad\qquad \text{(4)}$$

(2) was then converted into the tosylhydrazone (3); reduction with lithium hydride in refluxing toluene[2] afforded the diene (4). The overall yield of 40% compares with that obtained by the NBS process and has the advantage that the $\Delta^{4,6}$-diene is not obtained as a by-product.

[1]W. G. Dauben and D. S. Fullerton, *J. Org.*, **36**, 3277 (1971).
[2]L. Caglioti and P. Casselli, *Chim. Ind.* (*Milan*), 559 (1963).

**Lithium iodide**, **1**, 615–618.

***Cyclobutanone.*** Salaün and Conia[1] have recently published a convenient synthesis of cyclobutanone, in which the key step involves isomerization of an epoxide with lithium iodide functioning as a Lewis acid. The starting material is methallyl chloride (1), which is dehydrochlorinated with sodium amide–sodium $t$-butoxide to methylene-cyclopropane (2) in 60 % yield. Epoxidation of (2) with $p$-nitroperbenzoic acid in methylene chloride at $-10°$ gives oxaspiropentane (3) in essentially quantitative yield. This epoxide is converted on heat treatment alone into cyclobutanone (4, 40%) and other

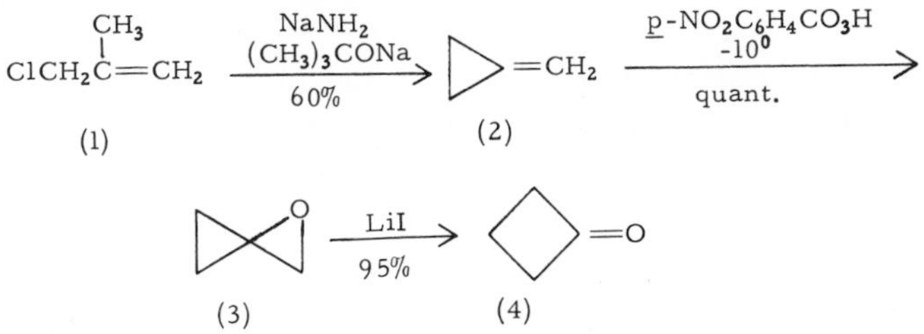

products. Addition of a catalytic amount of lithium iodide (LiI·1.5H$_2$O) produces an exothermic conversion of (3) to (4), which can be isolated in > 95% yield.

*Cleavage of alkyl aryl ethers.*[2] ArOR → ArOH. Alkyl aryl ethers are cleaved in high yield when treated with lithium iodide in dry 2,4,6-collidine. The reaction mixture becomes mildly basic as the cleavage proceeds but can be buffered with an acid such as benzoic acid. The reactions can also be carried out with LiI·3H$_2$O at 180–200° in the absence of solvent.

[1] J. R. Salaün and J. M. Conia, *Chem. Commun.*, 1579 (1971); J. R. Salaün, J. Champion, and J. M. Conia, *Org. Syn.*, submitted 1972.
[2] I. T. Harrison, *Chem. Commun.*, 616 (1969).

**Lithium iodide–Boron trifluoride.**

*Reduction of α-haloketones.*[1] Treatment of α-bromoketones with lithium iodide and boron trifluoride in ether or THF at room temperature affords the parent ketone in high yield. The hydrogen was eventually found to originate, presumably as water, in the lithium iodide (Alfa Inorganics, "anhydrous"). Only one exception to the reduction was observed, namely α-bromocamphor failed to react. The procedure is also effective for some α-chloroketones (phenacyl chloride, 2-chloropentanone), but fails for hindered chloroketones (α-chloronorbornanone).

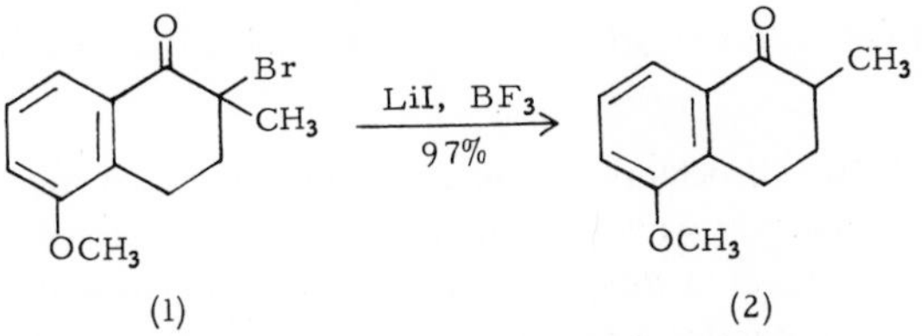

[1] J. M. Townsend and T. A. Spencer, *Tetrahedron Letters*, 137 (1971).

**Lithium N-isopropylcyclohexylamide (LiICA).**    Mol. wt. 147.18.

$$LiN{<}^{\displaystyle}_{\displaystyle CH(CH_3)_2}$$

The reagent is prepared by the reaction of *n*-butyllithium with N-isopropylcyclo-hexylamine (suppliers: Aldrich, Fluka, K and K, Koch-Light, MCB, P and B, Schu-chardt).

**Lithium ester enolates.**[1] The reagent reacts with a wide variety of esters at low temperatures in THF to give the corresponding lithium enolates:

$$HCCOOR + LiN{<}_{CH(CH_3)_2} \xrightarrow[-78^0]{THF} LiCCOOR + HN{<}_{CH(CH_3)_2}$$

A variety of other bases were tried unsuccessfully; the unique success of LiICA may be due partly to its high solubility in THF at low temperatures.

The ester enolates react at 20° in the presence of DMSO with a variety of organic halides to give good yields of the corresponding alkylated esters:

$$LiCCOOR + R'X \xrightarrow[50-95\%]{DMSO-THF \atop 20^0} R'-CCOOR + LiX$$

The ester enolates react with bromine or iodine at −78° to give α-halogenated esters in high yield:

$$LiCCOOR + Br_2(I_2) \xrightarrow[85-95\%]{THF, -78^0} Br(I)CCOOR + LiBr(I)$$

The method is superior to the usual procedure of refluxing the carboxylic ester with halogen and phosphorus trichloride.[2]

**Enolates of α,β-unsaturated esters.**[3] Addition of ethyl crotonate to a 0.50 *M* solution of LiICA in THF containing 20% (by volume) of HMPT[4] at −78° followed by

$$CH_3CH{=}CHCOOC_2H_5 \xrightarrow[87\%]{1)\ LiICA \atop 2)\ H_3O^+} CH_2{=}CHCH_2COOC_2H_5$$

quenching with dilute hydrochloric acid gives the nonconjugated ester, ethyl 3-buteno-ate, in 87% yield together with 13% of recovered ethyl crotonate. The procedure provides a simple, general method for conversion of α,β-unsaturated esters into the corresponding β,γ-unsaturated esters. The reaction provides a simple synthesis of the allenic ester methyl 2,3-butadienoate from methyl 2-butynoate.

$$CH_3C{\equiv}CCOOCH_3 \xrightarrow[\underset{60\%}{}]{\overset{LiICA}{H_3O^+}} CH_2{=}C{=}CHCOOCH_3$$

Alkylation of the lithium enolate solution of ethyl crotonate furnishes alkylated non-conjugated esters:

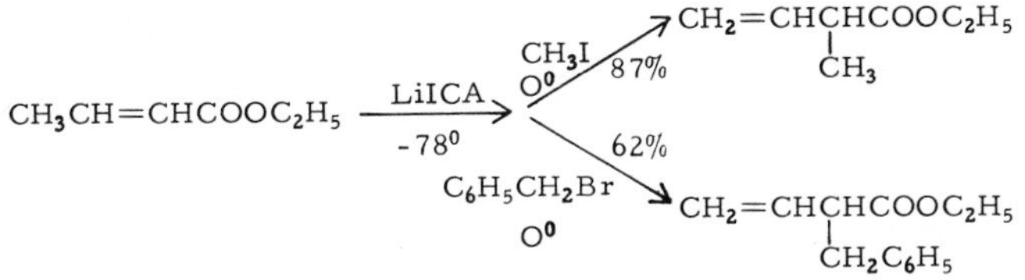

$$CH_3CH{=}CHCOOC_2H_5 \xrightarrow[-78^0]{LiICA}$$

$$\overset{CH_3I}{\underset{0^0}{}} 87\% \nearrow \underset{CH_3}{\overset{|}{CH_2{=}CHCHCOOC_2H_5}}$$

$$\underset{0^0}{C_6H_5CH_2Br} 62\% \searrow \underset{CH_2C_6H_5}{\overset{|}{CH_2{=}CHCHCOOC_2H_5}}$$

Addition of acetone to the solution of the lithium enolate of ethyl crotonate gives the nonconjugated β-hydroxy ester, ethyl 3-hydroxy-2-vinylisovalerate, in good yield.

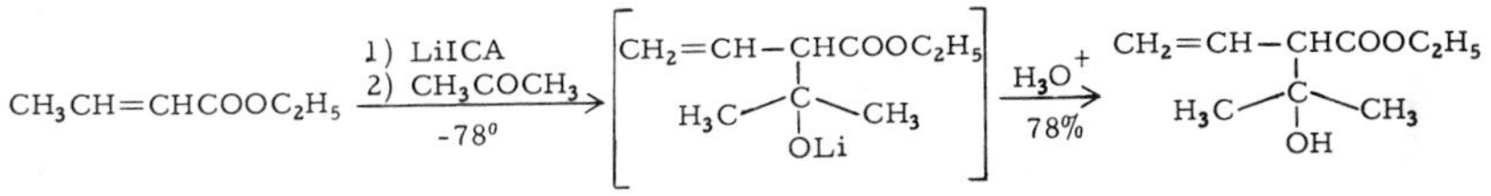

$$CH_3CH{=}CHCOOC_2H_5 \xrightarrow[-78^0]{\overset{1)\ LiICA}{2)\ CH_3COCH_3}} \left[ \underset{OLi}{\overset{CH_2{=}CH-CHCOOC_2H_5}{H_3C-C-CH_3}} \right] \xrightarrow[78\%]{H_3O^+} \underset{OH}{\overset{CH_2{=}CH-CHCOOC_2H_5}{H_3C-C-CH_3}}$$

**α-Methyl-γ-butyrolactone.** Posner and Loomis[5] have effected α-methylation of γ-butyrolactone in good yield by conversion to the lactone enolate by treatment with an excess of lithium N-isopropylcyclohexylamide or lithium diisopropylamide in THF

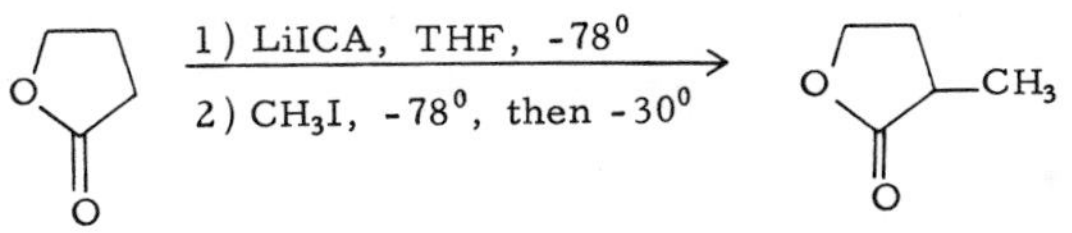

$$\xrightarrow[2)\ CH_3I,\ -78^0,\ then\ -30^0]{1)\ LiICA,\ THF,\ -78^0}$$

at −78°; an excess of methyl iodide is added to the enolate solution at −78° and then the reaction is allowed to warm to −30°.

**Claisen rearrangement of allyl esters.** Ireland and Mueller[6] report that lithium enolates of allyl esters rearrange rapidly at room temperature or slightly above to the corresponding γ,δ-unsaturated acids. Thus the allyl ester (1) is converted into the lithium enolate (2) by treatment with lithium isopropylcyclohexylamide in THF at −78°. The solution of (2) is then allowed to warm to 25° for 10 min. The γ,δ-unsaturated acid (3)

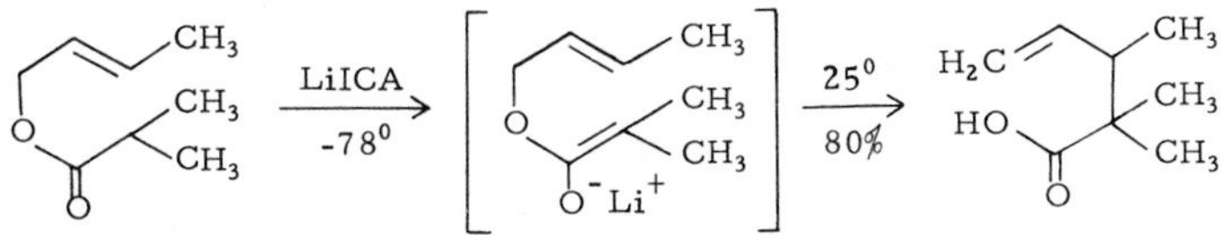

(1)                    (2)                    (3)

is obtained in 80% yield. In some cases the rearrangement proceeds slowly; in that case it is advantageous to quench the lithium enolate at $-78°$ with trimethylchlorosilane before allowing the reaction to warm to room temperature. This procedure was used

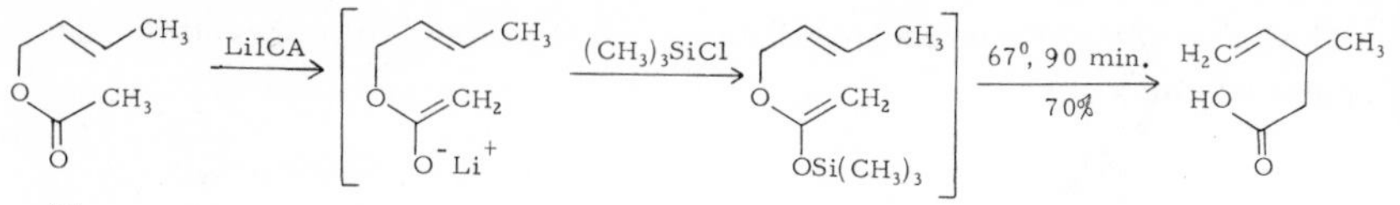

(4)                                                                                          (5)

with the acetate (4). The intermediate ketene acetal rearranges at $67°$ to (5) in 70% yield.

*Methylation of α,β-unsaturated ketones.*[7] Alkylation of α,β-unsaturated ketones usually proceeds almost exclusively at the α-position. However, methylation of anions of α,β-unsaturated ketones generated by treatment with lithium N-isopropylcyclohexylamide gives predominantly α'-methyl derivatives.

Examples:

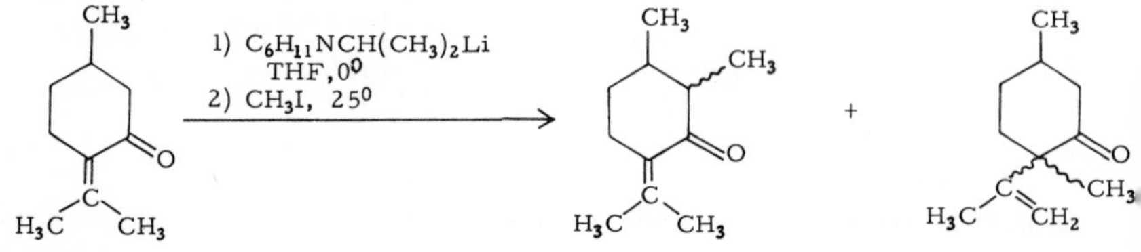

Pulegone                                      56%                          23%

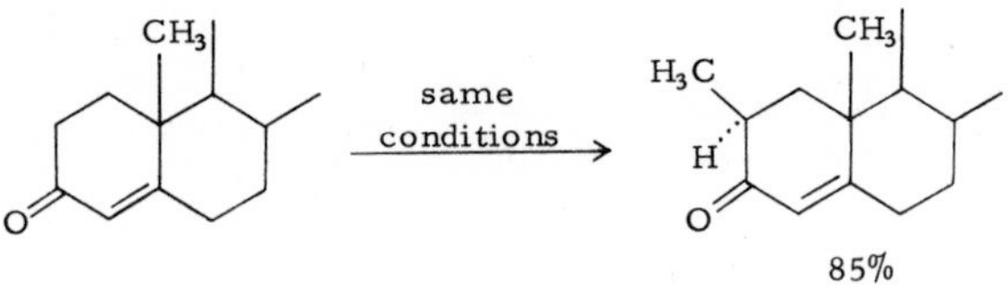

85%

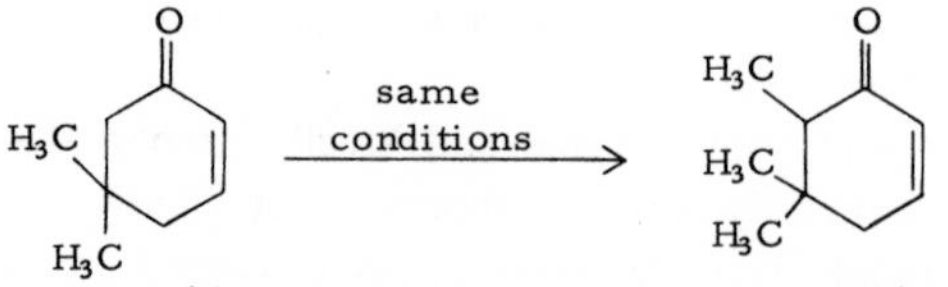

(1)                                      83%

Use of trityllithium (**1**, 1256; **2**, 454) as the base in the reactions above yields predominantly α-methyl derivatives. An anomalous product (2) is obtained with this

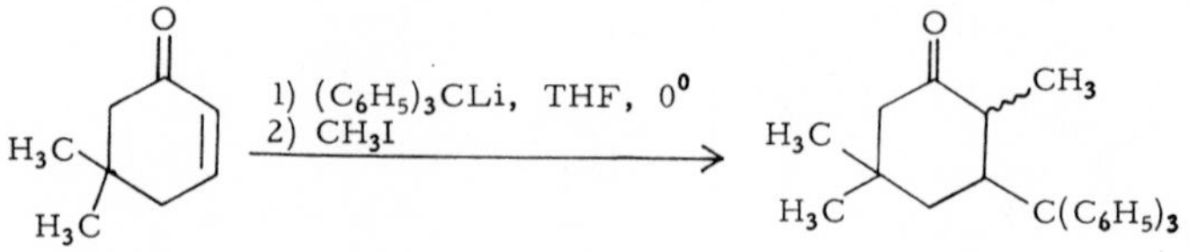

(1)                                                      (2)

base from 5,5-dimethylcyclohexene-2-one-1 (1).[8] Trityllithium also gives a product of conjugate addition (4) to ethyl crotonate (3) in THF solution at $-23°$.

$$CH_3CH=CHCOOC_2H_5 + (C_6H_5)_3CLi \xrightarrow[> 65\%]{\begin{array}{c}THF\\-23°\end{array}} (C_6H_5)_3CCHCH_2COOC_2H_5$$

(3)                                                                 $\overset{|}{CH_3}$

(4)

[1] M. W. Rathke and A. Lindert, *Am. Soc.*, **93**, 2318 (1971).
[2] H. T. Clarke and E. R. Taylor, *Org. Syn., Coll. Vol.*, **1**, 115 (1944).
[3] M. W. Rathke and D. Sullivan, *Tetrahedron Letters*, 4249 (1972).
[4] Much lower yields of the nonconjugated ester (23%) are obtained in the absence of HMPT.
[5] G. H. Posner and G. L. Loomis, *Chem. Commun.*, 892 (1972).
[6] R. E. Ireland and R. H. Mueller, *Am. Soc.*, **94**, 5897 (1972).
[7] R. A. Lee, C. McAndrews, K. M. Patel, and W. Reusch, *Tetrahedron Letters*, 965 (1973).
[8] R. A. Lee and W. Reusch, *ibid.*, 969 (1973).

**Lithium orthophosphate**, $Li_3PO_4$. Mol. wt. 115.80. Supplier: ROC/RIC.

*Preparation.* A solution of sodium orthophosphate dodecahydrate (570 g., ROC/RIC) in water is added with stirring to a solution of lithium hydroxide (126 g.) in $H_2O$. The resulting precipitate is filtered, washed with water, and then dispersed in water (1.5–2 l.) at 60° for 20 min. The solid material is filtered and washed. This leaching process is repeated four more times. The filter cake is dried for 16 hr. in an oven at 200°.

*Isomerization of cyclooctene oxide.* Isomerization of cyclooctene oxide (1) with lithium diethylamide gives the bicyclic alcohol (2) and 3-hydroxycyclooctene (3) in the approximate ratio of 4:1 (**2**, 247). The abnormal product (2) is obtained only with medium-ring epoxides; large-ring epoxides give exclusively allylic alcohols. Sheng[1] finds that cyclooctene oxide can be isomerized exclusively to 3-hydroxycyclooctene (3) if

(1)                     (2)                     (3)

potassium *t*-butoxide in DMSO is used as base (yield 46%). An even higher yield (70%) of (3) can be obtained by passing a solution of cyclooctene oxide in benzene through a column packed with lithium orthophosphate pellets at 180°.

[1] M. N. Sheng, *Synthesis*, 194 (1972).

**Lithium perhydro-9b-boraphenalylhydride**, **3**, 187.

*Specific reduction of E prostaglandins to $F_\alpha$ prostaglandins.* One useful reaction in prostaglandin research is the interconversion of primary[1] prostaglandins, for example, reduction of prostaglandin $E_2$ (1) to prostaglandin $F_{2\alpha}$ (2). Reduction of (1) with sodium

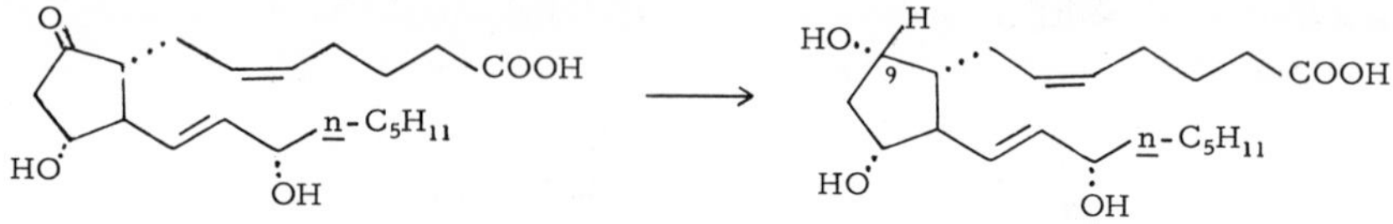

(1)                                                        (2)

borohydride proceeds nonselectively to give (2) and the $C_9$ epimer, $F_{2\beta}$, which is bio-
logically far less potent than $F_{2\alpha}$. Corey and Varma[2] now report that the desired reduction can be achieved stereospecifically (98.7% yield) with the bulky borohydride reagent
lithium perhydro-9b-boraphenalylhydride in THF at $-78°$.

[1]S. Bergström, *Science*, **157**, 382 (1967).
[2]E. J. Corey and R. K. Varma, *Am. Soc.*, **93**, 7319 (1971).

### Lithium 2,2,6,6-tetramethylpiperidide (LiTMP),

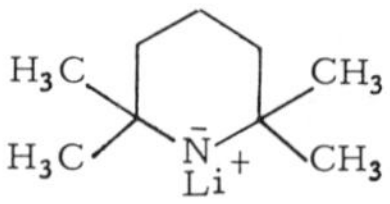

Mol. wt. 147.18.

(1)

The base is prepared by treatment of purified 2,2,6,6-tetramethylpiperidine (HTMP,
suppliers of crude material: Aldrich, Fluka, K and K) with either methyllithium or
*n*-butyllithium.

**Boron-stabilized carbanions.** Ordinarily bases coordinate with the boron atom of
organoboranes. However, Rathke and Kow[1] report that a highly hindered lithium
amide such as lithio-2,2,6,6-tetramethylpiperidine or lithio-*t*-butylneopentylamine[2]
can remove the α-proton from an organoborane to generate carbanions. Thus treatment of the boron compound B-methyl-9-borabicyclononane (1)[3] in benzene with the
former base for 12 hr. at room temperature followed by quenching with deuterium
oxide results in deuterium incorporation of 50% (equation I).

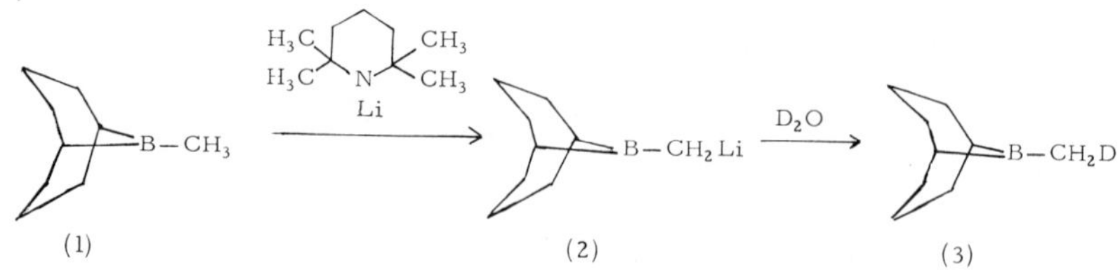

(1)                        (2)                        (3)

Rathke and Kow have reported one synthetic use of this new conversion. Thus
reaction of (2) with cyclohexanone gives methylenecyclohexane (4) in 55–65% yield
from (1):

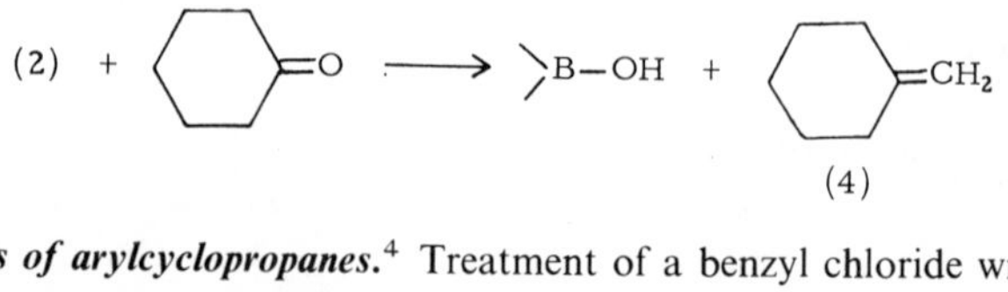

(4)

**Synthesis of arylcyclopropanes.**[4] Treatment of a benzyl chloride with LiTMP in
ether generates a phenylcarbene or carbenoid species, which reacts with an olefin

$$ArCH_2Cl \xrightarrow{\text{LiTMP}} "Ar\ddot{C}H" \xrightarrow{>=<} Ar-\!\!\triangleleft\!\!|$$

(used in large excess) to form an arylcyclopropane in yields in the range 50–80%. The cyclopropanes are formed by stereospecific *cis* addition; the less stable *syn*-isomer predominates over the *anti*-isomer. Thus 7-phenylnorcarane can be prepared in this way in 54% yield; the *syn/anti* ratio is 2.2. Olofson and Dougherty[4] examined the yield of 7-phenylnorcarane obtained using other $LiNR_2$ bases. Only a few bases proved to be as effective as LiTMP. These are lithium dicyclohexylamide [$LiN(Cy)_2$], LiN(t-amyl)*t*-Bu, and the lithium salt derived from N-(1-ethylcyclohexyl)-1,1,3,3-tetramethylbutyl-amine (LiETA). In these last two reagents, like LiTMP, the nitrogen is attached to two tertiary carbons.

By substitution of nonterminal acetylenes for the alkene, 3-arylcyclopropenes are obtained. Thus addition of LiTMP to an ether solution of $p$-$CH_3OC_6H_4CH_2Cl$ and 3-hexyne leads to formation of 1,2-diethyl-3-*p*-anisylcyclopropene (1) in 60% yield.

$$(\underline{p})\text{-}CH_3OC_6H_4-\!\!\triangleleft\!\!\bigg|\begin{smallmatrix} C_2H_5 \\ \\ C_2H_5 \end{smallmatrix}$$

(1)

*Benzynes.*[5] LiTMP is also an effective base for formation of benzynes from ArCl. Thus treatment of *o*-chloroanisole with LiTMP and $LiC{\equiv}CC_6H_5$ gave *m*-methoxytolane (1, tolane = diphenylacetylene) in 80% yield. Diphenyl sulfide was prepared in 93% yield from $C_6H_5Cl$, $C_6H_5SLi$, and LiTMP.

$$\begin{matrix} OCH_3 \\ \bigcirc\!\!-\!Cl \end{matrix} + LiC{\equiv}CC_6H_5 \xrightarrow[80\%]{\text{LiTMP}} C_6H_5C{\equiv}CC_6H_4OCH_3(\underline{m})$$

(1)

*Ester condensations.* Olofson and Dougherty[5] effected the synthesis of the $\beta$-keto ester (1) in 89% yield using LiTMP as base.

$$\begin{matrix} CH_3 \\ \diagup \\ CH_3 \end{matrix}\!\!CHCOOC_2H_5 \xrightarrow{\text{LiTMP}} \left[\begin{matrix} CH_3 \\ \diagdown \\ CH_3 \end{matrix}\!\!\bar{C}COOC_2H_5\right] \xrightarrow{C_6H_5COCl} C_6H_5CO\overset{\overset{\displaystyle CH_3}{|}}{\underset{\underset{\displaystyle CH_3}{|}}{C}}COOC_2H_5$$

(1)

[1] M. W. Rathke and R. Kow, *Am. Soc.*, **94**, 6854 (1972).
[2] G. F. Hennion and R. S. Hanzel, *ibid.*, **82**, 4908 (1960).
[3] H. C. Brown and M. M. Rogić, *ibid.*, **91**, 4304 (1969).
[4] R. A. Olofson and C. M. Dougherty, *ibid.*, **95**, 581 (1973).
[5] *Idem, ibid.*, **95**, 582 (1973).

**Lithium tri-*t*-butoxyaluminum hydride (Lithium aluminum tri-*t*-butoxyhydride), 1, 620–625; 2, 251–252; 3, 188.**

*Reduction of cyclic ketones.*[1] Competitive reductions of cyclic ketones of various types with lithium tri-*t*-butoxyaluminum hydride indicate that nonconjugated enones are less reactive than cyclic staturated ketones, but more reactive than conjugated enones. Steric effects do not appear to be important, since 3-ketocyclohexene (1) is less

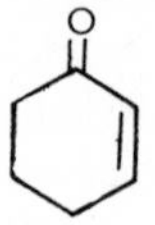

(1) 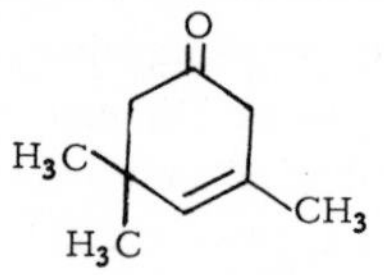                    (2)

reactive than $\beta$-isophorone (2, 3,5,5-trimethylcyclohexene-3-one-1). Evidently conjugation plays an important role in affecting reactivity. Lithium aluminum hydride is less selective than lithium tri-*t*-butoxyaluminum hydride in the reduction of $\alpha,\beta$-unsaturated and saturated six-membered ring ketones.[2]

*Reductive opening of tetrahydrofuran.*[3] The combination of lithium tri-*t*-butoxyaluminum hydride and triethylborane induces a rapid, essentially quantitative opening of the tetrahydrofurane ring:

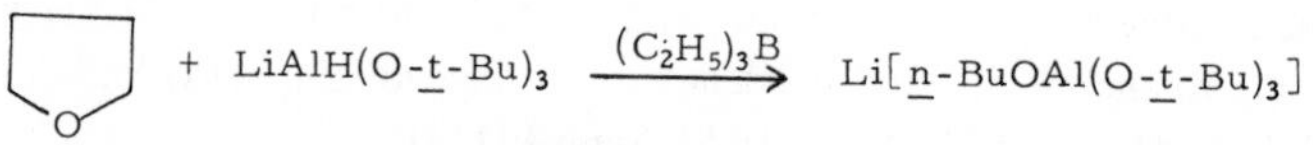

[1]H. Haubenstock and P. Quezada, *J. Org.*, **37**, 4067 (1972).
[2]H. Haubenstock, *ibid.*, **37**, 656 (1972).
[3]H. C. Brown, S. Krishnamurthy, and R. A. Coleman, *Am. Soc.*, **94**, 1750 (1972).

**Lithium tri-*sec*-butylborohydride, Li-*sec*-Bu₃BH. Mol. wt. 190.10.**

This reducing agent is prepared in quantitative yield by the reaction of lithium trimethoxyaluminum hydride (LTMA, **1**, 625; **2**, 252–253) with the highly hindered tri-*sec*-butylborane in THF at 25°. Aluminum methoxide is also formed, but since it is inert, the reaction mixture can be used directly.[1]

$$\text{LiAlH(OMe)}_3 + \underline{sec}\text{-Bu}_3\text{B} \xrightarrow{\text{THF}} \text{Li-}\underline{sec}\text{-Bu}_3\text{BH} + \text{Al(OMe)}_3$$

This new reagent is an active reducing agent and reduces cyclic and bicyclic ketones with superstereoselectivity.[1] Thus reduction of 2-methylcyclohexanone (1) gives *cis*-2-methylcyclohexanol in 99.3 % purity. Note that reduction with lithium trimethoxyaluminum hydride alone yields (2) in 69 % yield. Thus increasing the size of the alkyl substituents on boron enhances the stereoselectivity of the borohydride anion. Even

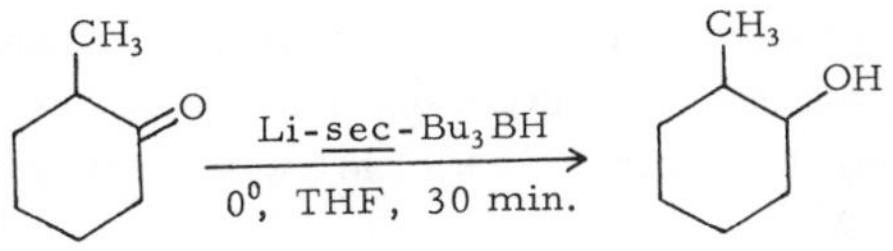

(1)                                          (2, 99.3%)

simple unhindered cyclohexanones with an alkyl substituent in the 3- or 4-position are reduced predominantly from the equatorial side. Thus 4-*t*-butylcyclohexanone (3) is reduced at −78° to *cis*-4-*t*-butylcyclohexanol (4) in 96.5% isomeric purity. Reduction of (3) with lithium perhydro-9b-boraphenalylhydride (PBPH, **3**, 187) gives (4) in

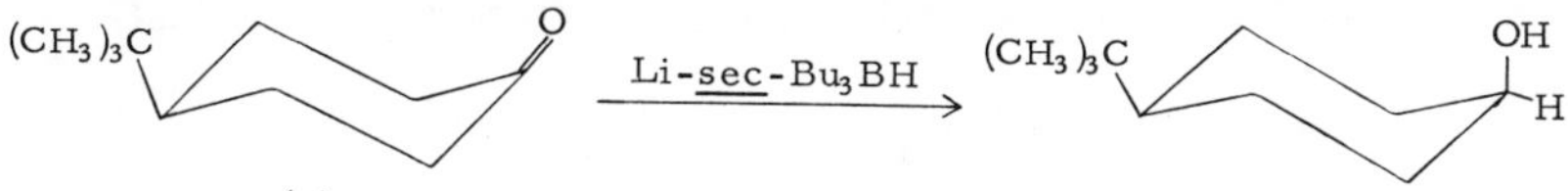

(3)                                          (4, 96.5%)

only 54% yield. Thus lithium tri-*sec*-butylborohydride appears to be the most stereoselective reagent known for reduction of cyclic and bicyclic ketones. It also has the advantage that it can be prepared *in situ* and utilized directly for reduction of ketones.

[1]H. C. Brown and S. Krishnamurthy, *Am. Soc.*, **94**, 7159 (1972).

**Lithium triethylborohydride**, $Li(C_2H_5)_3BH$. Mol. wt. 105.95. Supplier: Aldrich ("Super-Hydride").

The reagent is prepared in THF solution by the reaction of lithium hydride (**2**, 251) with triethylborane. It is stable indefinitely.

$$LiH + B(C_2H_5)_3 \xrightarrow[\quad 25°\quad]{THF} Li(C_2H_5)_3BH$$

*Reduction of alkyl halides.*[1] Allylic and benzylic chlorides and bromides are reduced by the reagent almost instantaneously at 25° ($S_N2$ displacement). Simple primary halides are completely reduced in 2 min. Even neopentyl bromide is reduced to neopentane (96% yield) in 3 hr. under reflux. Secondary cycloalkyl bromides are reduced at 25° in 24 hr. Even *exo*-2-bromonorbornane (1) can be reduced quantitatively. The

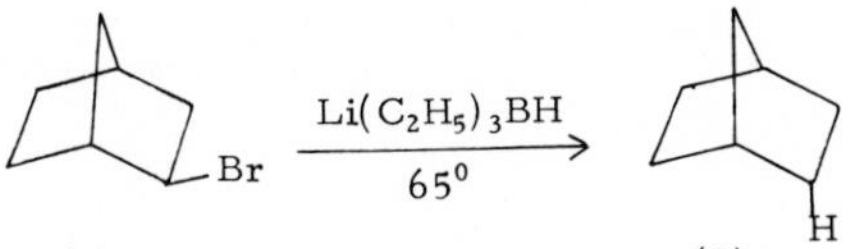

(1)                                          (2)

reduction of tertiary alkyl bromides is slow and results predominantly in elimination. The reagent is inert toward aryl halides.

The reduction with lithium triethylborohydride is far faster than that with sodium thiophenoxide, lithium aluminum hydride, or lithium borohydride.

Alkyl halides can also be reduced with lithium hydride in the presence of a catalytic quantity of triethylborane.

[1] H. C. Brown and S. Krishnamurthy, *Am. Soc.*, **95**, 1669 (1973).

## Lithium triethylcarboxide, $LiOC(C_2H_5)_3$. Mol. wt. 122.13.

The base is prepared from *n*-butyllithium and triethylcarbinol.

*Trialkylcarbinols from trialkylboranes.* Trialkylcarbinols have been prepared by the reaction of trialkylboranes with carbon monoxide in diglyme followed by oxidation with hydrogen peroxide (equation 1). See **2**, 60. The method has the disadvantage that at the temperature required isomerization of organoboranes can be significant.

$$(eq. 1) \quad \underline{n}\text{-Bu}_3B + CO \xrightarrow[125°, 8 \text{ hr.}]{Diglyme} [\quad] \xrightarrow{H_2O_2} \underline{n}\text{-Bu}_3COH$$
$$90\%$$

Brown *et al.*[1] now find that tri-*n*-butylcarbinol is obtained in essentially quantitative yield by the reaction of tri-*n*-butylborane with chlorodifluoromethane (or other trisubstituted methanes) under the influence of lithium triethylcarboxide (equation 2).

$$(eq. 2) \quad \underline{n}\text{-Bu}_3B + HCClF_2 \xrightarrow[65°, 1 \text{ hr.}]{THF, \ LiOC(C_2H_5)_3} [\quad] \xrightarrow{H_2O_2} \underline{n}\text{-Bu}_3COH$$
$$98\%$$

The new method has the advantage that the reaction proceeds rapidly at 65°, a temperature at which isomerization of organoboranes is not significant. Various alkoxides were examined; hindered alkoxides provided the best results. Of these lithium triethylcarboxide proved superior to potassium *t*-butoxide and potassium triethylcarboxide. Chlorodifluoromethane can be replaced by chloroform, in which case the yield is somewhat lower (85%) but the procedure is somewhat more convenient.

[1] H. C. Brown, B. A. Carlson, and R. H. Prager, *Am. Soc.*, **93**, 2070 (1971).

# M

**Magnesium, 1,** 627–629; **2,** 254; **3,** 189.

*Activated magnesium.* Rieke and Hudnall[1] have described a method for preparation of a highly active form of magnesium in a finely divided state suitable for the preparation of difficultly formed Grignard reagents. The process involves reduction of anhydrous magnesium chloride (Alfa Inorganics, K and K) with an alkali metal in an inert ethereal solvent under an inert atmosphere. Two combinations which have been found to be useful are sodium and diglyme and potassium and THF. The sodium or potassium salts which are formed need not be removed and the Grignard reagents can be prepared by addition of the alkyl or aryl halide directly to the suspension of powdered magnesium metal. For example, phenylmagnesium bromide has been prepared in over 60% yield from phenyl bromide and the activated magnesium in THF at $-78°$ in 30 min. Even the hitherto unknown phenylmagnesium fluoride has been prepared, albeit in only about 5% yield.

*Coupling.* Japanese chemists[2] have reported the synthesis of 8,8′-biheptafulvenyl (4) from the bromide (1). Treatment of (1) with magnesium in ether produced 1,2-di(3-cycloheptatrienyl)ethane (2) in 85% yield. Hydride abstraction with trityl fluoroborate (**1,** 1256–1258; **2,** 454; this volume) in methylene chloride gave the salt (3). Treatment

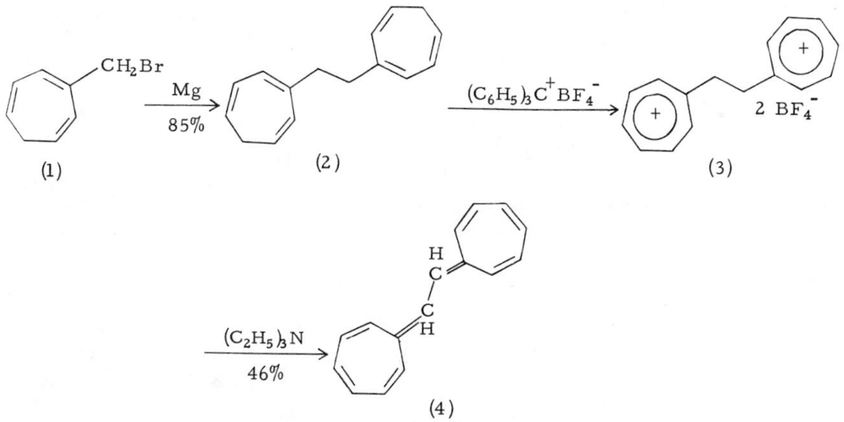

of (3) with triethylamine in methylene chloride effected cleavage of fluoroboric acid to give (4) in 46% yield.

*Allenes and cyclopropanone ketals.*[3] Magnesium in THF (or zinc in HMPT) reacts with $\alpha,\alpha'$-dibromo ketals (1) to give allenes (2) and cyclopropanone ketals (3). If zinc is used in place of magnesium, HMPT is necessary as solvent.

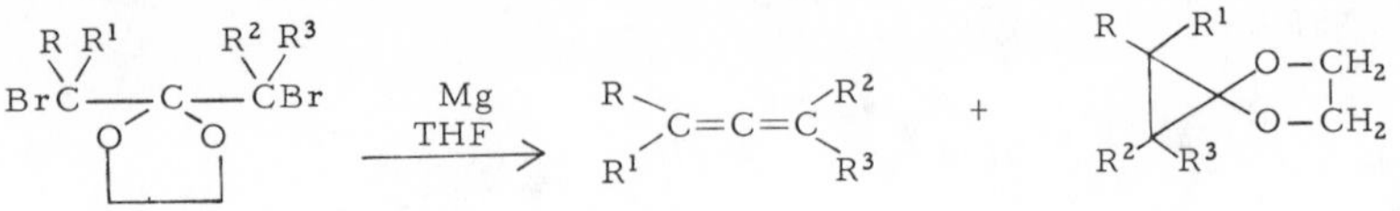

(1)                    (2, 20-40%)                    (3, 40-90%)

[1] R. D. Rieke and P. M. Hudnall, *Am. Soc.*, **94**, 7178 (1972).
[2] S. Kuroda, M. Oda, and Y. Kitahara, *Angew. Chem., internat. Ed.*, **12**, 76 (1973).
[3] G. Giusti and C. Morales, *Bull. soc.*, 382 (1973).

**Magnesium aluminum hydride**, $Mg(AlH_4)_2$. Mol. wt. 86.34.

The reagent is best prepared by the reaction of magnesium iodide and sodium aluminum hydride in THF; it is insoluble in this solvent (and also in ether) and separates as the solvate, $Mg(AlH_4)_2 \cdot THF$, in 58% yield.[1]

*Reduction.* An original report by a Czechoslovak group[2] indicated that organic compounds are reduced with difficulty and in poor yields by magnesium aluminum hydride. However, James[3] has reported recently that the reagent (suspended in ether) reduces aldehydes, ketones, acids, and oximes in high yield in a reasonable time (4–12 hr.). The reagent thus rivals lithium aluminum hydride in efficiency, but not in convenience.

[1] E. C. Ashby, R. D. Schwartz, and B. D. James, *Inorg. Chem.*, **9**, 325 (1970).
[2] J. Plešek and S. Heřmánek, *Coll. Czech.*, **31**, 3060 (1960).
[3] B. D. James, *Chem. Ind.*, 227 (1971).

**Maleic anhydride.** Mol. wt. 98.06, m.p. 52–54°. Suppliers: Aldrich, Eastman, Fluka.
Addition of thiolacetic acid: **1**, 13.
Addition to allene: **1**, 18.
Synthesis of naphthazarin: **1**, 1027–1028.
Reaction with phthaloyl chloride: **1**, 1290.
Diels–Alder addition to dienes: **1**, 181–182, 238, 781, 1149.

*Diels–Alder addition in the presence of chloranil[1]:*

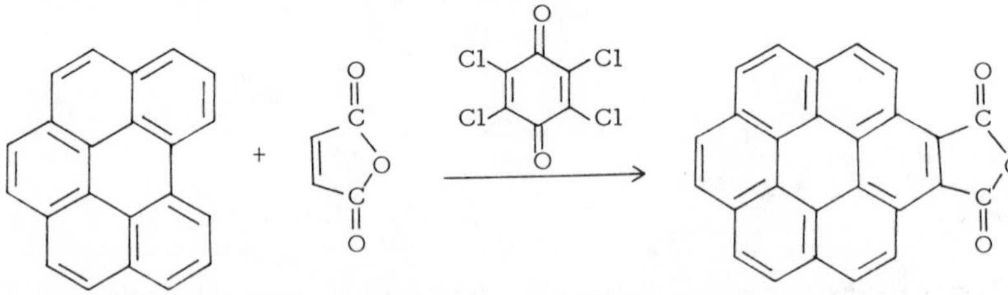

In the absence of a dehydrogenating agent, the intermediate Diels–Alder adduct could in some cases be isolated.

[1] M. Zander, *Ann.*, **723**, 27 (1969).

**Malononitrile**, $CH_2(CN)_2$. Mol. wt. 66.07, m.p. 32–34°, b.p. 220°. Suppliers: Aldrich, Eastman.

*Preparation*[1]:

1. $CNCH_2CONH_2 + PCl_5 \rightarrow CH_2(CN)_2 + POCl_3 + 2HCl$
2. $2CNCH_2CONH_2 + POCl_3 \rightarrow 2CH_2(CN)_2 + HPO_3 + 3HCl$

*Reaction with tetracyanoethylene* (**1**, 962, 1135).
*Bromination* (**1**, 1133).

| *t-Butylation*[2] with | Yield, % |
|---|---|
| $(CH_3)_3CBr/AlCl_3$ | 60 |
| $(CH_3)_3CCl/BF_3$ | 43–56 |
| $(CH_3)_3CBr/AgClO_4$ | 68 |
| $(CH_3)_2C{=}CH_2/HClO_4$ | 59 |

[1] B. B. Corson, R. W. Scott, and C. E. Vose, *Org. Syn., Coll. Vol.*, **2**, 379 (1943).
[2] P. Boldt, H. Militzer, W. Thielecke, and L. Schulz, *Ann.*, **718**, 101 (1968).

**Manganese dioxide**, **1**, 637–643; **2**, 257–263; **3**, 191–194.

*Oxidation* of 4,4-dimethyl-1,7-diphenyl-1,6-heptadiyne-3,5-diol (1) with active manganese dioxide (German firm of Merck)[1]:

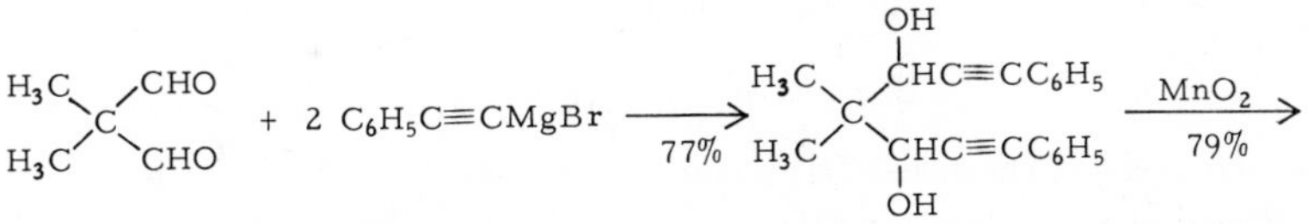

(1)

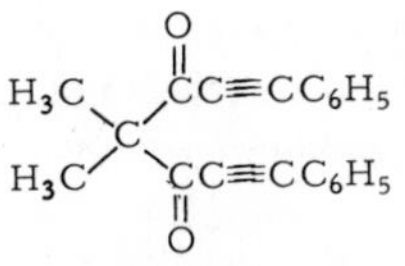

(2)

*Carboxylic amides.* Gilman[2] has extended the procedure of Corey *et al.* (**2**, 261) for oxidation of allylic alcohols to carboxylic esters to the preparation of carboxylic amides. Thus oxidation of aromatic and $\alpha,\beta$-unsaturated aldehydes with $MnO_2$ in the presence of sodium cyanide and an amine gives the corresponding carboxylic amides in high yield. In the absence of sodium cyanide, high yields of nitriles are formed from aldehydes, ammonia, and manganese dioxide.

*Oxidative coupling of aldehydes.*[3] Aliphatic aldehydes bearing an $\alpha$-hydrogen atom on treatment with activated manganese dioxide undergo $\alpha$-hydrogen abstraction and C–C and C–O dimerization. For example, isobutyraldehyde (1) is converted in about

80% yield into a mixture of 2,2,3,3-tetramethylsuccinaldehyde (2) and 2-methyl-2-(2'-methyl-1'-propenoxy)propionaldehyde (3) in the ratio 43:57.

(1)                    (2)                    (3)

*Dehydrogenation.* The final step in a total synthesis of the ergot alkaloid *dl*-isosetoclavine (2) involved dehydrogenation of (1). This step was carried out in 36% yield by means of activated manganese dioxide in chloroform at 25°.[4]

(1)                                    (2)

[1] E. Müller and A. Segnitz, *Ber.*, **106**, 35 (1973).
[2] N. W. Gilman, *Chem. Commun.*, 733 (1971).
[3] J. C. Leffingwell, *ibid.*, 357 (1970).
[4] E. C. Kornfeld and N. J. Bach, *Chem. Ind.*, 1233 (1971).

## Manganic acetate (Manganese(III) acetate), 2, 263–264.

*Preparation.*[1] Manganese(III) acetate can be prepared by permanganate oxidation of manganese(II) acetate tetrahydrate (suppliers: Alfa Inorganics, Fisher).

*Oxidation of toluenes.*[2] Toluenes are oxidized to benzyl acetates by the reagent. The reaction is markedly catalyzed by metal halides such as potassium bromide. The reaction is believed to proceed by way of a radical intermediate:

$$XC_6H_4CH_3 \xrightarrow[\text{HOAc}]{\text{Mn(OAc)}_3,\ \text{KBr}} XC_6H_4CH_2 \cdot \longrightarrow$$

$$XC_6H_4CH_2Br \longrightarrow XC_6H_4CH_2OAc$$

*Allylic oxidation.*[1] Cyclohexene is oxidized to cyclohex-2-enyl acetate in 75% yield by the reagent at 70° when catalyzed by potassium bromide (*ca.* 8 hr.). The reaction is slow in the absence of a catalyst. Norbornene is unreactive, but bicyclo[3.2.1]octene-2 is oxidized rapidly to about an equal mixture of *endo-* and *exo*-bicyclo[3.2.1]oct-3-ene-2-yl acetates. Acyclic olefins or cyclic olefins prone to allylic rearrangement give a complex mixture of allylic acetates.

[1] J. R. Gilmore and J. M. Mellor, *J. Chem. Soc.* (*C*), 2355 (1971).
[2] *Idem, Chem. Commun.*, 507 (1970).

**Mercuric acetate, 1**, 644–652; **2**, 265–267; **3**, 194–196.

*Oxymercuration* (**2**, 265–267; **3**, 194). Brown and Geoghegan[1] have investigated the relative reactivities of a number of olefins in the oxymercuration reaction in a 20:80 (v/v) mixture of water and THF. The following reactivity is observed: terminal disubstituted > terminal monosubstituted > internal disubstituted > internal trisubstituted > internal tetrasubstituted. Thus steric factors play a major role in the reactivity of olefins. Increased substitution on the double bond and increased steric hindrance at the site of hydroxyl or mercury substituent attachment decrease the rate of reaction. In the case of olefins of the type RCH=CHR′, *cis*-olefins are more reactive than the corresponding *trans*-olefins. Inclusion of the double bond in ring systems causes a moderate rate increase which varies somewhat with structure: cyclohexene > cyclopentene ≫ cyclooctene; norbornene ≫ bicyclo[2.2.2]-octene-2.

The results above suggested that selective monooxymercuration of dienes is possible, and indeed this has been realized.[2] In the case of symmetrical dienes such as 1,5-pentadiene, the yield of enol is lower than predicted for a statistical reaction (50% enol); the statistical value is approached in longer chain dienes. Yields can be raised by using mercuric trifluoroacetate (**2**, 195). In the case of unsymmetrical dienes selective hydration can be achieved. Thus limonene (1) can be converted to the enol (2) in 70% yield.

Oxymercuration of bicyclo[3.2.1]octa-2,6-diene (3) in acetic acid at 50° for 7 hr. gives *exo,exo*-6-acetoxymercuri-7-acetoxybicyclo[3.2.1]octene-2 (4) in 86% yield. The

reaction thus occurs more readily with the strained double bond of (3). This fact holds also for reaction of (3) with palladium acetate (**1**, 778; **2**, 303) in acetic acid at 25°, which leads to (6).[3]

Olefins can be converted into ketones by oxymercuration followed by transmetalation of the oxymercurial with palladium chloride. Unlike oxymercurials, the analogous palladium compounds decompose in solution to give ketones and palladium metal.[4]

$$CH_3(CH_2)_3CH=CH_2 \ + \ Hg(OAc)_2 \ \xrightarrow{H_2O} \ CH_3(CH_2)_3\underset{\underset{OH}{|}}{C}HCH_2HgOAc \ \xrightarrow{PdCl_2}$$

$$CH_3(CH_2)_3\underset{\underset{OH}{|}}{C}HCH_2PdCl_2^- \ \xrightarrow{84\%} \ CH_3(CH_2)_3COCH_3 \ + \ Pd^0 \ + \ HCl \ + \ Cl^-$$

Oxymercuration has been reviewed by Kitching.[5]

**Oxidation of steroidal alkenes.**[6] Mercuric acetate (2 eq.) reacts with $\Delta^{14}$-steroids (1) to give as the major product $\Delta^{8(14)}$-15-ones (2) rather than the expected allylic acetates, 16$\xi$-acetoxy-14-enes. The tetrasubstituted bond of $\Delta^{8(14)}$-cholestene is unreactive under the same conditions. $\Delta^2$-Cholestene (disubstituted) reacts only slowly to give $\Delta^1$-cholestene-3-one in 20% yield with 70% recovery of the starting material. This unexpected reaction of steroidal trisubstituted double bonds may proceed through an allylic acetoxylation followed by an oxidation step.

(1)                                             (2)

**Synthesis of tertiary amines.**[7] Mercuric acetate reacts with an enamine (1) in DMF by attack at the $\beta$-carbon atom to afford an iminium salt (2). Reduction of the salt with sodium borohydride gives the corresponding tertiary amine (3). Yields are in the range 50–90%.

$$R_2NCH=C\!\!<\ \xrightarrow{Hg(OAc)_2}\ R_2\overset{+}{N}=CH\underset{\underset{HgOAc}{|}}{C}\!\!<\ \ \overset{-}{O}Ac\ \xrightarrow{NaBH_4}\ R_2NCH_2CH\!\!<$$

(1)                          (2)                                  (3)

**Oxidative cleavage of cyclopropenes.**[8] 1,3,3-Trimethylcyclopropene (1) is converted into 1,1-diacetoxy-2,3-dimethyl-2-butene (2) by oxidation with mercuric acetate in methylene chloride at room temperature. Reaction of (1) with thallium triacetate or

(1)                                             (2)

lead tetraacetate gives, in addition to (2), 3,3-diacetoxy-2-methylpropene (3) and 2-acetoxy-4-methyl-1,3-pentadiene (4). The oxidations are believed to proceed by

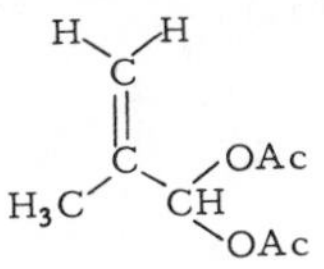

addition of the metal acetates followed by ring cleavage to give vinylcarbene–metal acetate complexes.

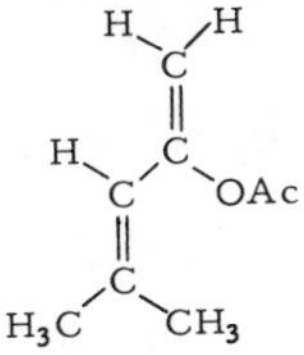

*Oxidative cyclization.* The final step in a synthesis of racemic ajmalicine (2) required oxidative cyclization of (1).[9] This reaction was carried out in 2.5% aqueous acetic acid with an excess of 1:1 mercuric acetate–ethylenediaminetetraacetic acid disodium salt (EDTA, 1, 373–374) followed by sodium borohydride reduction of iminium intermediates.

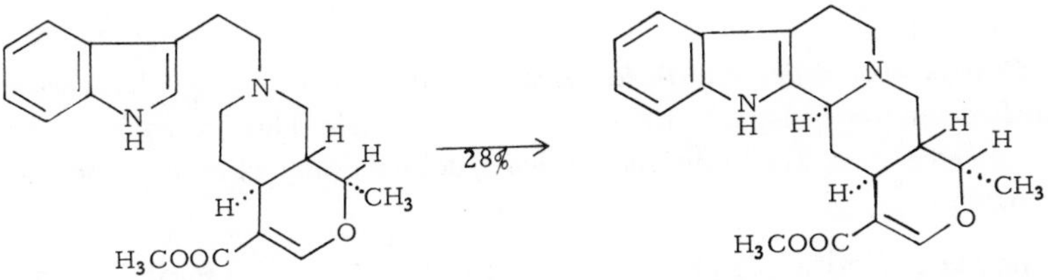

The same conditions were used by Stork and Guthikonda[10] for cyclization of the seco alcohol (3) to racemic yohimbine (4, 32% yield). Stork also notes that only 2 eq. of mercuric acetate is required since mercury, rather than mercurous acetate, is produced in the presence of EDTA.

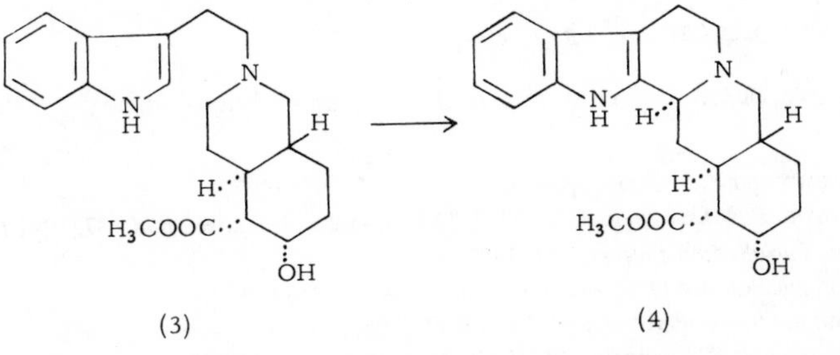

*Allylic oxidation of olefins.*[11] Oxidation of allylbenzene (1) with Hg(OAc)$_2$ in HOAc at reflux for 50 hr. gives metallic mercury (70%) and 72% of organic products.

$$C_6H_5CH_2CH{=}CH_2 \xrightarrow[72\%]{Hg(OAc)_2} C_6H_5CH{=}CHCH_2OAc \ + \ C_6H_5\underset{\underset{OAc}{|}}{C}HCH{=}CH_2 \ +$$

(1)                                    (2, 95%)                    (3, 0.5%)

$$C_6H_5CH_2CH(OAc)CH_2OAc \ + \ C_6H_5\overset{t}{CH}{=}CHCH_3$$

(4, 2%)                    (5, 0.1%)

The major product is cinnamyl acetate (2); α-phenylallyl acetate (3), 1,2-diacetoxy-3-phenylpropane (4), and *trans*-propenylbenzene (5) are formed in trace amounts. A similar distribution of allylic acetates is obtained from the solvolysis of cinnamyl-mercuric acetate, $C_6H_5CH{=}CHCH_2HgOAc$, which presumably is the intermediate in the oxidation of allylbenzene.

1-Butene is oxidized by $Hg(OAc)_2$ at 25–50° to give α-methylallyl acetate (6) as the exclusive product. This allylic acetate is also the major product of oxidation of *cis*- and *trans*-2-butene at moderate temperatures.

$$CH_3CH_2CH{=}CH_2 \xrightarrow{Hg(OAc)_2} CH_3CH(OAc)CH{=}CH_2$$

(6)

**Conversion of olefins into ethylene ketals.** Hunt and Rodeheaver[12] have described a method for conversion of olefins into ethylene ketals. Three steps are involved: 1) solvomercuration, 2) transmetalation, and 3) decomposition of the organopalladium intermediate.

1) $Hg(OAc)_2 \ + \ RCH{=}CH_2 \ + \ HOCH_2CH_2OH \xrightarrow{H^+} R\underset{\underset{OCH_2CH_2OH}{|}}{C}HCH_2HgOAc \ + \ HOAc$

2) $R\underset{\underset{OCH_2CH_2OH}{|}}{C}HCH_2HgOAc \ + \ PdCl_2 \longrightarrow R\underset{\underset{OCH_2CH_2OH}{|}}{C}HCH_2PdCl \ + \ HgOAcCl$

3) $R\underset{\underset{OCH_2CH_2OH}{|}}{C}HCH_2PdCl \longrightarrow \underset{R \quad CH_3}{O{\diagup}{\diagdown}O} \ + \ Pd \ + \ HCl$

For example, hexene-1 is converted in this way into 2-butyl-2-methyl-1,3-dioxolane in 77% yield.

[1]H. C. Brown and P. J. Geoghegan, Jr., *J. Org.*, **37**, 1937 (1972).
[2]H. C. Brown, P. J. Geoghegan, Jr., G. J. Lynch, and J. T. Kurek, *ibid.*, **37**, 1941 (1972).
[3]M. Sakai, *Tetrahedron Letters*, 347 (1973).
[4]G. T. Rodeheaver and D. F. Hunt, *Chem. Commun.*, 818 (1971).
[5]W. Kitching, *Organometal. Chem. Rev.*, **3**, 61 (1968).
[6]E. C. Blossey and P. Kucinski, *J.C.S. Chem. Commun.*, 56 (1973).
[7]R. D. Bach and D. K. Mitra, *Chem. Commun.*, 1433 (1971).
[8]T. Shirafuji and H. Nozaki, *Tetrahedron*, **29**, 77 (1973).

[9]J. Gutzwiller, G. Pizzolato, and M. Uskoković, *Am. Soc.*, **93**, 5907 (1971).
[10]G. Stork and R. N. Guthikonda, *ibid.*, **94**, 5109 (1972).
[11]Z. Rappoport, S. Winstein, and W. G. Young, *ibid.*, **94**, 2320 (1972).
[12]D. F. Hunt and G. T. Rodeheaver, *Tetrahedron Letters*, 3595 (1972).

## Mercuric azide, 3, 196.

*Caution:* Heathcock and Bach[1] report that a violent explosion occurred in the *in situ* generation of the azide when $Hg(N_3)_2$ precipitated out of solution at room temperature. The detonation was presumably caused by agitation by a magnetic stirring bar. They urge extreme caution when carrying out the azidomercuration reaction with olefins. Galle and Hassner[2] state that it is important to add the mercuric acetate to the solution of sodium azide in THF–water, and not vice versa.

*Cyclopropyl azides.*[2] Mercuric azide adds *syn* to cyclopropenes (1) to give *cis*-2-azidocyclopropylmercuric azide salt (2) in fair yield. Reductive demercuration with $NaBH_4$ gives cyclopropyl azides (3) in moderate yield.

[1]C. H. Heathcock and R. D. Bach, personal communication.
[2]J. E. Galle and A. Hassner, *Am. Soc.*, **94**, 3930 (1972).

## Mercuric oxide, 1, 655–658; 2, 267–268.

*Hunsdiecker–Cristol reaction.* (1, 657). A detailed procedure for the preparation of 1-bromo-3-chlorocyclobutane by the modified Hunsdiecker reaction has been published.[1] A 1-l. three-necked, round-bottomed flask wrapped with aluminum foil to exclude light and equipped with a mechanical stirrer, a reflux condenser, and an addition funnel, is charged with 37 g. (0.17 mole) of red mercuric oxide and 330 ml.

of carbon tetrachloride. To the flask is added 30.0 g. (0.22 mole) of 3-chlorocyclo-butanecarboxylic acid and the mixture is stirred and heated to reflux. With the heating bath maintained at about 120°, a solution of 40 g. (0.25 mole) of bromine in 180 ml. of carbon tetrachloride is added as rapidly as possible (4–7 min.) without loss of bromine from the condenser. After a short induction period, carbon dioxide is evolved at a rate of 150–200 bubbles per minute, as monitored by conducting the gas through rubber tubing from the condenser to a small amount of water where the bubbling can be observed (there is some loss of bromine). The solution is allowed to reflux until the rate of carbon dioxide evolution slows to about five bubbles per minute (25–30 min.). The mixture is then cooled in an ice bath, and the precipitate is filtered. Solvent is removed by distillation using a modified Claisen apparatus with a Vigreux column. Vacuum distillation of the residual oil gives 13–17 g. (35–46%) of 1-bromo-3-chloro-cyclobutane, b.p. 67–72° (45 mm.), $n_D^{23}$ 1.5065.

The degradation of carboxylic acids to alkyl halides using mercuric oxide and halogens involves the initial formation of the mercuric salt of the acid, followed by a normal Hunsdiecker reaction of the salt with halogen. The relative insensitivity of the reaction to water is a consequence of the solubility of the mercury salts in the solvent ($CCl_4$). There are two limitations: tertiary acids are not degraded, and use of iodine as the halogen frequently leads to the ester RCOOR as the major product. The yields in the modified reaction are usually lower than those obtained with the silver salt method.[2]

Cason and Walba[3] obtained the highest yields in the modified procedure when excess acid was used and the reaction conducted in refluxing $CCl_4$, with removal of water by azeotropic distillation. They were able to obtain tridecyl bromide from myristic acid in 90% yield.

*Oxidation of cyclooctane-1,2-dione bishydrazone* (1) with mercuric oxide yields mainly (2) along with a little cyclooctyne, triazoles, and their mercury derivatives.[4]

[1]G. M. Lampman and J. C. Aumiller, *Org. Syn.*, **51**, 106 (1971).
[2]N. J. Bunce, *J. Org.*, **37**, 664 (1972).
[3]J. Cason and D. M. Walba, *ibid.*, **37**, 669 (1972).
[4]E. Müller and H. Meier, *Ann.*, **716**, 11 (1968).

## Mercuric oxide–Bromine, HgO/Br₂.

*Bromination of alkanes.*[1] The combination of bromine and mercuric oxide is effective for free-radical bromination of alkanes:

$$2 \, RH + 2 \, Br_2 + \cdot HgO \longrightarrow 2 \, RBr + H_2O + HgBr_2$$

The combination reagent is significantly more reactive than elemental bromine or NBS, and should thus be useful where reactions with these two reagents are sluggish. The reaction appears to involve bromine monoxide as the reactive brominating agent. The new reaction is applicable to substitution of primary and secondary C–H bonds; yields for the most part are satisfactory.

$$2\ Br_2\ +\ HgO \longrightarrow HgBr_2\ +\ Br_2O$$

$$Br_2O\ +\ 2\ RH \longrightarrow 2\ RBr\ +\ H_2O$$

Note that mercuric oxide can be recovered from the spent mercury residue.[2]

[1] N. J. Bunce, *Canad. J. Chem.*, 3109 (1972).
[2] G. H. Cady, *Inorg. Syn.*, **5**, 156 (1957).

## Mercuric trifluoroacetate, 3, 195.

*Cyclization of isoprenoids.*[1] Acyclic isoprenoids can be cyclized in 30–60% yield by treatment with 1.2–2.0 eq. of mercuric trifluoroacetate in nitromethane or acetonitrile at −20° to 0°. After 5–30 min. the reaction mixture is treated with 2 eq. of $NaBH_4$ in 3 $M$ NaOH solution at 0°. An example is the cyclization of *trans*-geranylacetone (1) to 2,5,5,9-tetramethylhexahydrochromene (2).

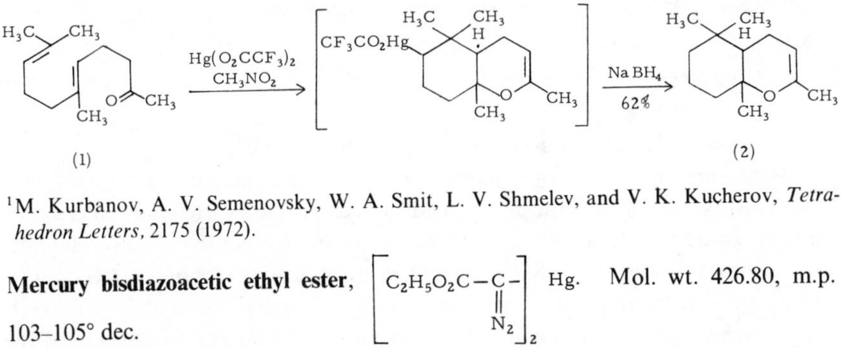

(1)    (2)

[1] M. Kurbanov, A. V. Semenovsky, W. A. Smit, L. V. Shmelev, and V. K. Kucherov, *Tetrahedron Letters*, 2175 (1972).

## Mercury bisdiazoacetic ethyl ester, $\left[ C_2H_5O_2C - \underset{\underset{N_2}{\|}}{C} - \right]_2 Hg.$    Mol. wt. 426.80, m.p. 103–105° dec.

*Preparation*[1] from ethyl diazoacetate and mercuric oxide and crystallization of the bright-yellow product from ether.

*Conversion to halodiazoacetic ethyl esters*[2]:

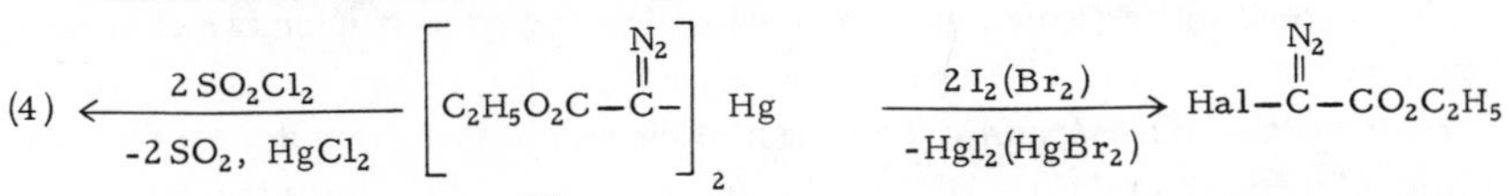

(2) Hal = I
(3) Hal = Br
(4) Hal = Cl

[1] E. Buchner, *Ber.*, **28**, 215 (1895).
[2] U. Schöllkopf, F. Gerhart, M. Reetz, H. Frasnelli, and H. Schumacher, *Ann.*, **716**, 204 (1968).

**Methanesulfonic acid, 1,** 666–667; **2,** 270.

*Cleavage of an epoxide.* As a first step in the synthesis of cyclobutanecarbox-aldehyde (D),[1] a solution of 48.0 g (0.50 mole) of methanesulfonic acid (b.p. 140°/0.2 mm.) in 500 ml. of anhydrous ether is stirred mechanically in an ice-water bath and a solution of 37 g. of *trans*-2-butene oxide in 500 ml. of anhydrous ether is added in 3–4 hr. After 6 hr. the cooling bath is removed and the mixture is stirred for an additional 12 hr.

$$
\begin{array}{c}
\square\text{—COCl} + \underset{\text{A}}{CH_3\overset{\displaystyle OH}{\overset{|}{C}H}-\overset{\displaystyle OSO_2CH_3}{\overset{|}{C}H}CH_3} \xrightarrow[\text{pyridine}]{0^0} \square\text{—}C\overset{\displaystyle O}{\underset{\displaystyle O}{\diagup}} \quad \underset{\text{B}}{CH_3\overset{|}{C}H-\overset{\displaystyle OSO_2CH_3}{\overset{|}{C}H}CH_3}
\end{array}
$$

NaBH$_4$ | pyridine, 115°

$$
\underset{\text{D}}{\square\text{—CHO}} \xleftarrow{H_3O^+} \underset{\text{C}}{\square\overset{\displaystyle H}{\underset{\displaystyle \overset{O}{|}\ \overset{O}{|}}{}}CH_3\overset{|}{C}H-\overset{|}{C}HCH_3}
$$

The ether is removed at 25° by means of a rotary evaporator and water aspirator to give 83–84 g. (99–100%) of *erythro*-2,3-butanediol monomesylate (A) as a clear, colorless, somewhat viscous liquid.

[1] M. R. Johnson and B. Rickborn, *Org. Syn.,* **51,** 11 (1971).

**Methanesulfonyl chloride (Mesyl chloride), 1,** 662–664; **2,** 268–269.

*Methanesulfonate esters (mesylates).*[1] Mesylates can be prepared in 85–95% yield by the reaction at 0–10° of an alcohol in methylene chloride (or cyclohexane or pentane) containing a 50% molar excess of triethylamine with a 10% excess of mesyl chloride, added over a period of 5–10 min. The reaction probably involves the sulfene $CH_2=SO_2$, formed by dehydrochlorination of mesyl chloride. Tertiary and neopentyl alcohols can be esterified by this procedure. Even reactive mesylates can be prepared if the temperature of the mesylate is kept at 0°.

Mesylates are less reactive toward solvolysis than the corresponding tosylates. Mesylates are better suited to reduction by lithium aluminum hydride than tosylates (**1,** 58) because the mesylate fragment is reduced to methyl mercaptan, which is easily removed.

*Elimination of iodohydrins.*[2] Treatment of the iodolactone (1) in dry pyridine with 1.3 eq. of mesyl chloride at −20° for 2 hr. and at 0° for 1.5 hr. affords the unsaturated lactone (2) in practically quantitative yield. The reaction provided a key step in the synthesis of A prostaglandins, which previously were available from the primary E prostaglandins by dehydration of the β-ketol unit.

The new elimination reaction may be related to the Cornforth procedure for conversion of epoxides to olefins (**1,** 881, 1116, 1279).

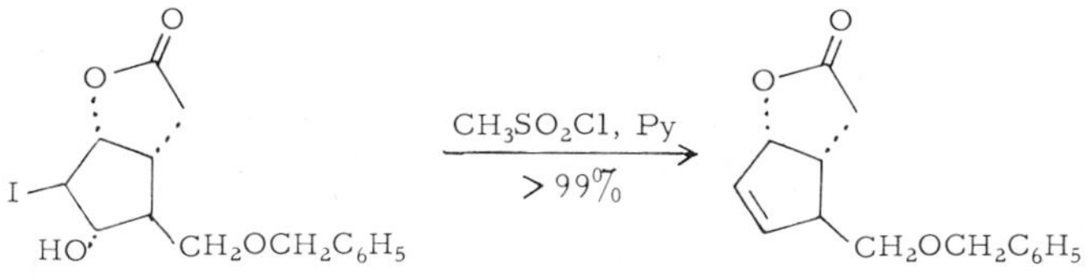

(1)             (2)

Crabbé and Guzmán have effected the same elimination with phosphoryl chloride in pyridine at room temperature (*see* **Phosphoryl chloride,** this volume).

*Allylic chlorides.*[3] Allylic alcohols (0.10 mole) are readily converted at 0° into allylic chlorides by treatment with mesyl chloride (0.11 mole) and a mixture of lithium chloride (0.10 mole), dry DMF, and *s*-collidine (0.11 mole). A nonallylic hydroxyl group, if present, is converted into the mesylate.

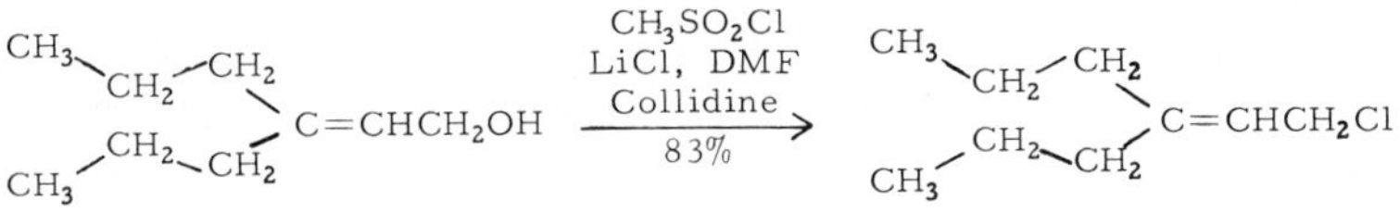

[1] R. K. Crossland and K. L. Servis, *J. Org.,* **35,** 3195 (1970).
[2] E. J. Corey and P. A. Grieco, *Tetrahedron Letters,* 107 (1972).
[3] E. W. Collington and A. I. Myers, *J. Org.,* **36,** 3044 (1971).

**2-Methoxyallyl bromide,** 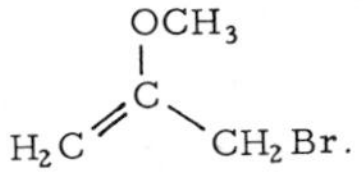    Mol. wt. 135.04, decomposes at room temperature.

*Preparation.*[1] The reagent is available by two different routes. It can be obtained in about 50% yield by the reaction of 2-methoxypropene[2] and NBS in $CCl_4$ (method A) or by pyrolysis of 1-bromo-2,2-dimethoxypropane (available from bromoacetone). The reagent (neat) decomposes at room temperature and is best stored in the dark as

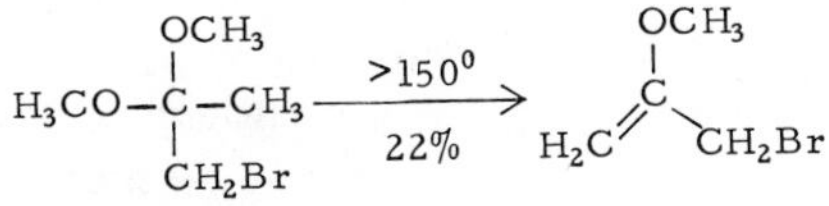

a 50% solution in $CCl_4$. Method A is preferred as a laboratory procedure; method B is adaptable to large-scale preparations.

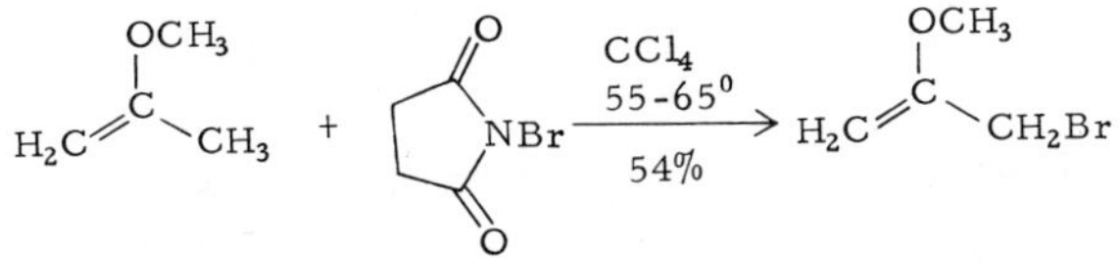

*Cycloadditions.*[3] Treatment of 2-methoxyallyl bromide with silver trifluoroacetate (**1,** 1018–1019) under carefully controlled conditions generates the 2-methoxyallyl cation (a), which undergoes cycloaddition with dienes. For example, the bromide in

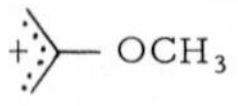

$$+\!\!\ddot{\rangle}\!\!-\!OCH_3$$

(a)

benzene is mixed with furane (1) and isopentane with solid $Na_2CO_3$ (buffer) suspended in the solution. Silver trifluoroacetate and $Na_2CO_3$ are added in small portions at room temperature over a period of 10 hr. and the resulting suspension is stirred in the dark with an efficient vibromixer (100 cps). The reaction is worked up by careful

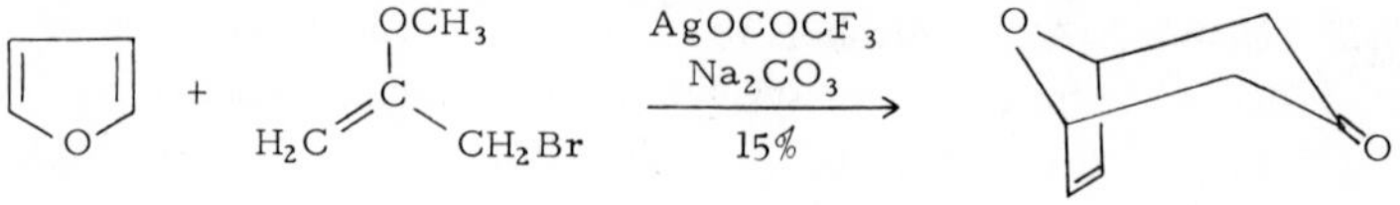

addition of nitric acid and filtration of the black silver residue. 8-Oxabicyclo[3.2.1]-octene-6-one-3 (2) is obtained in 15% yield. A similar reaction using cyclopentadiene gave bicyclo[3.2.1]octene-6-one-3 in 17% yield.

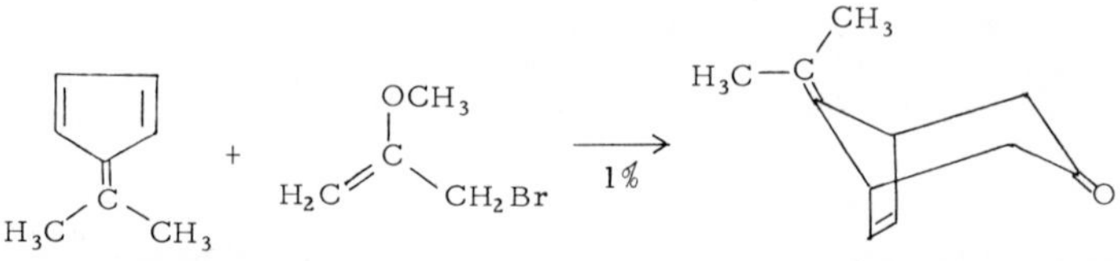

The reaction with 6,6-dimethylfulvene (3) led to 8-isopropylidenebicyclo[3.2.1]-octene-6-one-3 (4) in 1% yield.

[1] G. Greenwood and H. M. R. Hoffman, *J. Org.*, **37**, 611 (1972).
[2] G. Saucy and R. Marbet, *Helv.*, **50**, 1158 (1967).
[3] A. E. Hill, G. Greenwood, and H. M. R. Hoffmann, *Am. Soc.*, **95**, 1338 (1973).

**Methoxycarbenium hexafluoroantimonate,** $H_3C-O\!\!\overset{+}{=}\!\!CH_2$ $SbF_6{}^-$. Mol. wt. 134.01, m.p. 115–118°.

This stable alkoxycarbenium salt can be prepared[1] in about 90% yield by the reaction of either methoxyacetyl chloride or chloromethyl methyl ether with anhydrous hexafluoroantimonic acid[2] in sulfur dioxide or dichloromethane:

$$H_3COCH_2C\overset{O}{\underset{Cl}{\diagdown}} \xrightarrow[91\%]{HSbF_6} H_3C-O\overset{+}{=}CH_2SbF_6^- + CO + HCl$$

$$H_3COCH_2Cl \xrightarrow[89\%]{HSbF_6} H_3C-O\overset{+}{=}CH_2SbF_6^- + HCl$$

Methoxycarbenium ion is an ambident electrophile; reaction can occur at either C or O. Nucleophilic attack at the C atom results in methoxymethylation. Displacement of formaldehyde results in methylation of the nucleophile. The latter reaction is prevalent with aromatics. Thus the reaction of the reagent with benzene gives toluene in high yield.

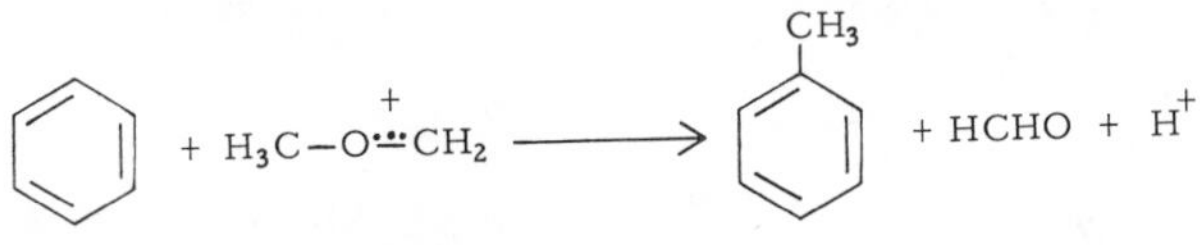

[1]G. A. Olah and J. J. Svoboda, *Synthesis*, 52 (1973).
[2]Suppliers: Aldrich, Alfa Inorganics, Cationics.

## 6-Methoxy-7-hydroxy-3,4-dihydroisoquinolinium methiodide (1). Mol. wt. 319.14, m.p. 216–218°.

**Preparation**[1]:

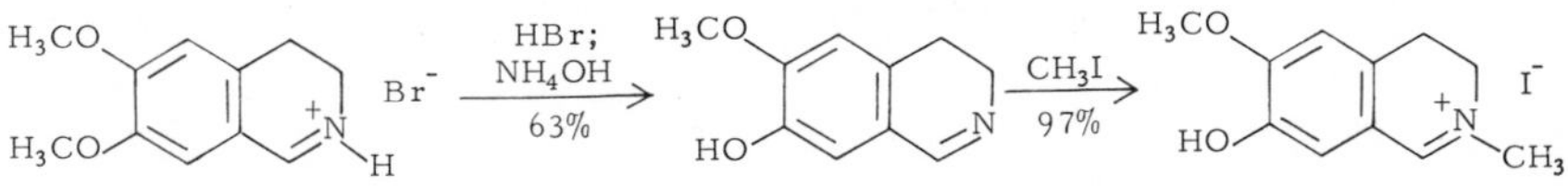

(1)

*Pschorr aporphine synthesis.* Kupchan *el al.*[2] have reported an improved synthesis of aporphines by using (1) for condensation with *o*-nitrotoluenes (2) in the presence of potassium *t*-butoxide in DMA to give 1-(2-nitrobenzyl)-6-methoxy-7-hydroxy-1,2,3,4-tetrahydroisoquinolines (3) in yields of 88–95%. These are hydrogenated to the

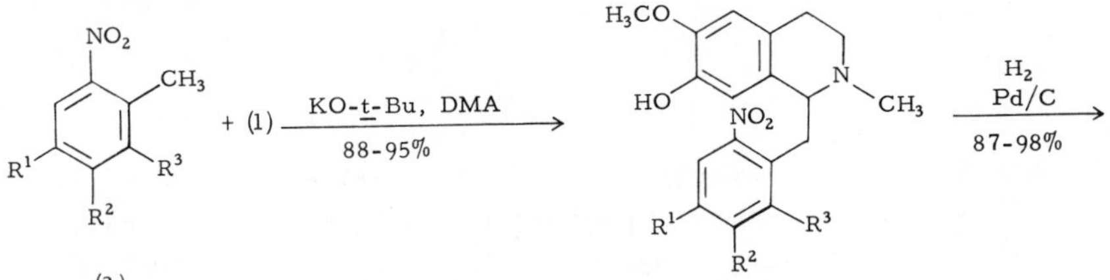

(2)

(3)

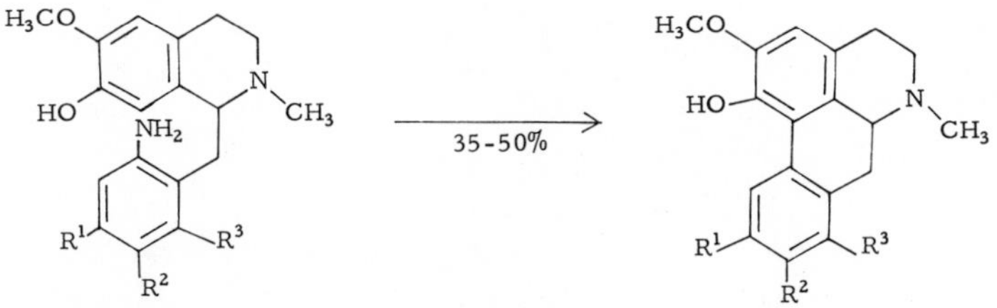

(4)

(5)

corresponding amino phenols (4), which are converted into 1-hydroxy aporphines (5) by the classical procedure of Pschorr: diazotization in 1:1 20% $H_2SO_4$–HOAc and cyclization with copper powder.

[1]A. Brossi, J. O'Brien, and S. Teitel, *Org. Prep. Proc.*, **2**, 281 (1970).
[2]S. M. Kupchan, V. Kameswaran, and J. W. A. Findlay, *J. Org.*, **38**, 405 (1973).

**3-Methoxyisoprene,** 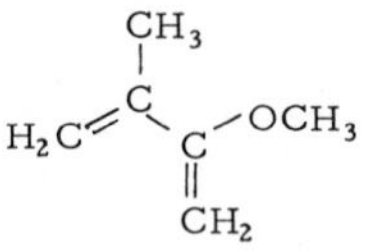 , **3**, 300–302. Mol. wt. 146.18, b.p. 93–95°.

*Preparation.*[1] The reagent (1) is prepared by the addition of methanol to 2-methyl-1-buten-3-yne using a catalyst prepared from red mercuric oxide, trichloroacetic acid, boron trifluoride etherate, and methanol.

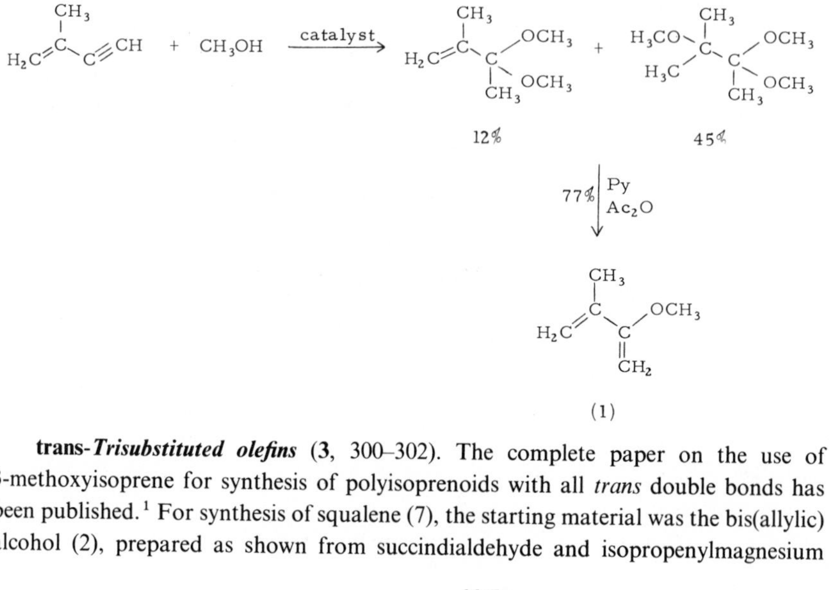

*trans-Trisubstituted olefins* (**3**, 300–302). The complete paper on the use of 3-methoxyisoprene for synthesis of polyisoprenoids with all *trans* double bonds has been published.[1] For synthesis of squalene (7), the starting material was the bis(allylic) alcohol (2), prepared as shown from succindialdehyde and isopropenylmagnesium

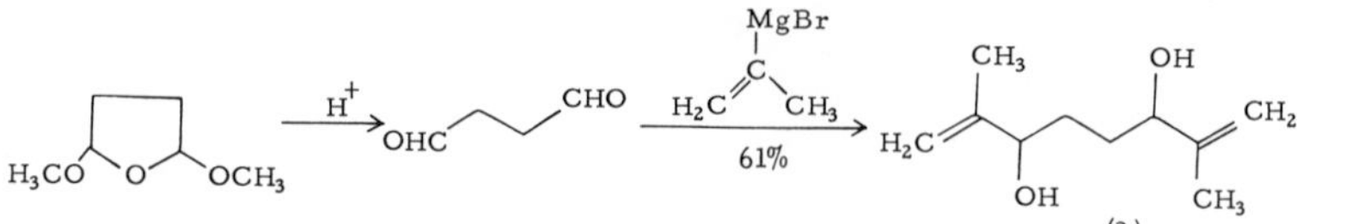

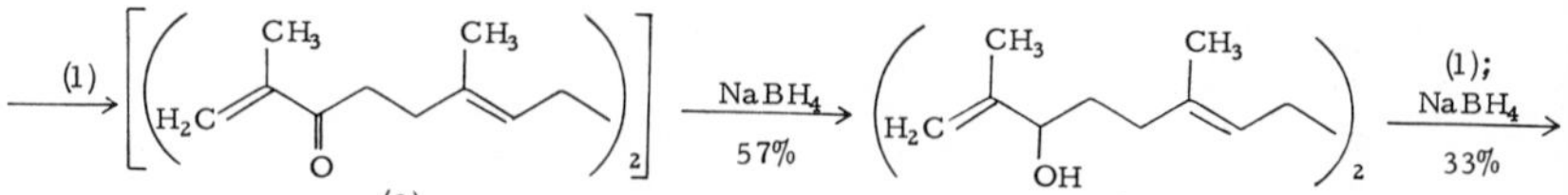

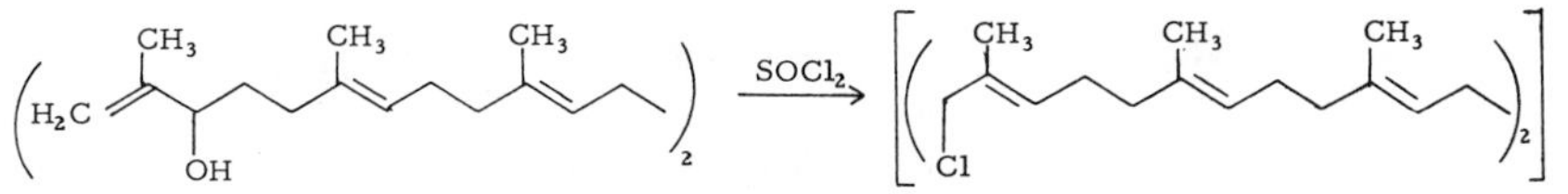

(5)    (6)

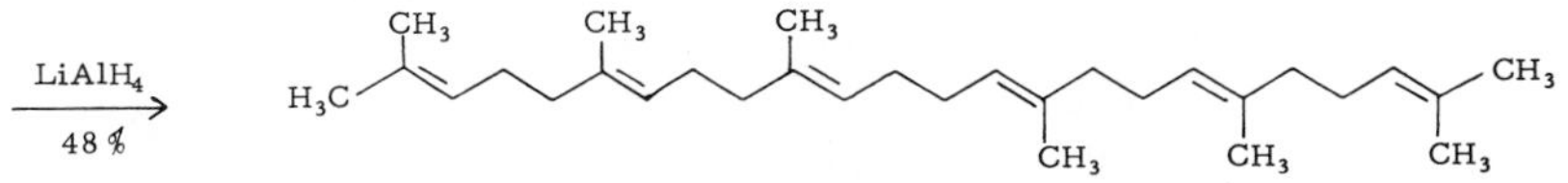

(7)

bromide.[2] This was treated with 5 molar eq. of 3-methoxyisoprene (1) in toluene containing small amounts of oxalic acid and hydroquinone at 140° for 24 hr. to give the $C_{20}$-tetraenedione (3), which was immediately reduced (NaBH$_4$) to the $C_{20}$-tetraenediol (4). Repetition of the two-step process gave the $C_{30}$-diol (5). Note that the yield in the Claisen rearrangement decreases as the length of the isoprenoid chain is increased. The $C_{30}$-diol (5) was rearranged to squalene (7) by treatment with thionyl chloride ($S_Ni'$ reaction) followed by reduction with lithium aluminum hydride. The squalene was essentially all-*trans* ($> 99\%$).

Faulkner and Petersen[1] also used the Claisen rearrangement for a synthesis of the optically active forms of *Cecropia* juvenile hormone (8). The stereoselectivity in this case was not so high as in the synthesis of squalene.

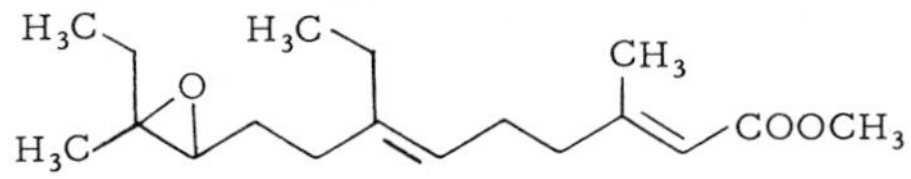

(8)

[1] D. J. Faulkner and M. R. Petersen, *Am. Soc.*, **95**, 553 (1973).
[2] 2-Bromopropene is available from Columbia Chemicals.

**Methoxymethyl methanesulfonate**, 3, 198 (note correction in name). Suppliers: Aldrich, WBL.

**2-Methoxy-6-methyl-1,4,3,5-oxathiadiazine 4,4-dioxide** (1),

Mol. wt. 162.12, m.p. 102–103° dec.

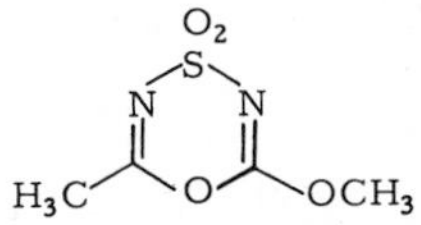

(1)

*Preparation.*[1] The reagent is prepared by the reaction of sodium hydride with N-carbomethoxysulfamoyl chloride (2, *see* **Methyl N-sulfonylurethane triethylamine complex**, this volume). This precursor is treated with sodium hydride and acetonitrile

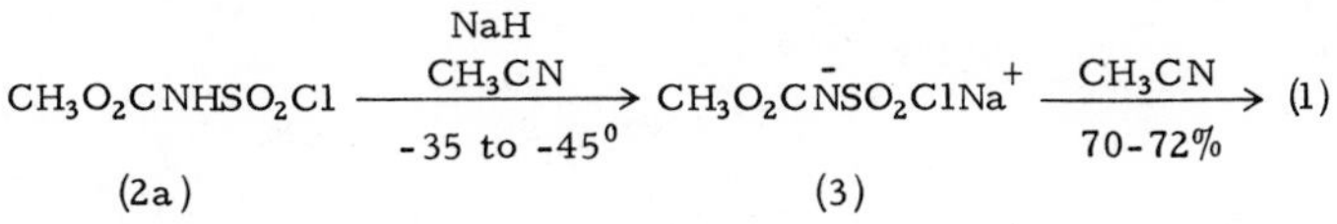

$$CH_3O_2CNHSO_2Cl \xrightarrow[-35\ to\ -45^0]{\underset{CH_3CN}{NaH}} CH_3O_2C\bar{N}SO_2ClNa^+ \xrightarrow[70-72\%]{CH_3CN} (1)$$

(2a)    (3)

at $-35$ to $-45°$. The solution is allowed to warm to room temperature; the sodium chloride which is formed is removed. The filtrate is then concentrated in a rotary evaporator and diluted with ether. The mixture is then cooled to $-30°$ overnight; (1) is collected by filtration in 70–72% yield.

*Methyl N-sulfonylurethane*, $CH_3O_2CN=SO_2$ (2). When (1) is heated in acetonitrile at $30°$, it undergoes [2 + 4] cycloreversion to give methyl N-sulfonylurethane (2) *in situ*. This reagent has also been obtained as the tetrahydrofurane complex from carbomethoxysulfamoyl chloride[2] as shown in the formulation.

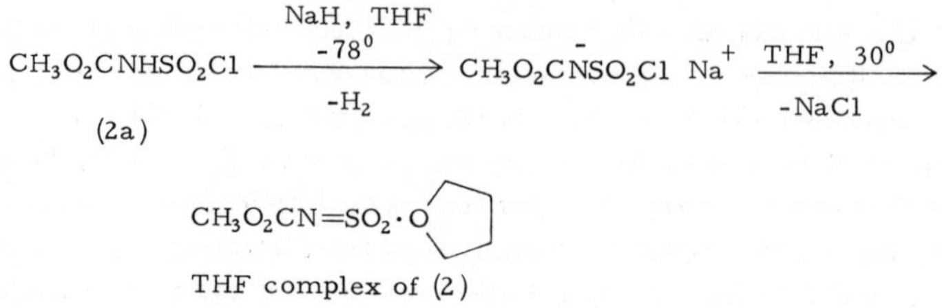

$$CH_3O_2CNHSO_2Cl \xrightarrow[\substack{-78° \\ -H_2}]{NaH, \ THF} CH_3O_2C\overset{-}{N}SO_2Cl \ Na^+ \xrightarrow[-NaCl]{THF, \ 30°}$$

(2a)

$$CH_3O_2CN=SO_2 \cdot O\ \langle \ \rangle$$

THF complex of (2)

The N-sulfonylurethane (2) readily undergoes [2 + 2] cycloadditions with olefins. Thus generation of (2) in acetonitrile in the presence of 1,1-diphenylethylene gives 2-carbomethoxy-3,3-diphenyl-1,2-thiazetidine 1,1-dioxide (3) and N-carbomethoxy-$\beta,\beta$-diphenylvinylsulfonamide (4).[1] Methyl N-sulfonylurethane (2) reacts with some

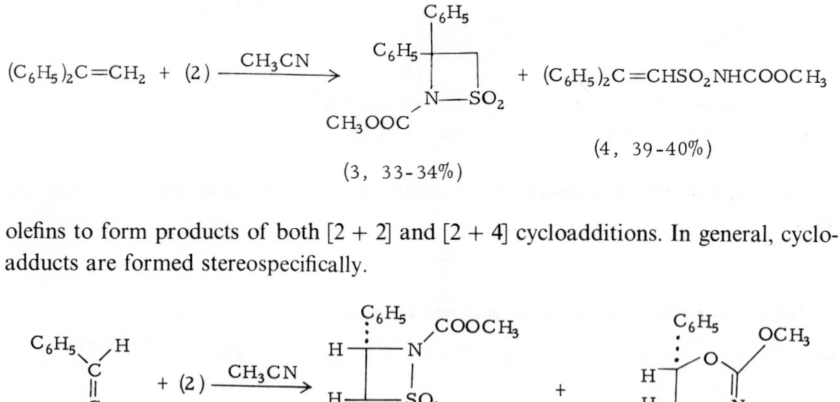

$$(C_6H_5)_2C=CH_2 + (2) \xrightarrow{CH_3CN} \underset{\substack{| \\ N-SO_2 \\ CH_3OOC}}{\overset{C_6H_5}{\underset{C_6H_5-}{|}}} + (C_6H_5)_2C=CHSO_2NHCOOCH_3$$

(3, 33–34%)            (4, 39–40%)

olefins to form products of both [2 + 2] and [2 + 4] cycloadditions. In general, cycloadducts are formed stereospecifically.

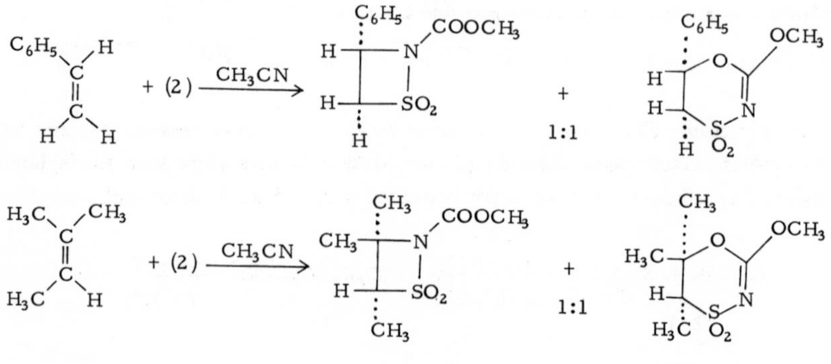

Addition to acetylenes proceeds exclusively by [2 + 4] cycloaddition. Thus phenylacetylene gives (5) exclusively.

(5)

Methyl N-sulfonylurethane (2) undergoes [2 + 2] cycloaddition to certain heteroatomic $\pi$-systems to give intermediates which spontaneously lose sulfur trioxide, in an overall "quasi"-Wittig reaction. Thus diphenylcyclopropenone (6) reacts with (2) in THF at 30° to give (7) in quantitative yield.

$^1$E. M. Burgess and W. M. Williams, *Org. Syn.*, submitted 1973.
$^2$*Idem, Am. Soc.*, **94**, 4386 (1972).

**Methyl *t*-butyl ether**, $CH_3OC(CH_3)_3$. Mol. wt. 88.15, b.p. 53–56°. Suppliers: Aldrich, Eastman. Ethyl *t*-butyl ether is also available from Aldrich and Eastman.

*Esterification.* Carboxylic acids react with alkyl *t*-butyl ethers in the presence of catalytic amounts of sulfuric acid or *p*-toluenesulfonic acid to form esters.[1] Yields are generally good to excellent. The reaction is carried out by heating the reactants either under reflux or on a steam bath for several minutes until the evolution of isobutylene ceases.

$$ROC(CH_3)_3 + R^1COOH \xrightarrow{H^+} R^1COOR + CH_2{=}C(CH_3)_2 + H_2O$$

The method has also been used to prepare alkyl glycosides.[2]

$^1$V. A. Derevitskaya, E. M. Klimov, and N. K. Kochetkov, *Tetrahedron Letters*, 4269 (1970).
$^2$*Idem, ibid.*, 4769 (1969).

**Methyl carbazate**, $NH_2NHCOOCH_3$. Mol. wt. 90.08, m.p. 70–72.5°. Supplier: Aldrich.

*Ketones → nitriles.*[1] Carbonyl compounds react in high yield with methyl carbazate (1) to give carbomethoxyhydrazones(2).[2] These are converted in nearly quantitative yield into methyl dialkyldiazenecarboxylates (4) by successive treatment with

$$R_1R_2C{=}O + NH_2NHCOOCH_3 \longrightarrow R_1R_2C{=}NNHCOOCH_3 \xrightarrow[CH_3OH]{HCN} R_1R_2C(CN)NHNHCOOCH_3$$

(1)                              (2)                                        (3)

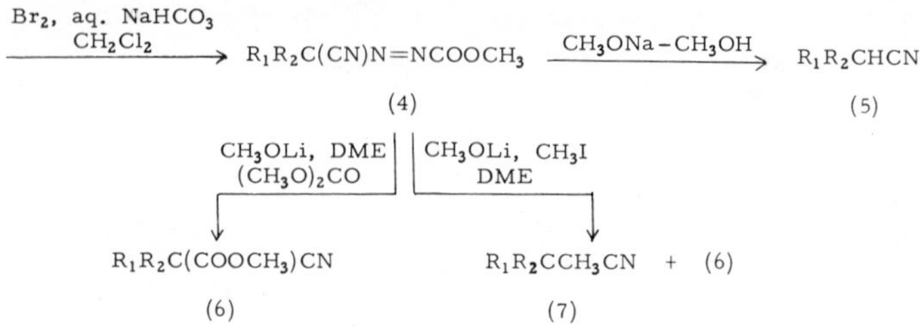

$$\xrightarrow[\text{CH}_2\text{Cl}_2]{\text{Br}_2, \text{ aq. NaHCO}_3} \text{R}_1\text{R}_2\text{C(CN)N=NCOOCH}_3 \xrightarrow{\text{CH}_3\text{ONa}-\text{CH}_3\text{OH}} \text{R}_1\text{R}_2\text{CHCN}$$

$$(4) \hspace{6cm} (5)$$

CH₃OLi, DME | CH₃OLi, CH₃I
(CH₃O)₂CO | DME

$$\text{R}_1\text{R}_2\text{C(COOCH}_3)\text{CN} \hspace{2cm} \text{R}_1\text{R}_2\text{CCH}_3\text{CN} \ + \ (6)$$

$$(6) \hspace{5cm} (7)$$

hydrogen cyanide and bromine. When the diazenes (4) are treated with sodium methoxide in methanol, nitrogen is evolved and nitriles (5) are obtained in high yield. If the diazenes are treated with lithium methoxide in DME containing dimethyl carbonate, cyano esters (6) are obtained together with trace amounts of nitriles (5). A third transformation of diazenes (4) is methylation by treatment with methyl iodide and lithium methoxide in DME to give α-methylnitriles (7). Catalytic amounts of base suffice for transformations (4) → (5) and (4) → (6), but stoichiometric quantities of base are required for the transformation (4) → (7).

[1]F. E. Ziegler and P. A. Wender, *Am. Soc.*, **93**, 4318 (1971).
[2]M. C. Chaco and N. Rabjohn, *J. Org.*, **27**, 2765 (1962).

## Methylcopper, CH₃Cu. Mol. wt. 102.59.

*Preparation.*[1] Methylcopper can be prepared from copper(I) iodide (4.4 mmole) in THF (40 ml.) and methyllithium (4 mmole) at −20°.

*Reaction with hindered carboxylic acid chlorides.* Israeli chemists[1] find that the very

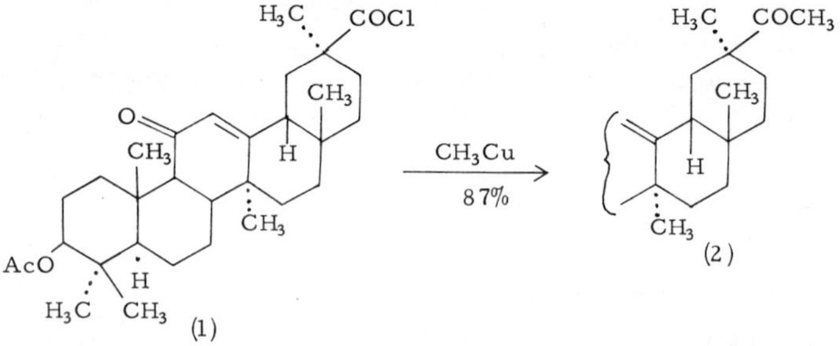

hindered acid chloride of 3-O-acetylglycyrrhetic acid (1) reacts readily with methylcopper to give the methyl ketone (2) in 87% yield. Reaction of (1) with methylmagnesium bromide at 0° gives (2) in only 25% yield. Reaction with dimethylcadmium gives even lower yields of the ketone. Reaction with dimethylcopperlithium gives about

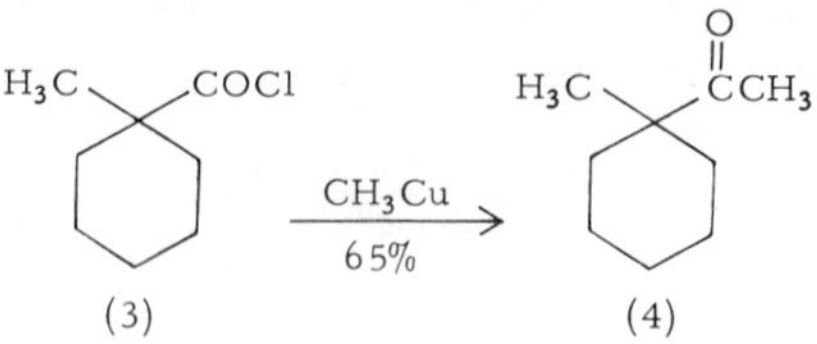

30–40% yields of (2). Methylcopper also reacts readily with 1-methylcyclohexane-carboxylic acid chloride (3) to give the corresponding methyl ketone (4)

[1]S. Rozen, I. Shahak, and E. D. Bergmann, *Synthesis*, 646 (1971).

**Methyl diazopropionate**, $H_3C-\overset{\overset{N_2}{\|}}{C}-COOCH_3$. Mol. wt. 114.11, b.p. 52°/18 mm., yellow.
*Preparation.*[1] The reagent can be prepared by several routes, as indicated.

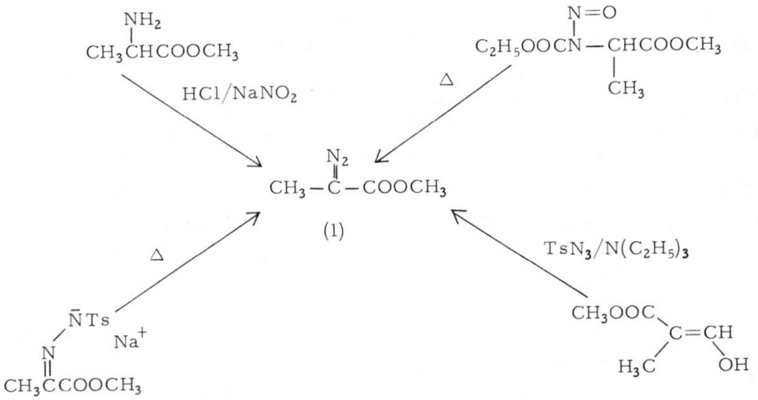

*Methylcarbomethoxycarbene.*[2] Alkylcarbenes are generally useless as cyclo-propane precursors since their singlet states rapidly undergo hydride shifts; the prevailing reaction is insertion into carbon–hydrogen bonds. For example, direct

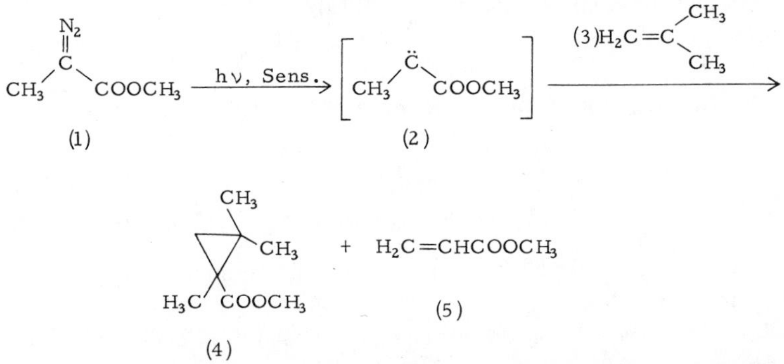

irradiation of methyl diazopropionate (1) gives methyl acrylate as the major product. However, benzophenone-sensitized decomposition of (1) gives triplet methylcarbo-methoxycarbene (2), which, in the presence of isobutylene (3), gives the cyclopropane derivative (4) in 72% yield. Only 3% of methyl acrylate (5) is formed. Photosensitized decomposition of (1) in the presence of *cis*- and *trans*-2-butene gave adducts in 50 and 46% yield, respectively.

(7, 33%)     (8, 11%)     (9, 3%)

Photosensitized irradiation of (6) in the presence of isobutylene gave (7) as the major product, but only in 33% yield.

[1] M. B. Sohn, M. Jones, Jr., M. E. Hendrick, R. R. Rando, and W. von E. Doering, *Tetrahedron Letters*, 53 (1972).
[2] M. B. Sohn and M. Jones, Jr., *Am. Soc.*, **94**, 8280 (1972).

## O-Methyldibenzofuranium fluoroborate (2). Mol. wt. 270.04.

The reagent is prepared by heating the diazonium salt (1) in anhydrous benzene. It is an unstable off-white solid. The precursor (1) is prepared from 2-methoxy-2'-nitrodiphenyl by reduction with hydrazine hydrate in the presence of palladium to give 2-amino-2'-methoxydiphenyl; this product is then diazotized with sodium nitrite in

(1)     (2)

the presence of aqueous fluoroboric acid to give the diazonium salt (1). The reagent can also be generated *in situ* by heating in methylene dichloride.[1]

*Methylation.*[1] This new fluoroborate is a powerful methylating agent comparable to methyl fluorosulfonate (**3**, 202; **4**, 339–340).

[1] A. J. Copson, H. Heaney, A. A. Logun, and R. P. Sharma, *Chem. Commun.*, 315 (1972).

## N-Methyl-N,N'-dicyclohexylcarbodiimidium iodiode. Mol. wt. 348.27, m.p. 111–113°.

(1)

The reagent (1) is prepared in 75% yield by prolonged heating of dicyclohexylcarbodiimide in methyl iodide.

*Alcohols → iodides.*[1] Aliphatic primary and secondary alcohols react with (1) in THF, benzene, or hexane at 35–50° to give the corresponding iodides, usually in high yield. Even sterically hindered alcohols react, although in somewhat low yields. The reaction proceeds with inversion; thus 3β-cholestanol is converted into 3α-iodocholestane in 84% yield. Cholesterol is converted into the previously unknown 3α-iodo-Δ⁵-cholestene (40% yield).

[1] R. Scheffold and E. Saladin, *Angew. Chem.*, *internat. Ed.*, **11**, 229 (1972).

**4,4′-Methylenebis(2,6-di-*t*-butylphenol)** ("**Ethyl antioxidant 702**"). Mol. wt. 424.6, m.p. 154°, b.p. 250°/10 mm. Solubility (wt. % at 20°):

| | | | | |
|---|---|---|---|---|
| Toluene | . . . . . . 32 | Isooctane | . . . . 3 |
| Benzene | . . . . . . 20 | Water | . . . . . Insoluble |
| Ethanol | . . . . . . 3.8 | 10% NaOH | . . . Insoluble |

Supplier: Ethyl Corporation.

Oxidation inhibitor in natural and synthetic rubbers, polyolefin plastics, resins, adhesives, petroleum oils, and waxes.

**Methylene chloride,. 1**, 676–677; **2**, 273.

*β-Tetralone synthesis.* Burckhalter and Campbell[1] prepared β-tetralones by mixing a phenylacetyl chloride and aluminum chloride in carbon disulfide and bubbling in ethylene for several hours at 0°. The resulting β-chloroethyl ketones need not be isolated but were converted into β-tetralones by treatment with hydrochloric acid.

Sims and co-workers[2] report that the procedure is markedly improved by use of methylene chloride rather than carbon disulfide; the yields are greatly improved and the reaction requires minutes rather than several hours. The ability of methylene chloride to dissolve the acid chloride–aluminum chloride complex and thus provide a homogeneous reaction medium may be responsible for the improved results.

[1] J. H. Burckhalter and J. R. Campbell, *J. Org.*, **26**, 4232 (1961).
[2] J. J. Sims, M. Cadogan, and L. H. Selman, *Tetrahedron Letters*, 951 (1971).

**Methylene chloride–*n*-Butyllithium.**

*Carbenoid synthesis of hydrocarbons.* Katz *et al.*[1,2] reported a simple synthesis of the previously rare valence isomer of benzene, benzvalene (3),[3] from cyclopentadiene (1). The diene is first converted into the cyclopentadienyl anion (2) by treatment with methyllithium in dimethyl ether; then a solution of methylene chloride and methyllithium in diethyl ether is added at −45°. Benzvalene (3) was obtained in 24% yield together with a trace (6.4%) of benzene.

*Caution:* Benzvalene is very explosive and has a vile smell.

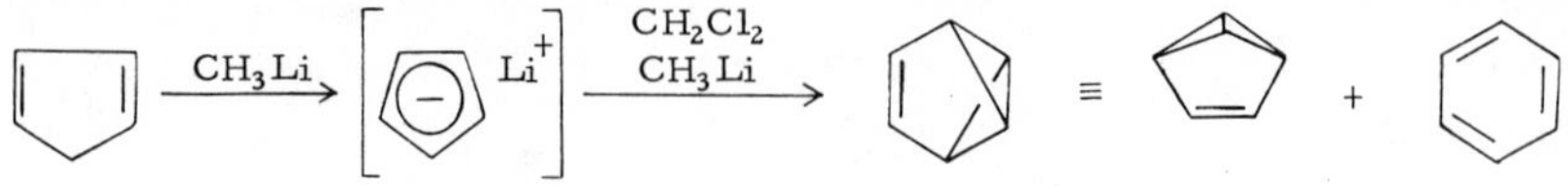

(1)                    (2)                                    (3)

The corresponding valence isomer of naphthalene, benzobenzvalene or naphthvalene (4), accompanied by naphthalene (5), was prepared in the same way, but in this case separation of (4) from (5) was not possible.

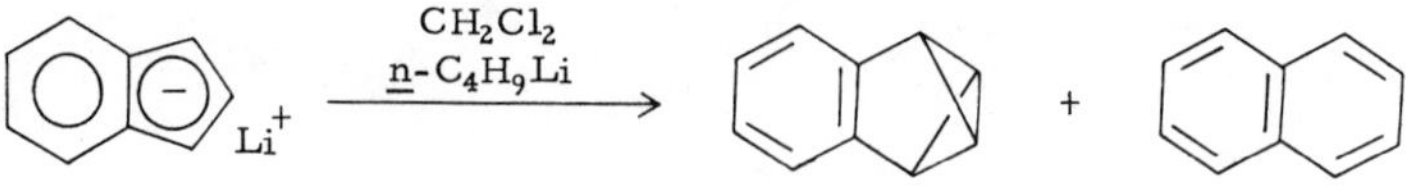

(4, 10%)                    (5, 18%)

Two laboratories[4,5] have reported the synthesis of isobullvalene (7, tricyclo-[5.3.0.0$^{2,10}$]deca-3,5,8-triene) in the same way from lithium cyclononatetraenide (6) by treatment with methylene chloride and *n*-butyllithium at a low temperature. The major product is isobullvalene (7), accompanied by the [10]annulene (8) and *cis*-9,10-

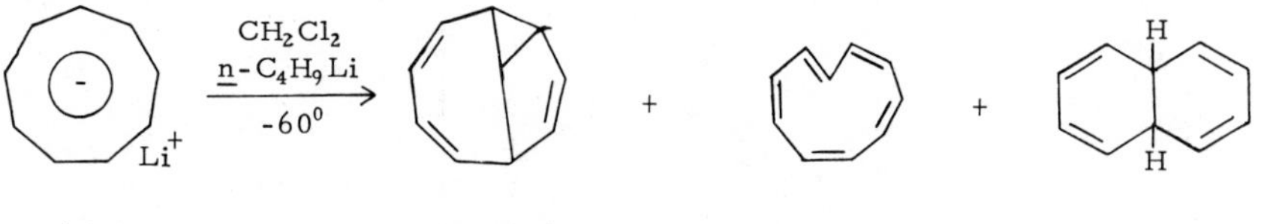

(6)                    (7, 55%)         (8, 14%)         (9, 14%)

(10)

dihydronaphthalene (9). The annulene (8) isomerizes at about $-20°$ to *trans*-9,10–dihydronaphthalene (10).

Two laboratories[5,6] simultaneously reported the synthesis of the valence isomer of pleiadiene (14), namely naphtho[1,8]tricyclo[4.1.0.0$^{2,1}$]heptene (13), by the reaction of the lithium anion of phenalene (11) with methylene chloride and *n*-butyllithium at low temperatures. A mixture of (13) and (14) is obtained in the ratio of 4:1. The new hydrocarbon (13, m.p. 76–78°) is surprisingly stable, but when heated at 150° in cyclohexane

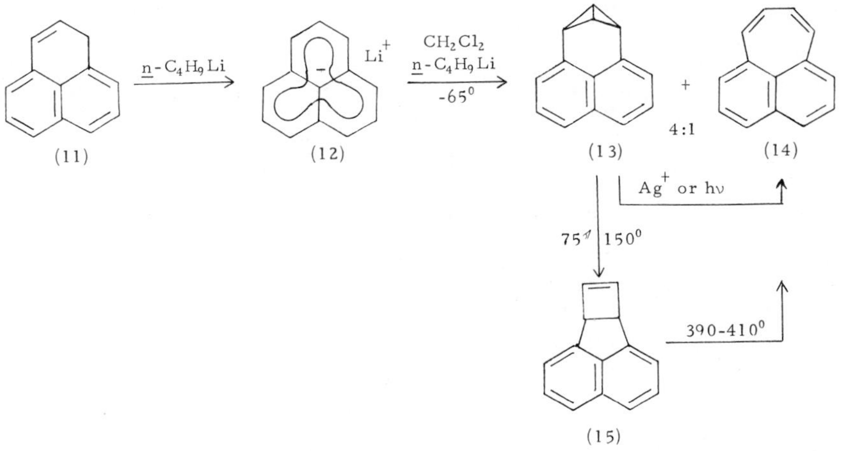

(11)    (12)    (13)    (14)

4:1

Ag⁺ or hν

75°  150°

390-410°

(15)

it yields the known naphthocyclobutene (15). It is isomerized by Ag(I) ion or by irradiation (2537 Å) to pleiadiene (14).

[1]T. J. Katz, E. J. Wang, and N. Acton, *Am. Soc.*, **93**, 3782 (1971).
[2]T. J. Katz, J. J. Cheung, and N. Acton, *ibid.*, **92**, 6643 (1970).
[3]This substance had previously been available in minute amounts by photolysis of benzene: K. E. Wilzbach, J. S. Ritscher, and L. Kaplan, *ibid.*, **89**, 1031 (1967); L. Kaplan and K. E. Wilzbach, *ibid.*, **90**, 3291 (1968); H. R. Ward and J. S. Wishnok, *ibid.*, **90**, 1085 (1968).
[4]K. Hojo, R. T. Seidner, and S. Masamune, *ibid.*, **92**, 6641 (1970).
[5]I. Murata and K. Nakasuji, *Tetrahedron Letters*, 47 (1973).
[6]R. M. Pagni and C. R. Watson, *ibid.*, 59 (1973).

**Methyl fluorosulfonate, 3**, 202. Suppliers: Aldrich (Magic Methyl), WBL.

*Alkylation.*[1] The ketone (1) undergoes base-catalyzed O-alkylation by the reagent to give (2; 7-methoxybicyclo[4.2.2]deca-2,4,7,9-tetraene) in 93–95% yield. Thus treatment of (1) with potassium *t*-butoxide (3 eq., 4 min.) and then with methyl fluorosulfonate (3 eq., 3 min.) in HMPT at 5° gives (2) in high yield. Significant C-alkylation is observed in less polar environments.

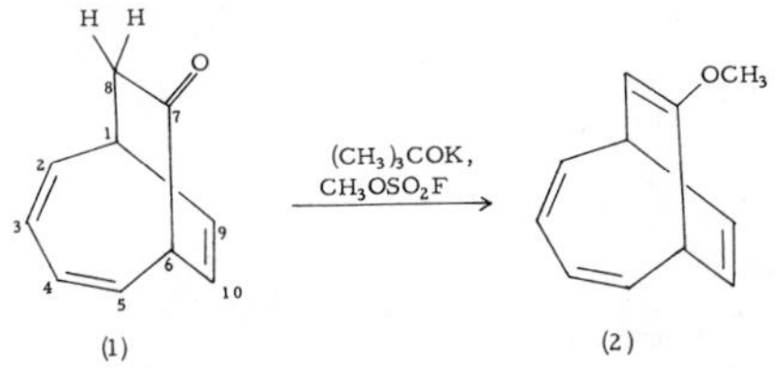

(1)    (2)

*Dethioacetalization.*[2] Dithioacetals (1) undergo facile methylation when treated with methyl fluorosulfonate in anhydrous benzene to give sulfonium salts (2). These are hydrolytically cleaved by sodium hydroxide to give the parent ketone (3). Overall yields are 60–90%.

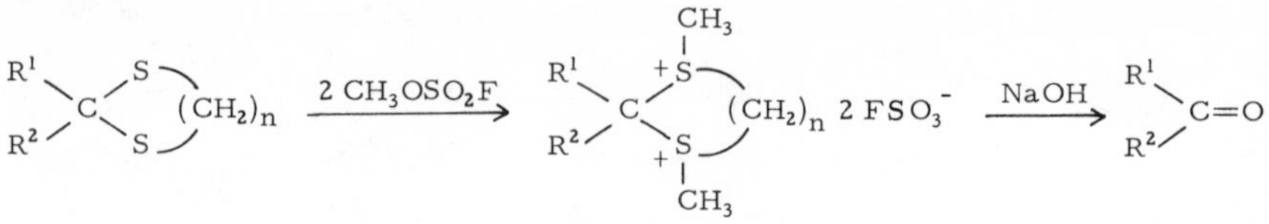

$$R^1 \underset{R^2}{\overset{}{>}}C\underset{S}{\overset{S}{<}}(CH_2)_n \quad \xrightarrow{2\ CH_3OSO_2F} \quad R^1 \underset{R^2}{\overset{}{>}}C\underset{\overset{+}{S}}{\overset{\overset{+}{S}}{<}}(CH_2)_n \ \ 2\ FSO_3^- \quad \xrightarrow{NaOH} \quad R^1 \underset{R^2}{\overset{}{>}}C=O$$

(with $CH_3$ groups on the sulfonium centers)

(1, n = 2 or 3)                    (2)                    (3)

*Methyl sulfonates.*[3] Phenols or phenol ethers when heated with the reagent at 100° for 12 hr. are converted into methyl sulfonates in moderate yield.

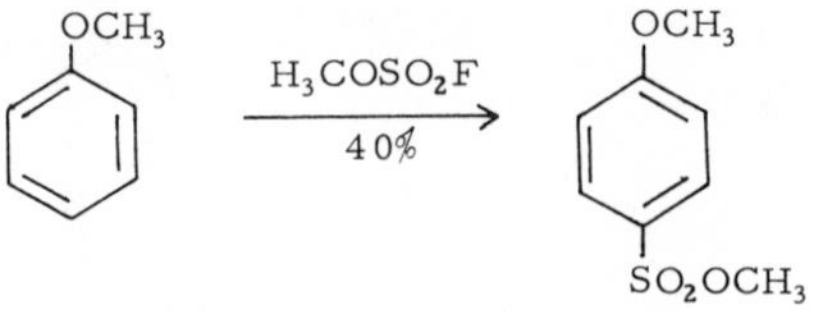

[1]J. B. Press and H. Shechter, *Tetrahedron Letters*, 2677 (1972).
[2]T.-L. Ho and C. M. Wong, *Synthesis*, 561 (1972).
[3]T. Kametani, K. Takahashi, and K. Ogasawara, *ibid.,* 473 (1972).

**Methylhydrazine,** $H_2NNHCH_3$. Mol. wt. 46.07, b.p. 87°, $n_D^{20}$ 1.4325. Suppliers: Aldrich, Columbia, Eastman, Fisher, Fluka, Howe and French, K and K, MCB, P and B, Sargent.

$\Delta^2$-**Pyrazoline-5-ones.**[1] Treatment of 2-bromo-1,3-diphenyl-1,3-propanedione (1) with methylhydrazine in ethanol at room temperature gives 1-methyl-3,4-diphenyl-$\Delta^2$-pyrazoline-5-one (2) in 56% yield. It is probably formed by a halohydrin rearrangement as indicated.

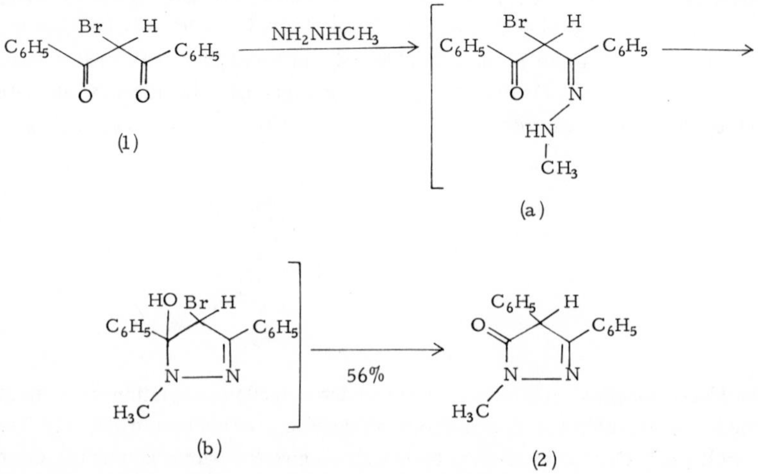

[1]M. J. Nye and W. P. Tang, *Canad. J. Chem.*, **51**, 338 (1973).

**Methyl iodide, 1,** 682–685; **2,** 274; **3,** 202.

*Thioacetals → carbonyl compounds.* Hydrolysis of thioacetals to the parent carbonyl compound often presents some difficulty. Fetizon and Jurion[1] have described two new methods for this conversion. In one method, a solution of the thioacetal in moist acetone is heated under reflux for several hours with excess methyl iodide. With

acid-sensitive compounds addition of a powdered carbonate is advisable for neutralization of the HI formed.

In the second method a solution of the thioacetal in liquid $SO_2$ is treated with excess methyl fluorosulfonate (**3,** 202; this volume).

[1]M. Fetizon and M. Jurion, *Chem. Commun.*, 382 (1972).

**Methyl isocyanate,** $CH_3N{=}C{=}O$. Mol. wt. 57.05, m.p. 37–39°. Supplier: Eastman.

*Conversion of aldoximes to nitriles.*[1] The dehydration of aldoximes to nitriles can be effected by reaction with methyl isocyanate in DMF in the presence of triethylamine to give an O-(methylcarbamoyl) aldoxime. The carbamate decomposes at 110–120° to give the nitrile and carbon dioxide and methylamine. Yields of nitriles are 65–99%.

[1]J. A. Albright and M. L. Alexander, *Org. Prep. Proc. Int.*, **4,** 215 (1972).

**Methyl methylthiomethyl sulfoxide,** $CH_3SOCH_2SCH_3$. Mol. wt. 125.23, b.p. 92–93°/2.5 mm. Supplier: Aldrich.

*Preparation.* The reagent is prepared in 78% yield by oxidation of formaldehyde dimethyl mercaptal with 30% hydrogen peroxide in acetic acid.[1]

*Aldehyde synthesis.*[2] The carbanion generated from this reagent (1) reacts with an alkyl halide to give an aldehyde dimethyl mercaptal S-oxide (2). This product is easily

hydrolyzed by a catalytic amount of sulfuric acid to give the corresponding aldehyde (3) and dimethyl disulfide. Yields of aldehyde are in the range 80–90%. An aldehyde that is sensitive to acid can be trapped as the acetal by treating (2) with sulfuric acid in the presence of triethyl orthoformate in ethanol.

*Phenylacetic acid derivatives.*[3] The reagent undergoes Knoevenagel-type condensation with benzaldehyde (Triton B) to give 1-methylsulfinyl-1-methylthio-2-phenylethylene (2). When (2) is treated with hydrochloric acid in ethanol at room temperature ethyl phenylacetate (3) is obtained in 78% yield.

$$C_6H_5CHO + CH_3SOCH_2SCH_3 \xrightarrow[\text{Triton B}]{\text{THF}} \underset{H}{\overset{C_6H_5}{>}}C=C\underset{SOCH_3}{\overset{SCH_3}{<}} \xrightarrow[78\%]{H^+, C_2H_5OH} C_6H_5CH_2COOC_2H_5$$

$$(1) \hspace{5cm} (2) \hspace{4cm} (3)$$

*α-Hydroxyaldehydes.*[4] The lithio derivative of the reagent reacts smoothly with a ketone (1) to afford an α-hydroxyaldehyde dimethyl mercaptal S-oxide (2), which on treatment with concentrated hydrochloric acid in THF at room temperature affords an α-hydroxyaldehyde (3). In the case of benzophenone ($R = R' = C_6H_5$) the product

$$\underset{R'}{\overset{R}{>}}C=O + CH_2\underset{SCH_3}{\overset{SOCH_3}{<}} \xrightarrow[\text{2) } H_2O]{\text{1) } \underline{n}\text{-BuLi}} R-\underset{R'}{\overset{OH}{\underset{|}{C}}}-CH\underset{SCH_3}{\overset{SOCH_3}{<}} \xrightarrow{H^+} R-\underset{R'}{\overset{OH}{\underset{|}{C}}}-CHO$$

$$(1) \hspace{5cm} (2) \hspace{4cm} (3)$$

(3, diphenylglycolic aldehyde) is obtained in 59% yield.

When $R' = H$ in (2) this procedure leads to isomerization. Thus phenylglycolic aldehyde dimethyl mercaptal S-oxide (2a) is converted into ω-hydroxyacetophenone (4). In this case a derivative of the desired α-hydroxyaldehyde can be obtained either by

$$C_6H_5\underset{H}{\overset{OH}{\underset{|}{\overset{|}{C}}}}-CH\underset{SCH_3}{\overset{SOCH_3}{<}} \xrightarrow{H^+} C_6H_5COCH_2OH$$

$$(2a) \hspace{5cm} (4)$$

protection of the hydroxyl group of (2a) as an ether or by protection of the newly formed formyl group as a dialkyl acetal.

[1] K. Ogura and G.-I. Tsuchihashi, *Bull. Chem. Soc. Japan*, **45**, 2203 (1972).
[2] *Idem, Tetrahedron Letters*, 3151 (1971).
[3] *Idem, ibid.*, 1383 (1972).
[4] *Idem, ibid.*, 2681 (1972).

## O-4-Methylphenyl chlorothioformate, $p$-$CH_3C_6H_4OC(S)Cl$. Mol. wt. 176.66, b.p. 52–53°/0.1 torr. Supplier: Lancaster.

The reagent (1) is prepared from thiophosgene in chloroform and $p$-cresol in aqueous sodium hydroxide.[1]

$$\underline{p}\text{-}CH_3C_6H_4OH + CSCl_2 \xrightarrow{\text{NaOH}} \underline{p}\text{-}CH_3C_6H_4OC(S)Cl$$

$$(1)$$

*Synthesis of olefins from sterically hindered alcohols.* Gerlach *et al.*[1] have reported a successful variation of the Chugaev reaction applicable to the preparation of temperature-sensitive olefins from sterically hindered alcohols. Thus *trans,trans*-spiro[4.4]nonane-1,6-diol (2) is converted into the thiocarbonate O-ester (3) by reaction with (1) in pyridine for 4 hr. at room temperature (80% yield). This derivative decomposes smoothly into the diene, 1,6-spiro[4.4]nonadiene (4), at a temperature of $\geqslant 135°$. In this case pyrolysis of the diacetate or of the di-S-methylxanthate failed to give the diene.

OH / OH (2) $\xrightarrow[80\%]{(1)}$ OC(S)OC$_6$H$_4$CH$_3$ / OC(S)OC$_6$H$_4$CH$_3$ (3) $\xrightarrow[68\%]{\geqslant 135°}$ (4) $+$ 2 CH$_3$C$_6$H$_4$OH $+$ 2 COS

[1]H. Gerlach, T. T. Huong, and W. Müller, *J.C.S. Chem. Comm.*, 1215 (1972).

### 3-Methyl-1-phenyl-3-phospholane oxide, 1, 695; 2, 279. Supplier: Eastman.
### Methyl N-sulfonylurethane triethylamine complex,

$$CH_3O_2C\bar{N}SO_2\overset{+}{N}(C_2H_5)_3\cdot(C_2H_5)_3\overset{+}{N}H\bar{C}l \quad (1).$$

Mol. wt. 173.59, m.p. 70–78° dec.

*Preparation.*[1,2] The reagent is prepared by the reaction of chlorosulfonyl isocyanate with methanol in benzene solution. Triethylamine is then added to the resulting N-carbomethoxysulfamoyl chloride.

$$ClSO_2NCO + CH_3OH \xrightarrow[88-92\%]{C_6H_6} CH_3O_2CNHSO_2Cl \xrightarrow[84-87\%]{2\ N(C_2H_5)_3} (1)$$

*Urethanes.*[1,2] The reagent (1) reacts with primary alcohols, for example 1-hexanol (2), to form the salt (3), which when heated at 95° for 1 hr. and then treated with water is converted into methyl-N-hexylcarbamate (4) in 55% yield. Benzyl alcohol is converted in the same way into methyl-N-benzylcarbamate (80% yield).

$$\underline{n}\text{-}C_6H_{13}OH + (1) \longrightarrow C_6H_{13}OSO_2\bar{N}CO_2CH_3\cdot\overset{+}{H}N(C_2H_5)_3 \xrightarrow[55\%]{\overset{\Delta}{H_2O}} \underline{n}\text{-}C_6H_{13}NHCOOCH_3$$

$$(2) \qquad\qquad (3) \qquad\qquad (4)$$

*Dehydration.*[1] Tertiary and secondary alcohols when treated with (1) are dehydrated to olefins. For example, 1,2-diphenylethanol (2) reacts with (1) in benzene or DMF to form 1,2-diphenylethyl-N-carbomethoxysulfamate triethylammonium salt (3); the salt

$$C_6H_5CH_2\underset{\underset{OH}{|}}{C}HC_6H_5 + (1) \longrightarrow C_6H_5CH_2\underset{\underset{OSO_2\bar{N}COOCH_3}{|}}{C}HC_6H_5 \xrightarrow[96\%]{\overset{50°}{H_2O}} \underset{H}{\overset{C_6H_5}{>}}C=C\underset{C_6H_5}{\overset{H}{<}}$$

$$(2) \qquad\qquad \overset{+}{H}N(C_2H_5)_3 \qquad\qquad\qquad (4)$$

$$(3)$$

is heated at 50° for 30 min., and then treated with water. *trans*-Stilbene (4) is obtained in 96% yield. The dehydration was shown to be a *cis* stereospecific elimination by use of deuterium-labeled derivatives of (2).

Other examples:

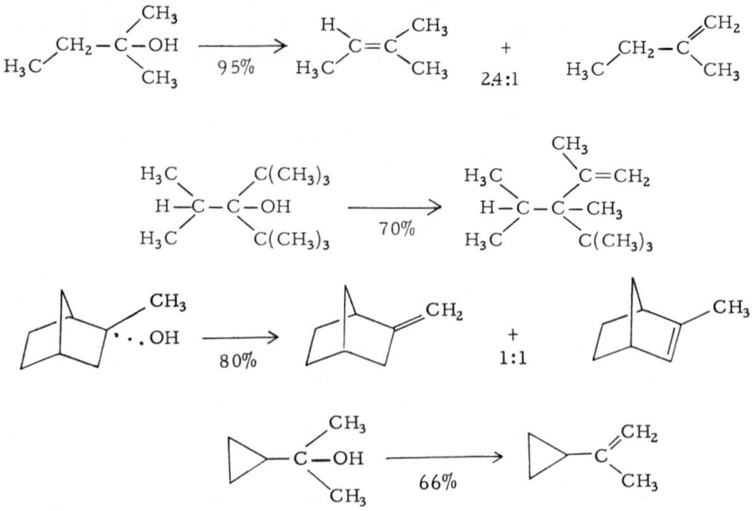

[1] E. M. Burgess, H. R. Penton, Jr., E. A. Taylor, and W. M. Williams, *J. Org.*, **38**, 26 (1973).
[2] *Idem, Org. Syn.*, submitted 1973.

**2-Methyl-3-thiazoline,** 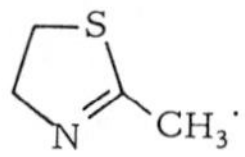 Mol. wt. 101.17, $n_D^{20}$ 1.5200. Suppliers: Aldrich,

Columbia Organic Chemicals.

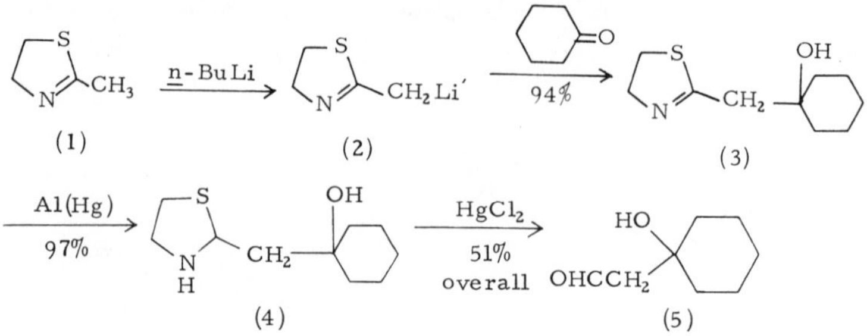

*Synthesis of aldehydes.* Meyers et al.[1] have described a method for synthesis of aldehydes similar to one using 2,4-dimethylthiazole (this volume). It is illustrated here for the synthesis of 2-(1-hydroxy)cyclohexylacetaldehyde (5). The thiazoline (1) is treated in THF under nitrogen at −78° with a slight excess of *n*-butyllithium. The resulting lithiothiazoline is then treated with cyclohexanone first at −78° and then at room temperature to give the elaborated thiazoline (3), which is reduced with aluminum amalgam to the thiazolidine (4). The crude product is then cleaved to the $\beta$-hydroxy-aldehyde (5) by treatment with mercuric chloride in aqueous acetonitrile at room temperature.

Note that the intermediate thiazolines (3) can be treated with a second electrophile. The main advantage of this route is that neutral conditions are used throughout.

[1]A. I. Meyers, R. Munavu, and J. Durandetta, *Tetrahedron Letters*, 3929 (1972).

**1-Methyl-3-*p*-tolyltriazene**, **1**, 696–697; **2**, 282. Suppliers: Aldrich, WBL.

**N-Methyl-N-tosylpyrrolidinium perchlorate,** ⟨structure⟩ $ClO_4^-$  Mol. wt. 326.79, m.p. 148–150°.

The reagent is prepared in 67.7% yield by the reaction of N-methylpyrrolidine with tosyl chloride in the presence of silver perchlorate in the dark.[1]

*Selective tosylation.*[1] The reagent selectively tosylates an amino group in the presence of a hydroxyl group:

$$\text{⟨structure⟩} \; ClO_4^- \; + \; {>}NH \xrightarrow{CH_2Cl_2} {>}NTs \; + \; \text{⟨structure⟩} \; ClO_4^-$$

[1]T. Oishi, K. Kamata, S. Kosuda, and Y. Ban, *Chem. Commun.*, 1148 (1972).

**R( − )-Mevalonolactone, 1**, 809–810.

*Isolation.* Tschesche *et al.*[1] found that mevaloside is the glucoside of R( − )-mevalonolactone and can be isolated in 1.75% yield. By enzymic hydrolysis the lactone is obtainable with ease from medlar leaves in yield of 0.63%.

[1]R. Tschesche, K. Struckmeyer, and G. Wulff, *Ber.*, **104**, 3567 (1971).

**Molecular sieves, 1**, 703–705; **2**, 286–287; **3**, 206.

*Ketimines and enamines.* Ketimines and enamines are prepared conveniently by the reaction of ketones with amines in the presence of molecular sieves (Linde 5A was used). Yields are 40–88%.[1]

*Acetals from aldehydes or ketones and secondary alcohols.* The synthesis of ketone di-*sec*-alkyl acetals ordinarily proceeds in low yield. Roelofsen and van Bekkum[2] report that such acetals can be prepared in 70–90% yield by use of type 5A molecular sieves to absorb selectively the water formed. *p*-Toluenesulfonic acid is used as the acid catalyst.

$$\text{>}C{=}O \; + \; 2 \; \underline{i}\text{-}C_3H_7OH \xrightleftharpoons{H^+} \; \text{>}C\begin{smallmatrix}OC_3H_7\text{-}\underline{i}\\OC_3H_7\text{-}\underline{i}\end{smallmatrix} \; + \; H_2O$$

[1]K. Taguchi and F. H. Westheimer, *J. Org.*, **36**, 1570 (1971).
[2]D. P. Roelofsen and H. van Bekkum, *Synthesis*, 419 (1972).

**Molybdenum hexacarbonyl, 2, 287; 3, 206–207.** Additional supplier: Climax Molybdenum Company.

*Reaction of ethers and acid halides.*[1] Acyclic ethers, when refluxed neat or in hexane or isooctane solution with acid chlorides in the presence of molybdenum hexacarbonyl, give esters, organic chlorides, and, in some instances, alkenes.

$$ROR + R'COCl \xrightarrow{Mo(CO)_6} RCl + R'COOR$$

Example:

$$\underline{n}\text{-}C_4H_9OC_4H_9\text{-}\underline{n} + C_6H_5COCl \longrightarrow C_6H_5COOC_4H_9\text{-}\underline{n}\,(73\%) + \underline{n}\text{-}C_4H_9Cl$$

Cyclic ethers under the same conditions give halo esters and/or elimination products:

$$+ CH_3COCl \xrightarrow[78\%]{Mo(CO)_6} CH_3COOCH_2CH_2CH_2CH_2Cl$$

$$+ CH_3COCl \xrightarrow[61\%]{Mo(CO)_6} CH_3COOCH_2CH=CHCH_2Cl$$

Of the various group VI metal carbonyls the catalytic order of effectiveness is $Mo(CO)_6 > W(CO)_6 > Cr(CO)_6$.

Complete retention of configuration is observed in the reaction of $3\beta$-ethoxy-$\Delta^5$-cholestene with acetyl chloride; $3\beta$-chloro-$\Delta^5$-cholestene is formed in high yield, accompanied by a small amount of $\Delta^{3,5}$-cholestadiene.

*Epoxidation of olefins* (**2**, 287). The procedure for epoxidation of olefins with *t*-butyl hydroperoxide catalyzed by molybdenum hexacarbonyl has been published.[2]

Kinetic data have been obtained from the reaction.[3] The mechanism is believed to involve 1) reversible complex formation between the catalyst and the hydroperoxide, 2) reversible inhibition by the coproduct alcohol, and 3) reaction of the hydroperoxide–molybdenum complex with the olefin to form the epoxide and by-product alcohol.

[1]H. Alper and C.-C. Huang, *J. Org.*, **38**, 64 (1973).
[2]M. N. Sheng and J. G. Zajacek, *ibid.*, **35**, 1839 (1970).
[3]T. N. Baker, III, G. J. Mains, M. N. Sheng, and J. C. Zajacek, *ibid.*, **38**, 1145 (1973).

**Monochloroborane–Ethyl etherate**, $BH_2Cl \cdot O(C_2H_5)_2$.

*Preparation.*[1] The reagent is prepared in ether solution by the reaction of lithium borohydride with boron trichloride.

$$LiBH_4 + BCl_3 + 2\,(C_2H_5)_2O \xrightarrow{0^0} 2\,BH_2Cl\text{--}O(C_2H_5)_2 + LiCl$$

*Hydroboration.* Monochloroborane in ethyl ether hydroborates a wide variety of olefins to give the corresponding dialkylchloroboranes in high yield:

$$2 \quad \text{(cyclopentene)} + BH_2Cl \xrightarrow[0^0,\ 1\ hr.]{(C_2H_5)_2O} \left(\text{(cyclopentyl)}\right)_2\!\!-BCl$$

In contrast, monochloroborane in THF gives a mixture of products: $R_3B$, $R_2BCl$, and $RBCl_2$.[2] Hydroboration–oxidation of alkenes with this reagent gives the anti-Markownikoff alcohols in greater than 99.5% isomeric purity. Thus norbornene is converted into *exo*-2-norbornanol in greater than 99.8% yield with this reagent. Styrene is converted into 2-phenylethanol (96% yield) and 1-phenylethanol (4%). The reagent thus exhibits a greater directive effect than diborane; it is almost comparable to disiamylborane (**1**, 57–59; **2**, 29; **3**, 22–23).

[1] H. C. Brown and N. Ravindran, *J. Org.*, **38**, 182 (1973).
[2] *Idem*, *Am. Soc.*, **94**, 2112 (1972).

# N

**Naphthalene–Lithium (Lithium naphthalenide), 2**, 288–289; **3**, 208.

*Reductive bisalkylation.* Treatment of (1) with 2 eq. of naphthalene–lithium (lithium naphthalenide) at −78° in THF affords dimethyl 1,4,6-cyclooctatriene-1,2-dicarboxylate (2) in 60% yield.[1]

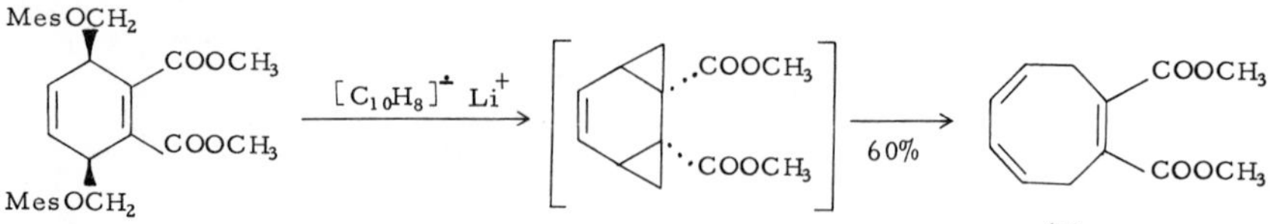

*Diisoprene dianion.*[2] When isoprene is added to a solution of naphthalene–lithium in THF, diisoprene dianion (1) is obtained. Addition of HMPT as cosolvent and then

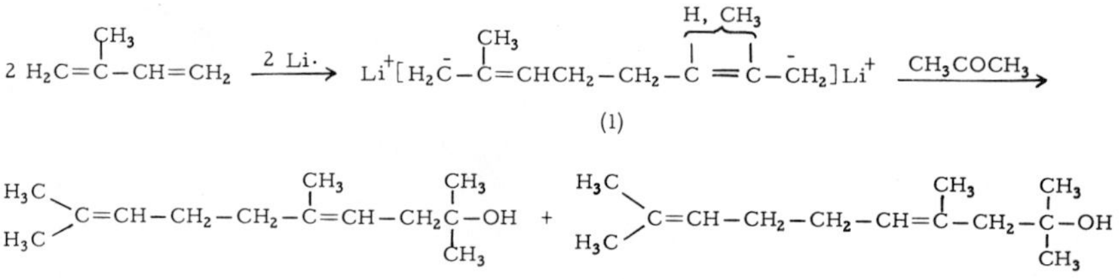

of acetone gives a mixture of monohydric alcohols (2) and (3). Use of HMPT as cosolvent is essential for the reaction.

*Reaction of α,β-unsaturated acids with ketones.*[3] Lithium naphthalenide in the presence of diethylamine reacts with α,β-unsaturated acids, for example crotonic acid (1), to give the α-anion of lithium crotonate. This anion reacts with ketones, for example cyclohexanone (2), to give δ-hydroxy acids, for example δ(1'-hydroxycyclohex-1'-yl)-crotonic acid (3).

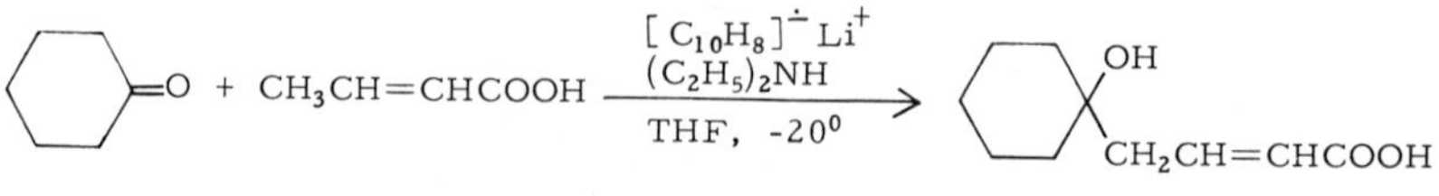

*Reaction with tetrahydrofurane.* Naphthalene–lithium (1) reacts with tetrahydrofurane at 65° to give the alcohols (2) and (3).[4]

(1)

(2, 92%)

(3, 8%)

*Reaction with alkyl and aryl halides.*[5] Alkyl and aryl halides are converted into organolithium reagents by reaction with naphthalene–lithium:

$$RX + 2C_{10}H_8 \overset{-}{} Li^+ \longrightarrow RLi + LiX + 2C_{10}H_8$$

Yields vary from 19 to 90% as determined by carbonation to give carboxylic acids. Naphthalene is separated from the products by sublimation or steam distillation.

[1] H. W. Whitlock, Jr., and P. F. Schatz, *Am. Soc.*, **93**, 3837 (1971).
[2] K. Suga and S. Watanabe, *Synthesis*, 91 (1971).
[3] S. Watanabe, K. Suga, T. Fujita, and K. Fujiyoshi, *Chem. Ind.*, 80 (1972); K. Suga, S. Watanabe, and T. Fujita, *Australian J. Chem.*, **25**, 2393 (1972).
[4] T. Fujita, K. Suga, and S. Watanabe, *Synthesis*, 630 (1972).
[5] C. G. Screttas, *Chem. Commun.*, 752 (1972).

**Naphthalene–Sodium (Sodium naphthalenide)**, **1**, 711–712; **2**, 289.

*Wurtz reaction.*[1] Bicyclo[1.1.1]pentane (2) can be prepared in 6.5% yield by the reaction of 3-bromomethylcyclobutyl bromide (1) with naphthalene–sodium in glyme under helium.[2]

(1)                    (2)

*Olefins from* vic-*dimethanesulfonates.*[3] Treatment of dimesylates of *vic*-diols with naphthalene–sodium or anthracene–sodium in THF or DME permits a rapid conversion to the corresponding alkene in high yield. The reaction is generally carried out by slow addition of a 0.3 $M$ solution of the anion radical to a degassed, stirred solution of the dimesylate until the intense color of the anion radical persists. The excess reagent is quenched with air or water, and the alkene isolated by usual techniques. The reaction is nonstereospecific, *cis*- and *trans*-alkenes being formed in close to the equilibrium ratio. Trimesitylborane–sodium (**3**, 309) can also be used, but in one case at least the yield of alkene was less than that obtained with an aromatic anion radical. The mesyl groups cannot be replaced by tosyl or brosyl groups; with these two latter groups, diols and some of the related epoxides are obtained.

**Alkylation of amines by conjugated olefins.**[4] Addition of isoprene to a solution of sodium naphthalenide in THF and diethylamine gives diethylisopentenylamine (1) in 70% yield.

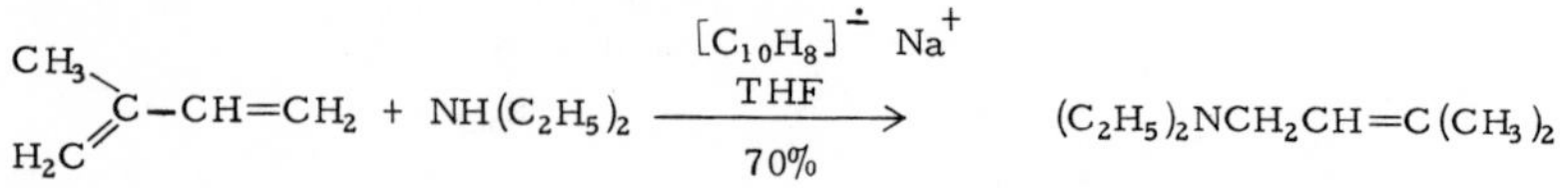

$$\underset{H_2C}{\overset{CH_3}{>}}C-CH=CH_2 \;+\; NH(C_2H_5)_2 \xrightarrow[70\%]{\underset{THF}{[C_{10}H_8]^{\underline{\cdot}}\;Na^+}} (C_2H_5)_2NCH_2CH=C(CH_3)_2$$

$$(1)$$

The yield is lower if lithium naphthalenide is used.[4]
Other examples:

$$C_6H_5CH=CH_2 \;+\; NH(C_2H_5)_2 \xrightarrow[95\%]{} C_6H_5CH_2CH_2N(C_2H_5)_2$$

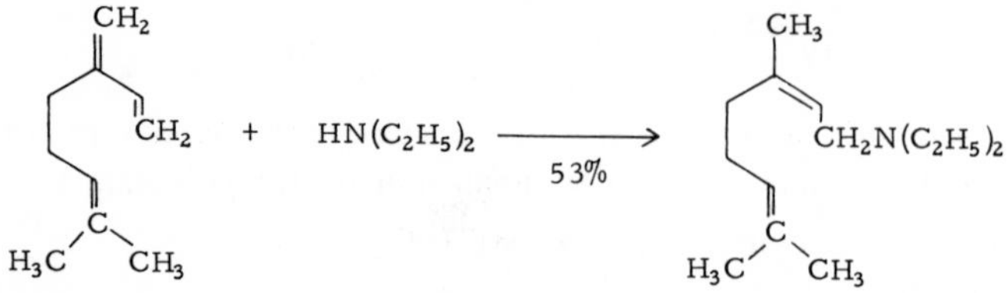

Myrcene

**Reaction with phenylacetonitrile.**[5] Sodium naphthalenide reacts with phenylaceto-nitrile to give products of electron or proton transfer, depending on the solvent. Electron transfer is particularly favored in THF–tetraglyme.

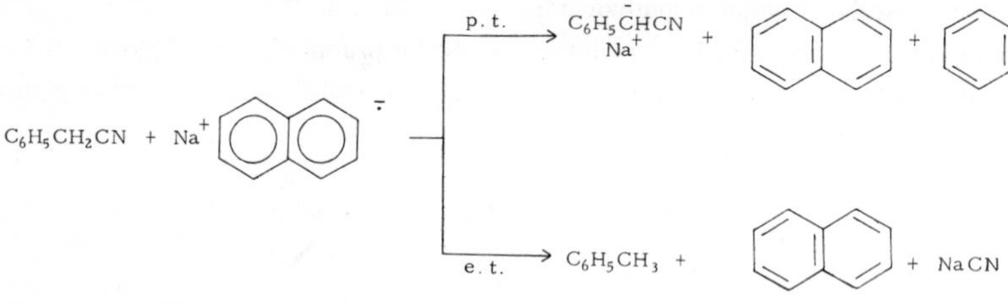

[1]G. M. Kosolapoff, *Org. React.*, **6**, 326 (1951).
[2]K. B. Wiberg and V. Z. Williams, Jr., *J. Org.*, **35**, 369 (1970).
[3]J. C. Carnahan, Jr. and W. D. Closson, *Tetrahedron Letters*, 3447 (1972).
[4]T. Fujita, K. Suga, and S. Watanabe, *Chem. Ind.*, 231 (1973).
[5]S. Bank and S. P. Thomas, *Tetrahedron Letters*, 305 (1973).

**Nickel–aluminum alloy (1:1 Raney nickel alloy)**, **1**, 718–720; **2**, 289. Additional supplier: British Drug Houses Ltd.

**Stephen reduction of an aromatic nitrile to the aldehyde.**[1] A 2-l. two-necked, round-bottomed flask fitted with a mechanical stirrer and a reflux condenser is charged with 40 g. of p-cyanobenzenesulfonamide, 600 ml. of 75% (v/v) formic acid, and 40 g. of

Raney nickel alloy. The stirred mixture is heated under reflux for 1 hr. The mixture is filtered with suction through a Büchner funnel coated with a filter aid such as Hyflo

Supercel, and the residue is washed with two 160-ml. portions of 95% ethanol. The combined filtrates are evaporated under reduced pressure with a rotary evaporator. The solid residue is heated in 400 ml. of boiling water and freed from a small amount of insoluble matter by decantation through a plug of glass wool placed in a filter funnel. The filtrate is chilled in an ice bath and the precipitate is collected by suction filtration, washed, and dried under vacuum.

The crude product weighs about 32 g., m.p. 112–114°. Recrystallization from ethanol with use of Norit affords 25.6–28.0 g. (62.9–68.8% yield) of *p*-formylbenzenesulfon-amide, m.p. 117–118°.

[1]T. van Es and B. Staskun, *Org. Syn.*, **51**, 20 (1971).

**Nickel boride**, $Ni_2B$, **1**, 720; **2**, 289–290; **3**, 208–210.

*Hydrogenation catalyst.* Russell and Hoy[1] have described a nickel boride catalyst which is useful for selective reduction of C=C bonds without hydrogenolysis of hydroxylic substituents or hydrogenation of carbonyl or epoxide groups. The black colloidal catalyst is prepared by reduction of nickel acetate in ethanol with 1.0 *M* sodium borohydride solution.

The catalyst is also recommended for hydrogenation of C=C bonds in substrates containing nitrogen functions. Amine and amide groups are not affected. Nitrile groups, however, are also reduced to primary amines.[2]

[1]T. W. Russell and R. C. Hoy, *J. Org.*, **36**, 2018 (1971).
[2]T. W. Russell, R. C. Hoy, and J. E. Cornelius, *ibid.*, **37**, 3552 (1972).

**Nickel bromide**, $NiBr_2$. Mol. wt. 218.52. Suppliers: Alfa Inorganics, ROC/RIC.

*Reaction of 1,3-dienes with active hydrogen compounds.* The catalyst resulting from the combination of nickel bromide and a dialkoxyphenylphosphine effects condensation of 1,3-dienes with active hydrogen compounds to give butenyl and octadienyl adducts.[1] Thus the reaction of morpholine (1, 0.05 mole) with butadiene (0.15 mole) in the presence of nickel bromide (1 mmole) and diisopropoxyphenylphosphine (1.1 mmole) at 100° for 30 min. gives a 79% conversion to the adducts (2), (3), (4), and (5).

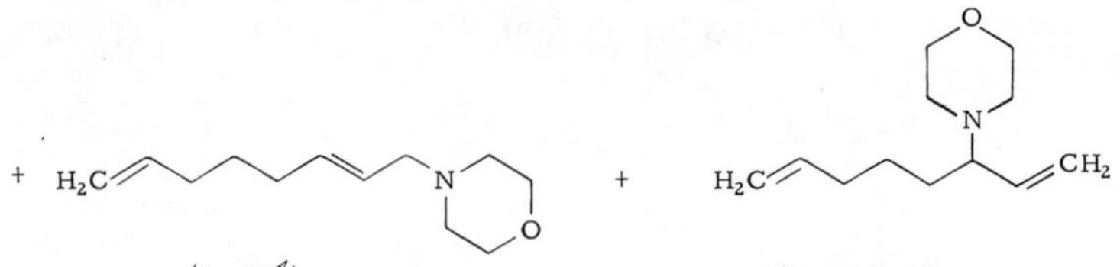

(4, 33%)                    (5, 22%)

Such condensation had been effected previously, but in lower yields, by a system comprising nickel(II) acetylacetonate, $Ni(C_5H_7O_2)_2$, diisopropoxyphenylphosphine, and sodium borohydride.[2] The borohydride in this case is essential to effect good conversions. The order of reactivity of nickel salts in the butadiene–amine–dialkoxyphenylphosphine system is

$$NiBr_2 \simeq NiCl_2 > Ni(OAc)_2 \simeq Ni(laurate)_2 \gg Ni(acac)_2$$

The order of reactivity of the dialkoxyphenylphosphines is

$$C_6H_5P[OCH(CH_3)_2]_2 > C_6H_5P(OC_2H_5)_2 > C_6H_5P(OCH_3)_2$$

Nickel bromide also is more effective than nickel(II) acetylacetonate in the condensation of benzyl methyl ketone with 1,3-butadiene. The reaction of this active methylene compound with 1,3-butadiene with the $NiBr_2$ system with sodium phenoxide

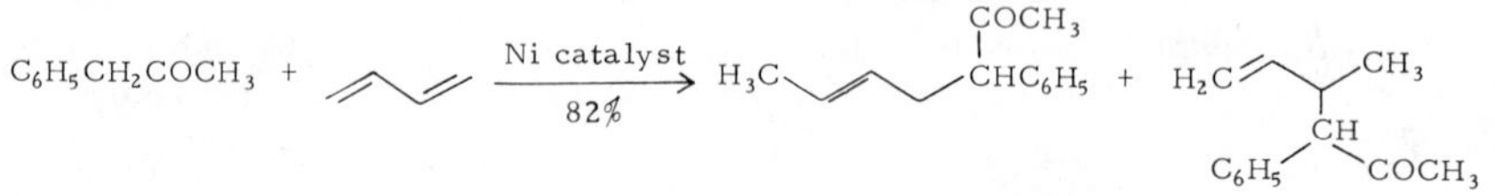

(6, 13%)                    (7, 8%)

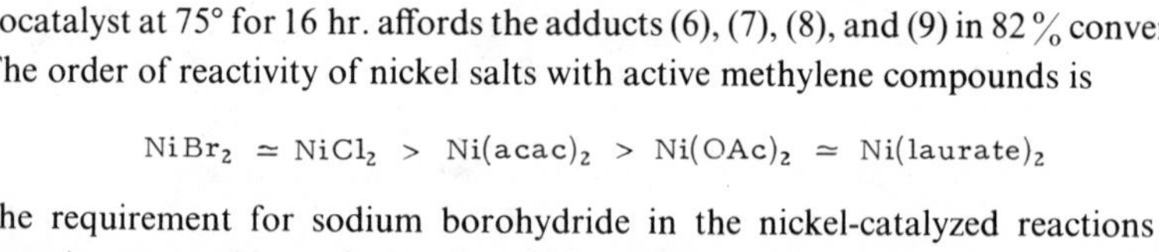

(8, 74%)                    (9, 5%)

as cocatalyst at 75° for 16 hr. affords the adducts (6), (7), (8), and (9) in 82% conversion. The order of reactivity of nickel salts with active methylene compounds is

$$NiBr_2 \simeq NiCl_2 > Ni(acac)_2 > Ni(OAc)_2 \simeq Ni(laurate)_2$$

The requirement for sodium borohydride in the nickel-catalyzed reactions with amines is presumably owing to its ability to prevent complex formation between $Ni(acac)_2$ and the amine. Even in cases where the borohydride is not essential, its addition results in some enhancement of the rate. The role of the phosphine is to reduce Ni(II) to Ni(0).

[1] R. Baker, A. H. Cook, and T. N. Smith, *Tetrahedron Letters*, 503 (1973).
[2] T. C. Shields and W. E. Walker, *Chem. Commun.*, 193 (1971); R. Baker, D. E. Halliday, and T. N. Smith, *ibid.*, 1583 (1971); *Idem. J. Organometal. Chem.*, **35**, C61 (1972).

**Nickel carbonyl, 1**, 720–723; **2**, 290–293; **3**, 210–212. Additional supplier: Matheson Gas Products.

*Caution:*   Nickel carbonyl is very toxic and should be handled with utmost care.[1]

**Review.**  The formation of carbon–carbon bonds via π-allylnickel compounds has been reviewed by Semmelhack.[2]

**Reaction of aryl halides with π-allynickel halides**: Methallylbenzene.[3] The yield of methallylbenzene is 67–72%.

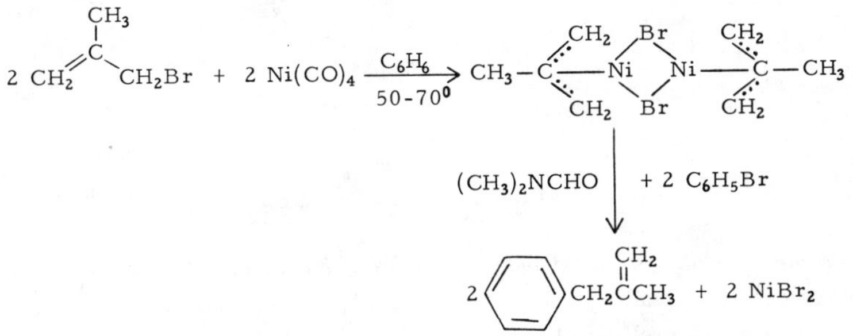

**Diaryl or dialkyl ketones.**[4] Organomercuric halides react with nickel carbonyl in DMF or THF at 60–70° (20 hr.) to give ketones in high yield. Thus phenylmercuric chloride or bromide is converted into benzophenone in 92–95% yield:

$$2\ C_6H_5HgCl(Br) + Ni(CO)_4 \xrightarrow[\text{92-95\%}]{\text{DMF}} C_6H_5\underset{O}{\overset{\parallel}{C}}C_6H_5 + 2\ Hg + NiCl_2(Br_2) + 3\ CO$$

Unsymmetrical diaryl ketones are obtained by the reaction of iodobenzene with nickel carbonyl in the presence of an organomercury halide:

$$C_6H_5I + Ni(CO)_4 \xrightarrow{-CO} [C_6H_5\underset{O\ I}{\overset{\parallel}{C}}Ni(CO)_2] \xrightarrow{ArHgCl} C_6H_5\underset{O}{\overset{\parallel}{C}}Ar$$

**Reaction with cyclooctyne.**[5] Conversion of cyclooctyne in ether gives a red nickel-containing complex (1), which decomposes at 300° to give the yellow ketone (2):

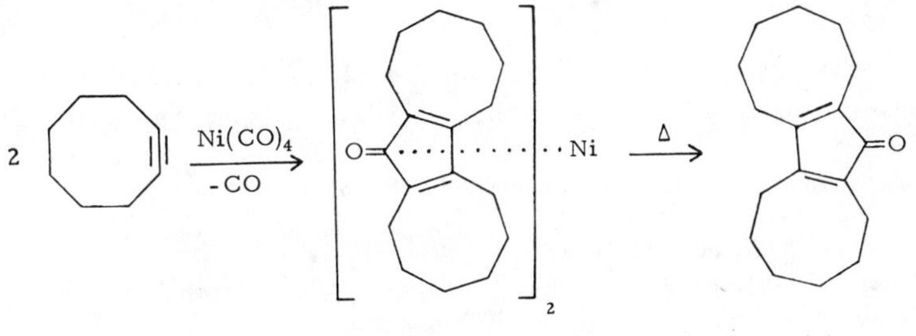

(1)                                (2)

Oxidation of the nickel carbonyl–cyclooctyne complex (3) with ceric ammonium nitrate (**1**, 120–121; **2**, 63–65; **3**, 44–45; this volume) followed by addition of tetraphenylcyclopentadienone (**1**, 1149–1150) affords (4) as a minor product. The major product is the trimeric cyclooctyne (5).

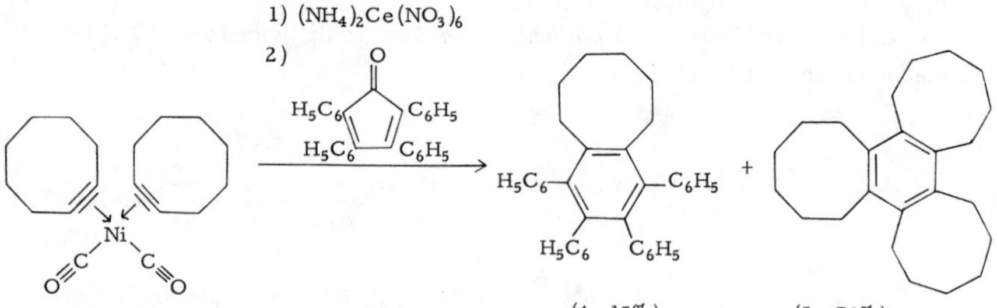

(3)    (4, 15%)    (5, 70%)

*Reaction of π-allylnickel bromide complexes with quinones.* Allylic bromides react with nickel carbonyl to afford π-allylnickel bromide complexes formulated as (1) in the case of allyl bromide itself (**2**, 291; **3**, 210). Hegedus *et al.*[6] have studied the reaction of

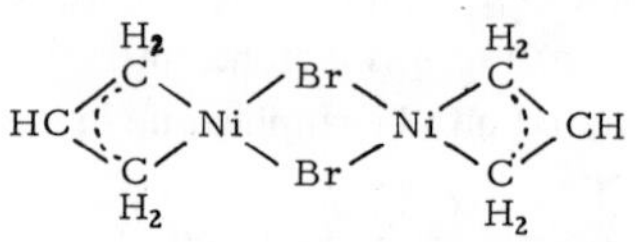

(1)

these complexes with quinones. *p*-Benzoquinone reacts with the complex of allyl bromide to give allylhydroquinone (2) in 58% yield. Other quinones, however, give

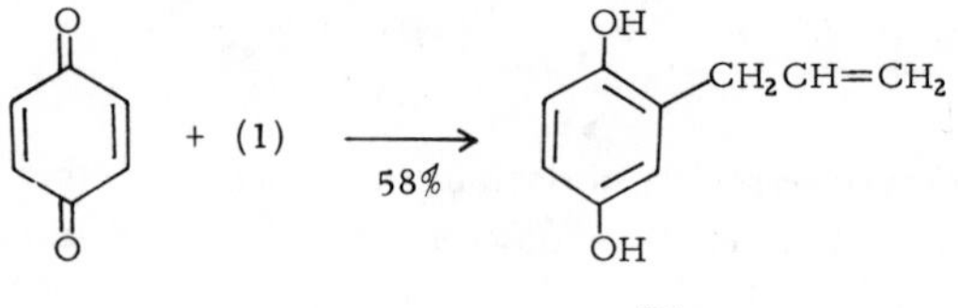

(2)

mainly allyl-substituted quinones. Thus the π-allyl complex derived from 2-methallyl bromide and nickel carbonyl reacts with 2-methylbenzoquinone to give (3) in 45% yield.

[1] "Nickel Carbonyl is Dangerous," Technical Bulletin, Matheson Co.
[2] M. F. Semmelhack, *Org. React.*, **19**, 115 (1972).
[3] M. F. Semmelhack and P. M. Helquist, *Org. Syn.*, **52**, 115 (1972).
[4] Y. Hirota, M. Ryang, and S. Tsutsumi, *Tetrahedron Letters*, 1531 (1971).
[5] G. Wittig and P. Fritze, *Ann.*, **712**, 79 (1968).
[6] L. S. Hegedus, E. L. Waterman, and J. Catlin, *Am. Soc.*, **94**, 7155 (1972).

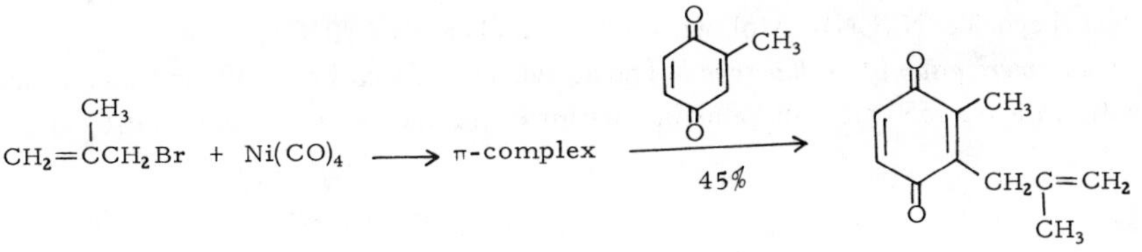

(3)

This new alkylation procedure has been used for a simple synthesis of plastoquinone-1 (4) in 61% yield from 2,3-dimethylbenzoquinone and 3-methyl-2-butenyl bromide. Co-

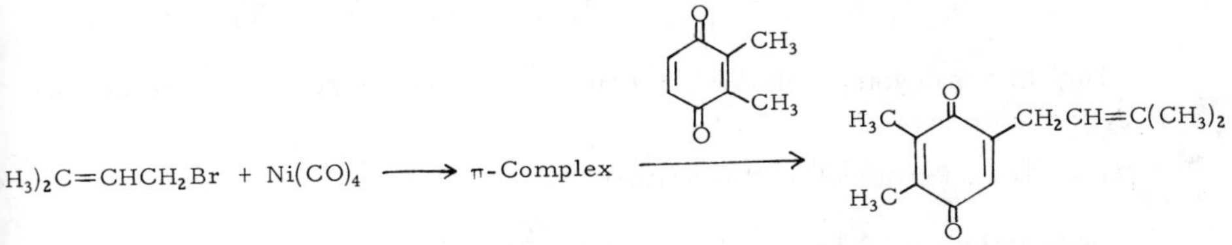

(4)

enzyme $Q_1$ (5) has been prepared in 40% yield from the reaction of 2,3-dimethoxy-5-methylbenzoquinone with the $\pi$-allyl complex of 3-methyl-2-butenyl bromide.

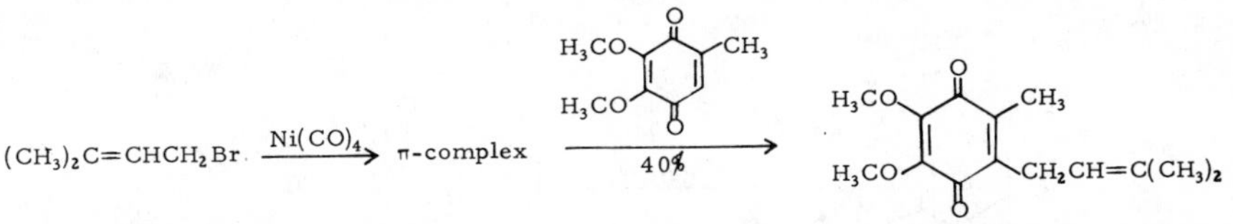

(5)

The major by-product of these syntheses is the hydroquinone of the starting quinone, which can be recovered and oxidized. Quinones with several unsubstituted positions are alkylated nonspecifically, but only products of monoalkylation are observed.

*Synthesis of macrolides.*[7] The intramolecular coupling reaction of allylic halides (**1**, 722–723; **2**, 290–292; **3**, 211) has now been shown to be applicable to the synthesis of macrocyclic lactones. Thus addition of the dibromo ester (1) to 6 eq. of nickel carbonyl in N-methylpyrrolidone effects cyclization to the macrolide (2) as the major product.

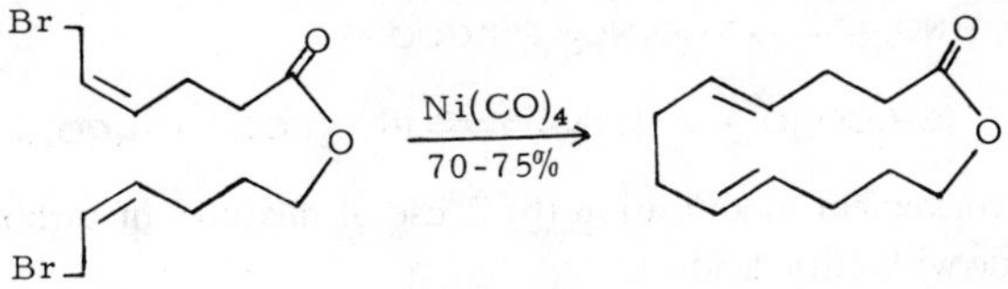

(1)                                 (2)

[7]E. J. Corey and H. A. Kirst, *ibid.*, **94**, 667 (1972).

**Nickel cyanide**, $Ni(CN)_2$. Mol. wt. 110.73. Supplier: ROC/RIC.

*Trimerization of cyclooctyne.*[1] The acetylenic hydrocarbon (540 mg.) was heated with catalyst (550 mg.) in refluxing tetrahydrofurane (40 ml.) under nitrogen for

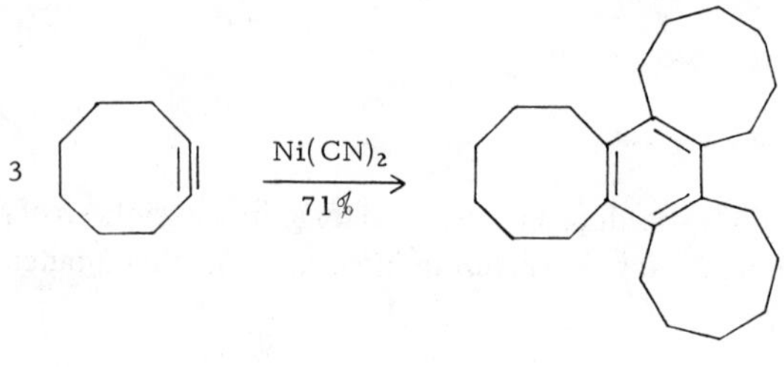

20 hr. Chromatography afforded 392 mg. of the pure trimer, octadecahydrotriscyclooctabenzene.

[1]G. Wittig and P. Fritze, *Ann.*, **712**, 79 (1968).

**Ninhydrin (Indane-1,2,3-trione)**. Mol. wt. 178.14, dec. 125°. Suppliers: Aldrich, Eastman, Fluka.

*Condensation with phosphate esters.*[1] The trione (1) reacts with trimethyl-, triethyl-, and triisopropylphosphite in anhydrous methylene chloride to give colorless 2:1 adducts regarded as cyclic, saturated pentahydroxyphosphoranes (2).

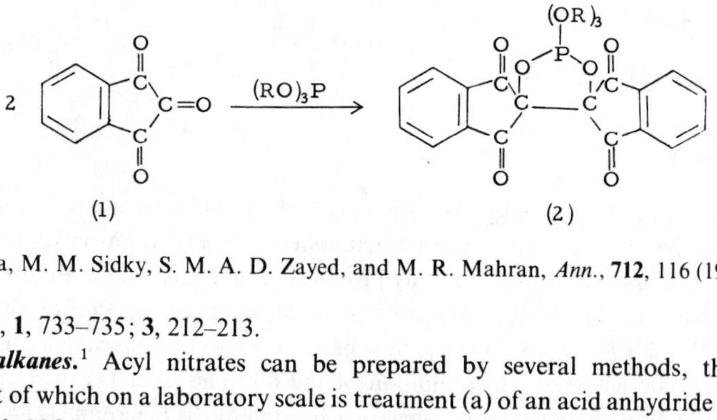

(1)                              (2)

[1]A. Mustafa, M. M. Sidky, S. M. A. D. Zayed, and M. R. Mahran, *Ann.*, **712**, 116 (1968).

**Nitric acid, 1**, 733–735; **3**, 212–213.

*Nitroalkanes.*[1] Acyl nitrates can be prepared by several methods, the most convenient of which on a laboratory scale is treatment (a) of an acid anhydride (tenfold excess) with 90% nitric acid (20°). The by-product acid can be reconverted into the

(a) $(RCO)_2O + HNO_3 \longrightarrow RCO_2NO_2 + RCOOH$

(b) $RCOOH + 2 (CH_3CO)_2O + 2 HNO_3 \rightleftharpoons RCO_2NO_2 + CH_3CO_2NO_2 + 3 CH_3COO\text{?}$

anhydride. A convenient modification (b) is use of mixtures of carboxylic acids and acetic anhydride with nitric acid.

Nitroalkanes can be obtained in satisfactory yields by heating the acyl nitrates at 270–300° (an inert solvent, acetonitrile or a nitroalkane, can be used). In a typical

$$RCO_2NO_2 \xrightarrow{270-300^0} RNO_2 + CO_2$$

experiment valeroyl nitrate affords 1-nitrobutane (56.5%), 1-butanol (20.0%), and *n*-butyl valerate (20.0%).

[1]G. B. Bachman and T. F. Bierman, *J. Org.*, **35**, 4229 (1970).

**2-Nitro-1-dimethylaminoethylene,** $(CH_3)_2NCH=CHNO_2$ (2).

*Nitrovinylation of aldehydes and ketones.*[1] Aldehydes and ketones containing a methylene group at the α-position react with 1-(dimethylamino)-2-nitroethylene (2) to give 2-*aci*-nitroethylidene derivatives (3). Ketones with two active methylene groups can react with 1 or with 2 eq. of (2). Treatment of derivatives of 4-*aci*-nitrocrotonaldehyde with nitroacetaldehyde produces the corresponding derivative of 1,3-dinitrobenzene.

$$\begin{array}{ccc} \underset{Y}{\overset{X}{>}}CH_2 + (CH_3)_2NCH=CHNO_2 & \xrightarrow[-(CH_3)_2NH]{C_2H_5OK} & \underset{Y}{\overset{X}{>}}C=CHCH=NO_2^-K^+ \\ (1) \qquad\qquad (2) & & (3) \end{array}$$

[1]R. Severin, P. Adhikary, E. Dehmel, and I. Eberhard, *Ber.,* **104**, 2856 (1971).

**Nitroethane,** $CH_3CH_2NO_2$. Mol. wt. 75.07, b.p. 112–114°. Suppliers: Aldrich, Baker, Columbia, Howe and French, MCB, Pfaltz and Bauer.

*Friedel–Crafts acylation.*[1] Substitution of nitroethane for the less volatile nitrobenzene in the Friedel–Crafts reaction of *p*-xylene with methylsuccinic anhydride

represents a major improvement; 3-(2,5-dimethylbenzoyl)-2-methylpropionic acid (3) can be obtained in 82% yield. This solvent also favors formation of (3) over (4).

[1]E. J. Eisenbraun, C. W. Hinman, J. M. Springer, J. W. Burnham, and T. S. Chou, *J. Org.*, **36**, 2480 (1971).

**Nitromethane,** **1**, 739–741.

*Spiro[4.4]nonatetraene* (5).[1] This elusive spiro compound has now been synthesized from the diacid chloride of diallylmalonic acid (1). This substance is treated in methylene chloride solution containing several mole equivalents of nitromethane with aluminum chloride at 25° to give, after hydrolysis with aqueous ammonium chloride, spiro[4.4]-nona-2,6-diene-1,5-dione (2) in 51% yield. Reduction of (2) with aluminum hydride

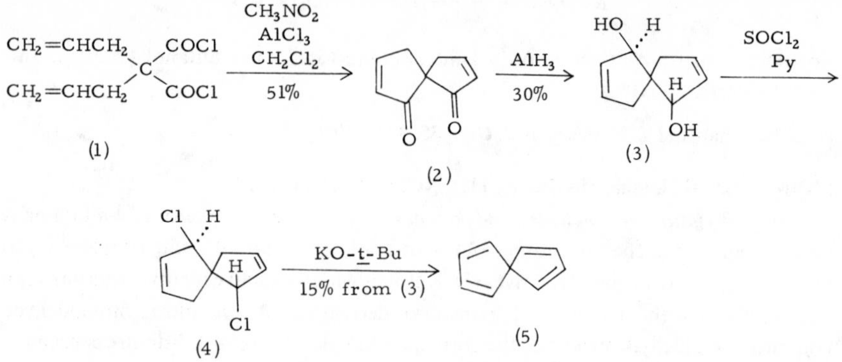

(1, 34–35; 2, 23–24; 3, 9–10) gave the diol (3), which was treated with thionyl chloride and pyridine in THF at 0° to give a mixture of dichlorides (4). Excess potassium *t*-butoxide was then added to a solution of (4) in tetraethylene glycol dimethyl ether. The solution was evaporated (0.1 torr) with a trap cooled to −196°. After a period of 1 hr., (5) was collected in the trap and isolated by glpc. The spiro compound rearranges at 65°

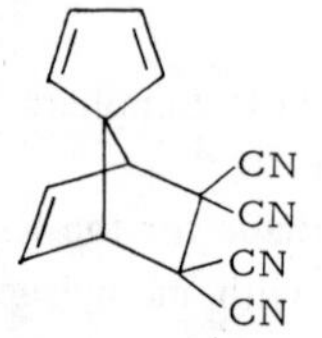

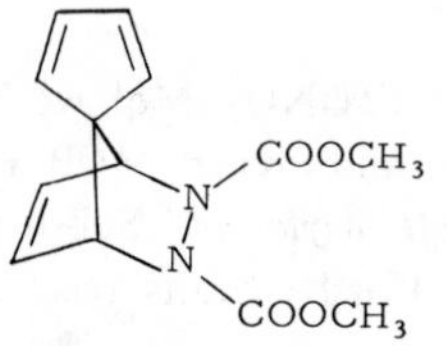

to indene (67 % yield) and high molecular weight products. It reacts rapidly with tetra-cyanoethylene and with dimethyl azodicarboxylate to give monoadducts (6) and (7), respectively.

[1] M. F. Semmelhack, J. S. Foos, and S. Katz, *Am. Soc.*, **94**, 8637 (1972).

**Nitronium tetrafluoroborate**, 1, 742. Supplier: Alfa Inorganics.

*Reaction with olefins.*[1] Olefins (primary, secondary, and tertiary) react with nitronium tetrafluoroborate in anhydrous acetonitrile at −15° to give products which on hydrolysis afford *vic*-nitroacetamide derivatives in 15–50% yield. The reaction is

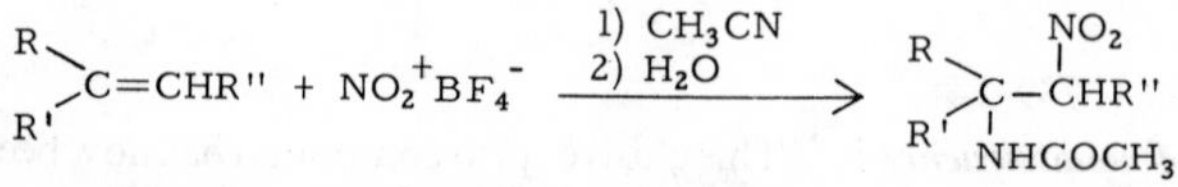

an extension of the Ritter reaction[2] for preparation of amines or amides by the reaction of olefins in an acidic medium with hydrogen cyanide or nitriles.

[1] M. L. Scheinbaum and M. Dines, *J. Org.*, **36**, 3641 (1971).
[2] J. J. Ritter and P. P. Minieri, *Am. Soc.*, **70**, 4045 (1948).

**4-(4-Nitrophenylazo)benzoyl chloride (NABS–Cl).**[1]  $O_2N\!-\!\langle\rangle\!-\!N\!=\!N\!-\!\langle\rangle\!-\!COCl.$

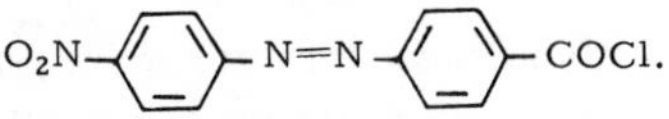

Mol. wt. 289.67, m.p. 163–165°. Suppliers: Aldrich, Schuchardt.

*Preparation:*

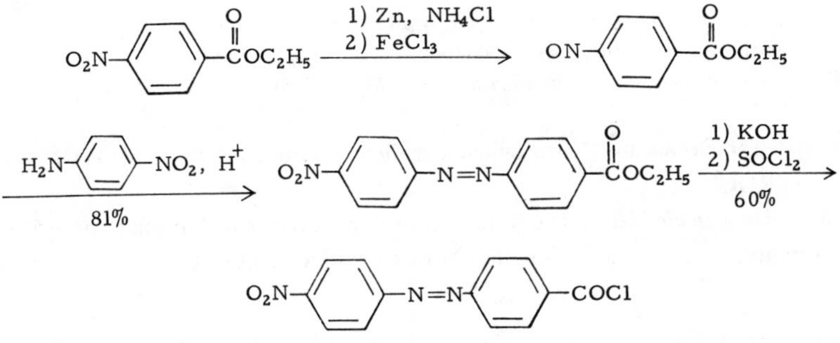

*Characterization of alcohols.*[2] NABS esters are generally solid, and the bright red-orange color makes them highly suitable for chromatographic purification. The molecular weight is easily determined from the ultraviolet absorption of the NABS chromophore.

[1] Abbreviation of German name.
[2] W. H. Nutting, R. A. Jewell, and H. Rapoport, *J. Org.*, **35**, 505 (1970).


*o*-**Nitrophenylsulfenyl chloride, 1, 745.**

*ω-Cyanoamino acids.* In a new synthesis of ω-cyanoamino acids, Chimiak and Pastuszak[1] found the *o*-nitrophenylsulfenyl group the most useful for protection of the amino group. Thus the NPS derivatives of L-asparagine (1, $n = 1$) and of L-glutamine (1, $n = 2$) were dehydrated with DCC to give (2). The protective group was removed by treatment with thioacetamide.[2] This reagent is available from Eastman.

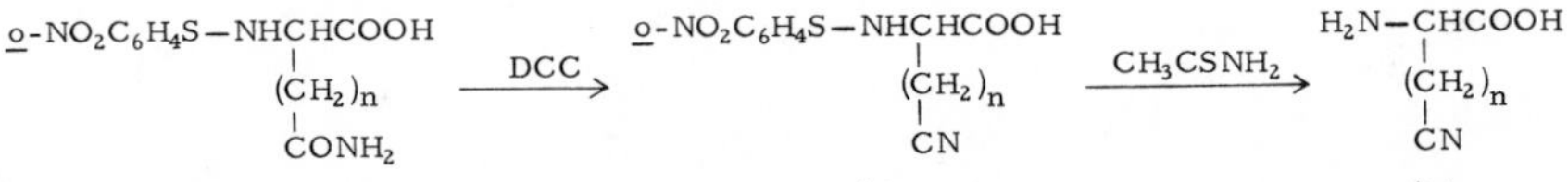

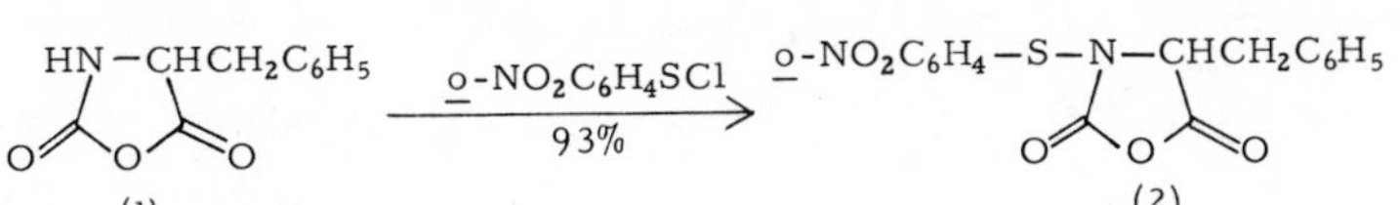

*N-Sulfenylamino acid N-carboxyanhydrides.*[3] The reagent reacts with an amino acid N-carboxyanhydride (NCA) to form *o*-nitrophenylsulfenyl (NPS) derivatives in high yield. Thus L-phenylalanine-NCA (1) reacts with the reagent in ethyl acetate or THF in the presence of triethylamine or 4-methylmorpholine to give NPS-L-phenyl-alanine-NCA (2) in 93% yield. The NPS–NCA amino acids are useful for peptide synthesis, since they are active esters and react without racemization. For example, the reaction of (2) with N-(trimethylsilyl)glycine trimethylsilyl ester in dioxane in the presence of ammonium sulfate ($CO_2$ is evolved) gives NPS-L-phenylalanylglycine in 94% yield.

[1]A. Chimiak and J. J. Pastuszak, *Chem. Ind.*, 427 (1971).
[2]W. Kessler and B. Iselin, *Helv.*, **49**, 1330 (1966).
[3]H. R. Kricheldorf, *Angew. Chem., internat. Ed.*, **12**, 73 (1973).

**Nitrosyl tetrafluoroborate (Nitrosonium tetrafluoroborate)**, $NOBF_4$. Mol. wt. 116.83. Supplier: ROC/RIC.

*Reaction with olefins.*[1]  Nitrosyl tetrafluoroborate reacts with primary or secondary olefins in acetonitrile to give 2-methyl-N-hydroxyimidazolium salts (1) at −15 to 0°.

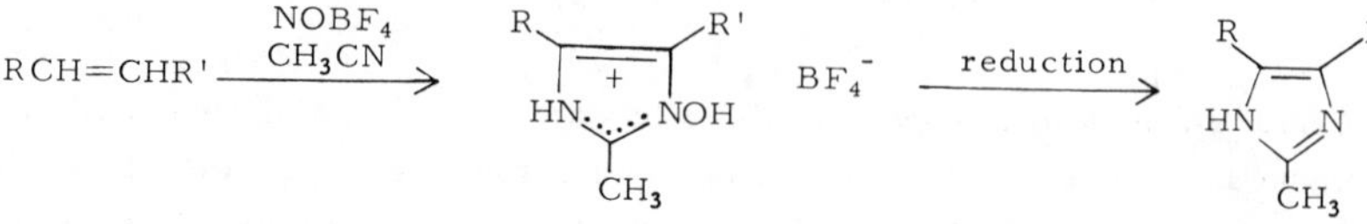

(1)                    (2)

These can be reduced by sodium bis-(2-methoxyethoxy)aluminum hydride (**3**, 260–261; this volume) to 2,4,5-trisubstituted imidazoles (3). Overall yields are in the range 50–80%. Benzonitrile can be substituted for acetonitrile.

[1]M. L. Scheinbaum and M. B. Dines, *Tetrahedron Letters*, 2205 (1971).

# O

**Osmium tetroxide–Potassium chlorate**, $KClO_3$. See **1**, 764; **2**, 301.

*Oxidation.* As one step in the synthesis of furaneol (3, a flavor principle of pine-apples and strawberries), Büchi et al.[1] wished to effect hydroxylation of 2,5-dimethyl-2,5-dimethoxy-2,5-dihydrofurane (1). They tried oxidation of (1 ; 15.8 mmole) with potassium chlorate (22.8 mmole) and osmium tetroxide (0.3 mmole) in aqueous tetrahydro-furane containing sodium bicarbonate at 30° for 63 hr.[2]; however, the expected diol

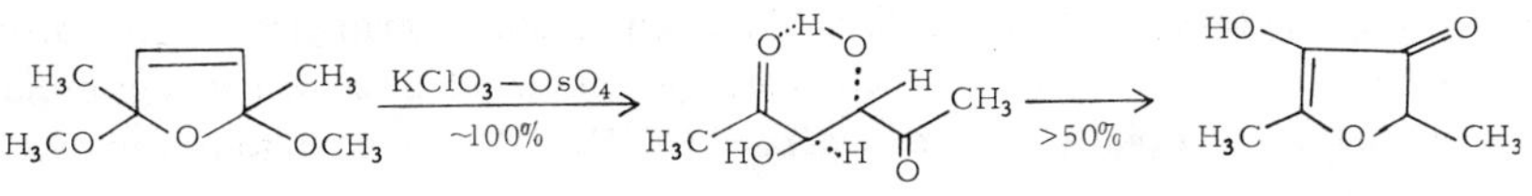

(1)                    (2)                    (3)

was obtained in only 10 % yield. Büchi reasoned that the low yield was due to hydrolysis. The oxidation was then performed in a more aqueous reaction medium without sodium bicarbonate. Under these conditions the dihydroxy ketone (2, *erythro* configuration) was obtained in nearly quantitative yield. Furaneol (3) was obtained in over 50 % yield by dehydration of (2) with either sodium bicarbonate or disodium hydrogen phosphate ($Na_2HPO_4$).

[1]G. Büchi, E. Demole, and A. F. Thomas, *J. Org.*, **38**, 123 (1973).
[2]H. Muxfeldt and G. Hardtmann, *Ann.*, **669**, 113 (1963).


**Oxalyl chloride**, **1**, 767–772; **2**, 301–302; **3**, 216–217.

*Isocyanatoquinones.* The first known member of this class, 2,5-dichloro-3,6-di-isocyanato-1,4-benzoquinone (2), was prepared by treating 2,5-diamino-3,6-dichloro-1,4-benzoquinone (1) with oxalyl chloride in butyl acetate and chlorobenzene.[1]

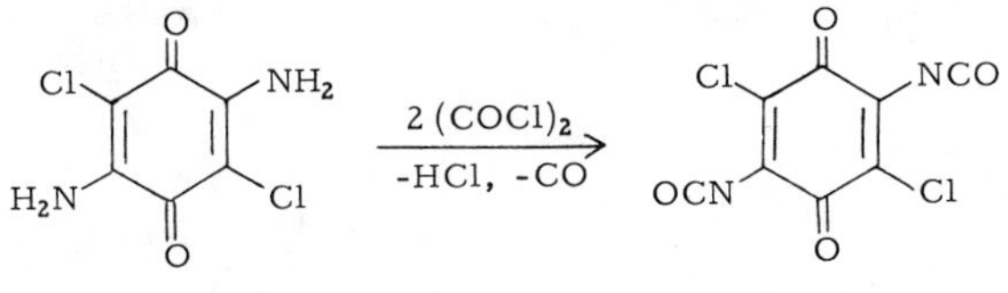

(1, 100 g.)                    (2, 50 g.)

[1]U. von Gizycki, *Angew. Chem., internat. Ed.*, **10**, 403 (1971).

**Oxygen, $O_2$.**

*Oxidation of trialkylboranes.*[1] Trialkylboranes react with oxygen[2] in THF to give alcohols in practically quantitative yield:

$$R_3B + 1\tfrac{1}{2}O_2 \longrightarrow [(RO)_3B] \xrightarrow{3 H_2O} 3 ROH + B(OH)_3$$

The reaction is rapid at first, but may require up to 75 min. for completion.

[1]H. C. Brown, M. M. Midland, and G. W. Kabalka, *Am. Soc.*, **93**, 1024 (1971).
[2]The oxygen is generated by the decomposition of hydrogen peroxide solution catalyzed by basic manganese dioxide in an automatic hydrogenator.

**Oxygen, singlet, $^1O_2$.**

*1,4-Cycloaddition.*[1] Photooxygenation of 1,1-diphenyl-2-methoxyethylene (1) in benzene in the presence of a sensitizer (dinaphthalenethiophene) yields as the major product the γ-lactone (5). The endoperoxide (2) is considered to be the first intermediate and indeed can be isolated and characterized (NMR, UV) if the photooxygenation is

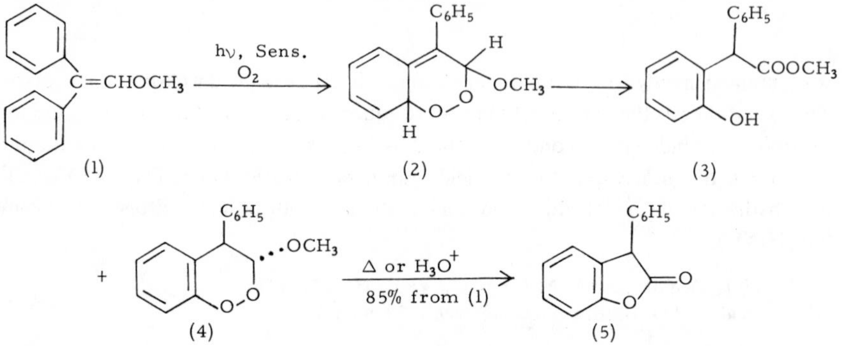

carried out at $-78°$. The methyl ester (3) and the peroxide (4) can be isolated and characterized. They are both converted into (5) by brief heating or treatment with acid. The reaction is thus an example of the Diels–Alder reaction with singlet oxygen.

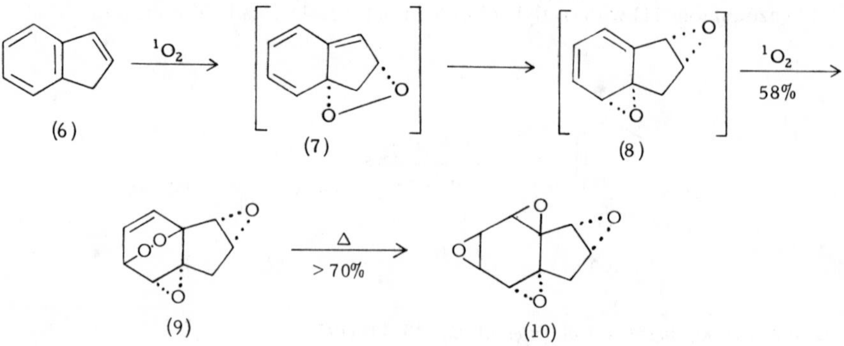

Photooxygenation of indene (6) at $-78°$ with a photosensitizer (Rose Bengal) gave as the major product (9), whose structure was assigned on the basis of IR and NMR spectra. Presumably the first step is a Diels–Alder addition to give the peroxide (7), which rearranges to the diepoxide (8). Addition of a second singlet oxygen molecule gives the observed product (9). When (9) is warmed it rearranges to two compounds, of which (10) is the major product. On the basis of the spectra it is assigned the tetra-epoxide structure (10).

*Reactions with steroidal monoolefins.*[2] Photooxygenation of $\Delta^2$-3-methyl-5$\alpha$-cholestene (1) with hematoporphyrin as sensitizer, followed by reduction of the initially formed hydroperoxides with methanolic sodium iodide, gives the two allylic alcohols (2) and (3). Similar oxidation of the isomeric $\Delta^2$-2-methyl-5$\alpha$-cholestene yields the

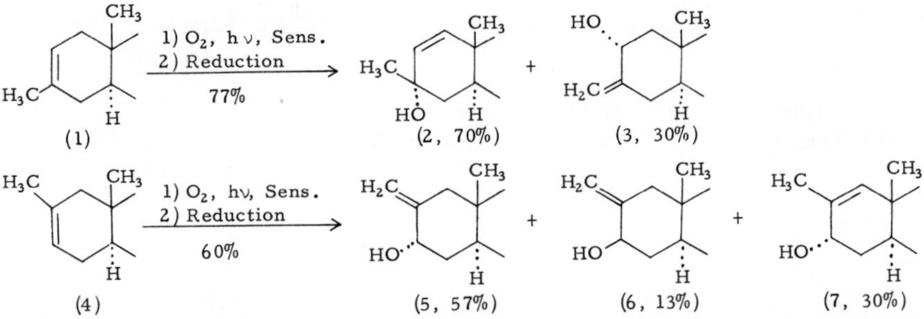

three allylic alcohols, (5)–(7). The preference for formation of $\alpha$-hydroperoxides is noteworthy. The results are best explained by a transition state that resembles the starting olefin more than it does the allylic hydroperoxide product.

[1]C. S. Foote, S. Mazur, P. A. Burns, and D. Lerdal, *Am. Soc.*, **95**, 586 (1973).
[2]A. Nickon, J. B. DiGiorgio, and P. J. L. Daniels, *J. Org.*, **38**, 533 (1973).

**Ozone, 1, 773–777.**

*Ozonization* of 2,6-bismethylenespiro[3.3]heptane (2a) and 2,10-bismethylene-trispiro[3.1.1.3.1.1]tridecane (4a) gives 2,6-dioxospiro[3.3]heptane (2b) and 2,10-dioxo-trispiro[3.1.1.3.1.1]tridecane (4b), respectively.[1]

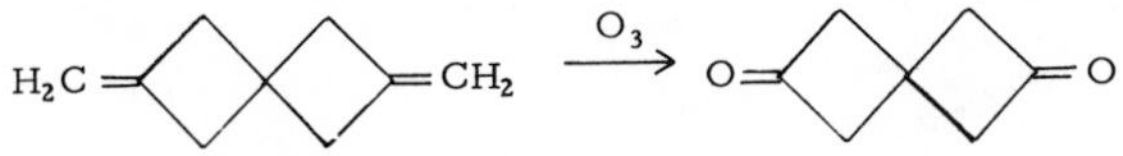

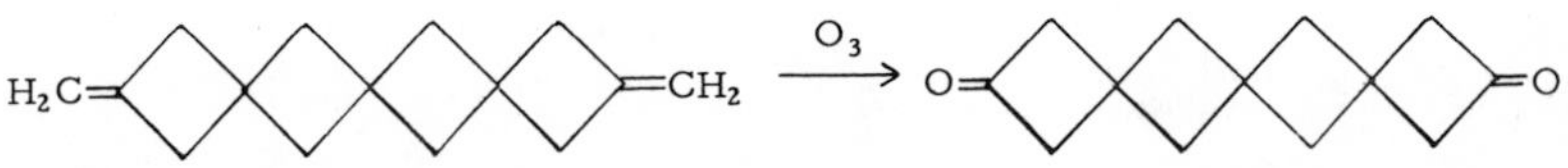

*Aldehydes → esters.* Ozone reacts in essentially quantitative yield with acetals to give the corresponding esters. Presumably the reaction proceeds by insertion of ozone into the C–H bond of the acetal.[2]

Examples:

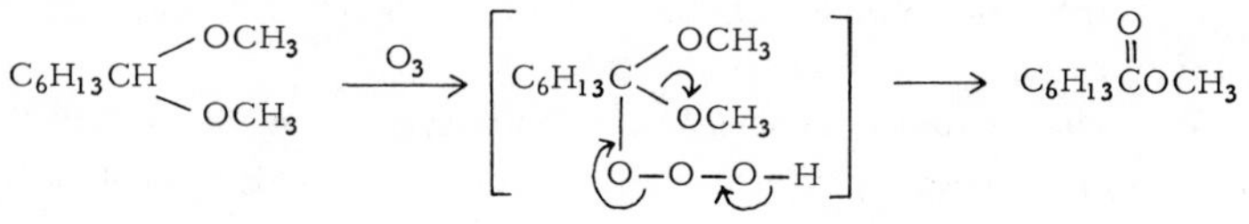

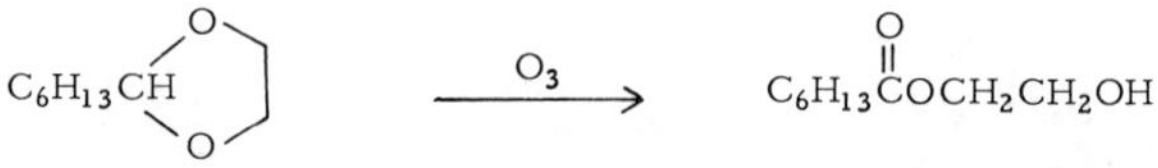

[1]E. Buchta and A. Kröniger, *Ann.*, **716**, 112 (1968).
[2]P. Deslongchamps and C. Moreau, *Canad. J. Chem.*, **49**, 2465 (1971).

# P

**Palladium acetate, 1**, 778; **2**, 303.

***π-Allylpalladium complexes.***[1] The enamine 1-piperidinocyclohexene (1) reacts with allyl phenoxide, palladium acetate, and triphenylphosphine in refluxing benzene to give, after hydrolysis, 2-allylcyclohexanone (2) and 2,6-diallylcyclohexanone (3), formed via a π-allylpalladium complex.

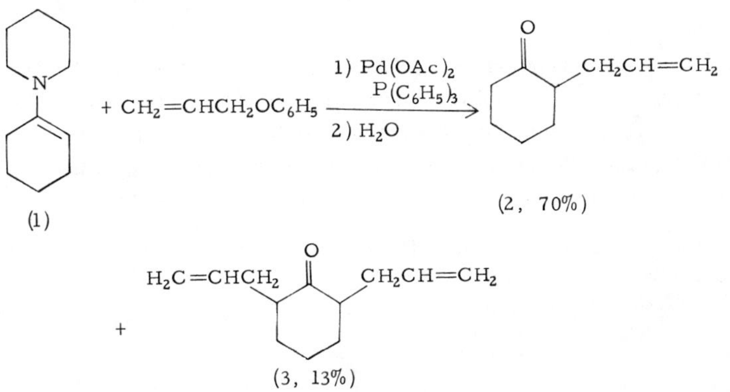

+ CH₂=CHCH₂OC₆H₅
1) Pd(OAc)₂
P(C₆H₅)₃
2) H₂O
CH₂CH=CH₂
(1)
(2, 70%)
+
H₂C=CHCH₂
CH₂CH=CH₂
(3, 13%)


Reaction of an enamine with this palladium catalyst with 1,4-diacetoxy-2-butene leads to a bicyclic ketone. Thus reaction of 1-piperidinocyclopentene (4) gives bicyclo-[4.2.1]non-3-ene-9-one (5) in 28% yield. This reaction involves formation of an intermediate allyl-substituted enamine, which can also form a π-allylpalladium complex.

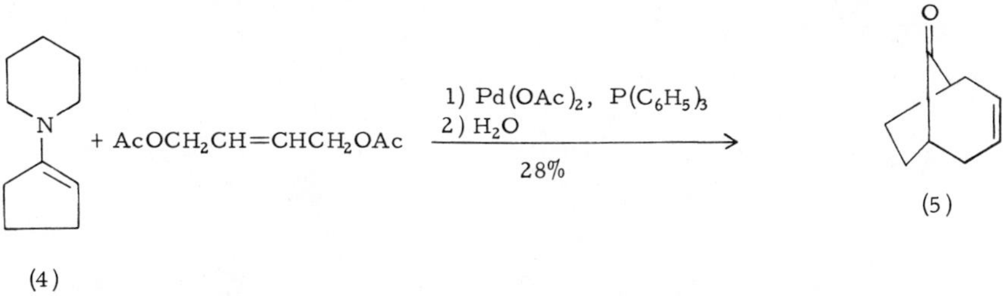

+ AcOCH₂CH=CHCH₂OAc
1) Pd(OAc)₂, P(C₆H₅)₃
2) H₂O
28%
(5)
(4)


[1] H. Onoue, I. Moritani, and S.-I. Murahashi, *Tetrahedron Letters*, 121 (1973).

**Palladium black, 1**, 778–782; **2**, 303.

***3β-Hydroxy-5α,8α-epidioxyandrostene-6-one-17*** (1) ***androstatriene-4,6,8,(14)-dione-3,17*** (3). Dodson *et al.*[1] reported in 1966 at an A.C.S. meeting that ergosterol epidioxide is converted in high yield into ergostatriene-4,6,8(14)-one-3 when stirred with a palladium catalyst in ethanol. The reaction has now been reported in detail in the androstane series.[2] Thus treatment of 3$\beta$-hydroxy-5$\alpha$,8$\alpha$-epidioxyandrostene-6-one-17 (1; 2.14 g.) in ethanol with palladium black (Engelhard) (10 g.) for 10 days affords

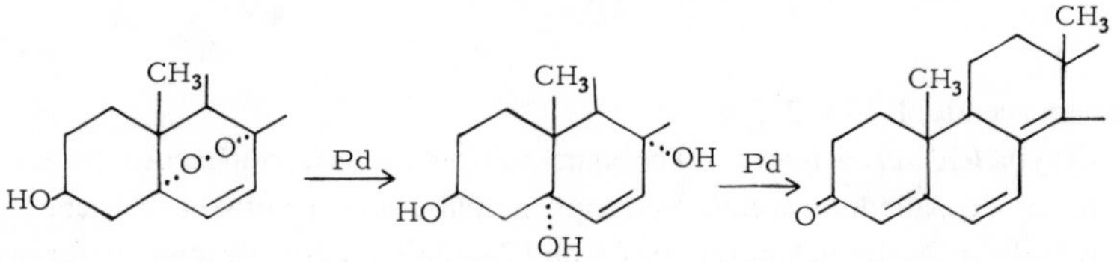

(1)                    (2)                    (3)

androstatriene-4,6,8(14)-dione-3,17 (3) in 60% yield (UV analysis). The trienedione is probably formed via the triol (2); indeed the triol can be obtained in 70% yield by treatment of (1) with palladium black (Fisher) in ethanol for 18 hr. at room temperature. Then (3) is presumably formed from (2) by catalyzed transfer of hydrogen to solvent resulting in oxidation of the 3-OH group followed by loss of water.

[1] R. M. Dodson, G. D. Valiaveedan, H. Ogasawara, and H. M. Tsuchiya, 151st National Meeting of the A.C.S., March, 1966, p. I-33.
[2] W. F. Johns, J. Org., **36**, 2391 (1971).

## 10% Palladium on calcium carbonate hydrogenation catalyst, 1, 778–782; 2, 303; 3, 218.

Hydrogenation of tridehydro[18]annulene (3) to [18]annulene (4) is carried out with a suspension of palladium catalyst in benzene in a hydrogenation apparatus. The yield of [18]annulene is 0.63% overall from 1,5-hexadiyne.[1]

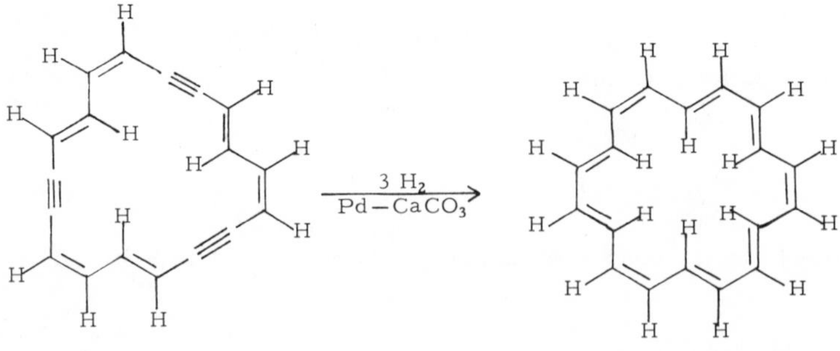

(3)                    (4)

[1] K. Stöckel and F. Sondheimer, Org. Syn., submitted 1971.

**Palladium-on-carbon**, **1**, 982; **2**, 303; **3**, 274.

*Dehydrogenation of hydroaromatic compounds.* Used in the last step of the synthesis of 3,4-benzpyrene from 1-oxo-1,2,3,4,5,6,7,8-octahydroanthracene[1]:

a: R = $CH_2CO_2H$
b: R = H

1) Zn-dust distn.
2) Pd/C

*Hydrogenation catalyst for Rosenmund reduction.*[2] Supplied by Engelhard Industries and dried in a vacuum oven at 115° for 48 hr. (becomes very pyrophoric). A pressure vessel, for example a 1.2-l. Hastelloy autoclave, is charged in order with 600 ml. of dry toluene, 25 g. (0.3 mole) of sodium acetate[3] (dried in a vacuum oven at 115° for 48 hr.), 3 g. of dry 10% palladium-on-carbon catalyst, 23 g. (0.1 mole) of 3,4,5-tri-methoxybenzoyl chloride (Aldrich), and 1 ml. of Quinoline S [*Org. Syn., Coll. Vol.*, **3**, 629 (1955)]. The pressure vessel is flushed with nitrogen, sealed, evacuated briefly, and

pressurized to 50 psi with hydrogen. The mixture is shaken with 50 psi of hydrogen for 1 hr. at room temperature, repressured if required, and heated at 35–40° for 2 hr. Agitation is continued overnight while the mixture cools to room temperature. The pressure is released and the vessel is opened. The mixture is filtered through 10 g. of Celite filter aid, and the insoluble material is washed with 25 ml. of toluene. The combined filtrates are washed with 25 ml. of 5% sodium carbonate solution and then with 25 ml. of water. The toluene solution is dried over anhydrous sodium sulfate and the filtered solution is concentrated at the reduced pressure of a water aspirator. Distillation of the residue through a 10-cm. Vigreux column with warm water circulating through the condenser to prevent crystallization of the distillate yields 12.8–16.2 g. (64–83%) of 3,4,5-trimethoxybenzaldehyde, b.p. 158–161°/7–8 mm., m.p. 74–75°.

[1] E. Buchta and R. Zöllner, *Ann.*, **716**, 102 (1968).
[2] A. I. Rachlin, H. Gurien, and D. P. Wagner, *Org. Syn.*, **51**, 8 (1971).
[3] This reagent serves as internal acceptor for hydrogen chloride.

**Palladium catalysts, 1**, 778–782; **2**, 303.

*Rosenmund reduction* (**1**, 975). Peters and Van Bekkum[1] have described a convenient modification of the Rosenmund reduction[2] of acid chlorides to aldehydes. The method is based on a procedure of Japanese chemists[3] which has received little attention. Palladium-on-charcoal is used as the catalyst, ethyldiisopropylamine as the hydrogen chloride acceptor, and acetone as the solvent (the hydrochloride of the amine is soluble in acetone). The reduction takes place at room temperature and atmospheric pressure.

*Amino acid synthesis.*[4] Phenylhydrazones of α-keto acids (1, prepared by reaction of a phenyldiazonium salt with a reactive methylene group, the Japp–Klingemann reaction[5]) are readily hydrogenated with 5 % Pd/C (also platinum) to give amino acids (2).

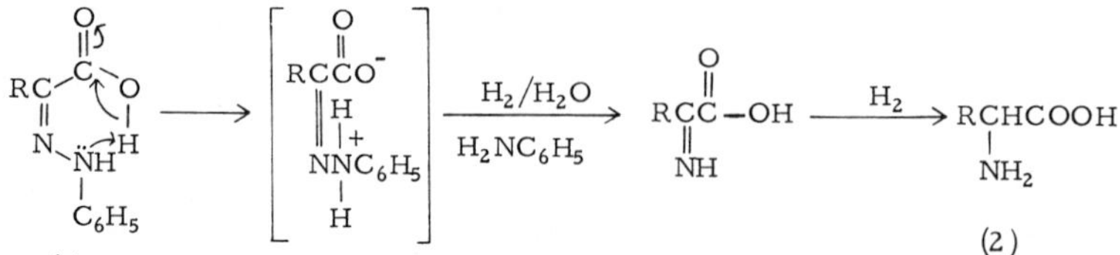

The reduction of (1) to (2) can also be effected with zinc dust in 75 % alcohol in the presence of mercuric chloride or calcium chloride.

*Isomerization of the diterpene A/B ring juncture.*[6] The steroidal type A/B ring juncture of diterpenes can be isomerized to the antipodal one by treatment with a 10 % palladium–charcoal catalyst[7] in refluxing triglyme. Thus methyl 5α,10α-podocarpa-8,11,13-triene-15-oate (1) can be isomerized in 83 % yield to methyl 5β,10α-podocarpa-8,11,13-triene-15-oate (2). The reaction was used in a synthesis of (−)-podocarpic acid from (+)-dehydroabietic acid.

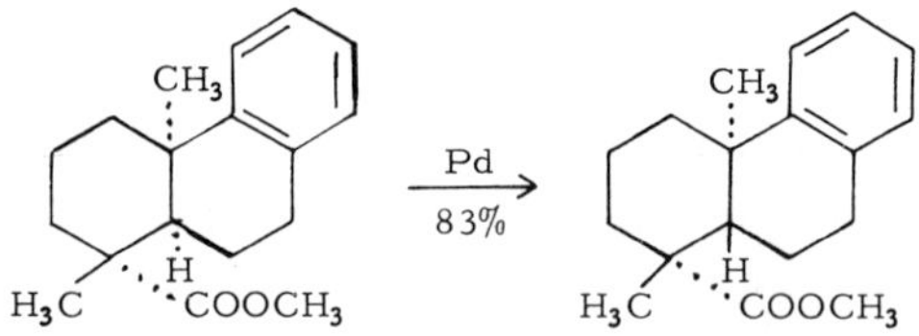

Treatment of estrone (3) in the same way gave moderate yields (48 %) of (+)-isoequilenin (4).

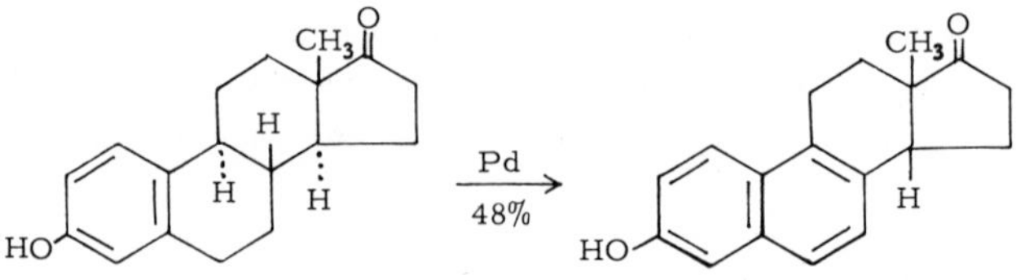

[1] J. A. Peters and H. Van Bekkum, *Rec. trav.*, **90**, 1323 (1971).
[2] E. Mosettig and R. Mozingo, *Org. React.*, **4**, 362 (1948).
[3] Y. Sakurai and Y. Tanabe, *J. Pharm. Soc. Japan*, **64**, 25 (1944).
[4] N. H. Khan and A. R. Kidwai, *J. Org.*, **38**, 822 (1973).
[5] R. F. Japp and F. Klingemann, *Ber.*, **20**, 2942, 3284, 3398 (1887).
[6] S. W. Pelletier, Y. Ichinohe, and D. L. Herald, Jr., *Tetrahedron Letters*, 4179 (1971).
[7] A. Ross, P. A. S. Smith, and A. S. Dreiding, *J. Org.*, **20**, 905 (1955); C. T. Mathew, G. S. Gupta, and P. C. Dutta, *Proc. Chem. Soc.*, 336 (1964).

**Palladium(II) chloride**, **1**, 782; **2**, 303–305.

*π-Allylpalladium complexes* (**2**, 304).[1] π-Allylpalladium complexes can be prepared from olefins in 80–100% yields by treatment either with palladium chloride in methylene chloride containing sodium carbonate or with palladium chloride and sodium chloride in acetic acid containing sodium acetate. Thus 2-methyl-l-octene (1)

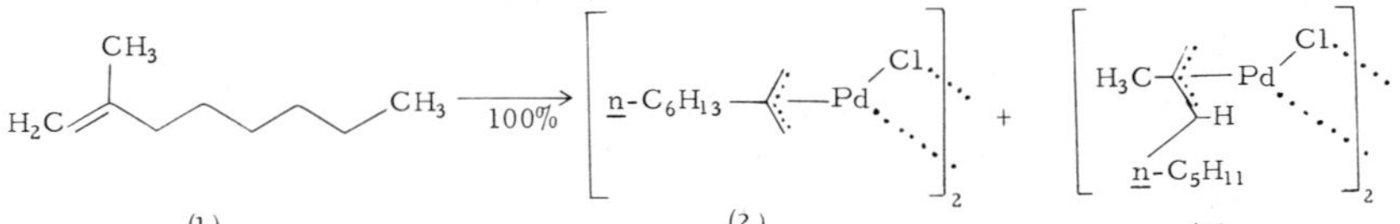

is converted quantitatively into the two π-allylpalladium complexes (2) and (3) in the ratio 1:1.6. These complexes can be alkylated by anions. For example, the reaction of the mixture of (2) and (3) with the anion of diethyl malonate in the presence of at least 4 eq. of triphenylphosphine[2] in THF or DMF proceeds in minutes at room temperature to give the mixture of alkylated olefins (4) and (5) in 63% yield.

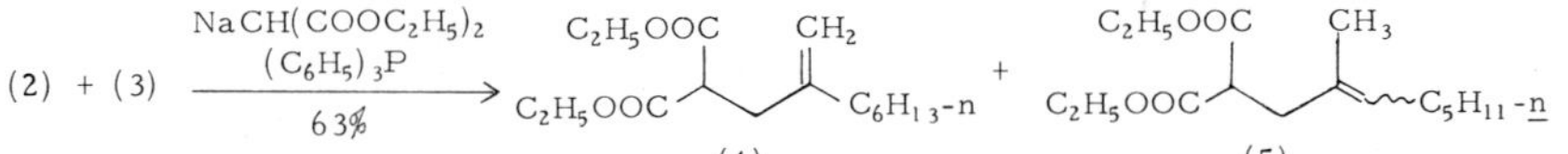

Homologation of a methyl group to an ethyl group was effected in the following way:

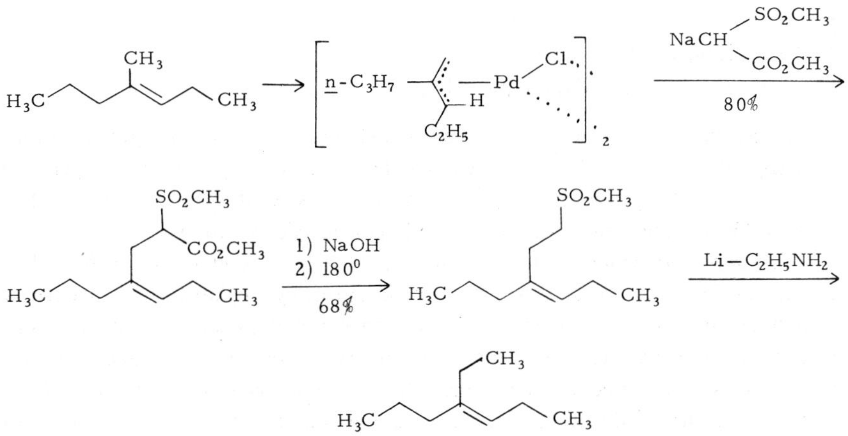

*Phenolic coupling.*[3] Palladium(II) chloride was used for the phenolic coupling of (1) to give (2). When (2) is heated for 2 hr. at 38° it is converted into carpanone, a lignan

(1)                              (2)                              (3)

obtained from the bark of the carpano tree. Note that (3) has five contiguous asymmetric centers.

[1] B. M. Trost and T. J. Fullerton, *Am. Soc.*, **95**, 292 (1973).
[2] No reaction occurs in the absence of triphenylphosphine. Bis(diphenyl)phosphinoethane can also be used as the palladium ligand.
[3] O. L. Chapman, M. R. Engel, J. P. Springer, and J. C. Clardy, *Am. Soc.*, **93**, 6696 (1971).

**Pentaerythrityl tetrabromide**, $C(CH_2Br)_4$. Mol. wt. 433.80, m.p. 158–160°.

*Preparation.*[1] Material supplied by Columbia Organic Chemicals Company, Inc., should be recrystallized from chloroform.

*Electrolytic reduction*[2] :

[1] H. L. Herzog, *Org. Syn. Coll. Vol.* **4**, 753 (1963).
[2] M. R. Rifi, *Org. Syn.*, **52**, 22 (1972).

**1,2,2,6,6-Pentamethylpiperidine (PMP)**, Mol. wt. 155.28, colorless liquid, $pK_a$ 11.25.

*Preparation.*[1] This base is prepared by methylation of 2,2,6,6-tetramethylpiperidine (Aldrich, Fluka, K and K) with methyl iodide in methanol at room temperature overnight; 1,2,2,6,6-pentamethylpiperidine hydriodide separates in 71 % yield. The free amine is liberated by treatment with 5 % sodium hydroxide solution.

*Exhaustive alkylation of amines.* Sommer *et al.*[1,2] have developed a new procedure by which primary and secondary amines are exhaustively alkylated to the quaternary stage in one step at room temperature. In the conventional procedure strong inorganic bases (sodium hydroxide, sodium carbonate) are used to bind the generated acid; high temperatures are required. In the new procedure a strong organic base is used; DMF or acetonitrile is the usual solvent, and the reaction is complete in a few hours. The most important requirements concern the organic base. It must have solubilities similar to those of the starting amines to permit a homogeneous reaction. It must be a stronger base than the starting amine, but must undergo alkylation slowly. In the first publication, 2,6-lutidine was found suitable for quaternization of aromatic amines in the $pK_a$ range of 3.86–5.34. In the latter publication 1,2,2,6,6-pentamethylpiperidine was shown to be suitable for exhaustive alkylation of a wide spectrum of amines. It is one of the strongest known organic bases; in addition the methyl groups produce severe steric hindrance to alkylation. In the general procedure, excess methyl iodide is added to a solution of the amine or amine base and a stoichiometric amount of PMP in an appropriate solvent and the reaction is allowed to proceed overnight at room temperature. The quaternary ammonium compound usually precipitates directly; if not, it can be caused to precipitate by addition of a nonpolar solvent (acetone, ethyl acetate, or ether). Yields of quaternary iodides are in the range 50–95%.

[1] H. Z. Sommer, H. I. Lipp, and L. L. Jackson, *J. Org.*, **36**, 824 (1971).
[2] H. Z. Sommer and L. L. Jackson, *ibid.*, **35**, 1558 (1970).

***trans*-3-Pentene-2-one, 2,** 306–307; **3,** 218–219.

*Preparation.* The preparation by Odom and Pinder (**3,** 219, ref. 1) has now been published.[1]

*Robinson annelation.* The Robinson annelation of unactivated cyclohexanones with *trans*-3-pentene-2-one under usual conditions proceeds in low yield, if at all. However, Scanio and Starrett[2] report that the sodium enolate of 2-methylcyclohexanone (1) in dioxane[3] reacts with 1 eq. of *trans*-3-pentene-2-one (2) to give *cis*-4,10-dimethyl-1(9)-octalone-2 (3) in 65% yield. If DMSO is used in place of dioxane, the

*trans* isomeric octalone (4) is obtained in 72% yield. Scanio and Starrett suggest that this significant solvent effect of the highly polar DMSO may result from a reversal in the usual order of steps in the annelation reaction; that is, aldol condensation may occur first followed by an intramolecular Michael reaction.

[1] H. C. Odom and A. R. Pinder, *Org. Syn.*, **51**, 115 (1971).
[2] C. J. V. Scanio and R. M. Starrett, *Am. Soc.*, **93**, 1539 (1971).
[3] Prepared by heating to reflux an equimolar mixture of NaH and (1) in dioxane for 3 hr.

**Peracetic acid, 1**, 785–791; **2**, 307–309; **3**, 219.

*Oxidation of vitamin A alcohol.* Oxidation of vitamin A alcohol (1) with peracetic

acid in THF at room temperature gives 11,12-epoxyvitamin A aldehyde (2) in 45%
yield. Probably steric factors inhibit attack at the 5,6- and 7,8-double bonds.[1]

[1] Y. Ogata, K. Tomizawa, and K. Takagi, *Tetrahedron*, **29**, 47 (1973).

### Peracetic acid–Boron trifluoride etherate.

*Baeyer–Villiger oxidation of macrocyclic ketones* (**1**, 463–464). Mookherjee *et al.*[1]
report that macrocyclic ketones are readily converted into the corresponding lactones
in good yield by treatment with 40% peracetic acid in the presence of boron trifluoride
etherate. Thus treatment of cyclododecanone (1) in chloroform with boron trifluoride

etherate (1 mole) and 40% peracetic acid (3 moles) at 25° in the dark for 2 weeks gave
cyclododecanolide (2) in 77.2% yield. The yield was lower when boron trifluoride
etherate was replaced by sulfuric acid (**1**, 788).

[1] B. D. Mookherjee, R. W. Trenkle, and R. R. Patel, *J. Org.*, **37**, 3846 (1972).

### Peracetic acid–Sulfuric acid.

*N-Oxidation of weakly basic N-heteroaromatic compounds.* N-Oxidation of poly-
halogenopyridines is difficult owing to the low basicity of the nitrogen atom. For
example, treatment of pentachloropyridine with trifluoroacetic acid and 90% hydrogen
peroxide under optimum conditions gives only a 20% yield of the 1-oxide. Chivers and
Suschitzky[1] now find that the oxidation can be effected in 95% yield by use of 90%
hydrogen peroxide, acetic acid, and concentrated sulfuric acid. The more expensive
trifluoroacetic acid can also be used but with no improvement in yield. No oxidation
occurs with use of 75% aqueous sulfuric acid. The function of the sulfuric acid is
probably to protonate the peracid and hence facilitate transfer of $OH^+$.

[1] G. E. Chivers and H. Suschitzky, *J. Chem. Soc.* (*C*), 2867 (1971).

**Performic acid**, **1**, 457–458.

*Cyclic carbonates.* Oxidation of atractyligenin methyl ester (1) with formic acid

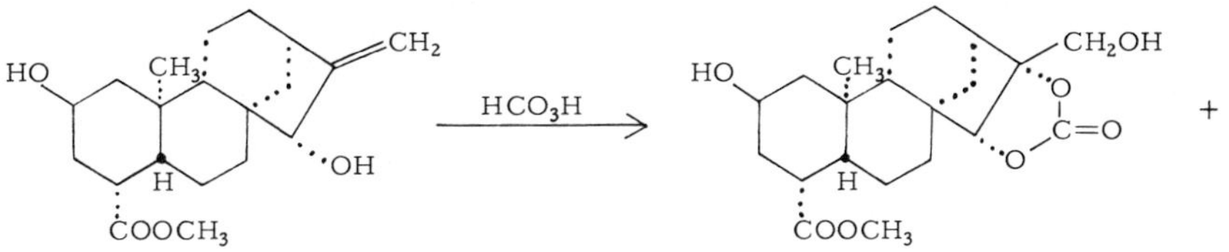

(99%) and hydrogen peroxide (36%) unexpectedly gives the three cyclic carbonates (2)–(4).

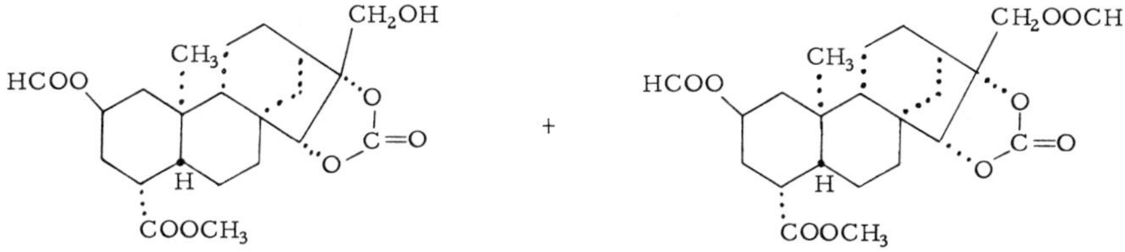

[1]F. Piozzi, G. Savona, and M. L. Marino, *Tetrahedron*, **29**, 621 (1973).

**Periodates**, **1**, 809–815; **2**, 311–313.

*Spiro-epoxy-2,4-cyclohexadienones.* Oxidation of *ortho*-hydroxymethylphenols having at least one bulky substituent by sodium periodate leads to spiro-epoxy-2,4-cyclohexadienones in good yield.[1] Thus oxidation of 2,4-di-*t*-butyl-6-hydroxymethyl-phenol (1) in methanol by a slight excess of sodium periodate in water gives (2) in 95% yield. In the absence of a bulky substituent the product undergoes a spontaneous

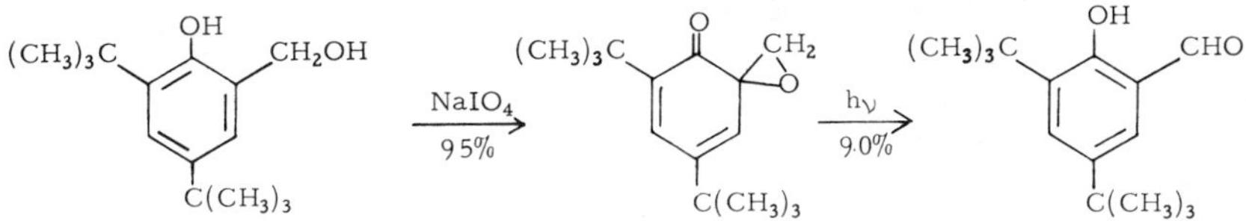

Diels–Alder reaction to give a dehydrodimer of the hydroxymethylphenol.[2] The dienone (2) when irradiated (high-pressure mercury lamp Phillips HPK 125W) at room temperature aromatizes in high yield to the substituted salicylaldehyde (3). The isomerization can also be effected, but in low yield, by heat treatment at 120°.

Oxidation of the carbinol (4) with periodate does not give the expected spiro-epoxy compound but the benzaldehyde acetal of 3,5-di-*t*-butylcatechol (5). This oxidative rearrangement is apparently limited to *o*-hydroxy-substituted diaryl- and triaryl-carbinols.[3]

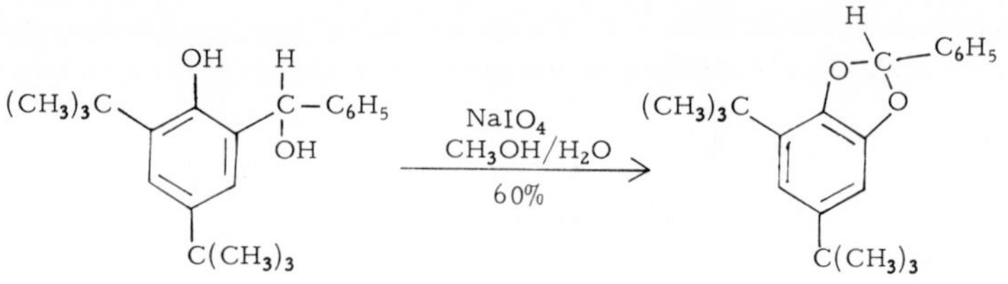

(4)                                    (5)

**Oxidative rearrangement of α-keto esters and amides.**[4] α-Keto esters and amides undergo oxidative rearrangement when treated with sodium periodiate at pH 7–9 to

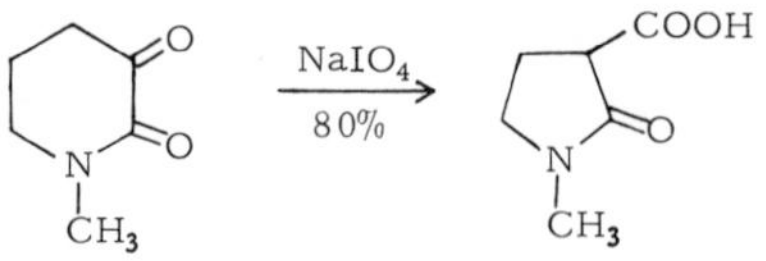

(1)                    (2)

give malonic acid derivatives. Thus 1-methyl-2,3-piperidinedione (1) is converted into 3-carboxy-1-methyl-2-pyrrolidinone (2) in 80% yield. Another example is the oxidation of N,N-dimethyl-2-ketobutanamide (3) to give N,N-dimethylmethylmalonamic acid (4, 69% yield). The original paper should be consulted for a suggested mechanism.

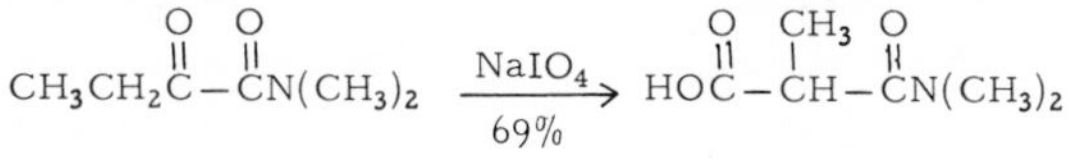

(3)                    (4)

[1] H.-D. Becker, T. Bremholt, and E. Adler, *Tetrahedron Letters*, 4205 (1972).
[2] E. Adler, S. Brasen, and H. Miyake, *Acta Chem. Scand.*, **25**, 2055 (1971).
[3] H.-D. Becker and T. Bremholt, *Tetrahedron Letters*, 197 (1973).
[4] M. L. Rueppel and H. Rapoport, *Am. Soc.*, **94**, 3877 (1972).

**Periodic acid**, **1**, 815–819; **2**, 313–315; **3**, 220.

**Oxidative cleavage of azines.**[1] Aromatic azines, for example benzalazine (1), are oxidatively cleaved to the parent aldehyde or ketone in high yield when treated with periodic acid at room temperature in glacial acetic acid or N,N-dimethylformamide (1:2 molar ratio of substrate to oxidant).

$$C_6H_5CH=N-N=CHC_6H_5 \xrightarrow[96\%]{H_5IO_6} 2\ C_6H_5CHO\ +\ N_2$$

(1)

**Methyl and ethyl glyoxylate.**[2] Methyl and ethyl glyoxylate can be prepared in about 80–85% yield by cleavage of the corresponding esters of *d*-tartaric acid in dry ether with periodic acid.

$$\begin{array}{c} \text{H} \\ | \\ \text{HO} - \text{C} - \text{COOR} \\ | \\ \text{HO} - \text{C} - \text{COOR} \\ | \\ \text{H} \end{array} \quad \xrightarrow{\text{H}_5\text{IO}_6, \ \text{ether}} \quad 2 \quad \begin{array}{c} \text{O} \\ \| \\ \text{H} \diagup \text{C} \diagdown \text{COOR} \end{array}$$

$$R = CH_3 \text{ or } C_2H_5$$

[1]A. J. Fatiadi, *Chem. Ind.*, 64 (1971).
[2]T. R. Kelly, T. E. Schmidt, and J. G. Haggerty, *Synthesis*, 544 (1972).

$$\overset{\text{OOH}}{\underset{|}{\phantom{.}}}$$

**Peroxybenzimidic acid,** $C_6H_5\overset{|}{C}=NH$, **1**, 469–470.

*Epoxidation.* Carlson and co-workers[1] have compared the relative rate of epoxidation of peroxybenzimidic acid with that of peracetic acid and note little correlation. They conclude that this peracid is relatively indiscriminate and is not the reagent of choice for selective epoxidation of polyunsaturated substrates.

[1]R. G. Carlson, N. S. Behn, and C. Cowles, *J. Org.*, **36**, 3832 (1971).

**Pertrifluoroacetic acid.**

*Pertrifluoroacetic acid–Boron trifluoride* (**1**, 826; **2**, 316). Hart[1] has reviewed oxidation with this system.

[1]H. Hart, *Accts. Chem. Res.*, **4**, 337 (1971).

**Phenanthrene–Sodium (Sodium phenanthrenide),** $[C_{14}H_{10}] \doteq Na^+$. Mol. wt. 153.18.

The reagent is prepared by adding 0.014 g.-atom of sodium to a solution of 0.015 mole of phenanthrene in 30 ml. of dry THF. The dark-green solution is stirred for 18 hr. at room temperature.

*Dechlorination.*[1] Treatment of a mixture of *cis*- and *trans,exo,exo*-3,4-dichlorotetracyclo[4.4.2.0$^{2,5}$.0$^{7,10}$]dodeca-8,11-diene, (1) and (2), with phenanthrene–sodium until the green color persists affords *exo,exo*-tetracyclo[4.4.2.0$^{2,5}$0$^{7,10}$]dodecatriene (3) in 69 % yield.

(1)                    (2)                    (3)

[1]L. A. Paquette and J. C. Stowell, *Am. Soc.*, **93**, 5735 (1971).

**Phenyl azide, 1**, 829–832; **2**, 45, 415, 443; **3**, 17.

*Addition reactions.*[1] Heterocyclic β-enamino esters of the dihydrofurane type

(1a, b) react with phenyl azide to give the triazoles (4a, b) with ring cleavage and migration of the ester group.

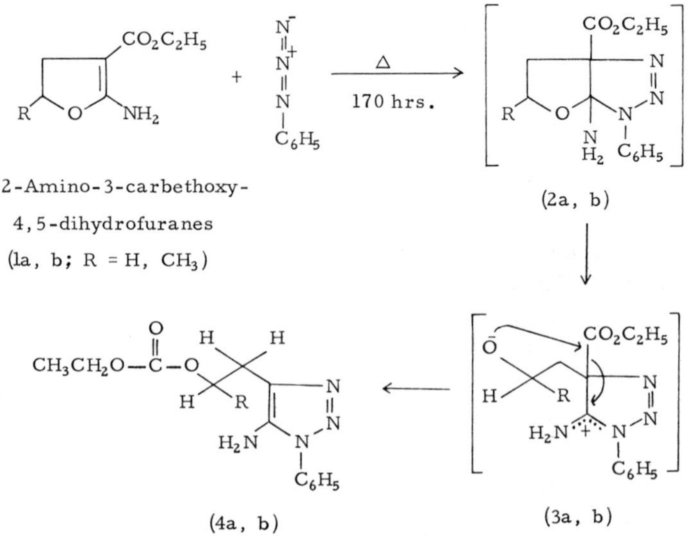

2-Amino-3-carbethoxy-
4,5-dihydrofuranes
(1a, b; R = H, CH$_3$)

(2a, b)

(4a, b)

(3a, b)

[1]H. Wamhoff and P. Sohár, *Ber.*, **104**, 3510 (1971).

**p-Phenylbenzoyl chloride (4-Diphenylcarbonyl chloride),** Mol. wt. 216.67, m.p. 110–112°. Supplier: Aldrich.

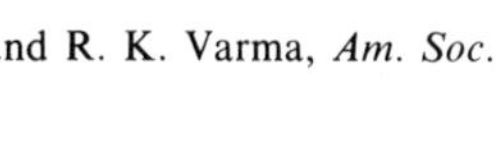

Corey *et al.*[1] used *p*-phenylbenzoyl esters in the course of synthesis of prostaglandins. The esters are crystalline and easily handled, characterized, and purified by chromatography.

[1]E. J. Corey, S. M. Albonico, U. Koelliker, T. K. Schaaf, and R. K. Varma, *Am. Soc.*, **93**, 1491 (1971).

**Phenyl(1-bromo-1-chloro-2,2,2-trifluoroethyl)mercury,** $C_6H_5HgCClBrCF_3$. Mol. wt. 474.14, m.p. 140–144°.

*Preparation*[1]:

$$C_6H_5HgCl + (CH_3)_3COK + CF_3CBrClH \xrightarrow[\phantom{xx}86\%\phantom{xx}]{THF, -10^0 \text{ to } 0^0} C_6H_5HgCClBrCF_3$$

*CF$_3$CCl Transfer.*[1] The reagent can be used in transfer reactions of $CF_3CCl$ at 130–140°:

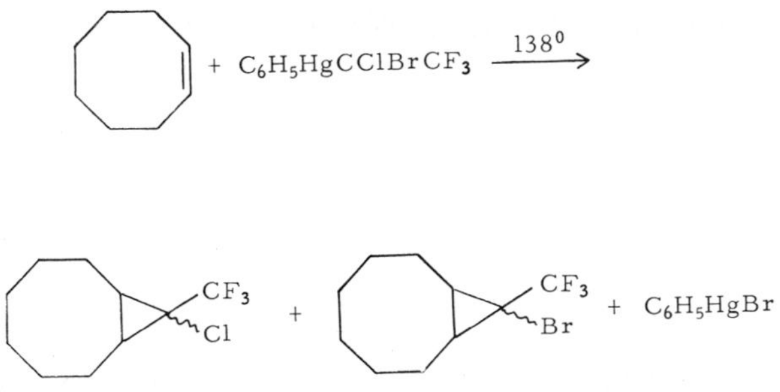

$$\text{+ } C_6H_5HgCClBrCF_3 \xrightarrow{138^0}$$

74%                    5%

[1]D. Seyferth and D. C. Mueller, *Am. Soc.*, **93**, 3714 (1971).

## 1-Phenyl-5-chlorotetrazole, 2, 319–320.

*Replacement of phenolic OH by H.* Details of the conversion of *p*-phenylphenol into diphenyl are recorded.[1]

[1]W. J. Musliner and J. W. Gates, Jr., *Org. Syn.*, **51**, 82 (1971).

## Phenyldichloroborane, $C_6H_5BCl_2$. Mol. wt. 158.83. Supplier: Ventron.

*Stereospecific synthesis of secondary amines.*[1] Phenyldichloroborane (1) reacts with an azide, for example cyclohexyl azide (2),[2] in benzene solution, first at 20° and

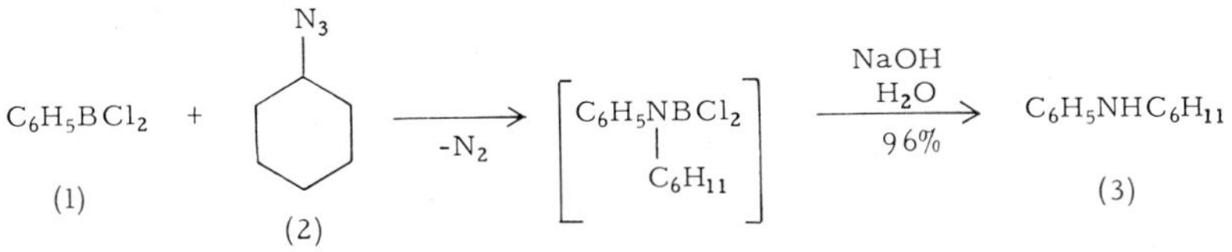

then at 80°, to give N-cyclohexylaniline (3) in 96% yield after alkaline treatment. The reaction is general for aryldichloroboranes.

Alkydichloroboranes[3] can be used equally well. Thus the reaction of cyclohexyldichloroborane with *n*-butyl azide gives N-*n*-butylcyclohexylamine in 95% yield.

The reaction is stereospecific. Thus *trans*-2-methylcyclopentyldichloroborane (4) reacts with cyclohexyl azide (2) to give N-(*trans*-2-methylcyclopentyl)cyclohexylamine (5).

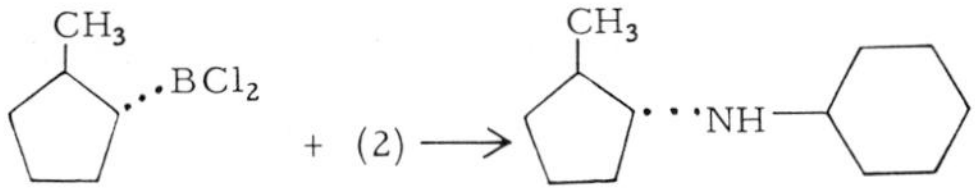

[1]H. C. Brown, M. M. Midland, A. B. Levy, *Am. Soc.*, **95**, 2394 (1973).
[2]Prepared from cyclohexyl bromide and excess sodium azide in DMF; see A. J. Parker, *J. Chem. Soc.*, 1328 (1961).
[3]H. C. Brown and A. B. Levy, *J. Organometal. Chem.*, **214**, 233 (1972); H. C. Brown and N. Ravindran, *Am. Soc.*, **95**, 2396 (1973).

**9-Phenyl-9-hydroxyanthrone (Tritylone alcohol).** Mol. wt. 286.31, m.p. 213–214°.

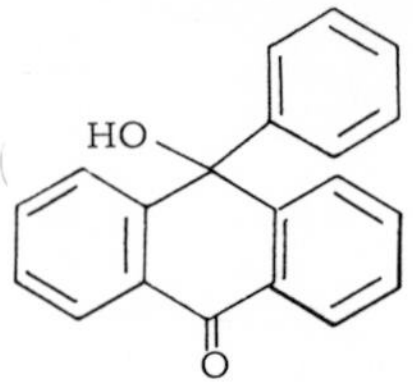

The reagent is prepared in 73% yield by oxidation of 9-phenylanthracene.

Tritylone alcohol reacts readily with alcohols to give tritylone ethers. Selective reaction of primary alcohols in the presence of secondary alcohols is possible. These new ethers are more stable to acid than trityl ethers. They can be cleaved by Wolff–Kishner reduction.[1]

[1]W. E. Barnett, L. L. Needham, and R. W. Powell, *Tetrahedron*, **28**, 419 (1972).

**Phenyl isocyanate, 1,** 843–844.

*Deoxygenation of phenols.*[1] Phenols are converted into urethanes by treatment with phenyl isocyanate using benzene as solvent for both the reaction and crystallization. Hydrogenolysis in acetic acid with Pd/C as catalyst gives arenes in yields of 20–80%. The variation in yield suggests that steric effects may be important.

$$ArOH + C_6H_5N=C=O \longrightarrow ArOCONHC_6H_5 \xrightarrow{H_2} ArH$$

[1]J. D. Weaver, E. J. Eisenbraun, and L. E. Harris, *Chem. Ind.*, 187 (1973).

**Phenylmethoxycarbenepentacarbonyltungsten(0)** (1). Mol. wt. 441.1, m.p. 59°.

*Preparation*[1]:

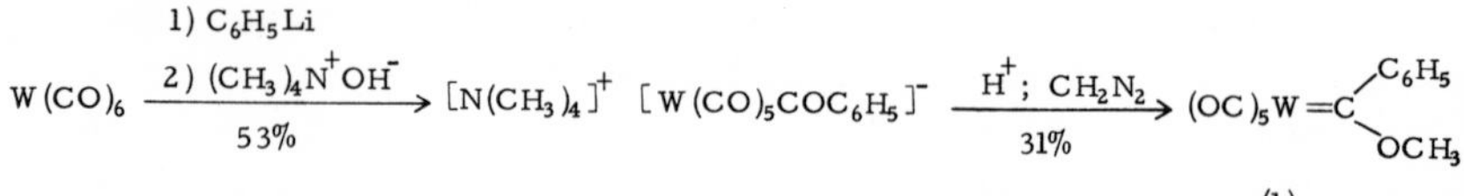

$$W(CO)_6 \xrightarrow[\phantom{xx}53\%\phantom{xx}]{\begin{array}{l}1)\ C_6H_5Li\\2)\ (CH_3)_4N^+OH^-\end{array}} [N(CH_3)_4]^+ [W(CO)_5COC_6H_5]^- \xrightarrow[31\%]{H^+;\ CH_2N_2} (OC)_5W=C\begin{array}{l}C_6H_5\\OCH_3\end{array}$$

(1)

*Vinyl ethers.*[2] This stable metal–carbene complex (1) reacts with Wittig reagents at room temperature to give vinyl ethers (2) in high yield and pentacarbonyltriphenylphosphinetungsten(0) (3).

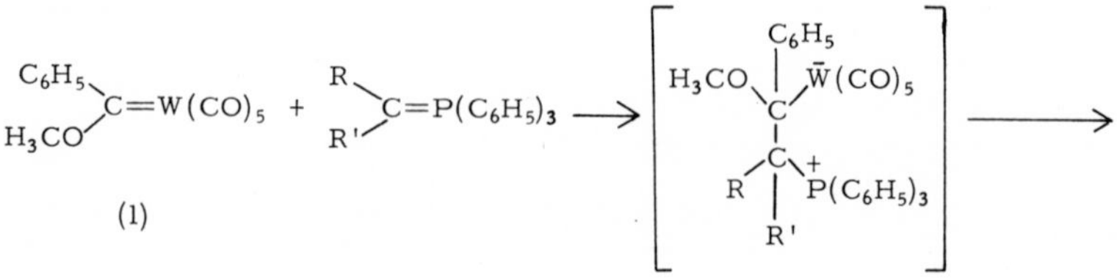

$$\underset{C_6H_5}{\overset{H_3CO}{\diagdown}}C=C\underset{R'}{\overset{R}{\diagup}} \quad + \quad (C_6H_5)_3PW(CO)_5$$

(2)                          (3)

[1] E. O. Fischer and A. Maasböl, *Angew. Chem., internat. Ed.*, **3**, 580 (1964); *Ber.*, **100**, 2445 (1967).
[2] C. P. Casey and T. J. Burkhardt, *Am. Soc.*, **94**, 6543 (1972).

**Phenylpalladium acetate**, generated *in situ* from phenylmercuric acetate (Eastman, mol. wt. 336.74) and palladium acetate (Engelhard Industries, mol. wt. 224.49).

*Synthesis of 2-methyl-3-phenylpropionaldehyde.*[1] A slurry of 33.6 g. (0.10 mole) of phenylmercuric acetate, 200 ml. of acetonitrile, and 14.4 g. (17 ml., 0.2 mole) of

$$C_6H_5HgOCOCH_3 + Pd(OCOCH_3)_2 \longrightarrow [C_6H_5PdOCOCH_3] + Hg(OCOCH_3)_2$$

$$[C_6H_5PdOCOCH_3] + CH_2=\underset{\overset{|}{CH_3}}{C}CH_2OH \longrightarrow \left[\underset{\overset{|}{PdOCOCH_3}}{C_6H_5CH_2\underset{\overset{|}{CH_3}}{C}CH_2OH}\right]$$

$$\longrightarrow C_6H_5CH_2\underset{\overset{|}{CH_3}}{C}HCHO + CH_3COOH + Pd$$

methallyl alcohol (Eastman) is stirred mechanically in a 500-ml. three-necked flask fitted with a condenser and a thermometer. The stirred slurry is cooled in an ice bath and 22.4 g. (0.1 mole) of powdered palladium acetate is added over 1 min. Stirring is continued with cooling for 1 hr. and then at room temperature overnight.

The black reaction mixture is diluted with 100 ml. of ether and poured onto 200 g. of ether-wet alumina (Woelm, activity grade 1) in a 45 × 2.5-cm. chromatographic column. The product is washed through the column with about 1 l. of ether. The brown eluate is concentrated by distilling the ether through a 45-cm. Vigreux column on a steam bath at atmospheric pressure. Then a slight vacuum is applied to remove most of the acetonitrile. When the volume reaches about 50 ml., the mixture is filtered into a 100–ml. distilling flask to free it from some precipitated palladium metal. Final distillation through a 10-cm. Vigreux column gives 9.4 g. (69%) of 2-methyl-3-phenyl-propionaldehyde of at least 95% purity, b.p. 77–80°/3 mm.

[1] R. F. Heck, *Am. Soc.*, **90**, 5526 (1968); *Org. Syn.*, **51**, 17 (1971).

**Phenylthiomethyllithium, 2**, 324.

*Ketone methylenation.*[1] The reagent reacts with ketones to yield adducts (2) which can be isolated in excellent yields by aqueous quench,[2] or converted into esters (3) by treatment with *n*-butyllithium and then an acylating reagent (acid chloride or acid anhydride). Reduction of the ester (3) with lithium in liquid ammonia affords the

$$\overset{R^1}{\underset{R^2}{}}C=O \quad \xrightarrow[\text{THF}]{C_6H_5SCH_2Li} \quad \overset{R^1}{\underset{R^2}{}}\overset{OH}{\underset{CH_2SC_6H_5}{C}} \quad \xrightarrow[\text{(CH}_3\text{CO)}_2\text{O}]{\underline{n}\text{-BuLi–THF}} \quad \overset{R^1}{\underset{R^2}{}}\overset{OCOCH_3}{\underset{CH_2SC_6H_5}{C}}$$

(2)                        (3)

$$\xrightarrow{\text{Li/NH}_3} \quad \overset{R^1}{\underset{R^2}{}}C=CH_2$$

(4)

exocyclic methylene compound (4). Yields are generally high. The isolation of the intermediate carbinol is generally not necessary. The reaction is even applicable to the highly hindered tricyclic ketone ($\pm$)-norzizanone (5).

(5)              (6)              (7)

The reaction has been used to convert $\Delta^5$-cholestenone-3 into 3-methylene-$\Delta^5$-cholestene in about 40% yield. Reaction of $\Delta^5$-cholestenone-3 with methylenetriphenylphosphorane gives impure 3-methylene-$\Delta^5$-cholestene in < 15% yield.

Two other novel transformations are possible with this reagent. Methyl decanoate (8) undergoes twofold addition with the reagent at $-25°$ to give bisphenylthiomethylcarbinol (9, 73% yield). Benzoylation and reduction (Li/NH$_3$) effects both vicinal elimination and allylic cleavage to give 2-methyl-1-undecene (10).

$$CH_3(CH_2)_8COOCH_3 \xrightarrow[73\%]{} CH_3(CH_2)_8\overset{OR}{\underset{}{C}}(CH_2SC_6H_5)_2 \longrightarrow CH_3(CH_2)_8C\overset{CH_2}{\underset{CH_3}{}}$$

(8)                    (9a, R = H, 73%)                    (10)
                       (9b, R = COC$_6$H$_5$, 82%)

Reaction of $\alpha$-$n$-butylthiomethylenecyclohexanone (11) with the reagent, followed by dehydration (10% HCl and mercuric chloride), gives the phenylthiomethylene aldehyde (12). Reduction and methylation of (12) gives the exocyclic $\beta,\gamma$-unsaturated aldehyde (13).

(11)                    (12)                    (13)

[1] R. L. Sowerby and R. M. Coates, *Am. Soc.*, **94**, 4758 (1972).
[2] E. J. Corey and D. Seebach, *J. Org.*, **31**, 4097 (1966).

**4-Phenyl-1,2,4-triazoline-3,5-dione (PTAD), 1,** 849–850; **2,** 324–326; **3,** 223–224.

*Preparation.* The preparation of Cookson *et al.* (**3,** 224, ref. 1) has now been published.[1]

4-Phenylurazole, the precursor, is now available from Oxford Organic Chemicals, 14 Lonsdale Rd., Oxford, OX2 7EW, U.K.

The definitive paper[2] on the use of the reagent for protection of steroidal ring B dienes (**3,** 223–224) has now been published.

*Cycloaddition reactions.* Methylenecyclopropanes (1) undergo very slow [2 + 2] cycloaddition of the exocyclic double bond with PTAD to give 1:1 adducts (2).[3]

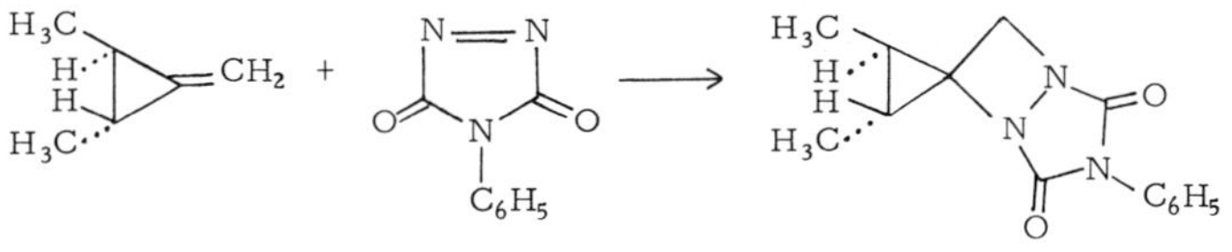

(1)                              (2)

On the other hand, alkenylidenecyclopropanes (3) react very rapidly with PTAD to give adducts (4) derived by [4 + 2] cycloaddition.[4]

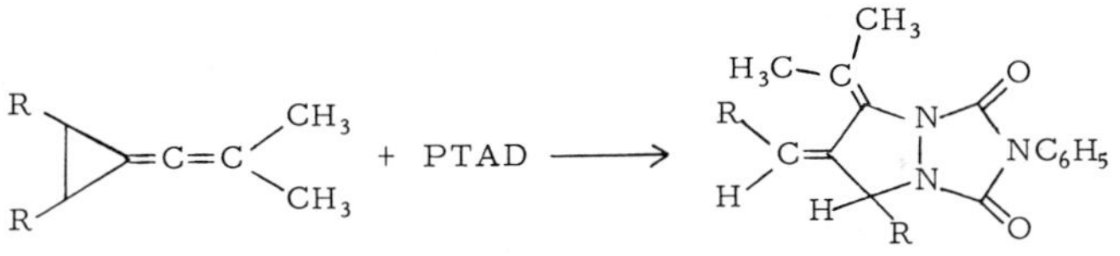

(3)                              (4)

The difference in the behavior of (1) and (3) is discussed in terms of the molecular orbital theory.

*Cycloaddition reactions with vinylcyclopropanes.*[5] α-Cyclopropylstyrene (1) does not react with maleic anhydride. However, it reacts with the more reactive dienophile PTAD to give the 2:1 adduct (4). This adduct is probably formed by cycloaddition of

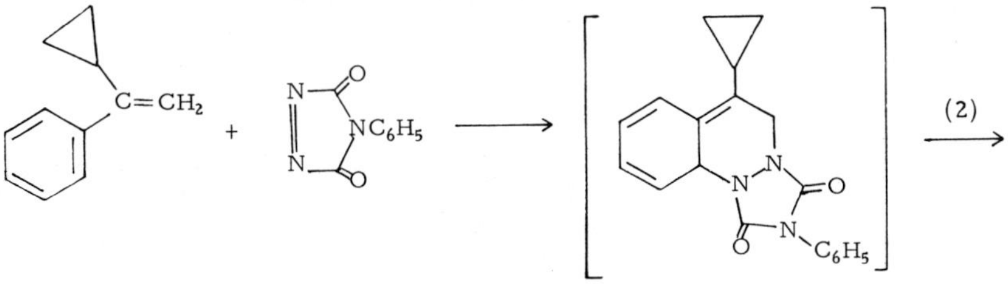

(1)                    (2)                              (3)

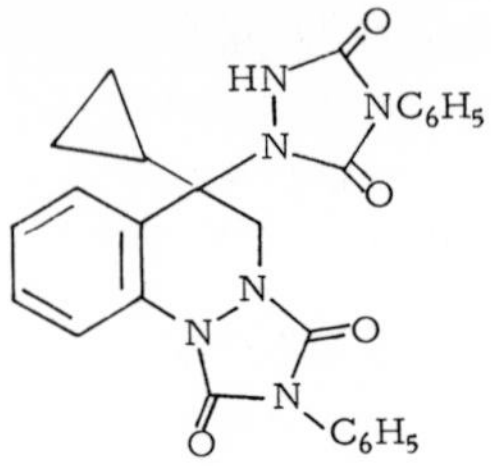

(4)

PTAD across the styrene chromophore of (1) to give (3), which reacts with PTAD by an ene reaction to regenerate the aromatic system.

*trans*-2-Phenyl-1-isopropenylcyclopropane (5) reacts rapidly with (2) (15 min. at 25°, $CH_2Cl_2$) to give only the ene product (6) in 91 % yield.

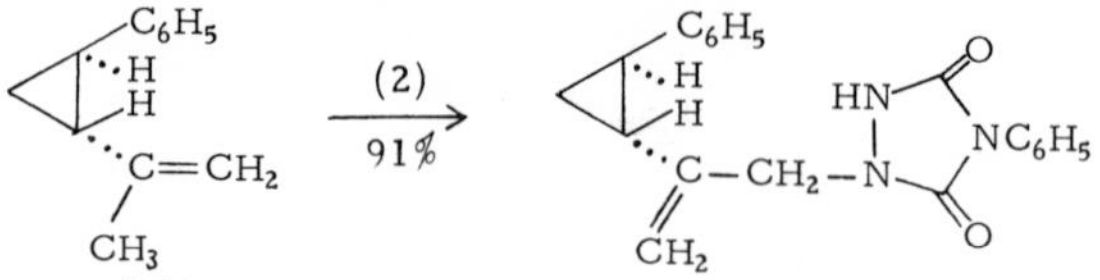

Vinylcyclopropane (7) reacts slowly with (2) to give the [2 + 2] cycloaddition product (8) in 87% yield.

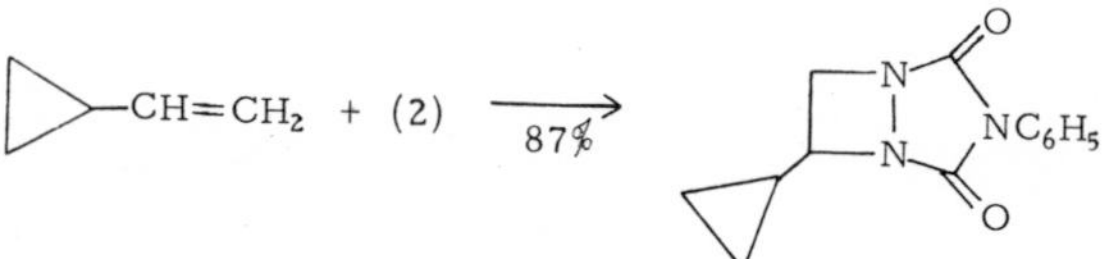

The vinylcyclopropanes (1) and (5) react with chlorosulfonyl isocyanate (CSI) (**1**, 117–118; **2**, 70; **3**, 51–53; this volume) in different ways. Thus (1) reacts with CSI ($CH_2Cl_2$, 0°) to give (9). In the case of (1) the cyclopropane ring is retained; in the reaction of (5) the cyclopropane ring is opened. A possible explanation for the contrasting behavior of (1) and (5) has been suggested.

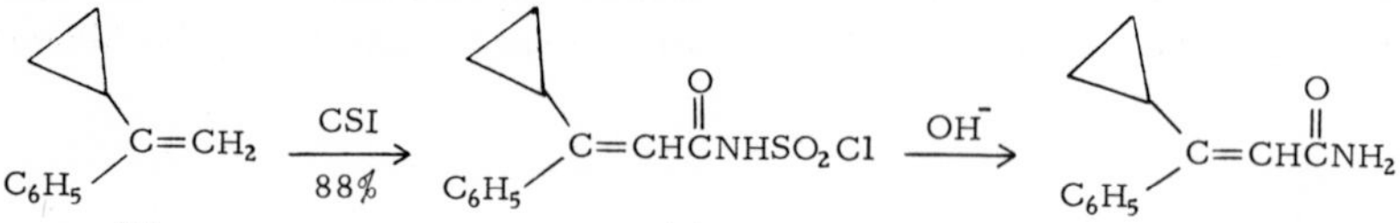

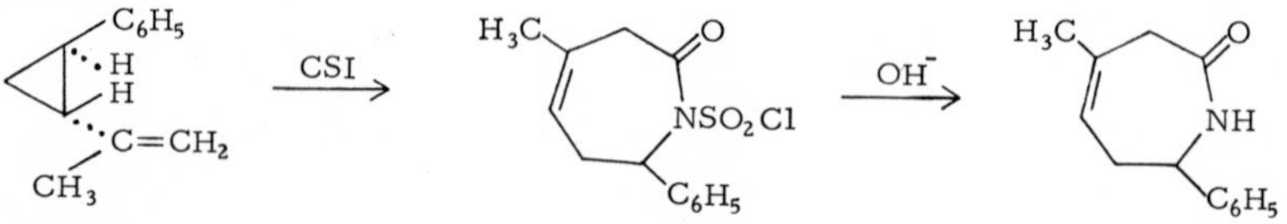

*Reaction with cyclooctatetraene.*[6] With most dienophiles, cyclooctatetraene (1) reacts through the quasiplanar diene system of the valence tautomer bicyclo[4.2.0]-octatriene (2). For example, tetracyanoethylene reacts in this way to give (3). However,

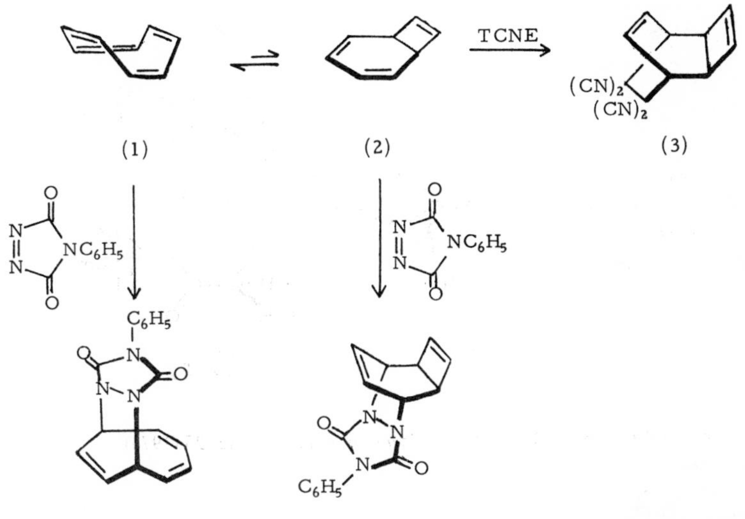

(1)          (2)                    (3)

(5)                    (4)

the more reactive dienophile 4-phenyl-1,2,4-triazoline-3,5-dione reacts not only with the valence tautomer (2) to give (4), but also undergoes 1,4-cycloaddition with (1) to give (5).

[1] R. C. Cookson, S. S. Gupte, I. D. R. Stevens, and C. T. Watts, *Org. Syn.*, **51**, 121 (1971).
[2] D. H. R. Barton, T. Shiori, and D. A. Widdowson, *J. Chem. Soc. (C)*, 1968 (1971).
[3] D. J. Pasto and A. F.-T. Chen, *Tetrahedron Letters*, 2995 (1972).
[4] D. J. Pasto, A. F. T. Chen, and G. Binsch, *Am. Soc.*, **95**, 1553 (1973).
[5] D. J. Pasto and A. F.-T. Chen, *Tetrahedron Letters*, 713 (1973).
[6] R. Huisgen, W. E. Konz, and U. Schnegg, *Angew. Chem., internat. Ed.*, **11**, 715 (1972).

## Phenyltrifluoromethylketene (1). Mol. wt. 220.15.

*Preparation.*[1] The reagent is prepared from α-trifluoromethylmandelic acid as shown:

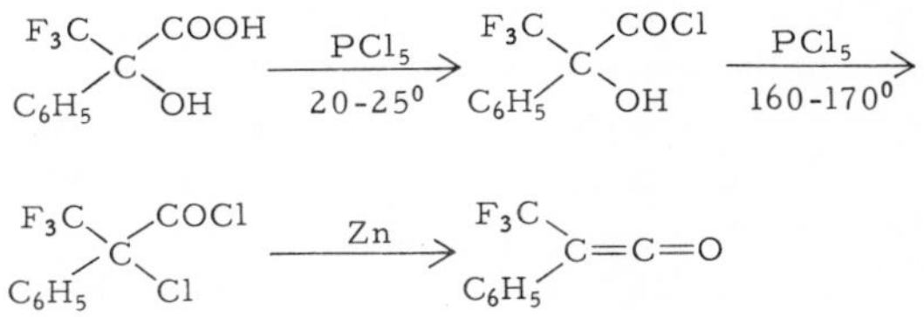

(1)

*Stereoselective acylation of racemic secondary alcohols.*[1] The reagent acylates a chiral secondary alcohol (2) to give the (S)-ester (3a) and the (R)-ester (3b). The product ratio of (3a) to (3b) was determined by [19]F-NMR. (S)-Diastereoisomers are preferentially formed.

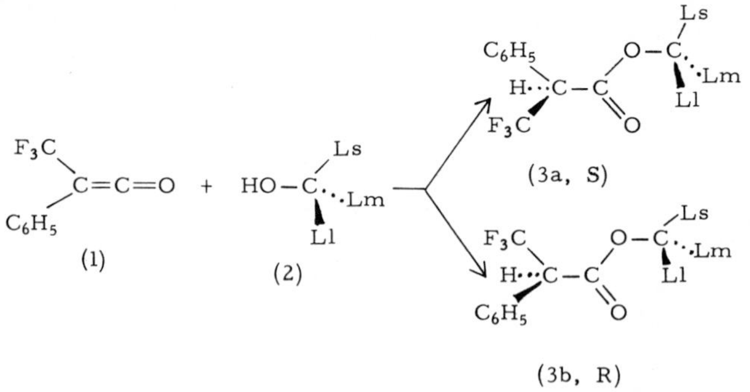

[1]E. Anders, E. Ruch, and I. Ugi, *Angew. Chem., internat. Ed.*, **12**, 25 (1973).

**Phenyl(trifluoromethyl)mercury, 3**, 224–225.

The definitive paper on the difluorocarbene transfer reagent phenyl(trifluoromethyl)mercury has been published.[1] The presently preferred method of preparation is as formulated.[1,2]

$$HgO + 2 CF_3COOH \longrightarrow Hg(O_2CCF_3)_2 + H_2O$$

$$Hg(O_2CCF_3)_2 \xrightarrow{\ 300°\ } CF_3HgO_2CCF_3$$

$$CF_3HgO_2CCF_3 + (C_6H_5)_2Hg \longrightarrow C_6H_5HgCF_3 + C_6H_5HgO_2CCF_3$$

$$C_6H_5HgO_2CCF_3 + NH_4^+Cl^- \xrightarrow{\ H_2O\ } C_6H_5HgCl + NH_4^+O_2CCF_3^-$$

Seyferth concludes that phenyl(trifluoromethyl)mercury is an excellent precursor for difluorocarbene. In the general procedure, 1 molar eq. of $C_6H_5HgCF_3$, 2.5–3.0 molar eq. of well-dried sodium iodide, and 3.0 molar eq. of the dried olefin are used ($N_2$). Benzene is distilled into the reaction flask directly from sodium benzophenone ketyl. The reaction mixture is stirred and heated at reflux under $N_2$ for 12–18 hr. Filtration removes phenylmercuric iodide and NaI and NaF; the *gem*-difluorocyclopropanes are isolated by distillation or by gas chromatography, usually in good yield.

Examples:

$$\text{(cyclohexene)} \xrightarrow{83\%} \text{(bicyclic } CF_2 \text{ adduct)}$$

$$\underset{H}{\overset{C_2H_5}{\diagdown}}C=C\underset{H}{\overset{C_2H_5}{\diagup}} \xrightarrow{93\%} \underset{H}{\overset{C_2H_5}{\diagup}}\triangle\underset{H}{\overset{C_2H_5}{\diagdown}} \, (F_2)$$

$$\underset{H}{\overset{C_2H_5}{\diagdown}}C=C\underset{C_2H_5}{\overset{H}{\diagup}} \xrightarrow{94\%} \underset{H}{\overset{C_2H_5}{\diagup}}\triangle\underset{C_2H_5}{\overset{H}{\diagdown}} \, (F_2)$$

$$\text{(furan)} \xrightarrow{67\%} \text{(}CF_2\text{ adduct)}$$

Some compounds that react reasonably well with dichlorocarbene or chlorofluoro-carbene do not react with the $C_6H_5HgCF_3/NaI$ reagent. Among these are cumene, *sym*-dichlorotetrafluoroacetone, and diethyl azodicarboxylate.

[1] D. Seyferth and S. P. Hopper, *J. Org.*, **37**, 4070 (1972).
[2] D. Seyferth, S. P. Hopper, and G. J. Murphy, *J. Organometal. Chem.*, **46**, 201 (1972).

## Phenyl(trihalomethyl)mercury, 1, 851–854; 2, 326–328; 3, 225.

*Review.* The chemistry of phenyl(trihalomethyl)mercury compounds has been reviewed by Seyferth.[1]

*Dichlorocarbene addition to azodicarboxylate esters.*[2] Phenyl(bromodichloro-methyl)mercury reacts with diethyl azodicarboxylate to give phenylmercuric bromide (98% yield) and a product (87% yield) which was found not to be the expected di-aziridine (1), but to have the structure (2). The same product is obtained by decarboxyla-tion of $CCl_3COONa$ in the presence of diethyl azodicarboxylate in refluxing 1,2-dimethoxyethane (69% yield).

$$\underset{Cl\diagup\diagdown Cl}{\overset{C_2H_5OOC}{\diagdown}}N\text{—}N\underset{}{\overset{COOC_2H_5}{\diagup}}C \qquad\qquad (C_2H_5OOC)_2NN=CCl_2$$

$$\quad\quad (1) \qquad\qquad\qquad\qquad\qquad\qquad (2)$$

[1] D. Seyferth, *Accts. Chem. Res.*, **5**, 65 (1972).
[2] D. Seyferth and H. Shih, *Am. Soc.*, **94**, 2508 (1972).

**Phenyltrimethylammonium perbromide (PTAB)**, **1**, 855; **2**, 328.
   *Preparation.*[1]

$$C_6H_5N(CH_3)_2 + (CH_3)_2SO_4 \longrightarrow C_6H_5\overset{+}{N}(CH_3)_3 \cdot CH_3SO_4^- \xrightarrow{\text{Br}_2, \text{HBr}} C_6H_5\overset{+}{N}(CH_3)_3 \cdot Br_3^- + CH_3SO_4H$$

In a 250-ml. wide-necked flask equipped with a thermometer, 26 ml. (0.2 mole) of freshly distilled dimethylaniline (b.p. 78°/13 mm.) is dissolved in 100 ml. of toluene.[2] The mixture is heated under magnetic stirring to about 40°, and 19 ml. (0.2 mole) of distilled dimethyl sulfate[3] is added through a funnel in about 20 min. After a few minutes the colorless sulfomethylate starts to crystallize. The temperature varies very little during the addition but rises slowly for 1 hr. until it reaches about 50°. One and a half hours after the end of the addition the mixture is heated on a steam bath for 1 hr. After cooling, the phenyltrimethylammonium sulfomethylate is collected, washed with 20 ml. of toluene, and dried under vacuum; yield 47.5 g. (96%).[4]

In a 125-ml. wide-necked flask, 10 g. (0.04 mole) of the sulfomethylate discussed above is dissolved in 10 ml. of a commercial aqueous solution of 48% hydrobromic acid diluted with 10 ml. of water, and 2.5 ml. of bromine is added dropwise through a funnel with magnetic stirring in about 20 min. An orange-yellow precipitate is formed immediately. Stirring at room temperature is continued for 5–6 hr. The product is collected on a filter, washed with water (about 10 ml.), and air-dried under a good hood. The crude product, amounting to 14.9 g., on crystallization from 25 ml. of acetic acid affords 14.2 g. (93%) of air-dried orange crystals, m.p. 115–116°.

[1]J. Jacques and A. Marquet, procedure submitted to *Org. Syn.*, 1971.
[2]Preferred to benzene because of its lower toxicity.
[3]All operations are carried out under a good hood. A slight deficiency of dimethyl sulfate allows the complete utilization of this toxic reagent.
[4]This product is slightly hygroscopic, but no special precautions are needed to handle it.

**Phosphonitrilic chloride (1,1,3,3,5,5-Hexachlorocyclophosphazatriene)** (1), **2**, 206–207. Mol. wt. 347.67, m.p. 112–114°.

(1)

Supplier: Hooker Chemical Company. Commercial material may be purified by crystallization from *n*-heptane,[1] or by distillation.[2]

*Dehydration of amides to nitriles.*[2] When a mixture of an amide and (1) in the ratio of 3:1 is heated at temperatures above 100°, nitriles are formed rapidly and in high yield (usually in the range 75–100%). The dehydration can also be carried out by refluxing the reactants in a solvent (usually chlorobenzene). In many cases, a ratio of amide to (1) of 6:1 is effective.

*Nitriles from aldoximes.*[3] Aliphatic, aromatic, and olefinic aldoximes (1 mole) are dehydrated by treatment with triethylamine (3 moles), and phosphonitrilic chloride (1 mole) in ether or THF at room temperature for 8–24 hr.; yields are in the range 70–95%.

Cyclohexanone oxime reacts with phosphonitrilic chloride to give (1) in 63% yield.

(1)

[1]H. R. Allcock and R. L. Kugel, *Am. Soc.*, **91**, 5452 (1969).
[2]J. C. Graham and D. H. Marr, *Canad. J. Chem.*, **50**, 3857 (1972).
[3]G. Rosini, G. Baccolini, and S. Cacchi, *J. Org.*, **33**, 1060 (1973).

**Phosphoric acid, 1, 860.**

*Rearrangement* of dl-p-hydroxy-α-(methylaminomethyl)benzyl alcohol (1, synephrin, Regis Chemical Company) to p-hydroxyphenylacetaldehyde.[1] A flask with an inserted thermometer is charged with 4.0 g. of (1) and 60 ml. of 85% phosphoric acid and immersed in an oil bath preheated to 170–180° and gently swirled until the resulting solution reaches 118°.[2] The flask is then removed from the bath for exactly 15 sec., during which time excess mineral oil can be wiped from its surface. The hot solution is

then immediately poured into 720 ml. of water at room temperature. In the procedure of Robbins[2] after 90 min. the solution is saturated with sodium chloride and the mixture is shaken with 120-, 70-, and 50-ml. portions of ether for at least 15 min. each. The resulting ether extract is combined with a second ether extract similarly prepared from another 4.0 g. of (1). The pooled extract is shaken for 10 min. with 40 ml. of sodium phosphate buffer (0.1 M, pH 7.6), prepared by adding 100 ml. of water to 13 ml. of a 0.2 M solution of $NaH_2PO_4$ and 87 ml. of a 0.2 M solution of $Na_2HPO_4 \cdot 7H_2O$. The ether layer is decanted, and the ether is removed by evaporation with a stream of dry nitrogen, leaving a yellow liquid containing the crude aldehyde. The procedure includes directions for generation of p-hydroxyphenylacetaldehyde from (3).

[1]J. H. Fellman, *Nature*, **182**, 311 (1958); *Arch. Biochem. Biophys.*, **85**, 345 (1959).
[2]Procedure of J. H. Robbins, *Org. Syn.*, submitted 1972.

**Phosphorus pentachloride, 1**, 866–870.

*Esters of halohydrins.*[1] Treatment of 2-carboxy-1,3-dioxolanes (1) with phosphorus pentachloride in methylene chloride at room temperature gives esters of 1,2-chlorohydrins (2) in about 80% yield. Similarly treatment of the 2-carboxy-1,3-dioxane (3) yields an ester of a 1,3-chlorohydrin (4). Reaction of (5; 2-carboxy-2-methyl-1,3-

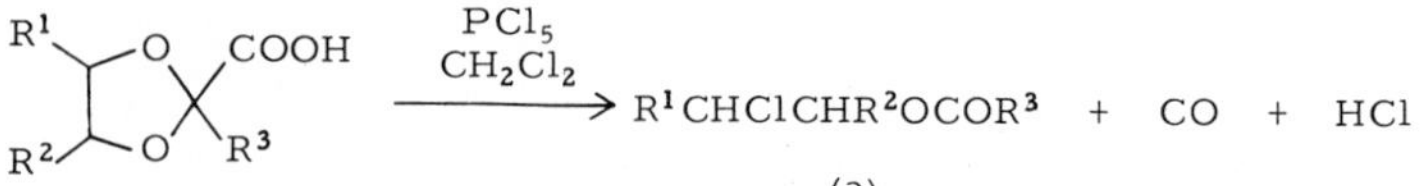

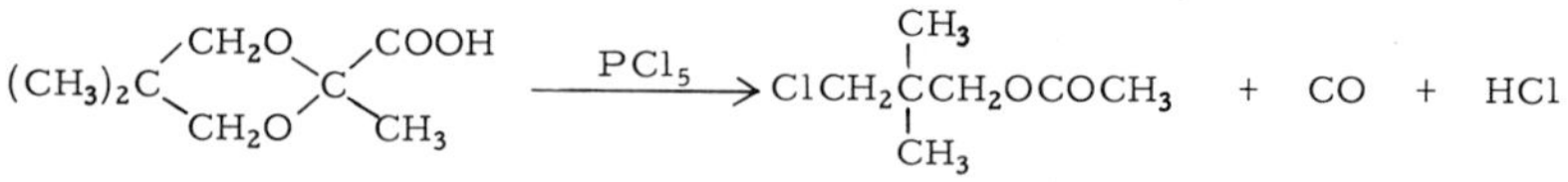

dioxapane) gives the ester of a 1,4-chlorohydrin (6). Thionyl chloride can also be used, but yields are lower.

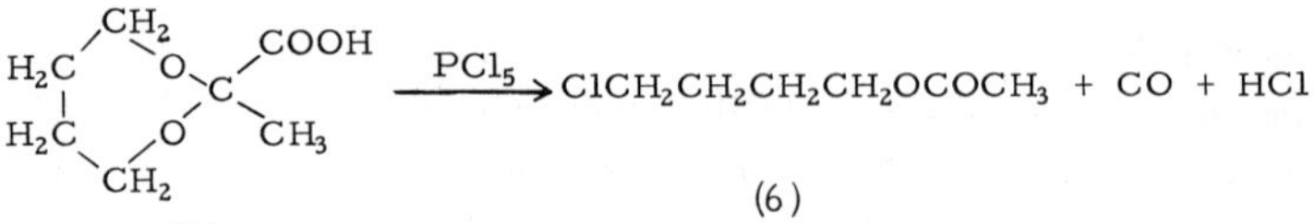

Evidence is presented that (8) is formed directly from (7) at −60°; it is converted into (9) at 0°.

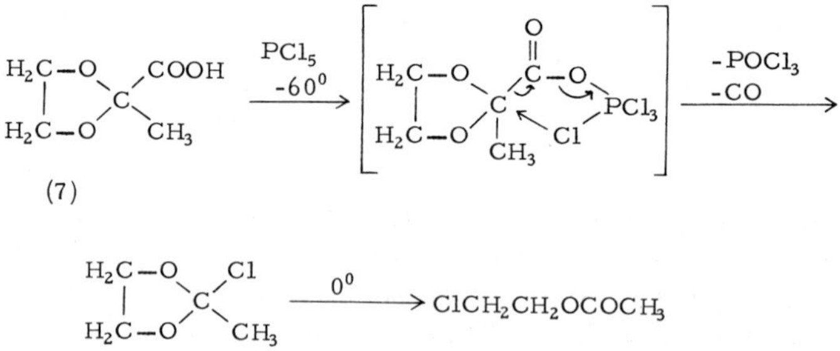

The reaction proceeds stereospecifically with inversion of configuration at the carbon–oxygen bond that is converted into a carbon–chlorine bond. Thus treatment of D-(−)-2-carboxy-2,4,5-trimethyl-1,3-dioxolane (10) yields L-(+)-*erythro*-3-chloro-2-butyl acetate (11). Treatment of (11) with KOH in ethylene glycol yields D-(+)-2,3-epoxybutane (12), a reaction known to proceed with inversion at the carbon–halogen bond.

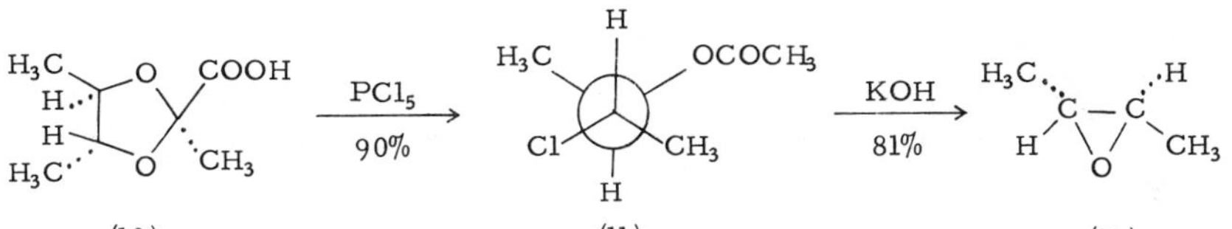

(10)    (11)    (12)

*O-Alkylbenzohydroximoyl chlorides.* O-Alkylbenzohydroximoyl chlorides (2) can be prepared in good yield by treatment of alkyl benzohydroxamates (1) with phosphorus

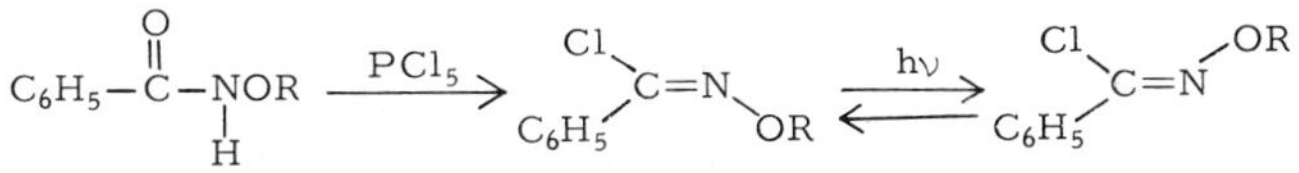

(1)    (2)    (3)

pentachloride.[2] Recent investigation[3] shows that (2) is obtained as the E isomer. On ultraviolet irradiation (2) is partially isomerized to the Z isomer (3).

*1-Chloroalkenes.*[4] Treatment of enols (1) in which $R^1$ and $R^2$ are electronegative substituents ($CN$, $COOC_2H_5$, $COCH_3$, $COC_6H_5$) with phosphorus pentachloride

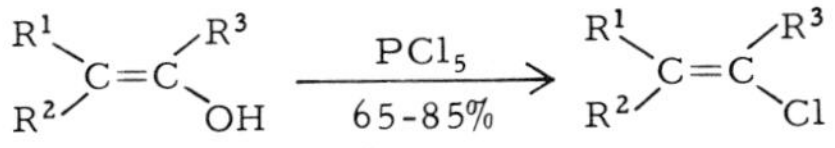

(1)    (2)

(reflux) gives 1-chloroalkenes (2). If the alkali metal salt of (1) is used, phosphoryl chloride (**1**, 876–882; **2**, 330–331; **3**, 228) is the reagent of choice (45–75% yield).

[1]M. S. Newman and C. H. Chen, *Am. Soc.*, **94**, 2149 (1972); *J. Org.*, **38**, 1173 (1973).
[2]J. E. Johnson, J. R. Springfield, J. S. Hwang, L. J. Hayes, W. C. Cunningham, and D. C. McClaugherty, *ibid.*, **36**, 284 (1971).
[3]J. E. Johnson, E. A. Nalley, and C. Weidig, *Am. Soc.*, **95**, 2051 (1973).
[4]K. Friedrich and H. K. Thieme, *Synthesis*, 111 (1973).

**Phosphorus pentasulfide, $P_4S_{10}$, 1, 870–871; 3, 226–228.**
    *Dehydrogenation–deoxygenation*[1]:

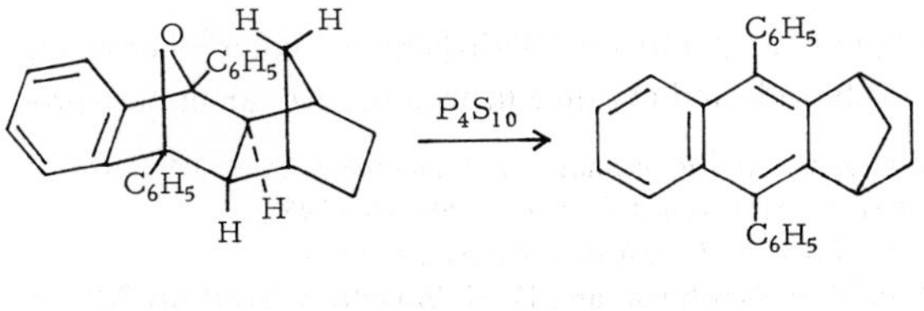

[1]M. P. Cava and F. M. Scheel, *J. Org.*, **32**, 1304 (1967).

**Phosphoryl chloride, 1,** 876–882; **2,** 330–331; **3,** 228.

*Quinazoline synthesis.*[1] When 3-nitroso-2-phenylindole (1) is heated with 3 moles of phosphoryl chloride (or tosyl chloride) in tetramethylene sulfone (**1,** 1144–1145,

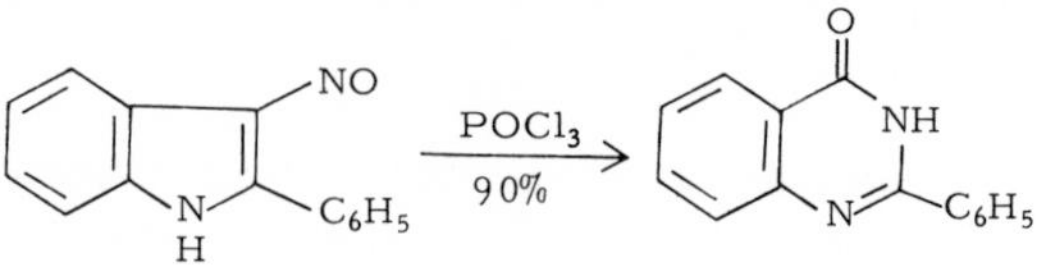

(1)                                             (2)

**2,** 402–403) at 200° for 1 hr. 2-phenylquinazoline-4(3H)-one (2) is formed in 90% yield. The reaction probably involves a Beckmann rearrangement.[2]

*Elimination of iodohydrins.*[3] Treatment of the iodohydrin (1) with freshly distilled phosphoryl chloride in pyridine first at 0° and then at room temperature (30 min.) affords the bicyclic olefin (2) in high yield. The reaction is generally applicable for olefin synthesis.

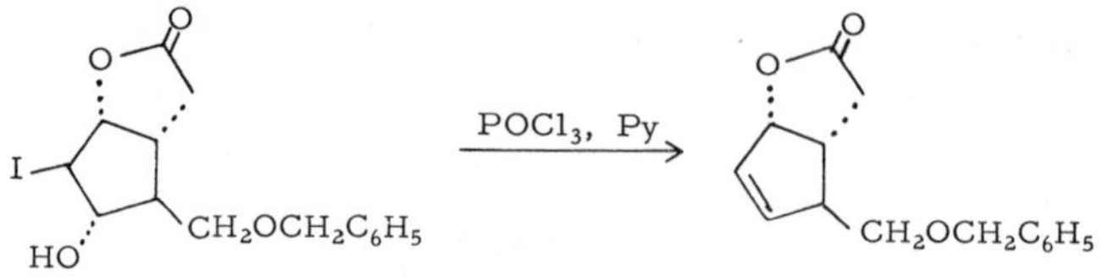

(1)                                             (2)

The same elimination has been effected by Corey and Grieco with mesyl chloride (*see* **Methanesulfonyl chloride,** this volume).

*Cholesteryl phosphorodichloridate.* Cholesteryl phosphorodichloridate (2) can be prepared in 87% yield by reaction of cholesterol (1) with phosphoryl chloride in the

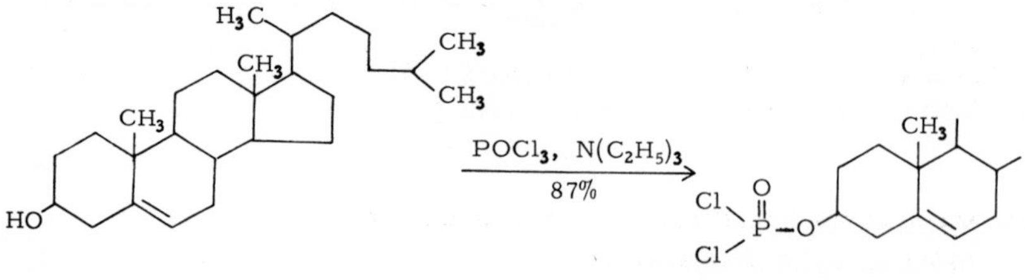

(1)                                             (2)

presence of an equimolar quantity of triethylamine in ether.[4] The reaction of cholesterol with phosphoryl chloride and pyridine in acetone gives an impure product.[5]

[1] F. Yoneda, M. Higuchi, and R. Nonaka, *Tetrahedron Letters*, 359 (1973).
[2] F. D. Popp and W. E. McEwen, *Chem. Rev.*, **58,** 370 (1958).
[3] P. Crabbé and A. Guzmán, *Tetrahedron Letters*, 115 (1972).
[4] R. J. W. Cremlyn, B. B. Dewhurst, and D. H. Wakeford, *Synthesis*, 648 (1971).
[5] H. Venner, *J. prakt. Chem.* [4], **12,** 59 (1960).

$CH_2Cl$

**4-Picolyl chloride hydrochloride,** ⟨structure⟩ · HCl . Mol. wt. 164.04, m.p. 160–163°.

Suppliers: Aldrich, Pfaltz and Bauer, ROC/RIC.

*Picolyl ester method of peptide synthesis.* Young *et al.*[1] have described a method of peptide synthesis in which the carboxyl terminal group is protected as the 4-picolyl

$$NH_2CHCOOCH_2-\overset{R}{|}$$

(1)

ester (1). These are prepared by the reaction of 4-picolyl chloride hydrochloride with the tetramethylguanidinium salt of the benzyloxycarbonyl amino acid in DMF at 90–100°. The N-protecting group can be removed from these derivatives with hydrogen bromide in acetic acid. 4-Picolyl esters can be cleaved by cold alkali, by catalytic hydrogenation, by electrolytic reduction, and by sodium in liquid ammonia. The new procedure provides an alternative to the solid-phase principle in which the carboxy terminal residue is esterified to a hydroxymethyl polystyrene resin. But it has the advantage that peptide synthesis proceeds in homogeneous solution. The weakly basic residue provides a handle by which the protected peptide product can be separated from the excess of acylating agents, coproducts, and byproducts. The separation is effected by adsorption on a cation exchanger or by extraction into an acidic aqueous phase. The new procedure was first used in a synthesis of the tetrapeptide L-leucyl-L-alanylglycyl-L-valine,[2] and has since been used in a synthesis of Val[5]-angiotensin-II (overall yield 38%)[2] and of bradykinin (overall yield 42%).[3]

[1]R. Camble, R. Garner, and G. T. Young, *J. Chem. Soc.* (*C*), 1911 (1969).
[2]R. Garner and G. T. Young, *ibid.*, 50 (1971).
[3]D. J. Schafer, G. T. Young, D. F. Elliott, and R. Wade, *ibid.*, 46 (1971).

**Picric acid, 1,** 884–885.

*Cleavage of hydrocarbon picrates.* Picrates of dimethylnaphthalenes are usually cleaved by absorption of the picrate on a column of basic alumina and elution of the hydrocarbon with benzene. Eisenbraun *et al.*[1] report that petroleum ether dissolves the arene but essentially no picric acid; the method reduces the amount of solvent and alumina required. In addition the picric acid can be recovered for reuse. The method is less suitable for picrates of more complex arenes which are not readily cleaved (fluoranthene picrate and pyrene picrate).

[1]E. J. Eisenbraun, T. E. Webb, J. W. Burnham, and L. E. Harris, *Org. Prep. Proc.*, 3, 249 (1971).

**Picryl chloride, 1,** 885.

*Condensation with N-phenacylquinolinium bromide* and deacylation to give 8,10-dinitroisoindolo[2.1-*a*]quinoline[1]:

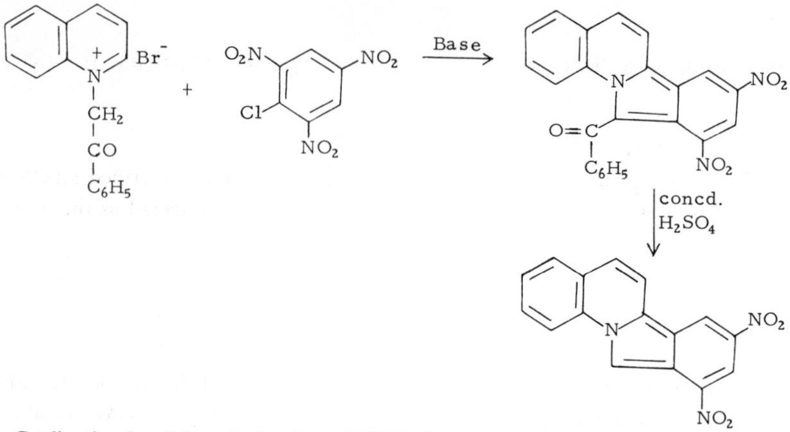

*Cyclization involving elimination of HNO$_2$.*[2] A 2-methylcycloimonium salt with a reactive N-methylene group condenses with picryl chloride at the 2-methyl group to give a 2-(2,4,6-trinitrobenzylidene)-substituted 1,2-dihydroheterocycle provided that the methylene group is only moderately activated. In the presence of a base such as piperidine the latter compounds form six-membered rings by loss of HNO$_2$.

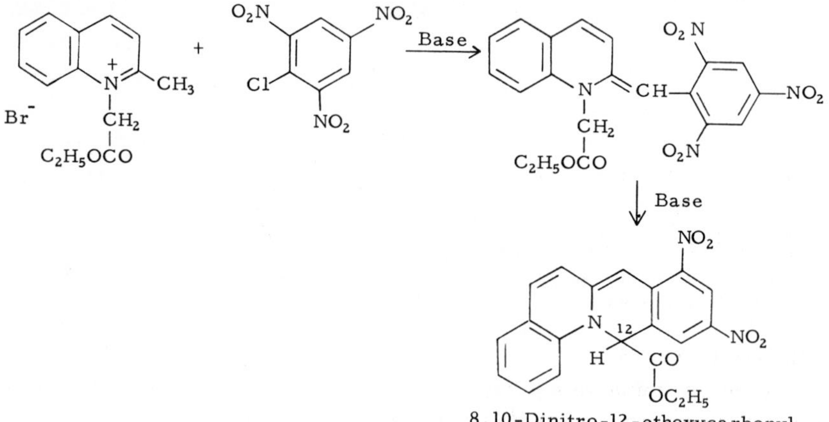

8,10-Dinitro-12-ethoxycarbonyl-12H-isoquinolino[2.3-a]quinoline

[1] D.-B. Reuschling and E. Kröhnke, *Ber.*, **104**, 2103 (1971).
[2] *Idem, ibid.*, **104**, 2110 (1971).

**Piperazine,** 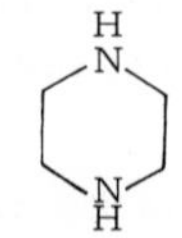
Mol. wt. 86.14, m.p. 108–110°. Suppliers: Aldrich, Baker, Columbia, Eastman, Fisher, Fluka, Howe and French, K and K, MCB, P and B, Sargent, Schuchardt.

*Lactonization of methyl o-formylbenzoate.* Piperazine is an unusually effective reagent for lactonization of methyl o-formylbenzoate (1) in either water[1] or dioxane[2]

(1)  (2)  (3)

solutions. It is much more effective than piperidine (**1**, 886–890; **2**, 332) or morpholine (**1**, 705–707).

[1]G. Dahlgren and D. M. Schell, *J. Org.*, **32**, 3200 (1967).
[2]G. H. Henderson and G. Dahlgren, *ibid.*, **38**, 754 (1973).

**Piperidine, 1, 886–890; 2, 332.**

*Correction* (**2**, 332). The reference should be D. E. Baldwin, V. M. Loeblich, and R. V. Lawrence, *J. Org.*, **23**, 25 (1958).

**Platinum–Alumina catalyst.**

*Adamantane synthesis.* Irish chemists[1] report a new catalyst that is much more effective than the so-called $AlBr_3$ sludge catalyst (**2**, 20) for isomerization of strained tricyclic hydrocarbons into adamantanes. The catalyst is prepared from alumina and chloroplatinic acid followed by sequential treatment with $H_2$, HCl, and $SOCl_2$. The catalyst can be reactivated by treatment with $O_2$ at 500° and then repetition of the $H_2$–HCl–$SOCl_2$ treatment.

[1]D. E. Johnston, M. A. McKervey, and J. J. Rooney, *Am. Soc.*, **93**, 2798 (1971).
[2]J. P. Gianetti and R. T. Sebulsky, *Ind. Eng. Chem., Prod. Res. Develop.*, **8**, 356 (1969).

**Polymethylhydrosiloxane (PMHS)**, $(CH_3)_3SiO(CH_3HSiO)_nSi(CH_3)_3$, where $n$ is $\sim 35$. PMHS is an easily handled, chemically inert liquid; it is made by Dow Corning and supplied by Aldrich. See **3**, 294.

*Reduction.*[1] PMHS and an organotin catalyst, bis(dibutylacetoxytin)oxide (DBATO),[2] in a refluxing protic solvent (usually ethanol) specifically reduce aldehydes and ketones to alcohols in high yield. One equivalent of siloxane hydride is required per mole of substrate with the solvent contributing a proton:

$$\underset{R_2}{\overset{R_1}{\diagdown}}C{=}O + \frac{1}{n}[CH_3HSiO]_n + ROH \xrightarrow{DBATO} \underset{R_2}{\overset{R_1}{\diagdown}}CHOH + \frac{1}{n}[CH_3Si(OR)O]_n$$

The active reducing agent in the system is probably dibutylacetoxytin hydride.
Examples:

$$(C_6H_5)_2C{=}O \longrightarrow (C_6H_5)_2CHOH \ (80\%)$$

$$C_6H_5CHO \longrightarrow C_6H_5CH_2OH \ (100\%)$$

$$CH_2{=}CHCOCH_3 \longrightarrow CH_2{=}CHCHOHCH_3 \ (65\%) + C_2H_5COCH_3 \ (35\%)$$

$$\underline{p}\text{-Benzoquinone} \longrightarrow \text{Hydroquinone} \ (81\%)$$

Under these conditions esters, carboxylic acids, lactones, amides, nitriles, alkyl halides, and nitro compounds are not reduced. Some of the groups are reduced, however, in refluxing 2-ethylhexanol (130°).

PMHS can also be used with a catalytic quantity of 5% palladium on charcoal to effect hydrogenation of terminal and *cis*-olefins, aromatic nitro compounds, and aromatic aldehydes. Ethanol (95%) containing one drop of concentrated HCl is used as solvent; the reaction is conducted at 40–60° (*caution:* the hydrogenations are sometimes exothermic). This method provides a convenient form of low-pressure hydrogenation.

Examples:

$$C_6H_5NO_2 \xrightarrow[\text{C}_2\text{H}_5\text{OH}]{\text{PMHS, Pd/C}} C_6H_5NH_2 \ (89\%)$$

$$C_6H_5CHO \longrightarrow C_6H_5CH_3 \ (84\%)$$

$$\text{1-Octene} \longrightarrow \underline{\text{n}}\text{-Octane} \ (88\%)$$

$$CH_2{=}CHCOCH_3 \longrightarrow C_2H_5COCH_3 \ (100\%)$$

[1] J. Lipowitz and S. A. Bowman, *J. Org.*, **38**, 162 (1973).
[2] DBATO is most conveniently prepared by the reaction of dibutyltin oxide with acetic acid; m.p. 54–57°, 80% yield.

**Polyphosphate ester (PPE)**, **1**, 892–894; **2**, 333–334; **3**, 229–231.

*Bischler–Napieralski reaction.*[1] Cyclization of the nitroamide (1) with the usual condensation reagents[2] failed, but was realized in moderate yield with polyphosphate ester prepared according to Cava (**3**, 229). The reaction was used in one step of a total synthesis of the thalictrum alkaloid adiantifoline.

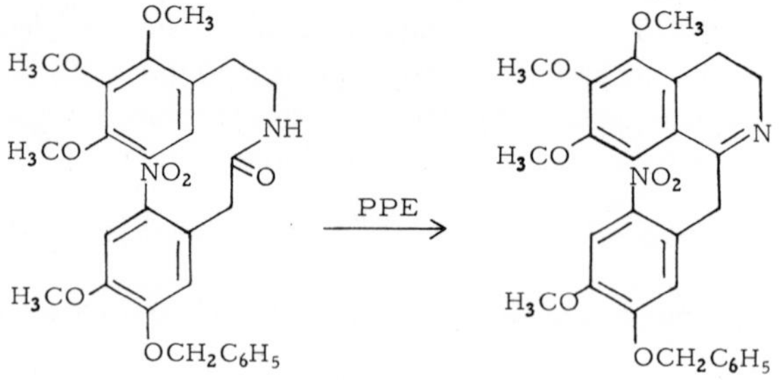

(1)                                     (2)

*1,3-Thiazine derivatives.* The cyanoamide (1)[3] condenses with an aromatic acid and PPE to form 5-carbamoyl-6-methylthio-1,3-thiazine-4-ones (2) in 27–90% yield. The reaction with aliphatic acids gives 5-cyano-6-methylthio-1-keto-1,3-thiazine-4-ones (3) in low yields.[4]

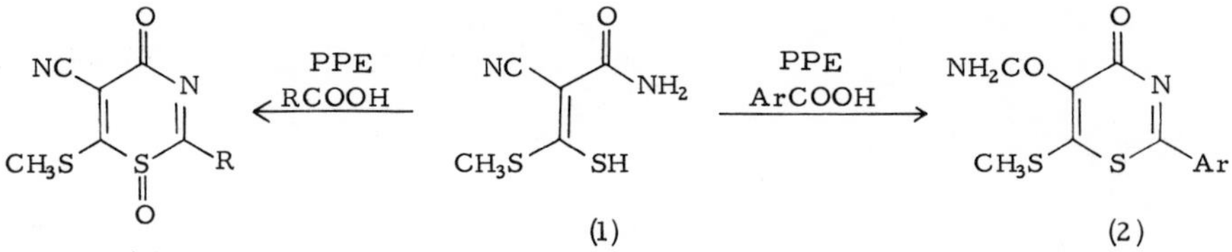

$$(3) \qquad (1) \qquad (2)$$

[1] R. W. Doskotch, J. D. Phillipson, A. B. Ray, and J. L. Beal, *J. Org.*, **36**, 2409 (1971).
[2] W. M. Whaley and T. R. Govindachari, *Org. React.*, **6**, 74 (1951).
[3] T. Takeshima, M. Yokoyama, N. Fukada, and M. Akano, *J. Org.*, **35**, 2438 (1970).
[4] M. Yokoyama, Y. Sawachi, and T. Isso, *ibid.*, **38**, 802 (1973).

**Polyphosphoric acid (PPA), 1**, 894–905; **2**, 334–336; **3**, 231–233.

*Cyclization of γ-arylbutyric acids.* Eisenbraun et al.[1] greatly prefer use of PPA to the acid chloride–AlCl$_3$ procedure for cyclization of 4-(2,5-dimethylphenyl)-2-methylbutyric acid (1) to 2,5,8-trimethyl-1-tetralone (2).

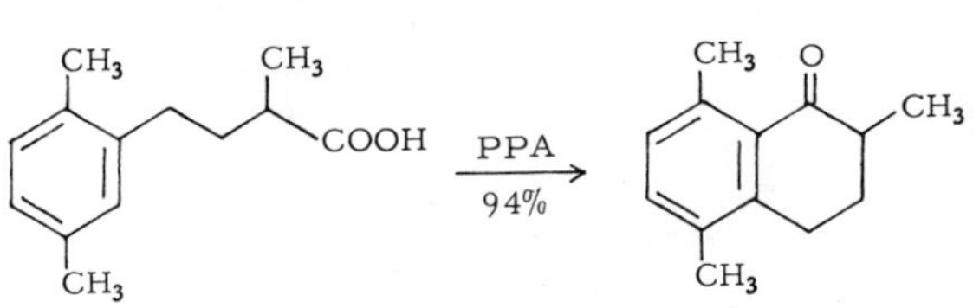

$$(1) \qquad\qquad (2)$$

*Aromatic aldazines and ketazines.*[2] Polyphosphoric acid is an excellent catalyst and solvent for production of azines from aromatic aldehydes and ketones in the presence of various carbonyl reagents, for example, hydrazine, its salts, semicarbazide hydrochloride, toluene p-sulfonohydrazide, and acid hydrazides. The reaction is usually complete at 100° within 15 min. The reaction is not useful in the case of aliphatic carbonyl compounds.

*Intramolecular acylations.* Usually a large excess of PPA is used for intramolecular acylations. Metz[3] in a detailed study of the cyclization of 2-phenoxybenzoic acid (1) to xanthone (2) found that a large excess of PPA is unnecessary and even leads to lower

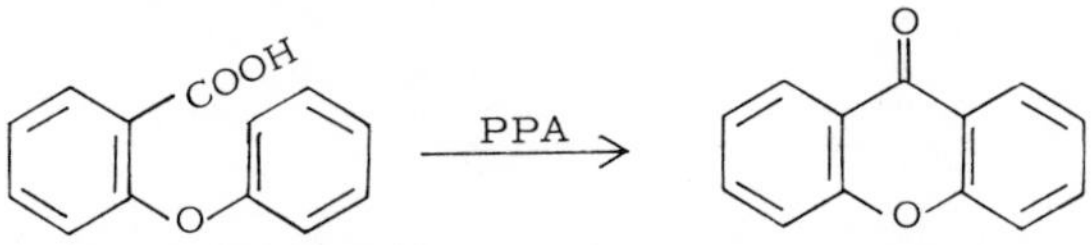

$$(1) \qquad\qquad (2)$$

yields. He recommends use of 350 g. of 84% PPA per mole of carboxylic acid with reaction times of 1–2 hr. at temperatures of 80–120°. Using these conditions (70°) he[4] was able to obtain a maximum yield of indanone-1 (4) from 3-phenylpropanoic acid (3).

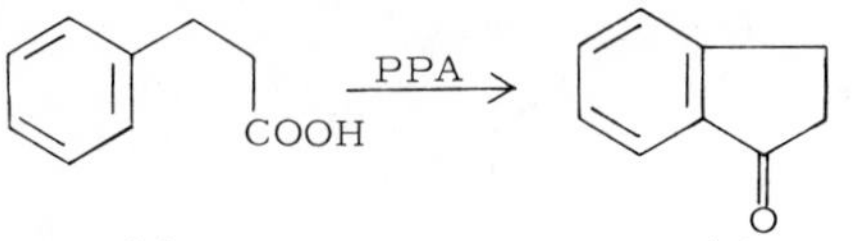

(3)                    (4)

**2-Aminoquinolines.**[5] Chloroformamidines (1) react with alkynyl Grignard reagents to afford the hitherto unknown propiolic acid amidines (2), which can be cyclized to 2-aminoquinolines (3) with PPA at temperatures of about 150°.

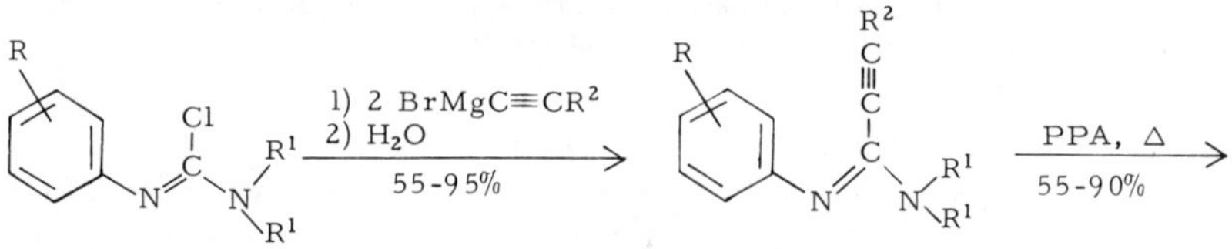

(1)                    (2)

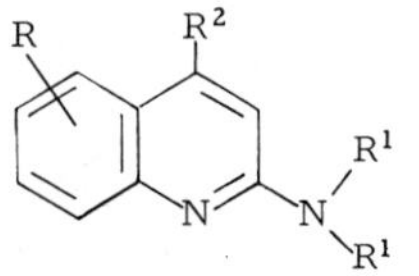

(3)

**Intramolecular Claisen condensation.** The synthesis of β-diketones by Claisen condensation[6] is ordinarily only possible when the product can be stabilized as the enolate. Gerlach and Müller[7] report that the nonenolizable β-diketone (2, bicyclo[2.2.2]octane-2,6-dione) can be obtained from 3-keto-1-cyclohexaneacetic acid (1) in 75% yield when a solution of (1) dissolved in glacial acetic acid is treated with polyphosphoric acid. The

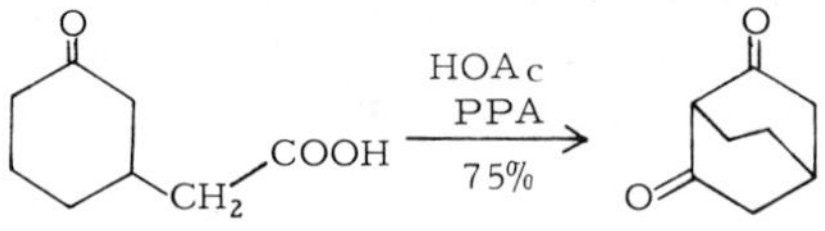

(1)                    (2)

yield is less if acetic acid is omitted; acetic anhydride cannot be used to replace acetic acid. Bartlett and Woods[8] were unable to prepare (2) by base-catalyzed condensation of methyl 3-keto-1-cyclohexaneacetate; they obtained the diketone, in low yield, by passing 3-keto-1-cyclohexaneacetic acid over manganese(II) oxide heated at 340°.

Under the same conditions, (4, spiro[4.4]nonane-1,6-dione) is obtained from (3) in 81% yield.

$$\text{(3)} \xrightarrow[\substack{\text{HOAc} \\ \text{PPA} \\ 81\%}]{} \text{(4)}$$

[1] E. J. Eisenbraun, C. W. Hinman, J. M. Springer, J. W. Burnham, and T. S. Chou, *J. Org.*, **36**, 2480 (1971).

[2] D. B. Mobbs and H. Suschitzky, *J. Chem. Soc. (C)*, 175 (1971).

[3] G. Metz, *Synthesis*, 612 (1972).

[4] *Idem, ibid.*, 614 (1972).

[5] W. Ried and P. Weidemann, *Ber.*, **104**, 3329 (1971).

[6] C. R. Hauser, F. W. Swamer, and J. T. Adams, *Org. React.*, **8**, 59 (1954).

[7] H. Gerlach and W. Müller, *Angew. Chem., internat. Ed.*, **11**, 1030 (1972).

[8] P. D. Bartlett and G. F. Woods, *Am. Soc.*, **62**, 2935 (1940).

**Potassium in graphite.** Potassium can be intercalated in graphite (Fisher Scientific) very easily by heating a mixture of the two elements, in the absence of air, at 70°. The stoichiometry can be adjusted to give $C_8K$, $C_{24}K$, $C_{36}K$, and $C_{48}K$.

*Reduction of ketones.*[1] Saturated and conjugated ketones can be reduced by the reagent to alcohols, probably by a mechanism similar to electrochemical reduction (as illustrated for acetophenone in scheme I). In some cases, pinacols are formed as well. Thus acetophenone is reduced to the alcohol (45 % yield) and the pinacol (45 % yield). Generally the alcohol is the predominant product. For example, benzophenone is reduced to benzhydrol in 98 % yield. $\alpha,\beta$-Unsaturated ketones are reduced to saturated alcohols. Reduction of camphor gives predominantly the *exo*-alcohol; note that reduction with sodium in alcohol or with potassium, in the presence of graphite (not intercalated), gives predominantly the *endo*-alcohol.

Scheme I:

[1] J.-M. Lalancette, G. Rollin, and P. Dumas, *Canad. J. Chem.*, **50**, 3058 (1972).

**Potassium amide, 1**, 907–909; **2**, 936; **3**, 232–233.

*Aminodehydroxylation.*[1] Rossi and Bunnett have reported a general method for conversion of phenols into anilines. The phenol is first converted into an aryl diethyl

$$ArOH \ + \ NaOH \ + \ (C_2H_5O)_2POCl \xrightarrow[80-90\%]{} ArOPO(OC_2H_5)_2$$

phosphate ester by reaction with diethyl phosphorochloridate (**1**, 248; **3**, 98) and NaOH. These esters are then reduced to anilines in good yield by treatment with potassium amide and potassium metal in liquid ammonia. This reaction is probably not applicable to nitro- or halogen-substituted phenols.

*Phenanthridine synthesis.*[2] Treatment of the haloanil (1) with potassium amide in liquid ammonia gives phenanthridine (2) in > 90% yield. A benzyne intermediate is

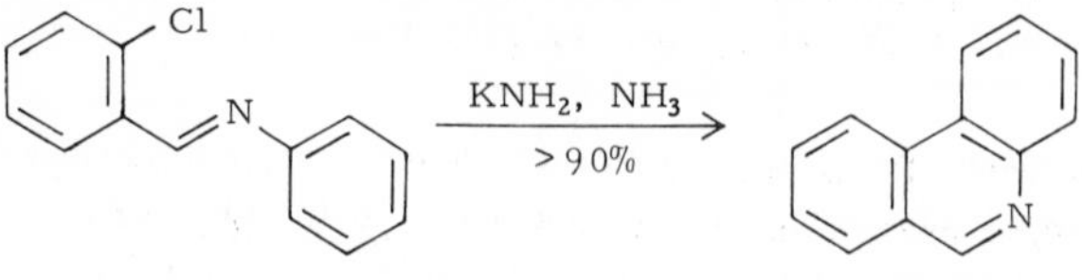

presumably involved. The cyclization succeeds in the presence of alkyl, alkoxy, dialkyl-amino, cyano, carboxy, halogen, and carbonyl groups, but fails in the case of nitro substituents.

*3,3-Dimethoxycyclopropene* (3). Baucom and Butler[3] have reported a relatively simple synthesis of 3,3-dimethoxycyclopropene (3) from 2,3-dichloropropene (1). This compound is converted into 1-bromo-3-chloro-2,2-dimethoxypropane (2) by treatment

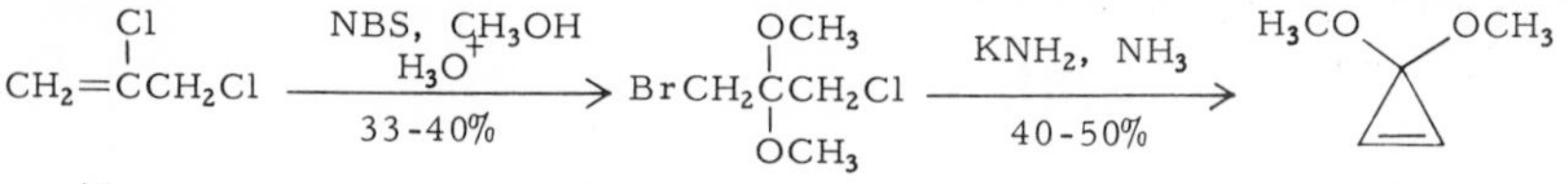

with NBS and methanol (sulfuric acid as catalyst). Cyclization of (2) to (3) was achieved by use of potassium amide in liquid ammonia.

[1]R. A. Rossi and J. F. Bunnett, *J. Org.*, **37**, 3570 (1972).
[2]S. V. Kessar, R. Gopal, and M. Singh, *Tetrahedron*, **29**, 167 (1973); S. V. Kessar, D. Pal, and M. Singh, *ibid.*, **29**, 177 (1973).
[3]K. B. Baucom and G. B. Butler, *J. Org.*, **37**, 1730 (1972).

**Potassium borohydride, KBH$_4$.** Mol. wt. 53.95. Supplier: Metal Hydrides, Inc., Beverly, Mass.

*Removal of the thiazolidine ring of a penicillin.*[1] Treatment of the sulfone (1),

(1, R = <u>o</u>-O$_2$NC$_6$H$_4$)
0.10 g.

(2)
8.6 mg.

derived from a penicillin, with excess potassium borohydride in cold aqueous 2-pro-panol gives the 2-azetidinone (2). The cleavage can also be effected with excess lithium tri-*t*-butoxyaluminum hydride (**1**, 620–625; **2**, 251–252; **3**, 188) in THF.

[1]J. C. Sheehan and C. A. Panetta, *J. Org.*, **38**, 940 (1973).

**Potassium *t*-butoxide**, **1**, 911–916; **2**, 338–339; **3**, 233–234.

*Dehydrobromination.* In a procedure for the preparation of tri-*t*-butylcyclo-propenyl fluoroborate (4), Ciabattoni *et al.*[1] prepared dineopentylketone (1) by the Grignard synthesis formulated, converted it to the $\alpha,\alpha'$-dibromide (2), and effected double dehydrobromination by treatment of this intermediate with potassium *t*-butoxide. The resulting cyclopropenone (3) was then brought into reaction with a solu-tion of commercial *t*-butyllithium in pentane. The mixture was quenched with water, and the pentane layer was washed, dried, and evaporated on a rotary evaporator. The resulting pale-yellow oil is taken up in ether and treated at 0° under rapid magnetic

$$(CH_3)_3CCH_2MgCl + (CH_3)_3CCH_2COCl \xrightarrow[75-77\%]{} MgCl_2 + (CH_3)_3CCH_2\overset{O}{\overset{\|}{C}}CH_2C(CH_3)_3$$
(1)

$$(CH_3)_3CCH_2\overset{O}{\overset{\|}{C}}CH_2C(CH_3)_3 + 2\ Br_2 \xrightarrow{85\%} (CH_3)_3CCHC\ CHC(CH_3)_3 + 2\ HBr$$
(1)    (2)

$$(CH_3)_3CCHC\ CHC(CH_3)_3 + 2\ (CH_3)_3COK \xrightarrow{79-83\%}$$
(2)

$$\xrightarrow{} \underset{(CH_3)_3C}{\overset{C=C}{}}\overset{O}{\overset{\|}{C}} + 2\ (CH_3)_3COH + 2\ KBr$$
(3)

$$\underset{(CH_3)_3C}{\overset{O}{\overset{\|}{C}}}\ C=C\ C(CH_3)_3 \xrightarrow[\substack{1)\ (CH_3)_3CLi \\ 2)\ H_2O \\ 3)\ HBF_4,\ Ac_2O,\ 68-79\%}]{}$$
(3)

$$\underset{(CH_3)_3C}{\overset{C(CH_3)_3}{}}\ C=C\ C(CH_3)_3 \quad BF_4^-$$
(4)

stirring with a freshly prepared 10% solution of fluoroboric acid in acetic anhydride (prepared under nitrogen at −40°). After the resulting suspension has been stirred for 20 min. the white precipitate is collected on a sintered glass funnel (medium porosity) under vacuum and washed thoroughly with three 75-ml. portions of ether. The product is then dissolved in a minimal amount of boiling acetone (*ca.* 300 ml.) and cooled in a freezer (*ca.* −25°) to afford 15.3 g. of tri-*t*-butylcyclopropenyl fluoroborate as white needles after filtration and washing with ether. Concentration of the mother liquors gives two further crops of pure product. The total yield is 24–28 g. (68–79%).

**Dehydrochlorination.** Dehydrochlorination of 1,6,7,7-tetrachloro-2,5-diphenyl-bicyclo[4.1.0]hept-3-ene (1) with a slight excess of potassium *t*-butoxide in THF gives

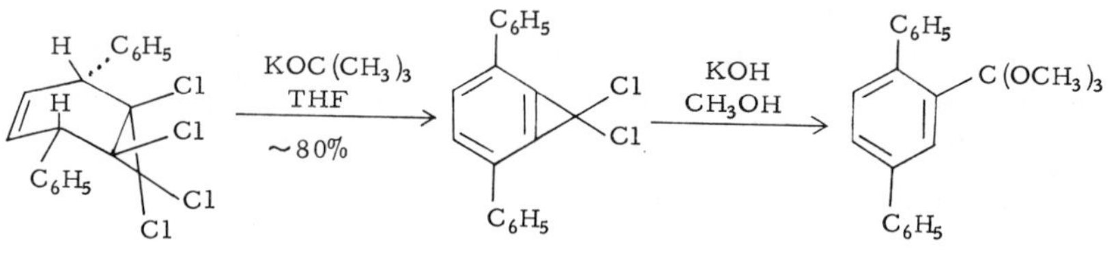

(1)                              (2)                              (3)

7,7-dichloro-2,5-diphenylbenzocyclopropene (2) in about 80% yield. Treatment of (1) with potassium hydroxide in methanol leads to the orthoester (3), evidently formed via

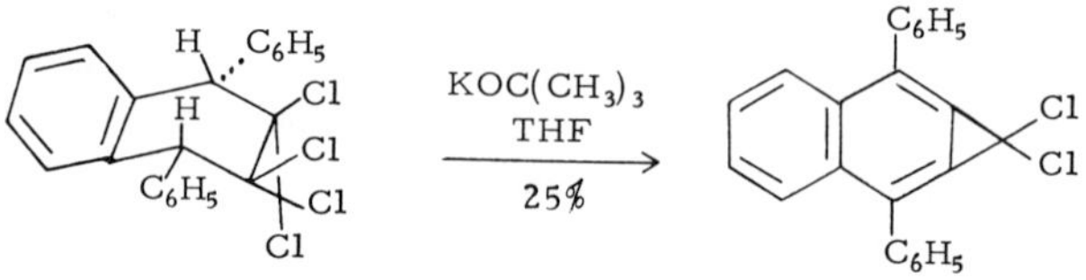

(4)                              (5)

(2) by a solvolytic ring cleavage.[2] Dehydrochlorination of the benzobicycloheptene (4) with potassium *t*-butoxide in THF gives 1,1-dichloro-2,7-diphenylcyclopropa[*b*]-naphthalene (5); in this case the yield is only 25%.[3]

2,3-Dichloro-2-norbornene (8) can be prepared by a three-step procedure.[4] A Diels–Alder reaction between cyclopentadiene (**1**, 181–182) and trichloroethylene (Firma Carl Roth, Karlsruhe, West Germany) gives 5,5,6-trichloro-2-norbornene (6); this is hydrogenated (5% Pd/C) to give trichloronorbornane (7). Dehydrochlorination of (7) with potassium *t*-butoxide in *t*-butanol gives (8). The overall yield is low, mainly

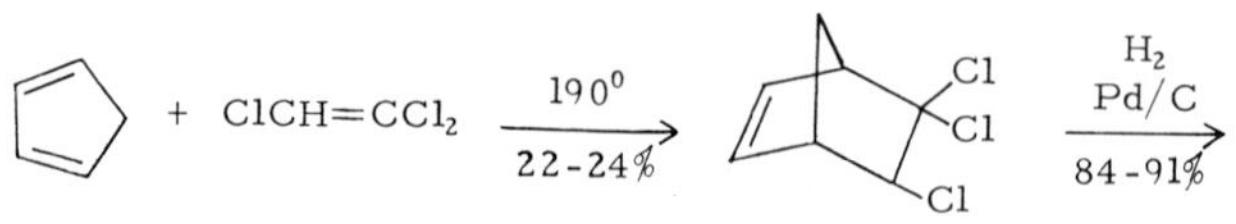

(6)

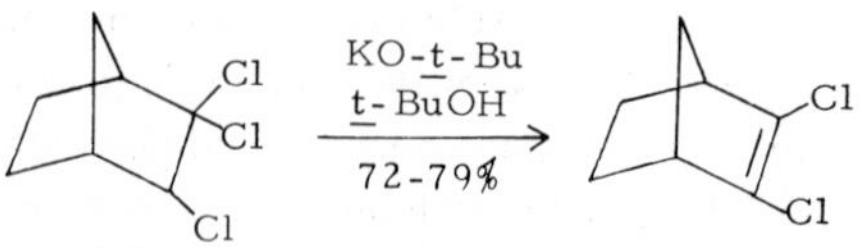

(7)                              (8)

owing to the fact that cyclopentadiene undergoes a further Diels–Alder reaction with (6).

Schoberth and Hanack[5] have reported a simple synthesis of cyclopropylacetylene (11) from methyl cyclopropyl ketone (9). The ketone is converted into the dichloride (10) by treatment with purified phosphorus pentachloride[6] in carbon tetrachloride, and this is then dehydrohalogenated with potassium *t*-butoxide in DMSO.

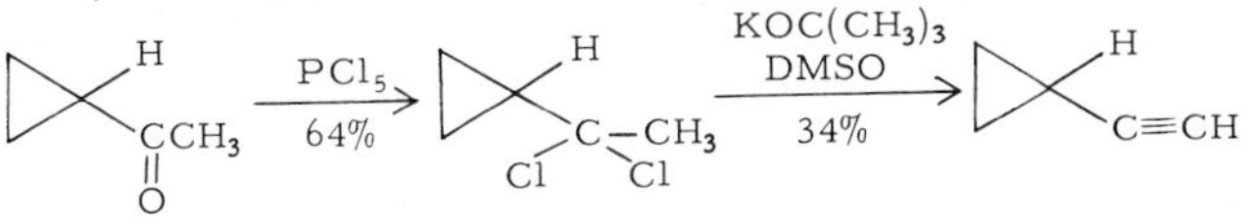

(9)                    (10)                    (11)

***Chloro-, bromo-, and iodocyclooctatetraene.***[7] The *cis*-7,8-dihalocycloocta-1,3,5-trienes which are formed by chlorination or bromination of cyclooctatetraene at $-60°$ in dichloromethane are dehydrohalogenated *in situ* by potassium *tert*-butoxide at $-45°$; yields of 85% of bromo- and 74–83% of chlorocyclooctatetraene are obtained. Iodocyclooctatetraene is accessible from cyclooctatetraenyllithium and iodine. Rational

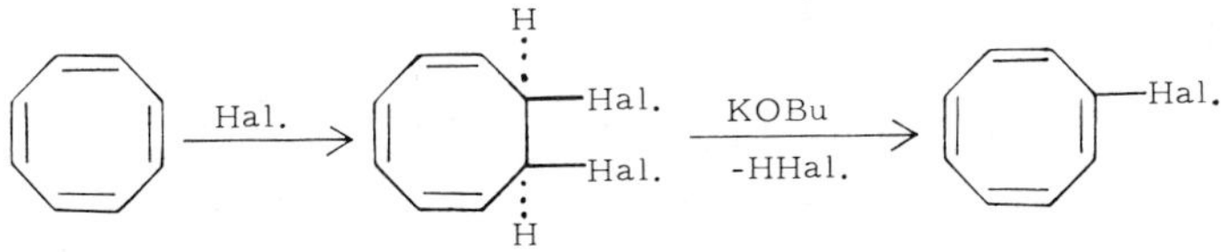

procedures are described for the preparation of methoxy-, phenoxy-, acetoxy-, methyl-, and phenylcyclooctatetraene.

***Cyclopropyl ketones.***[8] Cyclopropyl ketones can be obtained in 42–59% yields by heating esters of 3-hydroxypropylphosphonium salts with potassium *t*-butoxide in refluxing *t*-butanol. For example, 3-acetoxypropyltriphenylphosphonium bromide (1)[9] is converted into methyl cyclopropyl ketone (2) in 49% yield by way of the intermediates shown in the formulation.

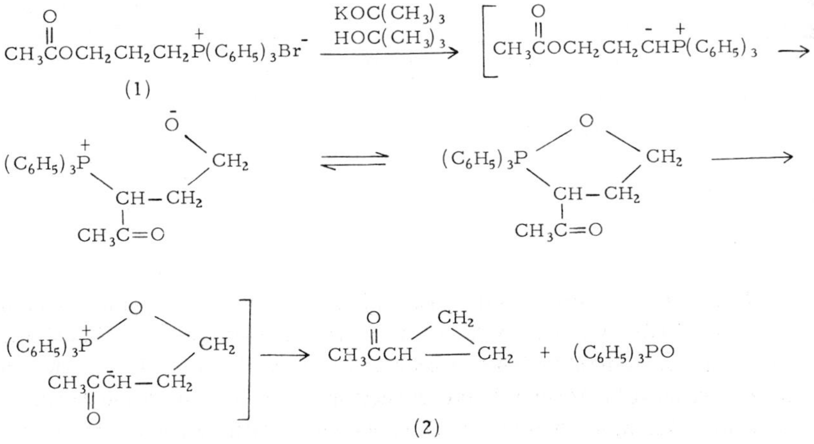

(1)

(2)

*Benzocyclopropene.*[10] Benzocyclopropene (3) can be prepared conveniently in two steps. 1,4-Cyclohexadiene (1) is treated with dichlorocarbene (generated from chloroform and potassium *t*-butoxide) to give 7,7-dichlorobicyclo[4.1.0]heptene-3 (2) in 41–44% yield. This intermediate is then treated with potassium *t*-butoxide and DMSO

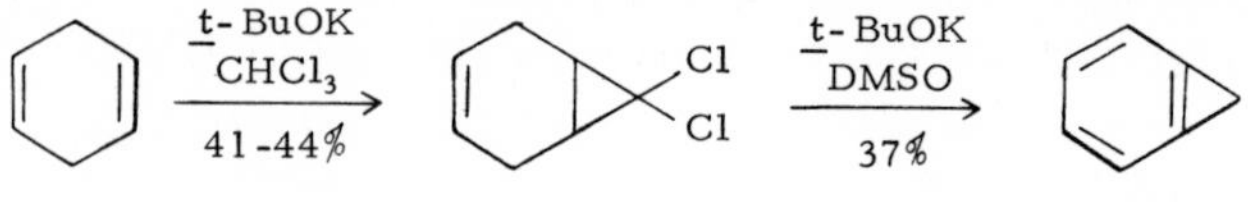

(1)                    (2)                    (3)

first at 15–20° and then at room temperature for 25 min. Benzocyclopropene (3) is obtained in 37% yield by a base-induced dehydrochlorination–isomerization reaction.

*Vinylalkylidenecyclopropanes.*[11] Vinylalkylidenecyclopropanes can be prepared from *gem*-dichlorocyclopropanes by dehydrochlorination with potassium *t*-butoxide in DMSO. The simplest example is the formation of vinylmethylenecyclopropane (2)

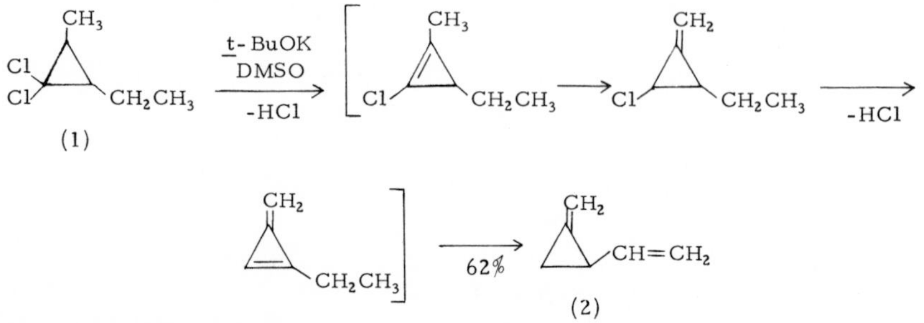

from 1,1-dichloro-2-ethyl-3-methylcyclopropane (1), possibly via the intermediates formulated.

*Butatrienes.*[12] Olefins can be converted into butatrienes by reaction with dichloro-carbene followed by treatment with potassium *t*-butoxide in *t*-butanol (or sodium meth-oxide in DMSO). Yields are in the range 50–90%.

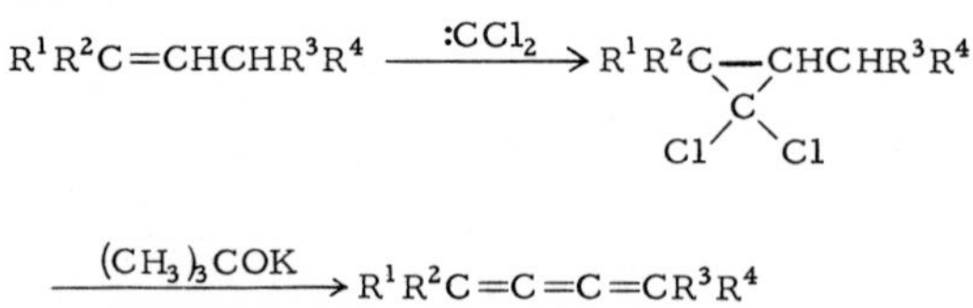

*9-Oxabicyclo[3.3.1]nonene-1 (3).*[13] This bridgehead alkene was synthesized in the following way. Hydroboration–oxidation of 1,5-cyclooctadiene gave *cis*-1,5-cycloocta-nediol, which was oxidized by Jones reagents to 1-hydroxy-9-oxabicyclo[3.3.1]-nonane (1) in 49% overall yield. Attempts to prepare derivatives of (1) usually led to derivatives of the hydroxy ketone (4); however, reaction of (1) with methanesulfonyl chloride and

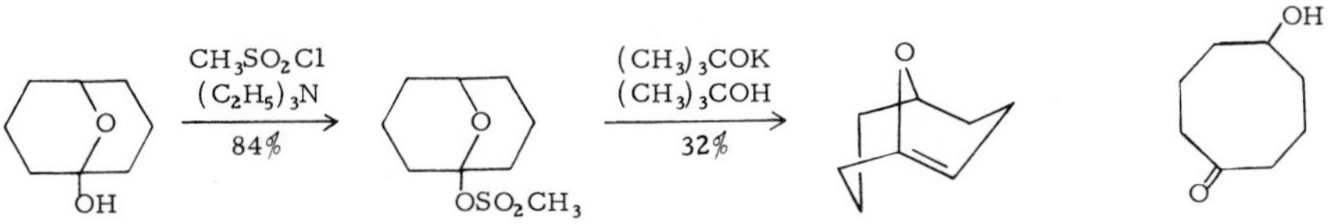

(1)                (2)                        (3)              (4)

triethylamine gave the mesylate (2) in good yield. The reaction of (2) with potassium
*t*-butoxide in *t*-butanol at 80° led to elimination of methanesulfonic acid and production
of (3) in 32% yield.

*Isomerization of unsaturated compounds* (**1**, 913–914; **2**, 336–337; **3**, 233). The
reaction of β-methallyl chloride (1) with sodamide in boiling dioxane or di-*n*-butyl
ether affords a mixture of methylenecyclopropane (2) and 1-methylcyclopropene (3).
Treatment of the mixture with potassium *t*-butoxide in DMSO affords methylenecyclopropane in a purity of 98.3–98.6% (glc analysis).[14]

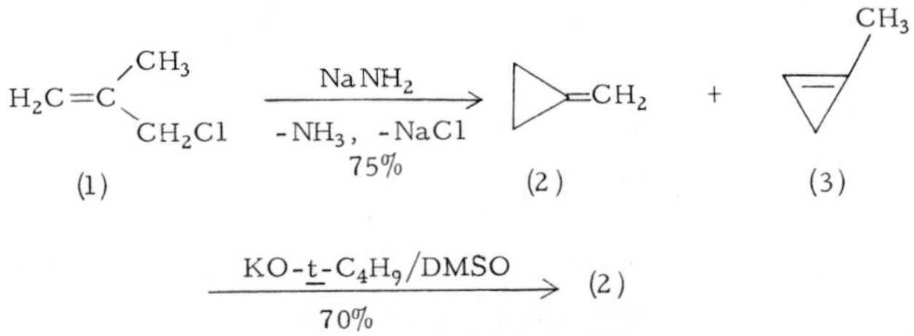

In the synthesis of [18]annulene,[15] the 35 g. of crude cyclooctadecadiyne (4) obtained
in the first step is dissolved in 800 ml. of benzene and heated to boiling on a water bath
in a 2-1. round-bottomed flask fitted with a reflux condenser and a calcium chloride
tube. A solution of potassium *t*-butoxide, prepared from 44 g. of potassium and 1 1.
of dry *t*-butyl alcohol by boiling under reflux under nitrogen until the metal dissolved,
is added and the mixture refluxed for 30 min. to isomerize three yne groups to diene
units. The workup is tedious and includes chromatography.

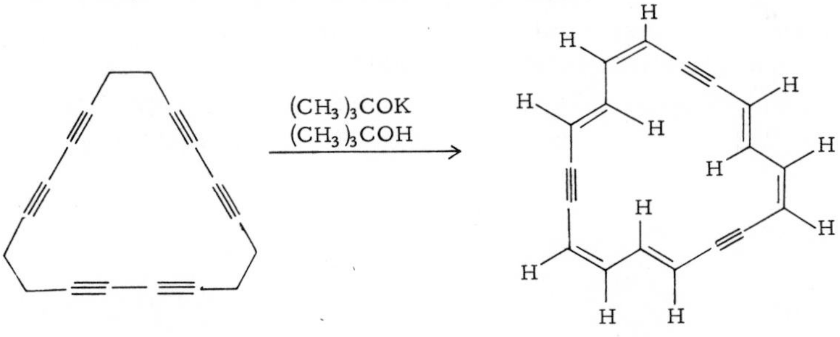

(4)                                    (5)

*Methylene transfer.*[16] Treatment of the sulfonium salt (1) with potassium *t*-butoxide
in 1,2-dimethoxyethane results in internal methylene transfer to give the cyclopropyl
compound (2) in > 95% yield.

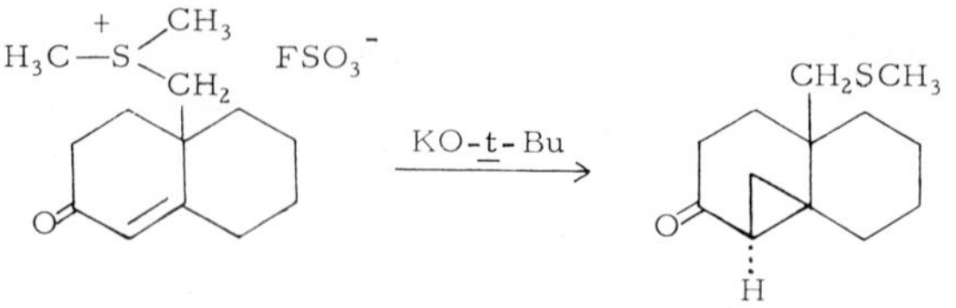

(1)                                         (2)

*Cyclopentene oxide and cyclohexene oxide.*[17] Treatment of *trans*-2-hydroxycyclo-pentylmercuric chloride (1a) and of *trans*-2-hydroxycyclohexylmercuric chloride (1b) with potassium *t*-butoxide in diglyme gives cyclopentene oxide (2a) and cyclohexene

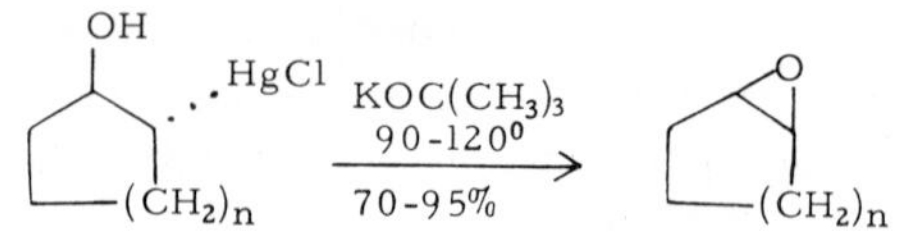

(1a, n = 1)                              (2a, n = 1)
(1b, n = 2)                              (2b, n = 2)

oxide (2b) in high yield. The corresponding cycloheptyl derivative (1c, $n = 3$) is converted mainly into cycloheptanone (70% yield).

*β-Hydroxydithiocinnamic acids.*[18] Potassium *t*-butoxide is the base of choice for synthesis of β-hydroxydithiocinnamic acids (1) by reaction of ketones with $CS_2$.

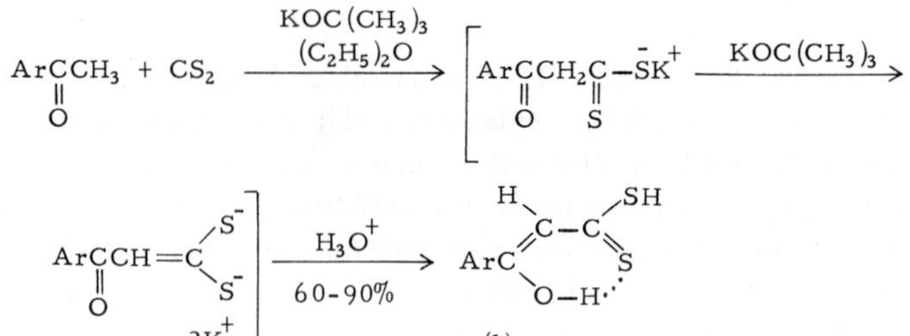

(1)

β-Hydroxydithiocinnamic acids (1) can be converted into the monosalt by treatment with tetra-*n*-butylammonium hydrogen sulfate[19] and $NaOCH_3$, which on alkylation gives the dithioesters (2) in high yield. Mercaptals (3) can be prepared in 70–88% yield by treatment of (2) with thallous ethoxide (**2**, 407–411) and an alkyl halide.

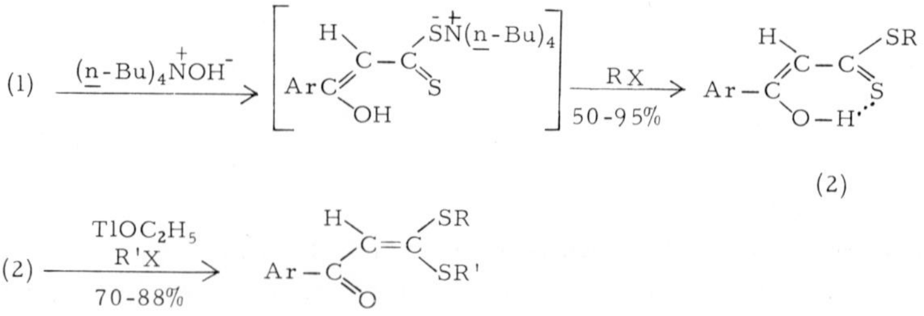

(2)

(3)

(4)                              (5)

When one of the substituents on sulfur in (3) is an allyl or a propargyl group, a thio-Claisen rearrangement is observed (4 → 5).

*Cleavage of nucleosides.*[20] Purine and pyrimidine bases can be cleaved from derivatives of adenosine, cytidine, and uridine by potassium *t*-butoxide in dioxane.

[1] J. Ciabattoni, E. C. Nathan, A. E. Feiring, and P. J. Kocienski, *Org. Syn.*, submitted 1971.
[2] B. Halton and P. J. Milsom, *Chem. Commun.*, 814 (1971).
[3] A. R. Browne and B. Halton, *J.C.S. Chem. Comm.*, 1341 (1972).
[4] G. Nagendrappa, *Org. Syn.*, submitted 1973.
[5] W. Schoberth and M. Hanack, *Synthesis*, 703 (1972).
[6] E. Ott, *Org. Syn., Coll. Vol.*, **2**, 529 (1943).
[7] J. Gasteiger, G. E. Gream, R. Huisgen, and W. E. Konz, *Ber.*, **104**, 2412 (1971).
[8] E. E. Schweizer and W. S. Creasy, *J. Org.*, **36**, 2379 (1971).
[9] Prepared in 97% yield by the reaction of 3-bromopropyltriphenylphosphonium bromide (Aldrich) with sodium acetate in 4:1 acetone–water.
[10] W. E. Billups, A. J. Blakeney, and W. Y. Chow, *Chem. Commun.*, 1461 (1971); procedure submitted to *Org. Syn.*, 1972.
[11] W. E. Billups, T. C. Shields, W. Y. Chow, and N. C. Deno, *J. Org.*, **37**, 3676 (1972).
[12] S. Kajigaeshi, N. Kuroda, G. Matsumoto, E. Wada, and A. Nagashima, *Tetrahedron Letters*, 4887 (1971).
[13] C. B. Quinn and J. R. Wiseman, *Am. Soc.*, **95**, 1342 (1973).
[14] R. Köster, S. Arora, and P. Binger, *Synthesis*, 322 (1971).
[15] K. Stökel and F. Sondheimer, *Org. Syn.*, submitted 1971.
[16] R. S. Matthews and T. E. Meteyer, *Chem. Commun.*, 1576 (1971).
[17] R. A. Kretchmer, R. A. Conrad, and E. D. Mihelich, *J. Org.*, **38**, 1251 (1973).
[18] F. C. V. Larsson and S.-O. Lawesson, *Tetrahedron*, **28**, 5341 (1972).
[19] Supplier: AB Biotec, Göteborg, Sweden.
[20] H. Follman, *Tetrahedron Letters*, 397 (1973).

**Potassium ferrate(VI)**, $K_2FeO_4$. Mol. wt. 198.05, purple crystals. The salt should be protected from moisture.

*Preparation.* Thompson *et al.*[1] recommend that the salt be prepared by sodium hypochlorite oxidation of ferric nitrate. A somewhat tedious purification procedure is necessary to obtain material of 92–96% purity in yields of 44–76%.

*Oxidation.*[2] Potassium ferrate (VI) is a reagent for selective oxidation of primary alcohols and amines to aldehydes and of secondary alcohols to ketones. Double bonds, aldehyde functions, tertiary hydroxyl groups, and tertiary amino groups are resistant to oxidation. The reaction is carried out at room temperature either in water or in aqueous solvents. In fact water is essential for oxidation. The reaction is carried out at an initial pH of 11.5; the final pH is 13.5. In a typical procedure $K_2FeO_4$ (0.002 mole) is added to the alcohol or amine (0.003 mole) in 10 ml. of water and the resulting mixture is shaken vigorously until the purple color of the ferrate(VI) disappears (1–90 min.) with precipitation of brown $Fe(OH)_3$. The product is isolated in the usual way by extraction with ether or benzene.

Examples:

$\underline{n}$-Heptanol $\longrightarrow$ $\underline{n}$-Heptaldehyde (30%)

Benzyl alcohol $\longrightarrow$ Benzaldehyde (80%)

Cyclohexanol $\longrightarrow$ Cyclohexanone (20-30%)

Cinnamyl alcohol $\longrightarrow$ Cinnamaldehyde (75%)

Benzylamine $\longrightarrow$ Benzaldehyde (70%)

1-Methylbenzylamine $\longrightarrow$ Acetophenone (70%)

In the case of carbohydrates, hydroxymethyl groups are selectively oxidized by potassium ferrate to aldehyde groups; secondary hydroxyl groups are not affected.[3]

[1]G. W. Thompson, L. T. Ockerman, and J. M. Schreyer, *Am. Soc.*, **73**, 1379 (1951); R. H. Wood, *ibid.*, **80**, 2038 (1958).

[2]R. J. Audette, J. W. Quail, and P. J. Smith, *Tetrahedron Letters*, 279 (1971).

[3]J. N. BeMiller, V. G. Kumari, and S. D. Darling, *Tetrahedron Letters*, 4143 (1972).

**Potassium ferricyanide, 1,** 929–933; **2,** 345.

*Oxidation of 2-amino-3,6-di-t-butylphenol* with the reagent at pH 7.2 affords 3,6-di-*t*-butylbenzoquinone-1,2 as a dark-blue crystallizate, m.p. 204°, characterized by UV and IR data.[1]

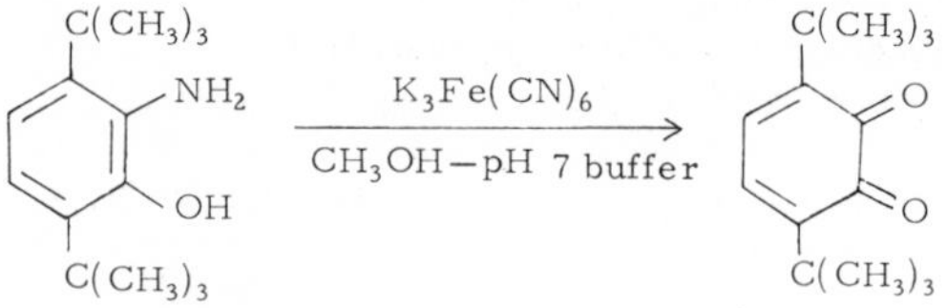

*Phenolic oxidative coupling.* Oxidation of the secondary amine (1) with potassium ferricyanide in the presence of sodium hydrogen carbonate in chloroform gives the dienone (2, norerythrinadienone).[2]

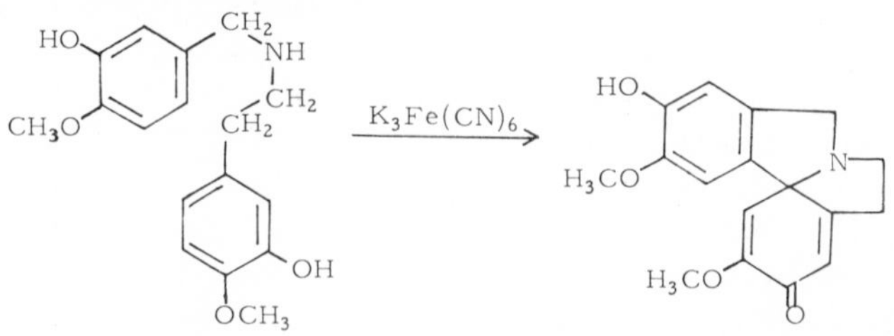

(1, 5 g.)                    (2, 270 mg.)

Oxidation of 2-ethoxycarbonylnorprotosinomenine (3) with potassium ferricyanide in the presence of ammonium acetate and ammonia (pH 9.2) affords the proerythrinadienone (4) in about 2 % yield. The product has a basic skeleton of the key intermediate in the probable biogenesis of *Erythrina* alkaloids.[3]

(3)                             (4)

[1]H. Brockmann and F. Seela, *Ber.*, **104**, 2751 (1971).
[2]T. Kametani, K. Takahashi, S. Shibuya, and K. Fukumoto, *J. Chem. Soc. (C)*, 1800 (1971).
[3]T. Kametani, R. Charubala, M. Ihara, M. Koizumi, and K. Fukumoto, *Chem. Commun.*, 289 (1971).

**Potassium hexachlorotungstate(IV)**, $K_2WCl_6$. Mol. wt. 474.86; red, crystalline salt.

The tungstate is prepared by heating tungsten hexachloride with potassium iodide in a sealed Carius tube for 3 days at 130°.[1]

*Olefin synthesis.* Sharpless and Flood[2] have described a new olefin synthesis by reductive deoxygenation of *vic*-diols (both *cis* and *trans*) by potassium hexachlorotungstate in refluxing THF. In a typical procedure the diol in THF under $N_2$ is converted into the dilithium salt by treatment with methyllithium (2 molar eq.) in diethyl ether. $K_2WCl_6$ (2 molar eq.) is then added and the heterogeneous mixture is heated under reflux for a period ranging from 4 hr. to 4 days, depending on the structure of the diol. Tetrasubstituted diols are reduced rapidly in moderate yield; mono-, di-, and trisubstituted diols are reduced more slowly, but still in moderate yield. The olefin is formed essentially by a *cis* elimination.

[1]C. D. Kennedy and R. D. Peacock, *J. Chem. Soc.*, 3392 (1963).
[2]K. B. Sharpless and T. C. Flood, *Chem. Commun.*, 370 (1972).

**Potassium hexamethyldisilazane [Potassium bis(trimethylsilyl)amide]**, $KN[Si(CH_3)_3]_2$. Mol. wt. 199.49.

*Preparation.*[1] The reagent is prepared by the reaction of potassium amide (**1**, 907; **2**, 936; **3**, 232–233) with hexamethyldisilazane in refluxing benzene $(N_2)$.[2]

*Haloketal cyclization.*[1] The ketal nitrile (1) is cyclized to the cyclohexanone derivative (2) by treatment with 1–3 eq. of potassium hexamethyldisilazane in benzene

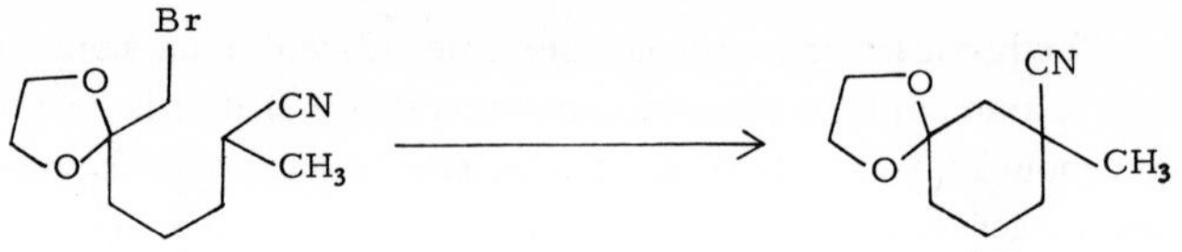

(1)                                          (2)

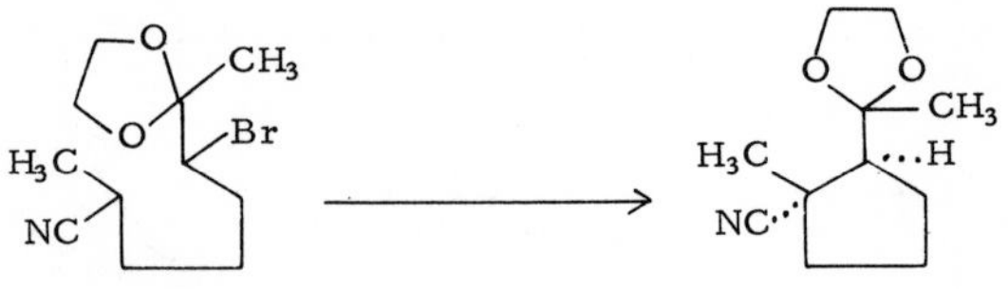

(3)                                          (4)

at room temperature (8–20 hr.) or under reflux (1–4 hr.). Even secondary α-haloketals can be cyclized [(3) → (4)].

The haloketal cyclization is most useful for synthesis of polycyclic systems. Thus (5) is converted into a *cis*-9-cyano-2-decalone (6), and (7) is converted into (8). The hydrindane (10) was prepared from (9).

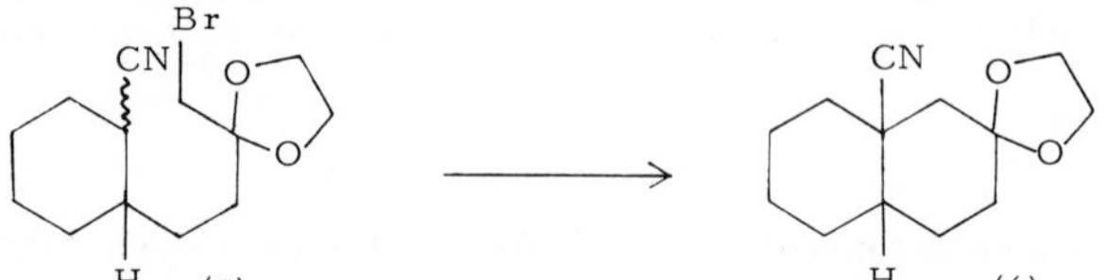

(5)                                          (6)

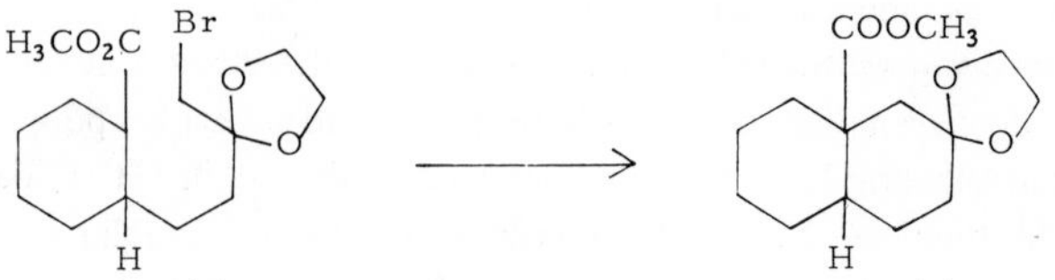

(7)                                          (8)

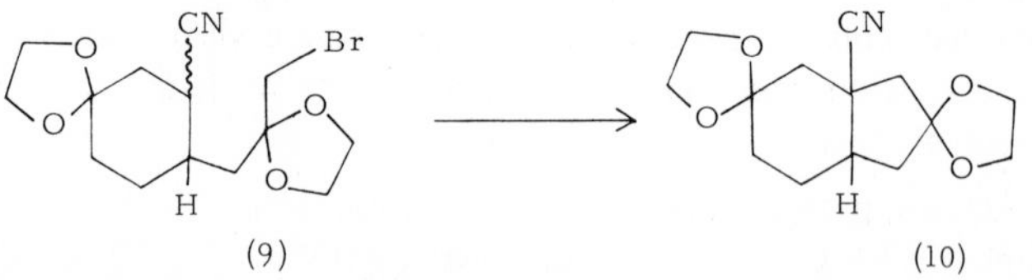

(9)                                          (10)

Two rings can be formed simultaneously. Thus (11) is converted into a mixture of (12) and (13) in the ratio of 95:5.

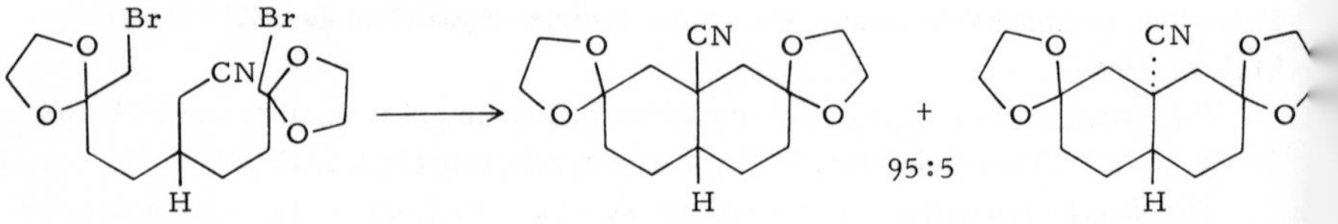

(11)                          (12)                          (13)

95:5

If potassium hexamethyldisilazane is replaced by lithium hexamethyldisilazane in the cyclization of (11), (13) is obtained as the predominant product (95% yield).[3] A possible explanation for the remarkable effect of the cation on this cyclization has been suggested.

[1]G. Stork, J. O. Gardner, R. K. Boeckman, Jr., and K. A. Parker, *Am. Soc.*, **95**, 2014 (1973).
[2]Compare procedure of V. Wannagat and H. Nidersprum, *Ber.*, **94**, 1540 (1961), for preparation of the corresponding sodium derivative.
[3]G. Stork and R. K. Boeckman, Jr., *Am. Soc.*, **95**, 2016 (1973).

**Potassium hydride, 1,** 935; **2,** 346.

*Caution:* Potassium hydride in oil reacts slowly with moist air; it is best handled under nitrogen or argon. Potassium hydride reacts vigorously with water. Residues should be disposed of in the same way as sodium residues.

*Kaliation of amines and feeble organic acids.*[1] Potassium hydride reacts rapidly at room temperature with amines to form potassium mono- and dialkylamides. These are exceptionally strong bases. Thus N-kalioethylenediamine (KEDA) aromatizes limonene

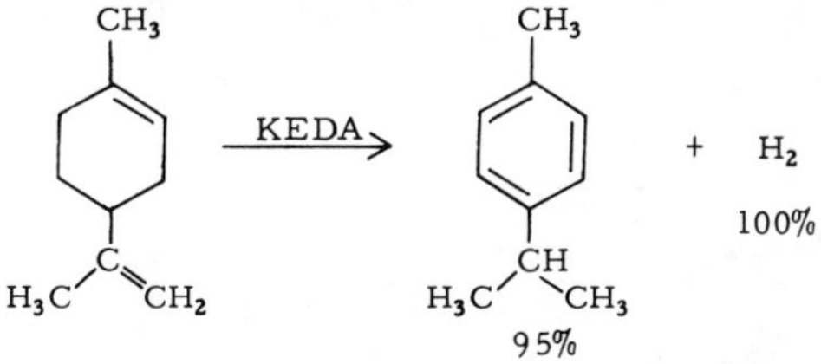

to *p*-cymene in 5 min. at 25°. In contrast the same reaction with lithioethylenediamine requires 2 hr. at 90°.

Potassium hydride reacts with DMSO at 25° to give dimsylpotassium; hydrogen evolution is quantitative in 8 min.

Hindered tertiary alcohols also react vigorously with potassium hydride with quantitative liberation of hydrogen.

[1]C. A. Brown, *Am. Soc.*, **95**, 982 (1973).

**Potassium hydroxide, 1,** 935–937; **2,** 346–347; **3,** 238.

*meso-Naphthodianthrone synthesis.*[1] Treatment of emodine (1) with 0.6 *N* KOH in a sealed tube (hydroquinone, $N_2$, 3 weeks at 110°) yields isohypericine (2) in 35%

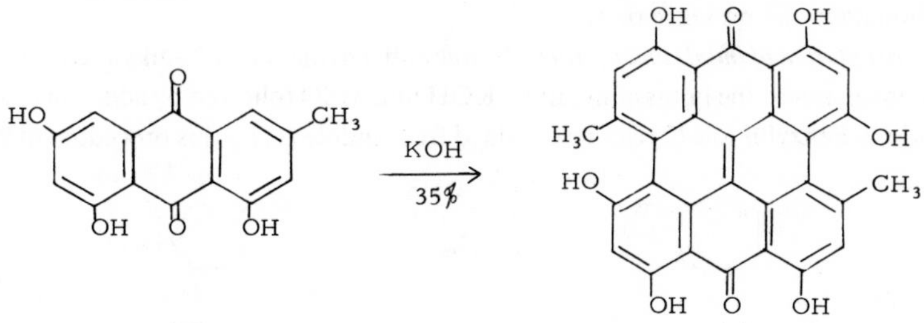

(1)                                    (2)

yield. The reaction is believed to involve Michael addition of an emodine anion to emodine, followed by a series of tautomerizations, condensations, and eliminations.

   *Dehydrohalogenation–hydrolysis.* Heathcock and Hassner used this reagent in ethanol in the completing step of the synthesis of an aziridine (4) from 1,2-dihydro-naphthalene (1)[2] via methyl (*trans*-2-iodo-1-tetralin)carbamate (3).[3]

$$AgNCO + I_2 \longrightarrow AgI + INCO$$

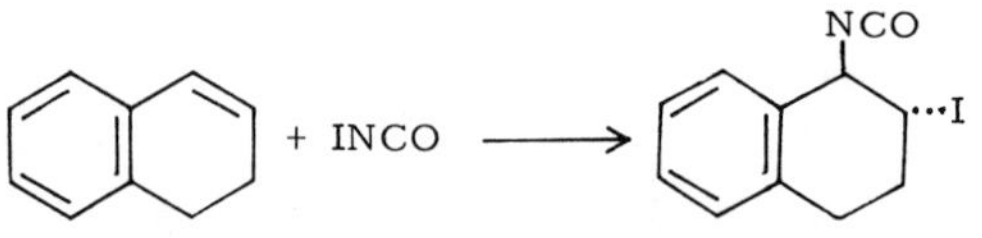

(1)    + INCO    →    (2)

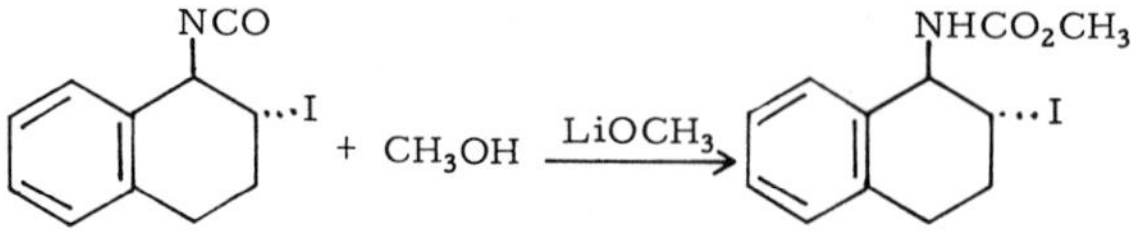

(2)    + CH$_3$OH    $\xrightarrow{\text{LiOCH}_3}$ 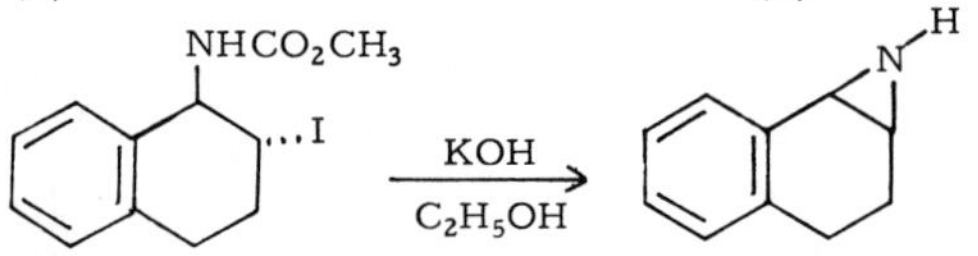    (3)

(3)    $\xrightarrow[\text{C}_2\text{H}_5\text{OH}]{\text{KOH}}$    (4)

   **Δ²-*Pyrazoline-5-ones*.**[4] Treatment of 5-hydroxy-3,5-diphenyl-Δ²-pyrazoline-4-one (1) with methanolic potassium hydroxide gives 4-hydroxy-3,4-diphenyl-Δ²-pyrazoline-5-one (2) in 76–80% yield. A pinacol rearrangement is probably involved. Compound (2)

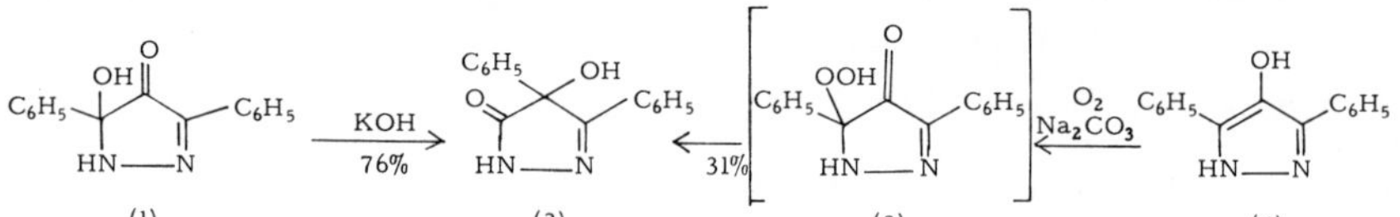

(1)    $\xrightarrow[76\%]{\text{KOH}}$    (2)    $\xleftarrow{31\%}$    (3)    $\xleftarrow[\text{Na}_2\text{CO}_3]{\text{O}_2}$    (4)

can also be obtained, in somewhat low yield, by boiling (4) in methanolic sodium carbonate in the presence of air.

   *Alkylation of indole and pyrrole.*[5] Indole and pyrrole can be N-alkylated in high yield by conversion to the potassium salt by KOH in DMSO followed by addition of an alkyl halide. 1-Benzylindole (2) can be obtained from indole (1) by this procedure in 92–97%

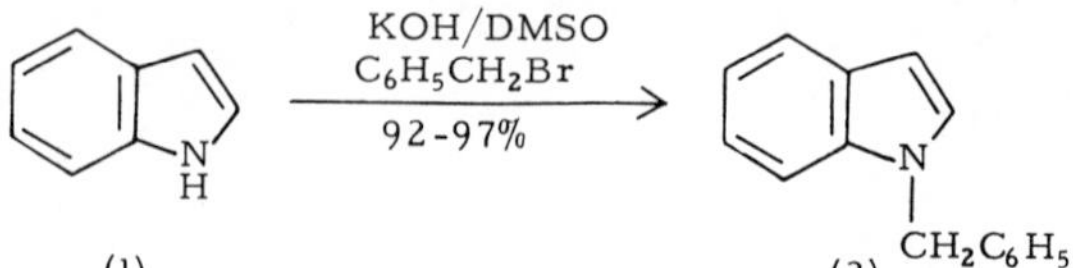

(1)    $\xrightarrow[92-97\%]{\substack{\text{KOH/DMSO} \\ \text{C}_6\text{H}_5\text{CH}_2\text{Br}}}$    (2)

yield. Yields are lower in the case of secondary alkyl halides; the reaction fails with tertiary halides.

[1]W. Steglich and R. Arnold, *Angew. Chem., internat. Ed.,* **12**, 79 (1973).
[2]C. H. Heathcock and A. Hassner, *Org. Syn.,* **51**, 53 (1971).
[3]*Idem, ibid.,* **51**, 112 (1971).
[4]M. J. Nye and W. P. Tang, *Canad. J. Chem.,* **51**, 338 (1973).
[5]H. Heaney, *Org. Syn.,* submitted 1973.

### Potassium iodide–Zinc copper couple.

The reductive elimination of vicinal methanesulfonyloxy groups in 2,3-O-dimesyl-α-D-glucopyranosides was originally carried out with sodium iodide and zinc dust.[1] Fraser-Reid and Boctor[2] report that the reaction is improved both in yield and in

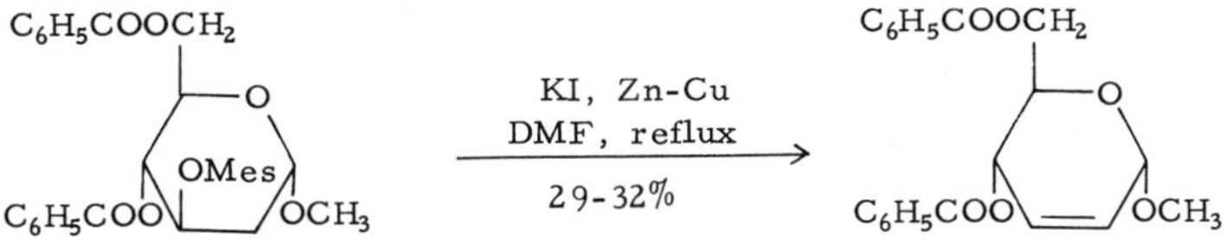

reproducibility by use of zinc–copper couple, as prepared in the Simmons–Smith reaction.[3] The reaction is carried out in refluxing DMF.

[1]R. S. Tipson and A. Cohen, *Carbohydrate Res.,* **1**, 338 (1965).
[2]B. Fraser-Reid and B. Boctor, *Canad. J. Chem.,* **47**, 393 (1969).
[3]R. D. Smith and H. E. Simmons, *Org. Syn.,* **41**, 72 (1961).

### Potassium (sodium) nitrosodisulfonate (Fremy's salt), 1, 940; 2, 347–348.

*Preparation:*[1]

$$NaNO_2 + 2\ SO_2 + NaHCO_3 \longrightarrow HON(SO_3Na)_2 + CO_2$$

$$HON(SO_3Na)_2 + OH^- \xrightarrow[\text{Stainless steel anode}]{-e^-} :\ddot{O}-\ddot{N}(SO_3Na)_2 + H_2O$$

The reagent can be used for preparation of 4,5-dimethyl-1,2-benzoquinone[2]:

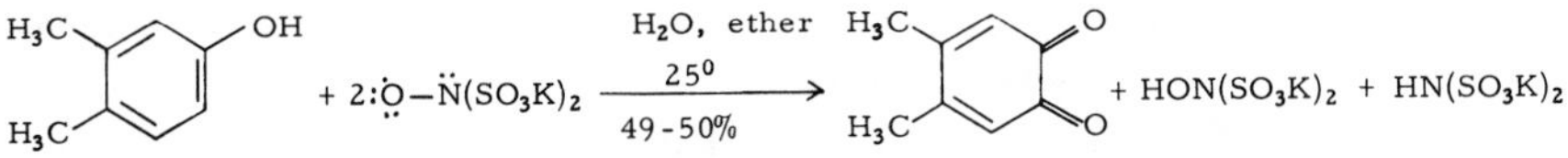

Oxidations with potassium nitrosodisulfonate have been reviewed in detail.[3]

[1]P. A. Wehrli and F. Pigott, *Org. Syn.,* **52**, 83 (1972).
[2]H.-J. Teuber, *ibid.,* **52**, 88 (1972).
[3]H. Zimmer, D. C. Lankin, and S. W. Horgan, *Chem. Rev.,* **71**, 229 (1971).

**Potassium osmate**, $K_2OsO_4$. Mol. wt. 332.40. Supplier: Alfa Inorganics. The reagent presents no problems as long as it is kept dry.

*Hydroxylation of olefins.* Lloyd *et al.*[1] have described two active catalyst solutions for hydroxylation of olefins using potassium osmate. One (catalyst A) is prepared using hydrogen peroxide as oxidant and acetic acid to neutralize the potassium osmate solution. In the other (catalyst B) sodium chlorate (**1**, 1056–1058) is used as oxidant.

$$\begin{array}{c} \diagdown \diagup \\ C \\ \| \\ C \\ \diagup \diagdown \end{array} \xrightarrow{\ K_2OsO_4,\ H_2O,\ HOAc,\ H_2O_2\ } \begin{array}{c} -\overset{|}{C}-OH- \\ -\overset{|}{C}-OH- \\ | \end{array}$$

$$\begin{array}{c} \diagdown \diagup \\ C \\ \| \\ C \\ \diagup \diagdown \end{array} \xrightarrow{\ K_2OsO_4,\ H_2O,\ HOAc,\ NaClO_3\ } \begin{array}{c} -\overset{|}{C}-OH- \\ -\overset{|}{C}-OH- \\ | \end{array}$$

Although catalyst B is simpler to prepare than catalyst A, it should not be used if the diol is to be distilled directly from the reaction mixture, since residual sodium chlorate can cause explosions when heated with an organic compound. Hydroxylation of allyl alcohol with catalyst A gave glycerol in 67% yield. *cis*-Cyclohexane-1,2-diol was obtained in 76% yield by hydroxylation of cyclohexene with catalyst B.

[1] W. D. Lloyd, B. J. Navarette, and M. F. Shaw, *Synthesis*, 610 (1972).

**Potassium perchromate**, $K_3CrO_8$. Mol. wt. 297.31.

*Singlet oxygen.*[1] Potassium perchromate can be generated when an aqueous solution of potassium chromate ($K_2CrO_4$, ROC/RIC) containing potassium hydroxide is added to an aqueous methanolic solution of 30% hydrogen peroxide. It is a convenient source of singlet oxygen. Thus generation of potassium perchromate in the presence of 2,3-dimethylbutene-2 gives rise to 2,3-dimethyl-3-hydroperoxybutene-1 in 35% yield.

$$\begin{array}{c} H_3C \\ \phantom{}\diagdown \\ H_3C \diagup \end{array} C=C \begin{array}{c} \diagup CH_3 \\ \phantom{} \\ \diagdown CH_3 \end{array} \xrightarrow[35\%]{K_3CrO_8} \begin{array}{c} H_3C \\ \diagdown \\ H_3C \diagup \end{array} \underset{\underset{OOH}{|}}{C} - C \begin{array}{c} \diagup CH_2 \\ \phantom{} \\ \diagdown CH_3 \end{array}$$

This is the product of sensitized photooxidation. The salt is a convenient source of singlet oxygen, which can be liberated at a controlled rate depending on the water content of the solvent and the temperature.

[1] J. W. Peters, J. H. Pitts, Jr., I. Rosenthal, and H. Fuhr, *Am. Soc.*, **94**, 4348 (1972).

**Potassium permanganate–Acetic anhydride.**

*α-Diketones from olefins.*[1] Olefins are oxidized directly to α-diketones by reaction with 3–4 molar eq. of potassium permanganate in acetic anhydride at temperatures below 10°. For example, cyclododecene (1) is oxidized to the dione (2); the keto acetate

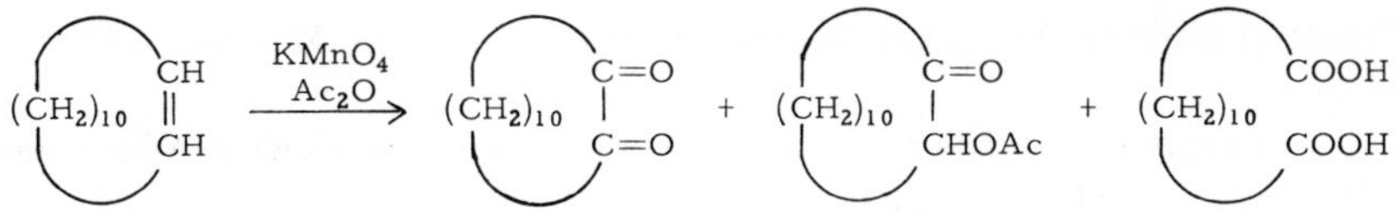

(3) and the dicarboxylic acid (4) are minor products. Since (3) can be hydrolyzed to the α-hydroxy ketone, which is then oxidized to (2) with cupric acetate, the total yield of (2) can be raised to 63%. The reaction can be carried out on a large scale [50 g. of (1)].

Another example is the oxidation of 1-decene (5):

$$CH_3(CH_2)_7CH=CH_2 \longrightarrow CH_3(CH_2)_7\overset{\overset{\displaystyle O}{\|}}{C}CH_2OAc + CH_3(CH_2)_7COOH$$

(5)                    (6) 36%              (7) 50%

Dimethoxyethane can be added as a diluent in cases where the substrate is not soluble in acetic anhydride at low temperatures.

[1] K. B. Sharpless, R. F. Lauer, O. Repič, A. Y. Teranishi, and D. R. Williams, *Am. Soc.*, **93**, 3303 (1971).

**1,3-Propanedithiol**, $HS(CH_2)_3SH$, **1**, 956. B.p. 169–170° (previous b.p. incorrect). Additional suppliers: Columbia, Fluka, K and K, MCB, Pfaltz and Bauer, Sargent, Schuchardt.

*Heterolytic fragmentation of 1,3-dithianyl tosylates.* Marshall and Belletire[1] have shown that a 1,3-dithiane anion can participate in a heterolytic fragmentation reaction.[2] Thus condensation of the aldehyde (1) with 1,3-propanedithiol gives the 1,3-dithiane (2),

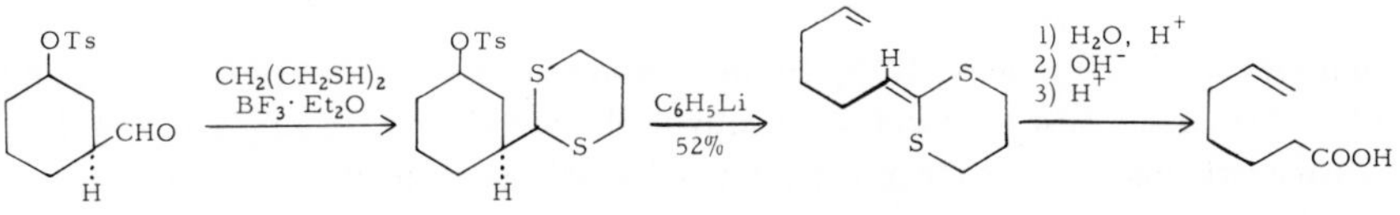

which on treatment with phenyllithium gives the ketene thioacetal (3) in 52% yield of 75% purity. This was converted into the acid (4) by acid hydrolysis with methanolic aqueous *p*-toluenesulfonic acid to give the thiol ester of (4); basic hydrolysis then afforded the acid (4).

[1] J. A. Marshall and J. L. Belletire, *Tetrahedron Letters*, 871 (1971).
[2] C. A. Grob and P. W. Schiess, *Angew. Chem., internat. Ed.*, **6**, 1 (1967).

**Propargyl aldehyde** (also called **Propiolaldehyde**), CH≡CCHO. Mol. wt. 54.05, b.p. 54–57°.

*Preparation* by oxidation of propargyl alcohol from the General Aniline and Film Corporation, Easton, Penn.[1]

*Reaction with o-phenylenediamine* to give dihydrodibenzotetraza[14]annulene[2]:

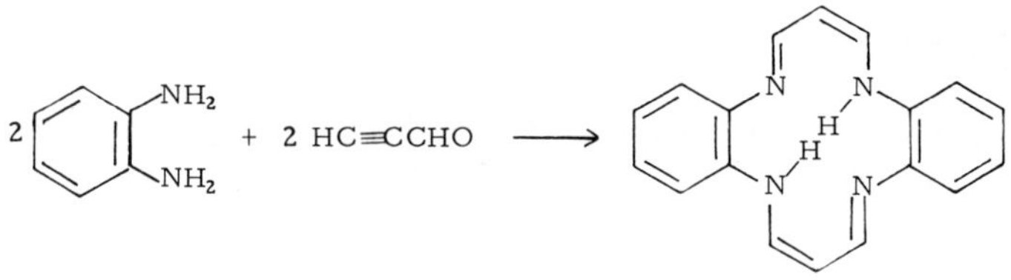

[1]J. C. Sauer, *Org. Syn. Coll. Vol.*, **4**, 813 (1962).
[2]H. Hiller, P. Dimroth, and H. Pfitzner, *Ann.*, **717**, 137 (1968).


**Pyrene.** Mol. wt. 202.24, m.p. 151° Suppliers: Coleman and Bell, Eastman, Matheson.

*Synthesis of 3,4-benzpyrene* (7).[1] Improvements in the succinic anhydride synthesis of this potent carcinogen have already raised the yield from the original 3% of Cook

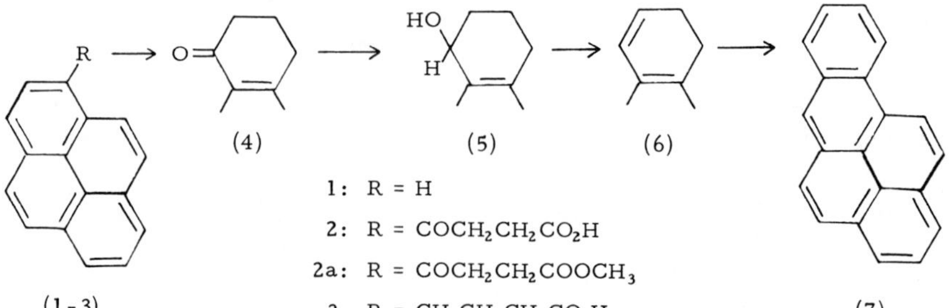

1:  R = H
2:  R = COCH$_2$CH$_2$CO$_2$H
2a:  R = COCH$_2$CH$_2$COOCH$_3$
3:  R = CH$_2$CH$_2$CH$_2$CO$_2$H

and Hewett (1933) to about 40%. Further improvements in the steps (2a) → (3) and (4) → (7) do not raise the yield but greatly facilitate the synthesis of (7) in 50-g. lots. Particularly important is the use of *n*-butanol as solvent in the Wolff-Kishner–Huang-Minlon reduction of (2a). The dehydrogenation of (6) to (7) was done with bromanil in xylene.

[1]H. Schlude, *Ber.*, **104**, 3995 (1971).


**Pyridine, 1**, 958–963; **2**, 349–351.

*Deacetoxylation.* In a synthesis of 11-desoxy-10α-hydroxyprostaglandins, Crabbé *et al.*[1] oxidized the alcohol (1) with Collins reagent (dipyridine chromium(VI) oxide, this volume) to give the diacetoxy aldehyde (2); treatment of this with pyridine for a few

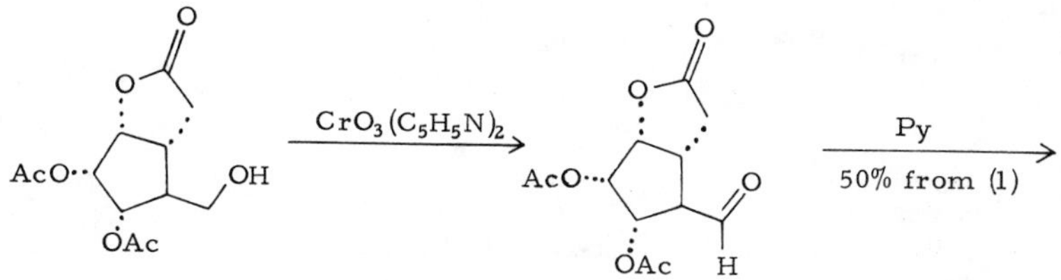

(1)   (2)

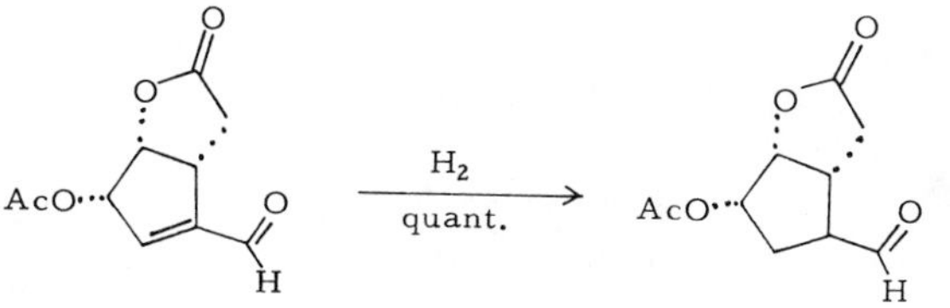

(3)    (4)

minutes affords the conjugated aldehyde (3) in 50% overall yield from (1). Hydrogenation of (3) gives (4) in quantitative yield.

[1]P. Crabbé, A. Cervantes, and M. C. Meana, *J.C.S. Chem. Comm.*, 119 (1973).

**Pyridine hydrochloride**, 1, 964–966; 2, 352–353; 3, 239–240.

*Xanthones.*[1] 2,2′-Dimethoxybenzophenones (1) are converted into xanthones (2)

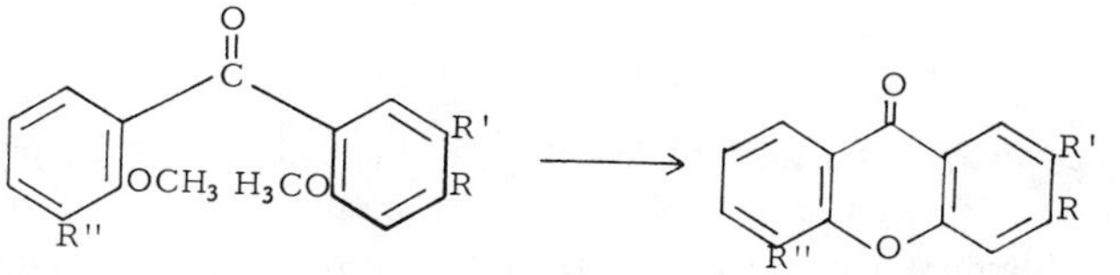

(1)    (2)

when heated for 24–48 hr. in a sufficient quantity of pyridine hydrochloride. Yields of 80–90% can be realized. The method is also applicable to synthesis of furanoxanthones, (3) → (4).

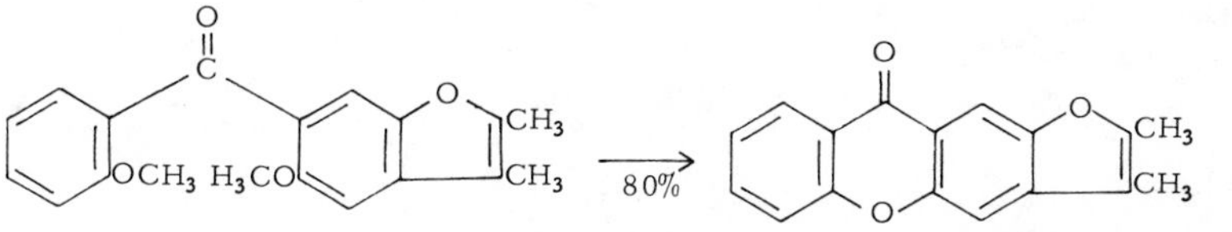

(3)    (4)

*Coumarins.*[2] Both *cis-* and *trans*-2-methoxycinnamic acids (1) are converted by treatment with warm pyridine hydrochloride into coumarins (2).

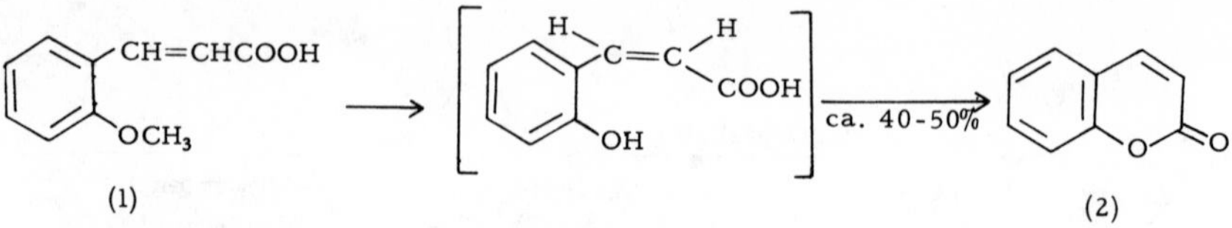

(1)                                              ca. 40–50%                    (2)

[1]R. Royer, J.-P. Lechartier, and P. Demerseman, *Bull. soc.*, 1707 (1971).
[2]R. Royer, B. Bodo, P. Demerseman, and J.-M. Clavel, *ibid.*, 2929 (1971).

**4-Pyrrolidinopyridine,**    Mol. wt. 148.21, m.p. 57°.

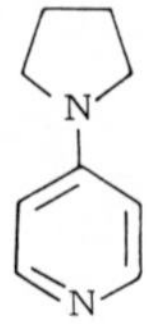

(1)

*Preparation*[1]:

67% overall

(1)

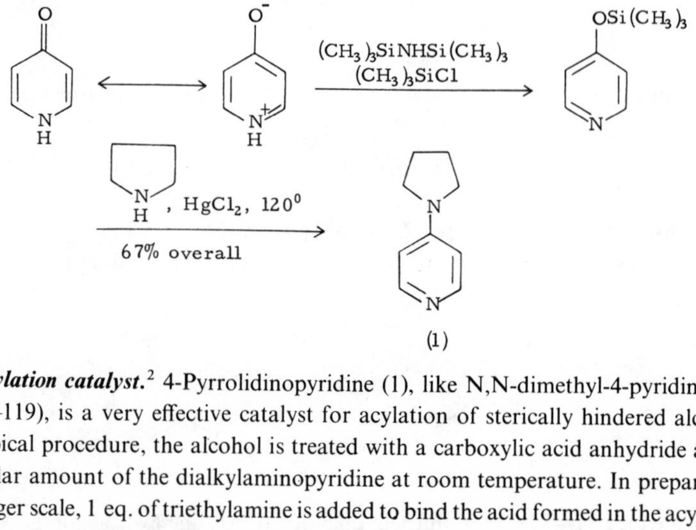

*Acylation catalyst.*[2] 4-Pyrrolidinopyridine (1), like N,N-dimethyl-4-pyridinamine (**3**, 118–119), is a very effective catalyst for acylation of sterically hindered alcohols. In a typical procedure, the alcohol is treated with a carboxylic acid anhydride and an equimolar amount of the dialkylaminopyridine at room temperature. In preparations on a larger scale, 1 eq. of triethylamine is added to bind the acid formed in the acylation. For example, methyl cholate is converted into the triacetate at room temperature within 2 hr. using 4-pyrrolidinopyridine as catalyst. Acetylation with $Ac_2O$/pyridine at room temperature affords only the 3,7-diacetate (the 12α-hydroxyl group is axial).[3] 4-Dialkylaminopyridines are particularly useful catalysts for acylation of acid-sensitive tertiary alcohols such as linalool (2), which can be acetylated in 80% yield by this new procedure.

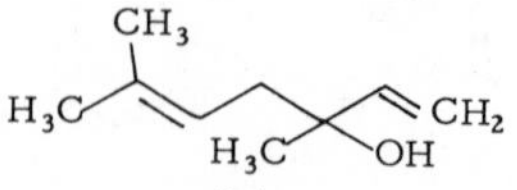

(2)

[1]H. Vorbrüggen, *Angew. Chem., internat. Ed.*, **11**, 305 (1972).
[2]G. Höfle and W. Steglich, *Synthesis*, 619 (1972).
[3]L. F. Fieser and S. Rajagopalan, *Am. Soc.*, **72**, 5530 (1950).

# Q

**Quinuclidine, 1**, 976.

The hydrochloride is available from Aldrich.

*Demethylation of 2-methoxytropone.*[1] 2-Methoxytropone (1) is demethylated to give (2) by a four-fold molar excess of quinuclidine in refluxing benzene in 60% yield. The reaction is slower and the yields are lower using DABCO (**2**, 99–101; this volume). A mercaptide in HMPT also effects demethylation.

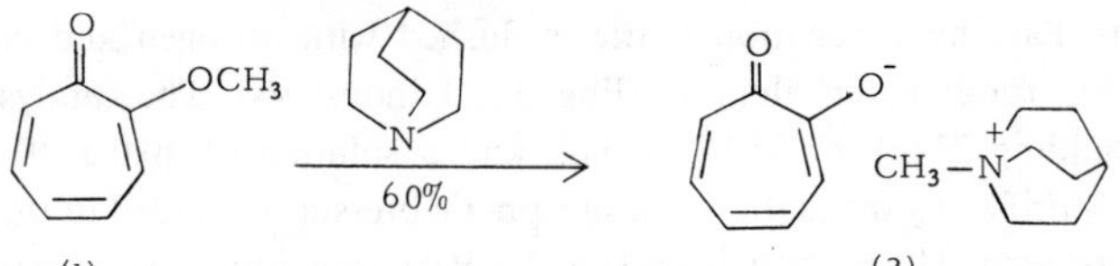

[1]G. Biggi, F. D. Cima, and F. Pietra, *Tetrahedron Letters*, 183 (1973).

# R

**Rhodium (5%) on alumina.**

*Hydrogenation of aromatic nuclei*: *cis,cis-1-decalol*.[1]

A 500-ml. Parr hydrogenation bottle is flushed with nitrogen and charged with 20.0 g. of 5% rhodium on alumina (Engelhard Industries). The catalyst is wet by cautiously adding 25 ml. of 95% ethanol, and a solution of 40.0 g. (0.28 mole) of 1-naphthol (purified by distillation at atmospheric pressure) is added to the bottle with 3 ml. of acetic acid. The mixture is shaken in a Parr apparatus at an initial pressure of 55–60 psi of hydrogen. The theoretical absorption of hydrogen is reached in about 12 hr. The catalyst is then removed by suction filtration and washed twice with 50-ml. portions of ethanol. The combined solutions are concentrated with a rotary evaporator to yield a viscous residue which is dissolved in 150 ml. of benzene. The solution is washed with 75 ml. of 10% sodium hydroxide solution and then with 75 ml. of water, dried over magnesium sulfate for 3 hr., and concentrated with a rotary evaporator to give 39–41 g. (94–97%) of a mixture of geometrical isomers of 1-decalol. *cis,cis*-1-Decalol may be isolated as a crystalline solid from the mixture by the addition of 15–20 ml. of heptane, followed by cooling. Recrystallization from a minimum amount of *n*-heptane affords 13–14 g. (30–33%) of *cis,cis*-1-decalol, m.p. 92–93°.

By the same method, and with 0.33 g. of catalyst per gram of reactant, hydroquinone and resorcinol afforded 90% and 85% of 1,4-cyclohexanediol and 1,3-cyclohexanediol, respectively, and 2-naphthol gave in 88% yield a mixture of 2-decalol geometrical isomers.

[1]A. I. Meyers, W. M. Beverung, and R. Gault, *Org. Syn.*, **51**, 103 (1971).

**Rhodium (5%) on carbon (Norit), 1, 982–983.**

*Dimerization and trimerization of norbornadiene*.[1] A 1-l. round-bottomed flask containing a magnetic stirring bar is fitted with a reflux condenser connected to a

mercury bubbler, from which the atmosphere in the flask can be evacuated and replaced with nitrogen and through which excess pressure in the flask can be vented to the air.

Norbornadiene (463 g., 5.03 moles), distilled from calcium hydride, and 5 % rhodium on carbon (2.5–4.5 g., 0.001–0.002 g.-atom of Rh, Engelhard Industries) are added and the mixture is stirred and refluxed for 1 day, and the rhodium on carbon is filtered. The filtrate is distilled at atmospheric pressure, yielding recovered norbornadiene and a distillation residue containing the reaction product. The procedure is repeated using the recovered norbornadiene, and the repetitions are continued until norbornadiene is consumed.

The initial treatments probably remove a catalyst poison from the norbornadiene. Thus while after the first or second reflux about 95 % of the norbornadiene is recovered, on a fifth reflux 120 g. of norbornadiene was completely oligomerized by 2.5 g. of 5 % rhodium on carbon in 10 hr. Norbornadiene that has been refluxed with Norit gives a higher yield than norbornadiene that has not, but the simplest procedure for effecting the oligomerization is to reflux norbornadiene with rhodium on carbon until the yield is high. Five such treatments using a total of 20 g. of 5 % rhodium on carbon may be required.

The distillation residues are combined and distilled (b.p. 70–73°/0.5 mm.), giving 326 g. (70 % yield) of norbornadiene dimer. The pot residue solidifies when cooled to room temperature and crystallization from 95 % ethanol gives 24.5 g. (5.3 %) of norbornadiene trimer, m.p. 176–178°.

Norbornadiene dimer is a colorless liquid consisting of a mixture of isomers (1, 84 %), (2, 12 %), and (3, 4 %). They can be separated by distillation through an annular Teflon spinning band column manufactured by Nester/Faust Corporation, Newark,

(1)　　　　　　　　　(2)　　　　　　　　　(3)

Delaware. Thus distillation of 73 g. of dimer yielded 50 g. of (1), b.p. 136°/25 mm., 98 % pure, and 5.0 g. of (2), b.p. 140°/22 mm., 97 % pure.

*Tetrahydrofurane-cis-2,5-dicarboxylic acid.*[2] Rhodium (5 %) on carbon is the catalyst of choice for hydrogenation of furane-2,5-dicarboxylic acid (1) to tetrahydrofurane-cis-2,5-dicarboxylic acid (2).

[1] J. J. Mrowca, R. J. Roth, N. Acton, and T. J. Katz, *Org. Syn.*, submitted 1972.
[2] J. A. Moore and J. E. Kelly, *Org. Prep. Proc.*, **4**, 289 (1972).

**Rhodium oxide**, $Rh_2O_3$. Mol. wt. 253.82. Suppliers: Alfa, Fluka, MCB, ROC/RIC, Schuchardt.

*Aminomethylation of alkenes.* The original procedure of Reppe[1] for aminomethylation of alkenes (equation 1) employed iron pentacarbonyl, $Fe(CO)_5$ (**1**, 519–520: **2**, 229–230; **3**, 167), as catalyst. Iqbal[2] now finds that rhodium oxide is a far more effective

(1)    $$\text{>C=C<} + 3\ CO + H_2O + HN< \xrightarrow{\text{cat.}} HC-\overset{|}{\underset{|}{C}}-CH_2N< + 2\ CO_2$$

catalyst. However, higher yields (*ca.* 90%) are obtained by use of a small amount of iron pentacarbonyl as cocatalyst.

*Imidazoles.*[3] α-Olefins react with carbon monoxide and concentrated aqueous ammonia in an autoclave (150°) to give 2,4,5-trialkylimidazoles in 50–60% yields. Thus ethylene gives 2,4,5-triethylimidazole in 52% yield:

$$CH_2=CH_2 + CO + NH_3 \xrightarrow[CH_3OH-H_2O]{Rh_2O_3} [CH_3CH_2\underset{O}{\overset{O}{C}}CHNHCCH_2CH_3]$$

$$\xrightarrow[52\%]{NH_3}$$

The ring carbons of the imidazole ring presumably arise from carbon monoxide. Conceivably rhodium oxide is converted into $HRh(CO)_3$, which functions as a carbonylation reagent; the oxide also functions as a reduction catalyst.

*Hydroformylation.* Swiss chemists[4] recommend rhodium(III) oxide as catalyst for hydroformylation of olefins. Thus they have prepared cyclohexanecarboxaldehyde (**2**) in 95% yield from cyclohexene (**1**). Lower yields were obtained using cobalt catalysts.

$$\text{(1)} + CO + H_2 \xrightarrow[95\%]{Rh_2O_3} \text{(2) CHO}$$

(1)                              (2)

[1] W. Reppe, *Experimentia*, **5**, 93 (1949).
[2] A. F. M. Iqbal, *Helv.*, **54**, 1440 (1971).
[3] Y. Iwashita and M. Sakuraba, *J. Org.*, **36**, 3927 (1971).
[4] P. Pino and C. Botteghi, *Org. Syn.*, submitted 1972.

**Ruthenium tetroxide**, **1**, 986–989; **2**, 357–359; **3**, 243–244.

*Oxidation of secondary alcohols in a two-phase system* (**3**, 243–244). Full details of the oxidation of hydroxylactones to ketolactones and of lactones to keto acids have been published.[1]

***Oxidation of alkynes.***[2] Oxidation of alkynes with ruthenium tetroxide leads to the corresponding diketones; terminal alkynes give only the corresponding acids.

Examples:

$$C_6H_5C\equiv CC_6H_5 \xrightarrow[\text{CCl}_4-\text{H}_2\text{O, }0^0]{\text{RuO}_2,\ \text{NaOCl}} C_6H_5CO-COC_6H_5 + C_6H_5COOH$$

$$\phantom{C_6H_5C\equiv CC_6H_5 \xrightarrow{RuO}} 83\% \phantom{C_6H_5CO-COC_6H_5 +} 7.5\%$$

$$C_3H_7C\equiv CC_3H_7 \xrightarrow{\hspace{2cm}} C_3H_7CO-COC_3H_7 + C_3H_7COOH$$

$$\phantom{C_3H_7C\equiv CC_3H_7 \xrightarrow{aaaa}} 58.5\% \phantom{C_3H_7CO-COC_3H_7 +} 40\%$$

$$C_6H_5C\equiv CH \xrightarrow{\hspace{2cm}} C_6H_5COOH$$

$$\phantom{C_6H_5C\equiv CH \xrightarrow{aaaa}} 66\%$$

[1] H. Gopal, T. Adams, and R. M. Moriarty, *Tetrahedron*, **28**, 4259 (1972).
[2] H. Gopal and A. J. Gordon, *Tetrahedron Letters*, 2941 (1971).

## Ruthenium trichloride hydrate, 2, 357; 3, 242–243.

***Selective hydrogenation of 1,5,9-cyclododecatriene to cyclododecene.*** A mixture of *trans,trans,trans*-1,5,9-cyclododecatriene (1) and *trans,trans,cis*-1,5,9-cyclododecatriene (2) in a 3:2 ratio can be homogeneously hydrogenated to cyclododecene

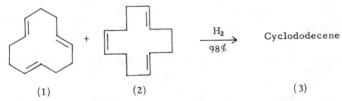

(1)          (2)          (3)

(3, *cis* and *trans*) in yields as high as 98% by homogeneous hydrogenation using $[(C_6H_5)_3P]_2(CO)_2RuCl_2$[1] as catalyst. The catalyst is more conveniently prepared *in situ*. Thus a mixture of ruthenium trichloride, triphenylphosphine, 4 psi of CO, the triene (1 + 2), and ethanol is stirred at 140° under hydrogen. Cyclododecene is again obtained in high yield along with cyclododecane (2%).[2]

[1] T. A. Stephenson and G. Wilkinson, *J. Inorg. Nucl. Chem.*, **28**, 945 (1966).
[2] D. R. Fahey, *J. Org.*, **38**, 80 (1973).

# S

**Selenium, 1**, 990–992.

    *Synthesis of ureas.*[1] Ureas can be synthesized by the reaction of aliphatic amines, carbon monoxide, and oxygen with selenium as a catalyst. For example, *n*-butylamine (0.1 mole) is dissolved in THF (100 ml.); amorphous selenium (0.005 g.-atom) is added and carbon monoxide is blown through the solution. Then oxygen (or air) is blown through. 1,3-Di-*n*-butylurea is obtained in stoichiometric yield. The reaction proceeds in two steps as formulated.

$$2\ RNH_2\ +\ Se\ +\ CO\ \longrightarrow\ (RNH_3)^+\ (RNH\!-\!\overset{\overset{O}{\|}}{C}\!-\!Se)^-\ \xrightarrow{O_2}\ (RNH)_2CO\ +\ H_2O\ +\ Se$$

    *Oxidation of 3,5-dialkylpyridines.*[2] 3,5-Dialkylpyridines are oxidized to pyridine-3,5-dicarboxylic acids in very good yields by sulfur trioxide or oleum when catalyzed by 1–4% selenium.

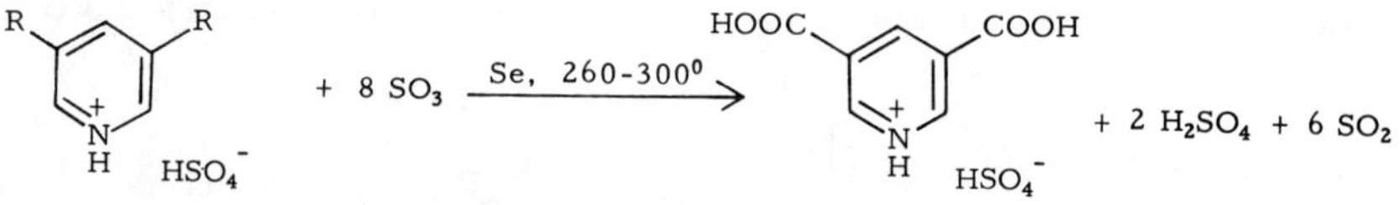

[1]N. Sonoda, T. Yasuhara, K. Kondo, T. Ikeda, S. Tsutsumi, *Am. Soc., ***93**, 6344 (1971).
[2]D. Dieterich, *Synthesis*, 631 (1972).

**Selenium dioxide, 1**, 992–1000; **2**, 360–362; **3**, 245–247.

    *Allylic oxidation.* The reaction was used by Meinwald and co-workers[1] in the first step of a synthesis of gyrinidal (9),[2] a norsesquiterpenoid from gyrinid beetles, from geranyl acetate (5).

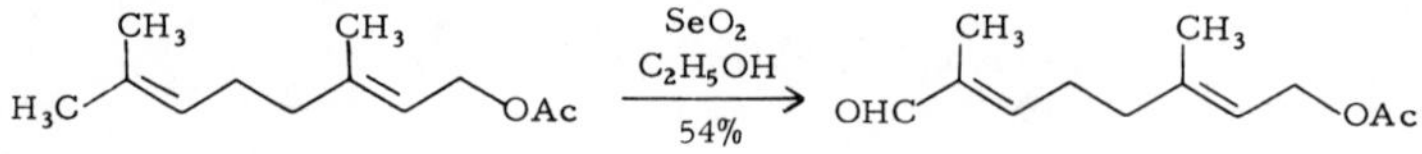

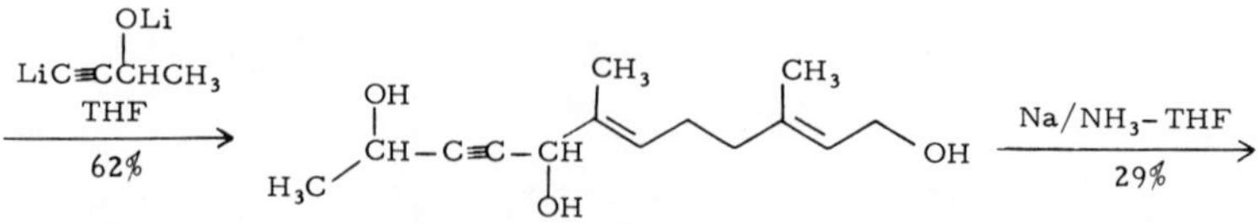

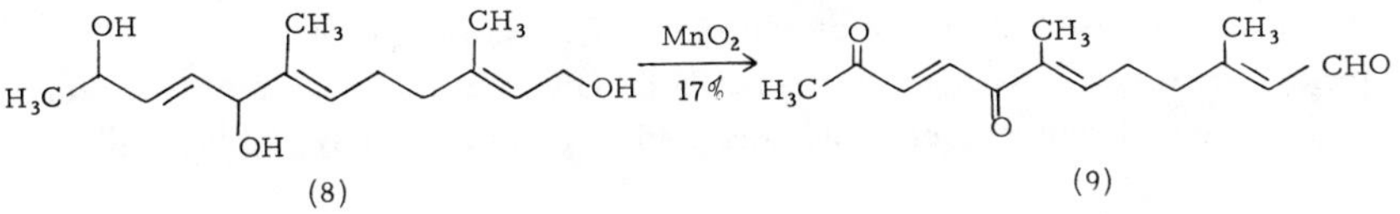

Oxidation of (5) with freshly sublimed selenium dioxide in refluxing 97% aqueous ethanol gave the aldehyde (6) in 54% yield. Addition of (6) to 10 molar eq. of the dilithio salt of 3-butyn-2-ol in THF gives the acetylenic triol (7), which is reduced to the triene (8) with excess sodium in liquid ammonia–THF and a trace of ethanol. The final step is oxidation of (8) with activated manganese dioxide (Winthrop Laboratories, New York City) at 0° in chloroform. An important feature of the synthesis is that the highly reactive 1,4-diene-3,6-dione system is generated in the final step.

A practically identical synthesis of gyrinidal (9) was reported simultaneously by Katzenellenbogen et al.[3] This group used geranyl mesitoate rather than the acetate as the starting material. The protective group was removed in the process of reduction of the triple bond by LiAlH$_4$ and sodium methoxide. These chemists also report a higher yield in the oxidation of (8) to (9) by manganese dioxide (40%).

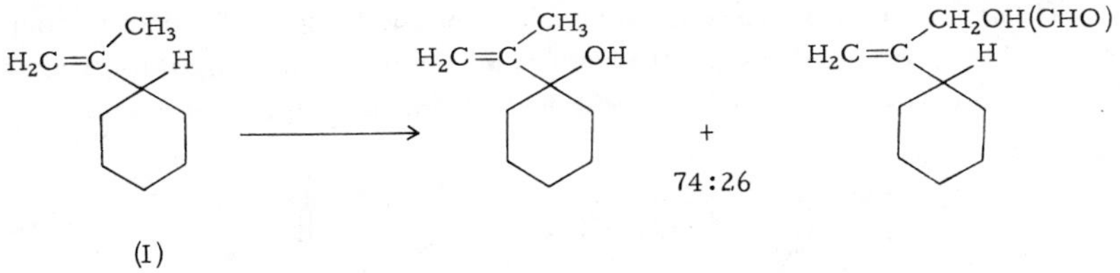

They also report that the oxidation of gem-dimethyl trisubstituted olefins with selenium dioxide in aqueous ethanol proceeds stereospecifically to give trans-alcohols

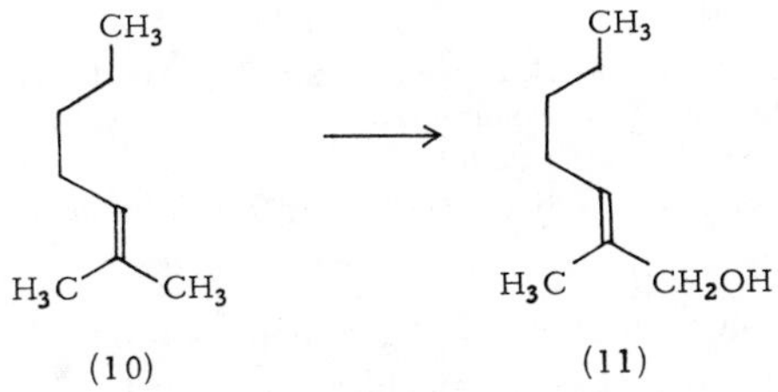

or trans-aldehydes. Thus oxidation of 2-methyl-2-heptene (10) gives trans-2-methyl-2-heptene-1-ol (11, > 98%).

The reaction was used for the oxidation of 2,7-dimethyl-2,6-octadiene (12) to 2,7-dimethyl-*trans,trans*-2,6-octadiene-1,8-dial (13). This is a tail-to-tail all-*trans* bifunctional isoprenoid synthetic unit and was used in a convenient synthesis of all-*trans* squalene.[5]

(12)                                              (13)

**Selective oxidation.** In a synthesis of *dl*-sirenin (3) Rapoport *et al.*[6] found that the oxidation of the unsaturated ester (1) with selenium dioxide in ethanol proved highly selective and gave a 55% yield of the *trans*-α,β-unsaturated aldehyde (2). This product was converted into *dl*-sirenin in 86% yield by reduction with mixed hydride.

(1)                                    (2)                                    (3)

**Flavones.** 2'-Hydroxychalcones are oxidized in good yield to flavones in boiling amyl alcohol or xylene.[7] This was the first example of use of the reagent for oxidative cyclization and is still one of the two best methods for the synthesis of flavones.[8]

[1]J. Meinwald, K. Opheim, and T. Eisner, *Tetrahedron Letters*, 281 (1973).
[2]The IUPAC name is (*E, E, E*)-3,7-dimethyl-8,11-diketo-2,6,9-dodecatrienal.
[3]C. H. Miller, J. A. Katzenellenbogen, and S. B. Bowlus, *Tetrahedron Letters*, 285 (1973).
[4]U. T. Bhalerao and H. Rapoport, *Am. Soc.*, **93**, 4835 (1971).
[5]*Idem, ibid.*, **93**, 5311 (1971).
[6]J. J. Plattner, U. T. Bhalerao, and H. Rapoport, *ibid.*, **91**, 4933 (1969).
[7]H. S. Mahal, H. S. Rai, and K. Venkataraman, *J. Chem. Soc.*, 866 (1935).
[8]Cf. G. Bargellini and G. B. Marini-Bettolo, in *Recent Progress in the Chemistry of Natural and Synthetic Colouring Matters*, T. S. Gore *et al.*, Eds., Academic Press, New York, 261 (1962).

**Silicon tetrachloride**, $SiCl_4$. Mol. wt. 169.92. Suppliers: K and K, MCB.

**Amides.**[1] Silicon tetrachloride is an efficient coupling reagent for preparation of amides from a carboxylic acid and an amine.

*Peptide synthesis.*[2] Dipeptides can be obtained in moderate yield by reaction of an N-protected amino acid with $SiCl_4$ in pyridine followed by addition of an amino acid ester. However, racemization is extensive and the Cb-protective group is not stable under the reaction conditions.

[1]T. H. Chan and L. T. L. Wong, *J. Org.*, **34**, 2766 (1969).
[2]*Idem, ibid.*, **36**, 850 (1971).

**Silver carbonate, 1**, 1005; **2**, 363.

*α-Acylaminoacrylic esters.* Oxidation of methyl N-acetylcysteinate (1) with a slight excess of silver carbonate in refluxing methanol for 1 hr. gives methyl α-acetamido-acrylate (2) in 78% yield. The oxidation is applicable to various cysteine and penicillamine derivatives.[1]

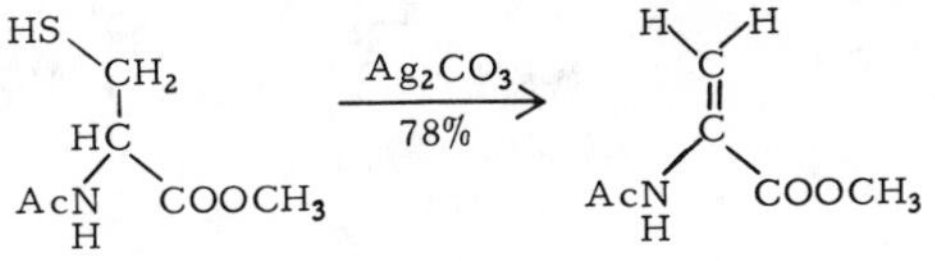

Oxidation of hydroquinones.* Schäfer *et al.*[2] used the combination of silver carbonate and magnesium sulfate in absolute benzene for oxidation of hydroquinones to quinones. Reported yields were 60–100%.

[1]D. Gravel, R. Gauthier, and C. Berse, *J.C.S. Chem. Commun.*, 1322 (1972).
[2]W. Schäfer, R. Leute, and H. Schlude, *Ber.*, **104**, 3211 (1971).

**Silver carbonate–Celite, 2**, 363; **3**, 247–249.

*Mechanism.* Fetizon *et al.*[1] have proposed a concerted mechanism for oxidations with silver carbonate on Celite in which the first step is a reversible absorption of the

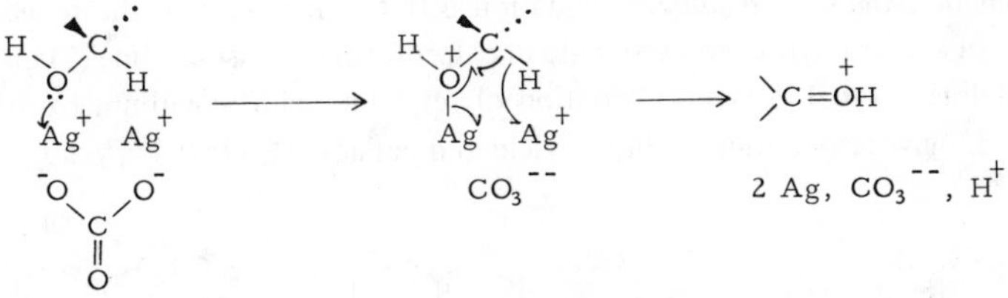

alcohol on the Celite. This facilitates coordination between one $Ag^+$ cation and the oxygen atom, which becomes positive. The second step involves a concerted transfer of electrons involving the second silver cation.

*Oxidation of a 1,3-amino alcohol.*[2] Oxidation of the alcohol (1) with silver carbonate–Celite in benzene does not give the anticipated saturated ketone, but approximately a 2:1 mixture of (2) and (3). This one-step procedure should be useful for preparation

(1) (2) (3)

of α-amino-α,β-unsaturated ketones. 2-Acetyl-1,4,5,6-tetrahydropyridine (2) is a constituent of the aroma of freshly baked bread.

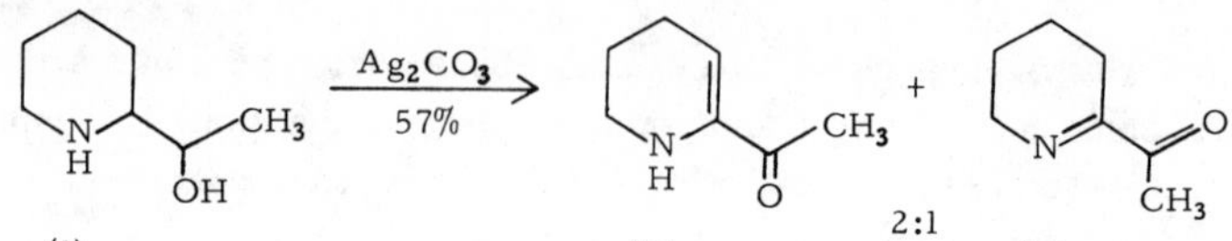

***Oxidation of carbohydrates.*** Fetizon and Moreau[3] have reported briefly on the oxidation of some hexoses. The oxidation of glucose and mannose is complex. However, the oxidation of galactose (1) is straightforward. If carried out in water at 80°, the

(1) (2) (3)

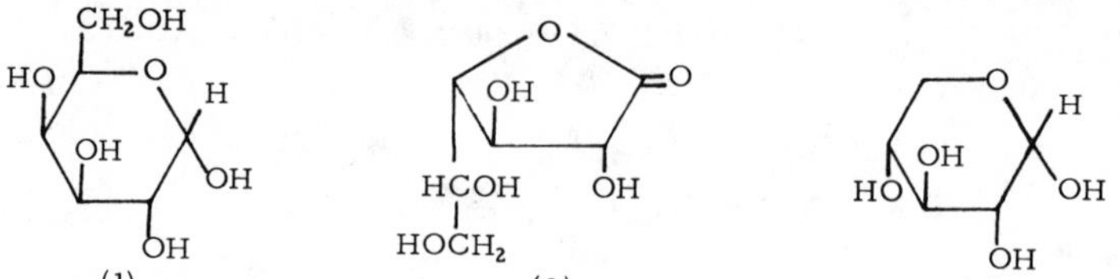

product is D-galactonolactone (2). When carried out in refluxing ethanol, the main product is D-lyxose (3), isolated in 36% yield.

***Oxidation of α-glycols.*** α-Glycols of the type (1) are selectively oxidized by Fetizon's reagent to α-ketols (2) in about 65–80% yield.[4] The α-ketols can be oxidized to α-diketones (3) by bismuth oxide in about 80% yield.

$$\underset{(1)}{CH_3CH{=}CH{-}\underset{OH}{CH}{-}\underset{OH}{CH}{-}CH{=}CHCH_3} \xrightarrow[80\%]{Ag_2CO_3} \underset{(2)}{CH_3CH{=}CH{-}CO{-}\underset{OH}{CH}{-}CH{=}CH{-}CH_3}$$

$$\xrightarrow{Bi_2O_3} \underset{(3)}{CH_3CH{=}CHCOCOCH{=}CHCH_3}$$

***Phenolic oxidative coupling.***[5] Oxidation of reticuline (1) with the reagent in the presence of sodium hydrogen carbonate in chloroform gives isoboldine (2) in 3% yield and pallidine (3) in 1% yield. Oxidation of reticuline with vanadium oxytrichloride (**3**, 331–332) gives isoboldine (2) in 1% yield and palladine (3) in 0.3% yield.

(1) (2) (3)

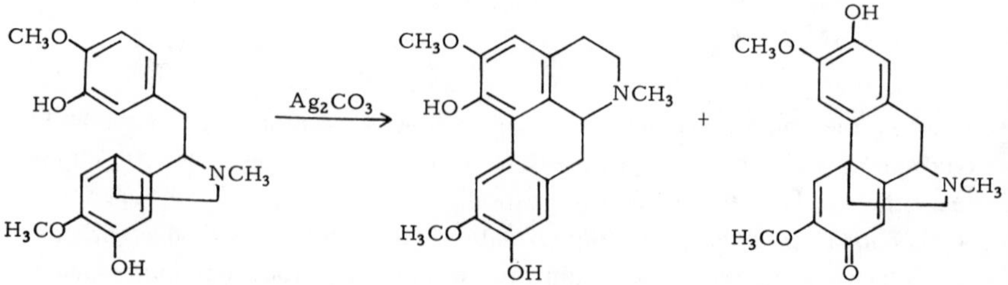

***C-Nitroso compounds.***[6] C-Nitroso compounds can be obtained in 55–95% yield by oxidation of hydroxylamines with silver carbonate–Celite. The reaction is carried out at or below room temperature in $CH_2Cl_2$ or $CFCl_3$. A practical advantage is that

$$RNHOH \longrightarrow RNO$$

coupling between unreacted hydroxylamine and the nitroso compound to give azoxy derivatives is not observed because the reaction is rapid. Also isomerization of primary and secondary nitroso compounds to oximes is suppressed.

***Fragmentation reactions.***[7] Silver carbonate on Celite (Fetizon's reagent) quantitatively cleaves steroidal 17-ethynyl alcohols. Thus treatment of 17α-ethynylestradiol methyl ether (1) with silver carbonate on Celite in toluene under reflux gives estrone methyl ether (2) in quantitative yield. In the same way, dehydroepiandrosterone (4) was obtained from (3).

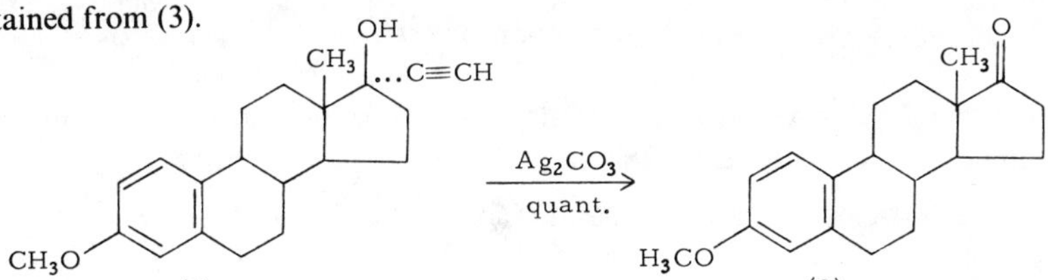

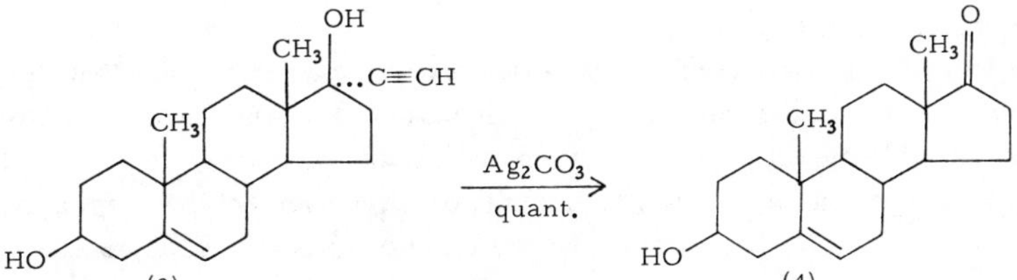

Fragmentation of steroidal cyanohydrins was also investigated. Treatment of pregnenolone acetate cyanohydrin (5) with the reagent in refluxing toluene led to pregnenolone acetate (6) in quantitative yield.

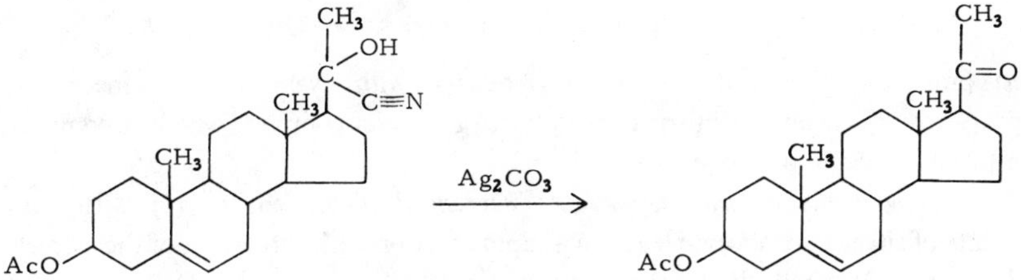

[1]M. Fetizon, M. Golfier, and P. Mourgues, *Tetrahedron Letters,* 4445(1972).
[2]G. Büchi and H. Wüest, *J. Org.,* **36,** 609 (1971).
[3]M. Fetizon and N. Moreau, *Compt. Rend.,* **275 (C),** 621 (1972).
[4]S. Thuan and J. Wiemann, *ibid.,* **272 (C),** 233 (1971).

[5]T. Kametani, A. Kozuka, and K. Fukumoto, *J. Chem. Soc.* (*C*), 1021 (1971).
[6]J. A. Maassen and Th. J. de Boer, *Rec. trav.*, **90**, 373 (1971).
[7]G. R. Lenz, *Chem. Commun.*, 468 (1972).

$$\overset{O}{\underset{\uparrow}{}}$$

**Silver diethyl phosphate,** $(C_2H_5O)_2POAg$. Mol. wt. 260.98.

The reagent is prepared by the reaction of diethyl phosphate (0.12 mole) and silver carbonate (0.05 mole) in ether at 25°; 82% yield.

**Peroxy acids.** Konen and Silbert[1] have developed a mild, general synthesis of peroxy acids entailing first acylation of silver diethyl phosphate in ether solution followed by perhydrolysis of the mixed anhydride with hydrogen peroxide (98%

$$RCOCl + (C_2H_5O)_2\overset{O}{\underset{\uparrow}{P}}OAg \xrightarrow[-AgCl]{} (C_2H_5O)_2\overset{O}{\underset{\uparrow}{P}}O\overset{O}{\overset{\parallel}{C}}R \xrightarrow{H_2O_2,\ H^+} RCO_3H + (C_2H_5O)_2\overset{O}{\underset{\uparrow}{P}}OH + H^+$$

concentration, FMC Corporation). Both steps proceed in high yield. Methanesulfonic acid was usually used as the acid catalyst.

[1]D. A. Konen and L. S. Silbert, *J. Org.*, **36**, 2162 (1971).

**Silver fluoroborate (Silver tetrafluoroborate),** **1**, 1015–1018; **2**, 365–366; **3**, 250–251. Additional supplier: WBL.

**Cycloaddition reactions of benzyne with cyclic olefins.** Benzyne reacts with cyclo-hexadiene to give, as the main products, hydrocarbons (1)–(4).[1] The first (1) arises by 2 + 4 cycloaddition, (2) and (3) arise by ene cycloadditions, and (4) is a result of 2 + 2 cycloaddition. Addition of catalytic amounts of silver fluoroborate exerts a marked effect; in this case (1) becomes almost the exclusive product. Silver ion, however,

(1, 48%)    (2, 18%)    (3, 21%)    (4, 13%)

exerts practically no effect on the reaction of benzyne with acyclic dienes.[2] A mechanism is proposed which involves formation of an $Ag^+$–olefin complex, which decomposes to a benzyne–olefin–$Ag^+$ complex.

**Rearrangements of tricyclo[3.2.0.0²,⁴]heptanes.**[3] Treatment of (1) with small amounts of silver fluoroborate leads to a rapid reaction with formation of the isomeric hydrocarbon (2). Similarly, (3) is converted into (4).

(1)                          (2)

This is the first report of regio- and stereoselective migration of methyl groups promoted by $Ag^+$.

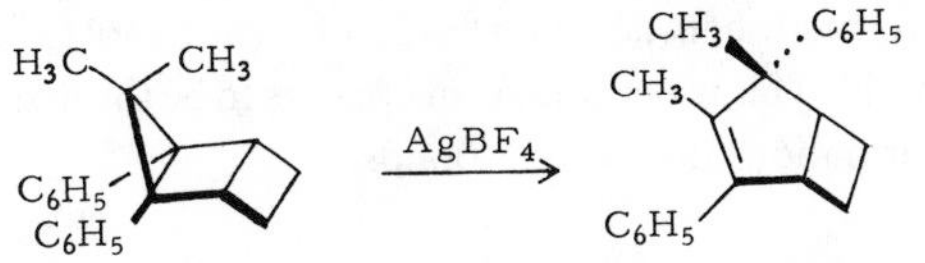

(3) (4)

[1]P. Crews and J. Beard, *J. Org.*, **38**, 522 (1973), and references cited therein.
[2]*Idem*, *ibid.*, **38**, 529 (1973).
[3]L. A. Paquette and L. M. Leichter, *Am. Soc.*, **94**, 3653 (1972).

**Silver nitrate, 1**, 1008–1011; **2**, 366–368; **3**, 252.

***Oxidation of vinylidenebisdialkylamines.***[1] Vinylidenebisdimethylamine (1) is oxidized by silver nitrate in acetonitrile to 1,1,4,4-tetrakis(dimethylamino)-2-butene-1,4-diylium nitrate (2), probably through the intermediates indicated.

$$2\ CH_2{=}C[N(CH_3)_2]_2 \xrightarrow[CH_3CN]{2\ AgNO_3} \left[ [(CH_3)_2N]_2\overset{+}{C}CH_2CH_2\overset{+}{C}[N(CH_3)_2]_2 \quad 2\ NO_3^- \right] \xrightarrow{(1)}$$

(1)

$$\left[ [(CH_3)_2N]_2C{=}CHCH{=}C[N(CH_3)_2]_2 + 2\ CH_3\overset{+}{C}[N(CH_3)_2]_2NO_3^- \right] \xrightarrow{2\ AgNO_3} [(CH_3)_2N]_2\overset{+}{C}CH{=}CH\overset{+}{C}[N(CH_3)_2]_2\ 2\ NO_3^-$$

(2)

Cyclization, rather than dimerization, can be achieved by this process in the case of dienetetramines. Thus oxidation of N,N,N′,N′,N″,N″,N‴,N‴-octamethyl-1,4-pentadiene-1,1,5,5-tetramine (3) affords in high yield a mixture of *cis*- and *trans*-cyclopropylenebis(dimethylaminomethylium) nitrate (4).

(3) (4)

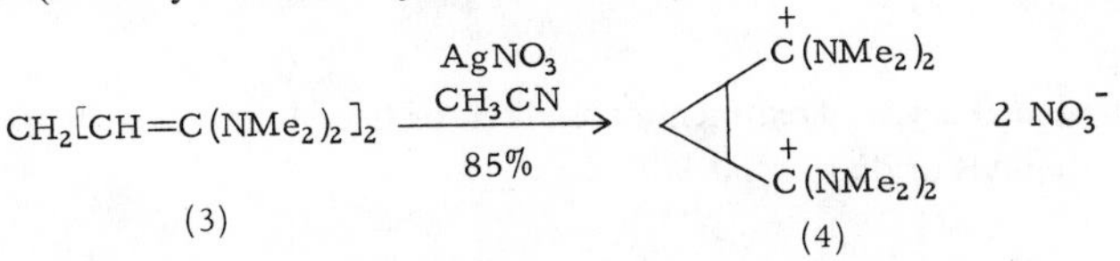

Even bicyclic ring systems can be synthesized by this method. Thus oxidation of (5), 1,3-bis[bis(dimethylamino)methylene]cyclohexane, leads to bicyclo[3.1.0]cyclohexane-1,5-ylenebis(dimethylaminomethylium) nitrate (6).

(5) (6)

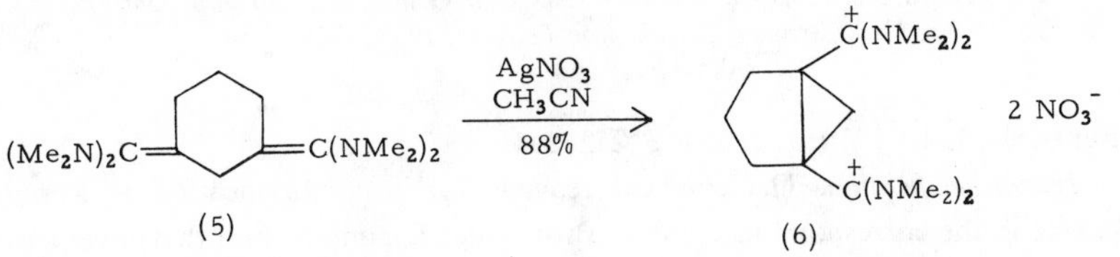

*Conversion of elemol to eudesmols.*[2] Thermolysis of elemol (1) at 210–220° results in partial isomerization to hedycaryol (2). Addition of a trace of silver nitrate to the thermolysis leads to formation, in addition to (2), of α-eudesmol (3) and β-eudesmol (4), both in about 0.8 % yield. This transformation appears to be the first instance of silver(I) ion-catalyzed transformations involving π-bonds.

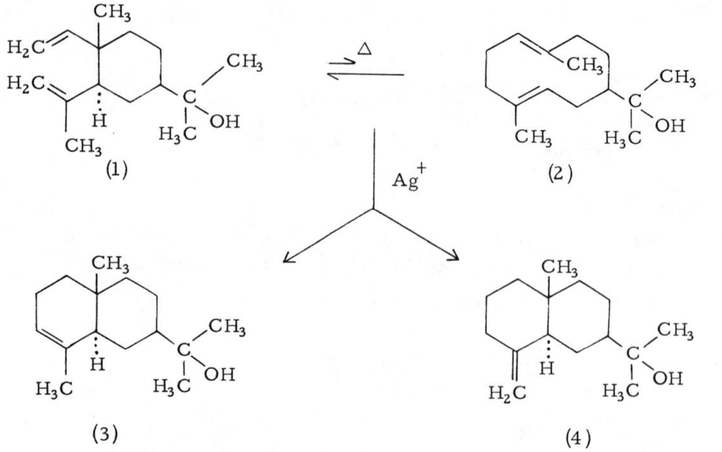

*β-Lactams.* Treatment of the N-chloro derivative of the carbinolamine (1) with excess silver nitrate in acetonitrile (3 hr.) affords the β-lactam (2).[3]

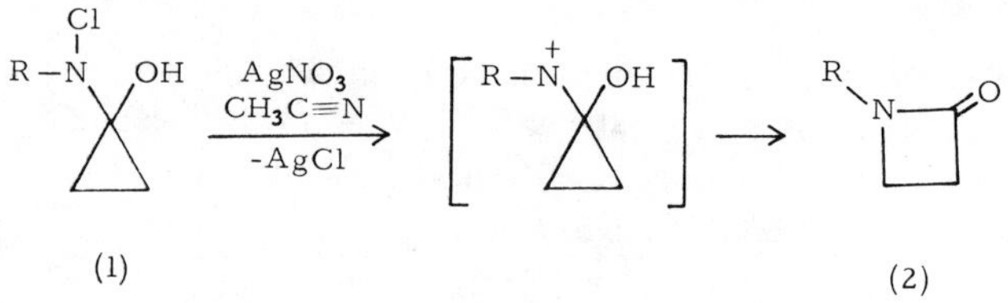

*Dialkyls and Diaryls.*[4] Treatment of trialkyl- or triarylboranes with alkaline silver nitrate gives dialkyls or diaryls (50–90% yield).

$$R_3B \xrightarrow{\text{AgNO}_3, \text{ OH}^-} R-R$$

[1]H. Weingarten and J. S. Wager, *J. Org.*, **35**, 1750 (1970).
[2]T. C. Jain and J. E. McCloskey, *Tetrahedron Letters*, 5139 (1972).
[3]H. H. Wasserman, H. W. Adickes, and O. Espejo de Ochoa, *Am. Soc.*, **93**, 5586 (1971).
[4]S. W. Breuer and F. A. Broster, *Tetrahedron Letters*, 2193 (1972).

**Silver oxide,** 1, 1011–1015; **2**, 368; **3**, 252–254.

*Transetherification.* The classical reagent for transetherification is a sodium alkoxide in the corresponding alcohol. Brown and Sugimoto[1] find that silver oxide is an excellent catalyst for transetherification of certain alkoxy-1,2,4,6,8-pentaaza-naphthalenes (1), alkoxynitropyrimidines (2), and related heterocyclic ethers by boiling alcohols. Silver acetate is less effective.

(1)                              (2)

*Oxidative methylation.* In an attempt to methylate 5-vanillylidenebarbituric acid (1) with silver oxide and methyl iodide in DMF by the Kuhn procedure (**1**, 683), Ethier and Neville[2] unexpectedly obtained the benzaldehyde (2) and tetramethylbarbituric

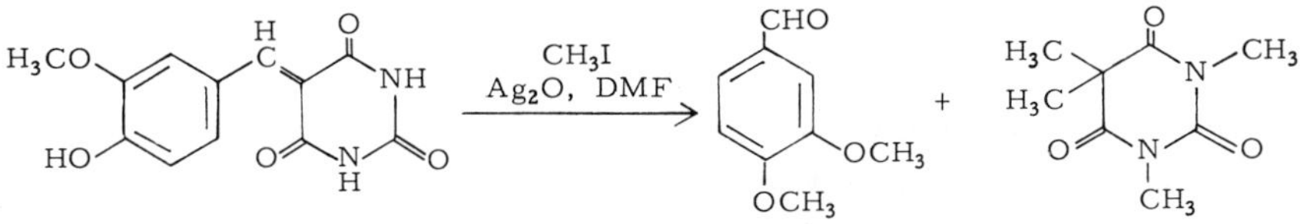

(1)                    (2, 80%)                (3, 45%)

acid (3). The reaction is probably promoted by interaction of silver ion with the olefinic double bond.

*Alkylation of 2-aryl-1,3,4-oxadiazoles.*[3] Alkylation of 2-aryl-1,3,4-oxadiazoles (1) by alkyl halides in acetonitrile in the presence of silver oxide gives O- and N-alkyl

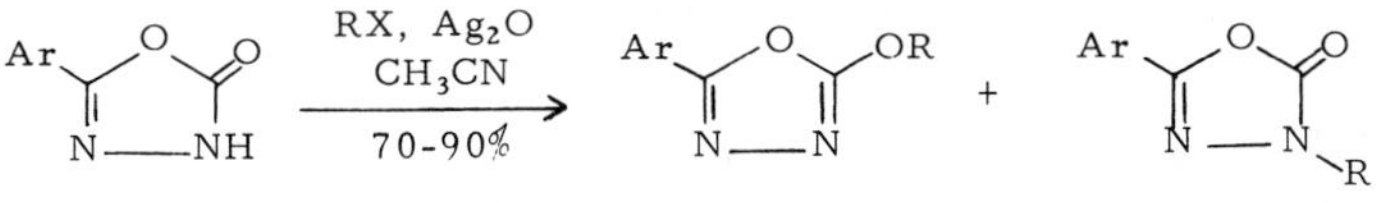

(1)                    (2)                    (3)

derivatives (2) and (3) in good yield. The products are readily separated by chromatography.

*Hydrolysis of thioketals.*[4] Thioketals are hydrolyzed to the parent ketone (75–85% yield) by treatment with a tenfold excess of silver oxide in refluxing aqueous methanol (1:10) for 16 hr. to 4 days.

[1] D. J. Brown and T. Sugimoto, *J. Chem. Soc. (C)*, 2661 (1970).
[2] J. C. Ethier and G. A. Neville, *Tetrahedron Letters*, 5297 (1972).
[3] M. Golfier and R. Milcent, *Bull. Soc.*, 254 (1973).
[4] D. Gravel, C. Vaziri, and S. Rahal, *J.C.S. Chem. Comm.*, 1323 (1972).

**Silver(II) oxide**, **2**, 369.

*Oxidative demethylation.* Rapoport *et al.*[1] examined the oxidation of the dimethyl ether of menaquinol-1 (1) with silver(II) oxide with the expectation of effecting oxidation

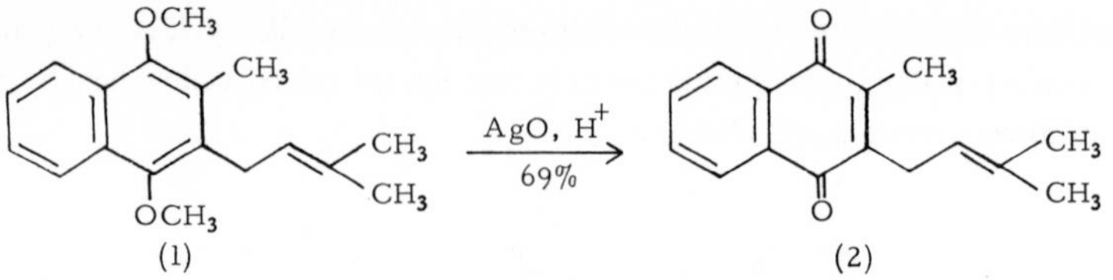

(1) → (2)    AgO, H$^+$  69%

of the allylic and benzylic methylene group at the $C^1$-position. Unexpectedly, menaquinone-1 (2) was the only product isolated. The reaction allows protection of quinones as the hydroquinone methyl ethers.

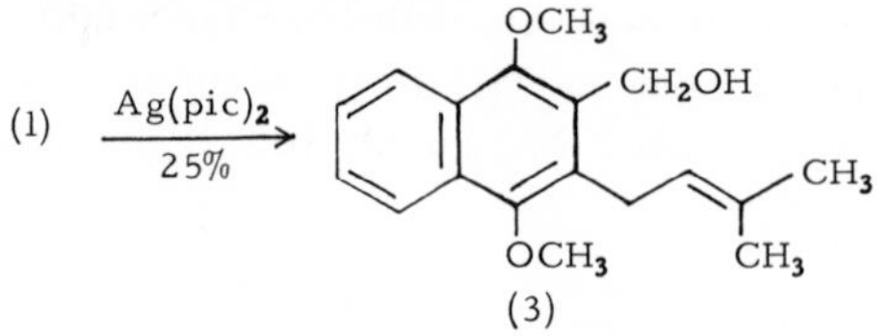

(1) $\xrightarrow[25\%]{\text{Ag(pic)}_2}$ (3)

Oxidation of (1) with silver(II) picolinate (3, 16) in DMSO effected oxidation of the 2-methyl group to give the benzyl alcohol (3) in 25% yield.

*Oxidative cleavage of hydroquinone ethers.*[2] The reagent oxidatively cleaves *p*-hydroquinone dimethyl ethers to *p*-quinones in good yield. A mineral acid (nitric or perchloric acids are most efficient) is required and the oxidant must be present in excess for high conversions. *o*-Quinones can be obtained in moderate yields.

*Miscellaneous oxidations.*[3] Some aromatic amines (*e.g.*, aniline, *p*-toluidine, *p*-chloroaniline) are oxidized by silver(II) oxide to the corresponding azo derivative in moderate yield. Hydroquinone is oxidized by AgO rapidly to *p*-benzoquinone in quantitative yield. Oxidation of benzil dihydrazone with AgO gives diphenylacetylene in 95% yield.

[1]C. D. Snyder, W. E. Bondinell, and H. Rapoport, *J. Org.*, **36**, 3951 (1971).
[2]C. D. Snyder and H. Rapoport, *Am. Soc.*, **94**, 227 (1972).
[3]B. Oritz, P. Villanueva, and F. Walls, *J. Org.*, **37**, 2748 (1972).

**Silver perchlorate, 2**, 369–370. Additional suppliers: Alfa Inorganics, Fluka, K and K, Pfaltz and Bauer, ROC/RIC, Sargent.

*Methanolysis of halogenocarbene adducts.* Silver perchlorate catalyzes the methanolysis of *exo*-8-bromo- and 8,8-dibromobicyclo[5.1.0]octanes, (1a) and (1b), to 3-methoxy- and 2-bromo-3-methoxy-*trans*-cyclooctenes (2a) and (2b), respectively in

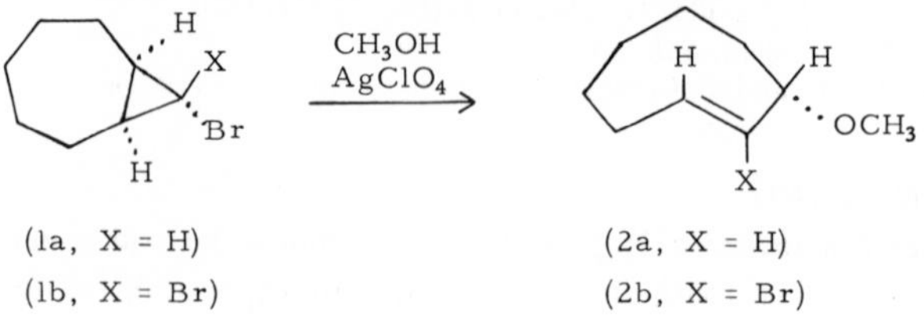

$\xrightarrow[\text{AgClO}_4]{\text{CH}_3\text{OH}}$

(1a, X = H)
(1b, X = Br)

(2a, X = H)
(2b, X = Br)

high yield.[1] Under the same conditions the corresponding bicyclo[5.1.0]octene-3 derivatives (3a) and (3b) are converted into 6-methoxy- and 5-bromo-6-methoxy-*cis*, *trans*-cyclooctadienes-1,4 (4a) and (4b), respectively, as the main products.[2]

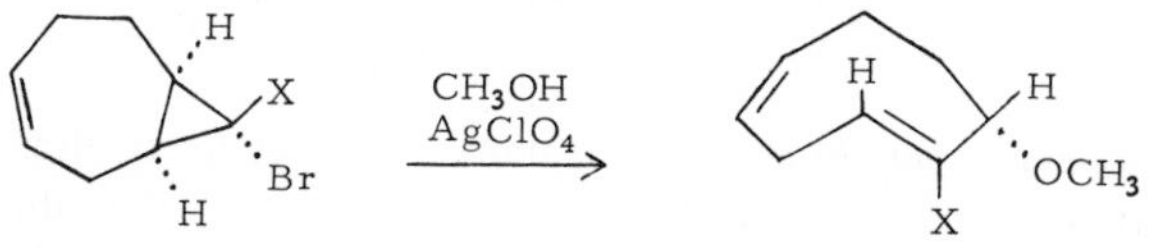

(3a, X = H)
(3b, X = Br)

(4a, X = H)
(4b, X = Br)

Under the same conditions, *exo*-8-bromobicyclo[5.1.0]octene-2 (5a) is converted into an approximately 1:1 mixture of two isomeric methoxycyclooctadienes, shown to be 3-methoxy-*cis*,*trans*-cyclooctadiene-1,4 (6a) and 5-methoxy-*cis*,*trans*-cyclooctadiene-1,3 (7a).[3] A similar result is obtained with 8,8-dibromobicyclo[5.1.0]octene-2 (5b), which

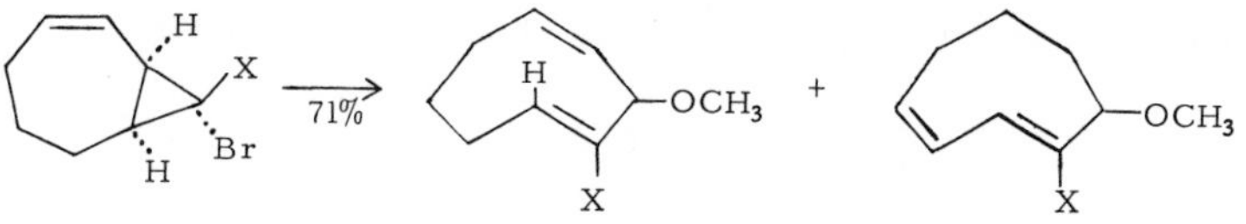

(5a, X = H)
(5b, X = Br)

(6a, X = H)
(6b, X = Br)

(7a, X = H)
(7b, X = Br)

gives (6b) and (7b). The products could arise by attack of methanol on both ends of a *trans*,*trans*-allylic cationic intermediate.

**Decarbonylation.**[4] Treatment of tricyclo[2.1.0.0²·⁵]pentane-3-ones with silver

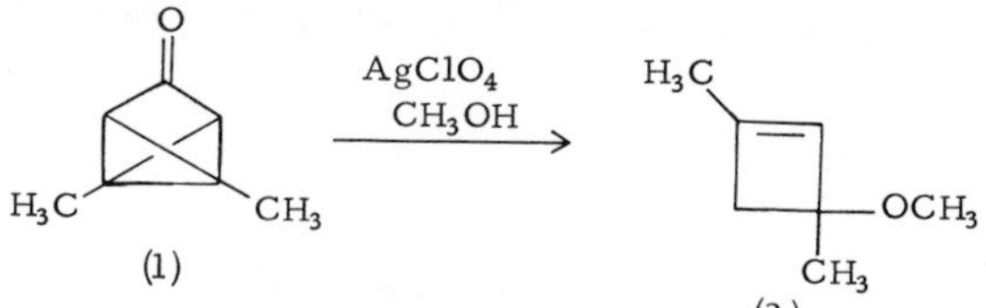

(1)

(2)

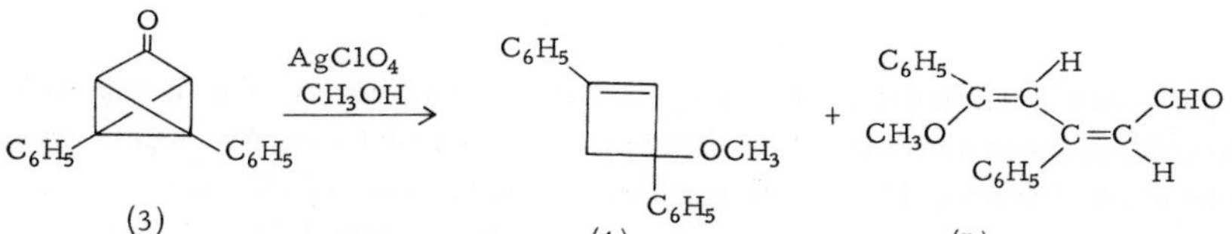

(3)

(4)

(5)

perchlorate in benzene at room temperature effects decarbonylation. Treatment of (1) with Ag⁺ in methanol yields (2); similarly (3) is converted into (4) and (5). The cation (6) is suggested as an intermediate.

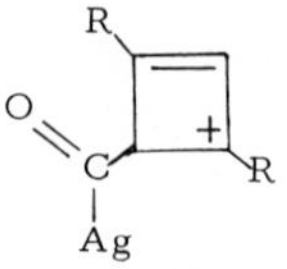

(6)

***Bicyclobutanes → butadienes.***[5] Bicyclobutane (1) upon treatment with silver perchlorate at room temperature rearranges to butadiene (2) in about 90% yield.

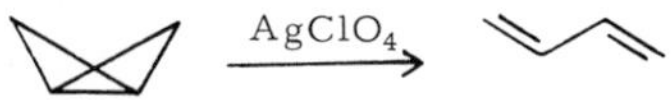

(1)    (2)

*exo,exo*-2,4-Dimethylbicyclobutane (3) and *endo,exo*-2,4-dimethylbicyclobutane (4) under the same conditions are converted into *trans,trans*- and *cis,trans*-2,4-hexadiene, (5) and (6), respectively. The reaction thus is largely, but not entirely, stereospecific and the stereochemistry is the reverse of that observed on thermolysis.

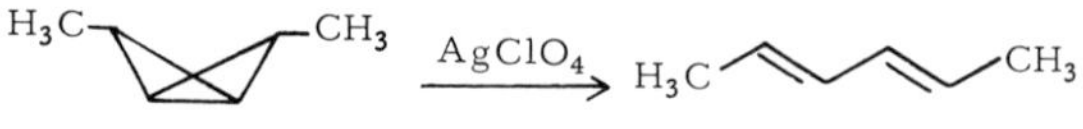

(3)    (5)

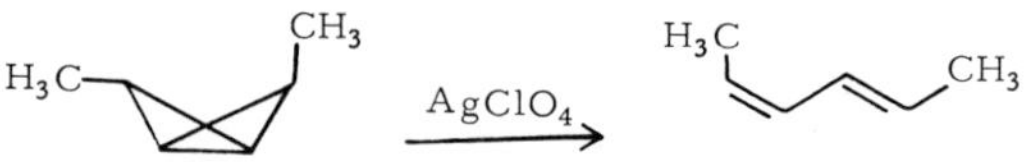

(4)    (6)

***Rearrangements*** (2, 250–251). When *endo*-2-methoxytricyclo[4.1.0.0³·⁷]heptane (1) is treated with 1–3 mole % of silver perchlorate in anhydrous benzene, it rearranges

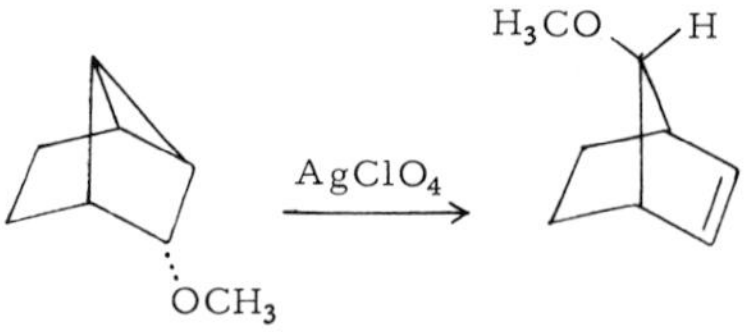

(1)    (2)

rapidly and quantitatively into *anti*-7-methoxynorbornene (2). The isomerization proceeds by initial ionization of the methoxyl group and not by any Ag⁺-strained bond interaction. Tricyclo[4.1.0.0³·⁷]heptane itself is stable to silver perchlorate.[6]

***Rearrangements of strained σ-bonds*** (*see* **Silver fluoroborate**, 3, 250). Paquette *et al.*[7] have now examined the Ag⁺-catalyzed rearrangement of the less strained *seco*-cubane derivatives. Treatment of the *endo,endo*-diester (1) with a solution of anhydrous silver perchlorate resulted in quantitative and stereospecific conversion to the *endo,endo*-tetracyclo[3.3.0.0²·⁸.0⁴·⁶]octane derivative (2). In like manner, the *endo,exo*-isomer (3) gave the isomer (4) of (2). Rearrangement of (5) gave the *exo,exo*-diester (6) as the only product. The three *seco*-cubyl esters rearrange at rates comparable to those of the more strained cubane systems. Clearly alleviation of strain is not a determining factor.

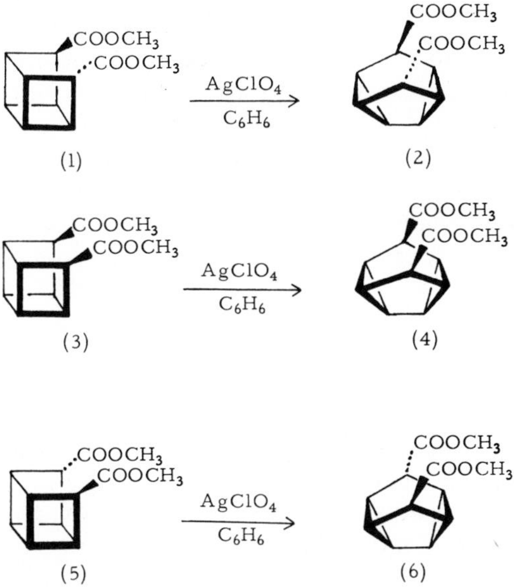

Isomerization of strained $\sigma$-bonds by Ag(I) ions has been reviewed by Paquette.[8]

[1] C. B. Reese and A. Shaw, *Am. Soc.*, **92**, 2566 (1970).
[2] M. S. Baird and C. B. Reese, *Chem. Commun.*, 1644 (1970).
[3] *Idem, Tetrahedron Letters*, 4637 (1971).
[4] H. Ona, M. Sakai, M. Suda, and S. Masamune, *J.C.S. Chem. Comm.*, 45 (1973).
[5] M. Sakai, H. Yamaguchi, H. H. Westberg, and S. Masamune, *Am. Soc.*, **93**, 1043 (1971).
[6] L. A. Paquette and G. Zon, *ibid.*, **94**, 5096 (1972).
[7] L. A. Paquette, R. S. Beckley, and T. McCreadie, *Tetrahedron Letters*, 775 (1971).
[8] L. A. Paquette, *Accts. Chem. Res.*, **4**, 280 (1971).

**Silver sulfate**, **1**, 1015; **3**, 254.
   *Hydrolysis of 1,3,5,7-tetrabromoadamantane[1] :*

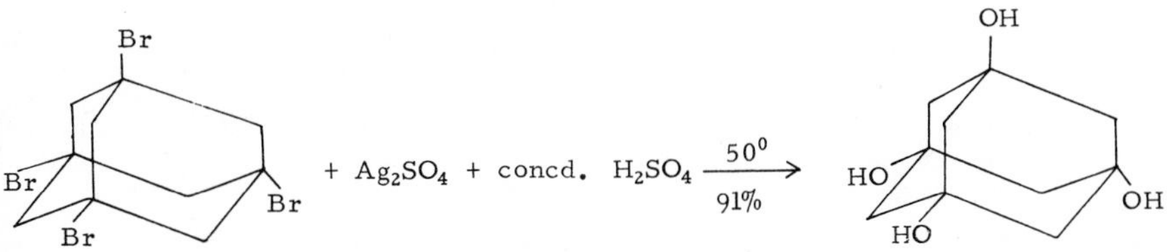

[1] H. Stetter and M. Krause, *Ann.*, **717**, 60 (1968).

**Simmons–Smith reagent, 1,** 1019–1022; **2,** 371–372; **3,** 255–258.

*Improved reaction.* Conia *et al.*[1] have reported two modifications of the Simmons–Smith reaction which gave improved yields. One is the use of a zinc–silver couple in place of the zinc–copper couple. This couple is prepared by adding granular zinc to a stirred hot solution of silver acetate in acetic acid. The mixture is stirred for 30 sec. and the zinc–silver couple formed is isolated by decantation and washed with acetic acid and ether. It is then stabilized by addition of a small amount of silver wool. The second improvement is that the reaction mixture is not subjected to acid hydrolysis. Instead an amine, for example pyridine, is added. This forms the insoluble complexes $ZnI_2 \cdot C_5H_5N$ and $ICH_2ZnI \cdot (C_5H_5N)_2$; the cyclopropane products are then isolated from the filtrate.

Using these improvements, Denis and Conia[2] converted bis(trimethylsiloxy)-2,3-butadiene (1) into the dicyclopropane (2) in 78 % yield. The bis-silyl ether was hydrolyzed

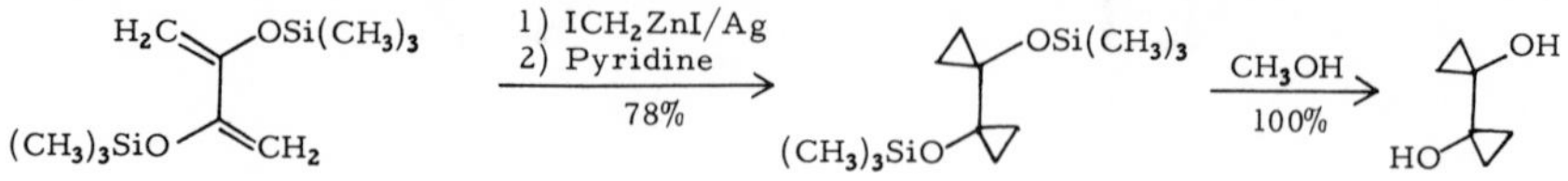

to the free diol (3) in quantitative yield by refluxing for 3 hr. in methanol. Hydrolysis could also be effected by refluxing (2) in acetone containing 2 eq. of water.

*Synthesis of methyl sterculate* (**3,** 255). Williams and Sgoutas[3] have shortened Gensler's synthesis of methyl sterculate (3) by direct decarbethoxylation of one

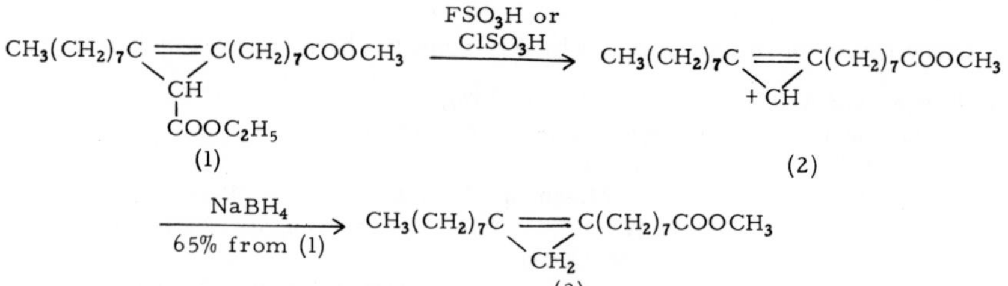

intermediate (1) with fluorosulfonic acid in methylene chloride at room temperature.

*Cyclopropanation of an allylic alcohol.*[4] The reaction of the allylic alcohol (1) with the Simmons–Smith reagent is extremely sensitive to dehydration to the diene (3). The best yields of (2) are obtained by careful control of the temperature (31–38°).

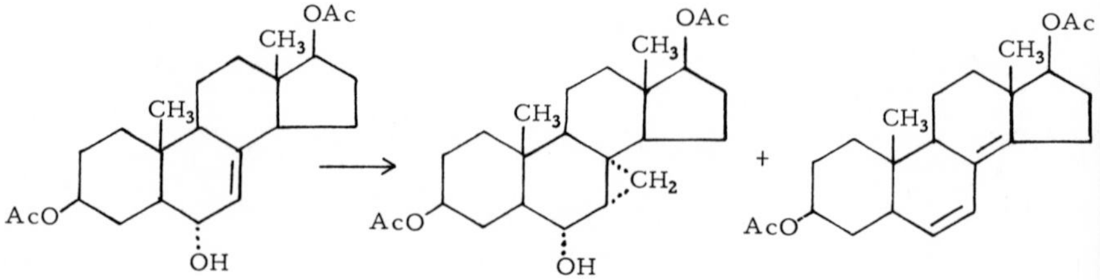

Under optimum conditions the desired product (2) can be obtained in 36% yield; yield of diene 11%.

*Reaction with cyclooctyne[5]:*

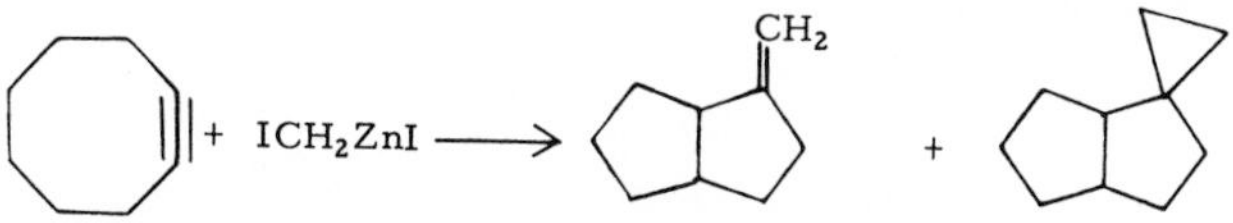

[1] J. M. Denis, C. Girard, and J. M. Conia, *Synthesis*, 549 (1972).
[2] J. M. Denis and J. M. Conia, *Tetrahedron Letters*, 4593 (1972).
[3] J. L. Williams and D. S. Sgoutas, *J. Org.*, **36**, 3064 (1971).
[4] W. G. Dauben and D. S. Fullerton, *ibid.*, **36**, 3277 (1971).
[5] G. Wittig and J. J. Hutchison, *Ann.*, **741**, 79 (1970).

**Sodium, 1**, 1022–1023.

*Cleavage of β-chloro cyclic ethers.* Tetrahydrofurfuryl chloride (1) is cleaved to 4-pentene-1-ol (2)[1] by powdered sodium in ether.[2]

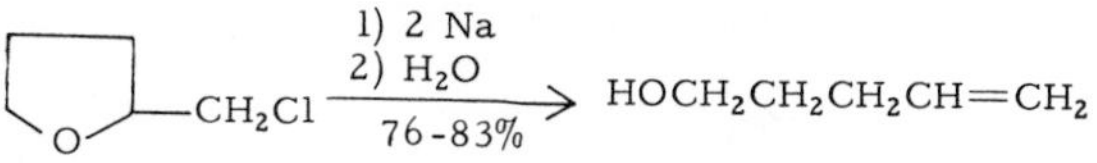

Detailed directions for cleavage of 3-chloro-2-methyltetrahydropyrane (3) to *trans*-4-hexene-1-ol (4) have been described.[3]

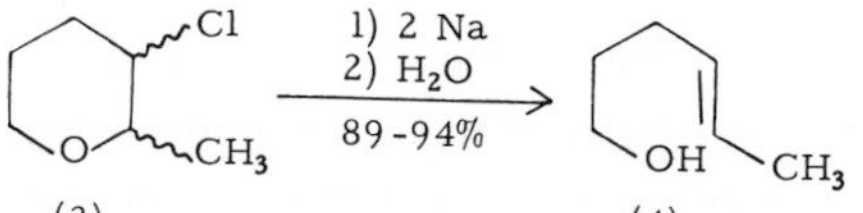

*Intramolecular Wurtz dehalogenation: bicyclo[1.1.0]butane[4]* (b.p. 8°). The reaction is carried out with stirring under nitrogen in refluxing dioxane that has been heated to reflux with the sodium ketyl prepared from 10 g. of benzophenone and 1 g. of sodium

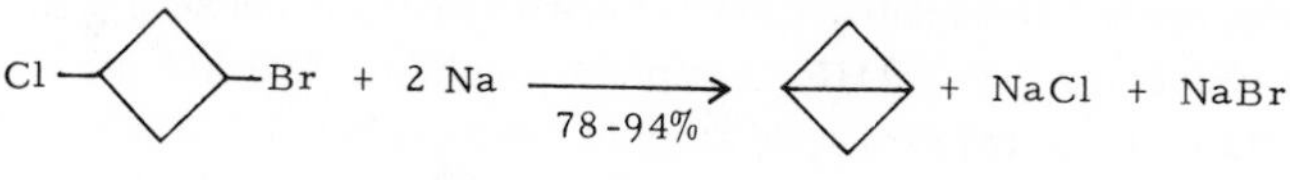

(and 2 l. of solvent) until a deep-blue solution results. The peroxide-free dioxane is distilled from the flask and used immediately. The apparatus employed and its operation are described in detail. The yield of bicyclobutane is 78–94%.

[1] L. A. Brooks and H. R. Synder, *Org. Syn.*, *Coll. Vol.*, **3**, 698 (1955).
[2] The powdered sodium is prepared under hot xylene with the aid of a Hershberg stirrer; the xylene is decanted and replaced by ether.
[3] R. Paul, O. Riobé, and M. Maumy, *Org. Syn.*, submitted 1973.
[4] G. M. Lampman and J. C. Aumiller, *ibid.*, **51**, 55 (1971).

**Sodium–Ammonia, 1,** 1041; **2,** 374–376; **3,** 259.

*Reduction of allenes* (**2,** 374–276). Moorthy and Devaprabhakara[1] have published a convenient synthesis of *cis,cis*-1,6-cycloundecadiene (4) from *cis,cis*-1,6-cyclodecadiene (1) as formulated.

(1)      $\xrightarrow[65\%]{CHBr_3,\ (CH_3)_3COK}$      (2)      $\xrightarrow[85\%]{CH_3Li}$

(3)      $\xrightarrow[72\%]{Na/NH_3}$      (4)

[1]S. N. Moorthy and D. Devaprabhakara, *Synthesis*, 612 (1972).

**Sodium aluminum chloride, 1,** 1027–1029; **2,** 372.

*1-Amino-4-hydroxyanthraquinone.*[1] Condensation of 4-aminophenol and its derivatives (1) with phthalic anhydride in an aluminum chloride–sodium chloride melt

(1)    +    $\xrightarrow[45\%]{1)\ AlCl_3-NaCl \quad 2)\ H_3O^+}$    (2)

(170–210°), followed by hydrolysis (2 *N* HCl), gives 1-amino-4 hydroxyanthraquinone (2) in rather low yields. Highest yields were obtained with the triacetyl derivative (1).

[1]V. P. Aggarwala, R. Gopal, and S. P. Garg, *J. Org.*, **38,** 1247 (1973).

**Sodium aluminum diethyl dihydride,** $NaAl(C_2H_5)_2H_2$. Mol. wt. 110.12. Supplier: Ethyl Corporation (trade name, OMH-1).

*Reducing agent.*[1] This relatively new reducing agent is available as a 25% solution in toluene containing 3–4% THF to improve solubility. The hydride is similar to $LiAlH_4$ in its ability to reduce a wide array of functional groups. Unlike $LiAlH_4$ it is soluble in aromatic hydrocarbons. It is also somewhat less expensive in terms of active hydrogen available.

[1]H. J. Sanders, *Chem. Eng. News*, June 19, 1972, p. 29.

**Sodium amalgam, 1,** 1030–1033; **2,** 373; **3,** 259.

*1-Ethyl-4-carbomethoxypyridinyl* (2). This stable radical is prepared more conveniently by reduction of 1-ethyl-4-carbomethoxypyridinium iodide (1) with 3% sodium amalgam in acetonitrile[1] than by zinc or magnesium.[2]

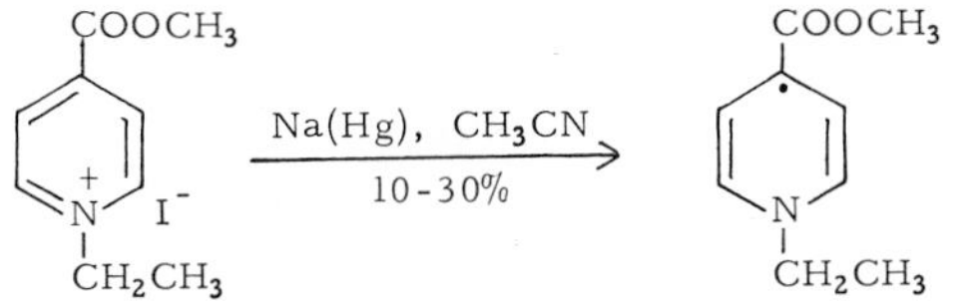

(1)                                    (2)

[1]E. M. Kosower and H. P. Waits, *Org. Prep. Proc. Int.*, **3**, 261 (1971).
[2]W. M. Schwarz, E. M. Kosower, and I. Shain, *Am. Soc.*, **83**, 3164 (1961).

**Sodium amide, 1,** 1034–1041; **2,** 373–374.

*N-Alkylanilines.*[1] Treatment of bromobenzene substituted in the *meta* position by electron-attracting groups [$OCH_3$, Cl, $N(CH_3)_2$] with sodium amide generates a benzyne derivative which reacts with a primary aliphatic amine to give the corresponding

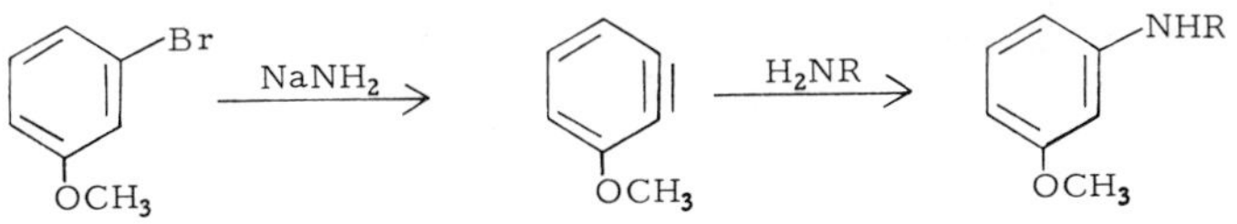

N-alkyl-*meta*-substituted aniline in 68–85 % yield. Less than 5 % of the *ortho*-isomer is obtained.

[1]E. R. Biehl, R. Patrizi, and P. C. Reeves, *J. Org.*, **36**, 3252 (1971).

**Sodium amide–Sodium *t*-butoxide.**

*Dehydrobromination.* β-Eliminations of HX to form olefins have generally been believed to proceed through an *anti* transition state. A recent review of bimolecular

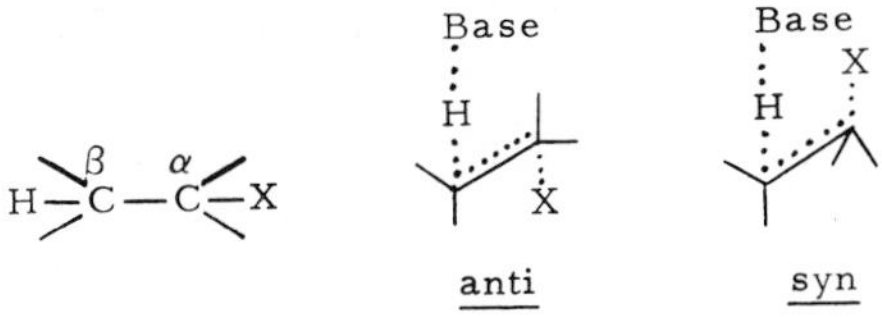

elimination by Sicher[1] notes that *syn* and *anti* eliminations can proceed side by side, particularly with those involving onium bases.

French chemists[2] report that treatment of *trans*-1,2-dibromocyclohexane (1) with the "complex base" sodium amide–sodium *t*-butoxide produces cyclohexene (2, 36 % yield) and 1-bromocyclohexene (3, 60 % yield). Use of either base (separately) leads

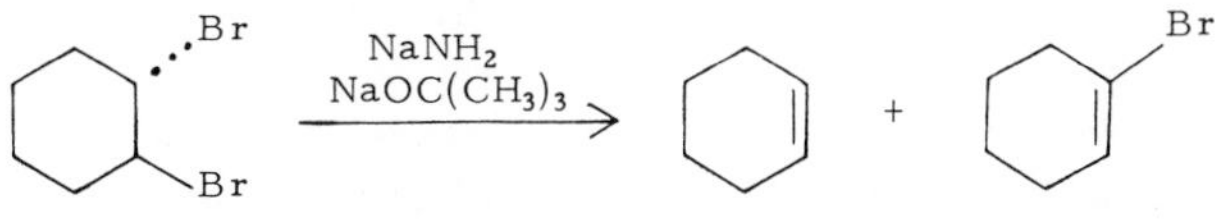

(1)                    (2, 36%)      (3, 60%)

mainly to recovery of the starting dibromide. Cyclohexene (2) is evidently formed by the expected *anti* elimination; 1-bromocyclohexene (3) can only be formed by *syn* elimination.

[1]J. Sicher, *Angew. Chem., internat. Ed.*, **11**, 200 (1972).
[2]P. Caubère and G. Coudert, *J.C.S. Chem. Comm.*, 1289 (1972).

**Sodium azide, 1**, 1041–1044; **2**, 376; **3**, 259–260.
   *Schmidt degradation.*[1,2]

$$\text{(CH}_2\text{)}_9 \underset{\substack{\text{CH} \\ \| \\ \text{C} \text{CO}_2\text{CH}_3}}{} \quad \xrightarrow[\text{H}_2\text{SO}_4, \ \text{CHCl}_3]{\text{NaN}_3} \quad \xrightarrow{\text{H}_2\text{O}} \quad \text{(CH}_2\text{)}_{10} \quad \text{C}=\text{O}$$

(3)                                                                          (4)

*Caution:*   Hydrazoic acid is toxic and has been reported to explode without apparent inducement.

In a well-ventilated hood, 600 ml. of concentrated sulfuric acid is cooled to 5° in a 3-l. three-necked flask equipped with an A. Thomas Magne-Matic Stirrer, Model 15, a 2-in. Teflon-coated, egg-shaped magnet, a water bath, an internal thermometer, a reflux condenser protected by a calcium chloride tube, and a rubber stopper.

A 191–196 g. batch of methyl cycloundecene-1-carboxylate (3) is added through a long-stemmed funnel to the slowly stirred acid. Stirring is then increased to a brisk rate in order to obtain a homogeneous solution; 500 ml. of chloroform is added, and the temperature of the resulting mixture brought to 35° by filling the water bath with warm water. To the well-stirred reaction mixture is added under protection of a safety shield during 50 min., 78 g. (1.2 moles) of sodium azide (Eastman; lumps should be broken up, contact with the skin avoided), while the reaction temperature is controlled to 35–40°. The reaction mixture is then cooled to 5°, poured onto 1 kg. of crushed ice, and transferred together with 1.5–2 l. of water to a 5-l. three-necked flask set up for steam distillation. The chloroform is distilled and the cycloundecanone is steam distilled with 3.5–4 kg. of steam. The steam distillate is extracted with the recovered chloroform and the extract filtered through anhydrous sodium sulfate. The aqueous phase is extracted once with 500 ml. of ether and the extract washed with saturated sodium chloride solution and filtered through anhydrous sodium sulfate. The organic phases are combined and the solvents evaporated under reduced pressure. Vacuum distillation of the residual oil affords 139–143 g. (83–85 %) of cycloundecanone (4) as a colorless or pale-yellow oil, b.p. 84–85°/2 mm., m.p. 16.2–16.6°.

*Modified Curtius reaction.* Kaiser and Weinstock[3] give a detailed procedure for the synthesis of 1-phenylcyclopentylamine from 1-phenylcyclopentanecarboxylic acid via the mixed carboxylic-carbonic anhydride.

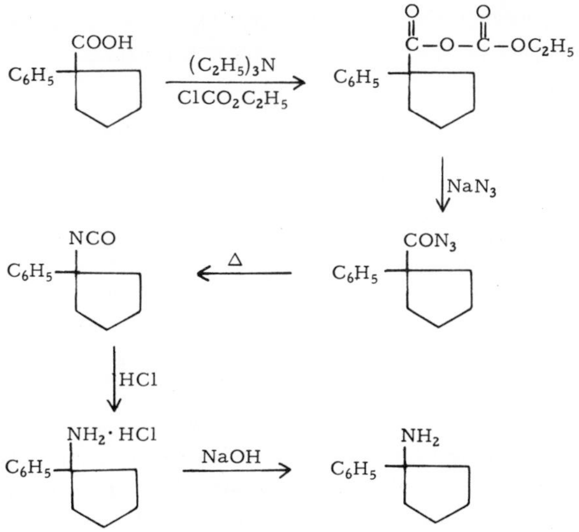

**1,2,3-Triazoles.**[4] *trans*-Arylnitroethylenes (*e.g.*, 1) on treatment with sodium azide

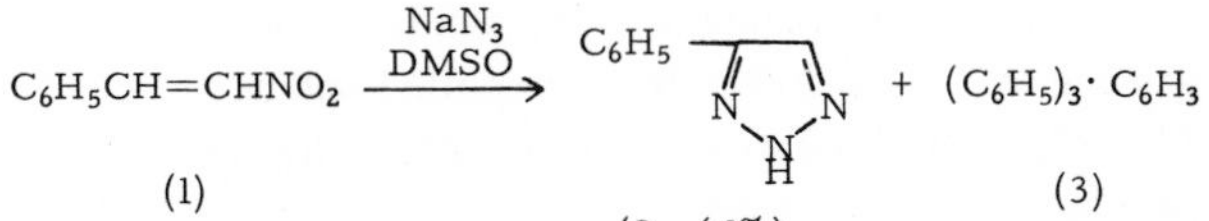

in DMSO or DMF give 1,2,3-triazoles (2) and, unexpectedly, *sym*-triarylbenzenes (3). The reaction is also applicable to α,β-unsaturated nitriles:

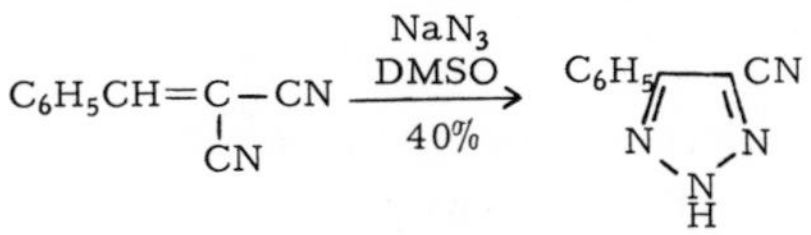

[1] E. W. Garbisch, Jr., and J. Wohllebe, *J. Org.*, **33**, 2157 (1968).
[2] *Idem., Org. Syn.*, submitted 1971.
[3] C. Kaiser and J. Weinstock, *ibid.*, **51**, 48 (1971).
[4] N. S. Zefirov, N. K. Chapovskaya, and V. V. Kolesnikov, *Chem. Commun.*, 1001 (1971).

**Sodium bis-(2-methoxyethoxy)aluminum hydride (SMEAH), 3, 260–261.**

*Alkylation.*[1] If benzophenone (1) is treated with SMEAH in *n*-propylbenzene at 162° for a short period of time, benzhydrol (2) is the major product (50%); diphenylmethane (3) is also obtained in 38.5% yield. Continued reaction decreases the amount of these two products and 1,1-diphenylethane (4) becomes the major product. Finally, on extension of the reaction to 3.5 hr., 2,2-diphenylpropane (5, 71% yield) and 1,1-diphenylcyclopropane (6, 11% yield) become the only products.

$(C_6H_5)_2C=O$         $(C_6H_5)_2CHOH$         $(C_6H_5)_2CH_2$         $C_6H_5\overset{\overset{\displaystyle CH_3}{|}}{C}HC_6H_5$

(1)                     (2)                     (3)                     (4)

$C_6H_5\overset{\overset{\displaystyle CH_3}{|}}{\underset{\underset{\displaystyle CH_3}{|}}{C}}C_6H_5$

$C_6H_5-\overset{\displaystyle C}{\underset{\displaystyle H_2C——CH_2}{|}}-C_6H_5$

(5)                                    (6)

Alkylation has never been encountered before with complex metal hydrides. The reaction probably proceeds by a free-radical mechanism by methyl radicals generated by fragmentation of the 2-methoxyethoxy groups. Note that the hydrogen atoms replaced are activated by phenyl groups. The reaction has been extended to fluorenone and anthraquinone and to other diarylmethanes.

*Reduction of 2-acetylnaphthalene* with the reagent in refluxing benzene yields the expected product, 1-(2-naphthyl)ethanol (97% yield). However, reduction with the hydride in *m*-xylene at 140° gives the bis-(2-naphthyl)butane (1, 27%) and 2-ethylnaphthalene (2, 16%). The formation of (1) is apparently the first example of the transformation of an alkyl aryl ketone into a diarylalkane by a metallic hydride.

(1)                     (2)

*Sulfonamides* are reductively cleaved by the reagent to regenerate the amines. Note that LiAlH$_4$ is not useful for this purpose.[3]

*Review.* Málek and Černý[4] have reviewed the preparation and reactions of alkoxyaluminohydrides.

[1] M. Černý and J. Málek, *Tetrahedron Letters*, 691 (1972).
[2] *Idem, Synthesis*, 53 (1973).
[3] E. H. Gold and E. Babad, *J. Org.*, **37**, 2208 (1972).
[4] J. Málek and M. Černý, *Synthesis*, 217 (1972).

**Sodium bistrimethylsilylamide, 1,** 1046–1047; **3,** 261–262.

*Monobromocyclopropanes.* Martel and Hiriart[1] report that monobromocarbene can be generated in fair yield by the reaction of methylene bromide with the strong base sodium bistrimethylsilylamide. This reaction constitutes a convenient synthesis of monobromocyclopropanes:

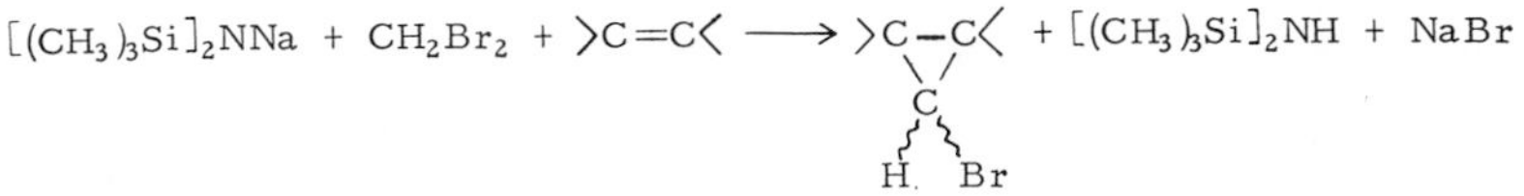
[(CH3)3Si]2NNa + CH2Br2 + >C=C< ⟶ >C–C< + [(CH3)3Si]2NH + NaBr
C
H Br

Heretofore the best method for synthesis of monobromocyclopropanes has been reduction of *gem*-dibromocyclopropanes with tributyltin hydride. The new procedure can also be used to prepare monochlorocyclopropanes by use of methylene chloride in place of methylene bromide.

[1]B. Martel and J. M. Hiriart, *Synthesis*, 201 (1972).

**Sodium borohydride, 1**, 1049–1055; **2**, 377–378; **3**, 262–264.

***Reduction of aromatic nitro compounds.*** In a continuation of a study of reductions with sodium borohydride in polar aprotic solvents (**3**, 262) Hutchins *et al.*[1] report that aromatic nitro compounds are reduced by the reagent in DMSO or sulfolane initially to azoxy compounds, which are subsequently reduced to azo compounds and amines. Electron-withdrawing groups facilitate both the initial and subsequent reductions. But electron-releasing groups slow the reduction of azoxy compounds to the extent that these reduction products can be obtained in reasonable yield.

***Nitroalkanes.***[2] Nitroalkanes can be readily prepared in 60–95% yield by reduction of 1-nitro-2-alkyl nitrates[3] with sodium borohydride in 95% ethanol.

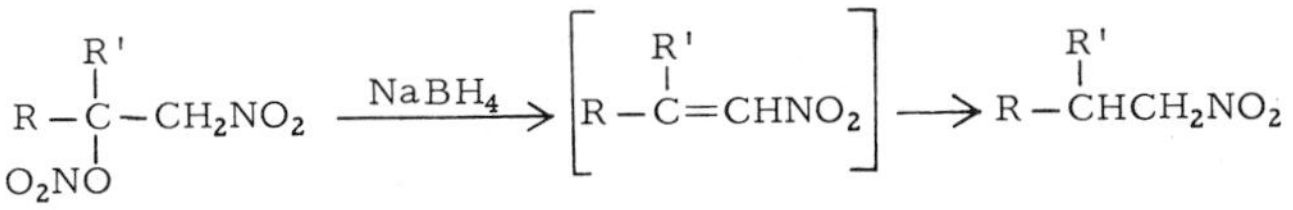
R'
R–C–CH2NO2
O2NO
NaBH4
R'
R–C=CHNO2
R'
R–CHCH2NO2

***Reduction of aldehydes and ketones under very mild conditions***[4] (**2**, 377–378):

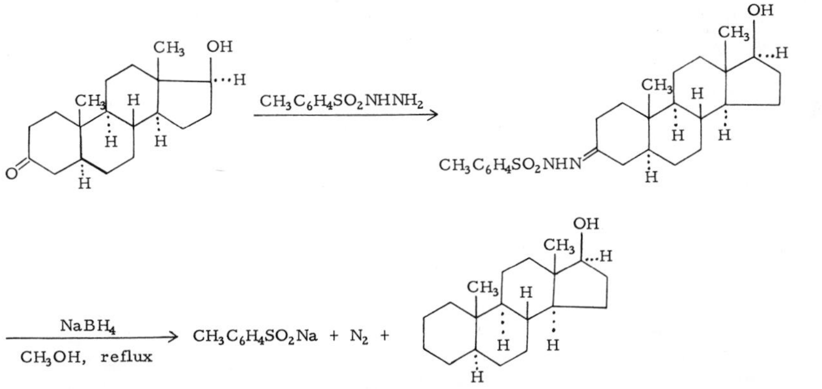
CH3 OH
H
CH3 H
H H
H
CH3C6H4SO2NHNH2
OH
CH3
H
CH3 H
H H
H
CH3C6H4SO2NHN
NaBH4
CH3OH, reflux
CH3C6H4SO2Na + N2 +
OH
CH3
H
CH3 H
H H
H

A mixture of 1 g. of 5α-androstane-17β-ol-3-one, 0.90 g. of tosylhydrazine, and 70 ml. of methanol is refluxed for 3 hr. and then cooled to room temperature. Sodium borohydride (2.5 g.) is added in small portions during 1 hr. and the mixture is refluxed for an additional 8 hr. The solvent is removed under reduced pressure and the residue washed in ether with water, aqueous sodium carbonate, 2 $M$ hydrochloric acid, and water. Chromatography affords 73–76% of pure 5α-androstane-17β-ol.[4]

*Reduction of tertiary halides.*[5] Tertiary halides are reduced to hydrocarbons by sodium borohydride in tetramethylene sulfone (sulfolane, **1**, 1144–1145; **2**, 402–403). The reaction proceeds by way of elimination, hydroboration, and protonolysis. Halides

lacking an α-hydrogen are also reduced; in this case the reaction probably occurs by initial ionization followed by hydride capture. Yields are in the range 45–90%.

*Sulfurated sodium borohydride* (**3**, 264). Supplier: Ventron.

The reduction of functional groups with sulfurated borohydrides has been reviewed by Lalancette *et al.*[6] The reagent is particularly useful for selective reductions. Thus selective reduction of a nitro, an oxime, or a nitrile group in the presence of an ester group is possible. In the case of steroidal ketones, the carbonyl group at $C_3$ is the most reactive site; selective reduction of a $C_3$-carbonyl group in the presence of other carbonyls at $C_{11}$, $C_{12}$, $C_{17}$, or $C_{20}$ is possible. The alcohol obtained is the equatorial isomer. Aldehydes can be reduced selectively in the presence of a carbonyl group if the molar ratio of $NaBH_2S_3$ is kept at a suitable value.

The reagent reduces aromatic nitro compounds to the corresponding amines in about 80% yield.[7] Ester, nitrile, ether, halide, or double bonds, if present, are not affected. Primary aliphatic nitro compounds are reduced to the corresponding nitrile in high yield. The reduction of amides gives only moderate yields of the corresponding amines.[8]

The reagent reacts with epoxides to give symmetrical bis-(2-hydroxyethyl) disulfides (1), which can be reduced by $LiAlH_4$ to 1,2-mercaptols (2).[8]

[1]R. O. Hutchins, D. W. Lamson, L. Rua, C. Milewski, and B. Maryanoff, *J. Org.*, **36**, 803 (1971).

[2]J. M. Larkin and K. L. Kreuz, *ibid.*, **36**, 2574 (1971).

[3]Prepared from 1-alkenes, nitrogen oxides, and oxygen: D. R. Lachowicz and K. L. Kreuz, U.S. Patent 3,282,983 (Nov. 1, 1966) [*C.A.*, **66**, 10577s (1967)].

[4]L. Caglioti, *Org. Syn.*, **52**, 122 (1972).

[5]R. O. Hutchins, R. J. Bertsch, and D. Hoke, *J. Org.*, **36**, 1568 (1971).

[6]J. M. Lalancette, A. Frêche, J. R. Brindle, and M. Laliberté, *Synthesis*, 526 (1972).

[7]J. M. Lalancette and J. R. Brindle, *Canad. J. Chem.*, **49**, 2990 (1971).

[8]J. M. Lalancette and A. Frêche, *ibid.*, **49**, 4047 (1971).

**Sodium chloride–Dimethyl sulfoxide.**

*Decarboalkylation.*[1]  Geminal diesters are decarboalkylated in 90–95% yield when heated at 140–180° for several hours with a slight excess of sodium chloride (table salt is satisfactory) in DMSO containing 2 moles of water per mole of substrate.

$$RR'C(COOC_2H_5)_2 \xrightarrow[\substack{90-95\%}]{\substack{NaCl-DMSO \\ 140-180^0}} RR'CHCOOC_2H_5 + CO_2 + CH_3CH_2OH$$

$\beta$-Keto esters and $\alpha$-cyano esters are also decarboalkylated in excellent yield by this procedure.

Examples:

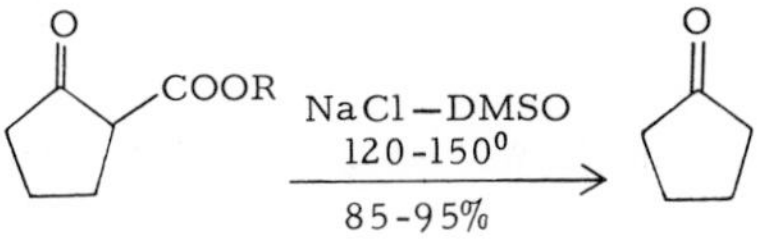

$$CH_3(CH_2)_3\underset{\underset{COOC_2H_5}{|}}{CH}\overset{\overset{O}{\|}}{C}CH_3 \xrightarrow{\substack{NaCl-DMSO \\ 153-165^0}} CH_3(CH_2)_4\overset{\overset{O}{\|}}{C}CH_3$$

$$NCCH_2COOC_2H_5 \xrightarrow{\substack{NaCl-DMSO \\ 135-165^0}} CH_3CN$$

$$CH_3(CH_2)_3\underset{\underset{COOC_2H_5}{|}}{CH}CN \xrightarrow{\substack{NaCl-DMSO \\ 152-168^0}} CH_3(CH_2)_4CN$$

Dimethylformamide (b.p. 153°) can be used in place of DMSO (b.p. 189°), but longer reaction times are necessary.

[1] A. P. Krapcho and A. J. Lovey, *Tetrahedron Letters*, 957 (1973).

**Sodium N-chloro-*p*-toluenesulfonamide (Chloramine-T),** $p$-$CH_3C_6H_4SO_2NClNa\cdot3H_2O$. Suppliers: Baker, Eastman, Fisher, Howe and French, K and K, MCB.

*Removal of ethylene hemithioacetal and -ketal protecting group,* 1,3-Oxathiolanes when treated with chloramine-T in water or ethanol under mild conditions give the corresponding aldehydes or ketones in good yields.[1] For example, 1,4-oxathiaspiro-[4.4]nonane (1) affords cyclopentanone (2) in 91% yield when treated with chloramine-T in 85% $CH_3OH-H_2O$ at 25° for 2 min.

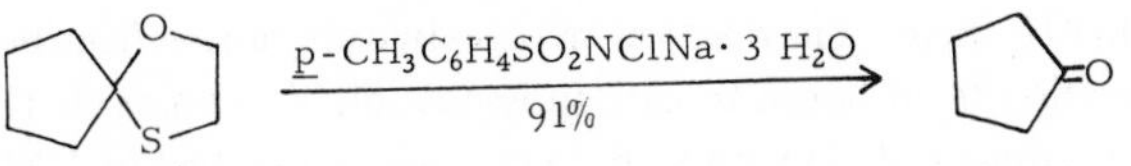

(1)                    (2)

The reagent also can be used for regeneration of ketones from 1,3-dithiolanes.[2] For example, spiro[1,3-dithiol-2,9'-fluorene] (3) is converted into fluorenone (4) in 86% yield by the reaction of chloramine-T in 80% ethanol.

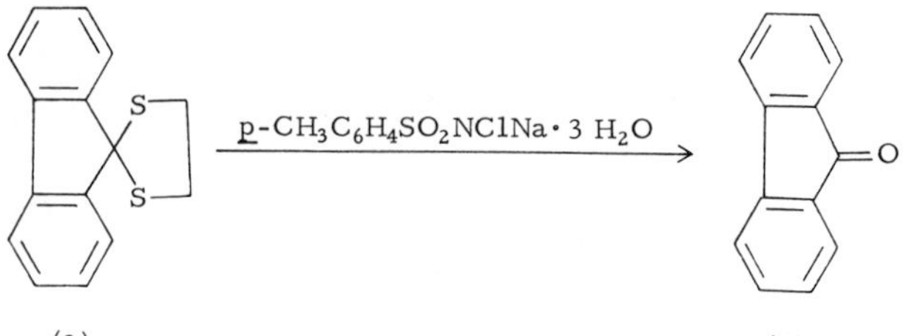

(3)                                                    (4)

[1] D. W. Emerson and H. Wynberg, *Tetrahedron Letters*, 3445 (1971).
[2] W. F. J. Huurdeman, H. Wynberg, and D. W. Emerson, *ibid.*, 3449 (1971).

**Sodium cyanide**, NaCN. Mol. wt. 49.02. Suppliers: Baker, Eastman, Fisher, Howe and French, MCB, E. Merck, Pfaltz and Bauer, Sargent.

*Cyanoboration.* Cyanoboration covers reactions in which there is migration from boron to carbon and which involve cyanoborate salts.

Hydroboration of 1-methylcyclopentene (1) with diborane gives tris-(*trans*-2-methyl-cyclopentyl)borane (2), which on oxidation gives *trans*-2-methylcyclopentanol (3).[1] Addition of sodium cyanide to a solution of (2), followed by treatment with trifluoro-acetic anhydride (TFAA) and then by oxidation with alkaline hydrogen peroxide, gives

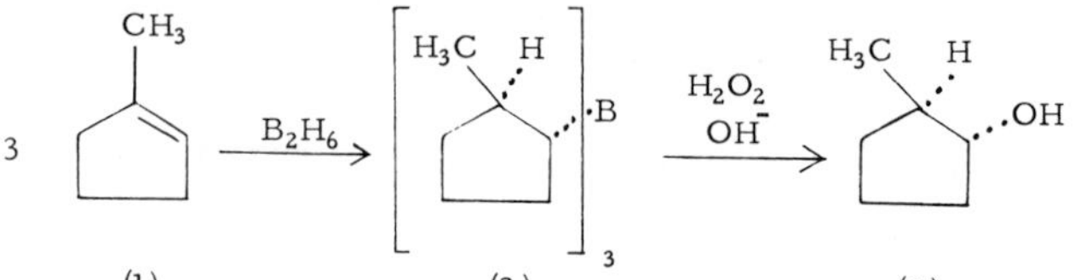

the ketone (4) in 82% yield.[2] The reaction involves two migrations from boron to carbon.

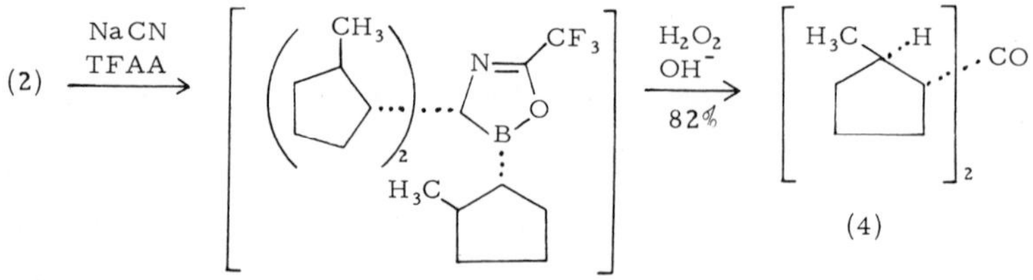

A third migration from boron to carbon is possible if pyridine or DMF is used as solvent. Thus treatment of (2) with sodium cyanide in pyridine or DMF followed by alkaline oxidation gives the carbinol (5).[3] Thus it is possible to induce a third migration of a bulky group from boron to carbon. The ease of migration of alkyl groups is in the order primary > secondary > tertiary for each of the migrations. Note that migrations in cyanoboration proceed with retention of configuration of the migrating group.

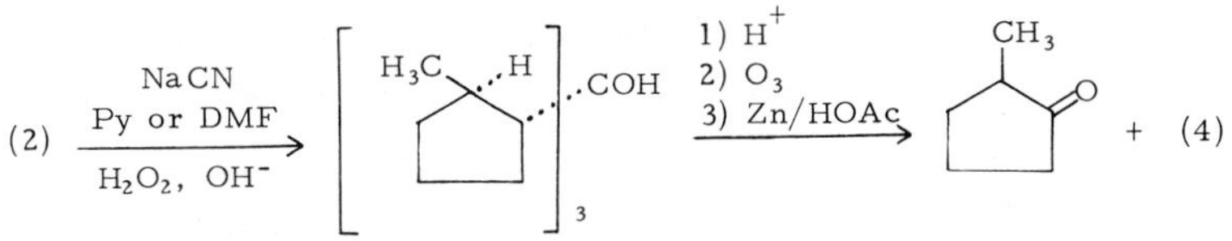

*Addition of aldehydes to activated double bonds.*[4] Aldehydes (1) react with α,β-unsaturated carboxylic esters (2a), ketones (2b), or nitriles under the catalytic influence of sodium cyanide to give γ-ketocarboxylic esters (3a), γ-diketones (3b), or γ-ketonitriles

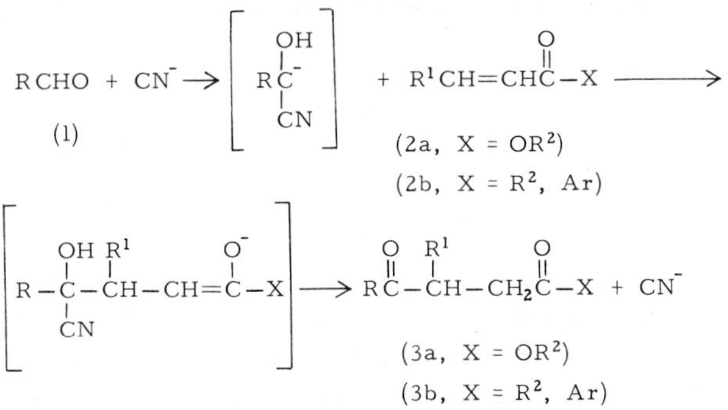

in yields of 30–90%. The reaction is carried out in polar solvents such as DMF or DMSO.

[1]H. C. Brown, M. M. Rogić, M. W. Rathke, and G. W. Kabalka, *Am. Soc.*, **91**, 2150 (1969).
[2]A. Pelter, M. G. Hutchings, and K. Smith, *Chem. Commun.*, 1529 (1970); 1048 (1971).
[3]*Idem, J.C.S. Chem. Commun.*, 186 (1973).
[4]H. Stetter and M. Schreckenberg, *Angew. Chem., internat. Ed.*, **12**, 81 (1973).

## Sodium cyanide–Dimethyl sulfoxide, 2, 381–382.

*Decarboalkylation* (**2**, 381–382). Definitive paper.[1]

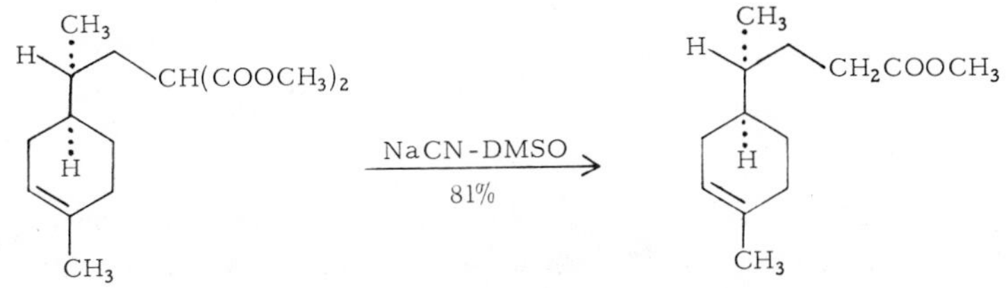

For an improved procedure *see* **Sodium chloride–Dimethyl sulfoxide**, this volume. van Tamelen and Anderson[2] effected decarbomethoxylation of (1) in 81% yield by treatment with sodium cyanide in DMSO solution at 130°. They note that milder conditions than those used originally (**2**, 381–382, ref. 1) are necessary for satisfactory yields.

[1] A. P. Krapcho and B. P. Mundy, *Tetrahedron*, **26**, 5437 (1970).
[2] E. E. van Tamelen and R. T. Anderson, *Am. Soc.*, **94**, 8225 (1972).

**Sodium cyanoborohydride**, $NaBH_3CN$. Mol. wt. 62.84, m.p. 240–242° dec. Suppliers: Aldrich, Alfa Inorganics.

Commercial material is suitable for most purposes, but can be purified by the method of Wade *et al.*[1]

*Reduction of alkyl halides and tosylates.*[2,3] Reduction with sodium cyanoborohydride in HMPT provides a rapid and selective removal of iodo, bromo, and tosyloxy groups in high yield. Thus 1-iododecane can be reduced in this way to *n*-decane in

$$CH_3(CH_2)_8CH_2I \xrightarrow[93\%]{NaBH_3CN,\ HMPT} CH_3(CH_2)_8CH_3$$

$$CH_3(CH_2)_{10}CH_2OTs \xrightarrow[61-80\%]{NaBH_3CN,\ HMPT} CH_3(CH_2)_{10}CH_3$$

93 % yield, and 1-dodecyl tosylate is reduced to *n*-dodecane in 61–80 % yield. Reduction of tosylates requires a large excess of sodium cyanoborohydride for satisfactory yields.

Since sodium cyanoborohydride is a very mild reducing agent, this method is particularly suitable for compounds containing reducible groups such as COOH, COOR, CN, $NO_2$, C=O. This method is thus more selective than the combination of sodium borohydride in DMF (**3**, 262).

Primary alcohols can be reduced to hydrocarbons in two steps; conversion into the iodide with triphenyl phosphite methiodide (methyltriphenoxyphosphonium iodide, **1**, 1249; **2**, 446[4]), followed by reduction with sodium cyanoborohydride in HMPT.

*Reductive amination.* In a procedure of Borch,[5] 4 g. of potassium hydroxide (pellets) is added in one portion to a magnetically stirred solution of 21 g. (0.25 mole) of dimethylamine hydrochloride in 150 ml. of methanol in a 500-ml. round-bottomed flask. Precipitation of potassium chloride begins immediately but does not interfere with the reaction. When the pellets are completely dissolved, 20 ml. (0.20 mole) of cyclohexanone is added in one portion. The resulting suspension is stirred at room temperature for

15 min., and then a solution of 4.75 g. (0.075 mole) of sodium cyanoborohydride in 50 ml. of methanol is added dropwise over 30 min. to the magnetically stirred suspension. After the addition is complete, the suspension is stirred for 30 min.

Potassium hydroxide (15 g.) is then added and the suspension is stirred until the pellets are completely dissolved. The reaction mixture is filtered by suction, and the volume of the filtrate is reduced to approximately 50 ml. on a rotary evaporator at a bath temperature maintained below 45°. Water (10 ml.) and saturated sodium chloride solution (25 ml.) are added to the residue and the layers are separated. The aqueous layer is extracted with ether and the organic layers are combined, and extracted with 6$M$ HCl. The acid layers are saturated with sodium chloride, and extracted with four portions of ether (gas chromatographic analysis shows the material extracted by ether to be cyclohexanol). The aqueous solution is stirred in an ice bath and brought to pH > 12 by cautious addition of potassium hydroxide pellets to the stirred solution. The layers are separated and the aqueous layer is extracted with two 40-ml. portions of ether. The combined organic layers are washed with 10-ml. of saturated sodium chloride solution, dried over potassium carbonate, and freed of ether on a rotary evaporator. This crude product is fractionated through a 6-in. Vigreux column. After the remaining solvent is removed, there is a forerun of 1–3 g., b.p. 144–155°, found by gas chromatography to be 80–85% N,N-dimethylcyclohexylamine and 15–20% cyclohexanol. Colorless N,N-dimethylcyclohexylamine (99.2% pure) then collected amounts to 15.7–17.5 g. (62–69%), b.p. 156–159°, $n_D^{25}$ 1.4521.

The procedure has been shown applicable to the reductive amination of cycloheptanone, cyclooctanone, acetophenone, norbornanone, and benzaldehyde. Amines used include ammonia, methylamine, dimethylamine, morpholine, aniline, and hydroxylamine.

*Selective reduction of aliphatic ketones and aldehydes to hydrocarbons.*[6] Aliphatic ketones and aldehydes can be reduced selectively in high yields to hydrocarbons with sodium cyanoborohydride and *p*-toluenesulfonylhydrazine in DMF–sulfolane containing *p*-toluenesulfonic acid at 100–105°. The prior preparation of tosylhydrazones is not necessary because carbonyl groups are reduced slowly by sodium cyanoborohydride. Maximum yields are obtained with a fourfold molar excess of NaBH$_3$CN. Yields are in the range 62–98%. Ester groups, if present, are not affected. Aromatic ketones are not reduced.

Examples:

$$CH_3CO(CH_2)_3COO(CH_2)_6CN \xrightarrow{75\%} CH_3CH_2(CH_2)_3COO(CH_2)_6CN$$

$$Cholestanone-3 \xrightarrow{88\%} Cholestane$$

$$4\text{-}\underline{t}\text{-Butylcyclohexanone} \xrightarrow{77\%} \underline{t}\text{-Butylcyclohexane}$$

$$CH_3(CH_2)_9CHO \xrightarrow{66\%} CH_3(CH_2)_9CH_3$$

Yields by this method are higher than those obtained by reduction of tosylhydrazones with NaBH$_4$ (**2**, 377–378); moreover NaBH$_3$CN is more selective than NaBH$_4$.

**Selective reductions.** Borch *et al.*[7] have recently reported a study of the reduction of various organic functional groups with sodium cyanoborohydride. Under neutral conditions, carbonyl groups are reduced to a negligible extent, but reduction is rapid at pH 3–4. Ketoximes are reduced smoothly at pH ~4 to the corresponding alkyl-hydroxylamines. Reduction of aldoximes results mainly in the dialkylhydroxylamine.

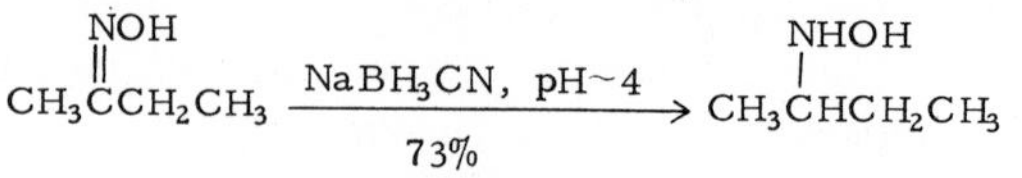

$$CH_3\overset{\overset{\displaystyle NOH}{\|}}{C}CH_2CH_3 \xrightarrow[73\%]{NaBH_3CN, \ pH\sim 4} CH_3\overset{\overset{\displaystyle NHOH}{|}}{C}HCH_2CH_3$$

$$C_6H_5CH=NOH \xrightarrow[60\%]{NaBH_3CN, \ pH\sim 4} (C_6H_5CH_2)_2NOH$$

Enamines are not reduced under neutral conditions, but are readily reduced in an acidic medium which generates an iminium salt, which is readily reducible:

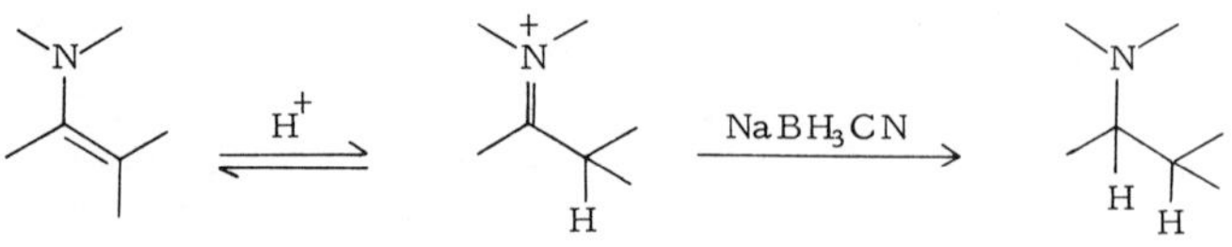

Yields for such reductions are in the range 50–85%. This ready reduction of iminium salts makes possible reductive amination of aldehydes and ketones at pH ~6:

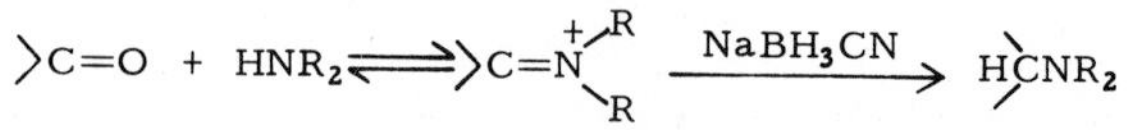

$$>C=O \ + \ HNR_2 \rightleftharpoons >C=\overset{+}{N}\overset{\nearrow R}{\searrow_R} \xrightarrow{NaBH_3CN} H\overset{|}{C}NR_2$$

Thus reaction of cyclohexanone, *n*-propylamine, and sodium cyanoborohydride in methanol at pH 6–8 at 25° for 24 hr. gives *n*-propylcyclohexylamine in 85% yield. The reaction is general for ammonia and primary and secondary amines; aromatic amines are somewhat sluggish. All aldehydes and relatively unhindered ketones can be reductively aminated. Yields are improved by use of 3A molecular sieves to absorb the water generated in the reaction. Note that reductive amination of substituted pyruvic acids with ammonia leads to α-amino acids. Thus alanine can be obtained from pyruvic acid in 50% yield. A pH of 7 is optimum for synthesis of α-amino acids.

Amide and nitrile groups are inert to sodium cyanoborohydride even at pH 2. Similarly, esters, acids, and lactones are inert to the reagent. Acid chlorides are reduced to the corresponding alcohols in THF.

**Methylation of amines.**[8] An aliphatic or aromatic amine ranging in basicity from $pK_a$ 10.66 to 2.47 can be reductively methylated by aqueous formaldehyde and $NaBH_3CN$ in acetonitrile. Yields range from 45 to 90%. The procedure is superior to the Clarke–Eschweiler method,[9] which can lead to complex mixtures.[10]

[1] R. C. Wade, E. A. Sullivan, J. R. Berschied, Jr., and K. F. Purcell, *Inorg. Chem.*, **9**, 2146 (1970).
[2] R. O. Hutchins, B. E. Maryanoff, and C. A. Milewski, *Chem. Commun.*, 1097 (1971).
[3] *Idem, Org. Syn.*, submitted 1972.

[4]This reagent can be purified by the procedure of J. Verheyden and J. G. Moffatt, *J. Org.*, **35**, 2319 (1970).

[5]R. F. Borch, *Org. Syn.*, **52**, 124 (1972).

[6]R. O. Hutchins, B. E. Maryanoff, and C. A. Milewski, *Am. Soc.*, **93**, 1793 (1971).

[7]R. F. Borch, M. D. Bernstein, and H. D. Durst, *ibid.*, **93**, 2897 (1971).

[8]R. F. Borch and A. I. Hassid, *J. Org.*, **37**, 1673 (1972).

[9]M. L. Moore, *Org. React.*, **5**, 301 (1949).

[10]S. H. Pine and B. L. Sanchez, *J. Org.*, **36**, 829 (1971).

**Sodium dihydrogen phosphate, monohydrate (Sodium phosphate, monobasic)**, $NaH_2PO_4 \cdot H_2O$. Mol. wt. 137.99. Suppliers: Baker, Fisher.

*Hydrolysis.*[1] The crude 1,5-dichloropentane-3-one obtained as a dark-brown oil (555–585 g.) by the aluminum chloride-catalyzed addition of 3-chloropropionyl chloride to ethylene in methylene chloride is hydrolyzed in two equal batches. For each batch a 3-l. three-necked, round-bottomed flask with a thermometer, reflux condenser, and 250-ml. dropping funnel is charged with 600 ml. of water, 600 ml. of dioxane, and 510 g. (3.70 moles) of sodium dihydrogen phosphate monohydrate and two or three No. 12

carborundum boiling stones and immersed in a sand bath which is maintained at 140–160°. When its contents are boiling vigorously at a temperature above 90°, the first batch of crude 1,5-dichloropentane-3-one is added dropwise over a period of 1 hr. The flask is then heated for a further 5 hr. so as to keep the reactants boiling vigorously at an internal temperature of 90–92°.

The reaction flask is allowed to cool and is then immersed in an ice-water bath. A solution of 160–180 g. (4–4.5 moles) of sodium hydroxide pellets in 500 ml. of water is added slowly with stirring until the pH rises to 5. After the addition of 400 ml. of ether, the products are transferred to a 5-l. separatory funnel and then shaken. The darkly colored organic upper layer is separated and the aqueous layer extracted with two 500-ml. portions of ether. The combined ether extracts are dried over anhydrous magnesium sulfate and concentrated to about 800 ml. under reduced pressure in a rotary evaporator. The concentrated extract is transferred to a 1-l. round-bottomed flask and fractionated through a 25-cm. Vigreux column under reduced pressure. A forerun of dioxane is collected and then tetrahydro-4H-pyrane-4-one distils as a colorless liquid, b.p. 59–60°/13 mm. The yield is 102–112 g. (50–55%) for the two steps, based on 3-chloropropionyl chloride.

[1]G. R. Owen and C. B. Reese, *Org. Syn.*, submitted 1972.

**Sodium ethoxide, 1,** 1065–1073.

*Cyclopropyl ketones.* Substituted cyclopropyl ketones can be obtained by base-induced isomerization of $\gamma,\delta$-epoxy ketones. Thus treatment of the epoxide (2) of

4-isopropylidenecyclohexanone (1) with sodium ethoxide in ethanol for 15 min. gave 7-hydroxysabina ketone (3), 1-(1'-hydroxy-1'-methyl)ethylbicyclo[3.1.0]hexane-4-one, in over 90% yield. (Use of 2 $N$ NaOH in refluxing ethanol is also effective.) The ketone (3) can be converted into a number of thujane derivatives.[1]

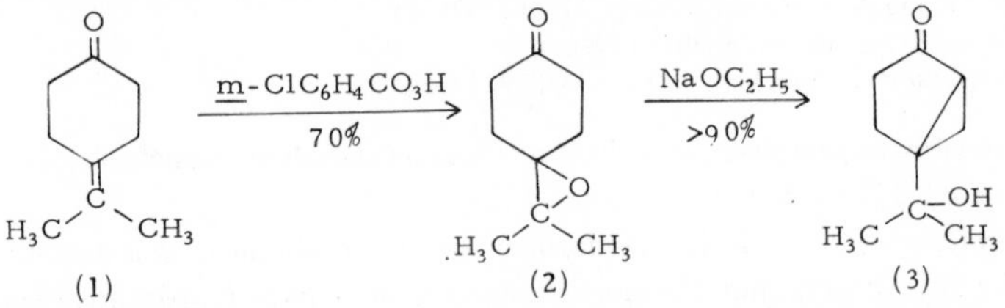

This isomerization reaction has been used in the synthesis of some norcarane derivatives.[2] Thus treatment of the epoxide (4) of karahanaenone[3] with sodium ethoxide in ethanol (reflux 15 min.) yielded stereospecifically one bicyclic hydroxy ketone (5; 5-endo-hydroxy-3,3,6-trimethylnorcarane-2-one) in over 80% yield.

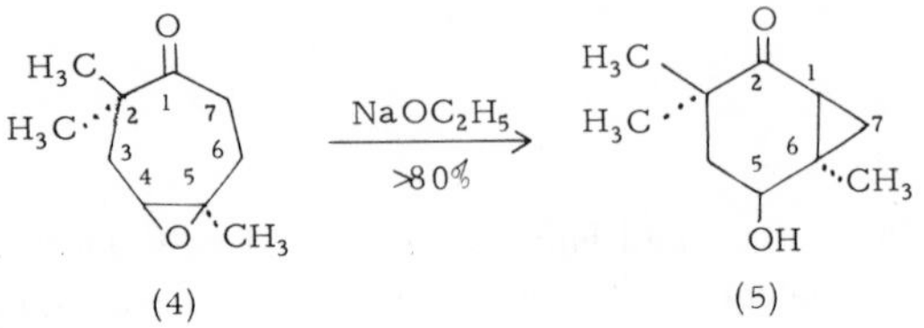

[1]Y. Gaoni, *Tetrahedron*, **28**, 5525 (1972).
[2]*Idem, ibid.*, **28**, 5533 (1972).
[3]E. Demole and P. Enggist, *Helv.*, **54**, 456 (1971).

**Sodium hydride, 1,** 1075–1081; **2,** 382–383.

*O-Alkylation of hydroquinones.*[1] The hydroquinone (1) can be O-alkylated in good yield by treatment with sodium hydride in DMSO and subsequent addition of an alkyl halide. Several other phenols have been alkylated by this procedure.[2]

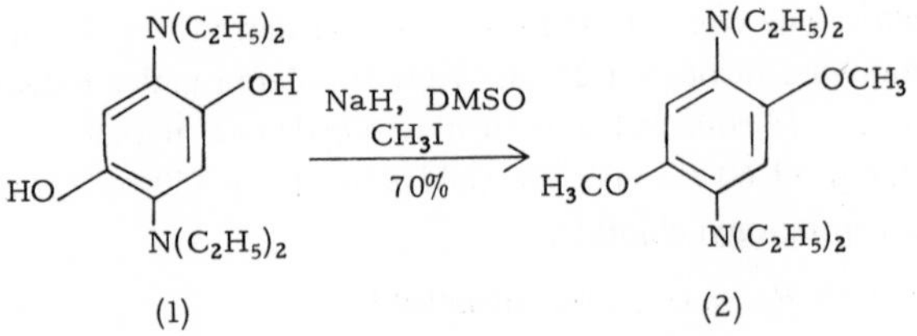

The same procedure is useful for N-alkylation of aromatic amines.[3] Thus N,N,N',N'-tetraethyl-1,4-diamino-2,5-dibromobenzene (4) was obtained in this way from 1,4-diamino-2,5-dibromobenzene (3) in 49% yield.

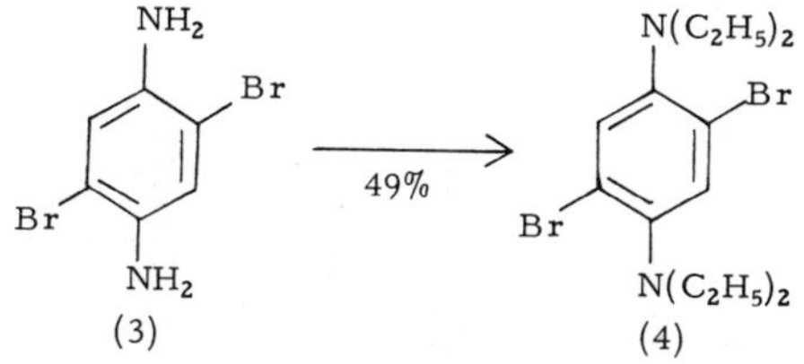

**Dimethyl cycloalk-1-ene-1,2-dicarboxylates.**[4] Dimethyl $\alpha,\alpha'$-dibromoalkanedicarboxylates (1, $n$ = 2–5) when treated with 2 eq. of sodium hydride in DMF undergo

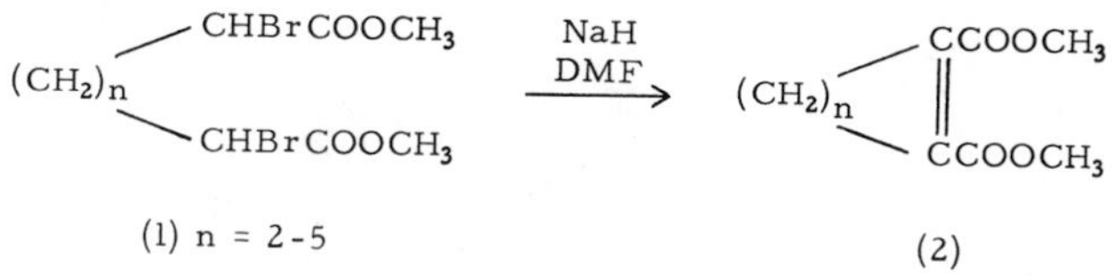

combined cyclization–elimination to give dimethyl cycloalk-1-ene-1,2-dicarboxylates (2). Treatment of dimethyl $\alpha,\alpha'$-dibromoglutarate (1, $n$ = 1) under the same conditions

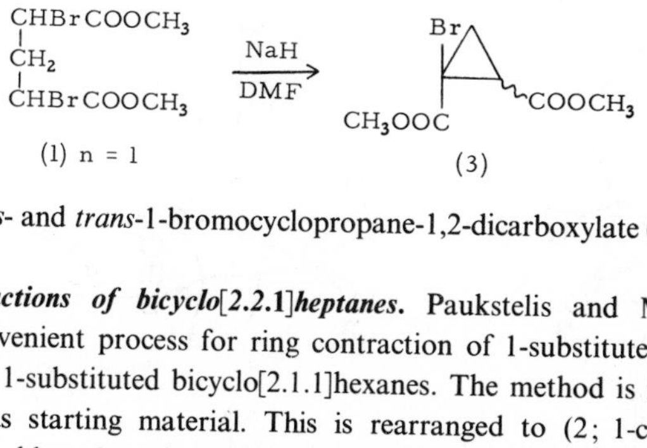

gives dimethyl *cis*- and *trans*-1-bromocyclopropane-1,2-dicarboxylate (3); the *cis–trans* ratio is 1:3.

**Ring contractions of bicyclo[2.2.1]heptanes.** Paukstelis and Macharia[5] have developed a convenient process for ring contraction of 1-substituted bicyclo[2.2.1]-heptane-3-ols to 1-substituted bicyclo[2.1.1]hexanes. The method is illustrated using *d*-camphor (1) as starting material. This is rearranged to (2; 1-chlorocamphene, IUPAC name, 1-chloro-2-methylene-3,3-dimethylbicyclo[2.2.1]heptane) by treatment with phosphorus trichloride and phosphorus pentachloride (reflux, 6 hr.) in 70% yield. Ozonolysis of (2) and direct reduction of the ozonide with sodium borohydride gives the epimeric alcohols (3) and (4) in the ratio of 5:1. Treatment of the mixture of (3) and (4) with sodium hydride (50% in oil) in DMF for a short time gives the aldehyde (5); a longer reaction period leads to the acid (6; 5,5-dimethylbicyclo[2.1.1]hexane-1-carboxylic acid) and the alcohol (7; 5,5-dimethylbicyclo[2.1.1]hexyl-1-methanol). The ring contraction is assumed to involve a pinacol-type *trans*-coplanar rearrangement.

**Reduction.** Benzophenone and fluorenone are reduced to benzhydrol and fluorenol, respectively, by sodium hydride. The best results (92–97% yields) are obtained with HMPT, DMF, or pyridine as solvent. Nonenolizable ketones, or ketones which enolize with difficulty, are reduced to the corresponding alcohols in very good yields. Side

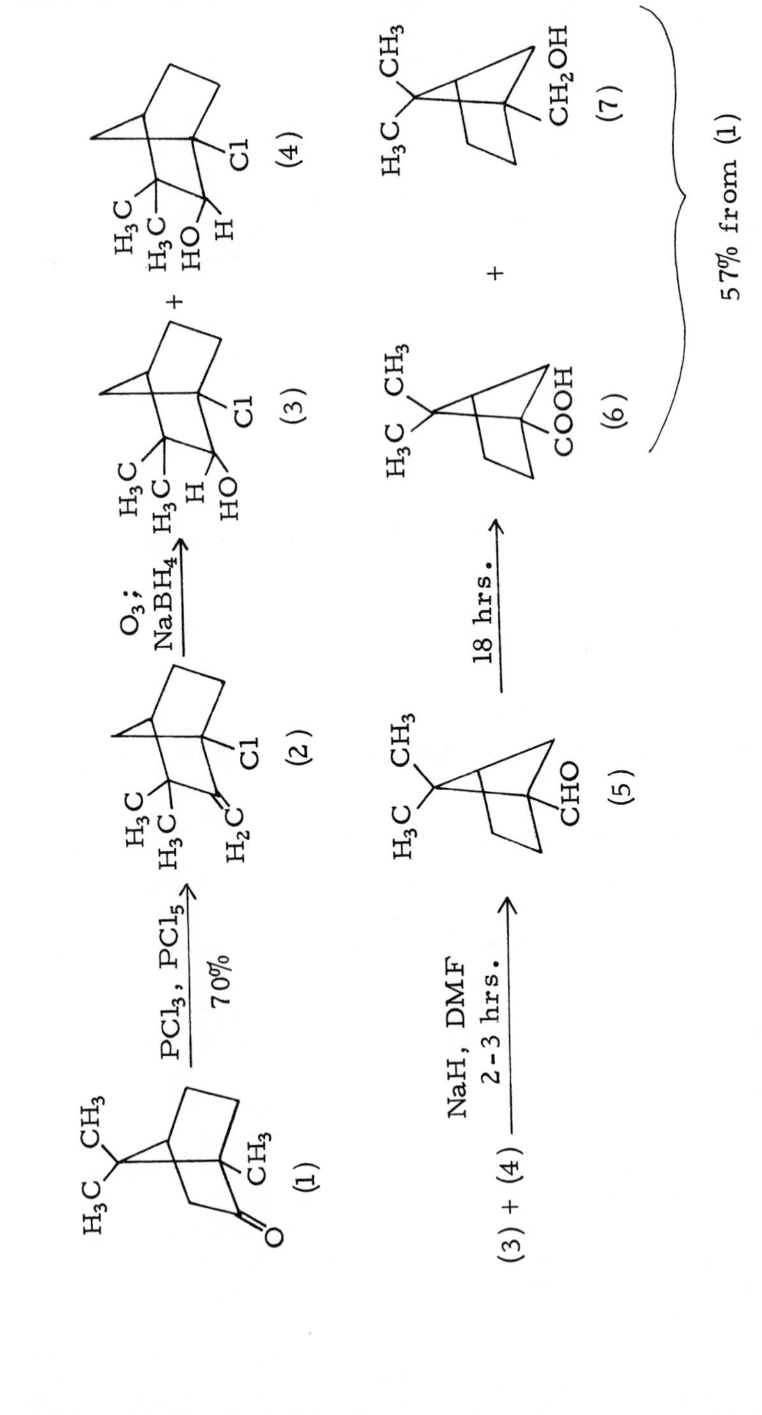
(1)
$PCl_3$, $PCl_5$
70%
(2)
$O_3$; $NaBH_4$
(3)
+
(4)
(3) + (4)
NaH, DMF
2-3 hrs.
(5)
18 hrs.
(6)
+
(7)
57% from (1)

reactions can occur in the reduction of aldehydes and esters. The reagent behaves as a base in the reaction with the oxides of styrene and of cyclohexene[7]:

$$C_6H_5CH \overset{}{\underset{O}{\diagdown\diagup}} CH_2 \quad \xrightarrow[40\%]{NaH/HMPT} \quad C_6H_5COCH_3$$

$$\xrightarrow[35\text{-}40\%]{NaH/HMPT}$$

*Methylation of phenols and alcohols.*[8] Phenols and alcohols are converted into methyl ethers by treatment with sodium hydride–methyl iodide in THF at room temperature. The reaction is even applicable to hindered phenols (2,6-di-*t*-butyl-*p*-cresol, 86% yield) and alcohols (triphenylcarbinol, 85% yield). However, acidic phenols (*p*-nitrophenol) are not methylated by this procedure even at 80°. This behavior contrasts to that of diazomethane, which readily methylates acidic phenols.

[1] T. Doornbos and J. Strating, *Syn. Commun.*, **1**, 175 (1971).
[2] T. Doornbos, personal communication.
[3] T. Doornbos and J. Strating, *Org. Prep. Proc.*, **1**, 287 (1969).
[4] R. N. McDonald and R. R. Reitz, *Chem. Commun.*, 90 (1971).
[5] J. V. Paukstelis and B. W. Macharia, *J. Org.*, **38**, 646 (1973).
[6] P. Caubère and J. Moreau, *Bull. Soc.*, 3270 (1971).
[7] *Idem, ibid.*, 3276 (1971).
[8] B. A. Stoochnoff and N. L. Benoiton, *Tetrahedron Letters*, 21 (1973).

## Sodium hydride–*t*-Butyl hypochlorite.

*Thiadiaziridine 1,1-dioxides.* Timberlake and Hodges[1] have prepared 2,3-di-*t*-butylthiadiaziridine 1,1-dioxide (2) in 46% yield by treatment of (1) with sodium

hydride–*t*-butyl hypochlorite in pentane. The three-membered heteroatomic ring system is surprisingly stable, but it is converted into the azoalkane (3) by oxidation with *t*-butyl hypochlorite. This product had been obtained previously from (1) by oxidation with sodium hypochlorite.[2] Pyrolysis of (2) in refluxing benzene also gives (3, *trans*-2,2′-dimethyl-2,2′-azobutane) in nearly quantitative yield. Increasing the size of the alkyl groups attached to nitrogen in the thiadiaziridine 1,1-dioxide leads to increased thermal stability, but results in decreasing stability in the corresponding azoalkane.

[1] J. W. Timberlake and M. L. Hodges, *Am. Soc.*, **95**, 634 (1973).
[2] R. Ohme and H. Preuschhof, *Ann.*, **713**, 74 (1968).

**Sodium hydride–Dimethyl sulfoxide, 1,** 1075–1081; **2,** 382–383.

*Amides.*[1] The reaction of equimolar mixtures of an ester, an amine, and sodium hydride (oil dispersion) in DMSO (or hexamethylphosphoric triamide, N,N-dimethylacetamide) provides a simple and convenient synthesis of amides. Yields are 68–94%.

[1]B. Singh, *Tetrahedron Letters*, 321 (1971).

**Sodium hypochlorite, 1,** 1084–1087; **2,** 67; **3,** 45, 243.

*Oxidation of N,N′-dialkylhydrazines.* The reagent is used in the preparation of azoethane by oxidation of N,N′-diethylhydrazine.[1]

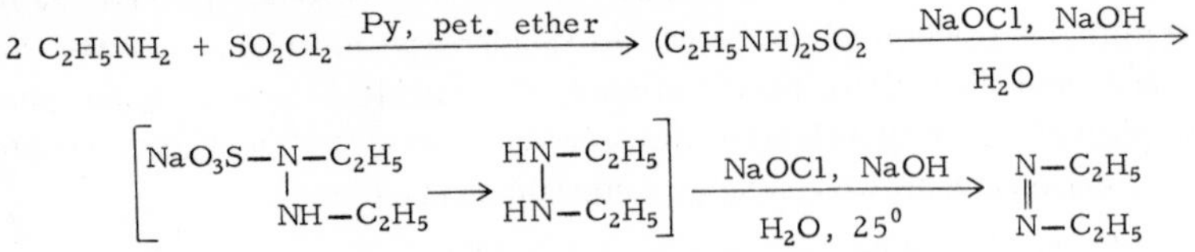

$$2 \; C_2H_5NH_2 + SO_2Cl_2 \xrightarrow{\text{Py, pet. ether}} (C_2H_5NH)_2SO_2 \xrightarrow[\text{H}_2\text{O}]{\text{NaOCl, NaOH}}$$

$$\begin{bmatrix} NaO_3S-N-C_2H_5 \\ \quad\quad | \\ \quad\; NH-C_2H_5 \end{bmatrix} \longrightarrow \begin{bmatrix} HN-C_2H_5 \\ | \\ HN-C_2H_5 \end{bmatrix} \xrightarrow[\text{H}_2\text{O, 25}^0]{\text{NaOCl, NaOH}} \begin{matrix} N-C_2H_5 \\ \| \\ N-C_2H_5 \end{matrix}$$

[1]R. Ohme, H. Preuschhof, and H.-U. Heyne, *Org. Syn.*, **52,** 11 (1972).

**Sodium iodide, 1,** 1087–1090; **2,** 384; **3,** 267.

*Reaction with tosylates* (cf. **1,** 1088). Conversion of methyl 2,3,6-tribenzoyl-4-tosyl-α-D-glucoside (1) into the iodide (2)[1]:

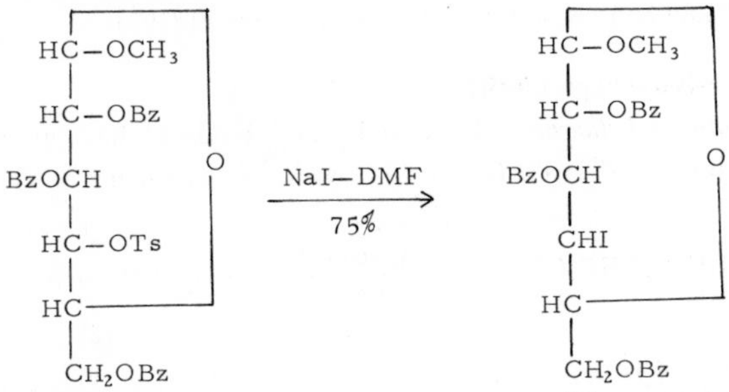

Conversion of methyl 2,3-dibenzoyl-4,6-ditosyl-α-D-glucoside (3) into methyl 2,3-dibenzoyl-4-tosyl-6-desoxy-6-iodo-α-D-glucoside (4)[2]:

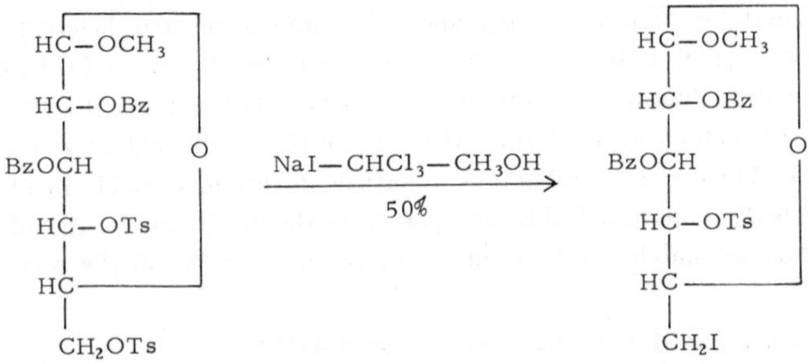

*Spiro[4.4]nona-1,3,7-triene* (5).[3] This interesting spiro compound has now been synthesized as shown in the formulation.

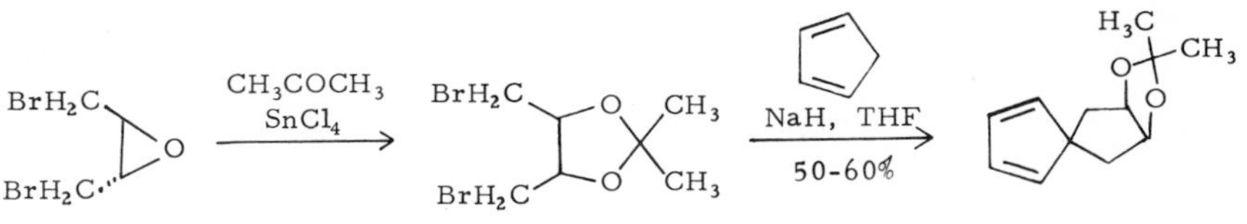

(1)          (2)                    (3)

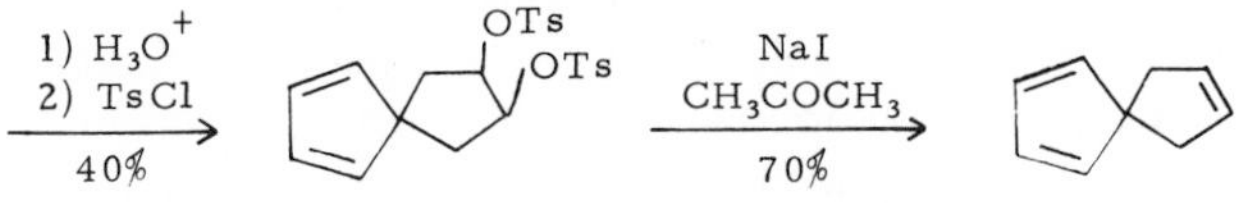

(4)                              (5)

[1]G. Siewert and O. Westphal, *Ann.*, **720**, 161 (1968).
[2]*Idem, ibid.*, **720**, 171 (1968).
[3]M. F. Semmelhack, J. S. Foos, and S. Katz, *Am. Soc.*, **94**, 8637 (1972).


**Sodium methoxide, 1,** 1091–1094; **2,** 385–386; **3,** 259–260.
   *Favorskii rearrangement* (2 → 3).[1]

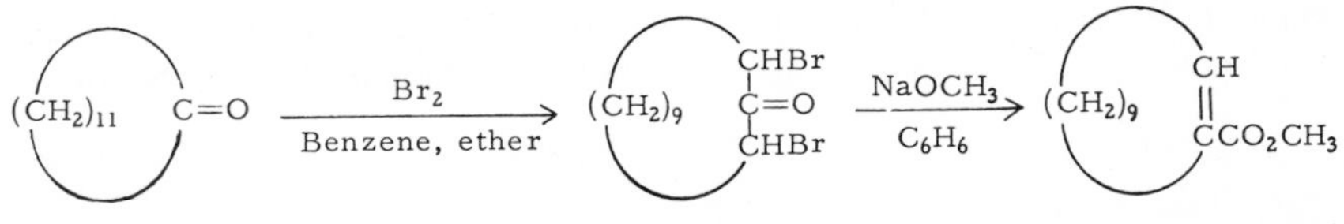

(1)                    (2)                        (3)

Cyclododecanone is converted into the α,α'-dibromide (2) in benzene–ether at 20–25°, the hydrogen bromide and ether are removed with a water aspirator, and the resulting solution of 2,12-dibromocyclododecanone (2) in benzene is stirred and treated during 30–40 min. with 125 g. (2.3 moles) of powdered sodium methoxide, added through the center neck of the flask, while the reaction temperature is maintained at 25–30° by adding ice to the water bath. Stirring is continued for 20 min. The reaction mixture is extracted successively with 500-ml. portions of water, 5% hydrochloric acid, and a concentrated aqueous solution of sodium chloride. The aqueous phases are successively extracted with 400 ml. of diethyl ether and discarded. The organic phases are filtered through anhydrous sodium sulfate, combined, and the solvents evaporated under reduced pressure. Distillation of the material boiling below 104° (0.4 mm.) through a short Vigreux column gives 191–196 g. (91–93%) of methyl cycloundecene-1-carboxylate as a pale-yellow oil, most of which distils at 83–87° (0.4 mm.).

   *Synthesis of alkyl aryl sulfides.*[2] *n*-Butyl 1-naphthyl sulfide is prepared by charging a 1-l. three-necked, round-bottomed flask equipped with a thermometer, a reflux condenser, an addition funnel, and a magnetic stirring bar with 586.7 g. (7.5 moles) of dimethyl sulfoxide and 225.5 g. (2.4 moles) of butanethiol. The solution is stirred and

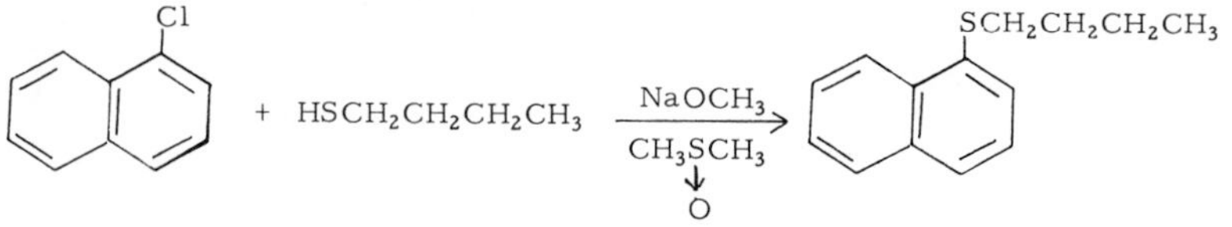

heated to 70° using a heating mantle, and 108 g. (2 moles) of sodium methoxide is slowly added. The reaction mixture is then heated to reflux temperature (115°). 1-Chloronaphthalene (81.3 g., 0.5 mole) is added rapidly through the addition funnel. The yellow reaction mixture is stirred and heated at reflux temperature for 45 hr. After about 8 hr. the mixture turns brown, a brown solid forms on the flask inner surface, and the reflux temperature drops to 98°.

The reaction mixture is then added to 500 ml. of cold water which has been saturated with sodium chloride. Extraction with ether and distillation gives a little by-product di-$n$-butyl disulfide and 75–86 g. (70–80%) of $n$-butyl 1-naphthyl sulfide, b.p. 115–130°/1 mm.

*Naphthalene 1,2-oxide.* Yagi and Jerina[3] have reported a new method for synthesis of arene oxides. The starting material for the preparation of naphthalene 1,2-oxide (5) is 1-hydroxy-2-bromotetralin, (1), which is acetylated with trifluoroacetic anhydride in chloroform to give (2) in 84% yield. This is converted into the dibromide (3) by treatment with NBS (note that tetralin 1,2-epoxide is unstable to bromination with NBS). The

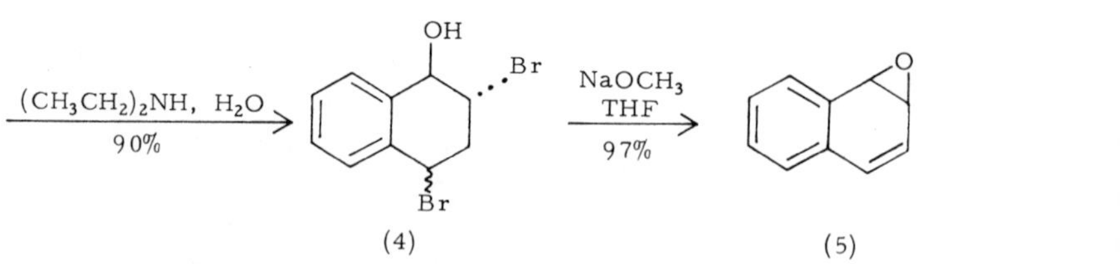

next step is removal of the blocking group by treatment with aqueous diethylamine in acetonitrile. Treatment of (4) with dry sodium methoxide in THF effects generation of the epoxide ring and dehydrobromination in one step to give naphthalene 1,2-oxide (5). The overall yield of (5) from (1) is 65%. In a previous preparation of (5) from (1) by a different route an overall yield of 14% was reported.[4]

The same sequence was also used to prepare the labile phenanthrene 3,4-oxide except that in this case sodium methoxide was replaced by 1,5-diazabicyclo[4.3.0]-nonene-5 (DBN, this volume).

[1]J. Wohllebe and E. W. Garbisch, Jr., *Org. Syn*, submitted, 1971.
[2]J. S. Bradshaw and E. Y. Chen, procedure submitted to *Org. Syn.*, 1972.
[3]H. Yagi and D. M. Jerina, *Am. Soc.*, **95**, 243 (1973).
[4]E. Vogel and F. G. Klärner, *Angew. Chem., internat. Ed.*, **7**, 374 (1968).

**Sodium N-methylanilide, 1**, 1095–1096.

*Thorpe–Ziegler cyclization.*[1] Doornbos and Strating used this reagent for effecting the cyclization of (1) to (2). Acyloin condensation of analogous esters failed in this case.

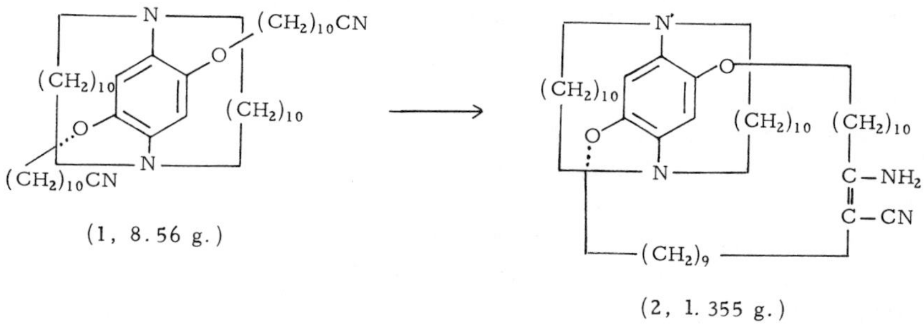

(1, 8.56 g.)

(2, 1.355 g.)

[1]T. Doornbos and J. Strating, *Syn. Commun.*, **1**, 193 (1971).

**Sodium nitrite, 1**, 1097–1101; **2**, 386–387.

*β-Chloropyruvaldoxime* (3). Taylor and Portnoy[1] have reported a convenient synthesis of β-chloropyruvaldoxime (3) by chlorination of diketene (1) to give γ-chloro-acetoacetyl chloride (2), followed by treatment of (2) with sodium nitrite in ether–water. The reaction involves hydrolysis, nitrosation, and decarboxylation. Condensation of

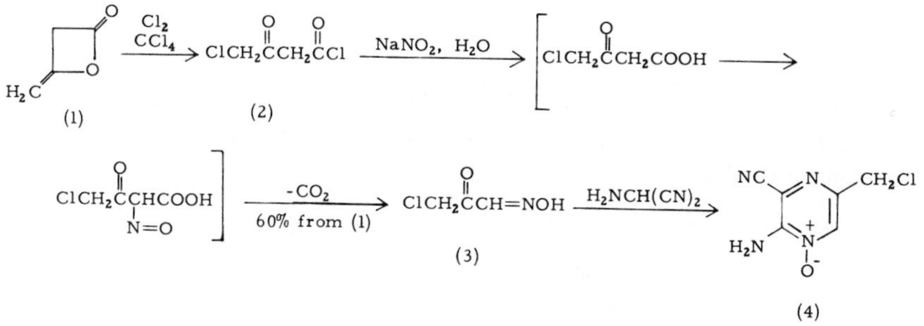

(3) with aminomalononitrile gives 2-amino-3-cyano-5-chloromethylpyrazine 1-oxide (4), a useful precursor to 6-substituted pteridines. β-Halopyruvaldoximes are highly lachrymatory; they are unstable at temperatures above $-20°$.

*3-Nitroisoxazoles.*[2] The reaction of acetylenic methylenebromides (1) with sodium nitrite gives 3-nitroisoxazoles (5) in 20–60% yield by way of the intermediates formulated.

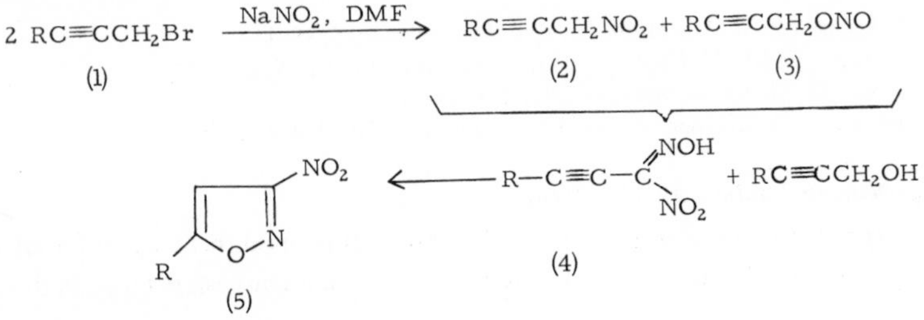

$$2 \ RC \equiv CCH_2Br \xrightarrow{NaNO_2, \ DMF} RC \equiv CCH_2NO_2 + RC \equiv CCH_2ONO$$

$$(1) \hspace{4cm} (2) \hspace{2cm} (3)$$

$$(5) \hspace{3cm} (4)$$

[1]E. C. Taylor and R. C. Portnoy, *J. Org.*, **38**, 806 (1973).
[2]S. Rossi and E. Duranti, *Tetrahedron Letters*, 485 (1973).

**Sodium sulfide,** $Na_2S \cdot 9H_2O$, **1**, 1104–1105.

*Olefin synthesis.* Kornblum *et al.*[1] have described a new synthesis of tetrasubstituted olefins from vicinal dinitro compounds. There are several methods for conversion of aliphatic and alicyclic nitro compounds into *vic*-dinitro compounds, the most useful of which is a procedure of Seigle and Hass.[2] A nitroparaffin is converted into the lithium salt (1) by treatment with lithium methoxide and this is then treated with bromine to form a bromonitro compound (2). Reaction of (2) with a second equivalent of the nitro-

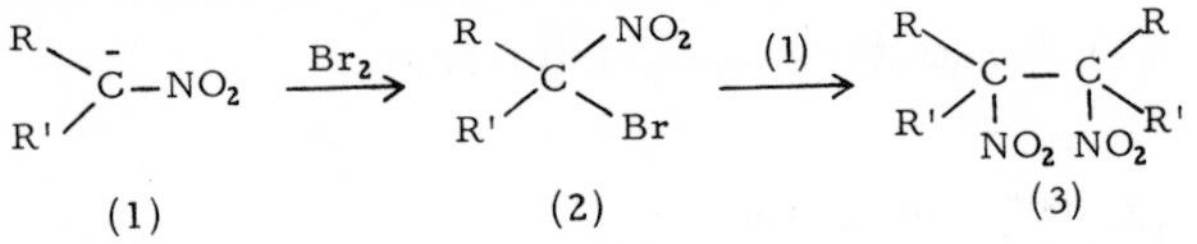

$$(1) \hspace{3cm} (2) \hspace{3cm} (3)$$

paraffin salt (1) gives the symmetrical vicinal dinitro compound (3) in 88–93% overall yield. Unsymmetrical vicinal dinitro compounds (4) are prepared from an α,α-dinitro compound as shown:

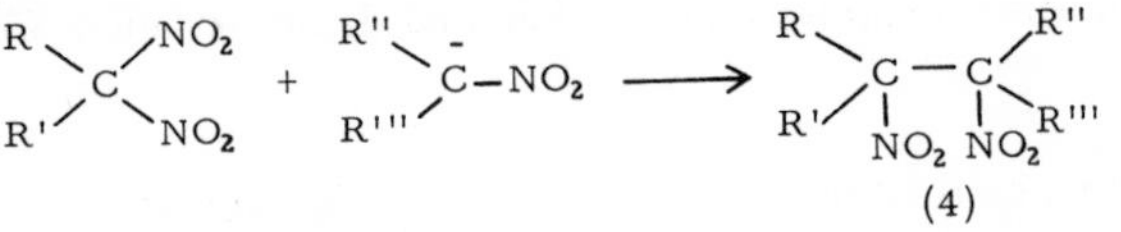

$$(4)$$

Treatment of dinitro compounds of type (3) or (4) with sodium sulfide in DMF under irradiation (20-W fluorescent light) produces olefins in high yield. Sodium thiophenoxide in HMPT is also very effective.

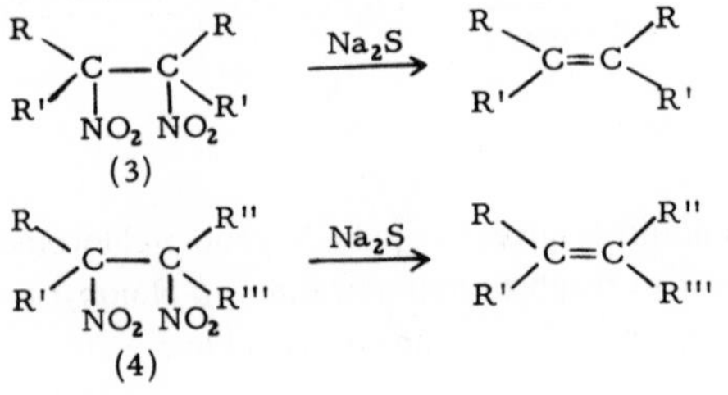

$$(3)$$

$$(4)$$

Examples:

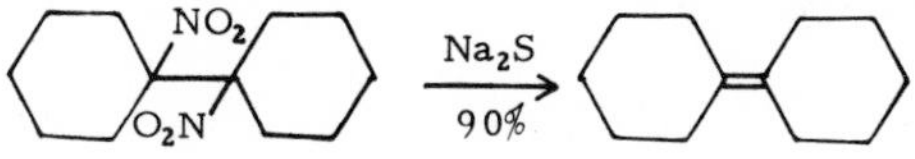

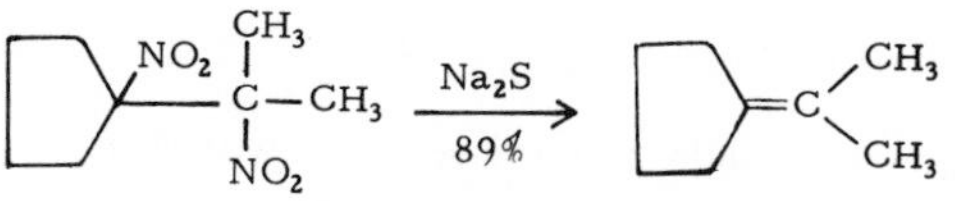

The new elimination process is noteworthy for high yields and absence of isomers. The reaction is evidently a chain process, since it is inhibited by di-*tert*-butyl nitroxide.

[1] N. Kornblum, S. D. Boyd, H. W. Pinnick, and R. G. Smith, *Am. Soc.*, **93**, 4316 (1971).
[2] L. W. Seigle and H. B. Hass, *J. Org.*, **5**, 100 (1940).

**Sodium tetraborate,** $Na_2B_4O_7$. Mol. wt. 201.27. Suppliers: Fisher, MCB, ROC/RIC, and others.

*N-Acetylneuraminic acid.*[1] The yield and stereoselectivity of the aldol condensation of 2-acetamido-2-desoxy-D-mannose (1) with oxalacetic acid (2) to give N-acetyl-neuraminic acid (3) are improved by adding sodium tetraborate to the reaction, which is carried out in aqueous solution at pH10. Thus in the absence of borate ion, yields of (3) of 11.5% have been reported, whereas addition of borate improves the yield to 21.6%. The borate ion inhibits the alkaline epimerization of various 2-acylamino-2-desoxy-aldoses.

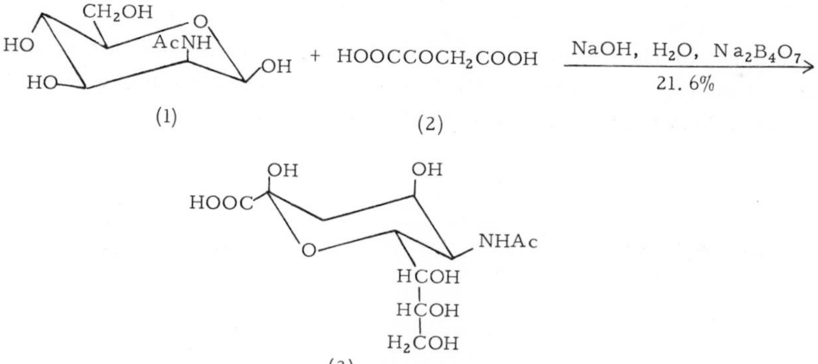

[1] M. J. How, M. D. A. Halford, M. Stacey, and E. Vickers, *Carbohydrate Res.*, **11**, 313 (1969).

**Sodium tetracarbonylferrate(-II) (Disodium tetracarbonylferrate),** 3, 267–268.

*Synthesis of aliphatic ketones.* Collman *et al.*[1] have described four similar but distinct syntheses of unsymmetrical ketones using $Na_2Fe(CO)_4$ as the reagent as outlined in (a)–(d).

(a) $RX + Na_2Fe(CO)_4 \longrightarrow RFe^-(CO)_4 \xrightarrow{R'X} R\overset{O}{\overset{\|}{C}}R'$

(b) $RX + Na_2Fe(CO)_4 + P(C_6H_5)_3 \longrightarrow R\overset{O}{\overset{\|}{C}}Fe^-(CO)_3[P(C_6H_5)_3] \xrightarrow{R'X} R\overset{O}{\overset{\|}{C}}R'$

(c) $RX + Na_2Fe(CO)_4 + CO \longrightarrow R\overset{O}{\overset{\|}{C}}Fe^-(CO)_4 \xrightarrow{R'X} R\overset{O}{\overset{\|}{C}}R'$

(d) $RCOCl + Na_2Fe(CO)_4 \longrightarrow R\overset{O}{\overset{\|}{C}}Fe^-(CO)_4 \xrightarrow{R'X} R\overset{O}{\overset{\|}{C}}R'$

In method (a), an alkyl halide (or tosylate) reacts with $[Fe(CO)_4]^{2-}$ in N-methyl-2-pyrrolidone to give an anionic iron alkyl (1), some of which have been isolated and characterized. Further reaction with another alkyl halide gives a complex tentatively formulated as (2), which may decompose to the ketone via the complex (3) formed by solvent-assisted migratory insertion. No alkyl coupling products are detected.

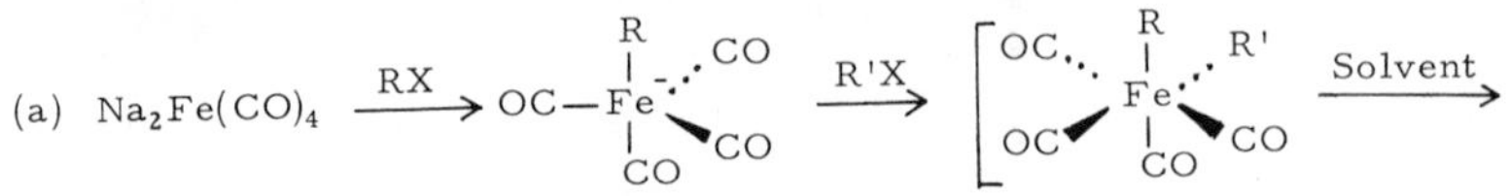

(a) $Na_2Fe(CO)_4 \xrightarrow{RX}$ (1) $\xrightarrow{R'X}$ (2) $\xrightarrow{Solvent}$

(1)   (2)

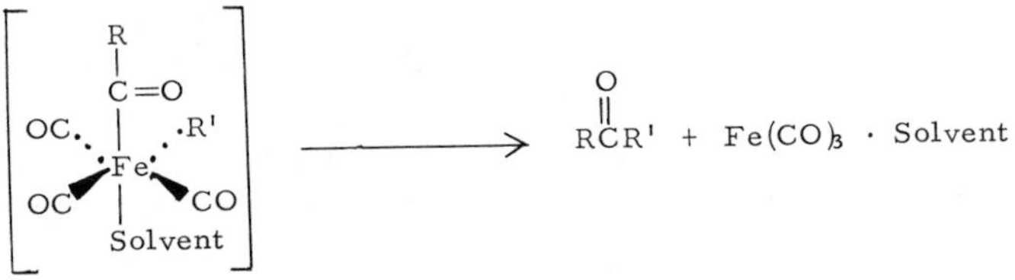

(3)

$\longrightarrow R\overset{O}{\overset{\|}{C}}R' + Fe(CO)_3 \cdot Solvent$

In method (b), (1) is again formed; the ligand triphenylphosphine then promotes migratory insertion to give the anionic iron acyl complex (4). Reaction with another alkyl halide gives a complex (5) corresponding to (3) above.

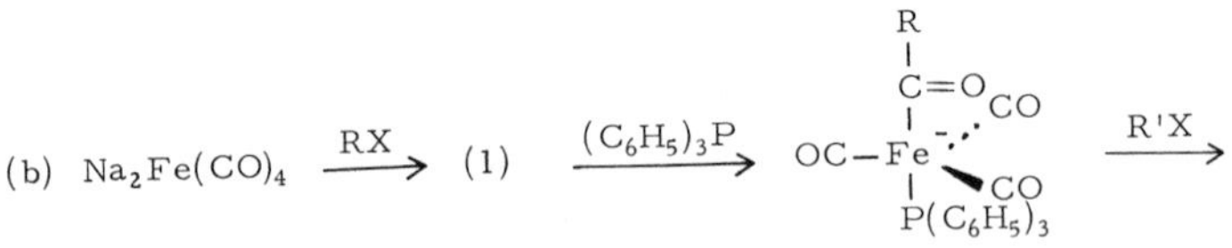

(b) $Na_2Fe(CO)_4 \xrightarrow{RX}$ (1) $\xrightarrow{(C_6H_5)_3P}$ (4) $\xrightarrow{R'X}$

(4)

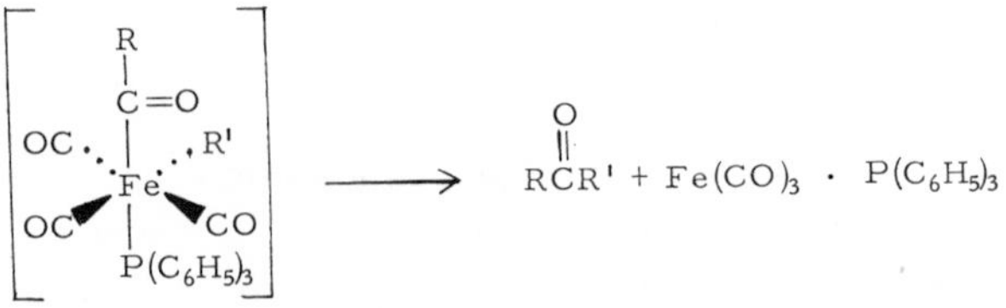

(5)

$\longrightarrow R\overset{O}{\overset{\|}{C}}R' + Fe(CO)_3 \cdot P(C_6H_5)_3$

Method (c) differs from method (b) only in that the ligand $P(C_6H_5)_3$ is replaced by CO. In the one case reported THF–HMPT (2:1) was used as solvent.

In method (d) the acyl iron complex (6) is formed directly from $Na_2Fe(CO)_4$ and RCOX. Reaction then affords (3), which decomposes to the ketone. Eight examples of this new method of synthesis have been reported; yields of isolated ketones are in the range 30–85%.

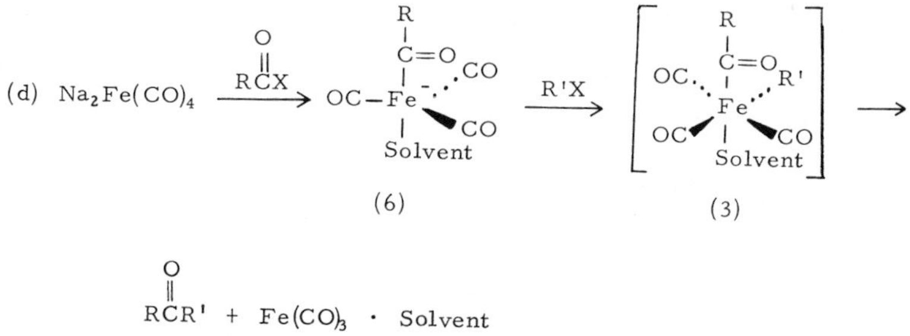

$$
\text{(d)} \quad Na_2Fe(CO)_4 \xrightarrow{RCX} (6) \xrightarrow{R'X} (3) \longrightarrow
$$

$$
\underset{\text{RCR'}}{\overset{O}{\parallel}} + Fe(CO)_3 \cdot \text{Solvent}
$$

**Synthesis of aliphatic carboxylic acids, esters, and amides.**[2] Sodium tetracarbonyl-ferrate(-II) (1) reacts with aliphatic halides and tosylates to give anionic alkyltetra-carbonyliron(0) complexes (2). In the presence of carbon monoxide these undergo

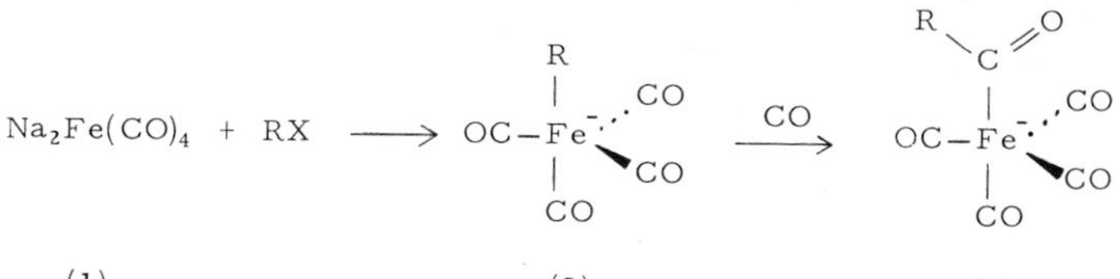

$$
Na_2Fe(CO)_4 + RX \longrightarrow (2) \xrightarrow{CO} (3)
$$

alkyl migration to give anionic acyl complexes (3). Oxidation ($O_2$ or sodium hypo-chlorite) of either (2) or (3) in THF or N-methyl-2-pyrrolidone (MP) followed by hydrolysis yields carboxylic acids. These are also formed by treatment of either (2) or (3) with iodine followed by hydrolysis (equations I and II). Treatment of either (2) or (3) with $I_2$ and an alcohol in THF, THF–HMPT, or MP affords esters (equation III).

$$
\text{I} \quad (2) \text{ or } (3) \xrightarrow{O_2 \text{ or } NaOCl} \xrightarrow{H_2O} RCOOH
$$

$$
\text{II} \quad (2) \text{ or } (3) \xrightarrow{I_2, \ H_2O} RCOOH
$$

$$
\text{III} \quad (2) \text{ or } (3) \xrightarrow{I_2, \ R'OH} RCOOR'
$$

$$
\text{IV} \quad (2) \text{ or } (3) \xrightarrow{I_2, \ R'R''NH} RCONR'R''
$$

Amides are formed by treatment of either (2) or (3) with $I_2$ and an amine in the same solvents (equation IV).

Yields (isolated) are about 80 % in the case of primary aliphatic halides and tosylates. Yields are less satisfactory in the case of secondary substrates owing to olefin-forming elimination reactions, which can be minimized by use of THF as solvent. Primary

$$ClCH_2(CH_2)_4CH_2Br \xrightarrow[84\%]{} ClCH_2(CH_2)_4CH_2COOH$$

bromides react with (1) in THF much faster than the corresponding chlorides. Thus 1-bromo-6-chlorohexane can be converted into 7-chloroheptanoic acid in 84 % yield.

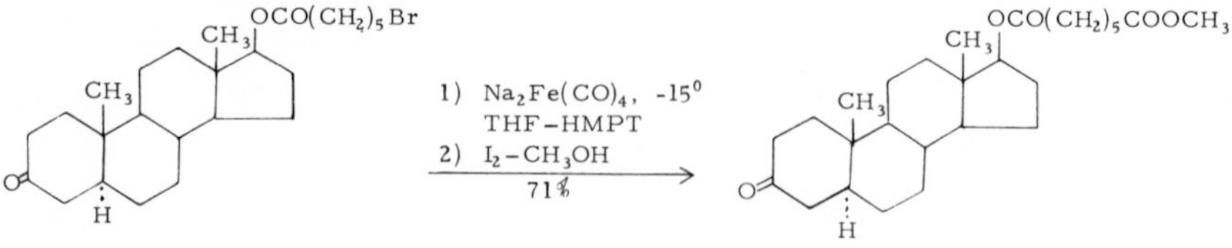

1) Na$_2$Fe(CO)$_4$, -15°
   THF–HMPT
2) I$_2$–CH$_3$OH
   71%

(4)                                                      (5)

Ester and carbonyl groups are not affected under these conditions; thus the bromo ketosteroid (4) was converted into the methyl ester (5) in 71 % yield.

**Aldehydes from acyl halides.**[3-5] Acyl halides, for example benzoyl chloride, are converted into aldehydes, for example benzaldehyde, by treatment with sodium tetra-carbonylferrate(-II) in the presence of triphenylphosphine and an acid. The aldehydes

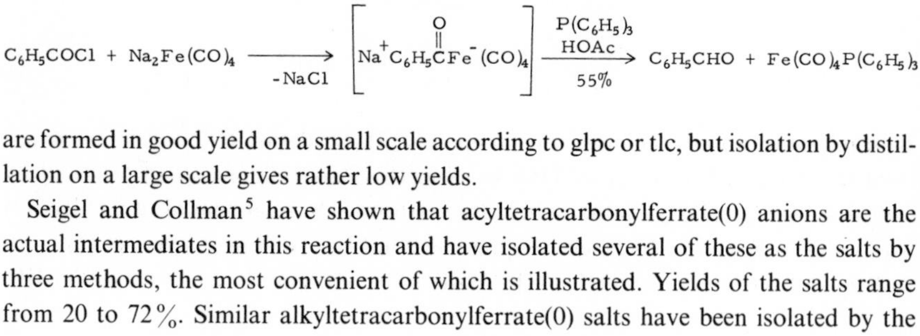

are formed in good yield on a small scale according to glpc or tlc, but isolation by distillation on a large scale gives rather low yields.

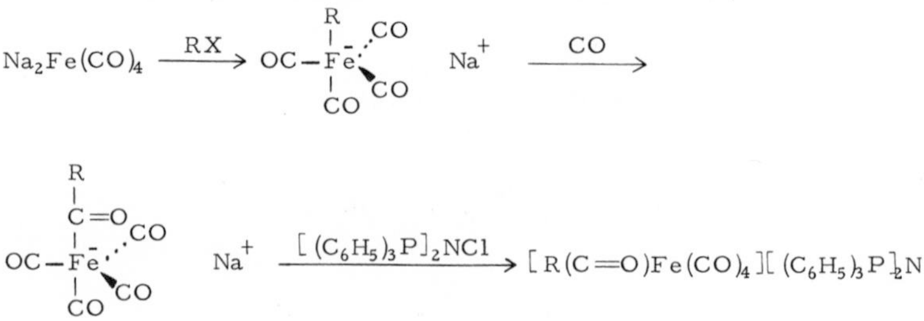

Seigel and Collman[5] have shown that acyltetracarbonylferrate(0) anions are the actual intermediates in this reaction and have isolated several of these as the salts by three methods, the most convenient of which is illustrated. Yields of the salts range from 20 to 72%. Similar alkyltetracarbonylferrate(0) salts have been isolated by the reaction of alkyl bromides with sodium tetracarbonylferrate followed by treatment with bis(triphenylphosphine)iminium chloride and shown to be intermediates in the conversion of aliphatic halides (tosylates) into ketones.

The reagent converts phthaloyl dichloride (1) into biphthalidylidene (2) in 23% yield.[6]

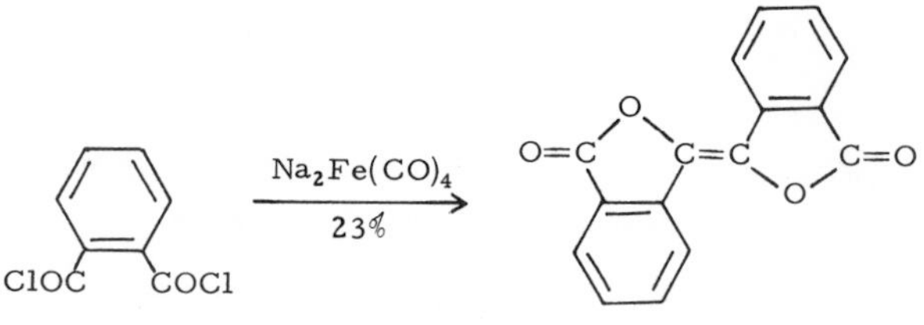

(1)                                    (2)

[1] J. P. Collman, S. R. Winter, and D. R. Clark, Am. Soc., **94**, 1788 (1972).
[2] J. P. Collman, S. R. Winter, and R. G. Komoto, ibid., **95**, 249 (1973).
[3] Y. Watanabe, T. Mitsudo, M. Tanaka, K. Yamamoto, T. Okajima, and Y. Takegami, Bull. Chem. Soc. Japan, **44**, 2569 (1971).
[4] Y. Watanabe, T. Mitsudo, and Y. Takegami, Org. Syn., submitted 1973.
[5] W. O. Seigel and J. P. Collman, Am. Soc., **94**, 2516 (1972).
[6] T. Mitsudo, Y. Watanabe, M. Tanaka, K. Yamamoto, and Y. Takegami, Bull. Chem. Soc. Japan, **45**, 305 (1972).

**Sodium thioethoxide**, $NaSC_2H_5$. Mol. wt. 84.13.

The reagent is generated in situ by the reaction of sodium hydride (60% oil dispersion) with ethanethiol in DMF.

*Cleavage of aryl methyl ethers.*[1] The reagent cleaves aryl methyl ethers in high yield. The reaction should be conducted in a hood since methyl ethyl sulfide (foul

$$ArOCH_3 \xrightarrow[\text{DMF}]{NaSC_2H_5} ArOH + CH_3SC_2H_5$$

smelling) is liberated. The method is particularly valuable for acid-sensitive ethers. Another important feature is that dimethyl ethers can be monodemethylated selectively. Thus orcinol dimethyl ether (1) can be converted into orcinol monomethyl ether (2) in 88–96% yield.

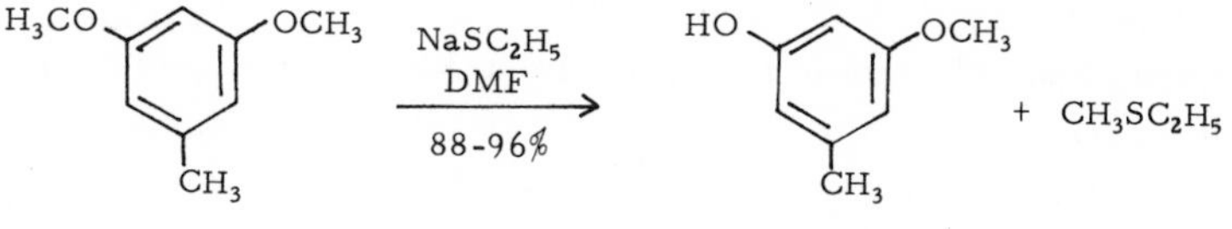

(1)                                    (2)

[1] G. I. Feutrill and R. N. Mirrington, Tetrahedron Letters, 1327 (1970); Org. Syn., submitted 1972.

**Sodium thiosulfate, 1**, 1107. Anhydrous salt is available from ROC/RIC.

*Debromination of* **vic-***dibromides.*[1] *vic*-Dibromides are debrominated, usually in high yield, by treatment with excess sodium thiosulfate in DMSO (60°, 8 hr.). For example, *erythro*-stilbene dibromide is converted into *trans*-stilbene (99% yield). The reaction is apparently *trans* stereospecific.

[1]K. M. Ibne-Rasa, A. R. Tahir, and A. Rahman, *Chem. Ind.*, 232 (1973).

**Solvents, 1**, 1109–1111.

*Solvents for Grignard reactions.* J. T. Baker supplies dry, peroxide-free ether and THF containing an activator for Grignard-reagent formation.

**Squaric acid (3,4-Dihydroxy-3-cyclobutene-1,2-dione).**
Mol. wt. 114.06, m.p. > 300°,
$pK_2$ 2.2. Supplier: Aldrich (96%).

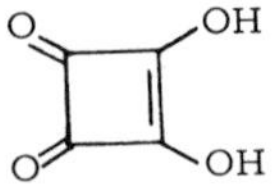

*Preparation* from 1,2-dichloro-3,3,4,4-tetrafluorocyclobutene (Penninsular Chemical Research Company, Inc.).[1]

*Conversion to a four-ring trimethine dye*[2]:

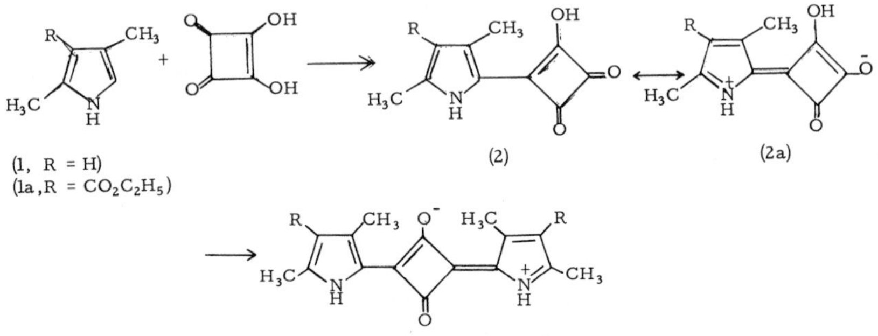

(1, R = H)
(1a, R = $CO_2C_2H_5$)

(2)

(2a)

(3, R = H; 3a, R = $CO_2C_2H_5$)

*Chlorination with dimethylformamide–thionyl chloride.* See **3**, 116.

[1]S. Cohen and S. G. Cohen, *Am. Soc.*, **88**, 1533 (1966).
[2]A. Treibs and K. Jacob, *Ann.*, **712**, 123 (1968).

**(+)-(R)-***trans***-β-Styryl *p*-tolyl sulfoxide.**
Mol. wt. 242.33, m.p. 81.5–82°, $\alpha D$ +164.3°.

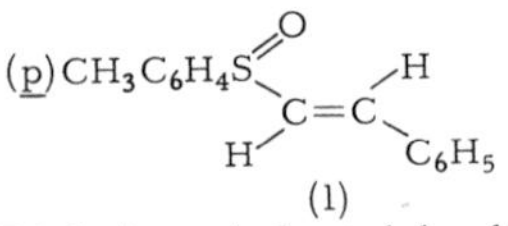

(1)

*Preparation.*[1] The reagent is prepared from (R)-(+)-methyl *p*-tolyl sulfoxide as shown.

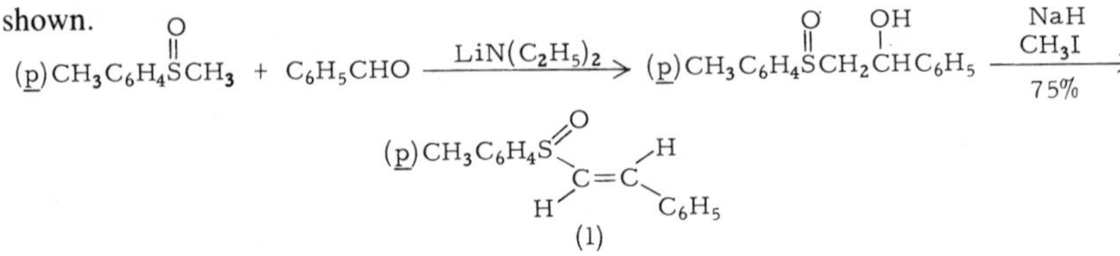

(1)

*Asymmetric synthesis by a Michael reaction.*[1] Japanese chemists report that Michael addition to α,β-unsaturated sulfoxides proceeds readily. Thus *p*-tolyl vinyl sulfoxide (2) reacts with diethyl malonate and ethyl acetoacetate in the presence of an equimolar amount of sodium ethoxide in ethanol to give the Michael adducts (3) and (4).

$$(p) CH_3C_6H_4\overset{\overset{O}{\|}}{S}CH=CH_2$$

(2)

$$\xrightarrow[61\%]{CH_2(COOC_2H_5)_2} (p) CH_3C_6H_4\overset{\overset{O}{\|}}{S}CH_2CH_2CH(COOC_2H_5)_2$$

(3)

$$\xrightarrow[71\%]{CH_3COCH_2COOC_2H_5} (p) CH_3C_6H_4\overset{\overset{O}{\|}}{S}CH_2CH_2\overset{\overset{\displaystyle COOC_2H_5}{\diagup}}{\underset{\underset{\displaystyle COCH_3}{\diagdown}}{CH}}$$

(4)

They then examined the Michael addition of the optically active α,β-unsaturated sulfoxide (1) with diethyl malonate with an equimolar amount of sodium ethoxide at 80° under argon. Diethyl 2-(*p*-tolylsulfinyl)-1-phenylethylmalonate (5) was obtained in

$$(1) + CH_2(COOC_2H_5)_2 \xrightarrow[82\%]{C_2H_5ONa} (p) CH_3C_6H_4\overset{\overset{O}{\|}}{S}CH_2\underset{\underset{\displaystyle C_6H_5}{|}}{CH}CH(COOC_2H_5)_2$$

(5a, 5b)

$$\longrightarrow (+)-(R)s-5a \xrightarrow[92\%]{\overset{Raney\ Ni}{C_2H_5OH}} C_6H_5\underset{\underset{\displaystyle CH_3}{|}}{CH}CH(COOC_2H_5)_2$$

αD + 93.1⁰

(-)-(6)

82% yield as a mixture of the diastereomers (5a) and (5b) in the ratio 4:1. The major isomer (5a) could be obtained by fractional crystallization in 51% overall yield from (1). Raney nickel desulfurization gave (−)-diethyl 1-phenylethylmalonate (6). This was converted by hydrolysis and decarboxylation into the known optically active (−)-3-phenylbutyric acid, $C_6H_5CH(CH_3)CH_2COOH$. The major isomer (5a) was accordingly obtained with at least 95% optical purity and has the (R)s-(R)c configuration (a).

$$(\underline{p})CH_3C_6H_4 \diagdown \underset{O}{\overset{}{\underset{\diagdown}{S}}} \diagup \overset{\displaystyle CH_2}{} \diagdown \underset{\underset{\displaystyle CH(COOC_2H_5)_2}{\overset{H}{}}}{C} \diagup \overset{\displaystyle C_6H_5}{}$$

(a)

[1]G. Tsuchihashi, S. Mitamura, S. Inoue, and K. Ogura, *Tetrahedron Letters*, 323 (1973).

**Succinic anhydride.** Mol. wt. 100.07, m.p. 118–120°. Suppliers: Aldrich, Eastman.

***Double succinoylation of indane***[1]**:**

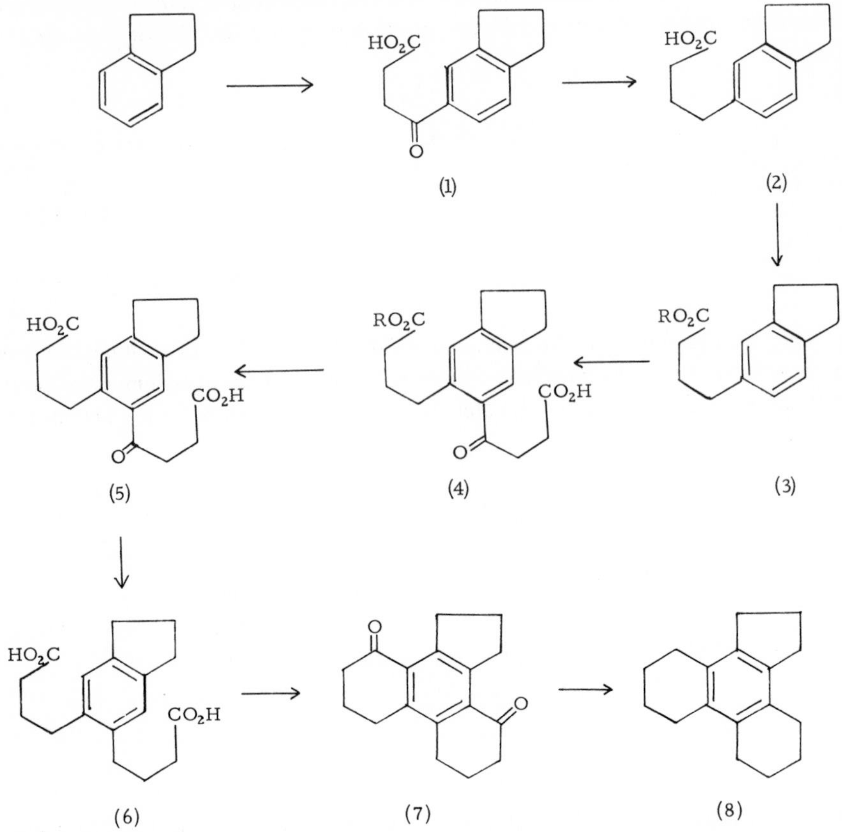
HO$_2$C
HO$_2$C
(1)
(2)
HO$_2$C
CO$_2$H
RO$_2$C
CO$_2$H
RO$_2$C
(5)
(4)
(3)
HO$_2$C
CO$_2$H
(6)
(7)
(8)

[1] A. U. Rahman and M. del Carmen Torre, *Ann.*, **718**, 136 (1968).

**3-Sulfolene,** **2**, 389–391.

***Ozonization and conversion to 4-H-1,4-thiazine 1,1-dioxide***[1]**:**

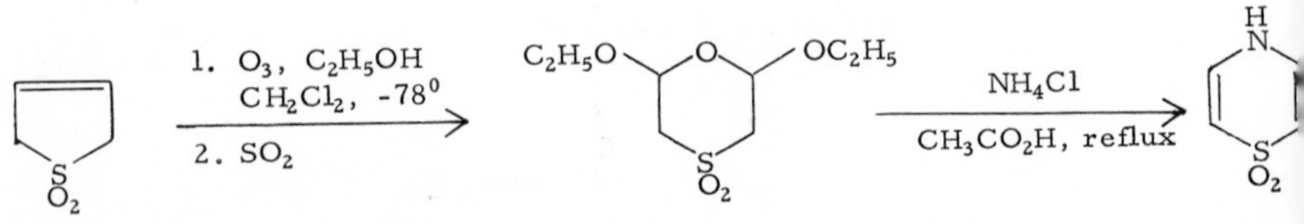
1. O$_3$, C$_2$H$_5$OH
CH$_2$Cl$_2$, -78°
2. SO$_2$
C$_2$H$_5$O
OC$_2$H$_5$
NH$_4$Cl
CH$_3$CO$_2$H, reflux

[1] W. E. Noland and R. D. DeMaster, *Org. Syn.*, **52**, 135 (1972).

*o*-Sulfoperbenzoic acid, 

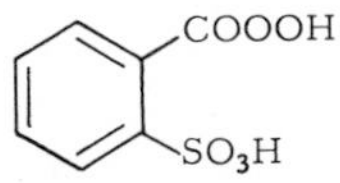

COOOH

SO₃H

. Mol. wt. 218.18.

The peracid is prepared in 65–75% yield by the reaction between *o*-sulfobenzoic anhydride and 30% hydrogen peroxide in acetone solution at −4 to 0°C. The solution is used directly for epoxidation of olefins. Since the reagent contains a strongly acidic group, the initially formed epoxide undergoes acid cleavage to a *trans*-diol unless a solid buffer such as sodium carbonate is present. The peracid can also be used for Baeyer–Villiger oxidation and for oxidation of heterocyclic *t*-amines to N-oxides.[1]

[1]J. M. Bachhawat and N. K. Mathur, *Tetrahedron Letters*, 691 (1971).

**Sulfur dichloride, 1,** 1121, 1122; **2,** 391–392.
Sulfur dichloride adds to norbornadiene (1) to give 2,8-dichloro-4-thiatricyclo-

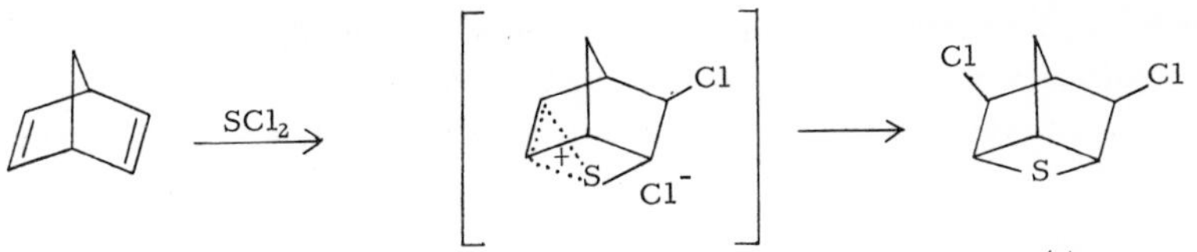

(1)                    (a)                    (2)

[3.2.1.0³·⁶]octane (2) in high yield.[1] The episulfonium salt (a) has been suggested as an intermediate.[2] Both the chlorine atoms of (2) can be replaced quantitatively by bromine by treatment of (2) with boron tribromide (**1,** 66–67; **2,** 33–34; **3,** 30–31).

[1]F. Lautenschlaeger, *J. Org.,* **31,** 1669 (1966); **34,** 3998 (1969).
[2]S. D. Ziman and B. M. Trost, *ibid.,* **38,** 649 (1973).

**Sulfur dioxide, 1,** 1122; **2,** 392.
    *Thiepin 1,1-dioxide* (4).[1] When sulfur dioxide is passed into a cold ethereal solution of vinyldiazomethane (1), nitrogen is evolved, and the sulfone 4,5-dihydrothiepin 1,1-dioxide (3) is obtained in 29% yield. The *cis*-divinyl episulfone (2), which can under-

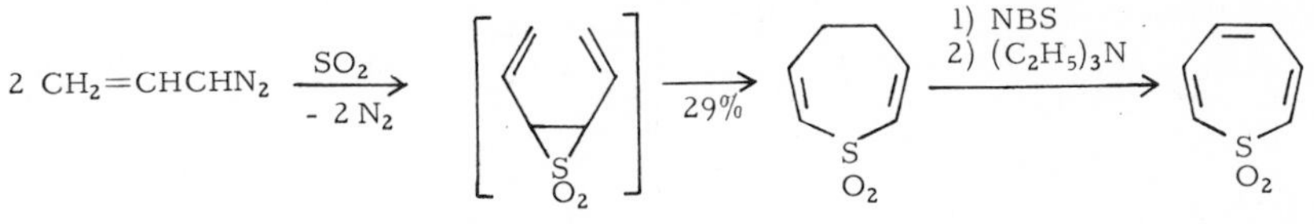

(1)              (2)              (3)              (4)

go rapid Cope rearrangement, is a probable intermediate. Thiepin 1,1-dioxide (4) is obtained by allylic bromination followed by dehydrobromination.

[1]L. A. Paquette and S. Maiorana, *Chem. Commun.,* 313 (1971).

**Sulfuric acid, $H_2SO_4$.**

**2-Oxa-adamantane.**[1] Treatment of bicyclo[3.3.1]nonane-2,6-diol (1) with 95%
sulfuric acid at room temperature affords 2-oxa-adamantane (2) in 35–40% yield.

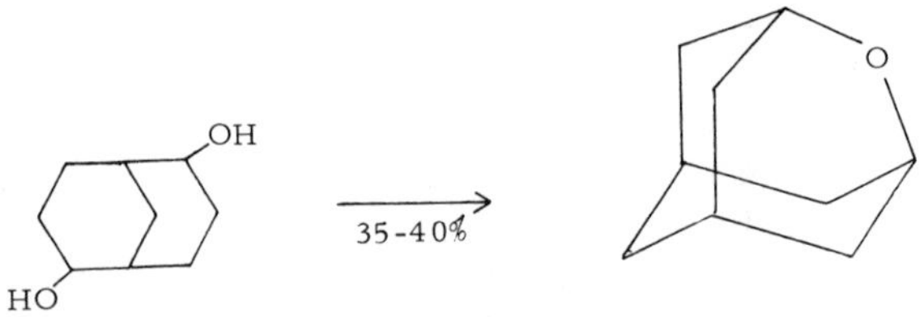

$$
\text{(1)} \xrightarrow{35-40\%} \text{(2)}
$$

The reaction involves formation of carbonium ions and 1,2-hydride shifts. Use of
more concentrated acid (8% oleum) or of 75% acid decreases the yield of (2) markedly.

*Ring contraction of 2-alkylidenecyclobutanols to cyclopropyl carbonyl compounds.*[2] 2-Methylenecyclobutanol (1) is converted quantitatively into 1-methyl-
cyclopropanecarboxaldehyde (2) when treated with 5% $H_2SO_4$ at 100° for 30 min.; the
rearrangement can be induced by heat treatment in a sealed tube at 245° for 4 hr. The

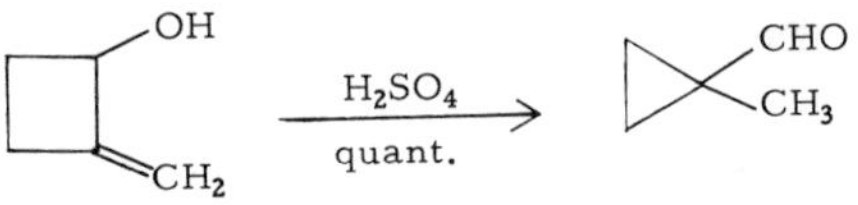

conditions for this ring contraction are more vigorous than those required for ring
expansion of 1-vinylcyclopropanol to 2-methylcyclobutanol (*see* **Hydrogen bromide,**
this volume).

*Hemiacetal formation.*[3] Treatment of the vinyl chloride (1), 2-(2'-methyl-3'-
chloroallyl)cyclohexanone, with concentrated sulfuric acid results in the formation
of the hemiacetal (2) in 85% yield. A chloronium ion (a) is postulated as the intermediate.

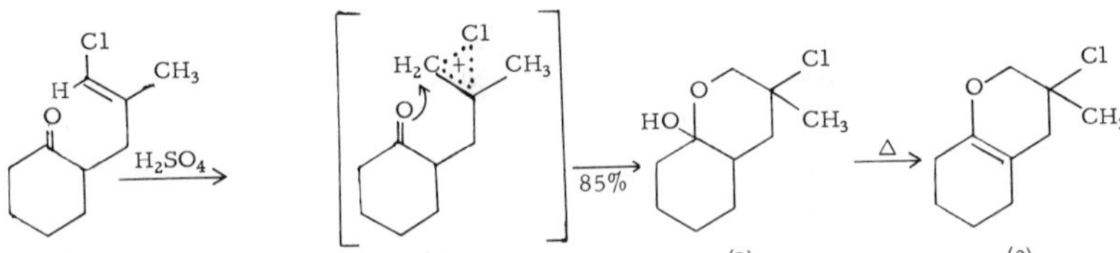

*Cyclization of citral to α-cyclocitral.* Citral (1, mixture of *cis*- and *trans*-isomers)
is converted into the enamine (2) by treatment with pyrrolidine in the presence of
molecular sieves 4A. Treatment of (2) with concentrated sulfuric acid and water
(10:1 by volume) affords α-cyclocitral (3) in 41% yield.[4]

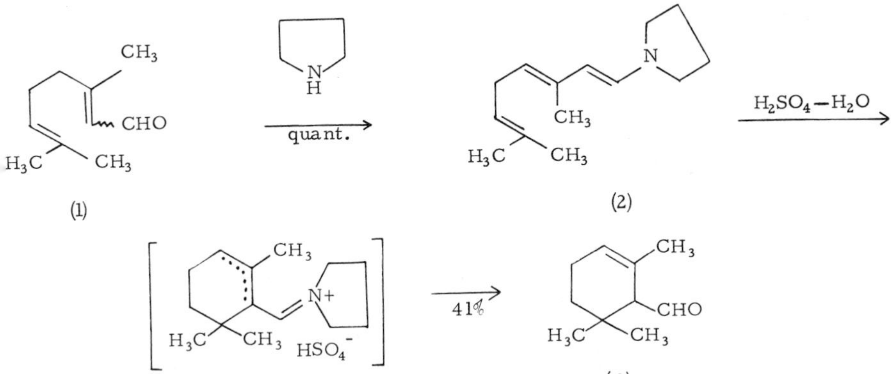

(1)   (2)

(3)

The method was used to synthesize (S)-(+)-α-cyclocitral in 12% optical yield by using an optically active pyrrolidine (4, prepared from L-proline). A higher optical yield, 27%, was obtained using (5).[5]

(4)   (5)

*Cyclization of (E)- and (Z)-4,8-dimethylnona-3,7-diene-2-one.* Acid-catalyzed cyclization of polyenes is now a fairly well-known reaction (3, 305).[6] For example cyclization of 6,10-dimethyl-3,5,10-undecatriene-2-one (1) with concentrated sulfuric acid gives β-ionone (2) in 85% yield.[7]

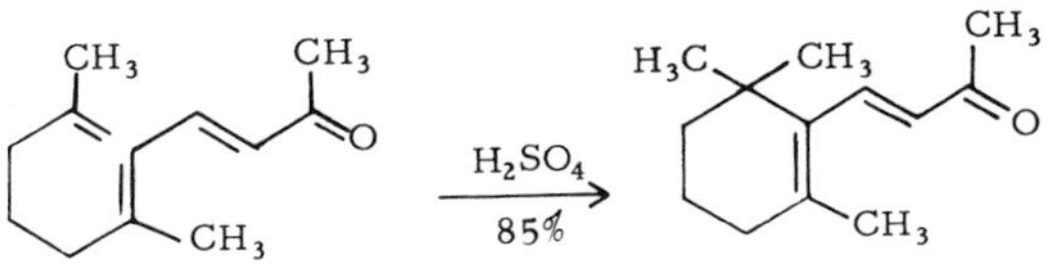

(1)   (2)

The cyclization of (E)- and (Z)-4,8-dimethylnona-3,7-diene-2-one (3) with 75% sulfuric acid gives (4; 1,3,5-trimethyl-2-oxabicyclo[3.3.1]nonene-3) and the tertiary alcohol (5) as the major products. Only minor amounts of the expected product (6; 1,1,3-trimethyl-2-acetylcyclohexene-3) are formed.

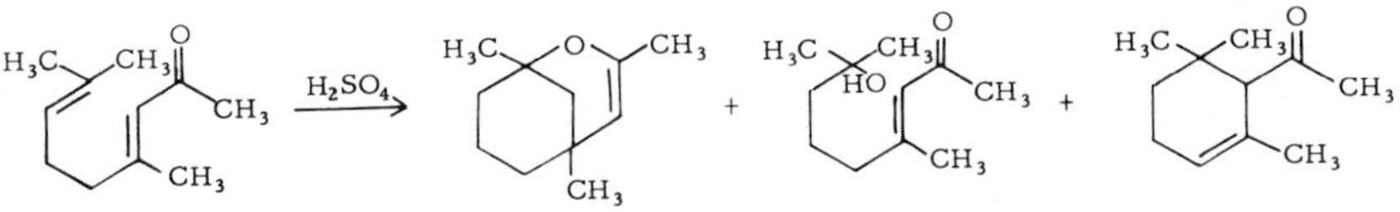

(3, 4.6 g)   (4, 1.5 g)   (5, 1.04 g)   (6, 0.35 g)

Büchi[8] suggests that the major product (4) is formed from the dienone (7).

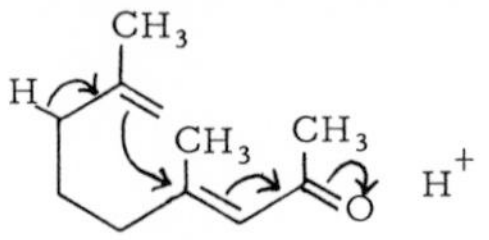

(7)

***Dethioacetalization.*** Dithioacetals are hydrolyzed by treatment with cold, concentrated sulfuric acid (10 min.).[9] This method was used in a synthesis of dihydrojasmone (4). Thus treatment of (2) with concentrated sulfuric acid yields (3). The vinylic

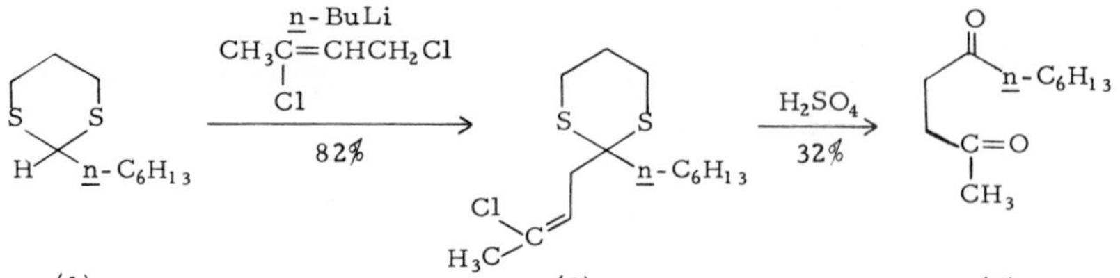

(1)                    (2)                    (3)

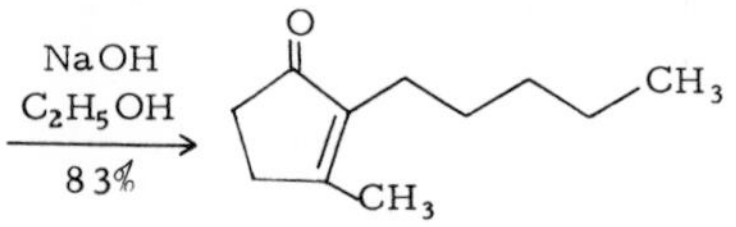

(4)

chloride is also hydrolyzed (Wichterle reaction).[10] The diketone (3) had previously been converted into (4) by treatment with sodium hydroxide in ethanol.[11]

Although the yield of (3) is rather low, dethioacetalization of simple ketones usually proceeds in high yield (90–98 %).

[1] N. V. Averina and N. S. Zefirov, *J.C.S. Chem. Comm.*, 197 (1973).
[2] J. P. Barnier, J. M. Denis, J. R. Salaün, and J. M. Conia, *ibid.*, 103 (1973).
[3] E. P. Woo, K. T. Mak, and H. N. C. Wong, *Tetrahedron Letters*, 585 (1973).
[4] S. Yamada, M. Shibasaki, and S. Terashima, *ibid.*, 377 (1973).
[5] *Idem, ibid.*, 381 (1973).
[6] W. S. Johnson, *Accts. Chem. Res.*, **1**, 1 (1968).
[7] W. Hoffmann, H. Pasedach, H. Pommer, and W. Reif, *Ann.*, **747**, 60 (1971).
[8] G. Büchi and W. Pickenhagen, *J. Org.*, **38**, 894 (1973).
[9] T.-L. Ho, H. C. Ho, and C. M. Wong, *Canad. J. Chem.*, **51**, 153 (1973).
[10] J. A. Marshall and D. J. Schaeffer, *J. Org.*, **30**, 3642 (1965).
[11] H. C. Ho, T.-L. Ho, and C. M. Wong, *Canad. J. Chem.*, **50**, 2718 (1972).

**Sulfur trioxide, 1,** 1125.

*Selective oxidations.*[1] The reaction of 2,3,4,5,6-pentachlorotoluene (1) with excess sulfur trioxide (stabilized) for 2–3 hr. at reflux temperature yields a substance with the composition $C_7H_3Cl_5 \cdot O_7S_2$ (2), and which on hydrolysis yields 2,3,4,5,6-pentachlorobenzyl alcohol (3) in 91 % yield. When the reaction is extended to 24 hr., pentachlorobenzaldehyde is obtained in nearly quantitative yield.

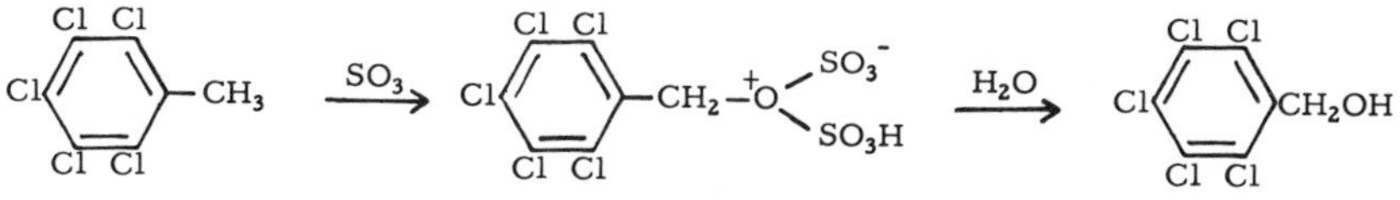

(1)                              (2)                              (3)

[1]V. Mark, L. Zengierski, V. A. Pattison, and L. E. Walker, *Am. Soc.*, **93**, 3538 (1971).

**Sulfur trioxide–Pyridine (Pyridinium-1-sulfonate), 1,** 1127–1128; **2,** 393–394; **3,** 275–276.

*Reduction* (3, 275–276). The reduction with sulfur trioxide–pyridine and then with LiAlH₄ was used in the last step of a synthesis of racemic α-*trans*-bergamotene (2), a natural sesquiterpene.[1]

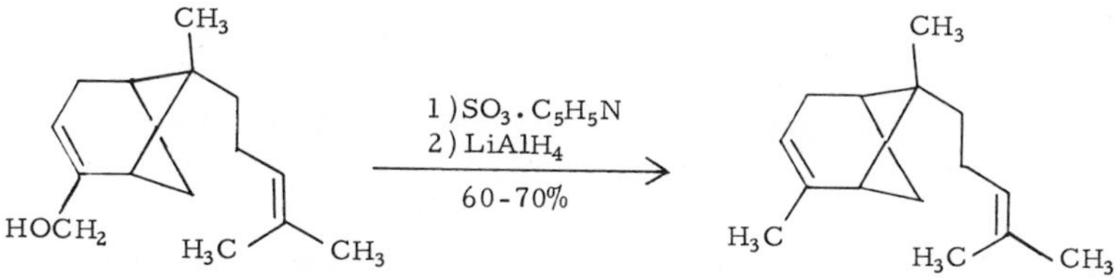

(1)                              (2)

This method was used for reduction of the ketal dienol (3) to the ketal diene (4); yield 70 %. The ketal diene (4) was used as an intermediate in a total synthesis of the sesquiterpene ($\pm$)-occidentalol (5).[2]

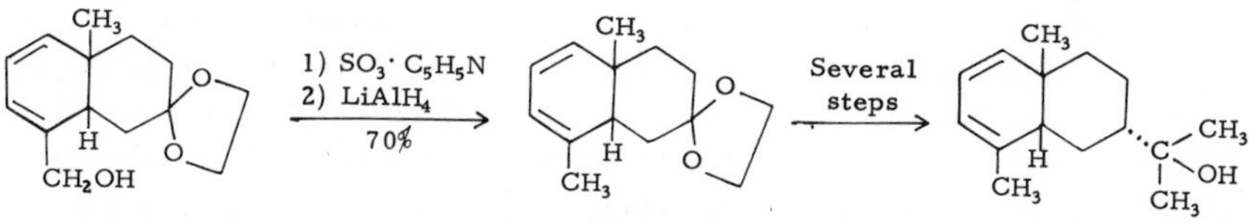

(3)                              (4)                              (5)

*Sulfonation of indole.*[3] Indole, 1-methyl-, 2-methyl-, and 1,2-dimethylindole are sulfonated by the reagent in refluxing pyridine at $C_3$. 3-Methylindole and 1,3-dimethylindole are sulfonated at $C_2$. 2,3-Dimethylindole does not react under these conditions.

[1]E. J. Corey, D. E. Cane, and L. Libit, *Am. Soc.*, **93**, 7016 (1971).
[2]D. S. Watt and E. J. Corey, *Tetrahedron Letters*, 4651 (1972).
[3]G. F. Smith and D. A. Taylor, *Tetrahedron*, **29**, 669 (1973).

**Sulfuryl chloride, 1,** 1128–1131; **2,** 394–395; **3,** 276.

*Free-radical chlorination of 1,1-cyclobutanedicarboxylic acid.*[1]

A 2-l. three-necked, round-bottomed flask equipped with a Trubore stirrer and paddle is charged with 172.8 g. (1.2 moles) of 1,1-cyclobutanedicarboxylic acid and 1500 ml. of benzene. The mixture is stirred and heated at reflux, and 200 ml. of benzene or benzene–water azeotrope is removed by distillation to ensure anhydrous conditions. The flask is then fitted with an addition funnel and a reflux condenser to which is attached a drying tube. Stirring and heating are continued and, over a 40-min. period, 170 g. (102 ml., 1.26 moles) of sulfuryl chloride (distilled before use) is added from the funnel while 4.0 g. of dibenzoyl peroxide is added simultaneously in small portions through the top of the condenser. There is a short induction period and then hydrogen chloride and sulfur dioxide are evolved. After the addition is complete, heating at reflux is maintained for 22 hr. The solid is dissolved after 1 hr., leaving a light-brown solution. After the heating period is complete, the benzene is removed by distillation and the residue heated to 190–210° for 45 min. in order to effect decarboxylation. The black residue is transferred to a small flask and distilled under vacuum through a 6-cm. Vigreux column. After a forerun of 25–30 g. (cyclobutanecarboxylic acid and 3-chloro-cyclobutanecarboxylic acid, b.p. 100–130°), 65–79 g. (40–49%) of *cis-* and *trans*-3-chlorocyclobutanecarboxylic acid is collected as a light-yellow liquid, b.p. 131–137° (15 mm.), $n_D^{24}$ 1.4790.

*α-Chlorination of sulfoxides* (**3,** 276). Tsuchihashi *et al.*[2] also recommend sulfuryl chloride for α-chlorination of sulfoxides, particularly for chlorination of hindered sulfoxides. They carried out the reaction in dichloromethane in the presence of pyridine.

Attempted chlorination of benzhydryl benzyl sulfoxide (1) with thionyl chloride in methylene chloride in the presence of either calcium oxide or pyridine resulted in cleavage to benzhydryl chloride (3) and α-toluenesulfinic acid (4). Phenylsulfine (5) is suggested as an intermediate to (4).[3]

$$(C_6H_5)_2CH\overset{\overset{\displaystyle O}{\|}}{S}CH_2C_6H_5 \quad \xrightarrow{SOCl_2}$$

(1)

$$\xrightarrow{CaO} (C_6H_5)_2CHCl \quad + \quad C_6H_5CH_2SO_2H$$

(3, 98%)          (4, 80%)

$$\xrightarrow{Py} (C_6H_5)_2CHCl \quad + \quad \text{unidentified product}$$

(3, 85%)

$$O{=}S{=}CHC_6H_5$$

(5)

*Chlorination of sulfones.*[4] Sulfones, unlike alkyl sulfides and sulfoxides, are chlorinated almost exclusively in the β-position. Thus tetramethylene sulfone (1,

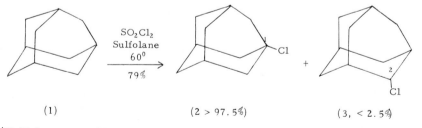

sulfolane) is converted into the β-chloro derivative (2) in almost quantitative yield.

*N-Monosubstituted amidosulfonyl chlorides.*[5] N-Monoalkyl amidosulfonyl chlorides are obtained conveniently and in high yield by the reaction of an alkylamine hydrochloride with sulfuryl chloride in refluxing acetonitrile in the presence of a Lewis acid such as antimony(V) chloride.

$$RNH_2 \cdot HCl \quad + \quad SO_2Cl_2 \quad \xrightarrow[-2\ HCl]{SbCl_5\ in\ CH_3CN} \quad RNH{-}SO_2{-}Cl$$

*Ionic chlorination of alkanes.*[6] Alkanes are readily chlorinated by sulfuryl chloride in sulfolane (1, 1144–1145; 2, 402–403). Thus adamantane (1) is converted almost exclusively into 1-chloroadamantane (2). Norbornane is converted under these conditions exclusively into 2-*exo*-chloronorbornane. *n*-Hexane is converted into 1-chlorohexane (20%), 2-chlorohexane (56%), and 3-chlorohexane (24%). The chlorination proceeds by an ionic mechanism.

$$\xrightarrow[\substack{Sulfolane \\ 60^0 \\ 79\%}]{SO_2Cl_2}$$

(1)                    (2 > 97.5%)                    (3, < 2.5%)

[1]G. M. Lampman and J. C. Aumiller, *Org. Syn.*, **51**, 73 (1971).

[2]G. Tsuchihashi, K. Ogura, S. Iriuchijima, and S. Tomisawa, *Synthesis*, 89 (1971).

[3]C. Y. Meyers and G. J. McCollum, *Tetrahedron Letters*, 289 (1973).

[4]I. Tabushi, Y. Tamaru, and Z. Yoshida, *ibid.*, 3893 (1971).

[5]G. Weiss and G. Schulze, *Ann.*, **729**, 40 (1969).

[6]I. Tabushi, Z. Yoshida, and Y. Tamaru, *Tetrahedron*, **29**, 81 (1973).

# T

**Tellurium tetrachloride**, $TeCl_4$. Mol. wt. 269.44. Suppliers: Alfa Inorganics, ROC/RIC. *Synthesis of bis-2,2'-biphenylenetellurium* (1)[1]:

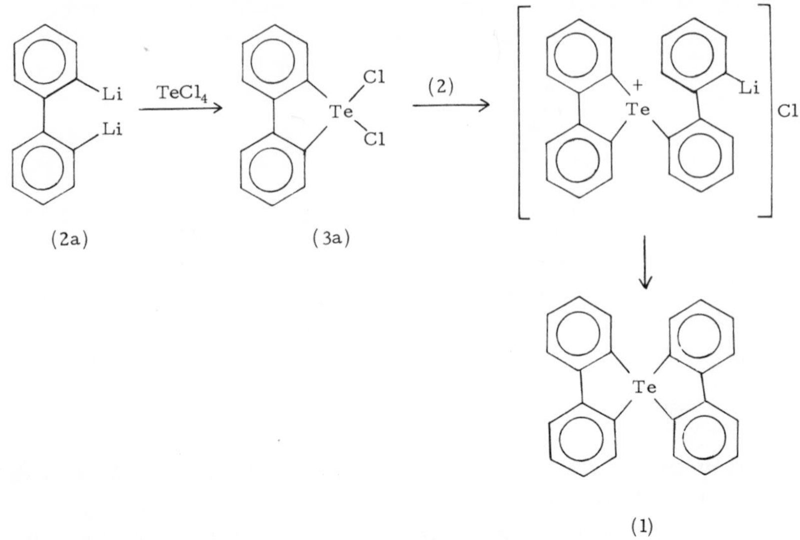

(2a)　　　　　　　　(3a)

(1)

[1] D. Hellwinkel and G. Fahrbach, *Ann.*, **712**, 1 (1968).

**2,4,4,6-Tetrabromocyclohexa-2,5-dienone**, 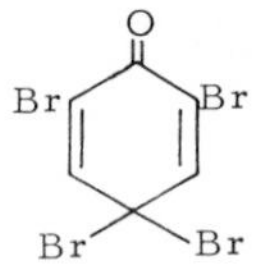 . Mol. wt. 409.44, m.p. 125–130° dec.

The reagent is prepared[1,2] in 80–90% yield by bromination of tribromophenol in acetic acid containing sodium acetate.

This reagent is useful for selective monobromination of phenols[3] and of aromatic amines,[2] predominantly in the *para* position. Methylene chloride and chloroform are used as solvents. Yields are generally higher than 90%.

A procedure has been submitted to *Organic Syntheses* for preparation of the reagent and its use for conversion of N,N-dimethyl-3-trifluoromethylaniline into 4-bromo-N,N-dimethyl-3-trifluoromethylaniline (82–90% yield).[4]

The reagent reacts readily with imidazole (1) to give a mixture of bromoimidazoles in which 4-bromoimidazole (2) predominates.[5] N-Methylimidazole is brominated to

(1)                              (2)

give 5-bromo-1-methylimidazole in 66% yield. Bromination of indole leads to 3-bromoindole in 88% yield.

[1]R. Benedikt, *Ann.*, **199**, 127 (1879).
[2]V. Caló, F. Ciminale, L. Lopez, and P. E. Todesco, *J. Chem. Soc.* (*C*), 3652 (1971).
[3]V. Caló, F. Ciminale, L. Lopez, G. Pesce, and P. E. Todesco, *Chim. Ind.*, **53**, 467 (1971).
[4]G. J. Fox, G. Hallas, J. D. Hepworth, and K. N. Paskins, procedure submitted to *Org. Syn.*, 1972.
[5]V. Caló, F. Ciminale, L. Lopez, F. Naso, and P. E. Todesco, *J.C.S. Perkin I*, 2567 (1972).

**Tetra-*n*-butylammonium bromide**, $[CH_3(CH_2)_3]_4N^+Br^-$. Mol. wt. 322.38, m.p. 100–102°. Suppliers: Eastman, Fluka.

*Dehydrobromination.* Lloyd and Parker[1] recommend tetra-*n*-butylammonium bromide in acetone containing 2,6-lutidine for dehydrobromination to give high proportions of the Saytzeff olefin and higher than thermodynamic proportions of the *trans*-olefin.

These conditions were used in the last step of a synthesis of 3,4-benzocyclobuta-[1,2-*b*]cycloheptatriene (2).[2] The yield, however, was only 20%.

1) NBS
2) $Bu_4N^+Br^-$,
lutidine, acetone
—————————→
20%

(1)                              (2)

[1]D. J. Lloyd and A. J. Parker, *Tetrahedron Letters*, 637 (1971).
[2]L. Lombardo and D. Wege, *ibid.*, 4859 (1972).

**Tetra-*n*-butylammonium fluoride**, $(CH_3CH_2CH_2CH_2)_4N^+F^-$. Mol. wt. 261.46.

*Preparation.*[1] The reagent can be prepared by neutralization of a 10% aqueous solution of tetra-*n*-butylammonium hydroxide with 48% hydrofluoric acid, concentration under reduced pressure, drying by azeotropic distillation under reduced pressure using 1:1 benzene–acetonitrile, and final drying at 30° and 0.5 mm. for 20 hr. For another preparation see Fowler *et al.*[2]

*Cleavage of silyl ethers.*[1] Tetra-*n*-butylammonium fluoride is a powerful agent for cleavage of silyl ethers.[3] It is particularly useful for cleavage of the recently introduced dimethyl-*t*-butylsilyl ethers (*see* **Dimethyl-*t*-butylchlorosilane**, this volume). The ethers are cleaved rapidly to alcohols by treatment with 2–3 eq. of the fluoride in THF at

25°. However, fluoride ion in THF is a sufficiently strong base that it cannot be used with substrates containing highly base-sensitive groups.

[1]E. J. Corey and A. Venkateswarlu, *Am. Soc.*, **94**, 6190 (1972).
[2]D. L. Fowler, W. V. Loebenstein, D. B. Pall, and C. A. Kraus, *ibid.*, **62**, 1140 (1940).
[3]E. J. Corey and B. B. Snider, *ibid.*, **94**, 2549 (1972).

**Tetra-*n*-butylammonium formate**, $(C_4H_9)_4\overset{+}{N}(O\overset{O}{\overset{\|}{C}}H)$. Mol. wt. 119.16.

The reagent is prepared by neutralization of an aqueous solution of tetra-*n*-butylammonium hydroxide with formic acid and evaporation to dryness.

*Epimerization of hydroxyl groups.* Epimerization of hydroxyl groups is often effected by $S_N2$ displacement of the tosylates of the alcohol with tetraethylammonium acetate (**1**, 1136–1137; **2**, 397) or with tetra-*n*-butylammonium acetate (**3**, 277). In connection with a study of configuration and biological activity in the prostaglandins, Corey and Terashima[1] examined the reaction of the tosylate of the model ($\pm$)-hydroxy-lactone (1) with 5.0 eq. of tetra-*n*-butylammonium acetate in acetone at 25° for 2 hr.

(1)    (2a) R = CH₃CO    (3)
       (2b) R = HCO

The product, obtained in 95% yield, consisted of a mixture of the desired (2a) and the cyclopentene (3) resulting from elimination in a ratio of 1.2:1.

In contrast, reaction of (1) with tetra-*n*-butylammonium formate (*see also* **Tetra-ethylammonium formate**, **1**, 1137–1138) under the same conditions gave the inverted formate (2b) and the cyclopentene (3) in a ratio of 3:1. The greater tendency for substitution at carbon by formate ion relative to acetate ion may be owing to smaller size or to decreased basicity.

Reaction of (1) with tetra-*n*-butylammonium oxalate (disalt) gave the unsaturated lactone (3) as the sole product, isolated in 82% yield. Oxalate ion thus appears to be an excellent nucleophile for effecting elimination under very mild conditions.

Corey and Terashima[2] used the $S_N2$ displacement with tetra-*n*-butylammonium formate in one step in a synthesis of *ent*-11,15-epi-PGE₂ (6). Thus reaction of (4) with 6.7 eq. of tetra-*n*-butylammonium formate in acetone at 25° for 16 hr. gave the inverted formate (5), which was transformed, by a sequence used previously, into (6).

In the prostaglandins, unlike most natural products, the optical antipodes of the natural forms approach the natural compounds in biological activity.

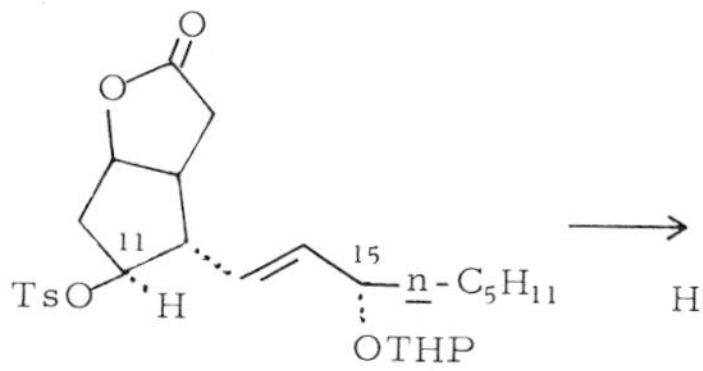

(4)

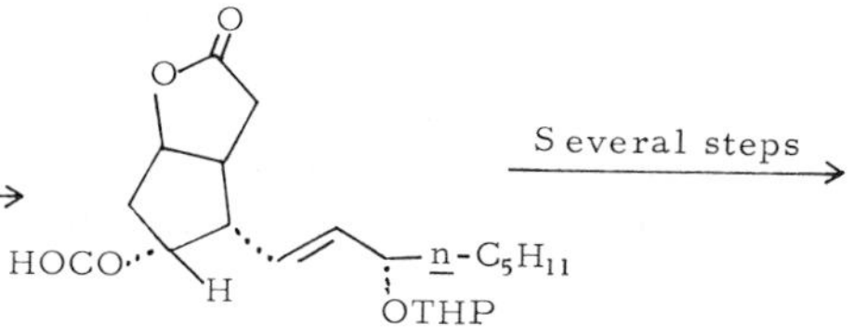

(5)

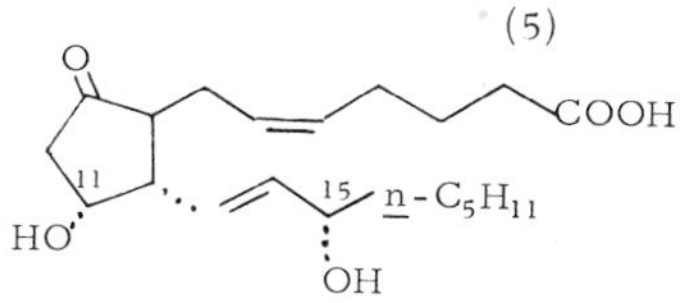

(6)

[1]E. J. Corey and S. Terashima, *Tetrahedron Letters*, 111 (1972).
[2]*Idem, J. Org.*, **37**, 3043 (1972).

**Tetrachloroethane–Nitrobenzene.**

*Succinoylation.*[1] The succinoylation of anisole with succinic anhydride to give 3-(*p*-methoxybenzoyl)propionic acid is carried out most conveniently in a mixture of tetrachloroethane and nitrobenzene (4:1).

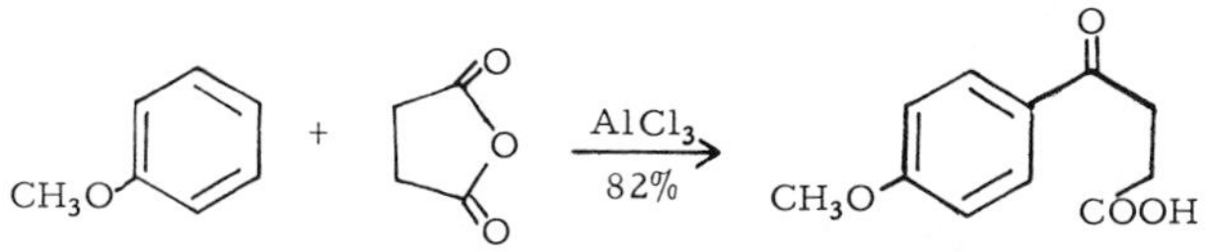

[1]Y. S. Rao and R. A. Kretchmer, *Org. Prep. Proc. Int.*, **3**, 177 (1971).

**Tetrakis(pyridine)copper(I) perchlorate,** $Cu^+(Py)_4ClO_4^-$. Mol. wt. 479.39.
*Preparation.*[1]

*Decomposition of aryl diazonium salts.* Cuprous oxide (**1**, 169–170) has been the catalyst of choice for homolytic decomposition of aryl diazonium salts.[2] It has the disadvantage of being effective only in an acidic medium. Lewin and Michl[3] have examined the effectiveness of various copper(I) perchlorates complexed witn hetero-cyclic amines. Of the various copper(I) salts, tetrakis(pyridine)copper(I) perchlorate is

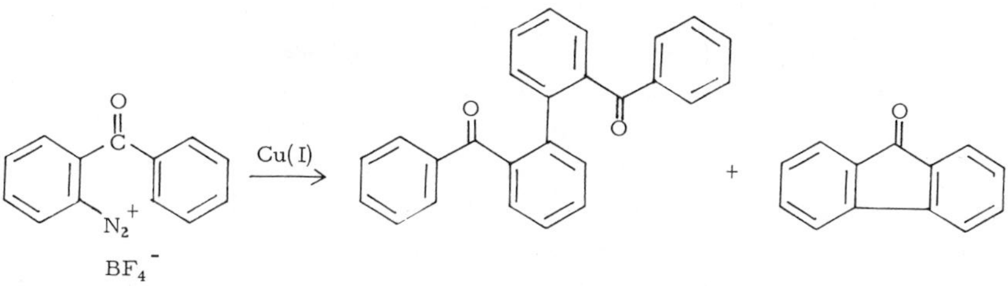

(1)                          (2)                          (3)

highly effective in a neutral medium. Tris(2-picoline)copper(I) perchlorate[4] is also effective in a neutral medium.

Decomposition of 2-diazobenzophenone tetrafluoroborate (1) in the presence of these Cu(I) complexes in a neutral medium gives as the major products 2,2'-dibenzoyldiphenyl (2, ~70% yield) and 9-fluorenone (3, ~30% yield). Decomposition of (1) with cuprous oxide at pH 1 gives as the major products benzophenone, 2-hydroxybenzophenone, and 9-fluorenone.

[1]K. L. Chen and R. T. Twansato, *Inorg. Nucl. Chem. Letters*, **4**, 499 (1968).
[2]A. H. Lewin, A. H. Dinwoodie, and T. Cohen, *Tetrahedron*, **22**, 1527 (1966); A. H. Lewin and T. Cohen. *J. Org.*, **32**, 3844 (1967).
[3]A. H. Lewin and R. J. Michl, *ibid.*, **38**, 1126 (1973).
[4]A. H. Lewin, R. J. Michl, P. Ganis, and U. Lepore, *J.C.S. Chem. Comm.*, 661 (1972).

**1,1,3,3-Tetramethylbutylisonitrile (TMBI), 3,** 279–280.

*Preparation.* The preparation has now been published.[1]

*Aldehydes from alkylmagnesium bromides.*[2] Aliphatic Grignard reagents react with 1 eq. of TMBI in THF to give aldimines (1), which on hydrolysis (oxalic acid)

$$RMgBr + C=N-\overset{\overset{\displaystyle CH_3}{|}}{\underset{\underset{\displaystyle CH_3}{|}}{C}}-CH_2-C(CH_3)_3 \longrightarrow RC\overset{\displaystyle N-\overset{\overset{\displaystyle CH_3}{|}}{C}-CH_2-C(CH_3)_3}{\underset{\displaystyle MgBr}{\diagdown}} \overset{H_3O^+}{\longrightarrow} RC\overset{\displaystyle O}{\underset{\displaystyle H}{\diagdown}}$$

(1)                                    (2)

yield aldehydes (2) in 48–67% yield. Carbonation of (1) provides the corresponding α-keto acid.

[1]H. M. Walborsky and G. E. Niznik, *J. Org.*, **37**, 187 (1972).
[2]H. M. Walborsky, W. H. Morrison, III, and G. E. Niznik, *Am. Soc.*, **92**, 6675 (1970).

**N,N,N',N'-Tetramethyldiamidophosphorochloridate,** $[(CH_3)_2N]_2POCl$. Mol. wt. 170.60, b.p. 98°/15 mm.

*Preparation.*[1] The reagent is prepared by the reaction of dimethylamine with phosphoryl chloride.

*Protection or reductive deoxygenation of alcohols and ketones.* Ireland *et al.*[2] have found that N,N,N',N'-tetramethylphosphorodiamidates (TMPDA derivatives) of alcohols and of ketone enolates are reduced in high yield by lithium–ethylamine. They are readily prepared by phosphorylation of alcoholate or enolate anions. The complete sequence is as follows. The alcoholate anion is simply prepared by treatment of an alcohol with a slight molar excess of *n*-butyllithium. The enolate anions of saturated ketones are prepared by treatment with lithium diisopropylamide. In the case of α,β-unsaturated ketones, lithium–ammonia reduction or conjugate organometallic addition is suitable. For phosphorylation of the anion a fivefold excess of N,N,N',N'-tetramethyldiamidophosphorochloridate in 4:1 dimethoxyethane (or THF)–N,N,-N',N'-tetramethylethylenediamine (TMEDA) is used. The reaction is complete after stirring for 1–2 hr. at 25°; yields of esters are high. Diethyl phosphorochloridate (**1**, 248; **3**, 98) can also be used and is preferred in the case of tertiary alcohols.

The reduction is carried out by addition of the derivative in THF and 2–4 eq. of $t$-butanol to an ice-cooled argon-protected solution of 10 eq. of lithium in dry ethylamine. It is not necessary to purify material at intermediate stages.

Examples:

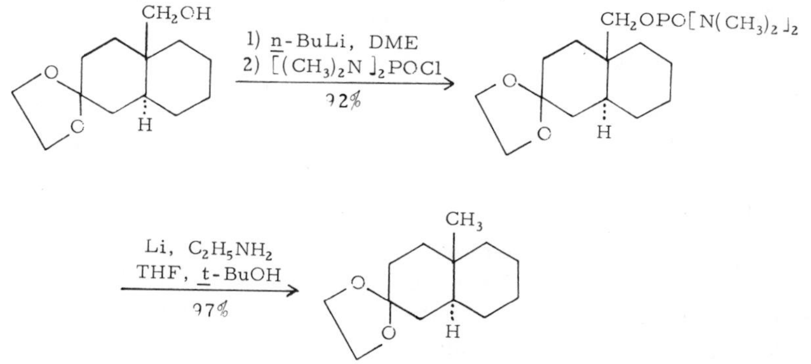

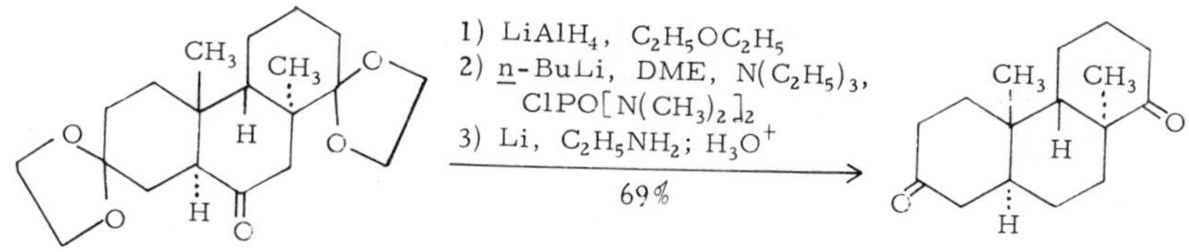

The TMPDA grouping is also a useful blocking group for hydroxyl functions; it is stable to $CH_3Li$, $LiAlH_4$, 1 $M$ $KOH–C_2H_5OH$, and 0.2 $M$ aqueous HCl–acetone. The alcohol group can be regenerated by treatment with 5 eq. of $n$-butyllithium in TMEDA for 30 min. at 25°.

[1] E. W. Crunden and R. F. Hudson, *J. Chem. Soc.*, 3591 (1962); H. G. Cook, J. D. Ilett, B. C. Saunders, G. J. Stacey, H. G. Watson, I. G. E. Wilding, and S. J. Woodcock, *ibid.*, 2921 (1949).

[2] R. E. Ireland, D. C. Muchmore, and U. Hengartner, *Am. Soc.*, **94**, 5098 (1972).

## 2,4,4,6-Tetramethyl-5,6-dihydro-1,3-(4H)-oxazine, 3, 280–284.

*Synthesis of aldehydes* (**3**, 280–282). Use of the reagent for synthesis of aldehydes has now been described in detail.[1]

The synthesis of 1-phenylcyclopentanecarboxaldehyde (**4**) from 2-benzyl-4,4,6-trimethyl-5,6-dihydro-1,3-(4H)-oxazine (**1**) has been published.[2]

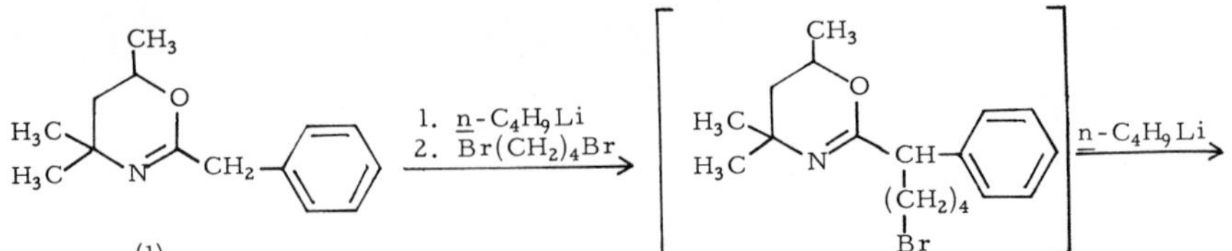

(1)

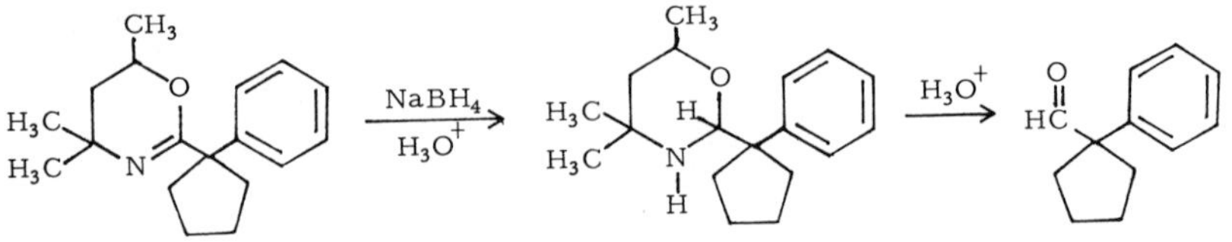

(2)          (3)          (4)

In a new variation for the two-carbon homologation of alkyl halides to give aldehydes,[3] 2,4,4,6-tetramethyl-5,6-dihydro-1,3-(4H)-oxazine is first converted into the
stable, crystalline methiodide salt (5) by treatment with 4.0 eq. of methyl iodide in the
dark for 20 hr. The salt (5) is treated with sodium hydride in dry DMF at room temperature. The alkyl halide is then added and the reaction mixture is stirred at 50–55° until

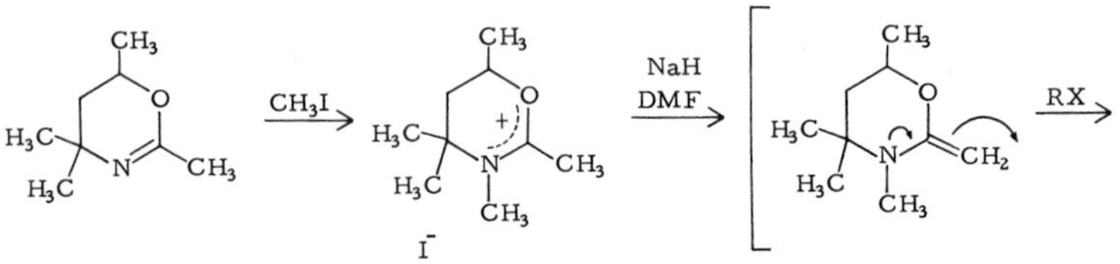

(5)

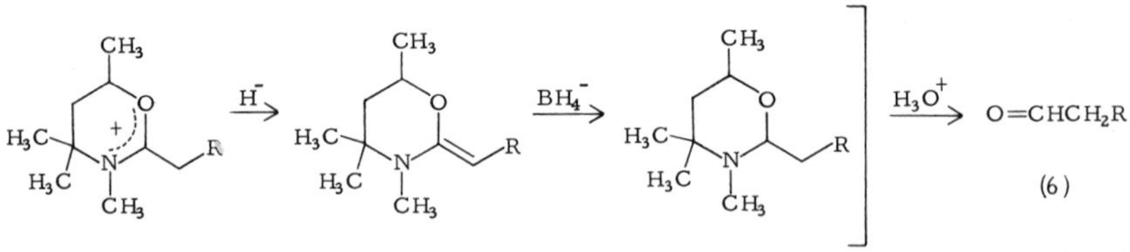

(6)

hydrogen evolution stops. The mixture is cooled and reduced with sodium borohydride
in absolute ethanol. The final step involves hydrolysis with aqueous oxalic acid. Yields
of aldehydes (6) are in the range 50–60%.

**Synthesis of ketones.** The dihydro-1,3-oxazine system itself is inert to Grignard
reagents. Meyers and Smith[4] now report that oxazines (1) form stable N-methyl
quaternary iodides (2) when treated with excess methyl iodide. These react with
Grignard reagents (2–2.5 eq.) at room temperatures to give an adduct (3), which on
hydrolysis with aqueous oxalic acid yields the ketone (4) and the amino alcohol (5).

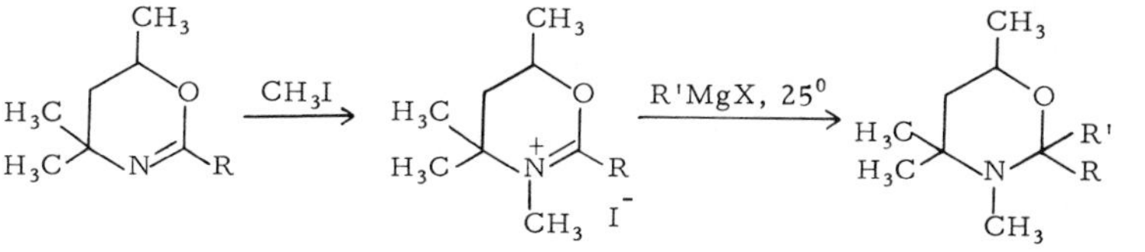

(1)                (2)                        (3)

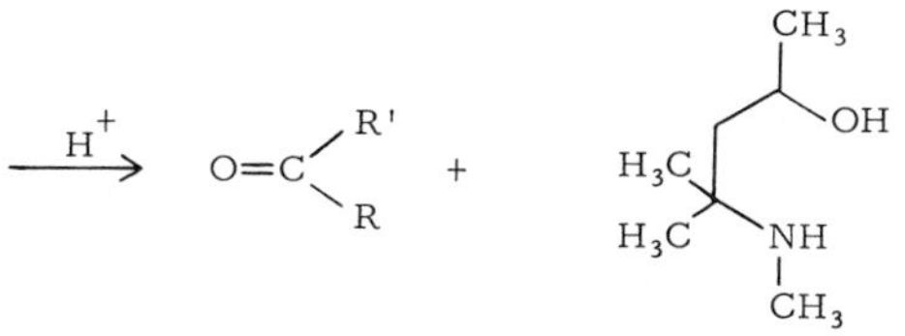

(4)                        (5)

Actually it is not necessary to prepare the N-methyl salts in a separate operation; they can be obtained in the same reaction vessel in which the Grignard addition is carried out. The yields of ketones by this procedure range from poor (12–20%) to good (50–85%). They are dependent on both the nature of R and on the Grignard reagent used.

Yields are low when R = CH$_3$ because of proton abstraction by the Grignard reagent to give the unstable ketene N,O-acetal (6). This acetal is best prepared *in situ* by

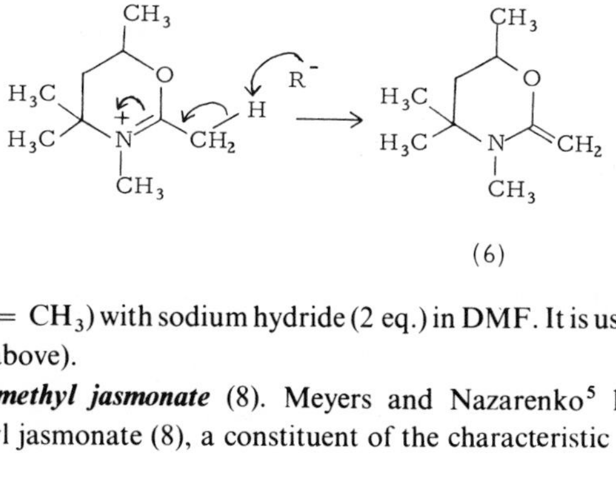

(6)

treatment of (1, R = CH$_3$) with sodium hydride (2 eq.) in DMF. It is useful for synthesis of aldehydes (see above).

*Synthesis of methyl jasmonate* (8). Meyers and Nazarenko[5] have reported a synthesis of methyl jasmonate (8), a constituent of the characteristic odor of jasmine,

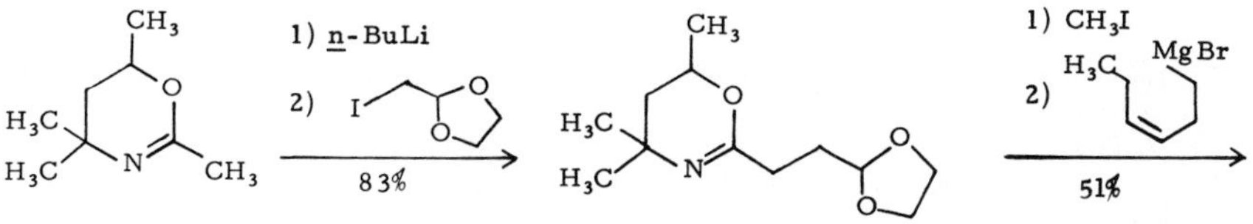

(1)                                    (2)

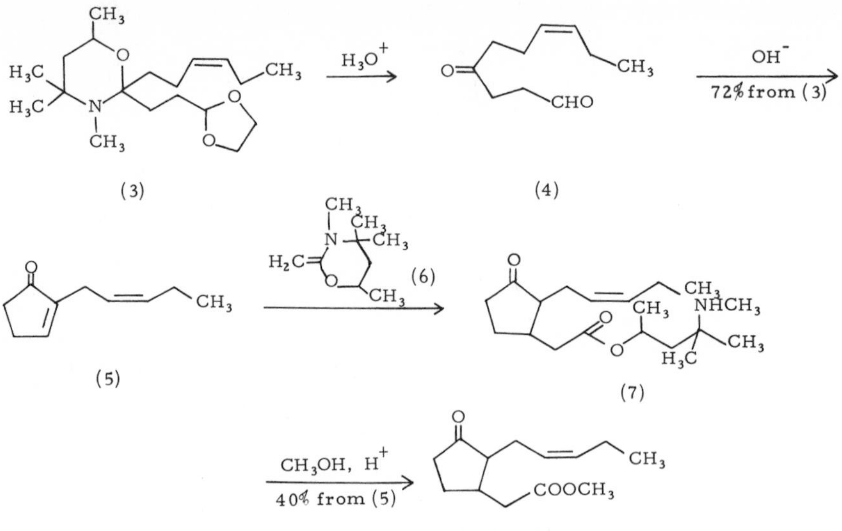

(3)

(4)

(5)

(6)

(7)

$CH_3OH, H^+$

$\overrightarrow{40\% \text{ from } (5)}$

(8)

starting with the oxazine (1). This is metalated as usual and then treated with 1.0 eq. of 2-iodomethyl-1,3-dioxolane to give the alkylated oxazine (2, 83% yield). This is converted into the methiodide salt (which need not be isolated) and this is treated with the Grignard reagent of cis-1-bromo-3-hexene in THF. The product (3) is heated in aqueous oxalic acid; both masking groups are removed to give the keto aldehyde (4) in good yield. The cyclopentenone (5) is formed when (4) is heated in 1% sodium hydroxide solution for 30 min. This is then treated with the ketene N,O-acetal (6) and the reaction quenched with $H_2O$. The resulting keto ester (7) is then transesterified with methanol (p-toluenesulfonic acid as catalyst) to give ($\pm$)-methyl jasmonate.

The synthesis presents two points of interest. It provides a method for synthesis of 2-alkylcyclopentenones. It also provides a method for addition of the unit $CH_2COOCH_3$ to electrophilic olefins. Thus the reaction of 2-cyclohexenone (9) with (6) followed by the steps outlined above leads to the keto ester (10) in 55% yield.

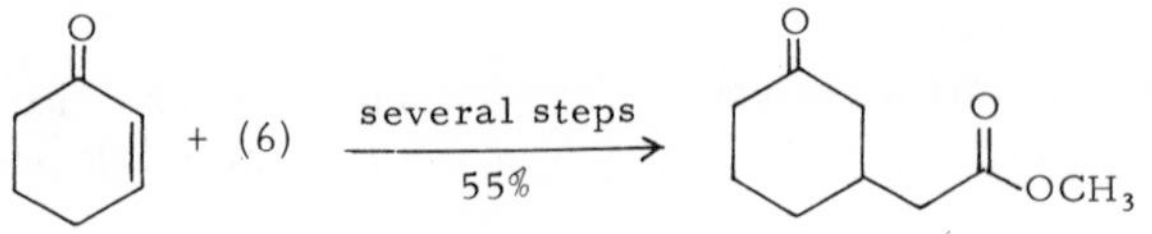

+ (6)    several steps    55%

(9)

(10)

[1]A. I. Meyers, A. Nabeya, H. W. Adickes, I. R. Politzer, G. R. Malone, A. C. Kovelesky, R. L. Nolen, and R. Portnoy, *J. Org.*, **38**, 36 (1973).
[2]I. R. Politzer and A. I. Meyers, *Org. Syn.*, **51**, 24 (1971).
[3]A. I. Meyers and N. Nazarenko, *Am. Soc.*, **94**, 3243 (1972).
[4]A. I. Meyers and E. M. Smith, *J. Org.*, **37**, 4289 (1972).
[5]A. I. Meyers and N. Nazarenko, *ibid.*, **38**, 175 (1973).

## N,N,N′,N′-Tetramethylethylenediamine (TMEDA), 2, 403; 3, 284–285.

*Dehydrogenation.* Dehydrogenation is frequently the final step in the synthesis of polycyclic aromatic hydrocarbons. This can be carried out efficiently with the *n*-butyl-lithium–TMEDA complex.[1] Thus treatment of *cis*-9-ethyl-10-methyl-9,10-dihydro-anthracene (1) with the complex in refluxing hexane generates the intensely purple dianion (a); the color is discharged by addition of cadmium(II) chloride to give 9-ethyl-10-methylanthracene (2) in 99% yield.

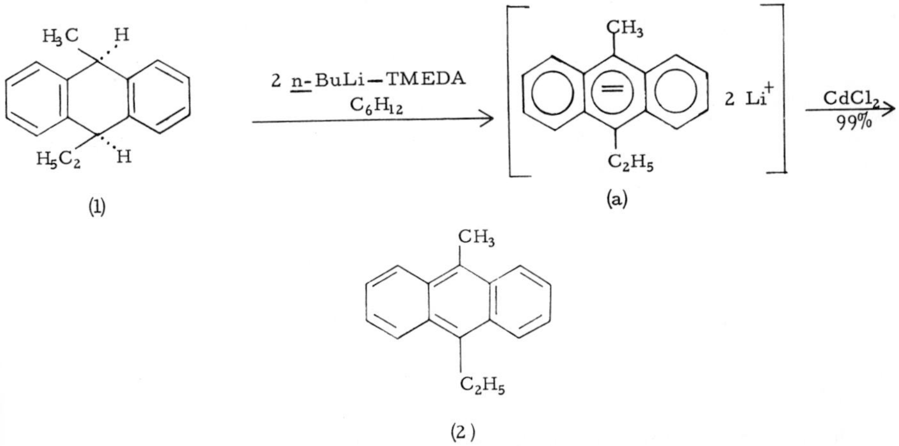

(1)    (a)

(2)

9,10-Dihydrophenanthrene (3) is dehydrogenated without added metal salt. In this case the intermediate 1:1 complex (b) is suggested.

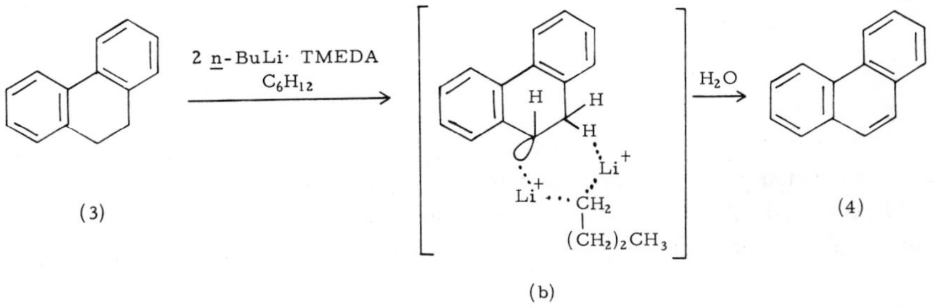

(3)    (b)    (4)

*Transmetalation activator* (2, 403). Metalations of certain secondary benzyl-amines occur more efficiently using TMEDA-activated *n*-butyllithium than in metala-tions with *n*-butyllithium alone. Thus N-methylbenzylamine undergoes dimetalation with *n*-butyllithium–TMEDA predominantly at the nitrogen and *o*-benzyl positions.[2]

*Metalation of olefins.*[3] Metalation of 1,1-dimethyl-2,2-diphenylethylene (1) with *n*-BuLi–TMEDA complex gives the anion (2) in good yield as shown by the reaction

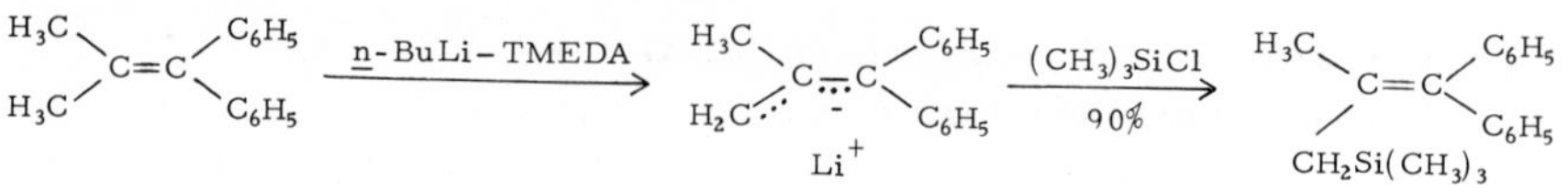

(1)                                    (2)                                         (3)

of (2) with trimethylchlorosilane to give (3). Reaction of (1) with *n*-BuLi in hexane–THF is much slower.

Reaction of α-cyclopropylstyrene (4) with the complex results in opening of the cyclopropane ring to give the salts (5) and (6).

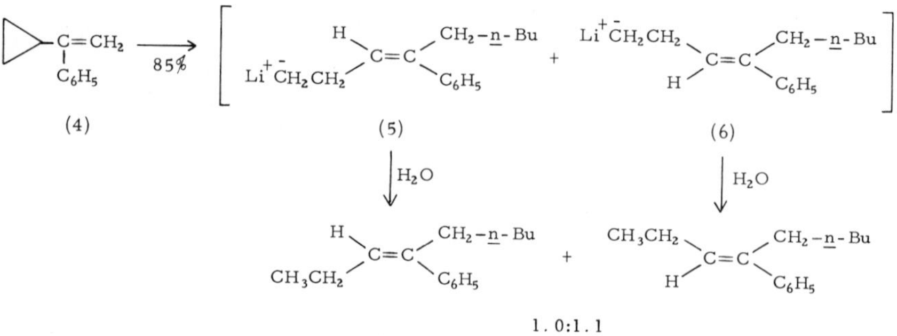

(4)                               (5)                                (6)

(7)                                                      (8)

1.0:1.1

Metalation of diphenylmethylenecyclobutane (9) proceeds equally well with *n*-BuLi in hexane–THF or with the *n*-Bu–TMEDA complex. In this case the crystalline salt (10) can be isolated.

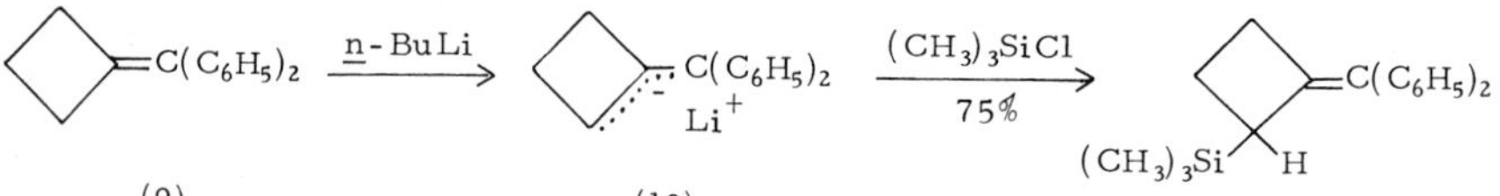

(9)                         (10)                                       (11)

**Metalation of limonene.** Crawford *et al.*[4] report that limonene (1) undergoes selective metalation at $C_{10}$ on treatment with the 1:1 complex of *n*-butyllithium and TMEDA to afford the 2-substituted allyllithium species represented by (2). The metalation is carried out by allowing a mixture of 2 eq. of limonene and 1 eq. of the complex

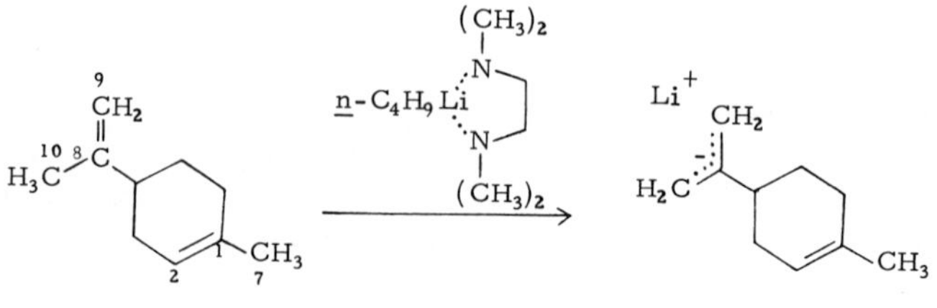

(1)                                                      (2)

(about 1.5 $M$ in $n$-hexane) to stand overnight at room temperature. Solutions of (2) are stable for several days when kept under nitrogen or argon.

Metalation can be conducted with both ( + )-(R)-limonene (3) and ( − )-(S)-limonene (4), both of which are commercially available.

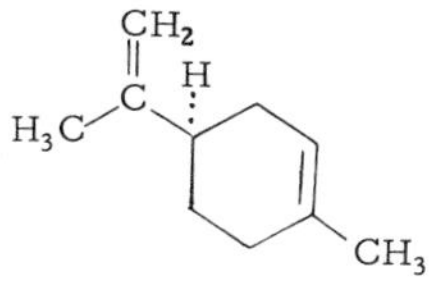

(3) 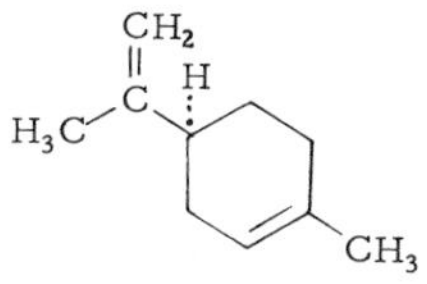                    (4)

This new selective metalation reaction is useful for synthesis of 10-substituted limonene derivatives. The excess limonene is readily separated from the product by fractional distillation. Thus metalation of ( + )-limonene (3) followed by reaction with carbon dioxide and then esterification affords the $\beta,\gamma$-unsaturated ester (5) in 19% yield. Isomerization with sodium methoxide in methanol converts (5) into the conjugated isomer (6).

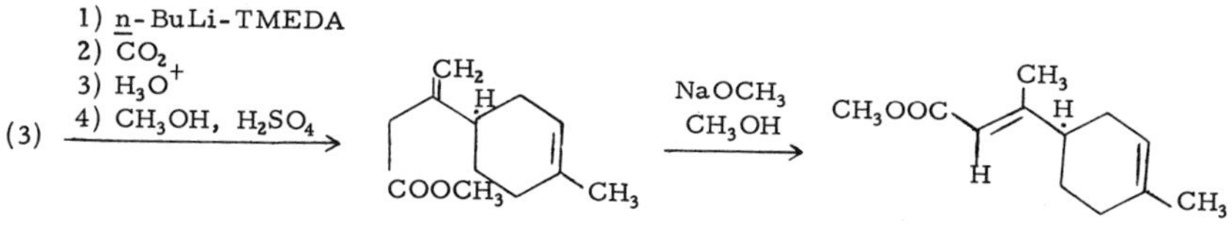

(5)                    (6)

This selective metalation has been used for the synthesis of several bisabolane sesquiterpenes. Thus the sesquiterpene ( − )-$\beta$-bisabolene (7) can be synthesized in one step from metalated ( − )-limonene by reaction with 1-bromo-3-methyl-2-butene. Two

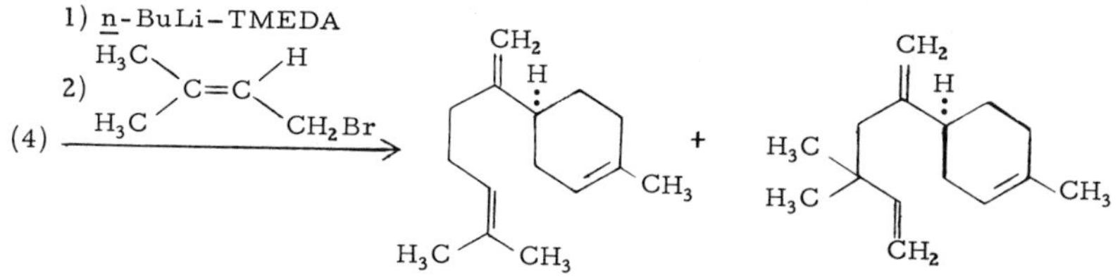

(7)        4:1        (8)

products, (7) and (8), are obtained in the ratio of 4:1; they are separable by spinning band distillation or by preparative glpc.

Crawford[5] has reported a synthesis of the monocyclic diterpene $ar$-artemisene (15), in which selective metalation is used in two of the four steps. Thus ( + )-limonene (3) is converted into the metalated derivative (9); reaction of (9) with paraformaldehyde provides the alcohol (10). This is converted into the aromatic alcohol (11) by treatment with N-lithioethylenediamine in refluxing ethylenediamine (1, 567–570). The alcohol is converted into the bromide (12) with phosphorus tribromide. Geraniolene (13) is selectively metalated with the $n$-butyllithium–TMEDA complex to give (14). Addition of (12) to the solution gives ($\pm$)-$ar$-artemisene (15).

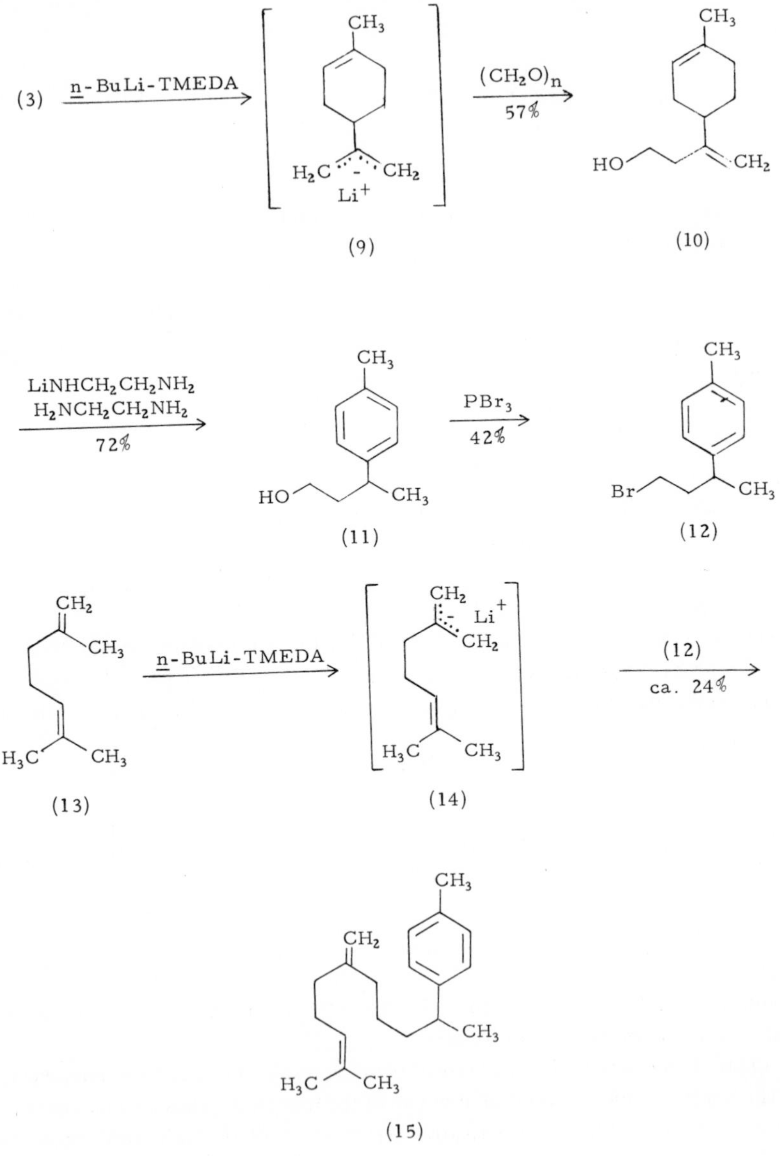

(3) →[n-BuLi-TMEDA]→ (9) →[(CH$_2$O)$_n$, 57%]→ (10)

→[LiNHCH$_2$CH$_2$NH$_2$ / H$_2$NCH$_2$CH$_2$NH$_2$, 72%]→ (11) →[PBr$_3$, 42%]→ (12)

(13) →[n-BuLi-TMEDA]→ (14) →[(12), ca. 24%]→

(15)

The procedure of Crawford *et al.* was used to effect allylic alkylation in a synthesis of the sesquiterpene ( − )-cryptomerione (19) from ( − )-carvone.[6] Thus allylic alkylation of the acetal of ( + )-dihydrocarvone (16) was effected by treatment with the complex of *n*-butyllithium and TMEDA followed by addition of 1-chloro-3-methylbutene-2. Dihydrocrytomerione (17) was obtained in 70% overall yield. Bromination of (17) with

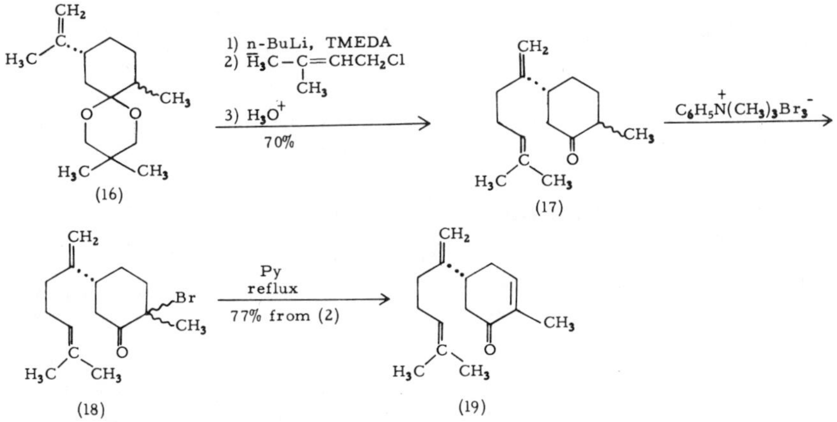

phenyltrimethylammonium perbromide (**1**, 855–856; **2**, 328) followed by dehydro-bromination gave (19) in good yield.

[1]R. G. Harvey, L. Nazareno, and H. Cho, *Am. Soc.*, **95**, 2376 (1973).
[2]R. E. Ludt and C. R. Hauser, *J. Org.*, **36**, 1607 (1971).
[3]E. Dunkelblum and S. Brenner, *Tetrahedron Letters*, 669 (1973).
[4]R. J. Crawford, W. F. Erman, and C. D. Broaddus, *Am. Soc.*, **94**, 4298 (1972).
[5]R. J. Crawford, *J. Org.*, **37**, 3543 (1972).
[6]G. L. Hodgson, D. F. MacSweeney, and T. Money, *J.C.S. Chem. Comm.*, 236 (1973).

## Tetramethylguanidine, 1, 1145.

*Michael additions of nitromethane.* Italian chemists[1] were able to effect Michael addition of nitromethane to $\alpha,\beta$-unsaturated carboxylic acid esters with catalysis by tetramethylguanidine. Yields of 1:1 adducts (1) are in the range 50–85% when $R^1$ or $R^2$ are alkyl or aryl groups. If $R^1$ and $R^2$ are H, 1:2 adducts (2) are also formed in substantial amounts.

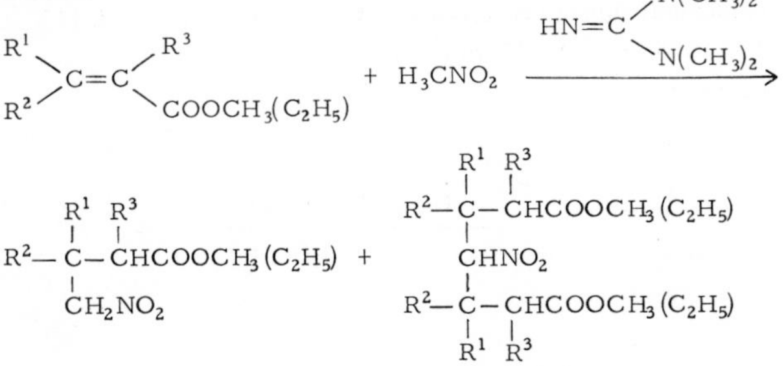

Michael addition of nitromethane with this catalyst to the ketone (3) was used to obtain the nitroketone (4) in 84% yield.[2] The reaction was the first step in a synthesis of (±)-11-desoxyprostaglandin $E_1$, $F_{1\alpha}$, and $F_{1\beta}$.

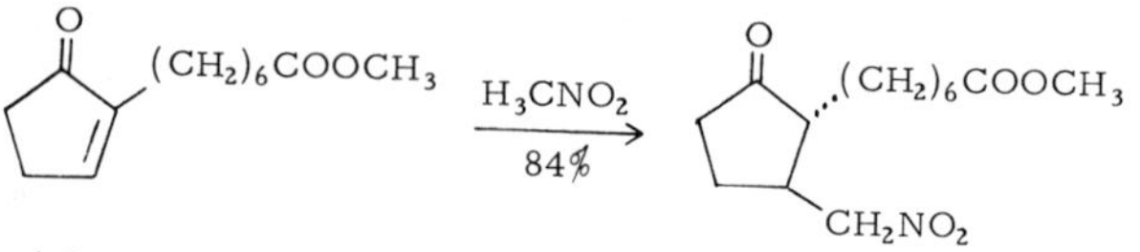

(3)                                    (4)

[1]G. P. Pollini, A. Barco, and G. DeGuili, *Synthesis*, 44 (1972).
[2]F. S. Alvarez and D. Wren, *Tetrahedron Letters*, 569 (1973).

**N,N,N',N'-Tetramethylsuccinamide,**  $CH_2CON(CH_3)_2$
Mol. wt. 152.23, m.p. 81–82°.       $CH_2CON(CH_3)_2$.

The reagent is prepared by the reaction of diethyl succinate and dimethylamine (sodium methoxide, Parr bomb, 100°, 24 hr.); yield 76%.

*1,4-Diketones.*[1] The reagent reacts with organolithium compounds at −78° to give 1,4-diketones. Yields vary, depending on the organolithium compound. Highest yields were obtained using 6-bromo-2-pyridyllithium, prepared from 2,6-dibromo-pyridine and *n*-butyllithium. Only a 4% yield of a 1,4-diketone was obtained with use of phenyllithium in the reaction above.

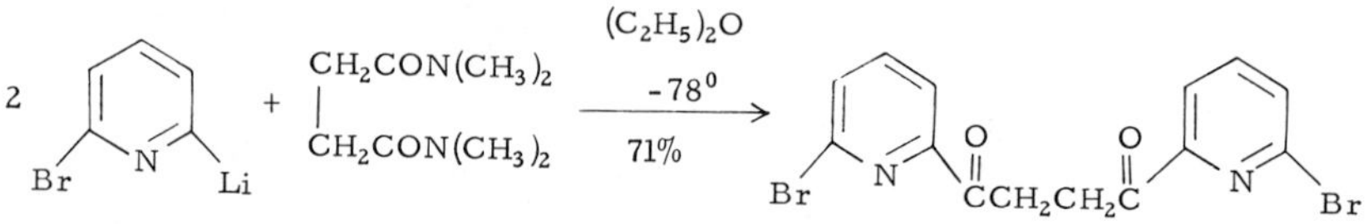

*1,5-Diketones.*[1] 1,5-Diketones are obtained in the same way by use of N,N,N',N'-tetramethylglutaramide instead of N,N,N',N'-tetramethylsuccinimide. Again, highest yields of 1,5-diketones were obtained by use of 6-bromo-2-pyridyllithium as the organolithium compound. In this case phenyllithium gave a reasonable yield of a 1,5-diketone (50%).

[1]D. C. Owsley, J. M. Nelke, and J. J. Bloomfield, *J. Org.*, **38**, 901 (1973).

**Tetraphenylcyclopentadienone (Tetracyclone), 1,** 1149; **2,** 345, 367; **3,** 111, 103, 139–140, 170–171, 176–177. Mol. wt. 384.45, m.p. 219°.
    *Preparation*[1]:

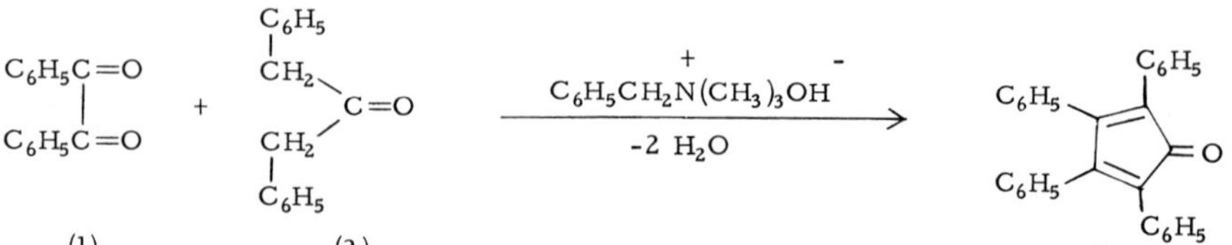

(1)            (2)                                              (3)

### Condensation with 1,2-dehydrocyclooctatetraene (5)[2]:

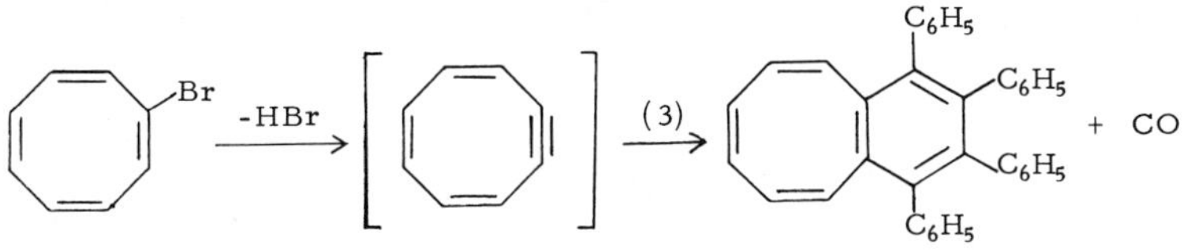

(4)     (5)     (6)

### Condensation with diphenylacetylene[3]:

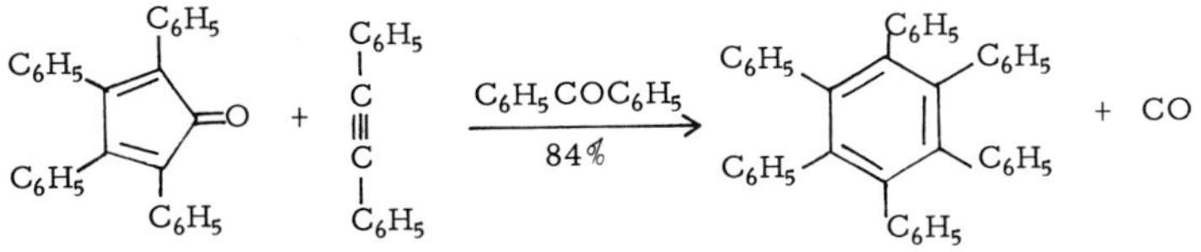

Hexaphenylbenzene

m. p. 454-456°

### Condensation with dehydrothiophene (2)[4]:

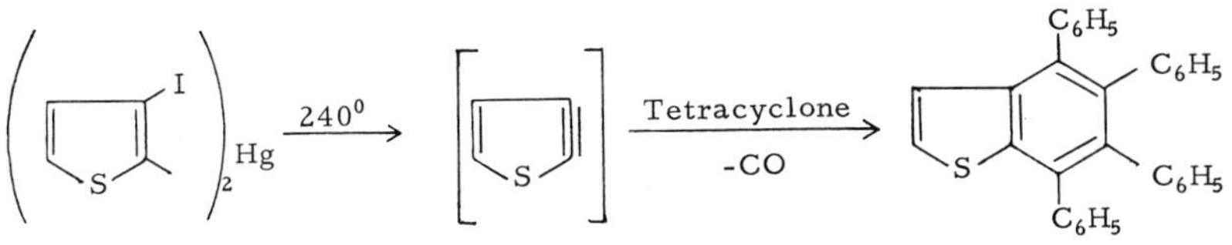

(1)     (2)     (3)

### Use to trap dehydrothiophene[4]:

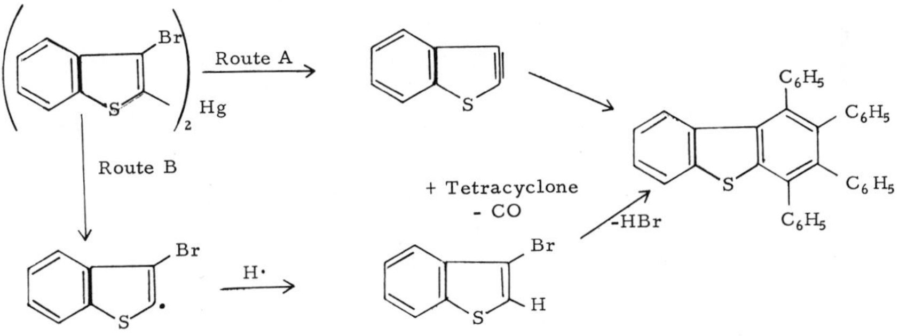

[1]L. F. Fieser, *Org. Syn.*, **46**, 44 (1966).
[2]A. Krebs and D. Byrd, *Ann.*, **707**, 66 (1967).
[3]L. F. Fieser, *Org. Exps.*, 2nd Ed., D. C. Heath and Co., Boston, 299 (1968); *Org. Syn.*, **46**, 44 (1966).
[4]G. Wittig and M. Rings, *Ann.*, **719**, 127 (1968).

**Thallium(III) acetate, 1,** 1150–1151; **2,** 406; **3,** 286.

*Aromatic bromides* (**3,** 286). The definitive paper on electrophilic aromatic bromination with bromine and thallium(III) acetate has been published.[1] The two most outstanding features are 1) monobromination is observed in almost all cases, and 2) exclusive *para* substitution is observed with almost all monosubstituted benzenes. Electron-withdrawing groups inhibit bromination of monosubstituted benzenes. It

$$(1) \quad Tl(OOCCH_3)_3 + 3ArH + 3Br_2 \longrightarrow 3ArBr + 3CH_3COOH + TlBr_3$$

can be concluded that the thallium(III) acetate/bromine reagent is an electrophile of low reactivity but high steric requirement. The stoichiometry of the reaction is shown in equation 1. Experimental results suggest that the reaction involves bromination by molecular bromine catalyzed by either thallium(III) bromide or thallium (III) acetate.

*α-Acyloxy carboxylic acids.*[2] The reaction of thallium(III) acetate with aliphatic carboxylic acids results in formation of α-acyloxy carboxylic acids, which can be readily hydrolyzed to α-hydroxy acids.

$$Tl(OCOCH_3)_3 + 3\ R^1R^2CHCOOH \xrightarrow[-3\ CH_3COOH]{} Tl(OCOCHR^1R^2)_3 \xrightarrow[-R^1R^2CHCOOH]{} \underset{\underset{OCOCHR^1R^2}{|}}{R^1R^2CCOO^-Tl^+}$$

[1] A. McKillop, D. Bromley, and E. C. Taylor, *J. Org.,* **37,** 88 (1972).
[2] E. C. Taylor, H. W. Altland, G. McGillivray, and A. McKillop, *Tetrahedron Letters,* 5285 (1970).

**Thallium(III) nitrate (TTN),** $Tl(NO_3)_3$. Mol. wt. 390.38. Suppliers: Aldrich, Alfa, Columbia, Fisher, Howe and French.

*Preparation.*[1] The salt is prepared by dissolving 50 g. of thallium(III) oxide in 150 ml. of warm concentrated nitric acid and cooling the pale-yellow solution to 0°. Thallium(III) nitrate trihydrate separates and is dried under vacuum over $P_2O_5$. The salt is stable if stored in tightly sealed bottles.

*Oxidation of olefins.*[2] Olefins are oxidized rapidly by thallium(III) nitrate in methanol at room temperature to aldehydes or ketones. Rearrangement is the exclusive or predominant pathway with olefins in which at least one of the substituted groups has a high migratory aptitude.

Examples:

$$\underset{R^2}{\overset{R^1}{>}}C=C\underset{R^4}{\overset{R^3}{<}} \xrightarrow{Tl(NO_3)_2} R^2-CO-\underset{\underset{R^4}{|}}{\overset{\overset{R^1}{|}}{C}}-R^3$$

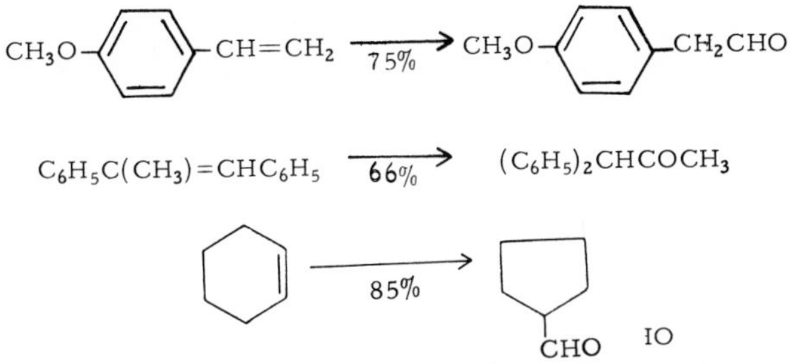

$$CH_3O\text{—}C_6H_4\text{—}CH=CH_2 \xrightarrow[75\%]{} CH_3O\text{—}C_6H_4\text{—}CH_2CHO$$

$$C_6H_5C(CH_3)=CHC_6H_5 \xrightarrow[66\%]{} (C_6H_5)_2CHCOCH_3$$

$$\text{cyclohexene} \xrightarrow[85\%]{} \text{cyclopentane-CHO} \quad \text{IO}$$

Cyclohexene is oxidized to cyclopentanecarboxaldehyde in 85% yield; in fact this TTN-induced ring contraction is probably the method of choice for preparation of this aldehyde. This reaction was used by Corey and Ravindranathan[3] as one step in a synthesis of 11-desoxyprostaglandins for the conversion of (1) into (2). However, in this case, the use of the original conditions led to complex mixtures containing only a

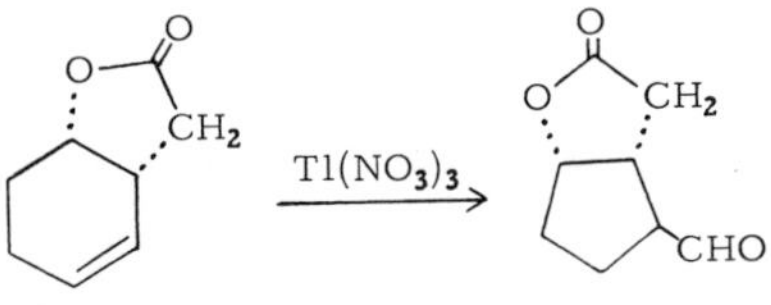

$$(1) \xrightarrow{Tl(NO_3)_3} (2)$$

(1)              (2)

moderate amount of (2). Satisfactory results can be obtained using 1.6 eq. of thallium(III) nitrate in water which is 0.5 $M$ in perchloric acid and 4.0 $M$ in sodium perchlorate at 25–28° for 0.5 hr.

*Chalcones → Benzils.*[2] Oxidation of chalcones with thallium(III) nitrate in an aqueous acid/glyme medium results in formation of benzils in about 50% yield. Three

$$ArCH=CHCOAr' \longrightarrow ArCOCOAr'$$

distinct oxidations are involved: oxidative rearrangement to the keto aldehyde (2), followed by cleavage to a desoxybenzoin (3); oxythallation to a benzoin (4); and finally oxidation to the benzil (5).

$$ArCH=CHCOAr' \longrightarrow Ar\overset{CHO}{\underset{|}{C}}HCOAr' \xrightarrow{H^+} ArCH_2COAr' \longrightarrow Ar\overset{OH}{\underset{|}{C}}HCOAr' \longrightarrow ArCO-COAr'$$

(1)      (2)      (3)      (4)      (5)

*2-Alkynoic esters.*[4] 5-Pyrazolones[5] are converted by treatment with 2 eq. of thallium(III) nitrate into 2-alkynoic esters. Yields are high and reaction conditions are extremely mild.

***Oxidation of acetylenes.***[6] Diarylacetylenes are converted into benzils in 45–95% yield by treatment with TTN in either aqueous acidic (perchloric acid) glyme or methanol:

$$\text{ArC}{\equiv}\text{CAr'} + 2\ \text{TTN} \xrightarrow{\text{H}_3\text{O}^+} \underset{\underset{\text{O}}{\|}}{\text{ArC}}-\underset{\underset{\text{O}}{\|}}{\text{CAr'}}$$

Dialkylacetylenes give acyloins in aqueous media and α-methoxyketones in methanol:

$$\text{RC}{\equiv}\text{CR'} + \text{TTN} \xrightarrow{\text{H}_2\text{O}} \underset{\underset{\text{OH}}{|}}{\text{RCH}}-\underset{\underset{\text{O}}{\|}}{\text{CR'}} + \underset{\underset{\text{O}}{\|}}{\text{RC}}-\underset{\underset{\text{OH}}{|}}{\text{CHR'}}$$

$$\text{RC}{\equiv}\text{CR'} + \text{TTN} \xrightarrow{\text{CH}_3\text{OH}} \underset{\underset{\text{OCH}_3}{|}}{\text{RCH}}-\underset{\underset{\text{O}}{\|}}{\text{CR'}} + \underset{\underset{\text{O}}{\|}}{\text{RC}}-\underset{\underset{\text{OCH}_3}{|}}{\text{CHR'}}$$

Terminal acetylenes undergo degradation to carboxylic acids containing one less carbon atom:

$$\text{RC}{\equiv}\text{CH} + 2\ \text{TTN} \xrightarrow[55-80\%]{\text{H}_3\text{O}^+} \text{RCOOH}$$

Alkylarylacetylenes undergo oxidative rearrangement in methanol to give methyl α-alkylaryl acetates in 80–98% yield:

$$\text{ArC}{\equiv}\text{CR} + \text{TTN} \xrightarrow[80-98\%]{\text{CH}_3\text{OH}} \text{Ar}\overset{\overset{\displaystyle R}{|}}{\text{C}}\text{HCOOCH}_3$$

***Oxidation of cyclohexanone.***[7] Oxidation of cyclohexanone with TTN in acetic acid at room temperature proceeds rapidly to give adipoin (5) in 84% yield. If the oxidation is performed at first at room temperature, the thallium(I) nitrate removed by filtration, and then the filtrate heated to about 40° for a few minutes, the product is cyclopentanecarboxylic acid (4), obtained in 84% yield. McKillop et al.[1] propose that both products are derived from a common precursor, an epoxy enol (3).

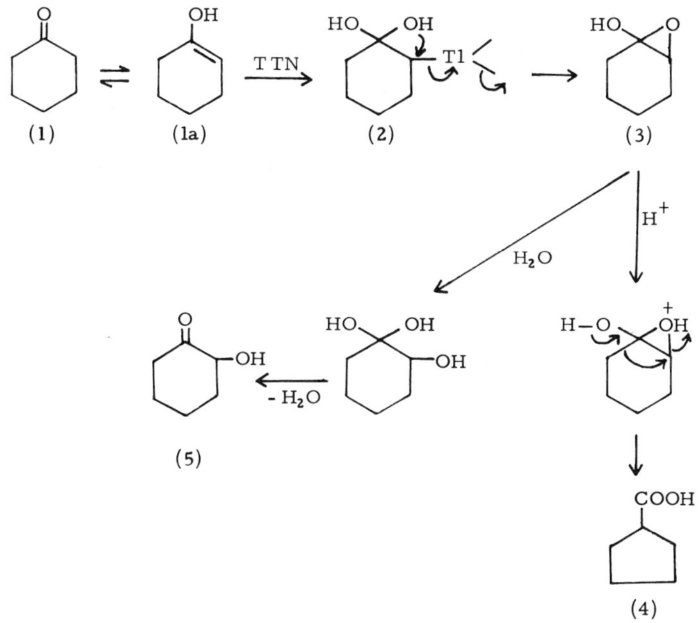

(1)   (1a)   (2)   (3)

(5)

(4)

***Carbamates.***[8] Thallium(III) nitrate reacts with isonitriles in methanol solution at room temperature within a few minutes to give carbamates in high yield (35–97%).

$$\text{RNC} + \text{Tl(NO}_3)_3 \xrightarrow[\text{2) H}_2\text{O}]{\text{1) CH}_3\text{OH}} \text{RNHCOOCH}_3 + \text{TlNO}_3 + 2\text{HNO}_3$$

Mercury(II) nitrate can also be used, but yields of carbamates are much lower. However, with this reagent alcohols other than methanol can be used. TTN appears to be stable only in methanolic solution.

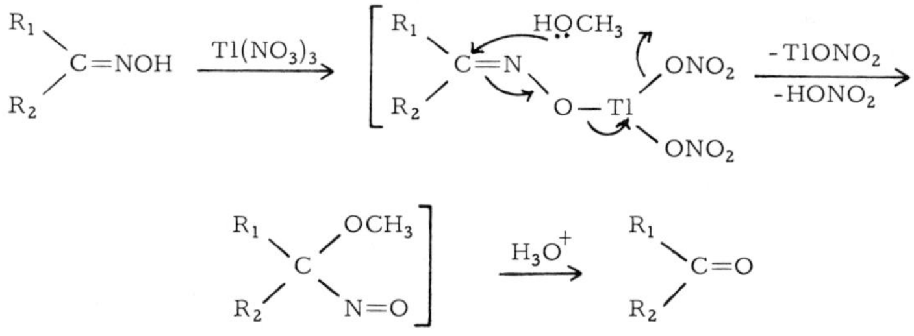

***Deoximation.***[9] Oximes are converted into aldehydes and ketones in yields of 70–95% when treated with TTN in methanol at room temperature. The reaction is complete within a few minutes. The precipitated thallium(I) nitrate is removed, the filtrate shaken with dilute sulfuric acid, and the aldehyde or ketone extracted with ether or chloroform. The process is formulated as shown. One limitation is that the method cannot be used with aryl aldehydes or ketones containing *ortho-* or *para*-hydroxyl groups or amino groups owing to concomitant oxidation of the aromatic ring to a quinone. However, the reaction is successful if the OH or $NH_2$ groups are protected by acetylation.

Semicarbazones and phenylhydrazones of carbonyl compounds react analogously, but the reaction in this case requires 1–5 min. for completion.

*Acetophenones → methyl phenylacetates*, $ArCOCH_3 → ArCH_2COOCH_3$.[10] Treatment of an acetophenone with thallium(III) nitrate in methanol containing a trace of perchloric acid at room temperature for 2–18 hr. results in formation of a methyl arylacetate in 60–95% yield. The mechanism suggested for this transformation is shown in the formulation. The reaction provides an alternative to the Willgerodt–Kindler reaction.[11]

$$ArCOCH_3 \underset{}{\overset{H^+}{\rightleftharpoons}} Ar-\underset{\overset{|}{OH}}{C}=CH_2 \xrightarrow[CH_3OH]{Tl(NO_3)_3} \left[ \underset{\overset{|}{OCH_3}}{H-O-C}-CH_2-Tl-ONO_2 \right]$$

$$\longrightarrow \underset{\overset{|}{OCH_3}}{O=C}-CH_2Ar \ + \ TlONO_2 \ + \ HONO_2$$

*Allenic esters.*[12] $\alpha$-Alkyl-$\beta$-keto esters (1)[13] can be converted into allenic esters (6) by treatment with 1 eq. of hydrazine, to form the 3,4-disubstituted 5-pyrazolone (2), and then with a solution of 2 eq. of TTN in methanol. Isolation of the intermediate pyrazolone is unnecessary. Overall yields of allenic esters are in the range 50–70%. Note that $R_3$ must be an alkyl group, otherwise $\alpha,\beta$-acetylenic esters are formed (see above). The conversion of 5-pyrazolones (2) into allenic esters (6) is explained by electrophilic

(1)          (2)          (6)

thallation of the enamine tautomer (2a) of the pyrazolone (2) followed by proton loss to give (4). Oxidation to (5) and solvolysis by methanol gives the allenic ester (6).

(2a)          (3)

$$(4) \xrightarrow{\text{TTN}} (5) \longrightarrow (6) + \text{TlONO}_2$$

*Oxidative cleavage of glycols.* McKillop *et al.*[14] have examined the oxidative cleavage of glycols with thallium(III) acetate (**1**, 1150–1151; **2**, 406; **3**, 286), thallium(III) trifluoroacetate (**3**, 286–289), and thallium(III) nitrate. The most efficient reagent is TTN in acetic acid, but cleavage occurs only with glycols containing vicinal

$$(1) \xrightarrow[61\%]{\text{TTN/HOAc}} 2 \;(2)$$

$$(3) \xrightarrow{91\%} 2 \;(4)$$

aromatic substituents such as (1). Glycol cleavage occurs in high yield with tetraaryl-substituted glycols (3). Tetraaryl-substituted glycols such as (3) are also cleaved smoothly by thallium(I) ethoxide (**2**, 407–411) in ethanol.

[1] A. McKillop, J. D. Hunt, E. C. Taylor, and F. Kienzle, *Tetrahedron Letters*, 5275 (1970).
[2] A. McKillop, B. P. Swann, and E. C. Taylor, *ibid.*, 5281 (1970).
[3] E. J. Corey and T. Ravindrathan, *ibid.*, 4753 (1971).
[4] E. C. Taylor, R. L. Robey, and A. McKillop, *Angew. Chem., internat. Ed.*, **11**, 48 (1972).
[5] 5-Pyrazolones are prepared in quantitative yield by the reaction of hydrazine with β-keto esters: R. H. Wiley and P. Wiley, in *The Chemistry of Heterocyclic Compounds*, A. Weissberger, Ed., Interscience, New York, Vol. **20** (1964).
[6] A. McKillop, O. H. Oldenziel, B. P. Swann, E. C. Taylor, and R. L. Robey, *Am. Soc.*, **95**, 1296 (1973).
[7] A. McKillop, J. D. Hunt, and E. C. Taylor, *J. Org.*, **37**, 3381 (1972).
[8] F. Kienzle, *Tetrahedron Letters*, 1771 (1972).
[9] A. McKillop, J. D. Hunt, R. D. Naylor, and E. C. Taylor, *Am. Soc.*, **93**, 4918 (1971).
[10] A. McKillop, B. P. Swann, and E. C. Taylor, *ibid.*, **93**, 4919 (1971).
[11] M. Carmack and M. A. Spielman, *Org. React.*, **3**, 83 (1947); F. Asinger, W. Schäfer, and K. Halcour, *Angew. Chem., internat. Ed.*, **3**, 19 (1964); R. Wegler, E. Kuhle, and W. Schäfer, *Newer Methods Prep. Org. Chem.*, **3**, 1 (1964).
[12] E. C. Taylor, R. L. Robey, and A. McKillop, *J. Org.*, **37**, 2797 (1972).
[13] For a recent preparation, see E. C. Taylor, G. H. Hawks, III, and A. McKillop, *Am. Soc.*, **90**, 2421 (1968).
[14] A. McKillop, R. A. Raphael, and E. C. Taylor, *J. Org.*, **37**, 4204 (1972).

**Thallium sulfate, $Tl_2(SO_4)_3$.** Mol. wt. 696.94. Suppliers: Alfa, Fisher, Fluka, K and K, E. Merck, P and B, Riedel de Haën, Schuchardt.

*Oxidation of cyclohexenes.*[1] 3-*t*-Butylcyclohexene (1) is oxidized by thallium sulfate to the *trans*-diaxial diol (2, 73 % yield), the *trans*-diequatorial diol (3, 10 % yield), and an unidentified product (18 % yield). Oxidation of 4-*t*-butylcyclohexene gives exclusively

the *trans*-diaxial 1,2-diol. The fact that only *trans*-diols are formed excludes an $S_N2$ substitution of thallium and suggests that the α-hydroxyl group participates in breaking the C–Tl bond of the intermediate organothallous compound.

[1] C. Freppel, R. Favier, J.-C. Richer, and M. Zador, *Canad. J. Chem.*, **49**, 2586 (1971).

**Thallium(III) trifluoroacetate (TTFA), 3, 286–289).**

*Aromatic iodides* (3, 287). The definitive paper on the synthesis of aromatic iodides by the reaction of arylthallium ditrifluoroacetates with potassium iodide has been published.[1] Four procedures have been developed. 1) Thallation is carried out as usual and then an aqueous solution of potassium iodide is added directly. 2) The intermediate arylthallium ditrifluoroacetate is isolated and then treated with potassium iodide. 3) For acid-sensitive substrates solid TTFA in acetonitrile is used for thallation. 4) These methods are unsuccessful with highly reactive compounds such as naphthalene and diphenyl. In such cases molecular iodine is used as the electrophilic reagent and TTFA is used as oxidant for the hydrogen iodide formed in the reaction.

The distribution rates for iodination of monosubstituted benzene derivatives have been reported.[2] Under conditions of thermodynamic control (elevated temperature), *meta* substitution is observed. Under conditions of kinetic control (room temperature), a significant preference for *para* substitution is observed for compounds containing *ortho–para*-directing substituent groups. *Ortho* substitution results when chelation of TTFA with the directing substituent permits intramolecular delivery of the electrophile. For example, methyl benzoate gives almost exclusively *ortho*-thallation (95 %).

*Oxidative phenol coupling.* Schwartz *et al.*[3] report that TTFA is an effective reagent for intramolecular oxidative phenol coupling. Thus oxidation of the diarylpropane derivative (1) with 1 molar eq. of TTFA in methylene chloride affords the

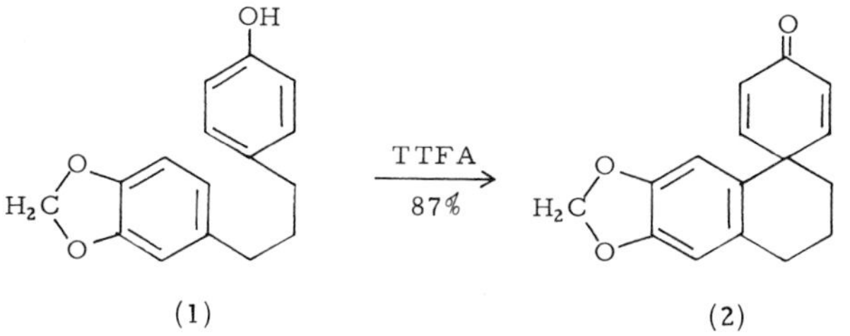

(1)                                                (2)

dienone (2) in 87% yield. The method was then extended to the system of the Amaryllidaceae alkaloids, but with a substantial decrease in yield. Thus oxidation of the norbelladine derivative (3) with TTFA gave only a 19% yield of the dienone (4). Hydrolysis of (4) with $Na_2CO_3$ in aqueous methanol afforded ($\pm$)-oxocrinine (95% yield).

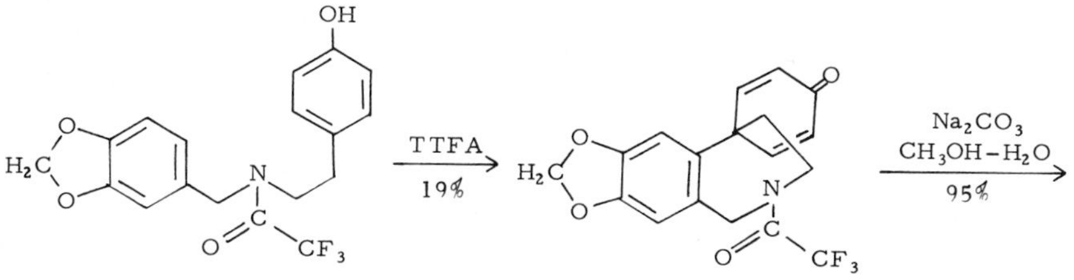

(3)                                          (4)

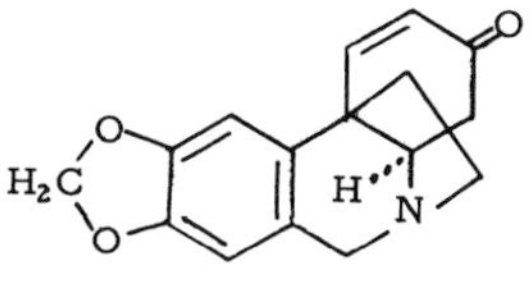

(5)

The new procedure was also used successfully for synthesis of ($\pm$)-O-methylandrocymbine (8, a colchicine precursor) from the phenethylisoquinoline (6). This was converted into the amine-borane (7) by treatment with diborane in THF–CHCl$_3$.[4] Oxidation of (7) with excess TTFA followed by removal of the blocking group gave (8) in 20% overall yield.

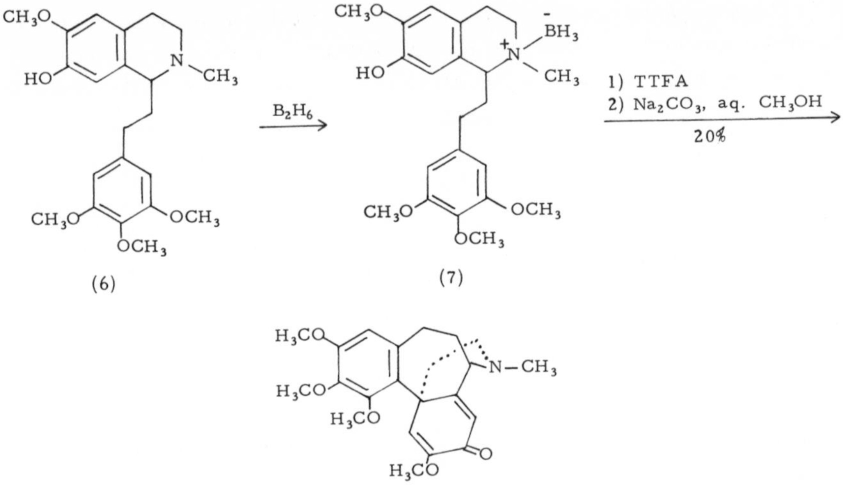

(6)                                        (7)

(8)

*1,6-Dioxa-6a-thiapentalene* (4).[5] Addition of thallium(III) trifluoroacetate in acetonitrile to a solution of 4*H*-pyrane-4-thione (1) in acetonitrile at room temperature gives the labile pyrylium salt (2), which on addition of water gives 1,6-dioxa-6a-thiapentalene (4) by sulfur–oxygen bond formation. The intermediate (3) is suggested.

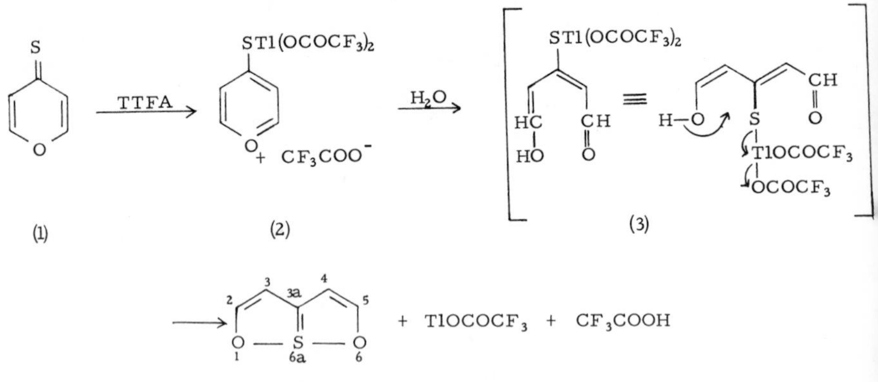

(1)            (2)                                    (3)

(4)

The driving force for the reaction is the energetically favorable Tl(III) → Tl(I) transformation.

*Oxidation to* **p**-*quinones* (**3**, 289). In the oxidation of 2,6-disubstituted-4-*t*-butylphenols to 2,6-disubstituted-*p*-benzoquinones, McKillop *et al.* (**3**, 289, ref. 6) postulated an intermediate 2,6-disubstituted-4-*t*-butyl-4-trifluoroacetoxycyclohexa-2,5-diene-1-one. A stable quinol trifluoroacetate of this type (2) has now been isolated from the reaction of estrone (1) with 2 eq. of TTFA.[6]

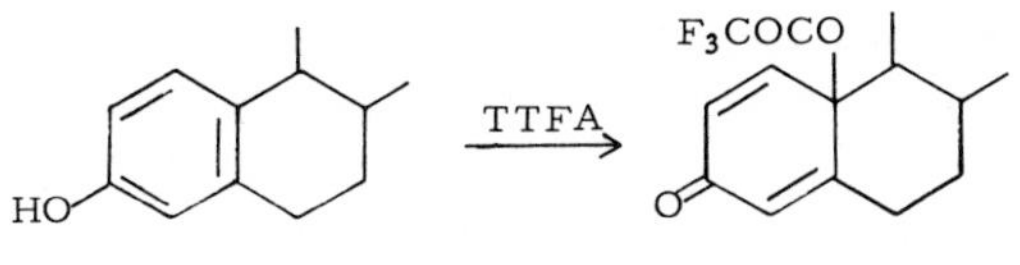

(1)          (2)

[1]A. McKillop, J. D. Hunt, M. J. Zelesko, J. S. Fowler, E. C. Taylor, G. McGillivray, and F. Kienzle, *Am. Soc.*, **93**, 4841 (1971).
[2]E. C. Taylor, F. Kienzle, R. L. Robey, A. McKillop, and J. D. Hunt, *ibid.*, **93**, 4845 (1971).
[3]M. A. Schwartz, B. F. Rose, and B. Vishnuvajjala, *ibid.*, **95**, 612 (1973).
[4]Oxidation of the free amine led to a myriad of products.
[5]D. H. Reid and R. G. Webster, *J.C.S. Chem. Comm.*, 1283 (1972).
[6]M. M. Coombs and M. B. Jones, *Chem. Ind.*, 169 (1972).

**Thallous ethoxide, 2,** 407–411.

*Alkylation of N-heterocyclic compounds.*[1] Pyrrole can be N-alkylated satisfactorily by conversion to pyrrylthallium(I) by reaction with thallous ethoxide. This is a relatively stable solid; the structure is uncertain. Reaction of this salt with an excess of an alkyl halide (or an alkyl tosylate) for 1–32 hr. at 20–60° gives N-alkylpyrroles in good yield. Alkylation of pyrrylmagnesium bromide gives isomeric 2- and 3-alkylpyrroles as the major products.

Carbazole (1), phenothiazine (2), and (to a lesser extent) dibenz[*b,f*]azepine (3) can be N-alkylated in high yields under mild conditions in the following way.[2] The heterocycle and a slight excess of thallous ethoxide dissolved in DMF–ether are stirred at

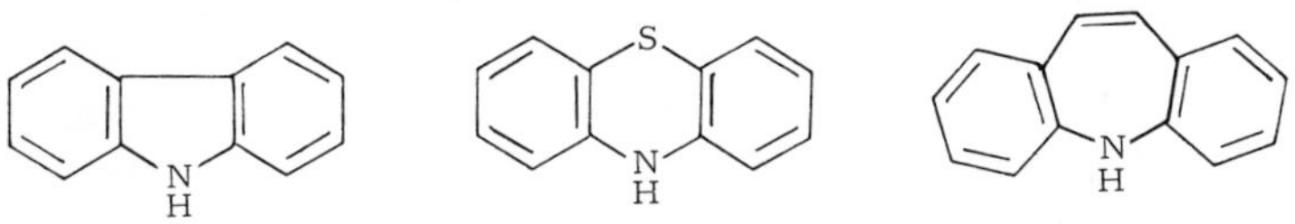

(1)          (2)          (3)

room temperature, and then treated with an excess of an alkyl iodide or bromide; the mixture is warmed briefly on a water bath.

N-Methylation of 2,3-dihydro-1,5-benzoxazepin-4(5*H*)-one (4) was achieved by treatment with thallous ethoxide followed by addition of methyl iodide.[3] However, in

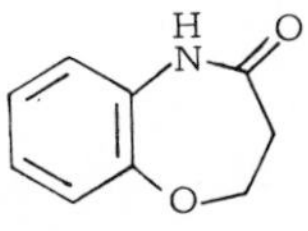

(4)

this case, alkylation was effected in higher yield by addition of sodium hydride in portions to a mixture of (4) and methyl iodide in dioxane. Note that (4) is readily cleaved by bases.

**Alkylation of cyclic 1,3-diketones.**[4] The reaction of cyclic 1,3-diketones with thallium(I) ethoxide, thallium(I) carbonate, or thallium(I) cyclopentadienide gives colorless, crystalline thallium(I) salts (2, 408). These salts can be alkylated in nonpolar solvents (benzene, DME) to give products of O-alkylation in high yield, as illustrated

for the case of dimedone (1). Yields are highest with the least reactive primary alkyl halides. Small quantities of C-dialkylated derivatives are obtained with methyl iodide or benzyl chloride.

[1]C. F. Candy and R. A. Jones, *J. Org.*, **36**, 3993 (1971).
[2]L. J. Kricka and A. Ledwith, *J.C.S. Perkin I*, 2292 (1972).
[3]D. Huckle, I. M. Lockhart, and M. Wright, *ibid.*, 2425 (1972).
[4]M. T. Pizzorno and S. M. Albonico, *Chem. Ind.*, 425 (1972).

**Thioacetamide,** $CH_3\overset{\underset{\|}{S}}{C}-NH_2$, Mol. wt. 75.13, m.p. 111–114°. Suppliers: Aldrich, Eastman, and others.

A general synthesis of thioamides from nitriles which utilizes thioacetamide as the source of hydrogen sulfide under acidic conditions is reported by Taylor and Zoltewicz.[1] The thioamides are formed in good yield and high purity and the method is equally applicable to aliphatic nitriles and to aromatic nitriles containing either electron-releasing, electron-withdrawing, or potentially reducible substituents. One equivalent of the nitrile is heated on a steam bath for 15–30 min. with 2 eq. of thioacetamide in

$$RCN + CH_3\overset{\underset{\|}{S}}{C}-NH_2 \xrightarrow[63-87\%]{HCl} R\overset{\underset{\|}{S}}{C}-NH_2 + CH_3CN$$

dimethylformamide saturated with dry hydrogen chloride. The reaction mixture is reduced to one-fourth its volume by distillation under water-aspirator pressure and aqueous sodium bicarbonate solution is added to neutralize excess acid. Filtration and cooling of the hot solution yields the thioamide. Yields are listed:

| Nitrile | Thioamide | Yield, % | M.P., °C |
|---|---|---|---|
| $p\text{-}O_2NC_6H_4CN$ | $p\text{-}O_2NC_6H_4CSNH_2$ | 83 | 158.5–159.5 |
| $p\text{-}CH_3OC_6H_4CN$ | $p\text{-}CH_3OC_6H_4CSNH_2$ | 87 | 148.5–149.5 |
| $CH_2(CN)_2$ | $CH_2(CSNH_2)_2$ | 63 | 211–212 |
| $NC(CH_2)_4CN$ | $H_2NCS(CH_2)_4CSNH_2$ | 78 | 178.5–179.5 |

The reaction is considered to involve the equilibrium (1), followed by its irreversible displacement to the right by removal of the low-boiling acetonitrile (2).

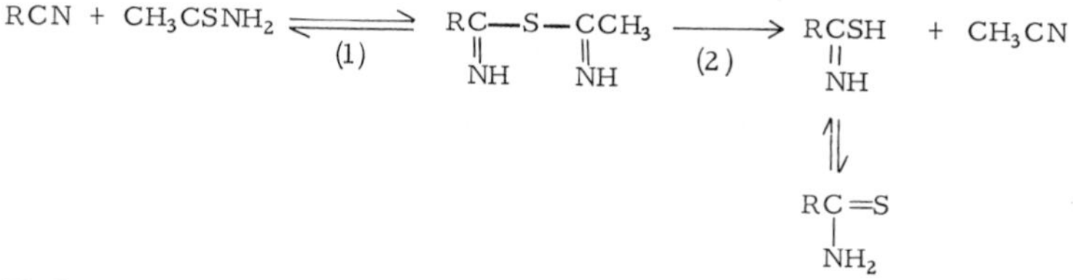

$$RCN + CH_3CSNH_2 \underset{(1)}{\rightleftharpoons} \underset{\overset{|}{NH}}{RC}-S-\underset{\overset{|}{NH}}{CCH_3} \xrightarrow{(2)} \underset{\overset{|}{NH}}{RCSH} + CH_3CN$$

$$RC=S \atop \underset{NH_2}{|}$$

[1]E. C. Taylor and J. A. Zoltewicz, *Am. Soc.*, **82**, 2656 (1960).

**4,4′-Thiobis(6-*t*-butyl-*o*-cresol) ("Ethyl" antioxidant 736).** Mol. wt. 358.5, m.p. 127°, b.p. 312°/40 mm. Solubility (wt. % at 20°)

| | | |
|---|---|---|
| Ethanol . . . . . . . . . . . 39 | Water . . . . . . . . . . . . Insoluble |
| Toluene . . . . . . . . . . . 8.8 | 10% NaOH . . . . . . . . < 0.002 |
| Isopentane . . . . . . . . . 0.2 | |

Supplier: Ethyl Corporation.

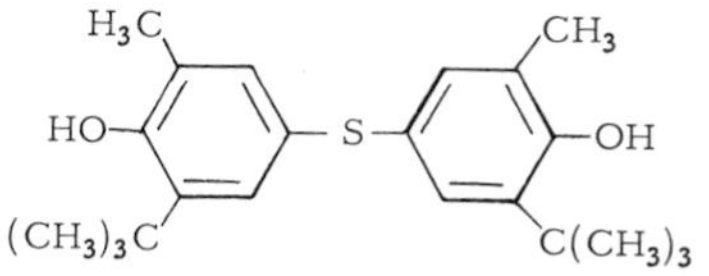

Oxidation inhibitor in natural and synthetic rubbers, polyolefin plastics, resins, adhesives, petroleum oils, and plastics.

**Thiocyanogen, 1,** 1152.

*Improved preparation.*[1] Thiocyanogen can be prepared in 85–90% yield by use of a two-solvent system (water and water-immiscible toluene) for the reaction between sodium thiocyanate and gaseous chlorine. The thiocyanogen formed in the aqueous phase is extracted into the toluene. The thiocyanogen solution can be stored at reduced temperatures.

[1]R. P. Welcher and P. R. Cutrufello, *J. Org.*, **37**, 4478 (1972).

**Thionyl chloride, 1,** 1158–1163; **2,** 412; **3,** 290.

*Isonitrile synthesis.* As the first step in a synthesis of *d*-2-methylbutan-1-al according to Walborsky *et al.*,[1] a 3-l. three-necked, round-bottomed flask fitted with a Hershberg stirrer, a 500-ml. pressure-equalizing addition funnel, and a nitrogen gas inlet tube is

$$\underset{\underset{CH_3}{|}}{\overset{\overset{CH_3}{|}}{CH_3-C}}-CH_2-\underset{\underset{CH_3}{|}}{\overset{\overset{CH_3}{|}}{C}}-NH-C\overset{O}{\underset{H}{\diagdown}} \xrightarrow{SOCl_2} \underset{\underset{CH_3}{|}}{\overset{\overset{CH_3}{|}}{CH_3-C}}-CH_2-\underset{\underset{CH_3}{|}}{\overset{\overset{CH_3}{|}}{C}}-\ddot{N}=\ddot{C}$$

$$\underset{\underset{CH_3}{|}}{\overset{\overset{CH_3}{|}}{CH_3-C}}-CH_2-\underset{\underset{CH_3}{|}}{\overset{\overset{CH_3}{|}}{C}}-\ddot{N}=\ddot{C} \quad + \quad CH_3\underset{\underset{Li}{|}}{CH}CH_2CH_3$$

$$\downarrow$$

$$\underset{\underset{CH_3}{|}}{\overset{\overset{CH_3}{|}}{CH_3-C}}-CH_2-\underset{\underset{CH_3}{|}}{\overset{\overset{CH_3}{|}}{C}}-N=\underset{\underset{\underset{CH_3}{|}}{CHCH_2CH_3}}{\overset{\overset{Li}{|}}{C}} \xrightarrow[\text{2. }H_3O^+]{\text{1. }D_2O} \quad \underset{D}{\overset{O}{\diagup}}\overset{\diagdown}{C}-\underset{\underset{CH_3}{|}}{CH}-CH_2CH_3$$

flame-dried and allowed to cool. The nitrogen inlet is replaced by a low-temperature thermometer, and the nitrogen line is attached to a Y-tube resting in the addition funnel. The flask is charged with 118 g. (0.75 mole) of N-(1,1,3,3-tetramethylbutylformamide) and 1500 ml. of dimethylformamide (b.p. 76–77°/1 mm.). Then 89 g. (55 ml., 0.75 mole) of thionyl chloride is mixed with 250 ml. of dimethylformamide (temperature rise of about 30°) and the mixture is placed in the addition funnel. The flask is immersed in a Dry Ice–acetone bath, and moderately fast stirring is commenced. When the temperature of the flask reaches $-50°$ the solution in the funnel is added at a rate such that the temperature ranges between $-55$ and $-50°$ (about 10 min.). After the addition is complete, the bath is removed momentarily to allow the reaction temperature to rise to $-35°$. The bath is then replaced and 159 g. (1.5 moles) of dry sodium carbonate (dried under vacuum at 130°) is added from a solid addition funnel under nitrogen over a 10-min. period. The mixture is then stirred overnight and poured into a 6-l. Erlenmeyer flask containing 3 l. of ice water (Hood). The reaction flask is rinsed with 300 ml. of pentane and sufficient water to dissolve the inorganic material that may be present, and the washings are added to the flask. The mixture is stirred vigorously for 5 min., and the layers are separated. The upper layer is washed twice with 100-ml. portions of water and then is dried over sodium sulfate. The solution is filtered and the pentane is removed by distillation. The crude tetramethylbutyl isonitrile is distilled through a 1.5 × 15-cm. Vigreux column. The yield of product of b.p. 55.5–56.5°/11 mm., $n_D^{30}$ 1.4178, $d^{25}$ 0.7944, is 86–90 g. (82–87%).

**4-Hydroxybenzenesulfonyl chloride.** Sodium 4-hydroxybenzenesulfonate (1) can be converted into 4-hydroxybenzenesulfonyl chloride (2) in 80–90% yield by treatment with

(1)    $\xrightarrow[80-90\%]{\begin{array}{c}SOCl_2\\DMF\end{array}}$    (2)

thionyl chloride containing a catalytic amount of DMF.[2] Other catalysts (*e.g.*, triphenyl-phosphine, HMPT) are less effective.

Phosphorus pentachloride cannot be used in place of $SOCl_2$; this reagent, however, can be used for preparation of 3,5-disubstituted-4-hydroxybenzenesulfonyl chlorides.[3]

*Acenaphtho[1,2-c]furazane* (2).[4] This furazane (2) can be obtained in 55 % yield by dehydration of acenaphthoquinone dioxime (1) with thionyl chloride in methylene chloride.

(1)                                    (2)

[1] G. E. Niznik, W. H. Morrison, III, and H. M. Walborsky, *Org. Syn.*, **51**, 31 (1971).
[2] R. W. Campbell and H. W. Hill, Jr., *J. Org.*, **38**, 1047 (1973).
[3] W. L. Hall, *ibid.*, **31**, 2672 (1966).
[4] A. J. Boulton and S. S. Mathur, *ibid.*, **38**, 1054 (1973).

## Thiophenol–Azobisisobutyronitrile.

*Isomerization of* **cis-** *to* **trans-*olefins*.** Sgoutas and Kummerow[1] reported in 1969 that *cis*-unsaturated fatty acid esters are isomerized to *trans*-esters when heated at 65° in a sealed vial with thiols or diphenylphosphine in the presence of azobisisobutyronitrile. The equilibrium mixture contains 75–80 % *trans* double bonds. There is no migration of the double bonds. Presumably addition of thiyl or phosphinyl radicals to the double bond is involved.

The method was used recently by Bhalerao and Rapoport.[2] Thus a *cis–trans* mixture of 2,3-dimethyl-3-octene (a, b; *cis–trans* ratio 85:15) when heated in a sealed tube at 65°

a                              b

for 8 hr. with thiophenol and azobisisobutyronitrile is converted into a *cis–trans* mixture in the ratio of 25:75. In this case use of selenium (220°, 1–2 hr.)[3] gave a complex mixture as indicated by glpc and NMR.

[1] D. S. Sgoutas and F. A. Kummerow, *Lipids*, **4**, 283 (1969).
[2] U. T. Bhalerao and H. Rapoport, *Am. Soc.*, **93**, 4835 (1971).
[3] J. C. Stowell, *J. Org.*, **35**, 244 (1970).

**Thiourea dioxide,** $\begin{array}{c} \overset{+}{H_2N} \\ \diagdown \\ \diagup \\ H_2N \end{array} C-SO_2^-$   Mol. wt. 108.13, m.p. 144°.

The reagent, also known as formamidinesulfinic acid, is prepared by oxidation of thiourea with hydrogen peroxide.[1]

**Reduction of ketones.**[2] Ketones are reduced to secondary alcohols by the reagent in yields of 75–100%.

$$\diagup C=O \ + \ \begin{array}{c} \overset{+}{H_2N} \\ \diagdown \\ \diagup \\ H_2N \end{array} C-SO_2^- \ + \ 2\,NaOH \ \longrightarrow \ \diagup CHOH \ + \ \begin{array}{c} H_2N \\ \diagdown \\ \diagup \\ H_2N \end{array} C=O \ + \ Na_2SO_3$$

In a typical experiment, fluorenone (0.01 mole) is dissolved in ethanol; an aqueous solution of sodium hydroxide (0.02 mole) and thiourea dioxide (0.01 mole) is added with stirring, and the mixture heated for 2 hr. at 90°. Fluorenol is obtained in 95.6% yield.

[1]E. B. Barnett, *J. Chem. Soc.*, **97**, 63 (1910).
[2]K. Nakagawa and K. Minami, *Tetrahedron Letters*, 343 (1972).

**Titanium(III) chloride,** TiCl₃, **2**, 415. Mol. wt. 154.27. Suppliers: Alfa Inorganics, Baker, Fisher, Howe and French, MCB, E. Merck, Pfaltz and Bauer, Riedel de Haën, ROC/RIC, Sargent.

**vic-Diols.** *vic*-Diols are obtained by interaction of carboxylic acids and alkyl-lithium compounds in 1,2-dimethoxyethane in the presence of titanium(III) chloride followed by aqueous workup.[1] Thus benzoic acid and methyllithium afford 2,3-diphenyl-butane-2,3-diol (33%), acetophenone (27%), and α-phenylpropiophenone (11%):

$$C_6H_5COOH \ + \ CH_3Li \ \xrightarrow[\text{2) Hydrolysis}]{\text{1) TiCl}_3} \ \underset{\overset{|}{OH}\ \ \overset{|}{OH}}{\overset{\overset{CH_3}{|}\ \ \overset{CH_3}{|}}{C_6H_5C-CC_6H_5}} \ + \ C_6H_5COCH_3 \ + \ \underset{}{\overset{\overset{CH_3}{|}}{C_6H_5CHCOC_6H_5}}$$

|  33% | 27% | 11% |

The diol is obtained as a mixture of the (±)- and *meso*-isomer in the ratio 2.4:1.

**Reduction.** Timms and Wildsmith[2] recently reported that oximes are rapidly reduced by aqueous titanium trichloride[3] to imines, which are readily hydrolyzed to

$$\underset{\overset{|}{NO_2}}{CH_3CH_2CHCH_2CH_2COCH_3} \ \xrightarrow{85\%} \ CH_3CH_2COCH_2CH_2COCH_3$$

(1)                              (2)

carbonyl compounds. Since oximes might be expected to be intermediates in the reduction of nitro compounds, McMurry and Melton[4] examined the reduction of 5-nitro-heptane-2-one (1) with aqueous titanium trichloride[5] and obtained 2,5-heptanedione (2) in 85% yield. Chromous chloride and vanadous chloride also accomplish this transformation, but in considerably lower yield (~25%). The reaction was used in a simple synthesis of *cis*-jasmone (6).

The method was used successfully for conversion of (7) to the aldehyde (9).[6] The precursor (8) is obtained by Michael addition to the steroidal 3-keto-1,4,6-triene of nitromethane in *t*-butanol in the presence of potassium *t*-butoxide.

[1]E. H. Axelrod, *Chem. Commun.*, 451 (1970).
[2]G. H. Timms and E. Wildsmith, *Tetrahedron Letters*, 195 (1971).
[3]A 15% aqueous solution of TiCl₃ containing ZnCl₂ and HCl (B. D. H. Chemicals Ltd.) was used.
[4]J. E. McMurry and J. Melton, *Am. Soc.*, **93**, 5309 (1971).
[5]The TiCl₃ must be taken from a full bottle and handled under an inert atmosphere.
[6]M. Kocór, M. Gumulka, and T. Cynkowski, *Tetrahedron Letters*, 4625 (1972).

**Titanium(IV) chloride, 1,** 1169–1171; **2,** 414–415; **3,** 291.

*Knoevenagel condensation.*[1] The combination of titanium tetrachloride and pyridine in THF is useful for Knoevenagel condensations.[2] For example, ethyl acetoacetate and ethyl nitroacetate undergo condensation readily under these conditions with aliphatic, aromatic, and heterocyclic aldehydes.

Diethyl malonate undergoes condensation with aliphatic and aromatic ketones to give satisfactory yields of α,β-unsaturated compounds.[3]

$$\underset{R^2}{\overset{R^1}{\diagdown}} C=O \ + \ H_2C(COOC_2H_5)_2 \ \xrightarrow[\underset{40-95\%}{}]{\overset{TiCl_4}{Py, \ THF}} \ \underset{R^2}{\overset{R^1}{\diagdown}} C=C(COOC_2H_5)_2$$

*Aromatic chlorination.*[4] Titanium tetrachloride in the presence of an oxidizing agent such as pertrifluoroacetic acid chlorinates a variety of aromatic substrates. Yields are high in the case of activated aromatics such as phenol; benzoic acid and nitrobenzene do not react.

*Selective* ortho *substitution of* o-*hydroxycarbonyl compounds.*[5] Bromination of compounds of the type (1) give exclusively 5-bromo derivatives. However, in the presence of titanium tetrachloride appreciable amounts of 3-bromo products are formed

(1)                    (2)

as well. A complex of type (2) is postulated as responsible for the selective *ortho* substitution. Similar results were obtained on formylation of compounds of type (1) with dichloromethyl methyl ether.

*Oximes* → *nitriles.*[6] Aldoximes are dehydrated to give nitriles by titanium tetra-chloride in the presence of pyridine and THF or dioxane. The reaction proceeds at room temperature in the case of aliphatic aldoximes, but a temperature of 80° is required for aromatic aldoximes. Yields are high (80–97%).

*Amides* → *nitriles.*[7] Primary acid amides are dehydrated to nitriles by treatment with titanium tetrachloride and a base (triethylamine, N-methylmorpholine) in THF at 0°. Yields range from 65 to 95%.

[1] J. R. Johnson, *Org. React.*, **1**, 226, 233 (1942).
[2] W. Lehnert, *Tetrahedron Letters*, 4723 (1970); *Tetrahedron*, **28**, 663 (1972).
[3] *Idem, ibid.*, **29**, 635 (1973).
[4] G. K. Chip and J. S. Grossert, *Canad. J. Chem.*, **50**, 1233 (1972).
[5] T. M. Cresp, M. V. Sargent, and J. A. Slix, *Chem. Commun.*, 214 (1972).
[6] W. Lehnert, *Tetrahedron Letters*, 559 (1971).
[7] *Idem, ibid.*, 1501 (1971).

*p*-Toluenesulfonic acid, 1, 1172–1178.

*Enol acetylation* (**1**, 1174–1178). As a first step in the synthesis of 3-*n*-butyl-2,4-pentanedione,[1] a mixture of 28.6 g. (0.25 mole) of 2-heptanone, 51.0 g. (0.50 mole) of acetic anhydride, and 1.9 g. (0.01 mole) of *p*-toluenesulfonic acid monohydrate contained in a stoppered 500-ml. round-bottomed flask equipped with a magnetic stirrer is stirred at room temperature for 30 min. Then 55 g. (0.43 mole) of the 1:1 boron trifluoride–acetic acid complex [*Reagents*, **1**, 69 (1967)] is added; some heat is evolved

$$\underline{n}\text{-}C_4H_9CH_2COCH_3 \ + \ (CH_3CO)_2O \ \xrightarrow{\underline{p}\text{-}CH_3C_6H_4SO_3H} \ \underline{n}\text{-}C_4H_9CH=\overset{\overset{\displaystyle OCOCH_3}{|}}{C}-CH_3$$

$$\xrightarrow[\ (CH_3CO)_2O\ ]{BF_3 \cdot CH_3COOH} \ \underline{n}\text{-}C_4H_9-\overset{\overset{\displaystyle CH_3}{|}}{\underset{\underset{\displaystyle CH_3}{|}}{C}}\underset{\diagdown}{\overset{\diagup}{\Big\langle}}\overset{C-O}{\underset{C=O}{\diagdown}}BF_2 \ \xrightarrow[\ H_2O, \ \triangle\ ]{CH_3CO_2Na} \ \underline{n}\text{-}C_4H_9CH(COCH_3)_2$$

during this addition. The resulting amber-colored solution is stirred in the stoppered flask at room temperature for 16–20 hr. and then a solution of 136 g. (1.00 mole) of sodium acetate trihydrate in 250 ml. of water is added. After the flask has been fitted with a reflux condenser, the reaction mixture is refluxed for 3 hr. and then cooled, and the product is extracted with three 100-ml. portions of petroleum ether (b.p. 30–60°). The combined extracts are washed with aqueous 5% sodium bicarbonate and with saturated sodium chloride solution. The petroleum ether solution is dried over Drierite, the solvent is removed with a rotary evaporator, and the residual oil is distilled. 3-*n*-Butyl-2,4-pentanedione is collected as a colorless liquid, b.p. 84–86°/6 mm. The yield is 25–30 g. (64–77%).

*Wagner–Meerwein rearrangement.* Treatment of (1) with TsOH in refluxing benzene

(1)                                         (2)

gives (2), formed by two Wagner–Meerwein shifts,[2] in quantitative yield.[3] In the same way, (3, mixture of isomers) is rearranged to (4; 3,4-dimethyltricyclo[3.3.3.0]undecane-2-one).

(3)                    (4)

*Hydrolysis–decarboxylation.* The *Organic Syntheses* procedure[4] for the conversion of diethyl 2,5-diketocyclohexane-1,4-dicarboxylate (1) to cyclohexane-1,4-dione (2) calls for treatment with water in a steel autoclave at 185–195° for 10–15 min. (yield 72–80%). The reaction can be carried out more conveniently and in higher yield by refluxing in aqueous ethylene glycol containing some *p*-toluenesulfonic acid.[5]

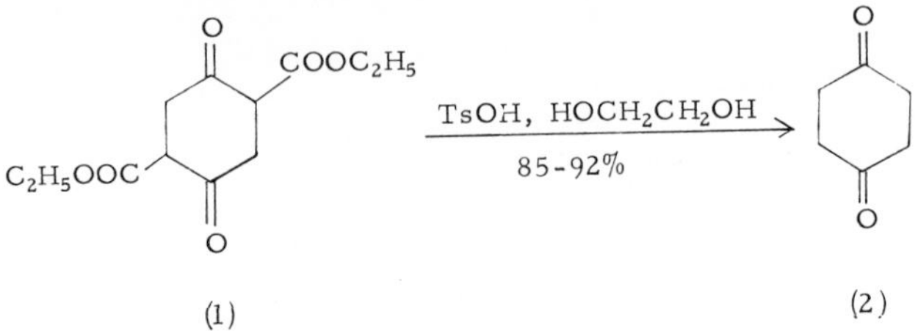

(1)                                         (2)

[1]C.-L. Mao and C. R. Hauser, *Org. Syn.*, **51**, 90 (1971).
[2]F. D. Popp and W. E. McEwen, *Chem. Rev.*, **58**, 375 (1958).
[3]N. P. Peet, R. L. Cargill, and D. F. Bushey, *J. Org.*, **38**, 1218 (1973).
[4]A. T. Nielsen and W. R. Carpenter, *Org. Syn.*, **45**, 25 (1965).
[5]S. A. Patwardhan and S. Dev, *Synthesis*, 427 (1971).

*p*-Toluenesulfonyl azide (Tosyl azide), **1**, 1178–1179; **2**, 415–417; **3**, 291–292. Additional supplier: WBL.

*Diazo group transfer.* Regitz[1] has reviewed diazo group transfer to active methylene compounds using tosyl azide.

*Reaction with tetramethoxyethylene.* In a reaction that is first order with respect to each of the reactants, tetramethoxyethylene (1) and tosyl azide (2) yield the imidocarbonate (3). These data, as well as cross-over experiments with isotopically labeled (1), exclude any dissociation of (1) into two molecules of dimethoxycarbene under the conditions applied.[2]

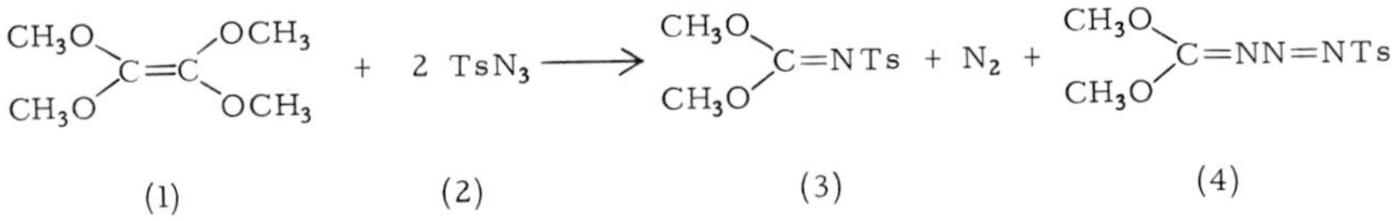

(1)              (2)           (3)              (4)

*α-Diazoketones* (**3**, 291–292). The procedure for the preparation of 2-diazocyclohexanone has been published.[3]

[1]M. Regitz, *Synthesis*, 351 (1972).
[2]R. W. Hoffmann, U. Bressel, J. Gehlhaus, H. Häuser, and G. Mühl, *Ber.*, **104**, 2611 (1971).
[3]M. Regitz, J. Rüter, and A. Liedhegener, *Org. Syn.*, **51**, 86 (1971).

*p*-Toluenesulfonyl chloride, **1**, 1179–1185; **3**, 292.

*Alkyl nitrones → N-alkylamides.*[1] The methyl nitrones (1) derived from $\Delta^4$-3-ketosteroids on treatment with *p*-toluenesulfonyl chloride in pyridine rearrange to

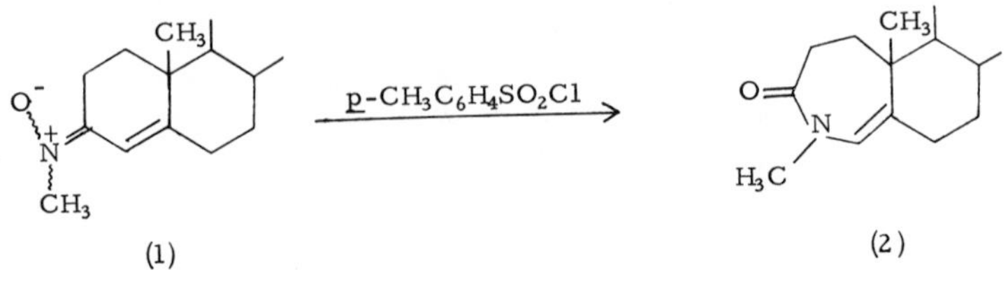

(1)                                         (2)

A-aza-A-homosteroids (2). The reaction is probably initiated by *p*-toluenesulfonylation of the nitrone followed by capture of a nucleophile at $C_3$. The resulting hydroxylamine O-tosylate then rearranges to (2). The rearrangement provides an alternative to the Beckmann rearrangement.

[1]D. H. R. Barton, M. J. Day, R. H. Hesse, and M. M. Pechet, *Chem. Commun.*, 945 (1971).

### *p*-Toluenesulfonylhydrazine, **1**, 1185–1187; **2**, 417–423; **3**, 293.

*New olefin synthesis* (**2**, 418–419). Shapiro and Duncan[1] report a detailed procedure for the preparation of *p*-toluenesulfonylhydrazine and its use for the synthesis of 2-bornene.

*Reaction with 2-cyclopentene-1-one and 2-cyclohexene-1-one*[2] :

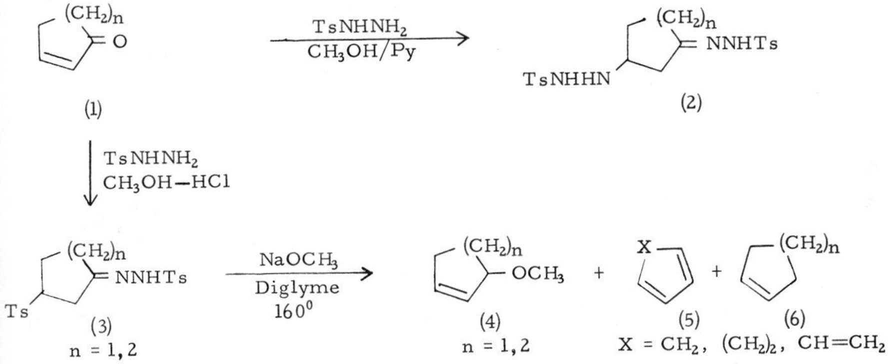

Carbenes are suggested as intermediates in the thermal reaction of (3) with sodium methoxide at 160°.

*Pyrolysis of lactone tosylhydrazone salts.* Agosta *et al.*[3] have developed a process that is formally a reversal of the Baeyer–Villiger oxidation, that is, a method for conversion of lactones into cyclic ketones. An example is the synthesis of the spiro[3.4]-octane-1-one (5). The starting material is the lactone (1); this is converted into the

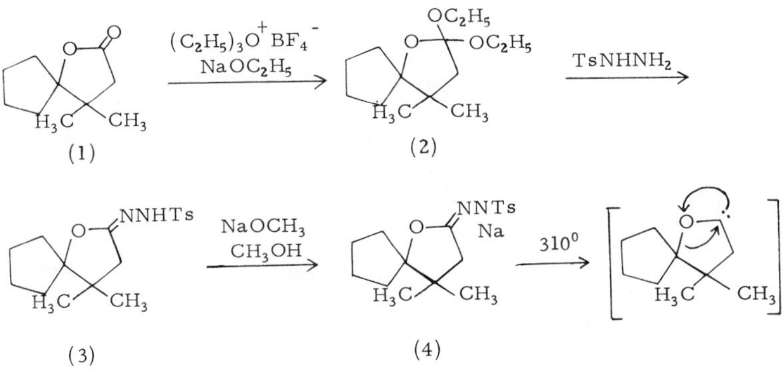

(5)

related *ortho*-lactone (2) by treatment with triethyloxonium fluoroborate (**1**, 1210–1212; **2**, 430–431; **3**, 303) and then with ethanol containing sodium ethoxide. Treatment of (2) with tosylhydrazine affords the lactone tosylhydrazone (3). This is converted into the sodium salt (4) by either sodium methoxide in methanol or sodium hydride in diethylene glycol diethyl ether. The dry sodium salt is then pyrolyzed at 310° to give the cyclo-butanone (5) in 42 % yield. This transformation presumably involves thermal elimination of nitrogen and *p*-toluenesulfinate anion to give an oxycarbene, which then undergoes rearrangement to (5).

Note that (5) can also be obtained in 35 % yield by irradiation for 125 min. of 1-cyclopentenyl isobutyl ketone (6).[4]

(6)

Foster and Agosta[5] have recently discussed possible mechanisms for the conversion of oxycarbenes to ring-contracted ketones.

**Ketones → Nitriles.**[6] Ketones can be converted into nitriles by way of the tosyl-hydrazones. Thus the tosylhydrazone of cyclohexanone (1) on treatment with potassium cyanide in methanol–acetic acid at room temperature followed by addition of 2 N

(1)                    (2)                    (3)

potassium hydroxide is converted into (2) in 85 % yield. When (2) is heated at 180° for 2 hr., it decomposes to give a nitrile (3) and N₂ and TsH in about 60 % yield.

[1] R. H. Shapiro and J. H. Duncan, *Org. Syn.*, **51**, 66 (1971).

[2] W. Kirmse and L. Ruetz, *Ann.*, **726**, 30 (1969); see also, *Idem, ibid.*, **726**, 36 (1968); W. Kirmse and G. Müncher, *ibid.*, **726**, 42 (1969).

[3] A. B. Smith, III, A. M. Roster, and W. C. Agosta, *Am. Soc.*, **94**, 5100 (1972); A. M. Foster and W. C. Agosta, *ibid.*, **94**, 5777 (1972).

[4] Irradiation was carried out with a Hanovia Model L mercury lamp in a quartz immersion well and a Corning uranium glass filter.

[5] A. M. Foster and W. C. Agosta, *Am. Soc.*, **95**, 608 (1973).

[6] S. Cacchi, L. Caglioti, and G. Paolucci, *Chem. Ind.*, 213 (1972).

***p*-Tolylsulfinylcarbanion lithium salt** (2). Mol. wt. 144.16.

*Preparation.* *p*-Tolylsulfinylcarbanion salt (2) is generated quantitatively from methyl *p*-tolyl sulfoxide (1) by lithium diethylamide (**1**, 610–611; **2**, 247–248) in THF; in this case *n*-butyllithium was less useful.

$$CH_3SC_6H_4CH_3(\underline{p}) \xrightarrow{\text{LiN}(C_2H_5)_2} Li^+\bar{C}H_2\overset{O}{\underset{\|}{S}}C_6H_4CH_3(\underline{p})$$

$$(1) \hspace{4cm} (2)$$

*Synthesis of optically active alcohols.*[1] For asymmetric synthesis of alcohols, the carbanion salt (2) is generated from optically active (R)-methyl *p*-tolyl sulfoxide (1).[2] The optically active carbanion reacts with benzaldehyde (3) to give a 1:1 diastereomeric mixture of 2-hydroxy-2-phenylethyl *p*-tolyl sulfoxides (4a) and (4b) in 84% yield. These can be separated by silica gel chromatography and fractional crystallizations to give (4a, $\alpha_D$ + 91.7°, 17% yield) and (4b, $\alpha_D$ + 202.8°, 15.5% yield). Raney nickel desulfurization of (4a) and (4b) gives (S)-(−)-1-phenylethanol (5a, $\alpha_D$ − 42.6°) and R-(+)-1-phenylethanol (5b, $\alpha_D$ +42.1°), respectively, in about 60% yield. Since the specific rotation of optically pure (5b) is $\alpha_D$ +43.5°, alcohols of high optical purity can be obtained in this way.

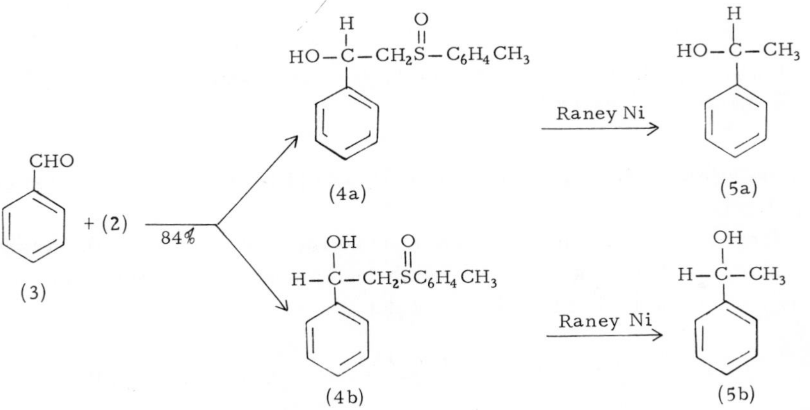

Reaction of optically active (2) with α-tetralone (6) gave a quantitative yield of a (1.8:1) diastereomeric mixture of 1-hydroxy-1-(*p*-tolylsulfinylmethyl)-1,2,3,4-tetra-hydronaphthalenes (7). The major isomer, $\alpha_D$ +77.6°, was isolated in 45.5% yield by chromatography. Desulfurization with Raney nickel gave (−)-1-hydroxy-1-methyl-1,2,3,4-tetrahydronaphthalene (8, $\alpha_D$ −31.0°) in 68% yield.

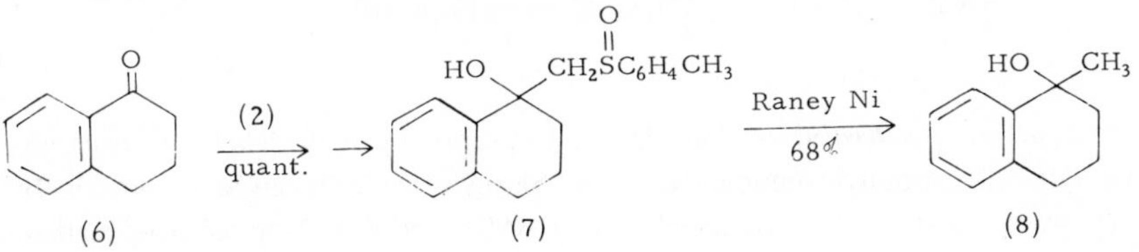

The reaction of optically active (2) with 1,2-epoxycyclohexane (9) in DME (reflux) led to a (1.5–2:1) mixture of (10a) and (10b) in 63.5% yield. These were separated by chromatography followed by fractional crystallization to give (10a, $\alpha_D$ +118.2°, 13.5% yield) and (10b, $\alpha_D$ + 225.3°, 13% yield). Desulfurization gave (R,R)-(−)-*trans*-2-methylcyclohexanol (11a) in 62% yield and (S,S)-(+)-*trans*-2-methylcyclo-hexanol (11b, $\alpha_D$ +40.7°, 77% yield). Since the specific rotation of pure (11b) is reported to be $\alpha_D$ + 42.9°,[2] alcohols of high optical purity can be achieved by use of carbanion (2).

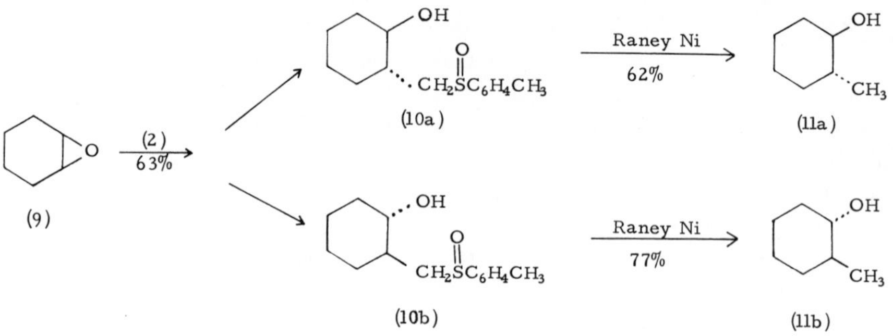

[1]G. Tsuchihashi, S. Iriuchijima, and M. Ishibashi, *Tetrahedron Letters*, 4605 (1972).
[2]K. K. Andersen, *ibid.*, 93 (1962).

**Tosylmethylisocyanide (TosMIC)**, $p\text{-}CH_3C_6H_4SO_2CH_2N{=}C$ (1). Mol. wt. 195.24, m.p. 116–117°.

*Preparation.* Dutch chemists[1] have reported two new syntheses of the reagent (equations I and II). Route II gives slightly higher yields of (1); route I, however, avoids use of evil-smelling volatile isocyanides.

$$\text{I.}\quad p\text{-}CH_3C_6H_4SCH_2NHCHO \xrightarrow{\ H_2O_2/Ac_2O\ } p\text{-}CH_3C_6H_4SO_2CH_2NHCHO$$

$$\xrightarrow[\substack{glyme,\ 0\text{-}10^0 \\ 80\%}]{\substack{POCl_3 \\ N(C_2H_5)_3}} p\text{-}CH_3C_6H_4SO_2CH_2N{=}C$$

$$(1)$$

$$\text{II.}\quad p\text{-}CH_3C_6H_4SO_2F \ + \ \underset{\underset{Li}{|}}{CH_2N{=}C} \xrightarrow{\ 87\%\ } (1)$$

*Synthesis of heterocycles.* Oxazoles can be prepared in high yield by the reaction of TosMIC with carbonyl compounds.[2] Thus 5-phenyloxazole (2) can be obtained in 91% yield by the reaction of benzaldehyde with TosMIC and $K_2CO_3$ in refluxing methanol

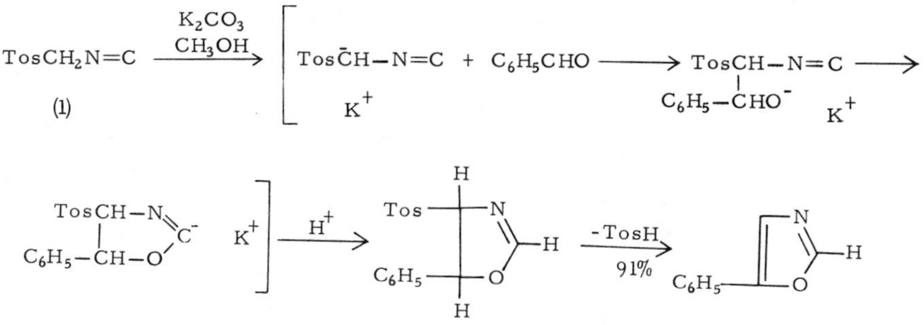

$$TosCH_2N{=}C \xrightarrow[CH_3OH]{K_2CO_3} \left[ Tos\bar{C}H{-}N{=}C \atop K^+ \right. + C_6H_5CHO \longrightarrow \left. TosCH{-}N{=}C \atop C_6H_5{-}\bar{C}HO \; K^+ \right] \longrightarrow$$

(1)

$$\left[ TosCH{-}N_{\diagdown C^-}^{\diagup} \atop C_6H_5{-}CH{-}O \; K^+ \right] \xrightarrow{H^+} \underset{\underset{H}{\overset{C_6H_5}{}}}{\overset{H}{\overset{Tos{-}N}{}}}{\diagdown}H \xrightarrow[91\%]{-TosH} \underset{C_6H_5{-}}{\overset{N}{}}{\diagdown}H$$

(2)

(2 hr.). If the reaction is carried out at 20°, the elimination of toluenesulfinic acid (TosH) can be avoided.

Acid chlorides or acid anhydrides can be used in place of an aldehyde. In this case, the oxazoles (3) are substituted by a tosyl substituent at the 4-position.

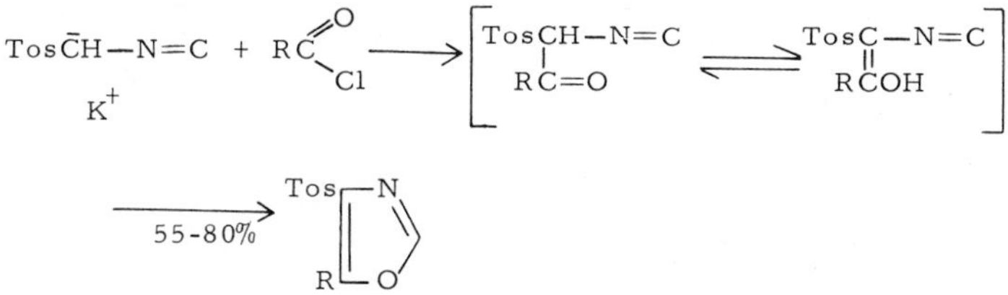

$$Tos\bar{C}H{-}N{=}C \atop K^+ + R\overset{O}{\underset{Cl}{C}} \longrightarrow \left[ TosCH{-}N{=}C \atop RC{=}O \rightleftharpoons TosC{-}N{=}C \atop RCOH \right]$$

$$\xrightarrow{55-80\%} \underset{R}{\overset{Tos{-}N}{}}{\diagdown}O$$

(3)

Tosyl-substituted imidazoles (5)[3] are available by the reaction of TosMIC (1) with imidoyl chlorides (4).[4] A mixture of 1:1 TosMIC and an imidoyl chloride (4) in glyme or THF is added to a stirred suspension of NaH in DMSO (20°, N₂).

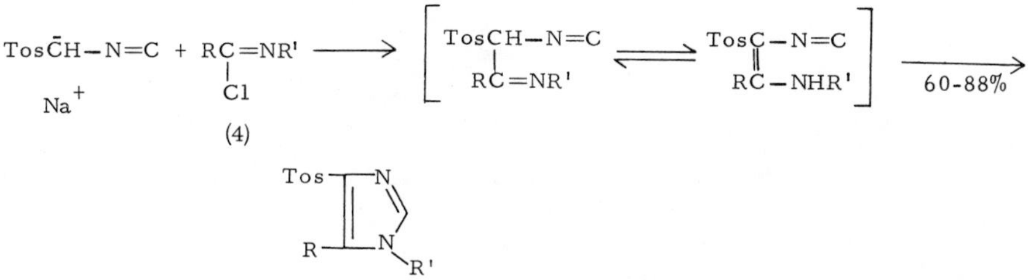

$$Tos\bar{C}H{-}N{=}C \atop Na^+ + RC{=}NR' \atop Cl \longrightarrow \left[ TosCH{-}N{=}C \atop RC{=}NR' \rightleftharpoons TosC{-}N{=}C \atop RC{-}NHR' \right] \xrightarrow{60-88\%}$$

(4)

$$\underset{R}{\overset{Tos{-}N}{}}\underset{N}{\diagdown}{}_{R'}$$

(5)

**p-Tolylthiomethylisocyanide.** The related reagent, p-tolylthiomethylisocyanide, $p\text{-}CH_3C_6H_4SCH_2N{=}C$ (1), mol. wt. 163.24, b.p. 60°/0.001 mm., has been prepared[5,6] by dehydration of the formamide (2) with $POCl_3$ in glyme and triethylamine.

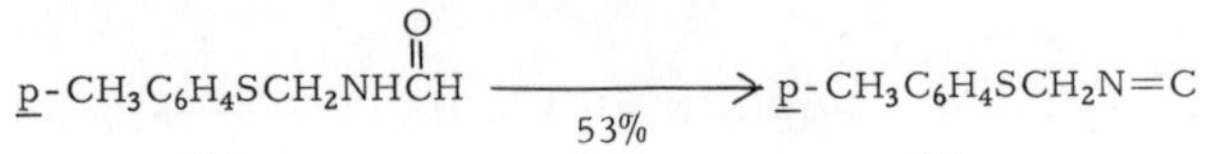

$$\underline{p}\text{-}CH_3C_6H_4SCH_2NH\overset{\overset{\displaystyle O}{\|}}{C}H \xrightarrow{\quad 53\% \quad} \underline{p}\text{-}CH_3C_6H_4SCH_2N=C$$

(2)                                    (1)

α-Lithio-*p*-tolylthiomethylisocyanide (3) prepared from (1) with *n*-BuLi in THF at −70° reacts with acetic anhydride in THF at temperatures rising from −70° to 20° (2 hr.) to give 5-methyl-4-*p*-tolylthiooxazole (4) in about 30% yield.

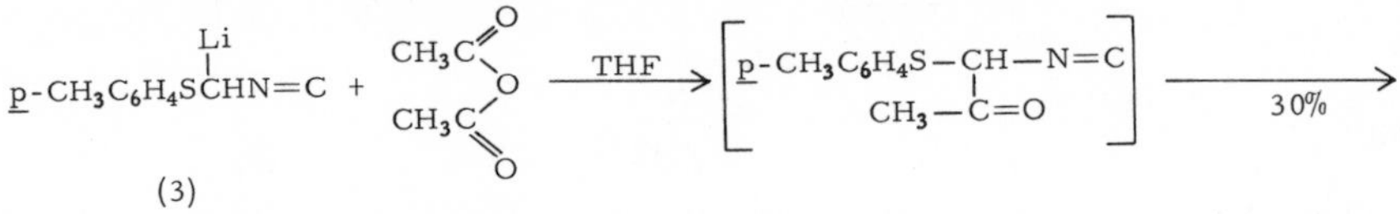

(3)

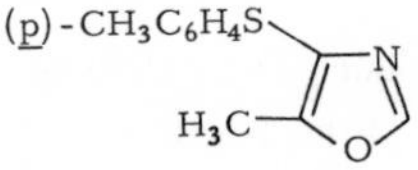

(4)

In the same way, (3) reacts with a ketone, for example acetone, to form the oxazoline

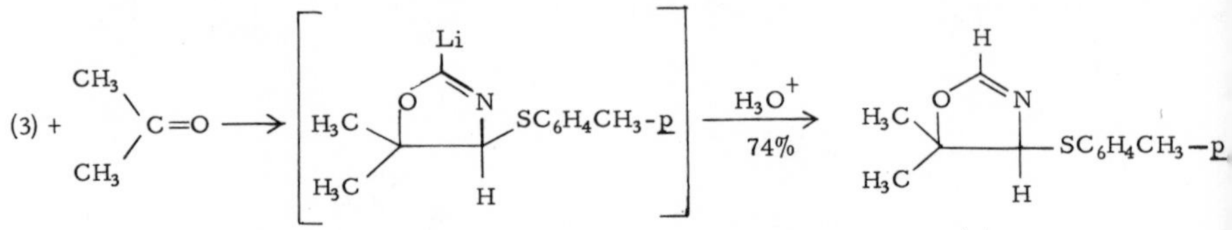

(5)

(5) in 74% yield. Reaction of (3) with methyl formate gives 4-*p*-tolylthiooxazole (6) in 52% yield.

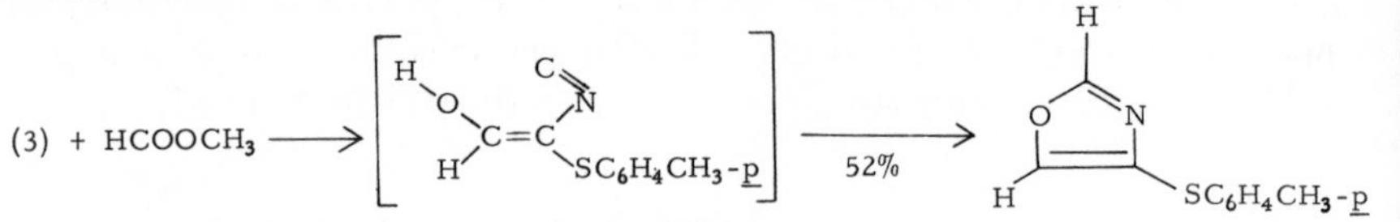

(6)

[1] A. M. van Leusen, G. J. M. Boerma, R. B. Helmholt, H. Siderius, and J. Strating, *Tetrahedron Letters*, 2367 (1972).
[2] A. M. van Leusen, B. E. Hoogenboom, and H. Siderius, *ibid.*, 2369 (1972).
[3] A. M. van Leusen and O. H. Oldenziel, *ibid.*, 2373 (1972).
[4] F. Cramer and U. Baer, *Ber.*, **93**, 1231 (1960).
[5] A. M. van Leusen and H. E. van Gennep, *Tetrahedron Letters*, 627 (1973).
[6] U. Schöllkopf and E. Blume, *ibid.*, 629 (1973).

**Tri-*n*-butylphosphine–Copper(I) iodide complex**, (*n*-Bu)$_3$PCuI. Mol. wt. 221.20, m.p. 75°.

*Preparation.*[1] The reagent is prepared in 96.7% yield by the reaction of copper(I) iodide (excess) with freshly distilled tri-*n*-butylphosphine.

*Synthesis of prostanoic acid skeleton.*[2] A key step in a total synthesis of prostaglandin E$_1$ (4, PGE$_1$) involved the condensation of (1) with 2 molar eq. of (2) in the presence of 1 molar eq. of tri-*n*-butylphosphine–copper(I) iodide complex.

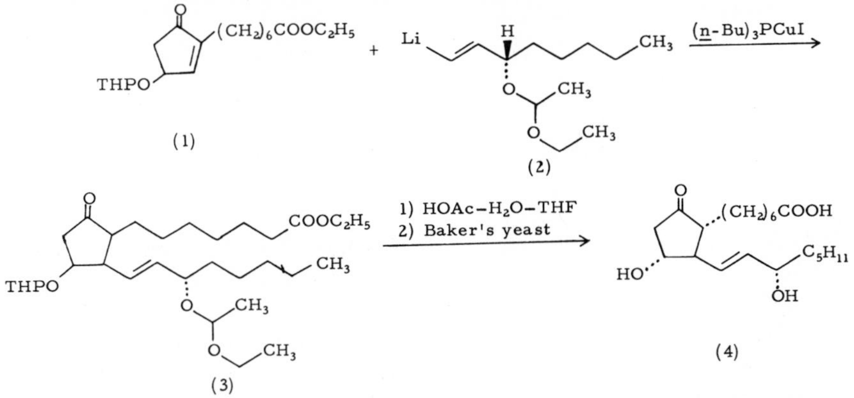

THPO—(CH$_2$)$_6$COOC$_2$H$_5$  (1)   +   Li—[...] H, CH$_3$, O—CH$_3$, O—CH$_3$  (2)   $\xrightarrow{(\underline{n}\text{-Bu})_3\text{PCuI}}$

THPO—[cyclopentanone]—COOC$_2$H$_5$, CH$_3$, O—CH$_3$, O—CH$_3$  (3)

$\xrightarrow[\text{2) Baker's yeast}]{\text{1) HOAc–H}_2\text{O–THF}}$

HO—[cyclopentanone]—(CH$_2$)$_6$COOH, C$_5$H$_{11}$, ÔH  (4)

Schaub and Weiss[3] effected conjugate addition of 3-*t*-butoxy-1-octylmagnesium bromide (6) to the cyclopentenone (5) in 35% yield by use of this complex to give, after deblocking with trifluoroacetic acid and saponification, the ketone (8). Reduction of (8) with lithium perhydro-9b-borophenalyl hydride (this volume) gave *rac*-11-desoxy-13-dihydroprostaglandin E$_1$.

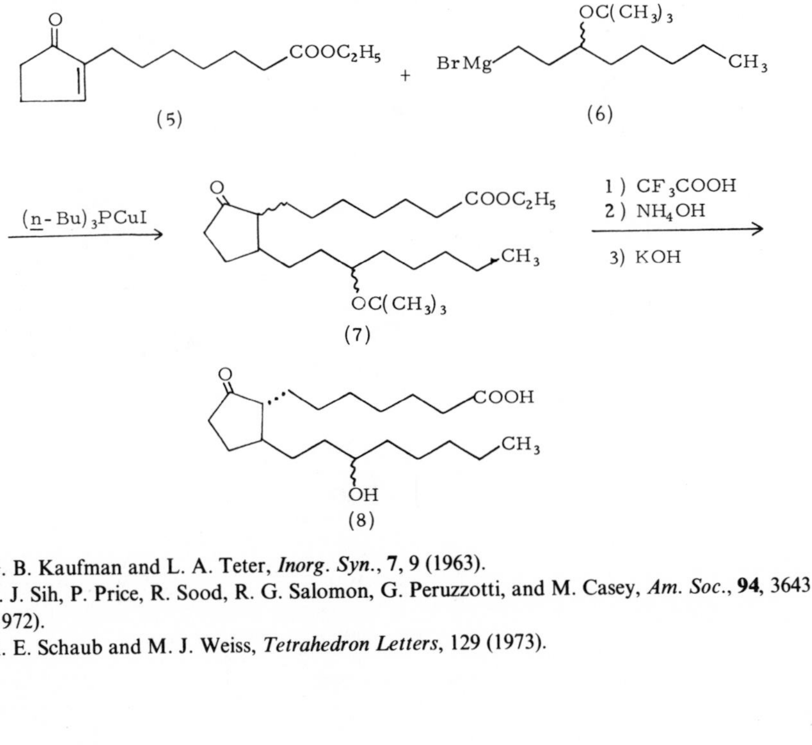

[cyclopentenone]—COOC$_2$H$_5$  (5)   +   BrMg—[...]—CH$_3$, OC(CH$_3$)$_3$  (6)

$\xrightarrow{(\underline{n}\text{-Bu})_3\text{PCuI}}$

[cyclopentanone]—COOC$_2$H$_5$, CH$_3$, ÓC(CH$_3$)$_3$  (7)

$\xrightarrow[\text{3) KOH}]{\substack{\text{1) CF}_3\text{COOH} \\ \text{2) NH}_4\text{OH}}}$

[cyclopentanone]—COOH, CH$_3$, ÔH  (8)

[1]G. B. Kaufman and L. A. Teter, *Inorg. Syn.*, **7**, 9 (1963).
[2]C. J. Sih, P. Price, R. Sood, R. G. Salomon, G. Peruzzotti, and M. Casey, *Am. Soc.*, **94**, 3643 (1972).
[3]R. E. Schaub and M. J. Weiss, *Tetrahedron Letters*, 129 (1973).

**Tri-*n*-butyltin hydride, 1,** 1192–1193; **2,** 424; **3,** 294.

*Reduction of halides under radical conditions.* Reduction of halides with tri-*n*-butyltin hydride under radical conditions is considered to proceed by the following chain reaction[1]:

$$\text{Initiator} + \text{Bu}_3\text{SnH} \longrightarrow \text{Bu}_3\text{Sn} \cdot$$

$$\text{Bu}_3\text{Sn} \cdot + \text{RX} \longrightarrow \text{R} \cdot + \text{Bu}_3\text{SnX}$$

$$\text{R} \cdot + \text{Bu}_3\text{SnH} \longrightarrow \text{RH} + \text{Bu}_3\text{Sn} \cdot$$

$$\text{R} \cdot + \text{Bu}_3\text{Sn} \cdot \longrightarrow \text{termination}$$

If the intermediate radical R· carries a hydrogen at the $\beta$-position, elimination of HX is often observed. Thus treatment of 2,3-dibromobutane with 2 eq. of Bu$_3$SnH gives butene-2 in 99 % yield and only 1 % of butane.[2]

Löffler[3] has examined the reduction of the adduct (2) of hydrogen bromide with bullvalene (1) with tri-*n*-butyltin hydride in benzene (80°) with azobisisobutyronitrile (AIBN, **1,** 45) as initiator. The reaction leads to five products: (3)–(7).

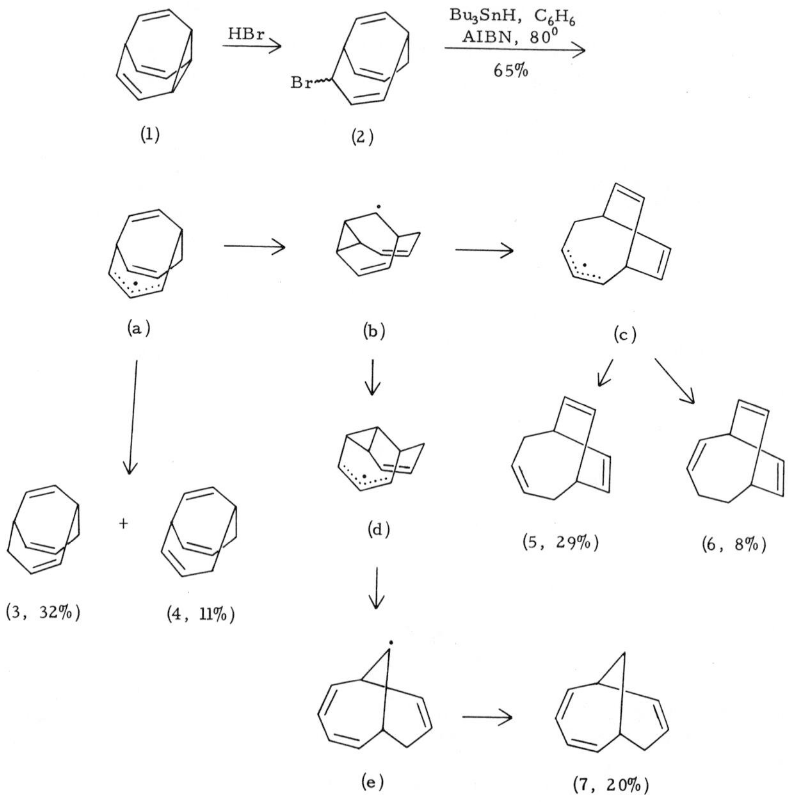

(1)    (2)

(a)    (b)    (c)

(3, 32%)    (4, 11%)    (d)

(5, 29%)    (6, 8%)

(e)    (7, 20%)

The products (5), (6), and (7) are believed to be formed by rearrangement of intermediate radicals (a), (b), and (c).

The reaction of the dibromide of bullvalene (8)[4] with $Bu_3SnH$ also gives an array of

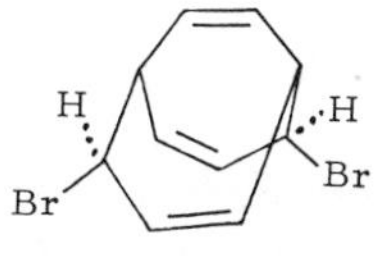

(8)                                     (9)

products. In this case the product of elimination (9) is formed in high yield (82%) with an increase in temperature and with decreasing concentration of the hydride reagent.

On the other hand, reaction of the dibromide (11) of semibullvalene (10)[5] with tri-*n*-butyltin hydride is relatively simple and gives as the main products the two dienes (12) and (13) in a ratio of 3:1.[6]

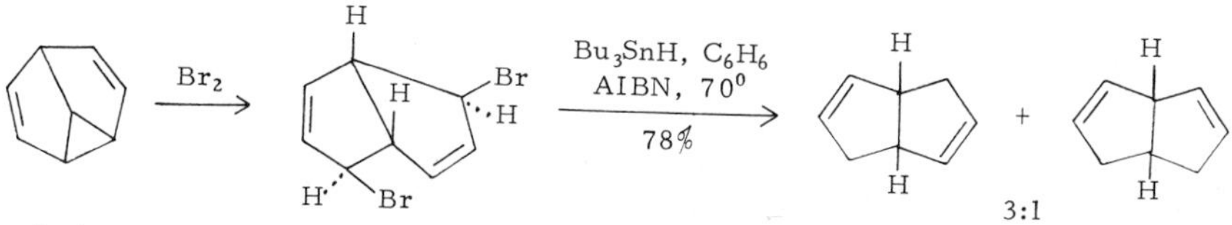

(10)              (11)                          (12)          (13)
                                                        3:1

*Cyclopropenone* (1). Breslow and Oda[7] have effected the preparation and isolation of pure recrystallized cyclopropenone (1, m.p. −29 to −28°) by reduction of tetrachlorocyclopropene in paraffin oil with tri-*n*-butyltin hydride under argon. A mixture

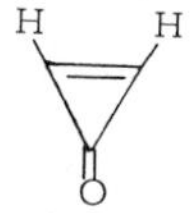

(1)

of trichloro-, dichloro-, and 3-chlorocyclopropene is obtained from which cyclopropenone can be obtained by treatment with $NaHCO_3$ and then with $Na_2SO_4$. The overall yield of the ketone is 41–46%.

**Ether synthesis from olefins and alcohols.** Grady and Chokski[8] have reported a two-step synthesis of ethers. An olefin (1) is treated with an anhydrous alcohol in the presence of NBS to give a bromo ether (2)[9] in about 50–60% yield. The bromo ether is then reduced with tri-*n*-butyltin hydride to give the ether (3) in yields of 70–90%.

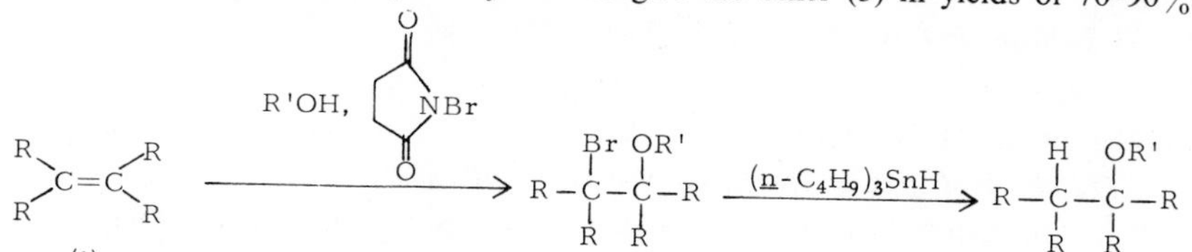

(1)                              (2)                        (3)

The reducing agent was prepared by the procedure of Grady and Kuivala (**3**, 294, ref. 5). This new synthesis of ethers avoids the strongly basic conditions of the classical Williamson synthesis.

*Intramolecular cyclization of unsaturated acyl chlorides.* Tributyltin hydride reduces acyl chlorides to aldehydes by a free-radical chain reaction involving an acyl radical (**1**, 1193). Čeković[10] now finds that treatment of acyl chlorides with a double bond in the 5- or 6-position with tributyltin hydride (azobisisobutyronitrile initiation) gives cyclohexanone derivatives. Thus 5-hexenoyl chloride (1) is converted into cyclohexanone (2), and citronelloyl chloride (3) is converted into menthone (4).

(1)    36%→    (2)

(3)    43%→    (4)

[1]H. G. Kuivila, *Accts. Chem. Res.*, **1**, 299 (1968).
[2]R. J. Strunk, P. M. DiGiacomo, K. Aso, and H. G. Kuivila, *Am. Soc.*, **92**, 2849 (1970).
[3]H.-P. Loffler, *Ber.*, **104**, 1981 (1971).
[4]Note that dibromination of bullvalene (1) proceeds stereoselectively to give the less stable *trans*-dibromide formed by *trans*-1,4-addition.
[5]Note that bromination of (10) proceeds stereoselectively by *cis,exo*-1,4-addition.
[6]L. A. Paquette, G. H. Birnberg, J. Clardy, and B. Parkinson, *J.C.S. Chem. Comm.*, 129 (1973).
[7]R. Breslow and M. Oda, *Am. Soc.*, **94**, 4787 (1972).
[8]G. L. Grady and S. K. Chokski, *Synthesis*, 483 (1972).
[9]K. L. Erickson and K. Kim, *J. Org.*, **36**, 2915 (1971).
[10]Z. Čeković, *Tetrahedron Letters*, 749 (1972).

**Trichloramine, 1,** 1193–1194; **2,** 424–425; **3,** 295.

   **vic-*Dichlorides.*** For a definitive paper on the chlorination of alkenes with trichloramine, see Field and Kovacic.[1]

[1]K. W. Field and P. Kovacic, *J. Org.*, **36**, 3566 (1971).

**Trichloroacetic acid, 1,** 1194; **2,** 425.

   *Hydrolysis of acetals of cyclopropanecarboxaldehyde.*[1] In a recent synthesis of cyclopropanecarboxaldehyde (2), the last step involved hydrolysis of acetals (1) of the aldehyde. Use of inorganic acids gave poor results. Best results were obtained by use of

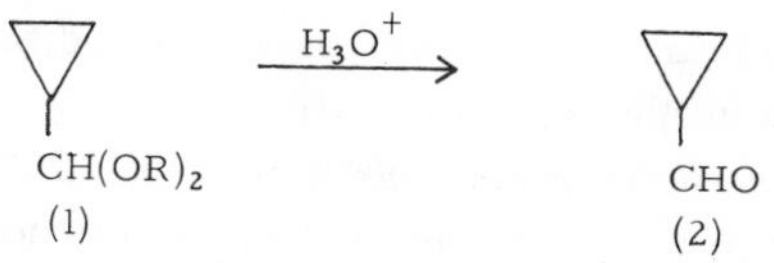

$$\text{CH(OR)}_2 \xrightarrow{\text{H}_3\text{O}^+} \text{CHO}$$

(1)                                         (2)

sufficient 1 $N$ solutions of trichloroacetic acid to form a homogeneous solution upon stirring.

[1]J. M. Stewart, C. Carlisle, K. Kem, and G. Lee, *J. Org.*, **35**, 2040 (1970).

**Trichloroacetyl chloride**, $CCl_3COCl$. Mol. wt. 181.85, b.p. 114–116°, $n_D^{20}$ 1.4689. Suppliers: Aldrich, Baker, Eastman, Fisher.

*Friedel-Crafts acetylation.*[1] The first step in the synthesis of ethyl pyrrole-2-carboxylate proceeds readily in ether solution without catalyst. Thus addition of 77 g. of

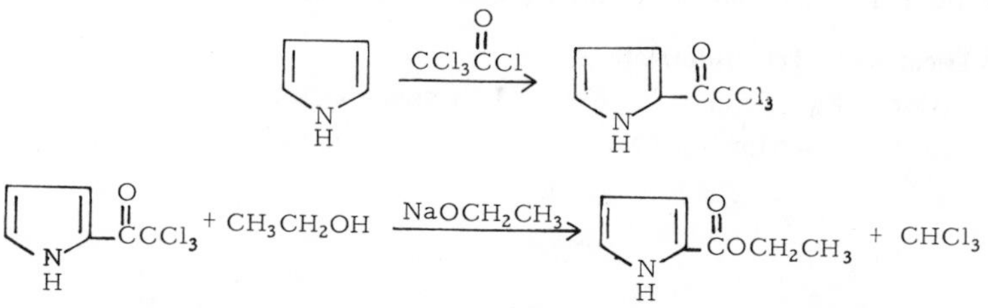

pyrrole in 640 ml. of anhydrous ether over 3 hr. to a stirred solution of 225 g. of trichloroacetyl chloride in 200 ml. of anhydrous ether develops enough heat to keep the mixture refluxing during the addition. The mixture is stirred for 1 hr., and then 100 g. (0.72 mole) of potassium carbonate in 300 ml. of water is added slowly. The organic phase is dried with magnesium sulfate, treated with 6 g. of Norit, and filtered. The solvent is removed by distillation on the steam bath, and the residue is dissolved in 225 ml. of hexane. The dark solution is cooled in ice, and the product crystallizes. The tan solid is collected and washed with 100 ml. of cold hexane, giving 189–196 g. (77–80%) of pyrrol-2-yl-trichloromethyl ketone, m.p. 73–75°.

[1]D. M. Bailey, R. E. Johnson, and N. F. Albertson, *Org. Syn.*, **51**, 100 (1971).

**Trichloroethanol, 3**, 295–296.

*Protection of aldehydes and ketones.*[1] Treatment of a dimethyl or diethyl acetal with 1.5 eq. of trichloroethanol in refluxing benzene under acid catalysis (*p*-TsOH) gives the mixed acetal; use of 4 eq. of the alcohol gives the di-2,2,2-trichloroethyl acetal.

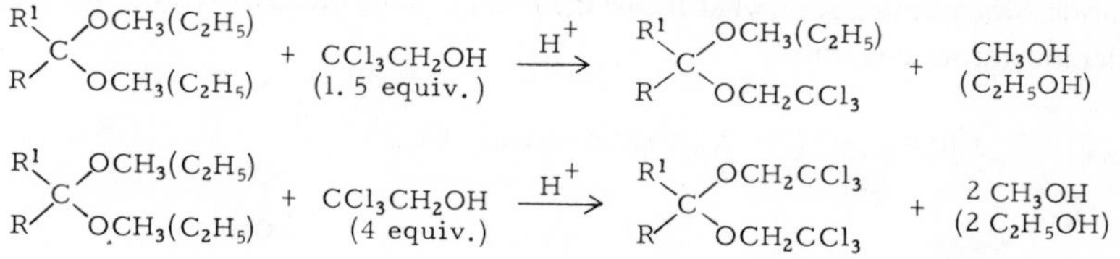

The carbonyl compound can be regenerated under nonacidic conditions by treatment with activated zinc dust in ethyl acetate or THF.

*Carboxyl protection during peptide synthesis.*[2] The 2,2,2-trichloroethyl esters of N-carbobenzoxyamino acids can be prepared by the reaction of the corresponding acid chlorides with trichloroethanol; the carbobenzoxy group is removed selectively with HBr–HOAc. The resulting esters are then coupled with N-carbobenzoxyamino acids or peptides with DCC in acetonitrile to give N-carbobenzoxypeptide trichloroethyl esters. The protecting group is removed selectively by reduction with zinc in acetic acid. Yields are fair to good (63–83%). Complete absence of isomerization remains to be proved, but the optical activity of the N-carbobenzoxypeptides agrees with values reported in the literature.

[1] J. L. Isidor and R. M. Carlson, *J. Org.*, **38**, 554 (1973).
[2] B. Marinier, Y. C. Kim, and J.-M. Navarre, *Canad. J. Chem.*, **51**, 208 (1973).

**β,β,β-Trichloroethyl chloroformate, 2, 426.**

A mixture of *endo*- and *exo*-bicyclo[3.3.1]nonanol-2-one-9, (1) and (2), can be separated by conversion to the trichloroethoxycarbonyl derivatives, separable by a

(1)                    (2)

combination of crystallization and chromatography.[1] The aldols can be regenerated by treatment with zinc dust in acetic acid without epimerization.

[1] D. Gravel, S. Rahal, and A. Regnault, *Canad. J. Chem.*, **50**, 3846 (1972).

**Trichloroisocyanuric acid (Cyanuric chloride), 2,** 426, 427; **3,** 297. Mol. wt. 184.41, m.p. 145–148°. Suppliers: Aldrich, Eastman, MCB, and others.

(1)

*Alkyl chlorides.*[1] Primary, secondary, and tertiary alcohols are converted into alkyl chlorides when heated somewhat below the boiling point with an excess of the reagent under anhydrous conditions.

$$3 \text{ ROH} + (1) \xrightarrow[30-90\%]{} 3 \text{ RCl} +$$

*Alkyl iodides.*[2] Alkyl iodides can be prepared in about 40–75% yield by heating the reagent with an excess of an alcohol and sodium iodide.

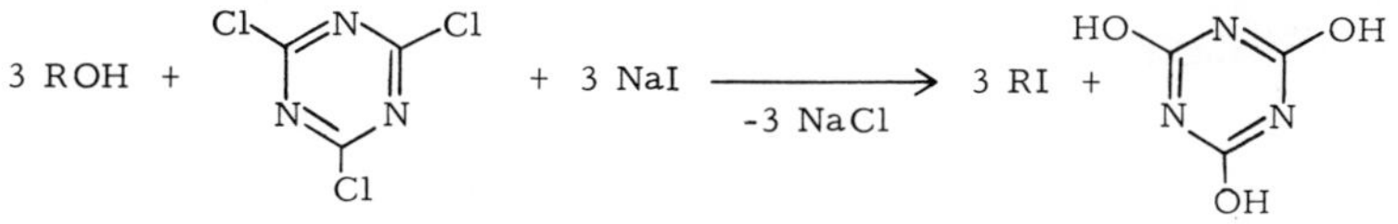

$$3 \text{ ROH} + \underset{\text{Cl}}{\underset{\big|}{\text{(triazine, Cl at positions)}}} + 3 \text{ NaI} \xrightarrow[-3 \text{ NaCl}]{} 3 \text{ RI} + \text{(triazine, OH)}$$

*Nitriles.*[3] Aldoximes are dehydrated to nitriles in good yield by treatment with trichloroisocyanuric acid in pyridine at room temperature.

[1] S. R. Sandler, *J. Org.*, **35**, 3967 (1970).
[2] *Idem, Chem. Ind.*, 1416 (1971).
[3] J. K. Chakrabarti and T. M. Hotten, *J.C.S. Chem. Comm.*, 1226 (1972).

**Trichloromethylisocyanide dichloride (TMI)**, $Cl_3CN{=}CCl_2$. Mol. wt. 215.31.

The reagent is prepared by high-temperature chlorination of methylisocyanide dichloride.[1]

*Review.*[2] The reagent effects cyclization in good yield of N-benzoylanthranilic

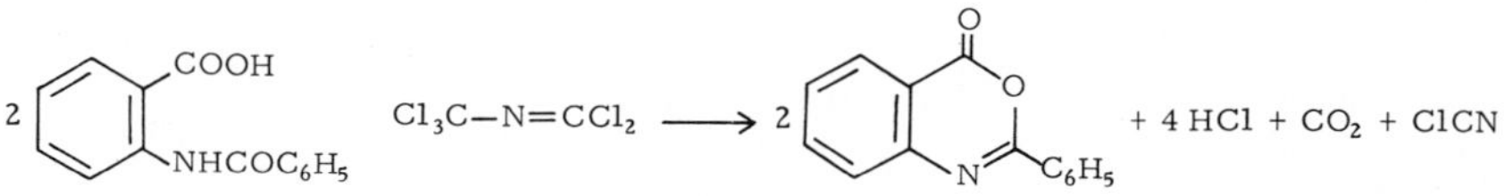

$$2 \text{ (anthranilic acid deriv.)} \quad Cl_3C-N{=}CCl_2 \longrightarrow 2 \text{ (oxazine product)} + 4 \text{ HCl} + CO_2 + ClCN$$

(1)

acid to 4-keto-2-phenylbenzo[*d*]-1,3-oxazine (1). Amides are dehydrated to nitriles, usually in good yield:

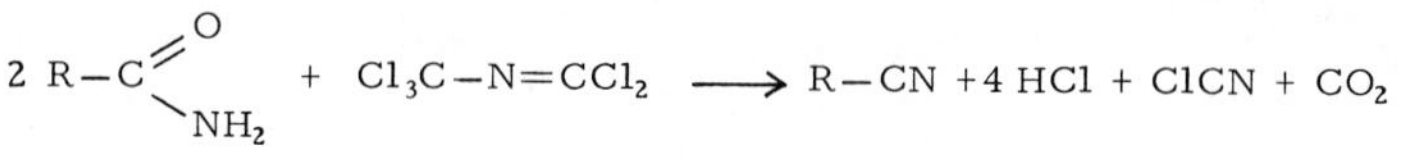

$$2 \text{ R-C}\begin{smallmatrix}O\\ \\NH_2\end{smallmatrix} + Cl_3C-N{=}CCl_2 \longrightarrow R-CN + 4 \text{ HCl} + ClCN + CO_2$$

Carboxylic acids and sulfonic acids are converted into acid chlorides in satisfactory yield:

$$2 \text{ RCOOH} + Cl_3C-N{=}CCl_2 \longrightarrow 2 \text{ RCOCl} + CO_2 + ClCN + 2 \text{ HCl}$$

The reagent is useful for synthesis of heterocycles, for example, 1,3,5-triazines, pyrimidines, quinazolines, 1,2,4-triazoles, and 1,2,4-oxadiazoles.

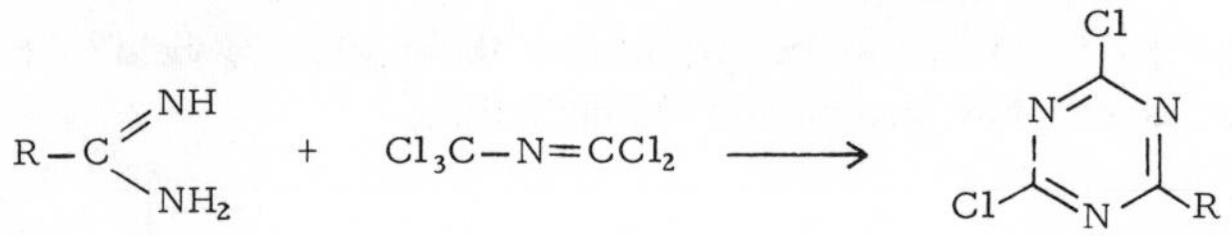
R—C
NH
NH₂
+
Cl₃C—N=CCl₂
Cl
N
N
Cl
N
R

1, 3, 5-Triazine

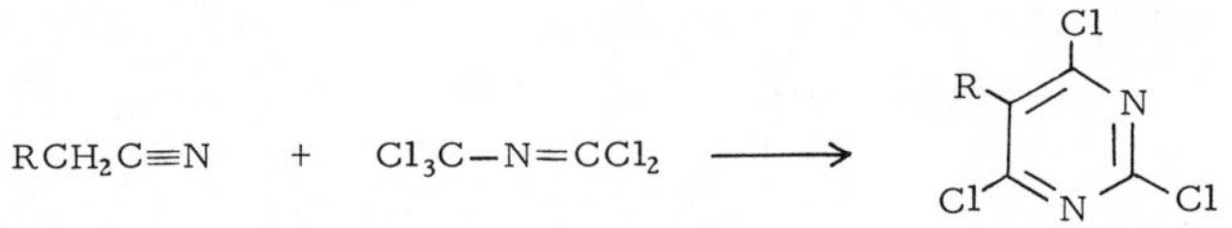
RCH₂C≡N
+
Cl₃C—N=CCl₂
Cl
R
N
Cl
N
Cl

Pyrimidine

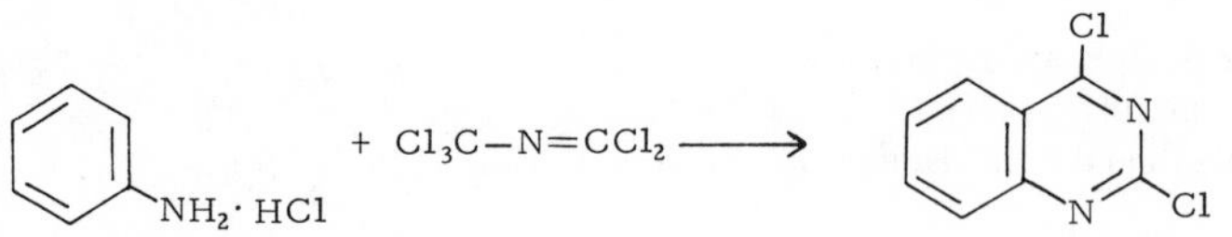
NH₂· HCl
+ Cl₃C—N=CCl₂
Cl
N
N
Cl

2, 4-Dichloroquinazoline

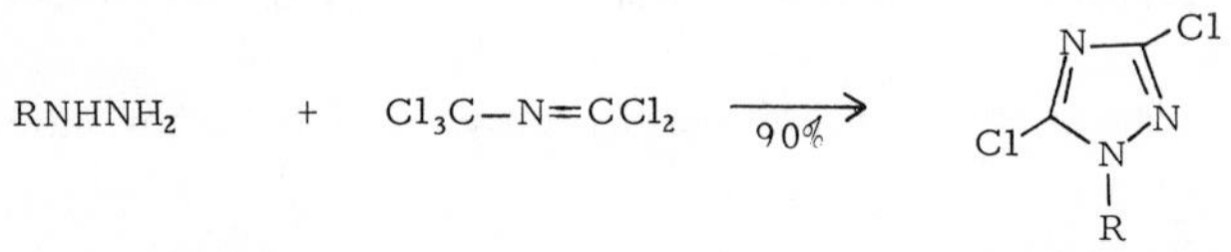
RNHNH₂
+
Cl₃C—N=CCl₂
90%
N
Cl
N
Cl
N
R

1, 2, 4-Triazole

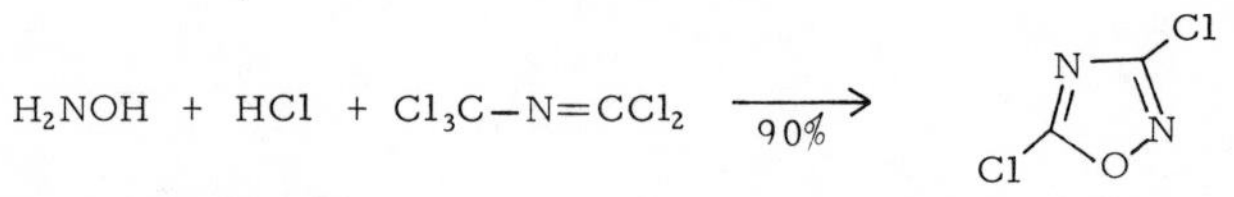
H₂NOH + HCl + Cl₃C—N=CCl₂
90%
Cl
N
N
Cl
O

3, 5-Dichloro-1, 2, 4-oxadiazole

[1]H. Holtschmidt, E. Degener, H.-G. Schmelzer, H. Tarnow, and W. Zecher, *Angew. Chem., internat. Ed.*, **7**, 856 (1968).

[2]K. Findeisen, K. Wagner, and H. Holtschmidt, *Synthesis*, 599 (1972).

**Trichloromethyllithium**, **1**, 1196; **2**, 119.

***Reaction with cyclooctyne***[1]**:**

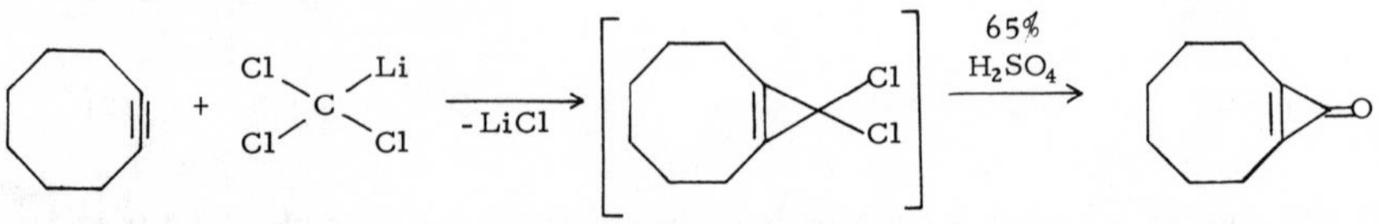
+
Cl
Li
C
Cl
Cl
-LiCl
Cl
Cl
65%
H₂SO₄

[1]G. Wittig and J. J. Hutchison, *Ann.*, **741**, 79 (1970).

**Trichlorosilane, 3,** 298–299. Additional supplier: Union Carbide Corporation.

The reagent can be purified by treatment with quinoline to remove hydrogen chloride followed by distillation, first at atmospheric pressure and then under vacuum.

Benkeser[1] has reviewed the chemistry of trichlorosilane–tertiary amine combinations.

*Methyl groups by reduction of aromatic carboxylic acids with trichlorosilane–tri-n-propylamine.*[2] In a hood well vented to permit open atmospheric transfer of trichlorosilane and to remove hydrogen chloride "off" gas produced, a 300-ml. three-necked,

round-bottomed flask is equipped with a magnetic stirrer, thermometer, glass stopper, and an efficient condenser to which is attached a nitrogen line with a gas bubbler. The glassware is thoroughly dried by flame or in an oven and the system is flushed with dry nitrogen and then charged with 19.8 g. (0.1 mole) of 2-diphenylcarboxylic acid (Aldrich, Fluka), 60 ml. (0.6 mole) of trichlorosilane, and 80 ml. of acetonitrile (dried over Matheson "Linde" type 4A molecular sieves). A dry inert atmosphere is maintained with a low nitrogen flow as the mixture is stirred and heated to reflux for 1 hr. (40–45°) or until gas evolution ceases and the carboxylic acid has dissolved. The solution is then cooled in a Dry Ice bath to at least 0°. The glass stopper is replaced with a 100-ml. pressure-equalized dropping funnel charged with 37.8 g. (0.264 mole) of tri-*n*-propylamine, which is emptied rapidly into the stirred solution. The ice bath is removed and the flask contents allowed to stir until the reaction ceases to be exothermic. Heat is supplied by a heating mantle to maintain reflux for 16 hr. (overnight) during which time the temperature rises to 70–75°.

The solution is poured rapidly into a 1-l. flask and treated with enough anhydrous diethyl ether to make a total volume of about 850 ml. and precipitate most of the tri-*n*-propylamine hydrochloride. After refrigeration in a sealed Erlenmeyer flask for 1 hr., the precipitate is collected on a 150-ml. Buchner funnel and washed with three 50-ml. portions of anhydrous ether. The clear yellow filtrate is placed in a 1-l. pressure-equalized dropping funnel and dropped at a constant rate through a 100-mm. Vigreux column fitted with a distillation head into a 300-ml. round-bottomed flask containing a magnetic stirring bar. The flask is stirred and heated until most of the ether has been distilled. The murky solution is then heated at 40° (80 mm.) to remove remaining volatiles. The Vigreux column, dropping funnel, and distilling head are replaced by a 100-ml. pressure-equalized dropping funnel charged with 100 ml. of methanol. The top of the funnel is left open for escape of hydrogen chloride as the methanol is added slowly to the oily flask contents. After vigorous boiling has ceased, the solution is heated to reflux for 1 hr. and then cooled in an ice bath and treated slowly with a solution of 56 g. (1 mole) of potassium hydroxide in 25 ml. of water and 50 ml. of methanol. The resulting mixture is refluxed for about 19 hr. (overnight) and then dissolved in 600 ml. of water and extracted with three 100-ml. portions of ether. The combined ether extracts are washed once with 50 ml. of 5 $N$ hydrochloric acid and then dried over anhydrous magnesium sulfate. Ether removal is effected by distillation as described previously and vacuum distillation of the residual liquid gives 12.5–13.4 g. (74–80%) of 2-methyl-diphenyl, b.p. 76–78°/0.5 mm., $n_D^{20}$ 1.5920 (literature: 1.5914).

Note that Ramakrishnan and Bickart[3] have isolated an anomalous product (2) in the reductive silylation of phenanthrene-3,4-dicarboxylic acid (1).

**Dialkyl ethers.**[4] Trichlorosilane can reduce alkyl aliphatic carboxylates under $\gamma$-irradiation to dialkyl ethers, in some cases in quantitative yield. The reaction is believed to proceed by a free-radical chain mechanism.[5]

$$R\,COOR' \xrightarrow[\text{Cl}_3\text{SiH}]{\gamma\,\text{ray}} R\,CH_2OR'$$

[1] R. A. Benkeser, *Accts. Chem. Res.*, **4**, 94 (1971).
[2] G. S. Li, D. F. Ehler, and R. A. Benkeser, *Org. Syn.*, submitted 1972; see also, R. A. Benkeser, K. M. Foley, J. M. Gaul, and G. S. Li, *Am. Soc.*, **92**, 3232 (1970).
[3] K. Ramakrishnan and P. Bickart, *J.C.S. Chem. Comm.*, 1338 (1972).
[4] J. Tsurugi, R. Nakao, and T. Fukumoto, *Am. Soc.*, **91**, 4587 (1969).
[5] Y. Nagata, T. Dohmaru, and J. Tsurugi, *J. Org.*, **38**, 795 (1973).

**Triethylaluminum**, **1**, 1197–1198; **2**, 427; **3**, 299.

**Hydrocyanation** (**2**, 427). Application of the Nagata procedure to the ketone (1) yielded 7-cyano-$\Delta^{1,9}$-octalin-2-one (2) in 48% yield.[1] No adduct was obtained using classical procedures.

Details of the reaction of 3$\beta$-acetoxy-$\Delta^5$-cholestene-7-one (**2**, 427) have been published.[2]

[1] R. A. Finnegan and P. L. Bachman, *J. Org.*, **36**, 3196 (1971).
[2] W. Nagata and M. Yoshioka, *Org. Syn.*, **52**, 100 (1972).

**Triethylamine, 1**, 1198–1203; **2**, 427–429.

*Dehydrohalogenation* (**1**, 1201–1202; **2**, 428). Diphenylketene is prepared by adding triethylamine dropwise to a stirred solution of diphenylacetyl chloride in ether under nitrogen, with ice cooling; distillation at 1 mm. gives diphenylketene as an orange oil in yield of 53–57%.[1]

$$(C_6H_5)_2CHCOCl \xrightarrow[53-57\%]{(C_2H_5)_3N} (C_6H_5)_2C{=}C{=}O \ + \ (C_2H_5)_3N \cdot HCl$$

[1] E. C. Taylor, A. McKillop, and G. H. Hawks, *Org. Syn.*, **52**, 36 (1972).

**Triethyl orthoacetate, 3**, 300–302.

*trans-Trisubstituted olefins.*[1] Trust and Ireland have described the preparation of ethyl *trans*-4-methyl-4,8-nonadienoate (2) by Claisen rearrangement of the allyl vinyl ether derived from 2-methyl-1,6-heptadiene-3-ol (1).

$$CH_2{=}CHCH_2CH_2\underset{\underset{\text{(1)}}{}}{\overset{OH\ \ CH_3}{\underset{|}{C}-\underset{|}{C}{=}CH_2}} \xrightarrow[83-88\%]{\begin{subarray}{c}CH_3C(OC_2H_5)_3\\ CH_3CH_2COOH\end{subarray}} CH_2{=}CHCH_2CH_2\overset{H}{\underset{\underset{\text{(2)}}{}}{{>}C{=}C{<}}}\overset{CH_2CH_2COOC_2H_5}{CH_3}$$

[1] R. I. Trust and R. E. Ireland, *Org. Syn.*, submitted 1972.

**Triethyl orthoformate, 1**, 1204–1209.

Simplified synthesis by the action of benzoyl chloride or ethyl chlorocarbonate with ethanol and formamide[1]:

$$OCHNH_2 \ + \ 3 \ C_2H_5OH \ + \ C_6H_5COCl \xrightarrow[40-44\%]{} HC(OC_2H_5)_3 \ + \ C_6H_5CO_2H \ + \ NH_4Cl$$

$$OCHNH_2 \ + \ 3 \ C_2H_5OH \ + \ ClCO_2C_2H_5 \xrightarrow[42\%]{} HC(OC_2H_5)_3 \ + \ CO_2 \ + \ NH_4Cl$$

[1] R. Ohme and E. Schmitz, *Ann.*, **716**, 207 (1968).

**Triethyloxonium fluoborate, 1**, 1210–1212; **2**, 430–431; **3**, 303.

*Amide acetals.*[1] Carboxamides (1) are alkylated in high yield (95–100%) by triethyloxonium fluoborate to give the corresponding carbonium fluoborates (2). As a rule, the reaction of (2) with sodium ethoxide at 0° affords an amide acetal (3).

$$RC\overset{O}{\underset{N(CH_3)_2}{<}} \xrightarrow[95-100\%]{(C_2H_5)_3O^+BF_4^-} RC\overset{+}{\underset{N(CH_3)_2}{<}}{\overset{OC_2H_5}{}} BF_4^- \xrightarrow[-NaBF_4]{NaOC_2H_5,\ 0^0} RC\overset{OC_2H_5}{\underset{N(CH_3)_2}{<}}{\overset{OC_2H_5}{}}$$

$$\text{(1)} \qquad\qquad\qquad \text{(2)} \qquad\qquad\qquad \text{(3)}$$

Yields are in the range 50–70%. The reaction of (3) with nucleophiles at 80° results in elimination of ethanol and formation of an aminomethylene compound. Thus the amide acetal (4) reacts with nitromethane to give (5) in 60% yield.

$$(C_2H_5O)_2CHC\underset{N(CH_3)_2}{\overset{OC_2H_5}{\overset{|}{\underset{|}{C}}}OC_2H_5} \xrightarrow[60\%]{CH_3NO_2} (C_2H_5O)_2CHC\underset{N(CH_3)_2}{\overset{|}{C}}=CHNO_2 + 2\ C_2H_5OH$$

(4)                                    (5)

***Azoxyalkanes.***[2] The reaction of alkane diazotates (probably *syn*) with triethyl-oxonium fluoroborate in $CH_2Cl_2$ suspensions gives azoxyalkanes in 50–60% yield:

$$\underset{N=N}{\overset{R}{\diagdown}}\diagup^{\bar{O}\ K^+} + (C_2H_5)_3\overset{+}{O}BF_4^- \xrightarrow{CH_2Cl_2} \underset{N=N}{\overset{R}{\diagdown}}\diagup^{\nearrow O}_{\diagdown C_2H_5} + KBF_4$$

Alternatively, azoxyalkanes can be prepared by the reaction of alkane diazotates with alkyl halides (preferably iodides) in hexamethylphosphoric triamide.

$$\underset{N=N}{\overset{R}{\diagdown}}\diagup^{\bar{O}\ K^+} + R'I \xrightarrow{HMPT} \underset{N=N}{\overset{R}{\diagdown}}\diagup^{\nearrow O}_{\diagdown R'} + KI$$

***Esterification.***[3] Hindered (as well as unhindered) carboxylic acids are easily esterified at room temperature by triethyloxonium fluoroborate in methylene chloride, the carboxylate anion being generated by addition of a bulky organic base such as diisopropylethylamine.[4] Yields are in the range 85–95%.

$$R-\overset{O}{\overset{\|}{C}}-O^- + CH_3CH_2-\overset{\diagup C_2H_5}{\underset{\diagdown C_2H_5}{\overset{+}{O}}} \longrightarrow R\overset{O}{\overset{\|}{C}}-OC_2H_5$$

***Cleavage of thioketals.***[5] Protection of ketones and aldehydes by conversion to thioketals is rarely used because thioketals are resistant to both acid- and base-catalyzed hydrolysis. Use of mercuric salts has been the most useful procedure known (**1**, 654; **2**, 182; **3**, 136). Japanese chemists now report that cleavage can be effected readily through alkylation with triethyloxonium fluoroborate. Thus alkylation of cyclo-hexanone ethylenethioketal (1) with the reagent affords the salt (2). Alkaline hydrolysis of (2) gives cyclohexanone in only 36% yield. However, if the salt (2) is shaken with 3% $CuSO_4$ solution in methylene chloride, cyclohexanone is obtained in 81% yield.

(1)                    (2)

81% ↓ CuSO₄

In an alternative procedure (1) is dialkylated to give the salt (3); this can be hydrolyzed with 10% NaOH solution to give cyclohexanone in 81% yield. Using this procedure,

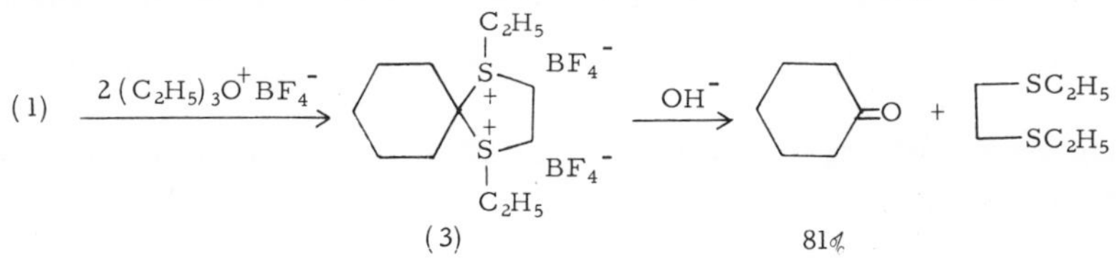

(3)          81%

the Japanese chemists were able to effect satisfactory cleavage of thioketals of a number of aldehydes and ketones.

[1]H. Bredereck, W. Kantlehner, and D. Schweizer, *Ber.*, **104**, 3475 (1971).
[2]R. A. Moss, M. J. Landon, K. M. Luchter, and A. Mamantov, *Am. Soc.*, **94**, 4392 (1972).
[3]D. J. Raber and P. Gariano, *Tetrahedron Letters*, 4741 (1971).
[4]S. Hünig and M. Kiessel, *Ber.*, **91**, 380 (1958); available from Aldrich.
[5]T. Oishi, K. Kamemoto, and Y. Ban, *Tetrahedron Letters*, 1085 (1972).

**Triethyl phosphite**, **1**, 1212–1216; **2**, 432–433; **3**, 304.

*Dechlorination.*[1] Triethyl phosphite is the reagent of choice for dechlorination of

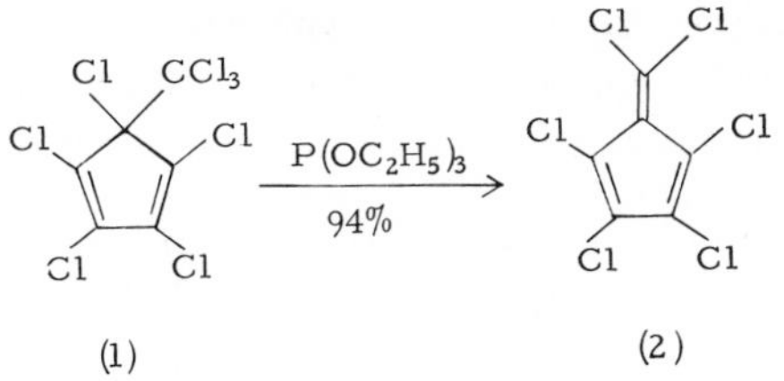

(1)                    (2)

(1) to hexachlorofulvene (2). Sodium iodide (98.2% yield) and stannous chloride (63% yield) are also effective.

*Deoxygenation of N-functions* (**1**, 1215–1216; **2**, 433). Brooke and co-workers[2] have presented evidence that the cyclization of (1) to the carbazole (2) by triethyl phosphite does not proceed by way of a nitrene as originally believed, but probably through the bipolar species (a).

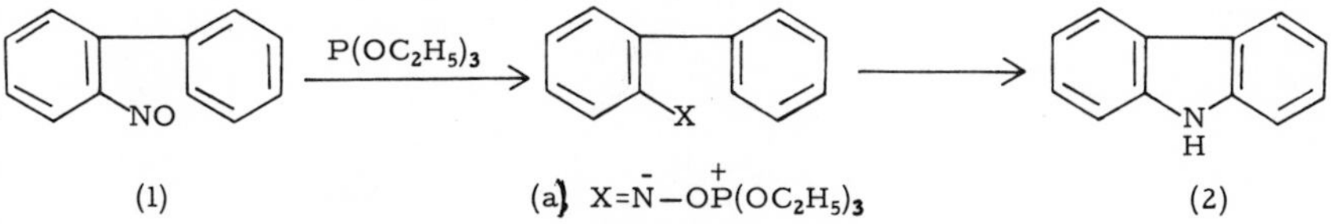

(1)              (a) $X = \overset{-}{N} - \overset{+}{O}P(OC_2H_5)_3$              (2)

[1]E. J. McBee, E. P. Wesseler, D. L. Crain, R. Hurnaus, and T. Hodgins, *J. Org.*, **37**, 683 (1972).
[2]P. K. Brooke, R. B. Herbert, and E. G. Holliman, *Tetrahedron Letters*, 761 (1973).

**Triethylsilane, 1,** 1218; **2,** 433; **3,** 304.

*Reduction of aldehydes and ketones to ethers.*[1] Carbonyl compounds are reduced to ethers by treatment with triethylsilane in an alcoholic acidic medium. Sulfuric,

$$R_2C=O + (C_2H_5)_3SiH + R'OH \xrightarrow[25-90\%]{} R_2CHOR' + (C_2H_5)_3SiOH$$

trifluoroacetic, and trichloroacetic acids are all effective catalysts. The following mechanism is proposed:

$$R_2C=O + H^+ \longrightarrow R_2C=OH^+$$

$$R_2C=OH^+ + R'OH \rightleftharpoons R_2C\begin{smallmatrix}OH\\ +\\ OR'\\ H\end{smallmatrix} \rightleftharpoons R_2C\begin{smallmatrix}\overset{+}{O}H_2\\ \\ OR'\end{smallmatrix} \rightleftharpoons R_2C=\overset{+}{O}R' + H_2O$$

$$R_2C=\overset{+}{O}R' + (C_2H_5)_3SiH + H_2O \longrightarrow R_2CHOR' + (C_2H_5)_3SiOH$$

[1]M. P. Doyle, D. J. DeBruyn, and D. A. Kooistra, *Am. Soc.,* **94,** 3659 (1972).

**Trifluoroacetic acid (TFA), 1,** 1219–1221; **2,** 433–434; **3,** 305–308.

*Dehydrogenation.* Andersen *et al.*[1] report that treatment of sesquiterpenes with trifluoroacetic acid leads to dehydrogenation with partial aromatization. The reaction is particularly suitable for production of tetralins. Thus treatment of the cadinene hydrocarbons (1)–(4) in *n*-decane solution for a few minutes with excess TFA at room

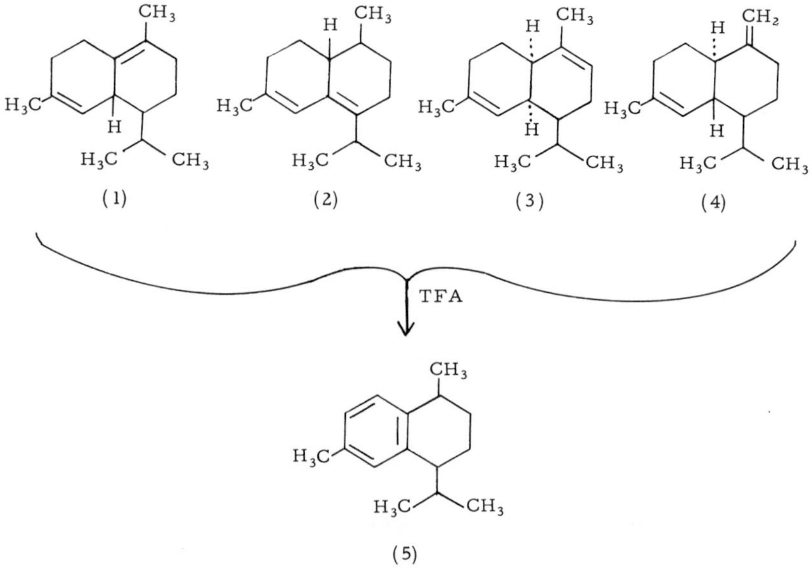

(1)    (2)    (3)    (4)

TFA

(5)

temperature results in formation of calamenene (5) in good yield. Limonene (6) requires hours for complete conversion to cymene (7).

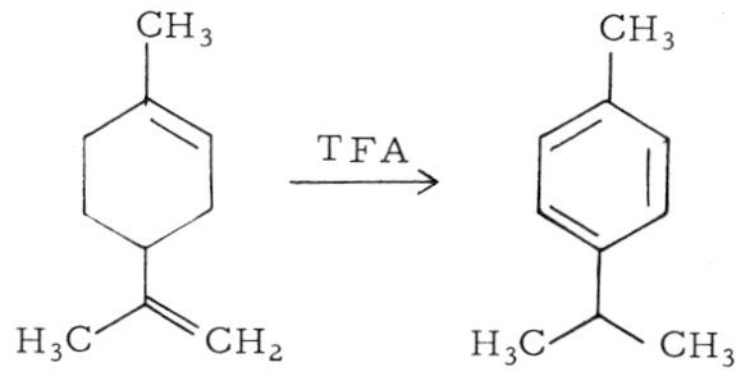

(6)            (7)

**Olefin cyclization** (3, 305–307). In an extension to the biogenetic-like cyclization of olefins, Johnson et al.[2] report that a suitably placed triple bond can participate in an olefinic cyclization to produce a *trans*-fused five-membered ring. Thus treatment of the trieynol (1) with methylene chloride containing 2% by weight of trifluoroacetic acid at −70° for 5 min. gives the tricyclic triene (2) in about 70% yield.

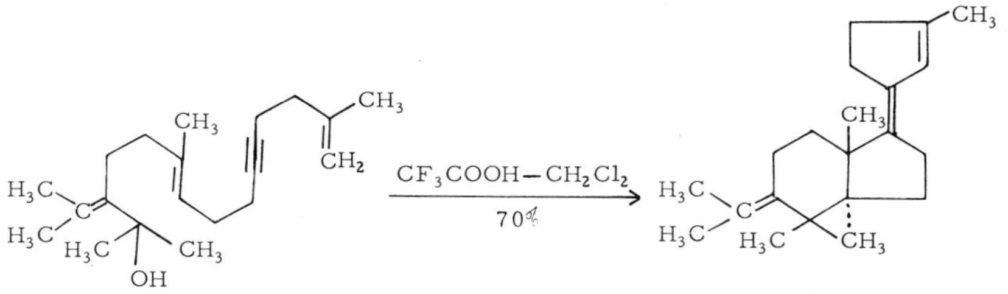

(1)                    (2)

This cyclization has obvious applications to the synthesis of steroids and indeed Johnson et al.[3] applied this reaction to a synthesis of *dl*-progesterone. The key step in the synthesis involves the cyclization of (3) to give (4). This reaction was carried out with trifluoroacetic acid as above, but ethylene carbonate was added to the reaction to trap the vinyl cation. After cyclization potassium carbonate was added to hydrolyze the enol complex. In this way (3) was converted into (4) in 71% yield. The tetracyclic ketone (4) was converted into progesterone (6) by ozonization followed by intramolecular aldol condensation. Note that (4) is a 5:1 mixture of the 17β- and 17α-epimeric ketones. The mixture was converted into (6) and then separated by fractional crystallization.

[1]N. H. Andersen, D. D. Syrdal, and C. Graham, *Tetrahedron Letters*, 903 (1972).
[2]W. S. Johnson, M. B. Gravestock, R. J. Parry, R. F. Myers, T. A. Bryson, and D. H. Miles, *Am. Soc.*, **93**, 4330 (1971).
[3]W. S. Johnson, M. B. Gravestock, and B. E. McCarry, *ibid.*, **93**, 4330 (1971).

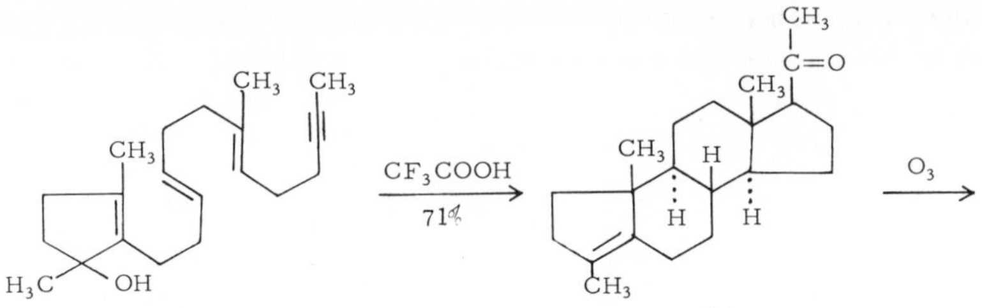

(3) → CF₃COOH / 71% → (4) → O₃

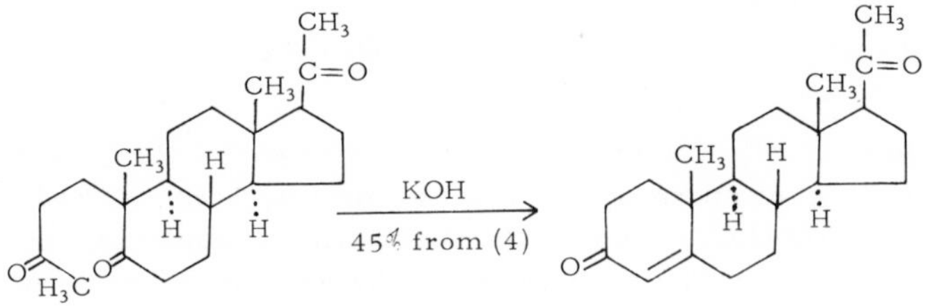

(5) → KOH / 45% from (4) → (6)

**ortho-*Claisen rearrangement*.**[4] Phenyl allyl ethers undergo Claisen rearrangement when dissolved in trifluoroacetic acid at room temperatures. The rate constants for several *para*-substituted compounds are about $10^5$ as great as those observed in the usual thermal rearrangement. The rearrangement is suitable for synthetic purposes. For example, 2,5-diallylhydroquinone can be isolated in about 70% yield by treatment of the diallyl ether of hydroquinone with trifluoroacetic acid for 1 hr. at 50°.

[4]U. Svanholm and V. D. Parker, *Chem. Commun.*, 645 (1972).

**Trifluoromethanesulfonic acid**, $CF_3SO_3H$. Mol. wt. 150.09. Suppliers: Fluka, WBL. Trifluoromethanesulfonic acid is available from the 3M Company under the name Fluorochemical acid FC-24. It is the strongest known acid and fumes upon exposure to air. Great care must be exercised in handling this acid.

*Vinyl trifluoromethanesulfonates (triflates)*.[1] Vinyl triflates are available by two procedures. In one, an acetylene is treated with the acid at low temperatures:

$$CH_3C{\equiv}CCH_3 + CF_3SO_3H \xrightarrow{55\%} CH_3CH{=}C(CH_3)OSO_2CF_3$$

$$\underline{cis\ and\ trans}$$

In the other, a ketone is treated with trifluoromethanesulfonic anhydride[2]:

$$(CH_3)_2CHCOCH_3 + (CF_3SO_2)_2O \xrightarrow{58\%} (CH_3)_2C{=}C(CH_3)OSO_2CF_3$$

Triflates have been used extensively in solvolysis studies.

[1]P. J. Stang, *Am. Soc.*, **91**, 4600 (1969); *Org. Syn.*, submitted 1972; T. E. Dueber *et al.*, *Angew. Chem., internat. Ed.*, **9**, 521 (1970); P. J. Stang and T. E. Dueber, *Org. Syn.*, submitted 1972.
[2]Prepared by distillation of the acid from a large excess of $P_2O_5$; b.p. 83–85°.

**Trifluoromethanesulfonic anhydride**, $(CF_3SO_2)_2O$. Mol. wt. 266.15, b.p. 84°.

*Preparation.*[1] To 32.1 g. of trifluoromethanesulfonic acid (3M Company) cooled to 0° is added in three parts 25 g. of $P_2O_5$, and the trifluoromethanesulfonic anhydride is distilled by gradually heating the reaction mixture to 110° (bath temperature) during 1 hr. The fraction boiling from 80–100° (760 mm.) is redistilled repeatedly from about

$$HC\equiv CH \;+\; Na \;+\; CH_2-CH_2 \xrightarrow[\;35\%\;]{Liq.\ NH_3} HC\equiv CCH_2CH_2OH$$

$$(1)$$

$$(1) \;+\; (CF_3SO_2)_2O \longrightarrow HC\equiv CCH_2CH_2OSO_2CF_3$$

$$(2)$$

$$(2) \xrightarrow[\;\;\;\;\;]{\begin{array}{l}1)\ CF_3CO_2H + CF_3CO_2Na\\ 2)\ H_2O\end{array}}$$

8 g. of $P_2O_5$ to remove traces of acid until it no longer fumes when a glass rod is dipped into the distillate and waved in the air. The yield of anhydride, b.p. 84°, is 25 g. (83%).

*Synthesis of cyclobutanone.*[2] The synthesis is accomplished in three steps, as formulated. In the last step the procedure calls for heating and stirring at 60–65° for 7–8 days.

[1]J. Burdon, I. Farazmand, M. Stacey, and J. C. Tatlow, *J. Chem. Soc.*, 2574 (1957).
[2]M. Hanack, T. Dehesch, K. Hummel, and A. Nierth, *Org. Syn.*, submitted 1971.

**Trifluoromethanesulfonic-carboxylic anhydrides,** $CF_3SO_2-O-C{\overset{O}{\underset{R}{\diagdown}}}$

*Preparation.*[1] The reagents are prepared in 80–100% yields by the reaction of the silver salt of trifluoromethanesulfonic acid with acyl halides:

$$CF_3SO_3Ag \;+\; RC{\overset{O}{\underset{Cl}{\diagdown}}} \xrightarrow[-AgCl]{} CF_3SO_2-O-C{\overset{O}{\underset{R}{\diagdown}}}$$

*Acylation.*[1] These anhydrides are powerful acylating agents. Thus nonactivated arenes such as benzene are readily acylated without addition of Friedel–Crafts catalysts:

$$C_6H_6 + CF_3SO_2-OC\overset{\displaystyle O}{\underset{\displaystyle C_6H_5}{\diagdown}} \quad \xrightarrow[90\%]{60°, \ 5 \ hrs.} \quad C_6H_5COC_6H_5$$

The trifluoromethanesulfonic acid can be recovered almost quantitatively as the barium salt.

The mixed anhydrides can also be formed *in situ* by the reaction of acyl halides with arenes catalyzed by trifluorosulfonic acid.[2]

$$RC\overset{\displaystyle O}{\underset{\displaystyle Cl}{\diagdown}} + \ \bigcirc\!-R' \ \xrightarrow{CF_3SO_3H} \ RC\overset{\displaystyle O}{\diagdown}\bigcirc\!-R'$$

[1]F. Effenberger and G. Epple, *Angew. Chem., internat. Ed.*, **11**, 299 (1972).
[2]*Idem, ibid.*, **11**, 300 (1972).

**Triiron dodecacarbonyl,** $Fe_3(CO)_{12}$. Mol. wt. 513.67. Supplier: Alfa Inorganics.

The reagent is supplied with about 10% by weight of methanol as stabilizer. It can be made methanol free by keeping the reagent at 0.5 mm. for at least 5 hr.

**Reductions.**[1] Methanolic solutions of triiron dodecacarbonyl specifically reduce

$$\underset{0.01\underline{M}}{ArNO_2} + \underset{0.01\underline{M}}{Fe_3(CO)_{12}} + \underset{(0.1\underline{M})}{CH_3OH} \ \xrightarrow[75-95\%]{benzene, \ reflux} \ ArNH_2$$

nitroaryls to primary amines in high yield. Various functional groups (C=C, C=O, COOR, NHAc) are not affected. No reduction occurs in the absence of methanol. Spectrographic evidence indicates that methanol and triiron dodecacarbonyl react to form the hydridoundecacarbonyltriferrate anion $[HFe_3(CO)_{11}{}^-]$. This is evidently the actual reducing agent.

Alper[2] has used this same system for reduction of the carbon–nitrogen double bond. Thus treatment of phthalazine (1) with $Fe_3(CO)_{12}$ in refluxing methanol–benzene for

$$\text{(structure with } N=N\text{)} \quad \xrightarrow[54\%]{Fe_3(CO)_{12} \\ CH_3OH, \ C_6H_6} \quad \text{(structure with } NH\text{)}$$

(1)                                    (2)

12–16 hr. gives 1,2-dihydrophthalazine (2) in 54% yield. The system was used successfully for reduction of a number of Schiff bases. For example, N-benzylideneaniline is reduced to N-benzylaniline in 88% yield.

[1]J. M. Landesberg, L. Katz, and C. Olsen, *J. Org.*, **37**, 930 (1972).
[2]H. Alper, *ibid.*, **37**, 3972 (1972).

**Triisobutylaluminum (TIBA)**, $Al[CH_2CH(CH_3)_2]_3$, 1, 260. Mol. wt. 198.32. Additional supplier: Ventron Corporation.

*Reduction of α-ketols.* Katzenellenbogen and Bowlus[1] studied the stereoselectivity

erythro                                                threo

of reduction of aliphatic α-ketols with a variety of aluminum hydride reagents. According to Cram's cyclic model,[2] the *erythro*-diol should be formed predominantly. The most selective reducing agent is TIBA. The decreased selectivity of aluminum hydride reagents is attributed to agglomeration. For example, diisobutylaluminum hydride (DIBAH, 1, 260–262), which exists as the trimer, displays slight selectivity.

[1]J. A. Katzenellenbogen and S. B. Bowlus, *J. Org.*, **38**, 627 (1973).
[2]D. J. Cram and D. R. Wilson, *Am. Soc.*, **85**, 1245 (1963).

**2,4,6-Triisopropylbenzenesulfonylhydrazine**,
Mol. wt. 310.46, m.p. 121–122° dec.

The reagent can be prepared[1] in about 80 % yield by the reaction of 2,4,6-triisopropyl-benzenesulfonyl chloride[2] in THF with hydrazine hydrate.

*Diimide.*[1] This hydrazine decomposes at temperatures between 35 and 65° to give diimide. It can be used for hydrogenation of olefins in good yield. Benzenesulfonyl-hydrazine[3] and *p*-toluenesulfonylhydrazine[4] have been used as precursors of diimide but much higher temperatures are required for generation of diimide from these precursors.

[1]N. J. Cusack, C. B. Reese, and B. Roozpeikar, *Chem. Commun.*, 1132 (1972).
[2]R. Lohrmann and H. G. Khorana, *Am. Soc.*, **88**, 829 (1966).
[3]S. Hünig, H.-R. Müller, and W. Thier, *Tetrahedron Letters*, 353 (1961).
[4]E. E. van Tamelen and R. S. Dewey, *Am. Soc.*, **83**, 3729 (1961).

**2,2,2-Trimethoxy-4,5-dimethyl-1,3-dioxaphospholene** (1), 2, 97–98.

*Phosphorylation.*[1] Treatment of (1) with phosgene at 0° yields (2), which is transformed upon heating at 120° for 2 hr. into a mixture of diastereomeric cyclic acyl phosphates (3a) and (3b) and the enediol phosphate (4). The isomers (3a) and (3b) can be separated, but the equilibrium mixture (30:70) can be used for phosphorylation of

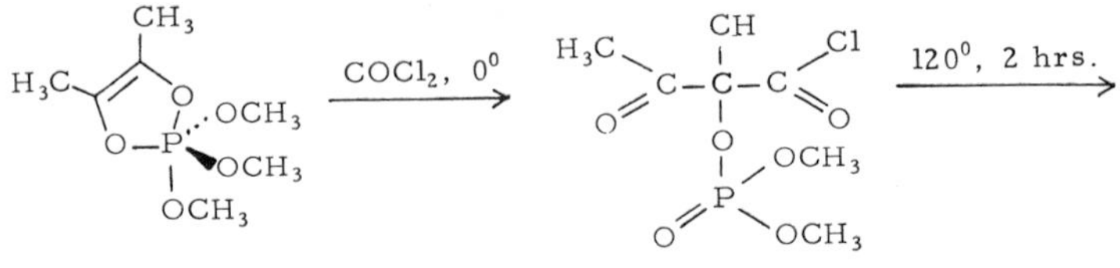

(1)         (2)

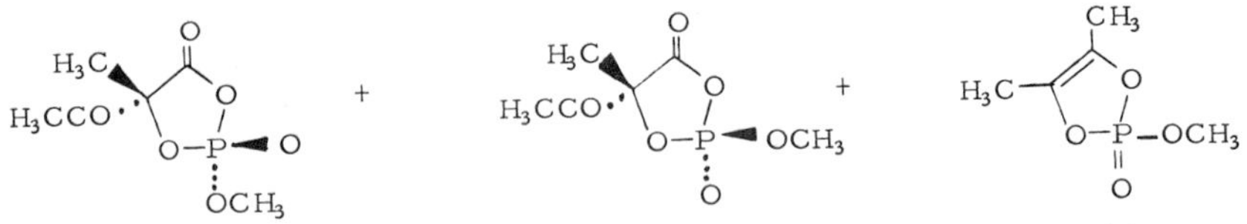

(3a)        (3b)        (4)

alcohols. Thus (3) reacts extremely rapidly with alcohols, with ring opening and decarboxylation, to form acetoin derivatives of type (5) [alkyl methyl (1-methyl-2-ketopropyl) phosphates]. These are very readily hydrolyzed to phosphate esters (6). The reactivity of (3) toward hydroxyl compounds is higher than that of any previously known organophosphorus compound, including (4).

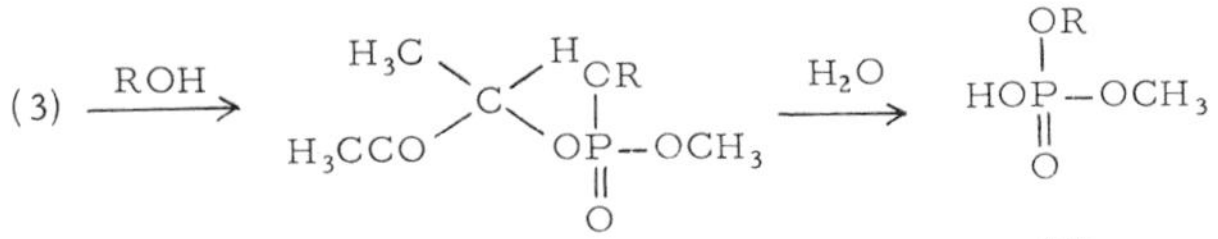

(5)         (6)

The acyl phosphate (3) also N-methylates pyridine to form the salt (7). Treatment of (7) with trimethyloxonium fluoroborate regenerates (3). Treatment of (7) with alcohols or phenols gives the salt (8), which is rapidly hydrolyzed to the phosphate ester (9).

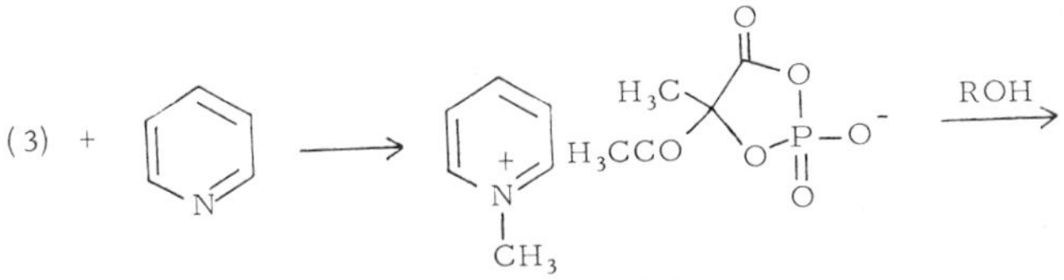

(7)

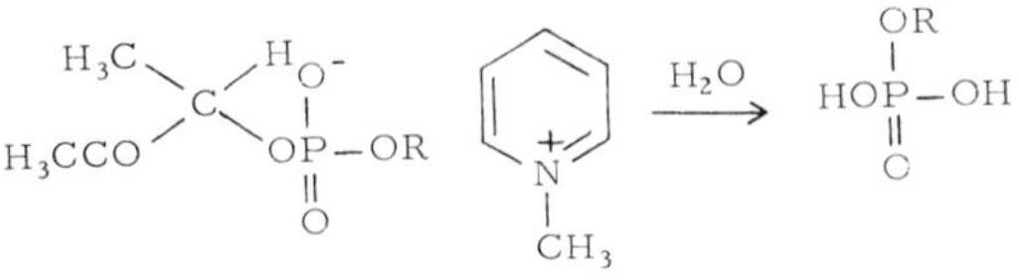

(8)         (9)

[1]F. Ramirez, S. Glaser, P. Stern, P. D. Gillespie, and I. Ugi, *Angew. Chem., internat. Ed.*, **12**, 66 (1973).

**Trimethylchlorosilane,** 1, 1232; **2,** 435–438; **3,** 310–312.

*Acyloin condensation* (**3,** 311–312). The method of converting bis(trimethylsiloxy)-cyclobutene derivatives into the corresponding 1,2-diketocyclobutanes by oxidative cleavage with bromine in an aprotic medium has since been shown to be a useful method of obtaining nonenolizable $\alpha$-diketones[1]:

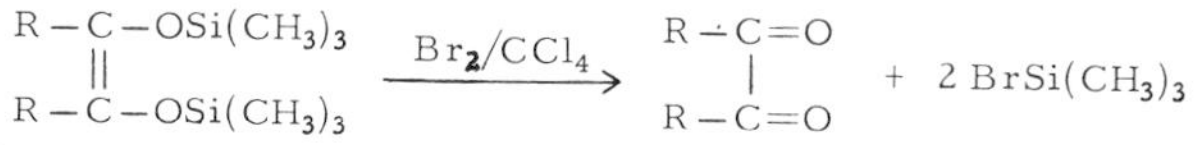

$$
\begin{array}{c}
R-C-OSi(CH_3)_3 \\
\| \\
R-C-OSi(CH_3)_3
\end{array}
\xrightarrow{\ Br_2/CCl_4\ }
\begin{array}{c}
R-C=O \\
| \\
R-C=O
\end{array}
+\ 2\,BrSi(CH_3)_3
$$

Chlorine and iodine can be used but bromine is the preferred reagent. Thus the reaction of (1) with bromine in carbon tetrachloride gives bipivaloyl (2) in 81 % yield.

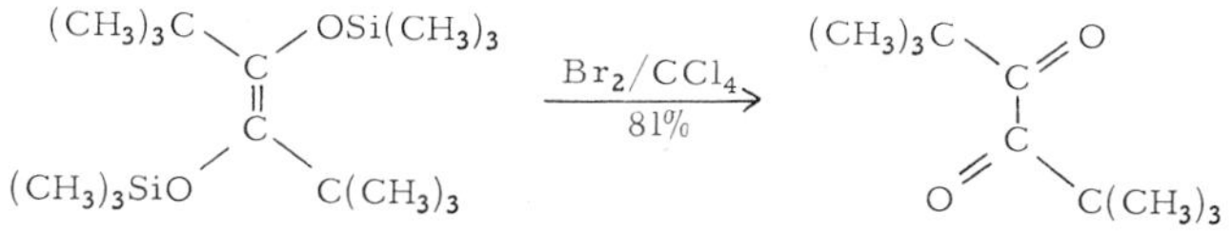

(1)                                   (2)

The reaction can also be used for preparation of cyclic nonenolizable $\alpha$-diketones. Thus (3) is converted into 1,2-diketo-3,3,7,7-tetramethylcycloheptane (4) in 98 % yield.

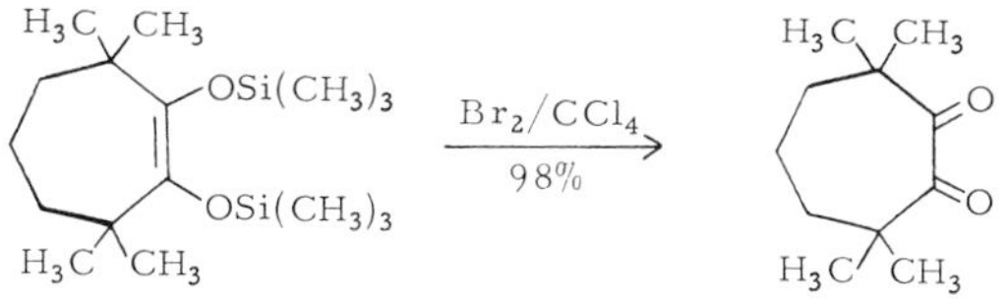

(3)                                   (4)

The reaction is not particularly useful for preparation of enolizable $\alpha$-diketones.[2] The main complication is that the liberated trimethylsilyl bromide reacts with the newly formed enolic function, thus giving rise to a new double bond to which bromine can be added. Thus several bromination products of the $\alpha$-diketones together with decomposition products are formed.

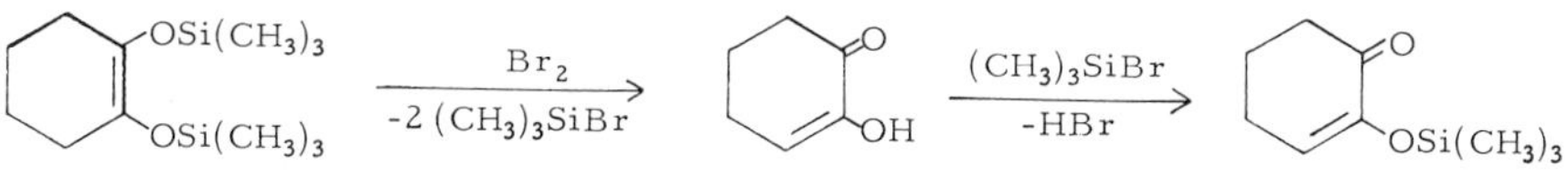

Kühlmann[3] has reviewed the reactions of carboxylic acid esters with sodium in the presence of trimethylchlorosilane.

$$
2\,RCOOR' \xrightarrow{\ 4\,Na\,+\,(CH_3)SiCl\ }
\begin{array}{c}
R-C-OSi(CH_3)_3 \\
\| \\
R-C-OSi(CH_3)_3
\end{array}
$$

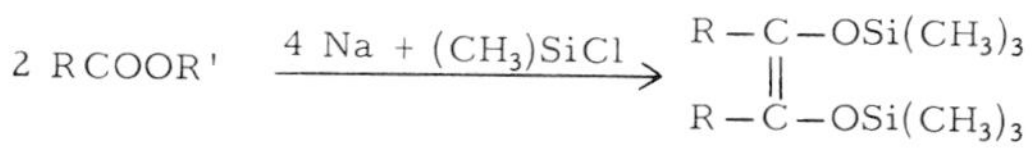

*β-Amino alcohols.* Parham and Roosevelt[4] have shown that trimethylsilyl enol ethers, prepared either by the procedure of Stork and Hendrlik (**2**, 436) or of House *et al.* (**3**, 310), can be converted into *β*-amino alcohols. Thus reaction of the trimethylsilyl enol ether of cyclohexanone (1) with anhydrous hydrogen cyanide containing a drop of sulfuric acid yielded α-cyanocyclohexyl trimethylsilyl ether (2) in 49% yield. Reduction of the nitrile group (lithium aluminum hydride) followed by hydrolysis of the ether group gave the *β*-amino alcohol in slightly impure form in 74.5% yield. This product was purified and characterized as the hydrochloride.

Parham and Roosevelt noted differences in the two methods for preparation of silyl enol ethers. Stork's method appears to be more suitable for hindered ketones, whereas House's method is better for less hindered, easily condensable ketones.

*Decarboxylation of disubstituted malonates.* In an attempt to effect acyloin-type reduction of disubstituted malonic esters, Ainsworth, Bloomfield, and co-workers[5] obtained instead in high yield ketene alkyl silyl acetals (1); an equivalent amount of

$$R^1R^2C(COOCH_3)_2 + Na + (CH_3)_3SiCl \longrightarrow R^1R^2C=C(OCH_3)OSi(CH_3)_3 + CH_3OSi(CH_3)_3 + CO$$

(1)

$$(1) + CH_3OH \longrightarrow R^1R^2CHCOOCH_3 + CH_3OSi(CH_3)_3$$

(2)

carbon monoxide is formed. Since the acetals (1) are readily hydrolyzed to the esters (2), the two-step process represents a convenient method for decarboxylation of disubstituted malonates.

*β-Keto acids.* A new synthesis of *β*-keto acids involves the reaction of dianions of carboxylic acids with esters.[6] The intermediates are trapped with trimethylchlorosilane and isolated as the trimethylsilyl esters. For example, isobutyric acid is converted into the dianion (1) by treatment with 2 eq. of lithium diisopropylamide in THF at 0°. Addition of 1 eq. of methyl pivalate (2) and an excess of trimethylchlorosilane yields the trimethylsilyl $\beta$-keto carboxylate (3) in 70% yield. The $\beta$-keto acid (4) is obtained in quantitative yield by solvolysis with methanol. Pyrolysis of (4) yields *tert*-butyl isopropyl ketone (5), again in quantitative yield.

**New synthesis of heptamethyldisilazane** (2).[7] The reaction of methylamine with trimethylchlorosilane (1) using *n*-pentane as solvent gave (2) in 85% yield.

$$2 \ (CH_3)_3SiCl \ + \ 3 \ CH_3NH_2 \ \longrightarrow \ (CH_3)_3Si-\underset{\underset{CH_3}{|}}{N}-Si(CH_3)_3 \ + \ 2 \ CH_3NH_3Cl$$

$$(1) \hspace{4cm} (2)$$

[1] J. Strating, S. Reiffers, and H. Wynberg, *Synthesis*, 209 (1971).
[2] *Idem, ibid.*, 211 (1971).
[3] R. Kühlmann, *ibid.*, 236 (1971).
[4] W. E. Parham and C. S. Roosevelt, *Tetrahedron Letters*, 923 (1971).
[5] Y.-N. Kuo, F. Chen, C. Ainsworth, and J. J. Bloomfield, *Chem. Commun.*, 136 (1971).
[6] Y.-N. Kuo, Y. A. Yahner, and C. Ainsworth, *Am. Soc.*, **93**, 6321 (1971).
[7] L. Birkofer and G. Schmidtberg, *Ber.*, **104**, 3831 (1971).

**Trimethylene dithiotosylate**, $TsS(CH_2)_3STs$. Mol. wt. 416.59, m.p. 67.5°.

*Preparation.* Potassium thiotosylate is prepared by the reaction of potassium hydrosulfide with tosyl chloride. This is then treated with trimethylene dibromide to give the reagent.

$$2 \ KHS \ + \ TsCl \ \longrightarrow \ TsSK \ + \ H_2S \ + \ KCl$$

$$2 \ TsSK \ + \ Br(CH_2)_3Br \ \longrightarrow \ TsS(CH_2)_3STs \ + \ 2 \ KBr$$

*Reaction with activated methylene groups.*[1] The reagent reacts with activated methylene groups to form a 2,2-disubstituted 1,3-dithiane (1) with elimination of 2 eq. of *p*-toluenesulfinic acid. The dithioketal group of (1), unlike the acetal group of analogous oxygen compounds, is remarkably acid stable, but dithianes can be readily converted back to methylene compounds by reduction with Raney nickel or hydrazine. The dithioketal group can also be converted to a carbonyl group by Hg(II)-catalyzed hydrolysis. The reagent is therefore useful for blocking active methylene groups.

$$TsS(CH_2)_3STs \ + \ XCH_2Y \ \xrightarrow[-2 \ TsH]{KOAc} \ \text{(dithiane)}$$

$$(1)$$

Carbonyl compounds that do not react with the reagent can be activated by conversion into an enamine or an hydroxymethylene derivative prior to the reaction. Thus cyclohexanone is converted into 2,2-trimethylenedithiocyclohexanone by way of the pyrrolidine enamine. The reagent was first used in an unambiguous synthesis of lanosterol from cholesterol for protection of the $C_2$-methylene group of $\Delta^4$-cholestene-3-one.[2]

Ethylene dithiotosylate, $TsS(CH_2)_2STs$, can be used in an analogous manner.

[1] R. B. Woodward, I. J. Pachter, and M. L. Scheinbaum, *J. Org.*, **36**, 1137 (1971); *Org. Syn.*, submitted 1972.

[2] R. B. Woodward, A. A. Patchett, D. H. R. Barton, D. A. J. Ives, and R. B. Kelley, *J. Chem. Soc.*, 1131 (1957).

**Trimethyl orthoformate**, $HC(OCH_3)_3$. Mol. wt. 106.12, b.p. 97–99°. Suppliers: Aldrich, Eastman.

For a simplified synthesis, *see* **Triethyl orthoformate** (this volume).

**N,4,4-Trimethyl-2-oxazolinium iodide**,       Mol. wt. 114.17. Supplier:

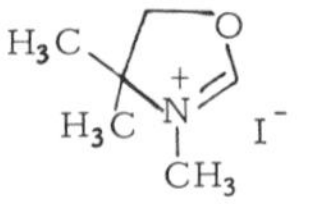

Columbia. The preparation of the reagent from 2-amino-2-methyl-1-propanol and formic acid is described in a note to a procedure for the synthesis of *o*-anisaldehyde[1]:

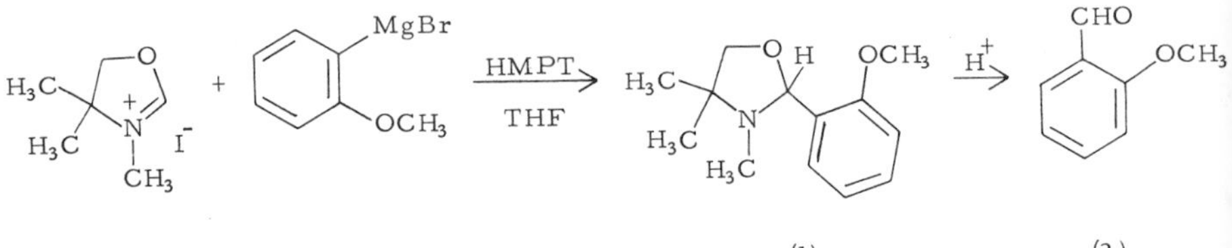

A 1-l. three-necked, round-bottomed flask equipped with a 500-ml. dropping funnel, a mechanical stirrer, and a nitrogen inlet tube is charged with 80 g. (0.33 mole) of N,4,4-trimethyl-2-oxoazolinium iodide. The system is flushed with nitrogen, and 150 ml. of tetrahydrofurane distilled from lithium aluminum hydride is added and the stirred suspension is cooled in an ice bath. Meanwhile, to a cooled solution of freshly prepared *o*-methoxyphenylmagnesium bromide (0.414 mole) from *o*-bromoanisole (77.5 g., 0.414 mole) and magnesium turnings (11 g., 0.458 g.-atom) is added 146 ml. (0.828 mole) of dry hexamethylphosphoramide (dried by distillation from calcium hydride). The resulting solution is blown over under a nitrogen atmosphere into the 500-ml. dropping funnel. The solution is slowly run into the cooled suspension where the methiodide salt dissolves. When the addition is complete, the reaction mixture is stirred at room temperature overnight.

The suspension is cooled to $-5°$, poured onto 600 ml. of ice water, and quickly acidified (pH 2–3) with cold 3 N hydrochloric acid. The acidic solution is rapidly extracted with 300 ml. of cold hexane and the extract discarded. The aqueous solution is then made basic by the addition of 20% sodium hydroxide solution (and ice, if needed). The suspension is extracted with three 1-l. portions of ether and the combined extracts are dried over potassium carbonate, filtered, and concentrated to give 75–85 g. of a pale-yellow mixture of oxazolidine (1) and hexamethylphosphoric triamide. The latter may be removed by elution of an ethereal solution through silica gel.

[1] R. S. Brinkmeyer. E. W. Collington, and A. I. Meyers, *Org. Syn.*, submitted 1971.

**Trimethyloxonium fluoroborate, 1,** 1232; **2,** 438; **3,** 314–315. Supplier: WBL.

*Methylation of 1,3-diazabicyclo[3.1.0]hex-3-enes.*[1] The 1,3-di-azabicyclo[3.1.0]hex-ene-3 (1) is selectively methylated by trimethyloxonium fluoroborate to form 2,2,3-trimethyl-4-phenyl-6-*p*-nitrophenyl-1-aza-3-azoniabicyclo[3.1.0]hexene-3   tetrafluoro-borate (2) in 93% yield. Reaction of (1) with *m*-chloroperbenzoic acid (**1,** 135–139;

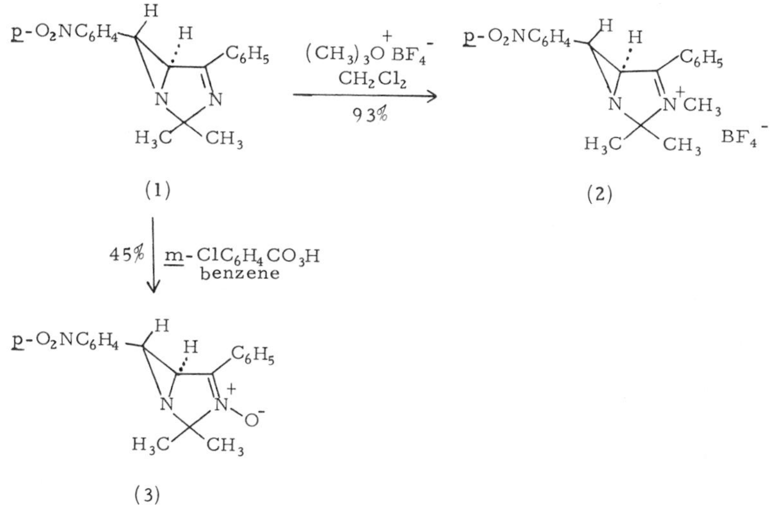

(1)    (2)

(3)

**2,** 68–69; **3,** 49–50) results in attack at the same position to give the N-oxide of (3; 2,2-dimethyl-4-phenyl-6-*p*-nitrophenyl-1,3-diazabicyclo[3.1.0]hexene-3).

[1] H. W. Heine, T. A. Newton, G. J. Blosick, K. C. Irving, C. Meyer, and G. B. Corcoran, III, *J. Org.*, **38,** 651 (1973).

**Trimethyl phosphite, 1,** 1233–1234; **2,** 439–441; **3,** 315–316.

*Corey–Winter olefin synthesis* (**1,** 1233–1234). Chong and Wiseman[1] were able to demonstrate the transient existence of bicyclo[3.2.1]octene-1 (2, a bridgehead alkene which violates Bredt's rule) by application of the Corey–Winter olefin synthesis. Thus treatment of the thionocarbonate (1) with triethyl phosphite at reflux (165°) for 24 hr. in the presence of 1,3-diphenylisobenzofurane (**1,** 342–343; **2,** 178–179) leads to the formation of two Diels–Alder adducts (3) and (4) derived from (2).

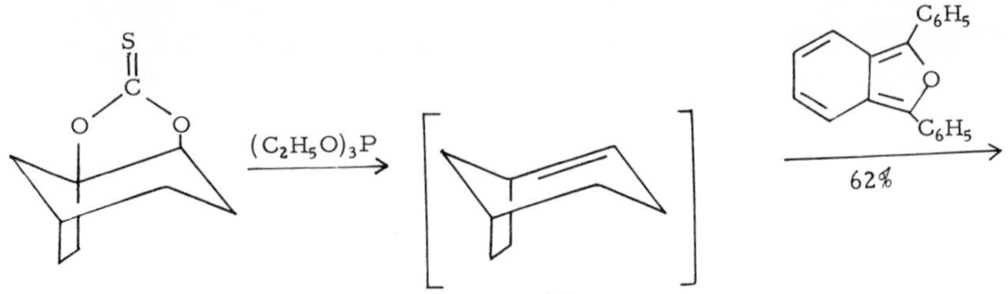

(1)                    (2)

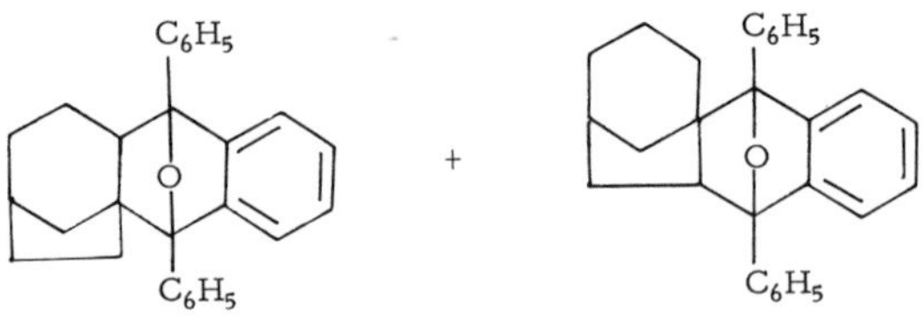

(3)                    (4)

*Esterification.* 1-Methylindole-2-carboxylic acid (1) can be converted into the methyl ester (2) in 94% yield by refluxing trimethyl phosphite.[2]

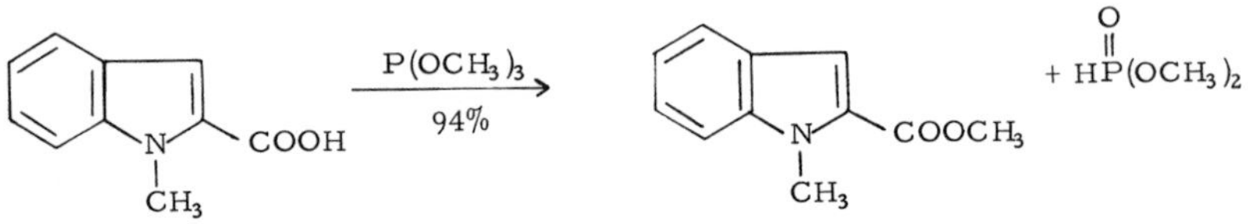

(1)                    (2)

[1] J. A. Chong and J. R. Wiseman, *Am. Soc.*, **94**, 8627 (1972).
[2] J. Szmuszkovicz, *Org. Prep. Proc. :nt.*, **4**, 51 (1972).

**Trimethylsilyl azide**, **1**, 1236; **3**, 316. Suppliers: PCR, WBL.

**Trimethylsilyl cyanide.** $(CH_3)_3SiCN$. Mol. wt. 81.15, b.p. 117°, $n_D^{26}$ 1.3883.

*Preparation.*[1] The reagent can be prepared in several ways:

$$(CH_3)_3SiCl + AgCN \xrightarrow[38\%]{} (CH_3)_3SiCN + AgCl$$

$$2 (CH_3)_3SiCl + K_2Hg(CN)_4 \xrightarrow[31.4\%]{} 2 (CH_3)_3SiCN + 2 KCl + Hg(CN)_2$$

$$(CH_3)_3SiNHSi(CH_3)_3 + 3 HCN \xrightarrow[45.5\%]{} 2 (CH_3)_3SiCN + NH_4CN$$

*Cyanosilylation of carbonyl compounds.*[2] Trimethylsilyl cyanide reacts readily with aldehydes (1, $R^2 = H$) to give cyanosilylated adducts (2, $R^2 = H$) in good yield.

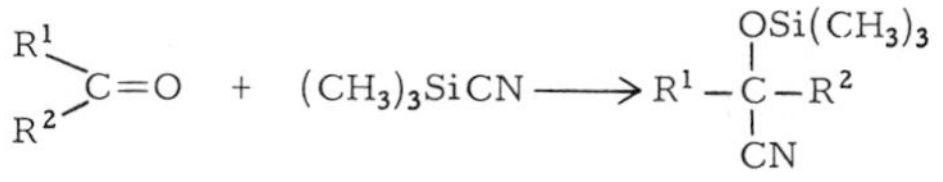

$$R^1 \atop R^2 \,{\Large{>}}\, C=O \quad + \quad (CH_3)_3SiCN \longrightarrow R^1 - \underset{\underset{CN}{|}}{\overset{\overset{OSi(CH_3)_3}{|}}{C}} - R^2$$

(1)                    (2)

Ketones react slowly under the same conditions but, in the presence of zinc iodide as catalyst, they also form the derivatives (2) at room temperature in excellent yields.

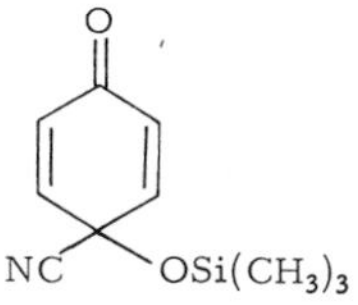

(3) 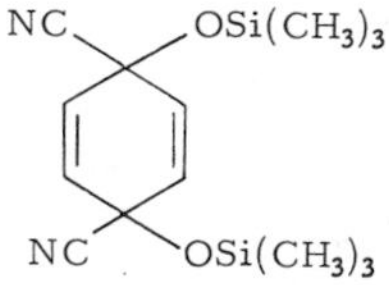                    (4)

$\alpha,\beta$-Unsaturated aldehydes and ketones give only the products of 1,2-addition. Benzoquinone gives the adducts (3) or (4) resulting from addition of either 1 or 2 eq. of the reagent. The adducts can be reduced by $LiAlH_4$ to $\beta$-amino alcohols.

The adducts can also be used as a means of protection of carbonyl groups. They are stable in aprotic media, but the carbonyl group is readily regenerated in dilute aqueous acid or base.

[1]T. A. Bither, W. H. Knoth, R. V. Lindsey, Jr., and W. H. Sharkey, Am. Soc., **80**, 4151 (1958).
[2]D. A. Evans, L. K. Truesdale, and G. L. Carroll, J.C.S. Chem. Comm., 55 (1973).

**Trimethylsilyldiazomethane**, $(CH_3)_3SiCHN_2$. Mol. wt. 114.23, b.p. 96°/775 mm.

*Preparation.*[1] This unusually stable diazoalkane can be prepared by the sequence shown.

$$(CH_3)_3SiCH_2Cl \xrightarrow{NH_3} (CH_3)_3SiCH_2NH_2 \xrightarrow{urea,\ HCl} (CH_3)_3SiCH_2NHCONH_2$$

m. p. 113-114°

$$\xrightarrow{HNO_2} (CH_3)_3SiCH_2N(NO)CONH_2 \xrightarrow[\underset{56\%}{C_6H_6}]{20\%\ KOH} (CH_3)_3SiCHN_2$$

(1)

*1,3-Dipolar addition.*[1] The reagent (1) in decalin solution undergoes 1,3-dipolar addition to acrylonitrile to give the 2-pyrazoline (2) in 73% yield.

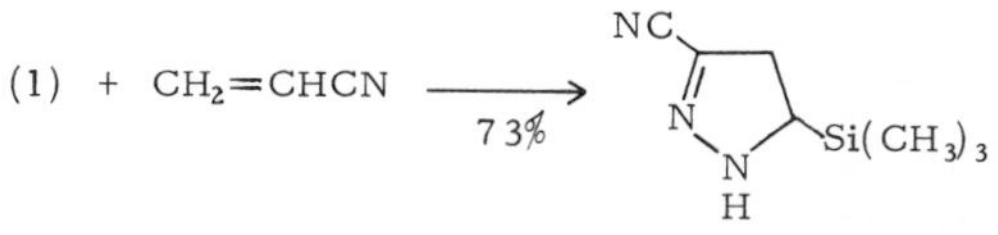

$$(1) \quad + \quad CH_2{=}CHCN \xrightarrow{73\%}$$

(2)

*Trimethylsilylcarbene.*[1] Addition of cuprous chloride to a solution of (1) in benzene in the presence of cyclohexene gives, as the major product, *anti*-7-trimethylsilylnorcarane (2, 65% yield).

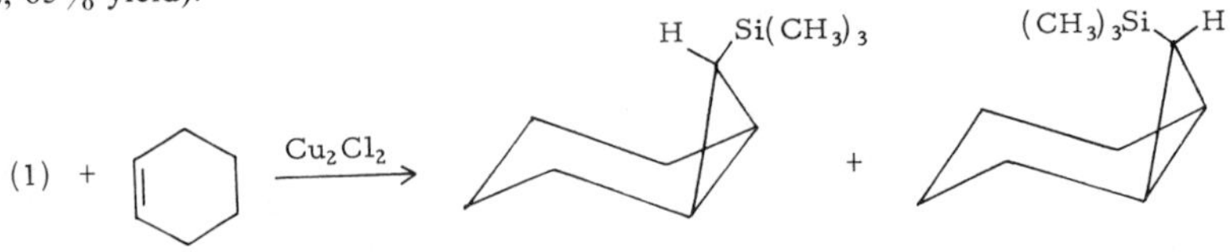

(1) +    Cu$_2$Cl$_2$ →

(2, 65%)            (3, 7%)

*Synthesis of acetylenes from carbonyl compounds.*[2] The reagent (1), or dimethyl-phosphonodiazomethane (2),[3] after treatment with *n*-butyllithium in THF at −78°, reacts with a carbonyl compound, for example benzophenone, to form an acetylene, diphenylacetylene.

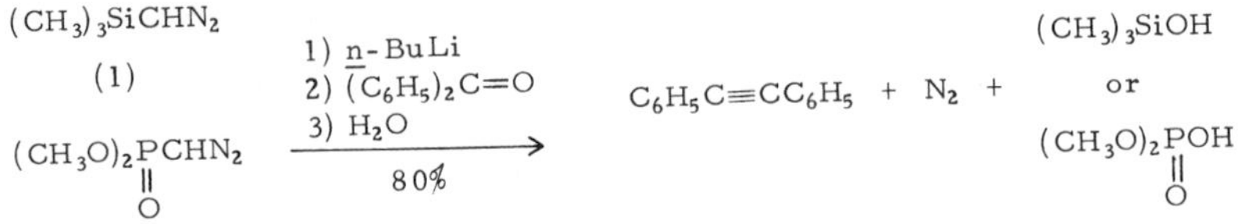

(CH$_3$)$_3$SiCHN$_2$
(1)

(CH$_3$O)$_2$PCHN$_2$
‖
O
(2)

1) *n*-BuLi
2) (C$_6$H$_5$)$_2$C=O
3) H$_2$O
————→
80%

C$_6$H$_5$C≡CC$_6$H$_5$ + N$_2$ +

(CH$_3$)$_3$SiOH
or
(CH$_3$O)$_2$POH
‖
O

Enolization is a competing reaction in the case of carbonyl compounds possessing α-hydrogen atoms. Thus acetophenone is converted into 1-phenylpropyne in only 16% yield with 50% recovery of the ketone. Aldehydes are converted into terminal acetylenes. For example, phenylacetaldehyde is converted into 3-phenylpropyne in 30% yield.

This reaction probably involves a Wolff rearrangement.[4]

[1] D. Seyferth, A. W. Dow, H. Menzel, and T. C. Flood, *Am. Soc.*, **90**, 1080 (1968).
[2] E. W. Colvin and B. J. Hamill, *J.C.S. Chem. Commun.*, 151 (1973).
[3] D. Seyferth, R. S. Marmor, and P. Hilbert, *J. Org.*, **36**, 1384 (1971).
[4] W. E. Bachman and W. S. Struve, *Org. React.*, **1**, 38 (1942).

## Trimethylsilyldiethylamine (TSiD), 3, 317.

*Conversion of F into E prostaglandins.* Yankee and Bundy[1] have reported that 15-methyl-PGF$_{2\alpha}$ methyl ester (1) can be silylated preferentially at the 11-position

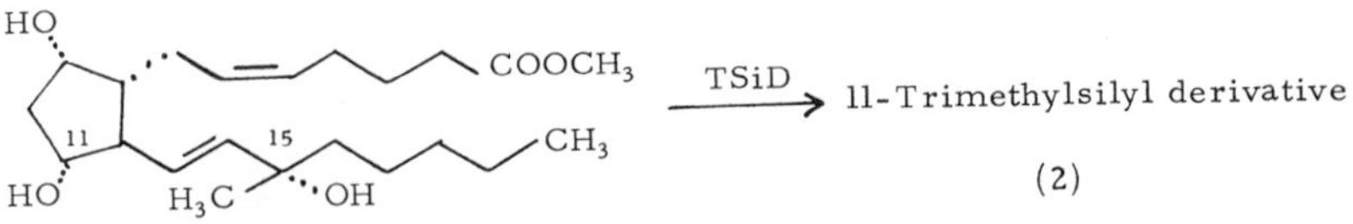

TSiD
————→ 11-Trimethylsilyl derivative

(2)

(1)

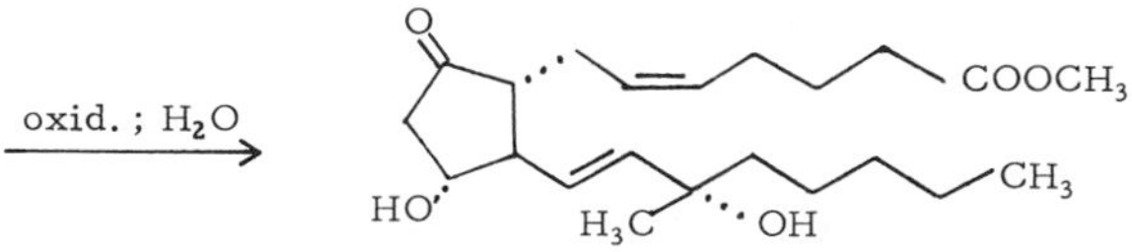

$$\xrightarrow{\text{oxid. ; } H_2O}$$

(3)

using TSiD. Oxidation of the resulting 11-trimethylsilyl derivative (2) with Collins reagent followed by hydrolysis of the TMS group with aqueous methanol containing a trace of acetic acid gives the methyl ester of 15-methyl $PGE_2$ (3) in $45\%$ yield from (1).

This method has now been shown to be general for conversion of F into E prostaglandins.[2] Thus the methyl ester of $PGF_{2\alpha}$ (4) on treatment with TSiD is disilylated at

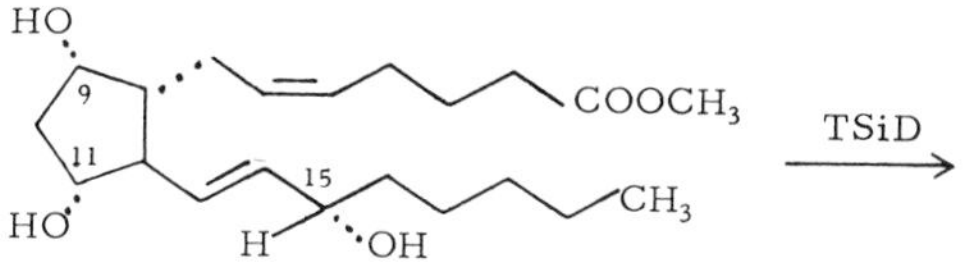

$$\xrightarrow{\text{TSiD}}$$  11, 15 - Ditrimethylsilyl derivative

(5)

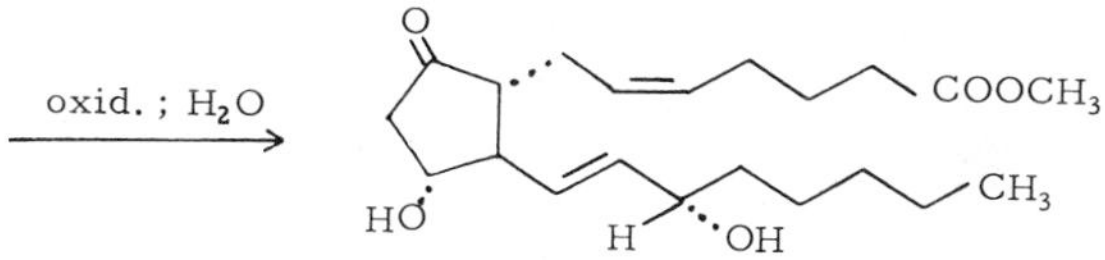

(4)

$$\xrightarrow{\text{oxid. ; } H_2O}$$

(6)

both $C_{11}$ and $C_{15}$, leaving the $C_9$-hydroxyl group free for oxidation. The methyl ester of $PGE_2$ (6) is obtained in this way in $35$–$50\%$ overall yield.

[1] E. W. Yankee and G. L. Bundy, *Am. Soc.*, **94**, 3651 (1972).
[2] E. W. Yankee, C. H. Lin, and J. Fried, *Chem. Commun.*, 1120 (1972).

**1,3,5-Trimethyl-2,4,6-tris[3,5-di-*tert*-butyl-4-hydroxybenzyl]benzene** ("Ethyl" antioxidant 330). Mol. wt. 775.2, m.p. 244°.
Solubility (wt. $\%$ at 18°)

| | | | |
|---|---|---|---|
| Methylene chloride | 31.9 | Methanol | 0.20 |
| Benzene | 20.0 | Isopropyl alcohol | 0.10 |
| Methylcyclohexane | 1.7 | Water | Insoluble |

Supplier: Ethyl Corporation.

A sterically hindered, temperature-stable, phenolic antioxidant. It is a highly effective, noncoloring, odorless stabilizer for plastics, resins, rubber, and waxes. It has exceptionally low volatility and is outstanding in applications requiring high processing temperatures.

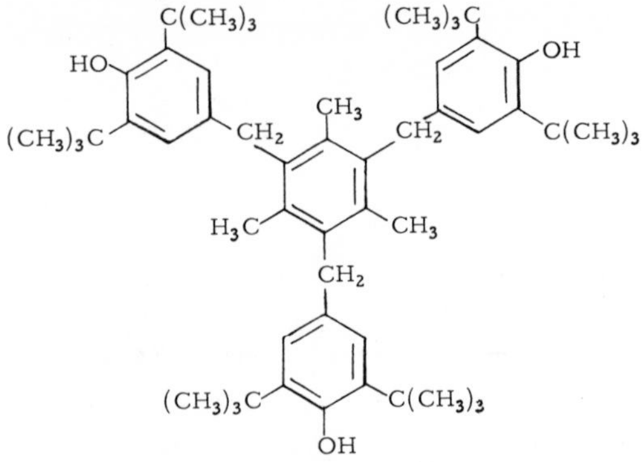

**1,2,3-Trioxo-2,3-dihydrophenalene** (1). Mol. wt. 210.1, red crystals, dec. 255–260°.

*Preparation* from 1,8-naphthalic anhydride in three steps with an overall yield of 60–65%.

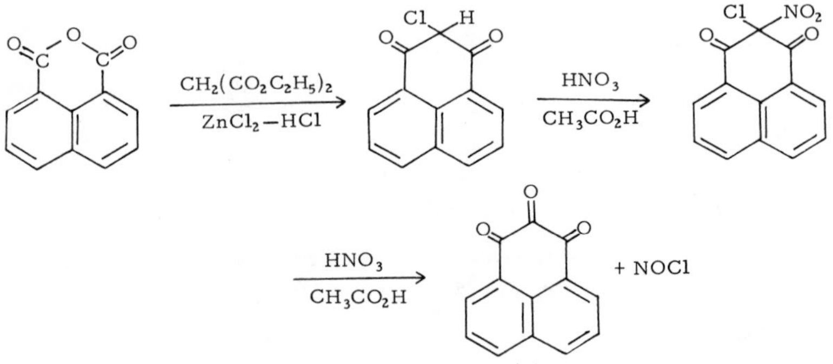

(1)

*Condensation with dibenzyl ketone:*

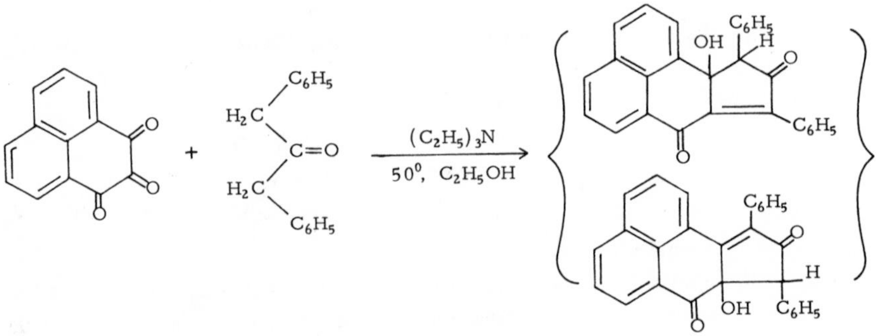

$$Ac_2O \longrightarrow$$

[1] W. Ried and M. L. Mehrotra, *Ann.*, **718**, 120 (1968).

**Triphenylarsine oxide**, $(C_6H_5)_3As{=}O$. Mol. wt. 332.21.

Anhydrous triphenylarsine oxide is obtained from commerical material (which contains water; suppliers: Columbia, Eastman, Fisher, K and K, P and B, Sargent, Schuchardt) by heating a sample under reflux with benzene under a Dean–Stark trap (*ca.* 5 hr.). The resulting solid is then dried at 135° (0.1 mm.).

*Reverse Wittig reactions.*[1] Dicyanoacetylene reacts with triphenylphosphine oxide in benzene at 160° to give triphenylphosphoranylideneoxalacetonitrile (1) in 78 % yield by way of the intermediates formulated.

$$NC-C{\equiv}C-CN \; + \; (C_6H_5)_3P{=}O \; \underset{\longleftarrow}{\longrightarrow} \; \left[ \begin{array}{ccc} NC-\overset{-}{C}{=}C-CN & \longrightarrow & NC-C{=}C-CN \\ \underset{(C_6H_5)_3\overset{+}{P}-O}{|} & \longleftarrow & \underset{(C_6H_5)_3\overset{|}{P}-O}{|\;\;|} \end{array} \right] \underset{78\%}{\longrightarrow}$$

$$(C_6H_5)_3P{=}C\begin{array}{l} {\diagup}CN \\ {\diagdown}COCN \end{array}$$

(1)

The reaction with triphenylphosphine oxide cannot be extended to other negatively substituted acetylenes. However, the reaction with triphenylarsine oxide proceeds much more readily, and this reagent reacts with methyl propiolate, dimethyl acetylene-dicarboxylate, ethyl phenylpropiolate, and hexafluoro-2-butyne to give products (2) in good yield (55–90 %).

$$R^1-C{\equiv}CR^2 \; + \; (C_6H_5)_3As{=}O \; \underset{\longleftarrow}{\longrightarrow} \; \left[ \begin{array}{ccc} (C_6H_5)_3\overset{+}{A}s-O & & (C_6H_5)_3As-O \\ \underset{R^1\overset{-}{C}{=}C-R^2}{|} & \longrightarrow & \underset{R^1C{=}C-R^2}{|\;\;|} \end{array} \right] \longrightarrow$$

$$(C_6H_5)_3As{=}C\begin{array}{l} {\diagup}COR^2 \\ {\diagdown}R^1 \end{array}$$

(2)

Triphenylstibine oxide, $(C_6H_5)_3Sb{=}O$, reacts with dicyanoacetylene at room temperature, but no pure products could be isolated. Reaction with methyl propiolate at 115° gives methyl phenylpropiolate in 40 % yield.

[1] E. Ciganek, *J. Org.*, **35**, 1725 (1970).

**Triphenylmethyl hexafluorophosphate**, $(C_6H_5)_3C^+PF_6^-$. Mol. wt. 388.28. Supplier: Ozark Mahoning Company.

*Synthesis of aldehydes and ketones.*[1] Trityl ethers[2] disproportionate to triphenyl-methane and aldehydes or ketones in the presence of small amounts of trityl salts: trityl hexafluorophosphate, trityl hexafluoroantimonate, $(C_6H_5)_3C^+AbF_6^-$, or trityl hexafluoroarsenate, $(C_6H_5)_3C^+AsF_6^-$.

$$(C_6H_5)_3COCH{\overset{R^1}{\underset{R^2}{\Big\langle}}} \longrightarrow (C_6H_5)_3CH + {\overset{R^1}{\underset{R^2}{\Big\rangle}}}C=O$$

In a typical experiment, trityl benzyl ether (5.0 mmole) is added to a stirred solution of the trityl salt (0.50 mmole) in acetonitrile. After 4 hr., water is added and the products isolated. Benzaldehyde and triphenylmethane are obtained in quantitative yields. The reaction thus provides a method for oxidation of alcohols to aldehydes or ketones.

The reaction probably involves a cationic chain reaction involving hydride transfer, and indeed the rate-limiting step in the oxidation of trityl benzyl ether was shown to be a chain-propagating step (equation I).

$$(I) \quad (C_6H_5)_3COCH_2C_6H_5 + (C_6H_5)_3C^+ \xrightarrow{k} (C_6H_5)_3C^+ + C_6H_5CHO + (C_6H_5)_3CH$$

[1]M. P. Doyle, D. J. DeBruyn, and D. J. Scholten, *J. Org.*, **38**, 625 (1973).
[2]Trityl ethers are conveniently prepared in high yield from trityl chloride (**1**, 1254–1256) and alcohols.

**Triphenylphosphine**, **1**, 1238–1247; **2**, 443–445; **3**, 317–320.

*Reduction of allocimene hydroperoxide (1) to ocimene alcohol (2)*[1]:

*Synthesis of polycyclic hydrocarbons by intramolecular C-alkylation of phosphine-alkylenes.*[2] An aromatic compound of type (1) can be converted by reaction with 1 mole of triphenylphosphine into a monophosphonium salt (2) which with base gives the corresponding phosphinealkene (5).

$$
\underset{(3)}{\overset{\displaystyle Ar}{\underset{\displaystyle \diagdown}{\overset{\displaystyle \diagup}{C}}} \overset{H}{\underset{}{\overset{|}{C}}} \overset{+}{-}P(C_6H_5)_3 \atop (CH_2)_n \overset{}{\diagdown} CH_2Br} \longrightarrow \left[ \underset{(4)}{Ar \overset{\overset{\displaystyle +}{P(C_6H_5)_3}}{\underset{(CH_2)_n}{\overset{|}{\underset{}{CH}}}} CH_2} \right] Br^- \xrightarrow{\text{Base}} \underset{(5)}{Ar \overset{P(C_6H_5)_3}{\underset{(CH_2)_n}{\overset{}{\underset{}{C}}}} CH_2}
$$

$$\downarrow \text{RCHO}$$

$$
\underset{(6)}{Ar \overset{\overset{\displaystyle HCR}{\overset{\|}{C}}}{\underset{(CH_2)_n}{}} CH_2}
$$

***Reaction with β-chlorovinyl ketones[3]:***

$$(C_6H_5)_3P + ClCH=CHCOCH_3 \longrightarrow [(C_6H_5)_3\overset{+}{P}-CH=CHCOCH_3]Cl^-$$

***Isocyanides.*[4]** In the previously preferred method of preparation of isocyanides, a formamide is dehydrated by the combination of phosphoryl chloride with either pyridine or potassium *t*-butoxide (**1**, 878). In a recently reported procedure the ternary

$$RNHCHO + (C_6H_5)_3P + CCl_4 + (C_2H_5)_3N \longrightarrow R-\overset{+}{N}\equiv\overset{-}{C}: + (C_6H_5)_3PO + HCCl_3 + (C_2H_5)_3N \cdot HCl$$

system triphenylphosphine–carbon tetrachloride–triethylamine is used for this purpose. Use of a 20% excess of triphenylphosphine gives yields of isocyanides of approximately 90%. 1,2-Dichloroethane, methylene chloride, or chloroform is used as solvent.

***Dibromoketene.*[5]** The reaction of trimethylsilyl tribromoacetate (1)[6] with triphenylphosphine in the presence of excess cyclopentadiene gives the cycloaddition product, 7,7-dibromobicyclo[3.2.0]hept-2-ene-6-one (2), in 89% yield.

$$\underset{(1)}{(CH_3)_3SiOCOCBr_3} + (C_6H_5)_3P + \left[\text{cyclopentadiene}\right] \longrightarrow$$

$$\underset{82\% \quad\quad 87\%}{(CH_3)_3SiBr + (C_6H_5)_3PO} + \left[\text{bicyclic structure with O and Br}\right]$$

$$(2, 89\%)$$

***Wittig-olefination.*[7]** The N-bridgehead bicyclic amides (1) and (2) react with (ethoxycarbonylmethylene)triphenylphosphorane or (cyanomethylene)triphenylphosphorane to form mono- and diolefins. The reaction takes place preferentially at a five-membered carbonyl group.

(1)        (3)        (4)

$$\begin{cases} R = CO_2C_2H_5 \\ R = CN \end{cases}$$

(2)                    (5)

***Olefin synthesis by double extrusion*** (**3**, 319–320). The first of a number of definitive papers on this new method of olefin synthesis has been published.[8]

[1] E. K. von Gustorf, F.-W. Grevels, and G. O. Schenck, *Ann.*, **719** 1 (1968).

[2] H. J. Bestmann, R. Härtl, and H. Häberlein, *ibid.*, **718**, 33 (1968).

[3] E. Zbiral and E. Werner, *ibid.*, **707**, 130 (1967).

[4] R. Appel, R. Kleinstück, and K.-D. Ziehn, *Angew. Chem., internat. Ed.*, **10**, 132 (1971).

[5] T. Okada and R. Okawara, *Tetrahedron Letters*, 2801 (1971).

[6] Prepared in 75% yield from the reaction of hexamethyldisilazane and tribromoacetic acid.

[7] W. Flitsch and B. Müter, *Ber.*, **104**, 2852 (1971).

[8] D. H. R. Barton and B. J. Willis, *J.C.S. Perkin I*, 305 (1972).

**Triphenylphosphine–Carbon tetrabromide** (*see also* **Triphenylphosphine–Carbon tetrachloride**).

$RCHO \rightarrow RC{\equiv}CH$. Corey and Fuchs[1] have effected the transformation $RCHO \rightarrow RC{\equiv}CH$ or $RC{\equiv}CR'$ in two steps:

$$RCHO \longrightarrow RCH{=}CBr_2 \longrightarrow RC{\equiv}CH \text{ or } RC{\equiv}CR'$$

(1)        (2)        (3)        (4)

The first step is carried out by either of two procedures. In one, the aldehyde (1 eq.) is added to a mixture of triphenylphosphine (4 eq.) and carbon tetrabromide (2 eq.) in methylene chloride at 0° with a reaction time of 5 min. In the second, preferred method, a reagent is prepared by the interaction of zinc dust (2 eq.), triphenylphosphine (2 eq.), and carbon tetrabromide (2 eq.) in methylene chloride at room temperature for 24–30 hr. The aldehyde (1 eq.) is added to this reagent and the reaction is allowed to proceed for 1–2 hr. The yields by the second method are 80–90%.

The dibromo olefins obtained in this way are converted into the lithium acetylide by treatment in THF with *n*-butyllithium or with powdered 1.5% lithium amalgam.

$$RCH=CBr_2 \xrightarrow[-78^\circ]{2BuLi} RC\equiv CLi \xrightarrow{H_2O} RC\equiv CH$$

(2) ... CO$_2$ ... (3)

$$RC\equiv CCOOH$$
(4)

Hydrolysis yields the terminal acetylene (3); carbonation yields the propargylic acid (4). Synthesis of a wide variety of acetylenes is possible by the reaction of the intermediate lithium acetylides with other electrophiles (alkyl halides, aldehydes, ketones).

[1]E. J. Corey and P. L. Fuchs, *Tetrahedron Letters*, 3769 (1972).

**Triphenylphosphine–Carbon tetrachloride**, 1, 1247; 2, 445; 3, 320.

*Reaction with enolizable ketones.*[1] Triphenylphosphine in carbon tetrachloride reacts with cyclohexanone under reflux to give (1) and (2) in the ratio of 93:7. Similar

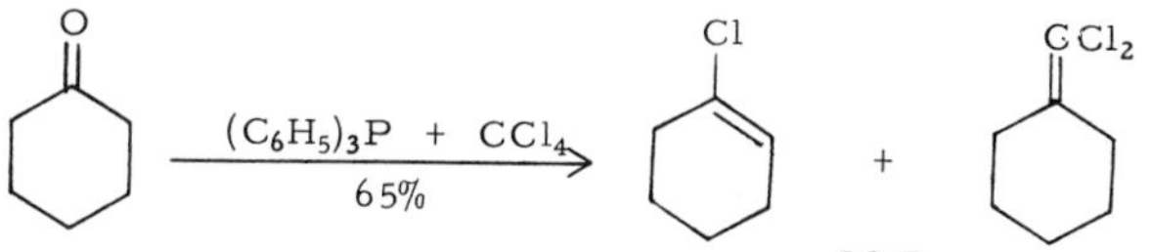

products were obtained in the case of cyclopentanone, but the product (4) corresponding to (2) was the major product. The dichloromethylene adducts (2) and (4) are probably formed via (5), which has been identified as a product of the reaction of triphenylphosphine with carbon tetrachloride.[2]

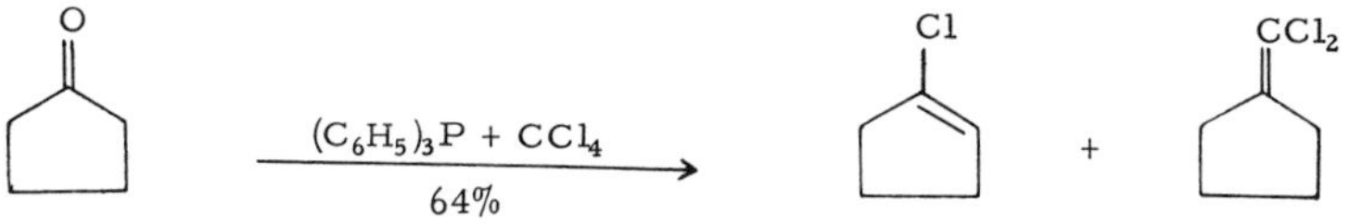

$$2 (C_6H_5)_3P + CCl_4 \longrightarrow (C_6H_5)_3P=CCl_2 + (C_6H_5)_3PCl_2$$
(5) ... (6)

*Reaction with epoxides.*[3] Epoxides react with triphenylphosphine and carbon tetrachloride (reflux under $N_2$ for 1–2 hr.) to give the corresponding *cis*-1,2-dichloroalkane and triphenylphosphine oxide. Thus cyclohexene oxide is converted into *cis*-1,2-dichlorocyclohexane in 80% yield. Only a trace of *trans*-1,2-dichlorocyclohexane is formed.

*Allylic chlorides.* Allylic alcohols can be converted into allylic chlorides with no, or slight, rearrangement by the triphenylphosphine–carbon tetrachloride reagent.[4]

$$CH_3CH=CHCH_2OH \xrightarrow{100\%} CH_3CH=CHCH_2Cl$$

$$CH_3CHOHCH=CH_2 \longrightarrow CH_3CHClCH=CH_2 + CH_3CH=CHCH_2Cl$$
89% ... 11%

The reagent is recommended for conversion of allylic alcohols into allylic chlorides without rearrangement. Thus geraniol (1) can be converted into geranyl chloride (2) in 75–81% yield.[5]

$$\text{(1)} \xrightarrow[75-81\%]{(C_6H_5)_3P/CCl_4} \text{(2)} + (C_6H_5)_3PO + HCCl_3$$

Carbon tetrachloride serves as both solvent and halogen source; it is usually desirable to employ a modest excess of triphenylphosphine. Hooz has suggested that the reaction proceeds through an $S_N2$ process.

$$(C_6H_5)_3P + CCl_4 \longrightarrow (C_6H_5)_3\overset{+}{P}Cl\ C\bar{C}l_3 \xrightarrow[-HCCl_3]{ROH} (C_6H_5)_3\overset{+}{P}Cl\ \bar{O}R \longrightarrow$$

$$(C_6H_5)_3\overset{+}{P}OR\ Cl^- \longrightarrow RCl + (C_6H_5)_3PO$$

*Carbodiimide synthesis.*[6] 1,3-Disubstituted ureas or thioureas react with triphenylphosphine, carbon tetrachloride, and triethylamine in methylene chloride at 40° to give carbodiimides in yields of 85–92%.

$$\underset{X = O,\ S}{\overset{\overset{X}{\underset{\|}{}}}{RNHCNHR'}} + (C_6H_5)_3P + CCl_4 + (C_2H_5)_3N \longrightarrow R-N=C=N-R' + (C_6H_5)_3P{=}X + HCCl_3 + (C_2H_5)_3N \cdot HCl$$

*Peptide synthesis.* Wieland and Seeliger[7] have used the combination of triphenylphosphine–carbon tetrachloride and triethylamine for coupling of Bocamino acids with amino acid esters to form peptides. However, extensive racemization is observed.

*Reaction with acylglycerols.*[8] The reaction of 1,3-distearoylglycerol (1) with $P(C_6H_5)_3$–$CCl_4$ under reflux gives 2-chlorodesoxy-1,3-distearoylglycerol (2) in 95%

yield. Normally nucleophilic substitution at $C_2$ in acylglycerols (glycerides) proceeds with migration of an acyloxy group at $C_1$ or at $C_3$ to $C_2$. The reaction occurs without inversion at $C_2$.

[1] N. S. Isaacs and D. Kirkpatrick, *Chem. Commun.*, 443 (1972).
[2] F. Ramirez, N. B. Desai, and N. McKelvie, *Am. Soc.*, **84**, 1745 (1962).
[3] N. S. Isaacs and D. Kirkpatrick, *Tetrahedron Letters*, 3869 (1972).
[4] E. W. Collington and A. I. M. Meyers, *J. Org.*, **36**, 3044 (1971); E. I. Snyder, *ibid.*, **37**, 1466 (1972).
[5] J. G. Calzada and J. Hooz, *Org. Syn.*, submitted 1973.
[6] R. Appel, R. Kleinstück, and K.-D. Ziehn, *Ber.*, **104**, 1335 (1971).
[7] T. Wieland and A. Seeliger, *ibid.*, **104**, 3992 (1971).
[8] R. Aneja, A. P. Davies, and J. A. Knaggs, *J.C.S. Chem. Comm.*, 110 1973).

**Triphenylphosphine–Diethyl azodicarboxylate**, **1**, 245–247.

*Selective phosphorylation of 5'-hydroxyl groups of nucleosides.*[1] Japanese chemists have effected selective phosphorylation of the 5'-hydroxyl group of the nucleosides

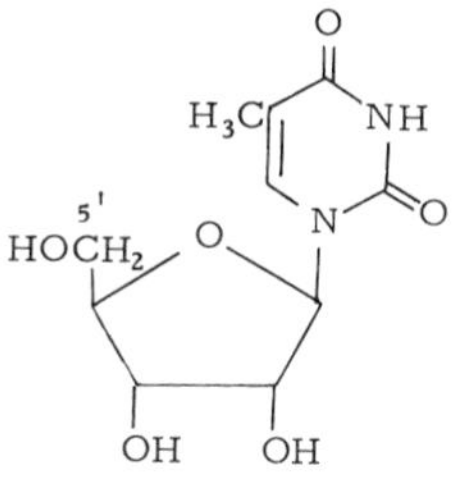

Thymidine 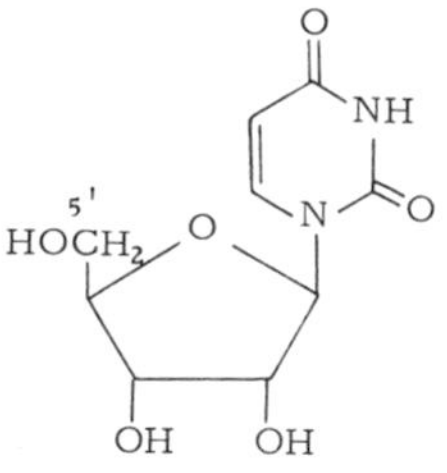                    Uridine

thymidine and uridine by reaction with diethyl azodicarboxylate, triphenylphosphine, and dibenzyl hydrogen phosphate. Quaternary phosphonium salts are considered to be involved. The selectivity for a primary hydroxyl group is ascribed to the bulkiness of the intermediate salt (a).

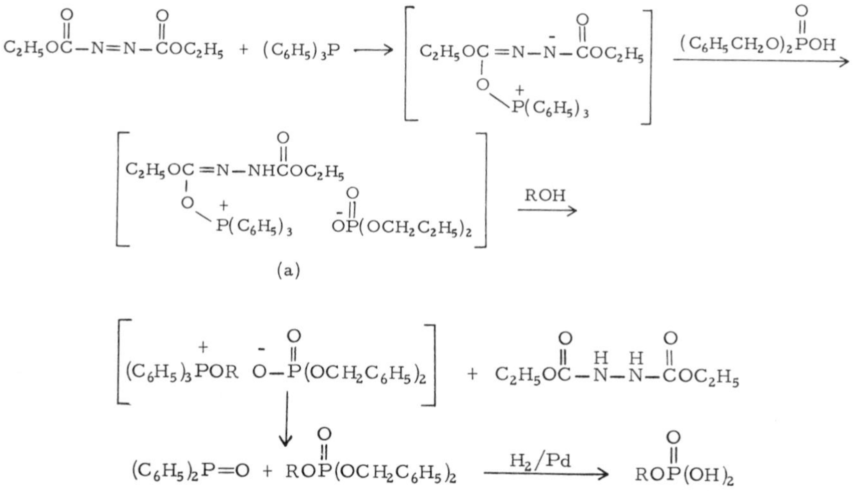

Thymidine 5'-phosphate was obtained by this procedure in 47% yield; uridine 5'-phosphate in 63% yield.

*Conversion of alcohols into amines.*[2] The reaction of phthalimide (1) with various alcohols and triphenylphosphine–diethyl azodicarboxylate gives N-alkylphthalimides (2) in yields of 60–90%. Since they are converted into amines by treatment with hydrazine hydrate (**1**, 442), the reaction provides a means of converting alcohols into amines.

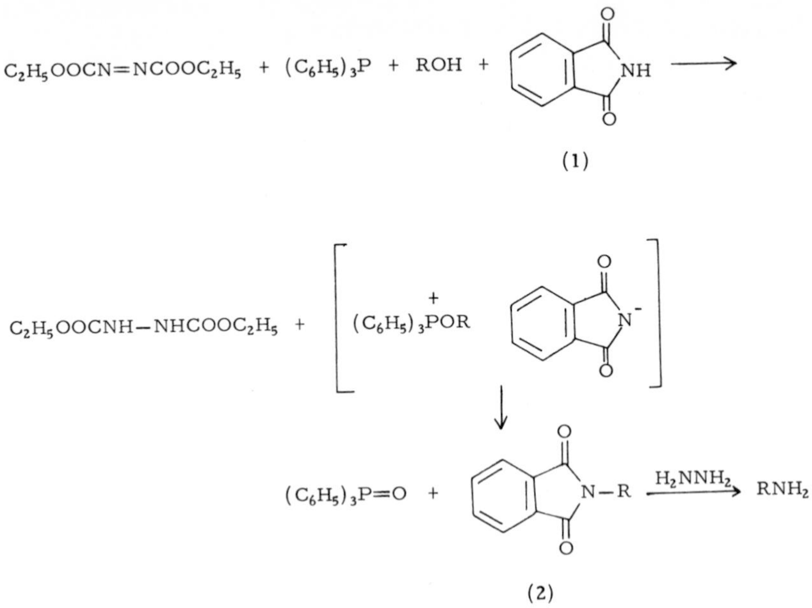

$$C_2H_5OOCN=NCOOC_2H_5 \;+\; (C_6H_5)_3P \;+\; ROH \;+\;$$

(1)

$$C_2H_5OOCNH-NHCOOC_2H_5 \;+\; \left[ (C_6H_5)_3\overset{+}{P}OR \right]$$

$$(C_6H_5)_3P=O \;+\; \underset{}{}N-R \xrightarrow{H_2NNH_2} RNH_2$$

(2)

Use of (S)-(+)-2-octanol of high optical activity in this reaction led to (R)-(−)-2-octylamine with high optical purity. The reaction thus proceeds with nearly complete inversion of configuration of the alkyl group.

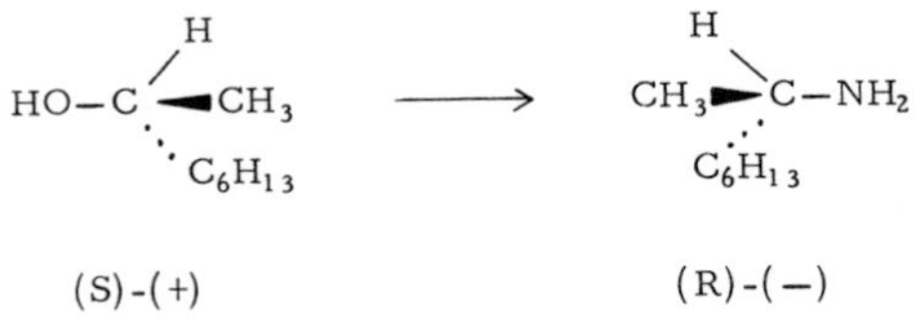

$$HO-C \overset{H}{\underset{C_6H_{13}}{\diagdown}} CH_3 \longrightarrow CH_3 \overset{H}{\underset{C_6H_{13}}{\diagdown}} C-NH_2$$

(S)-(+)                                    (R)-(−)

*Isocyanides.* Ugi *et al.*[3] investigated the synthesis of isocyanides (2) by dehydration of N-monosubstituted formamides (1) with triphenylphosphine–diethyl azodicarboxylate. In the 11 cases investigated, no isocyanide could be obtained in six of them. In the

$$RNHCHO \xrightarrow[-H_2O]{} RN=C:$$

(1)                                    (2)

other five cases, yields of isocyanides ranged from 25 to 58%. The method, however, is useful in preparation of isocyanides that racemize or decompose in a basic medium.

*Alkylation of active hydrogen compounds.* Japanese chemists[4] have reported a few examples of intermolecular dehydration between alcohols and active hydrogen

$$ROH + CH_2 \underset{Y}{\overset{X}{\diagup}} + C_2H_5O\overset{O}{\overset{\|}{C}}-N{=}N-\overset{O}{\overset{\|}{C}}OC_2H_5 + (C_6H_5)_3P \longrightarrow$$

$$X = Y = CN$$

$$X = CN; \quad Y = COOC_2H_5$$

$$X = CH_3CO; \quad Y = COOC_2H_5$$

$$R-CH\underset{Y}{\overset{X}{\diagup}} + C_2H_5O\overset{O}{\overset{\|}{C}}-\overset{H}{\overset{|}{N}}-\overset{H}{\overset{|}{N}}-\overset{O}{\overset{\|}{C}}OC_2H_5 + (C_6H_5)_3P{=}O$$

compounds using the combination of diethyl azodicarboxylate and triphenylphosphine. Yields are moderate. The reaction was extended to the alkylation of amides.

[1] O. Mitsunobu, K. Kato, and J. Kimura, *Am. Soc.*, **91**, 6510 (1969).
[2] O. Mitsunobu, M. Wada, and T. Sano, *ibid.*, **94**, 679 (1972).
[3] B. Beijer, E. von Hinrichs, and I. Ugi, *Angew. Chem.*, **11**, 929 (1972).
[4] M. Wada and O. Mitsunobu, *Tetrahedron Letters*, 1279 (1972).

### Triphenylphosphine–N-Halosuccinimide.

*OH → X.*[1] A carbohydrate hydroxyl group can be replaced by halogen (bromine, chlorine, iodine) by treatment in DMF with 2 eq. each of triphenylphosphine and an N-halosuccinimide. The by-products are succinimide and triphenylphosphine oxide. Yields are generally high. Primary hydroxyl groups can be selectively replaced in the presence of secondary hydroxyl groups.

[1] S. Hanessian, M. M. Ponpipom, and P. Lavallee, *Carbohydrate Res.*, **24**, 45 (1972).

### Triphenylphosphine dibromide, 1, 1247–1249; 2, 446; 3, 320–322.

*Carbodiimides.*[1] Triphenylphosphine dibromide reacts with N,N-disubstituted ureas in the presence of triethylamine to give carbodiimides.

$$[(C_6H_5)_3\overset{+}{P}Br]Br^- + R-\overset{H}{\overset{|}{N}}-\overset{O}{\overset{\|}{C}}-\overset{H}{\overset{|}{N}}-R' \longrightarrow \left[ R-\overset{(C_6H_5)_3PBr}{\overset{|}{\underset{H}{\overset{+}{N}}}}-\overset{O}{\overset{\|}{C}}-\overset{\phantom{H}}{\underset{H}{\overset{+}{N}}}-R \right] Br^- \longrightarrow R-N{=}C{=}N-R' + (C_6H_5)_3P{=}O$$

*Benzoins → Benzils.*[2] Benzoins are oxidized to benzils in high yield (75–98 %) by treatment with triphenylphosphine dibromide in acetonitrile.

$$Ar\overset{}{\underset{OH}{\overset{|}{C}}}H-\overset{O}{\overset{\|}{C}}Ar \xrightarrow{\ (C_6H_5)_3PBr_2,\ CH_3CN\ } Ar\overset{O}{\overset{\|}{C}}-\overset{O}{\overset{\|}{C}}Ar$$

[1] H. J. Bestmann, J. Lienert, and L. Mott, *Ann..* **718**, 24 (1968).
[2] T.-L. Ho, *Synthesis*, 697 (1972).

**Triphenylphosphine selenide**, $(C_6H_5)_3P{=}Se$. Mol. wt. 341.24, m.p. 183–184°. Supplier: Aldrich.

*Preparation*[1] The reagent is prepared by heating triphenylphosphine (3 moles) with selenium (1 mole).

*Deoxygenation of epoxides.*[2] The reagent, in combination with trifluoroacetic acid, converts epoxides into olefins at room temperature rapidly and stereospecifically with preservation of the stereochemistry about the C–C bond of the epoxide.

Examples:

$$1,2\text{-Epoxyoctane} \longrightarrow \text{Octene-1}\quad(71\%)$$

$$\underline{cis}\text{-Stilbene oxide} \longrightarrow \underline{cis}\text{-Stilbene}\;(\underline{ca.}\;71\%)$$

$$\text{Cyclohexene oxide} \longrightarrow \text{Cyclohexene}\;(53\%)$$

[1] A. Michaelis and H. v. Soden, *Ann.*, **229**, 295 (1885).
[2] D. L. J. Clive and C. V. Denyer, *J.C.S. Chem. Comm.*, 253 (1973).

**Triphenyl phosphite, 1,** 1249; **2,** 446; **3,** 322–323.

*Iodination procedure A.*[1] In this procedure of Rydon's, which is the simplest to perform and the best for the formation of iodides from sterically hindered alcohols, for example neopentyl alcohol, a 500-ml. two-necked. round-bottomed flask fitted with a reflux condenser carrying a calcium chloride drying tube is charged with 136 g.

$$A.\quad CH_3\underset{\underset{CH_3}{|}}{\overset{\overset{CH_3}{|}}{C}}CH_2OH + CH_3I + (C_6H_5O)_3P \longrightarrow CH_3\underset{\underset{CH_3}{|}}{\overset{\overset{CH_3}{|}}{C}}CH_2I + (C_6H_5O)_2POCH_3 + C_6H_5OH$$

(0.44 mole) of triphenyl phosphite, 35.2 g. (0.4 mole) of neopentyl alcohol, and 85 g. (37 ml., 0.60 mole) of methyl iodide, and a thermometer is inserted of sufficient length to extend into the liquid contents of the flask. The mixture is heated electrically under gentle reflux until the temperature of the liquid rises from an initial value of 75–80° to about 130° and the mixture darkens and begins to fume (about 24 hr.). It is necessary to adjust the heat imput as the reaction proceeds and the rate of refluxing diminishes. The reaction is conveniently monitored by infrared spectroscopy. As the reaction proceeds, a broad, strong band at 865 cm.$^{-1}$ with a shoulder at 880 cm.$^{-1}$ disappears and another broad, strong band at 945 cm.$^{-1}$ and a sharp, medium band at 1310 cm.$^{-1}$ appears. Distillation at reduced pressure, washing with water and then with cold 1 *N* sodium hydroxide solution to remove phenol, washing, drying, and redistillation affords 51–60 g. (64–75%) of neopentyl iodide, b.p. 54–55°/55 mm.

*Iodination procedure B.* This procedure is preferred for sensitive alcohols subject to ready elimination to give olefins, and is illustrated by the preparation of cyclohexyl iodide. A 500-ml. two-necked, round-bottomed flask fitted with a reflux condenser carrying a calcium chloride drying tube is charged with 124 g. (107 ml., 0.4 mole) of triphenyl phosphite and 85 g. (37 ml., 0.6 mole) of methyl iodide, and a thermometer is inserted of sufficient length to extend into the liquid contents of the flask. The mixture is

B.   $(C_6H_5O)_3P + CH_3I \longrightarrow [(C_6H_5O)_3PCH_3]I$

$[(C_6H_5O)_3PCH_3]I$ +     $\longrightarrow (C_6H_5O)_2POCH_3$ +     + $C_6H_5OH$

heated under gentle reflux by means of a heating mantle until the internal temperature has risen to about 120° and the mixture is dark and viscous. The flask is cooled and 40 g. (0.4 mole) of cyclohexanol is added to the oily methyltriphenoxyphosphorus iodide. The mixture is shaken gently until homogeneous and allowed to stand overnight at room temperature. The reaction may be followed by infrared spectroscopy. A strong, broad band at 1040 cm.$^{-1}$ disappears, and a similar band appears at 945 cm.$^{-1}$; the reaction appears to be complete after 6 hr. The mixture is distilled through a 13-cm. Vigreux column to give 62.5–63 g. (74–75%) of cyclohexyl iodide, b.p. 66–68°/12 mm., $n_D^{22}$ 1.5475.

*Steroidal amides.* Herz and Mantecón[2] have prepared amides from lithocholic acid 3-formate in satisfactory yields by the triphenyl phosphite method of Mitin and Glinskaya (**3**, 322–323).

[1]H. N. Rydon, *Org. Syn.*, **51**, 44 (1971).
[2]J. E. Herz and R. E. Mantecón, *Org. Prep. Proc. Int.*, **4**, 123 (1972).

**Triphenylphosphite methiodide (Methyltriphenoxyphosphonium iodide, MTPI)**, **1**, 1249–1250. Supplier: Eastman. Commercial product can be purified to give amber-colored material by washing with ethyl acetate.[1]

*Preparation.* Verheyden and Moffatt[1] have introduced some improvements in the original preparation of Landauer and Rydon (**1**, 1249–1250, ref. 1). Triphenyl phosphite (0.2 mole) and methyl iodide (0.26 mole) are placed in a 90° oil bath and the temperature of the bath is slowly raised to 125° over 8 hr. The pot temperature rises slowly from 70 to 85° and then rapidly to 115°. This temperature is maintained for 12–14 hr.; upon cooling and seeding, the mixture crystallizes to a solid brown mass. Dry ether is added; the resulting material is repeatedly washed by decantation with fresh dry ethyl acetate. The reagent is obtained as amber crystals in 90% yield. The reagent should be handled in a dry box under nitrogen.

*OH → I.* Verheyden and Moffatt[1] recommend DMF rather than the commonly used benzene as solvent for conversion of alcohols into iodides by this reagent. The reaction in this solvent occurs at room temperature and is usually rapid. Cholestanol is converted by the reagent in anhydrous DMF in 2 hr. at 25° into 3α-iodocholestane in 57% yield. The reaction of the reagent with alcohols is believed to proceed via

$$(C_6H_5O)_3\overset{+}{P}CH_3I^- \ + \ ROH \ \longrightarrow \ (C_6H_5O)_2\underset{\underset{CH_3}{|}}{\overset{+}{P}}OR \quad I^- \ \longrightarrow \ RI \ + \ (C_6H_5O)_2\overset{\overset{O}{\|}}{P}CH_3$$

nucleophilic attack on phosphorus with expulsion of phenol and formation of an alkoxy-phosphonium salt which then collapses to the alkyl iodide and diphenyl methylphos-phonate. This concerted mechanism accords with the observed inversion of configuration in the reaction.

Veryheyden and Moffatt[1] have used the reagent in DMF for iodination of primary hydroxyl groups of nucleosides. For example, the reaction of 2′,3′-O-isopropylide-neuridine (1a) with 2 eq. of methyltriphenoxyphosphonium iodide in DMF for 15 min. gives 5′-desoxy-5′-iodo-2′,3′-O-isopropylideneuridine (1b) in 96.5% yield. It is also possible to effect selective iodination of only the primary hydroxyl group. Thus brief

(1a, R = OH)
(1b, R = I)

(2a, R = OH)
(2b, R = I)

treatment of thymidine (2a) with 1.1 eq. of the reagent gives 5′-desoxy-5′-iodothymi-dine (2b) in 63% yield.

*Dehydration of secondary alcohols.* Hutchins *et al.*[2] attempted to convert *trans*-4-*tert*-butylcyclohexanol into the corresponding *cis*-iodide by treatment with MTPI in HMPT at room temperature. Instead, they obtained 4-*tert*-butylcyclohexene in 88% yield. They then found that secondary alcohols in general are dehydrated by treatment with a twofold excess of MTPI in HMPT at 25–75° for 0.25–25 hr. Primary alcohols are converted into the corresponding iodide in excellent yield under these conditions; tertiary alcohols are practically inert. Dehydration apparently involves initial conversion into the corresponding inverted iodide followed by dehydrohalogenation induced by iodide ion and HMPT (see **2**, 209–210). In most cases, the more stable Saytzeff alkene is formed in marked preference to the Hofmann alkene.

[1]J. P. H. Verheyden and J. G. Moffatt, *J. Org.*, **35**, 2319 (1970).
[2]R. O. Hutchins, M. G. Hutchins, and C. A. Milewski, *ibid.*, **37**, 4190 (1972).

**Triphenyl phosphite ozonide, 3**, 323–324.

*Oxidation of disulfides.*[1] A simple alkyl disulfide is oxidized by the reagent at $-78°$

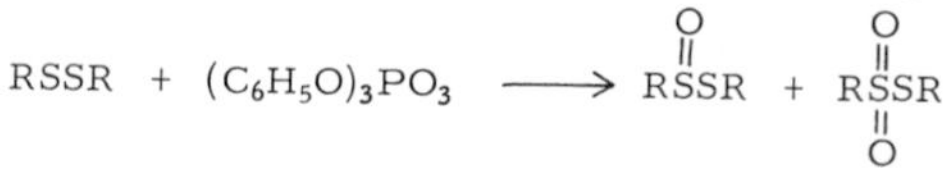

$$RSSR + (C_6H_5O)_3PO_3 \longrightarrow R\overset{\overset{O}{\|}}{S}SR + R\overset{\overset{O}{\|}}{\underset{\underset{O}{\|}}{S}}SR$$

initially to the thiolsulfinate and the thiolsulfonate in approximately 10:1 ratio. After standing at room temperature for 5–6 days the ratio becomes approximately 10:1 in favor of the thiolsulfonate. The oxidation may be due to singlet oxygen liberated from the ozonide, although the ozonide does not decompose at $-78°$.

[1]R. W. Murray, R. D. Smetana, and E. Block, *Tetrahedron Letters*, 299 (1971).

**Triphenyltin hydride, 1**, 1250–1251; **2**, 448; **3**, 324–325.

*Reduction* (**3**, 324–325). The $\alpha,\beta$-double bond of the ketone (1) can be selectively reduced by triphenyltin hydride in refluxing toluene (4 hr.) to (2) in 81 % yield.[1]

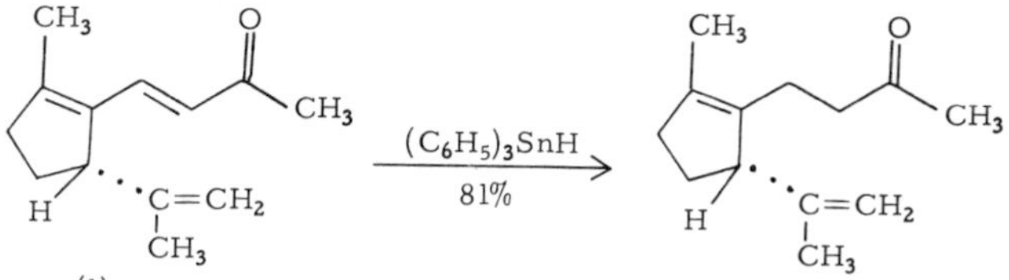

[1]M. Yamasaki, *J.C.S. Chem. Comm.*, 606 (1972).

**Tris(triphenylphosphine)chlororhodium, 1**, 1252; **2**, 448–453; **3**, 325–329. Additional supplier: Fluka.

*Decarbonylation* (**2**, 451; **3**, 327–329). The definitive paper by Walborsky and Allen[1] on the stereochemistry of the reaction has now been published. These chemists conclude that the reaction involves a radical pair:

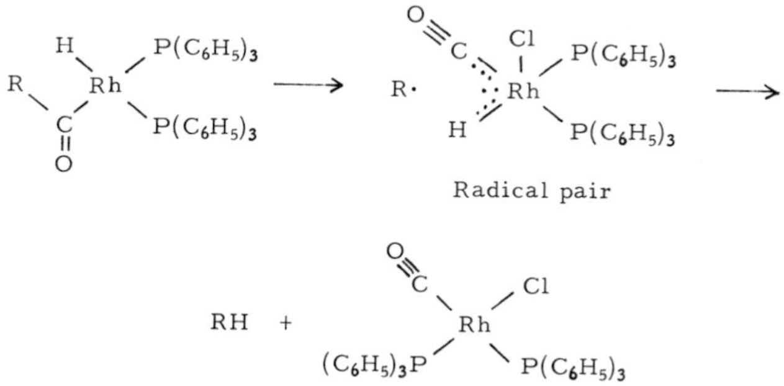

Radical pair

*Desulfonylation.*[2] Arenesulfonyl chlorides and bromides are catalytically desulfonylated by this rhodium complex by distillation (with or without nitrogen) or by heating in hexachlorobenzene. Yields are in the range 70–90%.

$$ArSO_2Cl(Br) \longrightarrow ArCl(Br) + SO_2$$

A number of other iridium, ruthenium, platinum, and palladium complexes were investigated, but highest yields were obtained with the rhodium complex. The desulfonylation is believed to proceed by formation of a metal–sulfinate complex and loss of $SO_2$.

*Cyclization.* Treatment of a 1-al-4-ene system (1) with the rhodium complex (1 eq.) in chloroform, benzene, or acetonitrile at room temperature yields a 2,3-dialkyl cyclopentanone (2) and a cyclopropane derivative (3) formed by decarbonylation. The

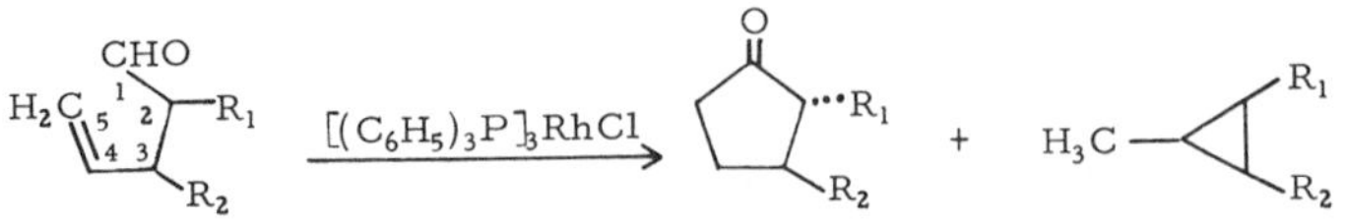

(1)                    (2)                    (3)

two products are obtained in essentially equal yields (20–35%). Treatment of (1) with $SnCl_4$ in nitromethane results in formation of only the cyclopentanone (2) (20–55% yield).[3]

A different type of cyclization by means of the rhodium complex is observed in the case of a compound containing a 1-al-6-ene system.[4] Thus reaction of (+)-citronellal (4) with 1 eq. of tris(triphenylphosphine)chlororhodium in freshly distilled chloroform at room temperature for 15 hr. results in the formation of (+)-neoisopulegol (5) and (−)-isopulegol (6) in the ratio of 3 : 1.

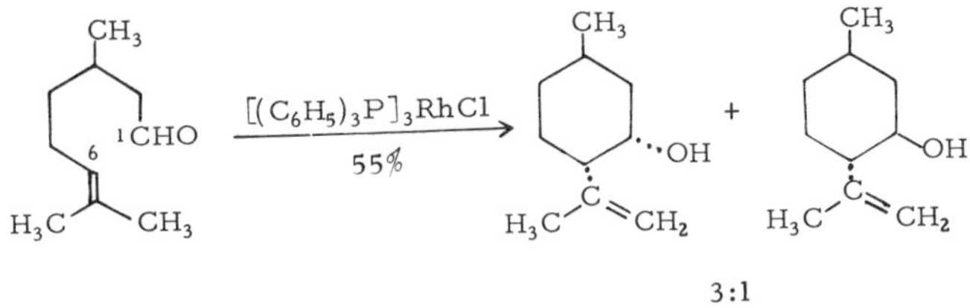

(4)                    (5)                    (6)

*Fluorenone from benzoic anhydride.*[5] Tris(triphenylphosphine)chlororhodium catalyzes the conversion of benzoic anhydride into fluorenone. The reaction proceeds at about 240°.

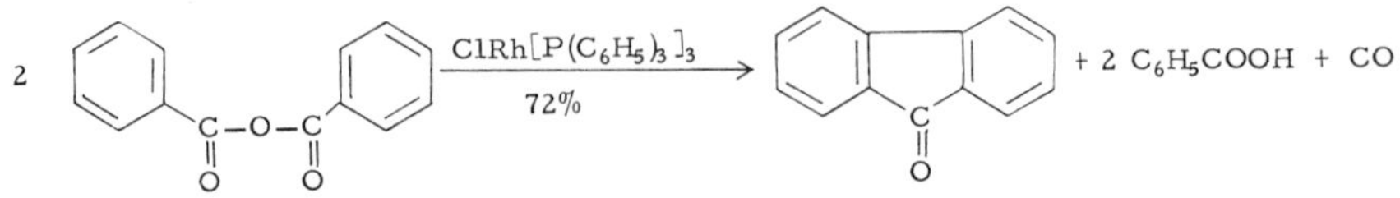

*Homogeneous catalytic hydrogenation of unhindered double bonds.*[6] To a 500-ml. two-necked flask with baffles containing a magnetic stirring bar and connected to an

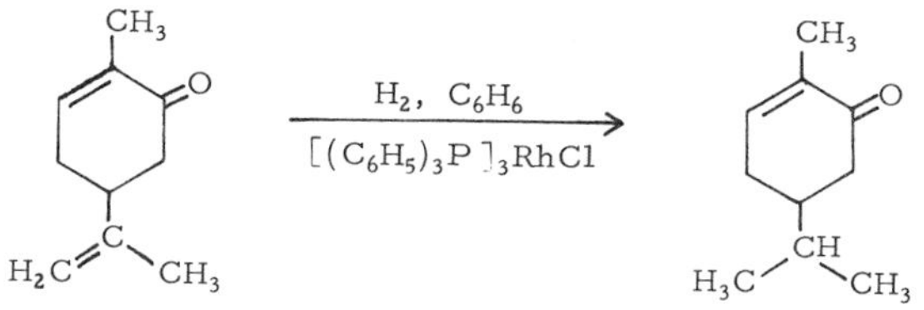

atmospheric pressure hydrogenation apparatus equipped with a graduated buret to measure the uptake of hydrogen, is added 0.9 g. of freshly prepared tris(triphenylphosphine)chlororhodium and 160 ml. of benzene distilled from calcium hydride. One neck is stoppered with a serum cap and the solution is stirred vigorously until homogeneous. Then the system is evacuated and filled with hydrogen. Through the serum cap 10 g. (0.066 moles) of freshly distilled carvone (b.p. 105–106°/14 mm.) is introduced with a syringe in the hydrogenation flask. The syringe is rinsed twice with 10 ml. of benzene and stirring is resumed. With fresh catalyst the uptake of hydrogen starts immediately and stops 3½ hr. later when the theoretical amount of hydrogen has been absorbed. The solution is filtered through a dry column (4-cm. diameter) of florisil (10/100 mesh). The column is washed with 300 ml. of ether and the combined solvent fractions are concentrated under reduced pressure. Vacuum distillation of the yellow residue through a Vigreux column (11 cm.) affords 9.1 g. (90%) of dihydrocarvone, b.p. 100–102°/14 mm.

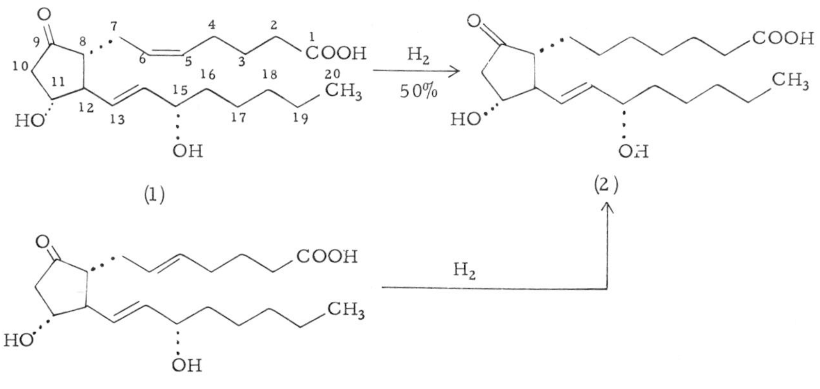

*Selective hydrogenation of prostaglandins.*[7] Hydrogenation of $PGE_2$ (1) with this catalyst in a mixture of benzene and acetone gives mainly $PGE_1$ (2, 50% yield). The observed selectivity of the 5,6-double bond over the 13,14-double bond is not entirely owing to the fact that one is *cis* and the other *trans*, since (3) is also hydrogenated to (2) as the major product. Similar selective hydrogenation of the 5,6-double bond of other prostaglandins was observed.

*Hydrogenation of allenes.*[8] Allenes can be hydrogenated to alkenes in moderate yield with this catalyst. The reaction is stereoselective; where isomerization is possible *cis*-alkenes are obtained. The least substituted double bond is reduced preferentially.

*Reduction of alkenes and alkynes.*[9] Alkenes and alkynes are reduced to alkanes by a mixture of formic acid and lithium formate at 40–60° in the presence of this catalyst. Triphenylphosphine complexes of ruthenium and iridium are also effective.

*Reduction of diazonium fluoroborates.*[10] Tris(triphenylphosphine)chlororhodium, $[(C_6H_5)_3P]_3RhCl$, and also $[(C_6H_5)_3P]_2RhCl(CO)$ catalyze the reduction of aryl diazonium fluoroborates by DMF (room temperature for 2 days or 80° for 1 day).

$$ArN_2^+BF_4^- \xrightarrow{\text{DMF, Rh complex}} ArH$$

*Alcoholysis of diarylsilanes.*[11] Tris(triphenylphosphine)chlororhodium and dichlorotris(triphenylphosphine)ruthenium are effective catalysts for alcoholysis of diarylsilanes:

$$R_2^1SiH_2 + R^2OH \longrightarrow R_2^1Si{\overset{\displaystyle H}{\underset{\displaystyle OR^2}{\big<}}} + H_2$$

They are also selective catalysts for hydrosilylation of carbonyl compounds:

$$R_2^1SiH_2 + {\overset{\displaystyle R^2}{\underset{\displaystyle R^3}{\big>}}}C{=}O \longrightarrow R_2^1SiOCH{\overset{\displaystyle R^2}{\underset{\displaystyle R^3}{\big<}}}$$

[1]H. M. Walborsky and L. E. Allen, *Am. Soc.*, **93**, 5465 (1971).
[2]J. Blum, *Tetrahedron Letters*, **35**, 3041 (1966); J. Blum and G. Scharf, *J. Org.*, **35**, 1895 (1970).
[3]K. Sakai, J. Ide, O. Oda, and N. Nakamura, *Tetrahedron Letters*, 1287 (1972).
[4]K. Sakai and O. Oda, *ibid.*, 4375 (1972).
[5]J. Blum and Z. Lipshes, *J. Org.*, **34**, 3076 (1969).
[6]R. E. Ireland and P. Bey, *Org. Syn.*, submitted 1972.
[7]F. H. Lincoln, W. P. Schneider, and J. E. Pike, *J. Org.*, **38**, 951 (1973).
[8]M. M. Bhagwat and D. Devaprabhakara, *Tetrahedron Letters*, 1391 (1972).
[9]M. E. Vol'pin, V. P. Kirkolev, V. O. Chernyshev, and I. S. Kolomnikov, *ibid.*, 4435 (1971).
[10]G. S. Marx, *J. Org.*, **36**, 1725 (1971).
[11]R. J. P. Corriu and J. J. E. Moreau, *J.C.S. Chem. Comm.*, 38 (1973).

**Tris(triphenylphosphine)chlororhodium–Triethylsilane.**

Triethylsilane, $(C_2H_5)_3SiH$, is supplied by K and K.

*Selective reduction of α,β-unsaturated carbonyl compounds.* Japanese chemists[1] have reported selective reduction of α,β-unsaturated carbonyl compounds in the terpene field by use of triethylsilane and the rhodium(I) complex as catalyst. Thus

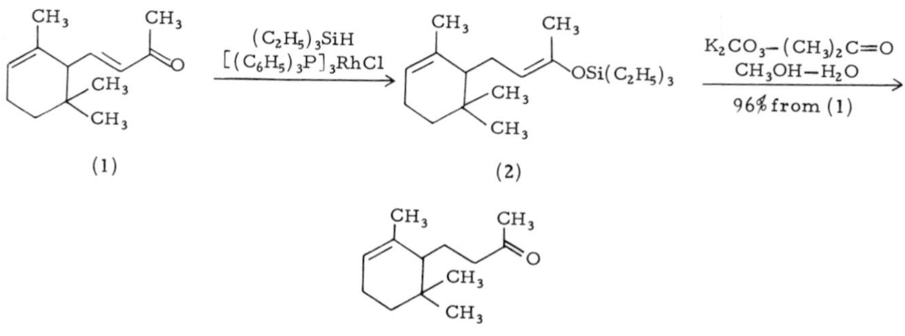

(1)

(2)

(3)

treatment of α-ionone (1) with a slight excess of triethylsilane in the presence of a catalytic amount of tris(triphenylphosphine)chlororhodium under nitrogen at 50° for 2 hr. yields the silyl enol ether (2); this was hydrolyzed by $K_2CO_3$–acetone–methanol–water to give dihydro-α-ionone (3) in 96% overall yield. In the same way citral (4) was reduced to citronellal (6) in 97% yield by way of the silyl enol ether (5). Note that under these conditions the rhodium(I) complex does not effect decarbonylation (2, 451; 3, 327–329).

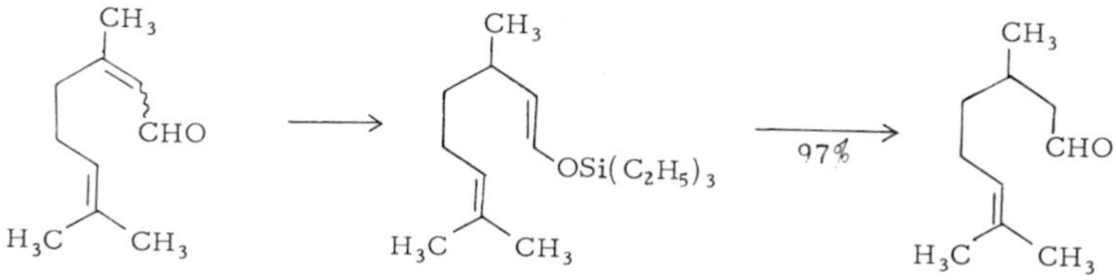

(4)

(5)

(6)

Application of the reaction to the conjugated dienone β-ionone (7) leads to dihydro-β-ionone (8) and an alcohol (9) in the ratio of 44:56 when triethylsilane is used. It was then found that the ratio of (8) to (9) is greatly affected by the hydrosilane employed. If phenyldimethylsilane is used, the ratio of (8) to (9) is 91:9. On the other hand, (9) can be obtained as the exclusive product by use of diethylsilane or diphenylsilane.

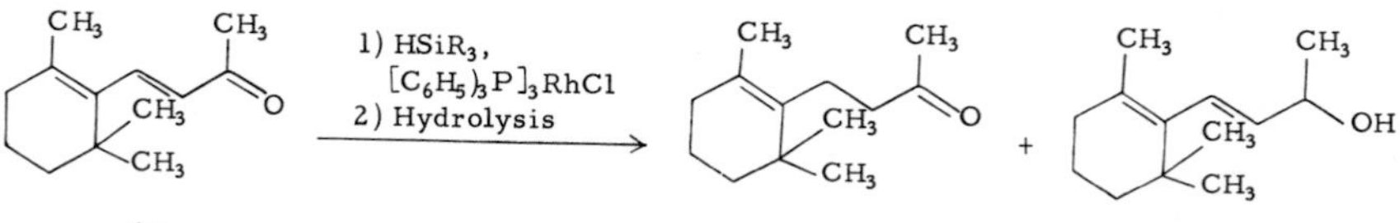

(7)

(8)

(9)

[1] I. Ojima, T. Kogura, and Y. Nagai, *Tetrahedron Letters*, 5085 (1972).

**Tris(triphenylphosphine)ruthenium dichloride,** $[(C_6H_5)_3P]_3RuCl_2$. Mol. wt. 959.44. Supplier: Strem.

*Homogeneous catalytic transfer–hydrogenation.* Sasson and Blum[1] introduced the use of this metal catalyst for transfer of hydrogen from primary alcohols to $\alpha,\beta$-unsaturated carbonyl compounds. Thus when benzyl alcohol is heated under nitrogen at 200° for 2 hr. with benzalacetone and the catalyst, benzaldehyde (90% yield) and 4-phenylbutane-2-one (92% yield) are formed. The reaction can be carried out in a

$$C_6H_5CH_2OH + C_6H_5CH{=}CHCOCH_3 \xrightarrow{[(C_6H_5)_3P]_3RuCl_2} C_6H_5CHO + C_6H_5CH_2CH_2COCH_3$$

solvent (boiling toluene, xylene, or mesitylene) but the reaction rate is reduced. Hydro-aromatics can replace alcohols as hydrogen donors, but require longer reaction periods and higher temperatures. The reaction is useful both for oxidation of alcohols and for reduction of $\alpha,\beta$-unsaturated ketones. The reaction does not proceed well in the case of aldehydes.

Regen and Whitesides[2] have used this system for direct oxidation of *vic*-diols to $\alpha$-diketones, a reaction which in the past has proved rather difficult.[3] Thus *cis*-1,2-cyclo-dodecanediol (1) can be oxidized to the diketone (2) in 53% yield. At low conversions

α-hydroxycyclododecanone can be detected in the reaction mixture. Thus the overall oxidation probably occurs in a two-stage process.

[1] Y. Sasson and J. Blum, *Tetrahedron Letters*, 2167 (1971).
[2] S. L. Regen and G. M. Whitesides, *J. Org.*, **37**, 1832 (1972).
[3] For example, see M. S. Newman and C. C. Davis, *ibid.*, **32**, 66 (1967).

*s*-**Trithiane, 3,** 329.

*Aldehyde synthesis.* The synthesis of *n*-pentadecanal from *s*-trithiane and 1-bromo-tetradecane (**3,** 329) has been published.[1] Purification is accomplished by extraction of 30 g. of commercial *s*-trithiane with 300 ml. of toluene.

[1] D. Seebach and A. K. Beck, *Org. Syn.*, **51**, 39 (1971).

**Trityl chloride, 1**, 1254–1256; **2**, 453–454.

*Acetates of chlorohydrins.* Newman and Chen[1] have published a convenient two-step procedure for conversion of 1,2- or 1,3-glycols into acetates of chlorohydrins. First the glycol is converted into the cyclic orthoester by reaction with trimethyl orthoacetate under acid catalysis (benzoic acid or chloroacetic acid). Yields are in the range 80–90%. The cyclic orthoesters are then treated with trityl chloride (1 eq.) in methylene chloride

$$RCHOHCHOHR \xrightarrow[-2\ CH_3OH]{\overset{CH_3C(OCH_3)_3}{\overset{H^+}{}}} \underset{RCH-CHR}{\overset{H_3C}{\underset{O}{\overset{|}{\underset{|}{}}}} \overset{OCH_3}{\underset{O}{\overset{C}{}}}} \xrightarrow{(C_6H_5)_3CCl}$$

$$\underset{RCH-CHR}{\overset{Cl\quad OCOCH_3}{\overset{|\quad\ \ |}{}}} \quad + \quad (C_6H_5)_3COCH_3$$

under reflux for 1–2 hr. Acetates of chlorohydrins are obtained in about 85–95% yield. Newman suggests that the trityl cation attacks the methoxy group of the orthoester to give the ambient cation (a), which reacts with trityl chloride or chloride ion to give the product. The method gives low yields in the case of 1,4-diols.

$$\overset{R}{\underset{R}{\overset{HC-O}{\underset{HC-O}{}}}}\overset{CH_3}{\underset{OCH_3}{\overset{C}{}}} \xrightarrow{(C_6H_5)_3C^+} \overset{R}{\underset{R}{\overset{HC-O}{\underset{HC-O}{}}}}\overset{}{\underset{}{}} \overset{+ \quad :C-CH_3}{} \quad + \quad (C_6H_5)_3COCH_3$$

(a)

$$\downarrow (C_6H_5)_3CCl$$

$$\underset{RCH-CHR}{\overset{Cl\quad OCOCH_3}{\overset{|\quad\ \ |}{}}} \quad + \quad (C_6H_5)_3C^+$$

[1]M. S. Newman and C. H. Chen, *Am. Soc.*, **95**, 278 (1973).

**Trityl fluoroborate, 1**, 1256–1258; **2**, 454. Supplier: WBL.

*Improved preparation.*[1] Trityl fluoroborate can be prepared conveniently by the reactions of triphenylmethyl chloride[2] with anhydrous tetrafluoroboric acid–dimethyl ether complex[3] in dry benzene:

$$(C_6H_5)_3CCl \ + \ HBF_4\cdot O(CH_3)_2 \xrightarrow[73\%]{C_6H_6} (C_6H_5)_3C^+BF_4^- \ + \ HCl \ + \ (CH_3)_2O$$

Trityl hexafluoroantimonate, $(C_6H_5)_3C^+SbF_6^-$, and trityl hexafluorophosphate, $(C_6H_5)_3C^+PF_6^-$, can be prepared in the same way using anhydrous fluoroantimonic acid and the ether complex of hexafluorophosphoric acid, respectively.

**Ergosterol acetate peroxide.**[4] Irradiation of ergosterol acetate (1) with a tungsten lamp (500 W) in air or pure oxygen in the presence of a catalytic amount of trityl fluoroborate gives ergosterol acetate peroxide (2) in quantitative yield in about 10 min. at a temperature of $-78°$. Photooxygenation at higher temperatures does not give clean products.

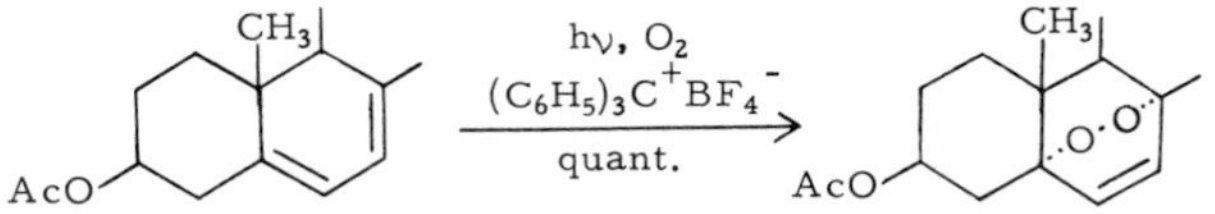

Under the same conditions ascaridole is formed from $\alpha$-terpinene in high yield.

The original paper should be consulted for a possible mechanism for this catalyzed photooxygenation.

**Oxidation of ketone acetals and ethers.**[5] Ketones can be regenerated from the ethylene acetal derivatives by treatment with trityl fluoroborate in dry dichloromethane ($N_2$) at room temperature. Thus the reaction of trityl fluoroborate with cyclohexanone ethylene acetal results in cyclohexanone (80%) and triphenylmethane. The reaction thus involves hydride transfer to the triphenyl carbonium ions. Triethyloxonium fluoroborate can also be used but is somewhat less effective than trityl fluoroborate.

The reaction with the acetal (1) gave cyclohexanone and benzoin (64% yield). This result and others suggested that the reaction would be useful for oxidation of a diol as

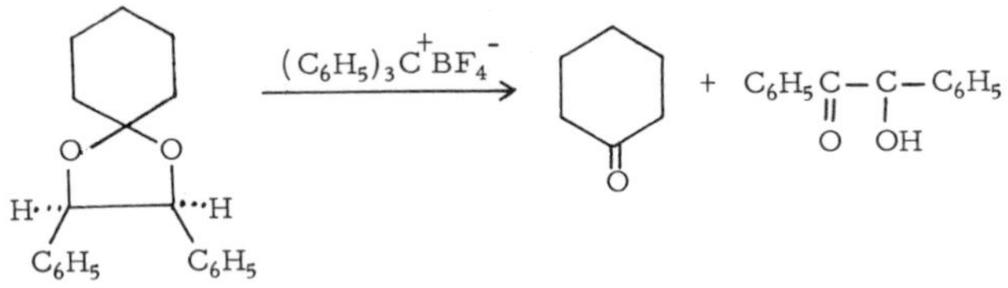

its acetonide or other acetal. Indeed the acetonide of $2\beta,3\beta$-dihydroxycholestane (2) on treatment with trityl fluoroborate gives $3\beta$-hydroxycholestane-2-one (3) in 79% yield.

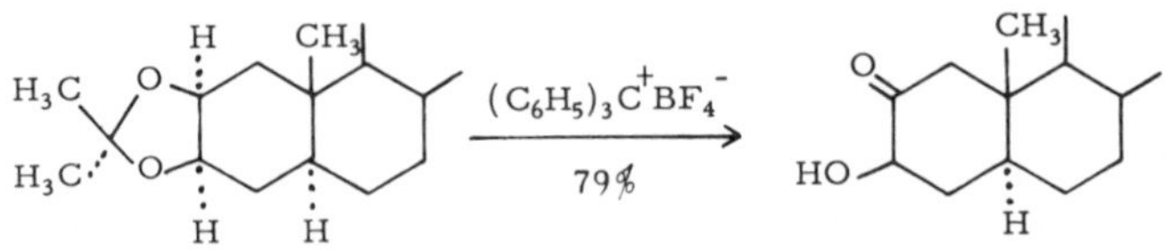

Various experimental evidence supports the following scheme for the oxidation:

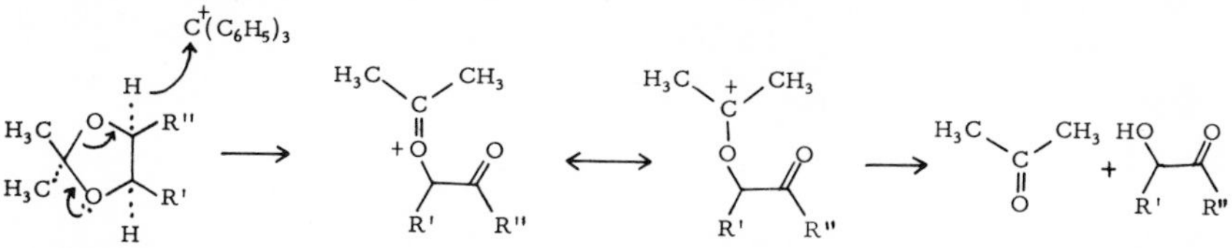

The reaction is useful in the preparation of keto-sugars. Thus the reaction of trityl fluoroborate with the acetonide (4) derived from D-mannitol gives the keto-sugar (5) in 65% yield.

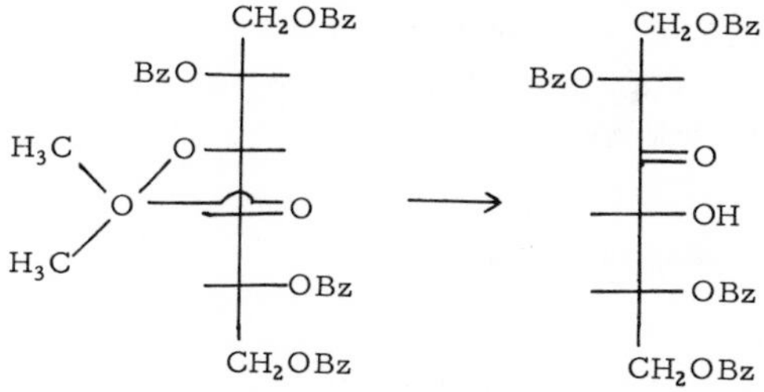

(4)                    (5)

Oxidation of benzyl ethers with trityl fluoroborate can be used either to prepare aromatic aldehydes or to cleave benzyl ethers of alcohols.
Examples:

$$C_6H_5CH_2OCH_3 \xrightarrow[75\%]{} C_6H_5CHO$$

$$C_6H_5CH_2O\cdot \text{Cholesterol} \longrightarrow \text{Cholesterol} + \Delta^4\text{-Cholestenone}$$

$$\phantom{C_6H_5CH_2O\cdot \text{Cholesterol} \longrightarrow} 60\% \phantom{xxxxxxx} 20\%$$

[1] G. A. Olah, J. J. Svoboda, and J. A. Olah, *Synthesis*, 544 (1972).
[2] W. E. Bachman, *Org. Syn.*, *Coll. Vol.*, **3**, 842 (1955).
[3] Suppliers: Aldrich; Cationics, Inc., 653 Alpha Drive, Cleveland, Ohio, 44143.
[4] D. H. R. Barton, G. Leclerc, P. D. Magnus, and I. D. Menzies, *ibid.*, 447 (1972).
[5] D. H. R. Barton, P. D. Magnus, G. Smith, G. Streckert, and D. Zurr, *J.C.S. Perkin I*, 542 (1972).

**Trityl hexachloroantimonate**, **3**, 330. Supplier: Aldrich.

**Trityllithium**, **1**, 1256; **2**, 454.

*Metal enolates as protecting groups for ketones.* Barton *et al.*[1] effected selective reduction of the 11-keto function of prednisone BMD (1) by conversion of the 3-keto group into the metal enolate (2) by trityllithium. Reduction of the 11-keto group *in situ*

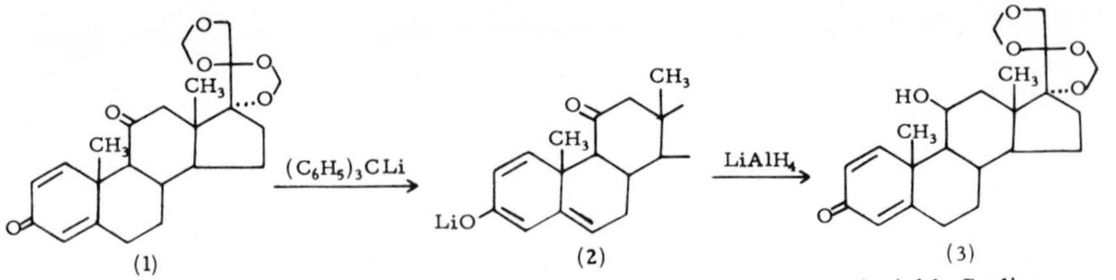

(1)          (2)          (3)

by lithium aluminum hydride affords prednisolone BMD in good yield. Sodium or lithium bistrimethylsilylamide[2] can also be used, but yields are lower. The visible color change attending protonolysis of trityllithium is an added advantage of this reagent. This method of selective protection was shown to be applicable to several steroidal ketones. The reactions were generally carried out at $-78°$ under argon in THF with 4–10 hydride eq. of lithium aluminum hydride per mole of substrate.

[1]D. H. R. Barton, R. H. Hesse, M. M. Pechet, and C. Wiltshire, *Chem. Commun.*, 1017 (1972).
[2]U. Wannagat and H. Niederprüm, *Ber*, **94**, 1540 (1961).

**Tropylium salts** (2).
*Preparation by hydrogen exchange[1]:*

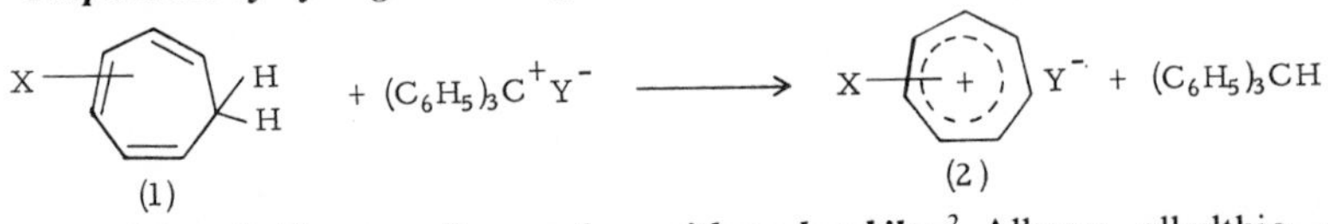

(1)          (2)

*Reactivity of chlorotropylium cations with nucleophiles.*[2] Alkoxy-, alkylthio-, and N-alkyl-N-arylaminotropylium salts are prepared by reaction of chlorotropylium salts with alcohols, thiols, and N-alkyl-N-arylamines, respectively. With dimethylamine or with benzenesulfonamides chlorotropylium salts rearrange to yield bis(dimethylamino)phenylmethane or N-benzylidenesulfonamides. It is postulated that the formation of chlorocycloheptatrienes (1A, 1B, 1C) is kinetically controlled, whereas substituted tropylium ions (3) are formed under thermodynamic control.

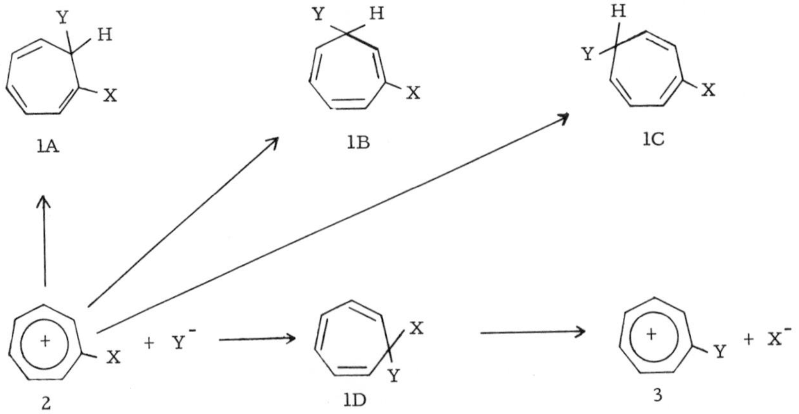

1A          1B          1C

2          1D          3

[1]H. J. Dauben, Jr., F. A. Gadecki, K. M. Harmon, and D. L. Pearson, *Am. Soc.*, **79**, 4557 (1957).
[2]B. Föhlisch and E. Haug, *Ber.*, **104**, 2324 (1971).

**Tungsten hexachloride**, $WCl_6$. Mol. wt. 396.66. Suppliers: Alfa Inorganics, Pressure Chemical Company, ROC/RIC.

*Deoxygenation.* Sharpless *et al.*[1] have prepared a series of lower valent tungsten halides from tungsten hexachloride as shown:

$$WCl_6 \; + \; 2 \; RLi \; \xrightarrow{\text{THF}} \; \text{Reagent I}$$

$$WCl_6 \; + \; 3 \; RLi \; \xrightarrow{\text{THF}} \; \text{Reagent II}$$

$$WCl_6 \; + \; 4 \; RLi \; \xrightarrow{\text{THF}} \; \text{Reagent III}$$

$$WCl_6 \; + \; 2 \; Li \; (\text{dispersion}) \xrightarrow{\text{THF}} \text{Reagent IV}$$

$$WCl_6 \; + \; 3 \; LiI \; \xrightarrow[\text{in vacuo}]{130^0} \text{Reagent V}$$

$$WCl_6 \; + \; 2 LiI \; \xrightarrow[\text{in vacuo}]{130^0} \text{Reagent VI}$$

These reagents (particularly reagent I) effect direct reductive coupling of aldehydes and ketones to olefins. For example, benzaldehyde can be converted into stilbene in 70–76% yield by treatment with excess reagent I.

$$2 \; C_6H_5CHO \; \xrightarrow[70-76\%]{\text{Reagent I}} C_6H_5CH{=}CHC_6H_5$$

In a typical experiment $WCl_6$ (0.8 mmole) is added to 10 ml. of THF cooled to $-78°$, followed by addition of *n*-butyllithium (1.6 mmole). The mixture is allowed to warm to room temperature and then the carbonyl compound (0.2 mmole) is added. After standing for 6 hr., the reaction is quenched with 20% NaOH solution, and the olefin extracted with ether.

Yields are highest in the case of aromatic aldehydes and ketones; application of the reaction to butanone-2, for example, gives only a 10% yield of olefin. Reagent I is more effective than the other tungsten reagents (II–VI) for the coupling reaction.

The new tungsten reagents (I–VI) are also effective for deoxygenation of epoxides to the parent olefin. Reagents V and VI are the cheapest and most convenient reagents for this purpose; however, reagents I–III serve well for small- and medium-scale reactions. In a typical experiment $WCl_6$ (0.15 mole) is added to THF (420 ml.) cooled to $-62°$; *n*-butyllithium (0.30 mole) is added slowly at this temperature. The mixture is allowed to warm slowly to room temperature. The oxide (0.08 mole) is added. A rapid exothermic reduction takes place. After 0.5 hr. the product is poured into an aqueous solution containing sodium tartrate and NaOH, and the olefin extracted with hexane.

Examples:

$$\text{trans-Cyclododecene oxide} \xrightarrow[97-98\%]{\substack{\text{Reagents} \\ \text{I, IV, VI}}} \underset{(95\% \; \underline{\text{trans}})}{\text{trans-Cyclododecene}}$$

$$\underline{\text{cis}}\text{-Cyclododecene oxide} \xrightarrow[70-90\%]{\text{Reagent I}} \underset{(94-98\% \; \underline{\text{cis}})}{\underline{\text{cis}}\text{-Cyclododecene}}$$

$$\underline{\text{trans}}\text{-Stilbene oxide} \xrightarrow[86\%]{\text{Reagent I}} \underset{(\overline{100\%} \; \underline{\text{trans}})}{\text{trans-Stilbene}}$$

Only limited stereoselectivity is observed in reduction of acyclic di- and trisubstituted epoxides.

Selective reduction of stigmasteryl acetate bisepoxide to 22,23-epoxystigmasteryl acetate in 83% yield can be carried out using a limited amount of reagent I (0.3 hr., room temperature).

See also **Potassium hexachlorotungstate(IV)**, this volume.

*Olefin metathesis.* Tungsten hexachloride with $C_2H_5AlCl_2$[2] or *n*-butyllithium[3] as cocatalyst has been used for olefin metathesis. Lithium aluminum hydride has been shown to be an effective cocatalyst and has the merit of availability and stability to air.[4] Thus treatment of heptene-3 in chlorobenzene with $WCl_6$–$LiAlH_4$ for 3 hr. yields heptene-3 (39%), octene-4 (23%), and hexene-3 (18.5%). A nonene, a pentene, and a butene were also formed in 10% yield.

[1] K. B. Sharpless, M. A. Umbreit, M. T. Nieh, and T. C. Flood, *Am. Soc.*, **94**, 6538 (1972).
[2] N. Calderon, W. A. Judy, E. A. Ofstead, K. W. Scott, and J. P. Wood, *ibid.*, **90**, 4132 (1968).
[3] H. R. Menapace and J. L. Wang, *J. Org.*, **33**, 3749 (1968).
[4] S. A. Matlin and P. G. Sammes, *J.C.S. Chem. Comm.*, 174 (1973).

# U

## Urushibara hydrogenation catalysts.

*Reviews.*[1-3] The catalysts were developed mainly by Y. Urushibara. They are prepared by precipitation of the catalyst metal (nickel, cobalt, or iron) from an aqueous solution of its salt (usually the chloride) by zinc dust or granular aluminum. The precipitated metal is then digested by a base or an acid (usually sodium hydroxide or acetic acid). The Urushibara catalysts are comparable to Raney catalysts. U-Catalysts can be used for hydrogenation of alkynes and alkenes to alkanes, of carbonyl compounds to alcohols, of aromatic nitro compounds to amines. They can also be used as dehydrogenation catalysts. Thus stigmasterol can be dehydrogenated to the corresponding $\Delta^4$-3-ketone, with cyclohexanone serving as the hydrogen acceptor. U-Catalysts have also been used to effect reductive desulfurization.

[1] *Urushibara Catalysts*, University of Tokyo Press, Tokyo, Japan (1971).
[2] *New Hydrogenating Catalysts*, Halsted Press, Wiley, New York (1972).
[3] K. Hata, I. Motoyama, and K. Sakai, *Org. Prep. Proc. Int.*, **4**, 180 (1972).

# V

**Vinylmagnesium chloride**, $CH_2$=CHMgCl.
*Preparation* (**1**, 418). H. Normant, *J. Org.*, **22**, 1602 (1957).

*1,6-Diketones.*[1] Vinylmagnesium chloride, in the presence of cuprous chloride as catalyst, reacts with a fatty acid or ester to give a 1,6-diketone, usually in rather low yield (5–55%). Thus the reaction as applied to *n*-valeric acid (1) gives *n*-tetradecane-5,10-dione (2) in 53% yield.

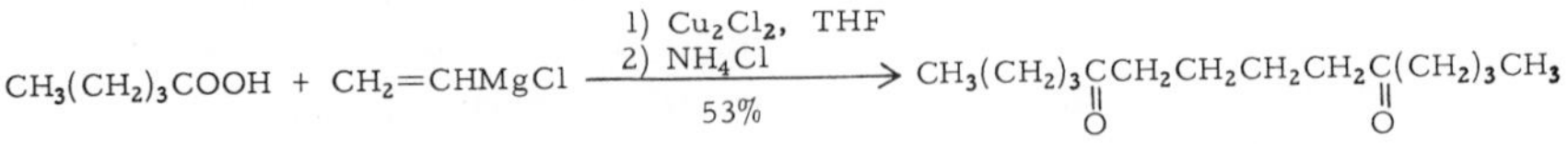

$$CH_3(CH_2)_3COOH + CH_2=CHMgCl \xrightarrow[53\%]{\substack{1)\ Cu_2Cl_2,\ THF \\ 2)\ NH_4Cl}} CH_3(CH_2)_3\underset{O}{\overset{\parallel}{C}}CH_2CH_2CH_2CH_2\underset{O}{\overset{\parallel}{C}}(CH_2)_3CH_3$$

(1)                                                   (2)

*Reaction with carboxylic acids.*[2] Vinylmagnesium chloride reacts with benzoic

$$C_6H_5COOH + 2\ CH_2=CHMgCl \xrightarrow{85\%} C_6H_5\underset{O}{\overset{\parallel}{C}}CH_2CH_2CH=CH_2$$

acid in THF to give 5-keto-5-phenyl-1-pentene. The reagent reacts with *n*-valeric acid to give a mixture of *n*-butyldivinylcarbinol and 5-keto-1-nonene.

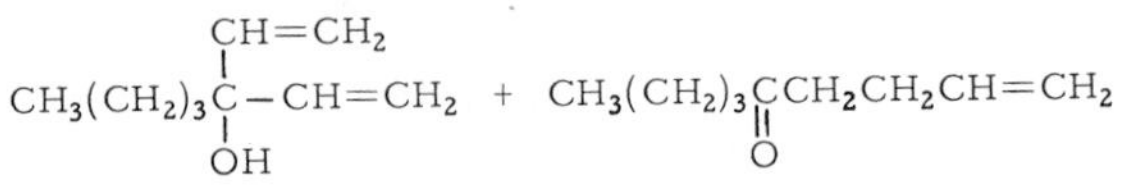

$$CH_3(CH_2)_3\underset{OH}{\overset{\overset{\displaystyle CH=CH_2}{|}}{C}}-CH=CH_2 + CH_3(CH_2)_3\underset{O}{\overset{\parallel}{C}}CH_2CH_2CH=CH_2$$

(35%)                             (30%)

[1]S. Watanabe, K. Suga, T. Fujita, and Y. Takahashi, *Canad. J. Chem.*, **50**, 2786 (1972).
[2]S. Watanabe, K. Suga, and Y. Yamaguchi, *J. Appl. Chem. Biotechnol.*, **22**, 43 (1972).

**Vinyl triphenylphosphonium bromide**, **1**, 1274–1275; **2**, 456–457; **3**, 333. Supplier: Aldrich.

# W

**Wittig reagents.** Aldrich supplies 15 Wittig reagents.
**Wittig reaction.**
   *Mechanism.* Smith and Trippett[1] have presented evidence that the Wittig reaction

$$HO^- \searrow$$
$$(C_6H_5)_3\overset{+}{P}-\overset{|}{C}R'$$
$$\|$$
$$CR_2$$

(a)

does not proceed through a vinyl phosphonium salt (a) as suggested by Schweizer.[2] Thus the accepted mechanism involving a four-membered ring is probably correct.

$$(C_6H_5)_3P{=}CHR + R'_2C{=}O \longrightarrow (C_6H_5)_3\overset{+}{P}-CHR \longrightarrow$$
$$\overset{|}{O}-\overset{-}{C}R'_2$$

$$(C_6H_5)_3P-CHR \longrightarrow RCH{=}CR'_2 + (C_6H_5)_3P{=}O$$
$$O-CR'_2$$

   *Solid-phase Wittig synthesis.* Two groups[3,4] have reported solid-phase Wittig synthesis. The resin is built up from styrene and the phosphine (1) with a small amount of 1,4-divinylbenzene for cross-linking. Alkylation and treatment with base gives a

resin bearing ylid groups (2). The Wittig reaction occurs when a carbonyl compound is added.

[1]D. J. H. Smith and S. Trippett, *Chem. Commun.*, 191 (1972).
[2]E. E. Schweizer, O. M. Crouse, T. Minami, and A. T. Wehman, *ibid.*, 1000 (1971).
[3]F. Camps, J. Castells, J. Font, and F. Vela, *Tetrahedron Letters*, 1715 (1971).
[4]S. V. McKinley and J. W. Rakshys, Jr., *Chem. Commun.*, 134 (1972).

# Z

Zinc, **1**, 1276–1284; **2**, 459–462; **3**, 334–337.

**Decyanation.** Swiss chemists[1] effected decyanation of the cyanamide (1), derived from lysergic acid, with zinc dust in 80% acetic acid (3 hr., 100°). The conversion can

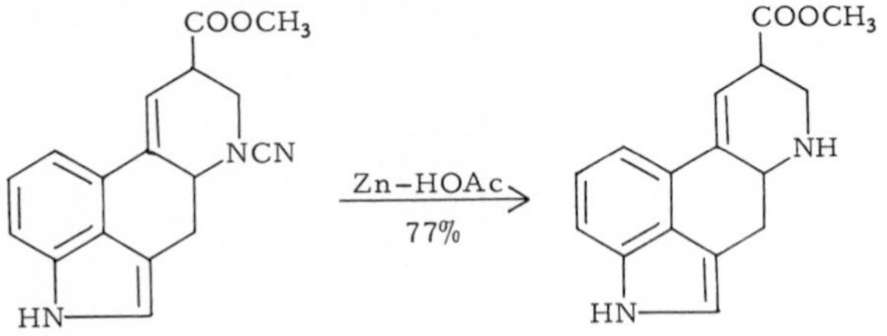

(1)                                    (2)

also be effected with hydrogen and Raney nickel. In this case use of $LiAlH_4$ is not possible because of the easy reducibility of the $\beta,\gamma$-unsaturated ester function.

Stork[2] used zinc in acetic acid for decyanation of (3) to give (4), which was used as one unit in a total synthesis of ($\pm$)-yohimbine (5).

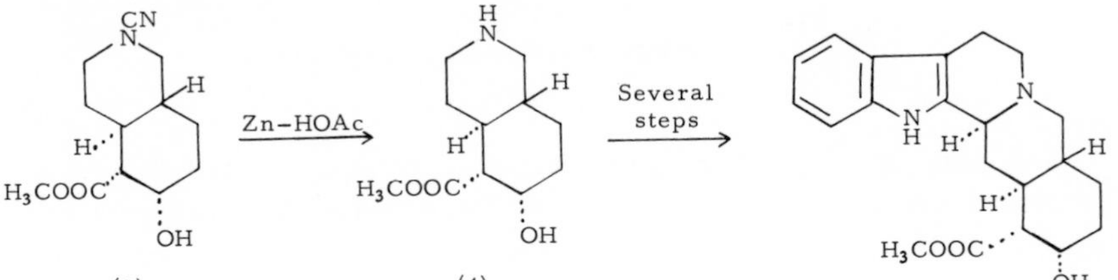

(3)                    (4)                                              (5)

**Reductive cleavage of allylic alcohols, ethers, or acetates by amalgamated zinc and HCl.**[3,4] Amalgamated zinc is prepared by treatment of zinc powder with an aqueous solution of $HgCl_2$. The resulting amalgam is washed with distilled water, ethanol, and finally dry ether.

Allylic alcohols, ethers, or acetates can be reduced to olefins by treatment with a large excess of amalgamated zinc and hydrogen chloride in ether. The new method differs from usual procedures (*e.g.*, sodium in liquid ammonia) in that the product is predominately, or exclusively, the less stable isomer. Thus 1-methylcyclohexene-3-ol (1)

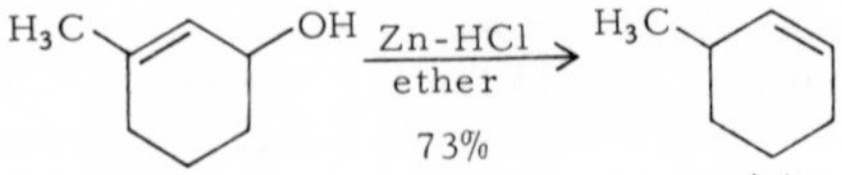

(1)                                    (2)

is converted into 3-methylcyclohexene (2) in 73 % yield. Isomeric allylic alcohols such as (3) and (4) give the same mixture of olefins.

$$
\left.\begin{array}{c}
\text{C}_6\text{H}_5\text{CH}=\text{CH}_2\text{CH}_2\text{OH} \\
(3) \\
\text{C}_6\text{H}_5\text{CHOHCH}=\text{CH}_2 \\
(4)
\end{array}\right\} \longrightarrow \underset{80\%}{\text{C}_6\text{H}_5\text{CH}_2\text{CH}=\text{CH}_2} + \underset{17\%}{\text{C}_6\text{H}_5\text{CH}=\text{CHCH}_3}
$$

**Acyloins.**[5] α-Diketones are reduced to acyloins in high yield by treatment with zinc

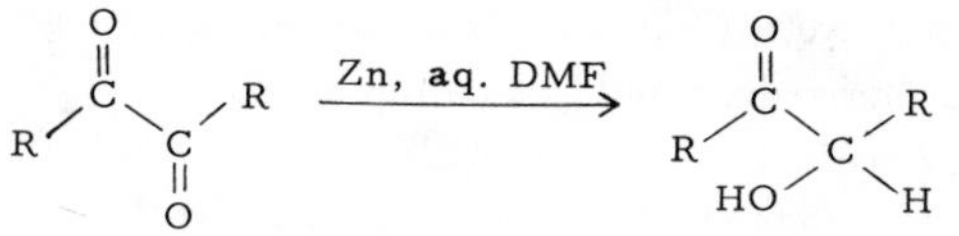

dust in refluxing aqueous DMF (N$_2$). The method affords a preparative route to olefins from nonenolizable α-diketones as shown in the example.

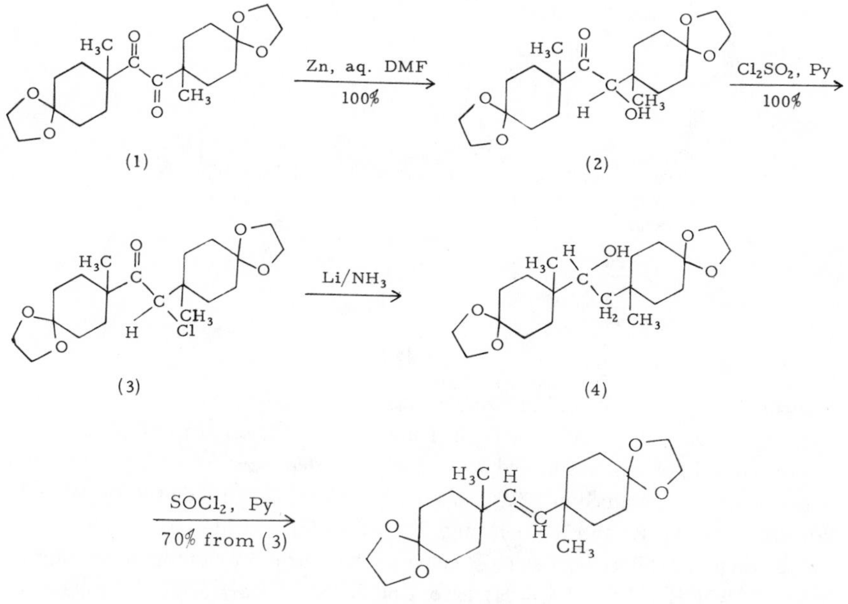

**Synthesis of pregnane and corticoid side chains.**[6] The 17α-ethynylcarbinol acetate (1) is reduced by zinc dust in refluxing diglyme, a process which is accompanied by rearrangement and elimination, to give the allenyl steroid (2) in 86 % yield. The desired corticoid side chain (3) can be formed from (2) by reaction with osmium tetroxide–

pyridine in benzene followed by cleavage of the osmate ester with sodium sulfite and potassium hydrogen carbonate.

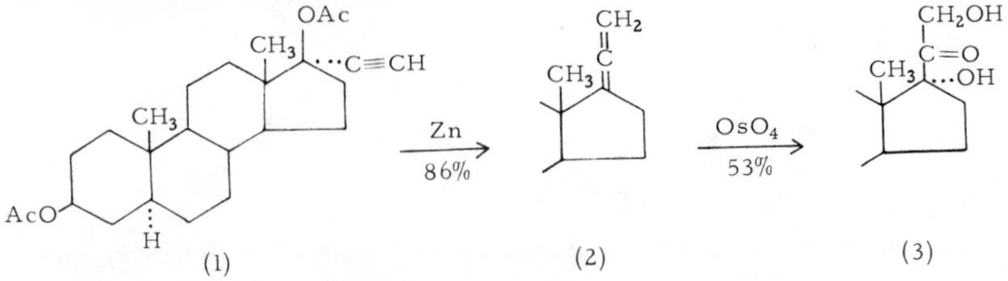

Peracid oxidation converts (2) into a 17α-hydroxy-20-ketopregnane. A 17,20-allene oxide is a probable intermediate in the oxidation.

*γ,δ-Unsaturated acids.* Baldwin and Walker[7] have reported a new route to γ,δ-unsaturated acids involving the Claisen rearrangement.[8] α-Bromo esters (1), prepared by esterification of allylic alcohols with α-bromo acid bromides,[9] are converted by a

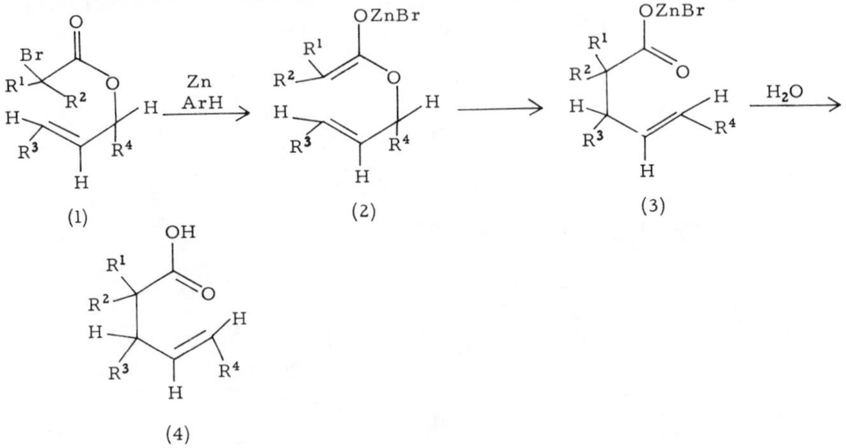

Reformatsky-type[10] reaction into zinc enolates (2) by treatment with zinc dust in a refluxing aromatic hydrocarbon solvent. These undergo Claisen rearrangement to zinc carboxylates (3) of γ,δ-unsaturated acids. Yields are depressed by α-alkyl substitution of (1); thus (3) is obtained in 100% yield when $R^1 = R^2 = CH_3$ and $R^3 = R^4 = H$. When $R^1 = CH_3$, $R^2$ and $R^4 = H$, and $R^3 = C_6H_5$, the yield of (3) is only 16%.

The reaction exhibits high stereoselectivity; thus the α-bromopropionate ester (1) ($R^1 = R^4 = CH_3$; $R^2 = R^3 = H$) gave 2-methylhex-4-enoic acid in a *trans-to-cis*

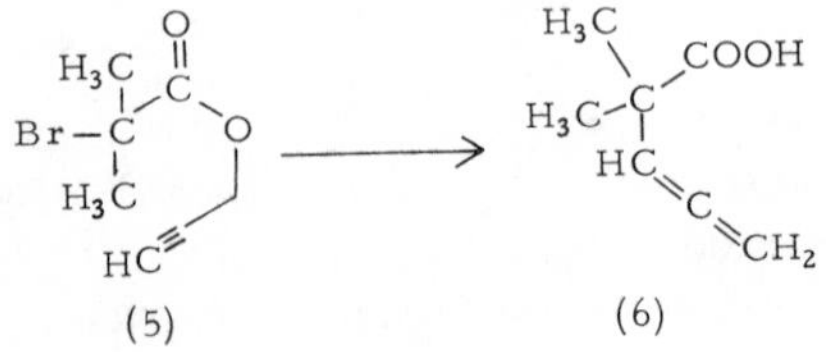

ratio of 93:7. The reaction is also applicable to acetylenic esters. Thus (5) is transformed by this procedure into the allenic acid (6).

[1]T. Fehr, P. A. Stadler, and A. Hofmann, *Helv.*, **53**, 2197 (1970).

[2]G. Stork and R. Nath Guthikonda, *Am. Soc.*, **94**, 5109 (1972).

[3]I. Elphimoff-Felkin and P. Sarda, *Tetrahedron Letters*, 725 (1972).

[4]*Idem, Org. Syn.*, submitted 1972.

[5]W. Kreiser, *Ann.*, **745**, 164 (1971).

[6]M. Biollaz, W. Haefliger, E. Verlade, P. Crabbé, and J. H. Fried, *Chem. Commun.*, 1322 (1971).

[7]J. E. Baldwin and J. A. Walker, *J.C.S. Chem. Comm.*, 117 (1973).

[8]D. S. Tarbell, *Org. React.*, **2**, 1 (1944); A. Jefferson and F. Scheinmann, *Quart. Rev.*, **22**, 391 (1968).

[9]C. W. Smith and D. G. Norton, *Org. Syn., Coll. Vol.*, **4**, 348 (1963).

[10]R. L. Shriner, *Org. React.*, **1**, 1 (1942); D. G. M. Diaper and A. Kuksis, *Chem. Rev.*, **59**, 89 (1959).

## Zinc–Acid anhydride–Catalyst.

*Reductive acetylation.* Thiele[1] established that reduction of benzil with zinc dust in acetic anhydride–sulfuric acid involves 1,4-addition of hydrogen to the α-diketone grouping and acetylation of the resulting enediol to give a mixture of the geometrical isomers (1) and (2). Thiele and later investigators isolated the more soluble, lower melting *cis*-stilbene-α,β-diol diacetate (2) in only impure form, m.p. about 110°. Separation by chromatography is not feasible, because the two isomers have the same degree of adsorbability on alumina. However, it is feasible by a process of fractional crystallization described in a student manual[2] to isolate both isomers in a pure condition. In

$$C_6H_5-\overset{\overset{O}{\|}}{C}-\overset{\overset{O}{\|}}{C}-C_6H_5 \xrightarrow[\text{1,4-addition}]{\text{2 HA}} \left[C_6H_5-\overset{\overset{OH}{|}}{C}=\overset{\overset{OH}{|}}{C}-C_6H_5\right] \xrightarrow{\text{2 Ac}_2\text{O, H}^+}$$

(1, trans)       (2, cis)
M. p. 155°       M. p. 119°
EtOH 271 mμ (ε = 23,400)       EtOH 265 mμ (ε = 12,800)

the method prescribed for the preparation of the isomer mixture, hydrochloric acid is substituted for sulfuric acid because the latter became reduced to sulfur and to hydrogen sulfide. If acetyl chloride (2 ml.) is substituted for the hydrochloric acid–acetic anhydride mixture, the *cis*-isomer (2) is the sole product.

One satisfactory procedure for preparing the diacetyl derivative of an air-sensitive reduction product of a quinone[3] consists in suspending the quinone (3) in acetic anhydride (5–6 ml./g.), adding one part of zinc dust and 0.2 part of fused, powdered sodium acetate, and warming the mixture gently until the colored material has disappeared and the supernatant liquid is colorless or pale yellow, depending on the purity of the starting material. After short boiling to complete the reaction, acetic acid is added to dissolve

$$\text{(3)} \xrightarrow{\text{Ac}_2\text{O, Zn, NaOAc}} \text{(4)}$$

(3)                                      (4)

the product and a part of the zinc acetate which has separated, the solution is filtered at the boiling point, the residue is washed with hot solvent, and the total filtrate is boiled under reflux and treated cautiously with sufficient water to hydrolyze the excess acetic anhydride and then to produce a saturated solution. The colorless solution on cooling deposits 0.6 g. of pure 2-methyl-1,4-naphthohydroquinone diacetate, m.p. 112–113°. Tetramethylammonium bromide (0.2 part) can be used in place of sodium acetate and is an even more effective catalyst.[3]

Earlier procedures specified the use of pyridine as catalyst[4] but this has the disadvantage that the reaction mixture tends to acquire a yellow color owing to a side reaction between pyridine, acetic anhydride, and zinc. The difficulty is completely obviated by use of the tertiary aliphatic amine, triethylamine, as in a procedure for the preparation of 2-methyl-1,4-naphthohydroquinone diacetate.[5]

Evidence that the synthetic substance (5), bicyclo[4.4.1]undeca-3,5,8-10-tetraene-2,7-dione, can be described as a quinone, is that treatment with zinc dust and acetic anhydride in the presence of pyridine leads readily to reductive acetylation that affords the aromatic 2,7-diacetoxy-1,6-methanol[10]annulene (6).[6]

$$\text{(5)} \xrightarrow{\text{Zn + (CH}_3\text{CO)}_2\text{O}} \text{(6)}$$

(5)                                      (6)

*Reductive benzoylation* has been accomplished by adding 1 ml. of benzoyl chloride dropwise to a cooled mixture of 0.5 g. of 2-methyl-1,4-naphthoquinone, 0.5 g. of zinc dust, and 3 ml. of pyridine, while cooling in ice.[4b] Reaction occurred at once, and after 10 min. at room temperature the pale-yellow mixture was heated to boiling, cooled,

$$\text{Zn + Py + 2 C}_6\text{H}_5\text{COCl}$$

and the product was extracted with ether–alcohol. After washing with dilute acid and with soda solution and concentrating the dried solution to a small volume, the dibenzoate separated as colorless needles, m.p. 178–179.5°. Two recrystallizations from alcohol (35 ml.) raised the melting point to 180–180.5°.

[1]J. Thiele, *Ann.*, **306**, 142 (1899).
[2]L. F. Fieser, *Org. Exp.*, 2nd Ed., D. C. Heath and Co., Boston, 221 (1968).
[3]*Idem, Experiments in Organic Chemistry*, 2nd Ed., D. C. Heath and Co., Boston, 399 (1941).
[4](a) *Idem, Am. Soc.*, **61**, 3473 (1939); (b) L. F. Fieser, W. P. Campbell, E. M. Fry, and M. D. Gates, Jr., *ibid.*, **61**, 3216 (1929).
[5]See **1**, 1202 (1967).
[6]E. Vogel, E. Lohmar, W. A. Böll, B. Söhngen, K. Müllen, and H. Günther, *Angew. Chem., internat. Ed.*, **10**, 398 (1971).

**Zinc carbonate**, $ZnCO_3$. Mol. wt. 125.39. Suppliers: Baker, Fisher, Howe and French, MCB, E. Merck, Riedel de Haën, Sargent.

*Preparation.* Zinc carbonate is precipitated from 0.1 $M$ zinc sulfate solution by addition of 0.1 $M$ sodium carbonate solution and dried under vacuum.

*Coumarin synthesis.* The Pschorr coumarin synthesis[1] involves the reaction of phenols with $\beta$-keto esters in the presence of a condensing agent (usually sulfuric acid). In a synthesis of the coumarin mold metabolite aflatoxin M$_1$ (3), Büchi and Weinreb[2] found that the usual conditions could not be employed because the phenol (1) is exceptionally sensitive to acidic reagents. However, they effected the desired condensation of

(1) and (2) with both zinc carbonate and magnesium carbonate in methylene chloride. On the other hand, sodium and potassium carbonate proved ineffective. The new method was shown to be effective for synthesis of related aflatoxins.

[1]S. Sethna and R. Phadke, *Org. React.*, **7**, 1 (1953).
[2]G. Büchi and S. M. Weinreb, *Am. Soc.*, **93**, 746 (1971).

**Errata for Volume 3**

Page 73, the correct formula for **Diazoethane** is $CH_3CHN_2$.

Page 187, line 9 up, *change* 2-methylcyclopentane *to* 2-methylcyclopentanone.

Page 319, *change* **Olefin synthesis by double extension** *to* **Olefin synthesis by double extrusion.**

Page 347, *change* Pfalts and Bauer *to* Pfaltz and Bauer.

Page 349, under ALKYLATION, *read* Hexamethylphosphoric triamide.

Page 350, *change* to DAKIN–WEST REACTION: N,N-Dimethyl-4-pyridinamine.

Page 352, *change* to OXIDATION OF SULFIDES TO SULFOXIDES.

Page 355, second column, line 11, *change* carbonium monoxide *to* carbon monoxide.

Pages 379–380, these pages should be reversed.


ADDITIONAL SUPPLIERS:

Bio-rad, 32nd and Griffin, Richmond, Calif.

Cationics, Inc., 653 Alpha Drive, Cleveland, Ohio 44143.

Kings Laboratories, Route 1, Box 107, Blythwood, South Carolina 29016.

Lancaster Synthesis Ltd., St. Leonard's House, St. Leonardgate, Lancaster, England.

Orgmet, Inc., 300 Neck Road, Haverhill, Mass. 01830.

PCR, Inc., P.O. Box 1466, Gainesville, Florida 32601

Polysciences, Inc., Paul Valley Industrial Park, Warrington, Pa., 18976.

E. H. Sargent and Co., 4647 West Foster Ave., Chicago, Ill. 60630.

Strem Chemicals, Inc., 150 Andover St., Danvers, Mass.

Willow Brook Laboratories, Inc., P.O. Box 526, Waukesha, Wis. 53186.

# INDEX OF REAGENTS ACCORDING TO TYPES

ACETALS: Ion-exchange resins
ACETYLATION: Bismuth triacetate. Ketene
ACIDS, INORGANIC: Boric acid. Chloroiridic acid. Hydrobromic acid. Hydrochloric acid. Hydrogen bromide. Hydrogen chloride. Hydrogen fluoride. Hydrogen fluoride–Boron trifluoride. Ion-exchange resins. Nitric acid. Periodic acid. Phosphoric acid. Polyphosphoric acid. Sulfuric acid.
ACIDS, ORGANIC: Acetyl sulfuric acid. d-(+)-Camphor-10-sulfonic acid. Formic acid. Methanesulfonic acid. Picric acid. p-Toluenesulfonic acid. Trichloroacetic acid. Trifluoroacetic acid. Trifluoromethanesulfonic acid.
ACYLATION: Phenyltrifluoromethylketene. 4-Pyrrolidinopyridine. Trifluoromethanesulfonic-carboxylic anhydrides.
ACYLATION, INTRAMOLECULAR: Polyphosphoric acid.
ACYLOIN CONDENSATION: Trimethylchlorosilane.
ALDOL CONDENSATION: t-Butylmagnesium chloride. Lithio propylidene-t-butylamine. Sodium tetraborate.
ALKYLATION: Benzyltriethylammonium chloride. n-Butyl mercaptan. Dialkylcarbonium fluoroborates. Diethoxycarbonium fluoroborate. Dimethylcopperlithium. Ethyl vinyl ether. Dimethylformamide dimethyl acetal. Dimethyl sulfide. Dimethylsulfonium methylide. Ethyl chloroformate. Hexamethylphosphoric triamide. Lithium diisopropylamide. Lithium N-isopropylcyclohexylamide. O-Methyldibenzofuranium fluoroborate. Methyl fluorosulfonate. Naththalene–Sodium. Palladium(II) chloride. 1,2,2,6,6-Pentamethylpiperidine. Potassium hydroxide. Silver oxide. Sodium bis-2-methoxyethoxyaluminum hydride. Sodium hydride. Thallous ethoxide.

ALKYLATION OF QUINONES: Nickel carbonyl.
ALLYLIC OXIDATION: Manganic acetate. Mercuric acetate. Selenium dioxide.
π-ALLYLNICKEL HALIDES: Bis-(1,5-cyclooctadiene)nickel(0).
π-ALLYLPALLADIUM COMPLEXES: Palladium acetate. Palladium(II) chloride.
AMIDE-CLAISEN REARRANGEMENT: N,N-Dimethylacetamide dimethyl acetal.
AMIDES, HYDROLYSIS: α-Chloro-N-cyclohexylpropanaldonitrone.
AMINOALKENYLATION: 1-Chloro-N,N,2-trimethylpropenylamine.
AMINODEHYDROXYLATION: Diethyl phosphorochloridate. 4-Chloro-2-phenylquinazoline. Potassium amide.
AMINOMETHYLATION OF ALKYNES: Rhodium oxide.
ANNELATION: Dimethyl acetylenedicarboxylate. Diphenylsulfonium cyclopropylide. trans-3-Pentene-2-one.
AROMATIC HYDROXYLATION: Hydrogen peroxide-Aluminium chloride.
ARYL DIAZONIUM SALTS, DECOMPOSITION: Tetrakis(pyridine)copper(I) perchlorate.
ASYMMETRIC SYNTHESIS: Diborane. Diisopinocamphenylborane. Hydrogen cyanide. (−) and (+)-2,3-O-Isopropylidene-2,3-dihydroxy-1,4-bis(diphenylphosphino)-butane. (+)-(R)-trans-β-Styryl-p-tolyl sulfoxide. p-Tolylsulfinylcarbonion lithium salt.

BAEYER VILLIGER OXIDATION: Carbomethoxyperbenzoic acid. m-Chloroperbenzoic acid. Hydrogen peroxide, acidic. Peracetic acid-BF$_3$·etherate. o-Sulfoperbenzoic acid.
BASES: Ascarite. Barium hydroxide. 1,8-Bis(dimethylamino)naphthalene. n-Butylamine. 1,5-Diazabicyclo[4.3.0]nonene-5 (DBN). 1,4-Diazabicyclo[2.2.2]octane-

(triphenylphosphine)chlororhodium.
DETHIOACETALIZATION: Methyl fluoro-
sulfonate. Sulfuric acid.
DEUTERATION: Gallium oxide.
DIAZO TRANSFER: Copper powder (salts).
p-Toluenesulfonyl azide.
DIELS-ALDER CATALYST: Aluminium
chloride.
DIELS-ALDER REACTIONS: Bis(triphenylphosphine)dicyanonickel(0). 2-Chloro-
acrylonitrile. 2-Chloroacrylyl chloride.
Cyclopropene. Dichloromaleic anhydride.
Dicyanoacetylene. "Indanocyclone." Oxygen, singlet. 4-Phenyl-1,2,4-triazoline-3,5-
dione.
DIIMIDE: 2,4,6-Triisopropylbenzenesulfo-
nyl hydrazide.
DIMERIZATION: Dibromobistriphenyl-
phosphine cobalt(II).
DUFF REACTION: Hexamethylenetetramine.

ELIMINATION OF IODOHYDRINS. Mesyl
chloride. Phosphoryl chloride.
ENOL ACETYLATION: p-Toluenesulfonic
acid.
ENOLATES: Lithium diisopropylamide.
Lithium N-isopropylcyclohexylamide.
EPIMERIZATION: Tetra-n-butylammonium
formate.
EPOXIDATION: N-Bromosuccinimide. Carbomethoxyperbenzoic acid. m-Chloroperbenzoic acid. Diperoxo-oxohexamethyl-
phosphoramidomolybdenium(VI). Molyb-
denium hexacarbonyl. Peractic acid. Peroxybenzimidic acid. o-Sulfoperbenzoic
acid.
ESCHWEILER-CLARKE REACTION:
Formaldehyde.
ESTER CONDENSATIONS: Lithium 2,2,-
6,6-tetramethylpiperidide.
ESTERIFICATION: Boric acid. Boron trifluoride etherate. Hexamethylphosphoric
triamide. Ion-exchange resins. Methyl
t-butyl ether. Triethyloxonium fluoroborate. Trimethyl phosphite.
ETHER CLEAVAGE: Aluminum chloride.
ETHERIFICATION: Dimethyl phosphite.
ETHYLENE TRANSFER: (Dimethylamino)-
phenyl-(2-phenylvinyl)-oxosulfonium

fluoroborate.

FAVORSKII REARRANGEMENT: Sodium
methoxide.
FLUORINATION: Bis(fluoroxyl)difluoro-
methane. Fluoroxytrifluoromethane.
FRAGMENTATION: Diethyl peroxide.
1,3-Propanedithiol. Silver carbonate-
Celite.
FRIEDEL-CRAFTS ACYLATION: Nitro-
ethane. Tetrachloroethane–nitrobenzene.
Trichloroacetyl chloride.

GEMINAL ALKYLATION: Diphenylsulfo-
nium cyclopropylide.
GLYCOL CLEAVAGE: Thallium trinitrate.
GRIGNARD TYPE SYNTHESIS: Lithium.

HALODECARBOXYLATION: Lead tetraacetate.
HALOGENATION: Ethanolamine. Hexamethylphosphoric triamide.
HOMOCONJUGATE ADDITION: Di-meth-
ylcopperlithium. Divinylcopperlithium.
HOMOLOGATION: 1,3-Butadiene monoxide. Ethyl diazoacetate. Isocyanomethyllithium.
HYDRIDE ABSTRACTION: Sodium methylsulfinylmethylide. Tropylium salts.
HYDROALUMINATION: Diisobutylaluminum hydride.
HYDROBORATION: Bis(3,6-dimethyl)bore-
pane. Bis(trans-2-methylcyclohexyl)bo-
rane. Catecholborane. Dicyclohexylborane. Diisopinocamphenylborane. 2,3-
Dimethyl-2-butylborane. Dimethyl sulfide–Borane. Monochloroborane–Diethyl
etherate.
HYDROCYANATION: Diethylaluminum
cyanide. Triethylaluminum.
HYDROFORMYLATION: Rhodium(III)
oxide.
HYDROGENATION CATALYSTS: Bis-
(pyridine)dimethylformamidedichlororho-
dium borohydride. Iron pentacarbonyl.
Lindlar catalyst. Nickel boride. Palladi-
um-on-calcium carbonate. Rhodium-on-
alumina. Rhodium-on-carbon. Ratheni-
um trichloride hydrate. Triron dodecacarbonyl. Tris(triphenylphosphine)chloro-

rhodium. Urushibara catalysts.
HYDROLYSIS: Sodium dihydrogen phosphate, monohydrate.
HYDROSILYLATION: Tris(triphenylphosphine)chlororhodium.
HYDROXYLATION: Hydrogen peroxide–Aluminum chloride. Potassium osmate.
HYDROXYMETHYLATION: Ethyl formate. Lithium diisopropylamide.
HYDROXYPROPYLATION: Ethyl 3-bromopropyl acetaldehyde acetal.

INTRAMOLECULAR CYCLIZATION OF UNSATURATED ARYL CHLORIDES: Tributyltin hydride.
IODINATION: Triphenylphosphite.
ISOMERIZATION: Potassium t-butoxide.
ACETYLENES AND ALLENES: Hydrogen fluoride-Boron trifluoride.
DIENES: Dichloromaleic anhydride.
DITERPENES: Palladium catalysts.
OLEFINS: Cuprous chloride. Lithium diphenylphosphide. Lithium orthophosphate. Potassium t-butoxide. Thiophenol–Azobutyronitrile.
OXASPIROPENTANE: Lithium iodide.

KALIATION: Potassium hydride.
KETALS: Ion-exchange resins.
KNOEVENAGEL CONDENSATION: Titanium tetrachloride.

LACTONES: Ceric acetate.
LACTONIZATION: Piperazine.
LITHIUM ESTER ENOLATES: Lithium N-isopropylcyclohexylamine.

MANNICH REAGENT: Dimethyl(methylene)ammonium iodide.
MEERWEIN-PONNDORF REDUCTION: Aluminum isopropoxide.
MESYLATES: Mesyl chloride.
METALLATION: n-Butyllithium. Cuprous tert-butoxide. N,N,N',N'-Tetramethylethylene diamine.
METATHESIS OF OLEFINS: Tungsten hexachloride.
METHANOLYSIS: Silver perchlorate.
METHOXYLATION: t-Butyl hypochlorite
METHOXYMETHYLATION: Chloromethyl

methyl ether. Methoxycarbenium hexafluoroantimonate. n-Butyl mercaptan.
Methoxycarbenium hexafluoroacetate.
METHYLATION: Trimethyloxonium fluoroborate.
N-METHYLATION: Formaldehyde. Sodium cyanoborohydride.
METHYLENATION OF KETONES: Phenylthiomethyllithium
METHYLENE TRANSFER: Potassium t-butoxide.
METHYLSULFONATION: Methyl fluorosulfonate.
MICHAEL ADDITION: Dimethylcopperlithium. Dimethyloxonium methylide.
trans-β-Styryl-p-tolyl sulfoxide. Tetramethylguanidine.

NITRATION: Dinitrogen tetroxide.
NITROVINYLATION: 2-Nitro-1-dimethylaminoethylene.

OLEFIN ADDITIONS: Bis(1,5-Cyclooctadiene)nickel(0). N-Bromoacetamide.
Lead (IV) acetate azides. Nitrosyl tetrafluoroborate. Sulfur dichloride.
ORGANOMETALLIC REAGENTS: N,N-Bisbromomagnesiumaniline. Bis-(1,5-cyclooctadiene)nickel(0). Bis(dibutylacetoxytin)oxide (see Polymethylhydrosiloxane). 1,3-Bis(methylthio)allyllithium. Bis-(pyridine)dimethylformamido-dichlororhodium borohydride. Bis(triphenylphosphine)dicyanonickel(0). 3-Butenyltri-n-butyltin. t-Butylmagnesium chloride. Di-n-butylcopperlithium. Dichlorobix(benzonitrile)palladium. Dichloromethyllithium. N,N-Dichlorourethane. Dicobaltoctacarbonyl. Di(cobalttetracarbonyl)zinc.
Diethylaluminum chloride. Diethylzinc–Iodoform. Diethylzinc–Methylene iodide.
Diiron nonacarbonyl. Diisobutylaluminum hydride. 2,2'-Dilithiumbiphenyl. Dimethylcopperlithium. Divinylcopperlithium.
Dodecacarbonyltriiron. Ethoxycarbonylmethylcopper. Hydridotris-(triphenylphosphine)carbonylrhodium. Iron(III)-acetylacetonate. Iron pentacarbonyl. Isocyanomethyllithium. Lead acetate. Lead acetate azide. Lead tetraacetate. Lead

tetra(trifluoroacetate). Lithio propylidene-t-butylimine. 2-Lithio-2-trimethylsilyl-1,3-dithiane. Lithium bis(trimethylsilyl)amide. Lithium t-butoxide. Lithium diethylamide. Lithium diisopropylamide. Lithium dimethylcarbamoylnickel tricarbonylate. Lithium diphenylphosphide. Lithium N-isopropylcyclohexylamide. Lithium perhydro-9b-boraphenalyl hydride. Lithium 2,2,6,6-tetramethylpiperidide. Lithium tri-t-butoxyaluminum hydride. Lithium triethylcarboxide. Mercuric acetate. Mercuric trifluoroacetate. Mercury bisdiazoacetic ethyl ester. Methylcopper. Methyllithium. Naphthalene–Lithium. Naphthalene–Sodium. Nickel carbonyl. Palladium(II) acetate. Palldium-(II) chloride. Phenanthrene–Sodium. Phenyl(1-bromo-1-chloro-2,2,2-trifluoromethyl) mercury. Phenylmethoxycarbenepentacarbonyltungsten(O). Phenylpalladium acetate. Phenylthiomethyllithium. Phenyl(trifluoromethyl)mercury. Phenyltrihalomethyl)mercury. Silver diethyl phosphate. Sodium aluminum diethyl hydride. Sodium bis-(2-methoxyethoxy)-aluminum hydride. Sodium N-methylanilide. Sodium tetracarbonylferrate(II). Tetrakis(pyridine)copper(I) perchlorate. Tri-n-butylphosphinecopper iodide complex. Tri-n-butyltin hydride. Triethylaluminum. Triirondodecacarbonyl. Triisobutylaluminum. Triphenylarsine oxide. Triphenyltin hydride. Tris(triphenylphosphine)chlororhodium. Tris(triphenylphosphine)chlororhodium–Triethylsilane. Tris(triphenylphosphine)ruthenium dichloride. Vinylmagnesium chloride.

OXIDATION CATALYST: Selenium.

OXIDATION, REAGENTS: Dimethylsulfoxide–Acetic anhydride. t-Amyl hydroperoxide. N-Bromosuccinimide. Ceric ammonium nitrate. Chloramine. o-Chloranil. 1-Chlorobenzotriazole. N-Chlorosuccinimide–Dimethyl Sulfide. Chromic acid. Chromic anhydride. Chromyl chloride. Cobalt(II) acetate. Cupric acetate monohydrate. Cupric nitrate–Pyridine complex. 2,3-Dichloro-5,6-dicyano-1,4-benzoquinone. Dicyclohexyl-18-crown-

6–Potassium permanganate. Dimethyl sulfide–Chlorine. Dimethyl sulfoxide. Dimethyl sulfoxide–Chlorine. Dimethylsulfoxide–Sulfur trioxide. Dipyridine chromium(VI) oxide. Iodine. Iodine–Potassium iodide. Iodine tris(trifluoroacetate). Iodosobenzene diacetate. Isoamyl nitrite. Lead tetraacetate. Manganese dioxide. Mercuric acetate. Mercuric oxide. Osmium tetroxide–Potassium chlorate. Ozone. Periodic acid. Pertrifluoroacetic acid. Potassium ferrate. Potassium ferricyanide. Potassium nitrosodisulfonate. Ruthenium tetroxide. Selenium dioxide. Silver carbonate. Silver carbonate–Celite. Silver nitrate. Silver oxide. Silver(II) oxide. Sodium hypochlorite. Sulfur trioxide. Thallium(III) nitrate. Thallium sulfate. Thallium(III) trifluoroacetate. Triphenyl phosphite ozonide. Triphenylphosphine dibromide. Trityl fluoroborate.

OXIDATIVE CLEAVAGE OF CYCLOPROPENES: Mercuric acetate.

OXIDATIVE COUPLING: Cuprous chloride. Ferric chloride–Dimethylformamide. Manganese dioxide. Palladium(II) chloride. Potassium ferricyanide. Silver carbonate–Celite. Thallium(III) trifluoroacetate.

OXIDATIVE CYCLIZATION: Mercuric acetate.

OXIDATIVE DECARBOXYLATION: Lead tetraacetate. Copper carbonate, basic.

OXIDATIVE METHYLATION: Silver oxide. Silver(II) oxide.

OXIDATIVE REARRANGEMENT: Cupric chloride. Periodates.

OXYGEN, SINGLET: Potassium perchromate.

OXYMERCURATION: Mercuric acetate.

OZONIZATION: 3-Sulfolene.

PEPTIDE SYNTHESIS: Diphenylphosphoryl azide. N-Ethoxycarbonyl-2-ethoxy-1,3-dihydroquinoline. o-Nitrophenylsulfenyl chloride. 2-Picolyl chloride hydrochloride. Silicon tetrachloride. Triphenylphosphine. Triphenylphosphine–Carbon tetrachloride. Triphenyl phosphite.

PROTECTION OF NITROGEN: t-Butyl azidoformate. Butyloxycarbonylfluoride.

Lithium–*t*-Butanol–Tetrahydrofurane.
REDUCTIVE DEHALOGINATION: Lithium aluminum hydride.
REDUCTIVE DEOXYGENATION OF ALCOHOLS AND KETONES: N,N,N$^1$,N$^1$-Tetramethyldiamidophosphorochloridate.
REFORMATSKY REACTION: Ethyl trichloroacetate. Lithium diisopropylamide.
REVERSE WITTIG REACTION: Triphenylarsine oxide.
RITTER REACTION: Nitronium tetrafluoroborate.
ROSENMUND REDUCTION: Palladium catalysts.

SCHMIDT DEGRADATION: Sodium azide.
SILYLATION: Trimethylsilyldiethylamine.
SIMMONS-SMITH REAGENT, *which see.*
SOMMELET REACTION: Hexamethylenetetramine.
SPIROALKYLATION: Diphenylsulfonium cyclopropylide.
STEPHEN REDUCTION: Nickel-aluminum alloy.
STEVENS REARRANGEMENT-ELIMINATION: Dialkylcarbonium fluoroborates.
SUCCINOYLATION: Succinic anhydride.
SULFONATION: Sulfur trioxide–Pyridine.
SYNTHESIS OF: ACETYLENE: Trimethylsilyldiazomethane.
   ACIDS, ESTERS, AND AMIDES: Sodium tetracarbonylferrate(-II).
   ACETYLENES: Triphenylphosphine–Carbon tetrachloride.
   γ,δ-ACETYLENIC KETONES: Diethylaluminum chloride.
   ADAMANTANE-2-CARBOXYLIC ACID: Dimethyloxosulfonium methylide.
   ALCOHOLS: Bis(2,6-dimethyl)borepane. N-Bromosuccinimide. *n*-Butyldithium. Oxygen. *p*-Tolylsulfinylcarbanion lithium salt.
   ALDEHYDES: 1,3-Bis(methylthio)allyllithium. 2,4-Dimethylthiazole. Ethyl vinyl ether. Methyl methylthiomethyl sulfoxide. 2-Methyl-3-thiazoline. Nickel-aluminum alloy. Sodium tetracarbonylferrate(-II). 1,1,3,3-Tetramethylbutyl isocyanide. 2,4,4,6-Tetramethyl-5,6-dihydro-1,3-(4H)-oxazine. Thallium(III) nitrate.

N,4,4-Trimethyl-2-oxazolinium iodide. *s*-Trithiane.
   α-ALKYLACRYLIC ACIDS: Lithium diisopropylamide.
   N-ALKYLAMIDES: *p*-Toluenesulfonyl chloride.
   N-ALKYLANILINES: Sodium amide.
   ALKYL ARYLSULFIDES: Sodium methoxide.
   O-ALKYLBENZOHYDROXIMOYL CHLORIDES: Phosphorus pentachloride.
   ALKYL BROMIDES: Dimethylbromosulfonium bromide.
   ALKYL CHLORIDES: Molybdenum hexacarbonyl. Trichloroisocyanuric acid.
   α-ALKYLHYDRACRYLIC ACIDS: Lithium diisopropylamide.
   ALKYL IODIDES: N-Methyl-N,N'-dicyclohexylcarbodiimidium chloride. Trichloroisocyanuric acid.
   ALKYL NITRATES: Dinitrogen tetroxide.
   ALKYLSILANES: Dichlorosilane.
   2-ALKYNOIC ESTERS: Thallium (III) nitrate.
   ALKYNYLDIETHYL ALANES: Diethylaluminum chloride.
   ALLENES: Diisopinocamphenylborane. Lithium N-isopropylcyclohexylamine. Magnesium.
   ALLENIC ACIDS: Diethyl(2-chloro-1,1,2-trifluoroethyl)amine.
   ALLYLIC ALCOHOLS: Lithium aluminum hydride–Sodium methoxide.
   *trans*-ALLYLIC ALCOHOLS: Diisobutylaluminum hydride.
   ALLYLIC CHLORIDES: N-Chlorosuccinimide–Dimethyl sulfide. Methanesulfonyl chloride. Triphenylphosphine–Carbon tetrachloride.
   AMIDE ACETALS: Triethyloxonium fluoroborate.
   AMIDES: N-Ethoxycarbonyl-2-ethoxy-1,2-dihydroquinoline (EEDQ). Silicon tetrachloride. Sodium hydride-Dimethylsulfoxide. Triphenyl phosphite.
   AMIDINES: Hexamethylphosphoric triamide.
   AMINES: Phenyldichloroborane. Triphenylphosphine-Diethyl azodicarboxylate.
   AMINO ACIDS: Hydrogen cyanide. (–)

and (+)-2,3,-O-Isopropylidene-2,3-dihy-
droxy-1,4-bis(diphenylphosphine)butane.
Palladium catalysts.
β-AMINO ALCOHOLS: Trimethylchloro-
silane.
AMINOANTHRAQUINONES: N,N-Di-
methylformamide.
ANILINES: 4-Chloro-2-phenylquinazoline.
ANNULENES: Propargyl aldehyde.
ANTHRONES: Pyridene hydrochloride.
APORPHINES: 6-Methoxy-7-hydroxy-3,4-
dihydroisoquinolinium methiodide.
ARYLACETIC ACIDS: Ceric acetate.
ARYLACETYLENES: n-Butylamine.
Iodoethynyl(trimethyl)silane.
ARYL BROMIDES: Thallium triacetate.
ARYLCYCLOPROPANES AND ARYL-
CYCLOPROPENES: Lithium 2,2,6,6-tetra-
methylpiperidide.
ARYL IODIDES: Cupric chloride. Thal-
lium(III) trifluoroacetate.
AZINES: Polyphosphoric acid.
AZIRIDINES: Benzenesulfonyl azide.
AZOALKANES: t-Butyl hypochlorite.
AZOXYALKANES: Triethyloxonium flu-
oroborate.
BENZILS: Thallium(III) nitrate.
BENZOCYCLOPROPENE: Potassium t-
butoxide.
3,4-BENZPYRENE: Pyrene.
BENZYLAMINES: Isovaline.
BENZYLIC HALIDES: N-Chlorosuccin-
imide–Dimethyl sulfide.
BICYCLIC ETHERS: Formic acid.
BIURET TRICHLORIDES: N-Dichloro-
methylene-N,N-dimethylammonium chlo-
ride.
BRANCHED SUGARS: 1,3-Dithiane.
α-BROMO ACIDS AND ESTERS: Barium
hydroxide.
BROMOHYDRINS: N-Bromoacetamide.
γ-BUTYROLACTONES: Hydrogen peroxide, basic.
CARBAMATES: Cupric dimethoxide.
Thallium(III) nitrate.
CARBODIIMIDES: Diethyl azodicarbo-
ylate: Triphenylphosphine–Carbon tetrachloride.
CARBONATES: Bis(triphenylphosphine)-
nickel(O).

CARBOXYLIC ACIDS: Isocyanomethyl-
p-tolyl sulfone.
CARBOXYLIC ACID DIMETHYLAMIDES:
Hexamethylphosphorous triamide.
2-CHLOROALKENES: Phosphorus penta-
chloride.
CHLOROHYDRINS: Trityl chloride.
α-CHLOROKETONES: Chromyl chloride.
β-CHLOROSULFONES: Cupric chloride.
α-CHLOROSULFOXIDES: Diazomethane.
COUMARINS: Pyridine hydrochloride.
Zinc carbonate.
ω-CYANOAMINO ACIDS: o-Nitrophenylsulfenyl chloride.
CYANOHYDRINS: Diethylaluminum cyanide.
CYCLIC CARBONATES: Performic acid.
CYCLIC DISULFIDES: Lead acetate.
CYCLIC KETONES: p-Toluenesulfonyl-
hydrazine.
CYCLOALKENE OXIDES: Potassium t-
butoxide.
CYCLOBUTANE-1,2-DIONE: Bromine.
CYCLOBUTANEDIONES: Chlorotrifluo-
roethylene.
CYCLOBUTANONES: 1-Chloro-N,N,2-
trimethylpropenylamine. Hydrogen bro-
mide. Lithium iodide. Trifluoromethane-
sulfonic anhydride.
CYCLOPENTANONES: Diphenylsulfonium cyclopropylide.
Δ³-CYCLOPENTENONES: 1,3-Dithieni-
um fluoroborate.
CYCLOPROPANONE KETALS: Magne-
sium.
CYCLOPROPANES: Methyl diazo propi-
onate. Copper(I)oxide–t-Butyl isonitrile.
CYCLOPROPENONE: Tri-n-butyltin hydride.
CYCLOPROPYL AZIDES: Mercuric azide.
CYCLOPROPYLCARBINYL COM-
POUNDS: 3-Butenyltri-n-butyltin.
CYCLOPROPYL KETONES: Potassium t-
butoxide. Sodium ethoxide. Sulfuric acid.
1,3-DEHYDROADAMANTANES: n-Butyllithium.
gem-DIALKYLALKANES: Dimethylcopperlithium.
DIALKYLS AND DIARYLS: Silver nitrate.
DIAMANTANE: Aluminium chloride.

1,3,2,4-DIAZADIPHOSPHETIDINES: Hexamethylphosphoric triamide.

DIAZOACETOPHENONE: Diazomethane.

α-DIAZOKETONES: *p*-Toluenesulfonyl azide.

DIBENZO-1,4-DIAZEPINE: Hydroxylamine-O-sulfonic acid.

1,3-DIENES: Dicyclohexylborane. Divinylcopperlithium.

DIHYDRO-3-FURANONES: 2,2-Dimethoxymethyl-1,3-dithiane.

DIHYDRO-γ-PYRONES: 2,2-Dimethoxymethyl-1,3-dithiane.

*vic*-DIOLS: Titanium(III) chloride.

α-DIKETONES: Potassium permanganate–Acetic anhydride. Ruthenium tetroxide.

1,3-DIKETONES: sodium cyanide.

1,4-DIKETONES: Copper(I) acetylacetonate. N,N,N′,N′-Tetramethylsuccinimide.

1,5-DIKETONES: N,N,N¹,N¹-Tetramethylsuccinamide.

1,6-DIKETONES: Vinylmagnesium chloride.

3,3-DIMETHOXYCYCLOPROPENE: Potassium amide.

2-DIMETHYLAMINOBENZO-1,3-DIOXOLE: Dimethylformamide–Thionyl chloride.

*trans*-1,2-DIOLS: Diborane.

1,3-DIOLS: Diborane.

1,6-DIOXA-6a-THIAPENTALENE: Thallium(III) trifluoroacetate.

ENAMINES: Molecular sieves.

ENOL PHOSPHATES: Dimethylcopperlithium.

EPISULFIDES: Iodine thiocyanate.

EPOXIDES: Dimethyloxosulfonium methylide. Hexamethylphosphorous triamide.

EPOXYOLEFINS: Hexamethylphosphorous triamide.

ETHERS: Aluminum chloride. Molybdenum hexacarbonyl. Sodium hydride. Sodium methylsulfinylmethylide. Tri-*n*-butyltin hydride. Trichlorosilane.

N-ETHYLBENZYLAMINE: Diethoxycarbonium fluoroborate.

ETHYLENE KETALS: Mercuric acetate.

α,β-ETHYLENIC SULFONES: *n*-Butyllithium.

1-FORMYLAMINO-1-DIETHYLPHOS-PHONYL ALKENES: Isocyanomethanephosphonic acid diethyl ester.

FURANES: Dimethylsulfonium methylide.

β-GLUCOPYRANOSIDES: Boron trifluoride etherate.

GLYCIDIC ESTERS: Lithium bis(trimethylsilyl)amide.

GLYCIDIC NITRILES: Benzyltriethylammonium chloride.

1,2-GLYCOLS: Dimethyl sulfoxide.

HALOHYDRIN ESTERS: Phosphorus pentachloride.

α-HALOKETONES: Dichloromethyllithium.

HETEROCYCLES: Copper(I) phenylacetylide. Dichlorobis(benzonitrile)palladium. N-Dichloromethylene-N,N-dimethylammonium chloride. Diiminosuccinonitrile. Dimethyl acetylenedicarboxylate. Dipotassium cyanodithioimidocarbonate. Ethoxycarbonyl isothiocyanate. Ethyldiisopropylamine. Ethylene oxide. Hydrogen fluoride. Isocyanomethane-phosphoric acid diethyl ester. Lead tetraacetate. Lithium aluminium hydride. Methylhydrazine. Phosphoryl chloride. Polyphosphate ester. Polyphosphoric acid. Potassium amide. Potassium hydroxide. Tolythiomethyl isocyanide. Tosylmethyl isocyanide. Trichloromethylisocyanide dichloride. Trimethylsilyldiazomethane.

HYDRINDANES: Acetic anhydride–Zinc chloride.

*vic*-HYDROPEROXY ALCOHOLS: Hydrogen peroxide.

HYDROXAMIC ACIDS: Diperoxo-oxohexamethylphosphoramidomolybdenum-(VI).

α-HYDROXY ACIDS: Methyl methylthiomethyl sulfoxide. THALLIUM(III) acetate.

β-HYDROXY ACIDS: Lithium diisopropylamide. Naphthalene-Lithium.

δ-HYDROXYCROTONIC ACIDS: Naphthalene–Lithium.

β-HYDROXYDITHIOCINNAMIC ACIDS: Potassium *t*-butoxide.

IMIDAZOLES: Rhodium oxide.

INDOLES: Ethylenediamine.

IODIDES: Triphenylphosphite methiodide.

ISOCYANATES: Cyanogen bromide.

Sodium iodide.
β-TETRALONES: Methylene chloride.
THIADIAZIRIDINE 1,1-DIOXIDES: Sodium hydride–*t*-Butyl hypochlorite.
THIOAMIDES: Thioacetamide.
TRIALKYLCARBINOLS: Carbon monoxide. Lithium triethylcarboxide.
TRIAZOLES: Ethyl azidoformate. Sodium azide.
TRITYL ETHERS: Triphenylmethyl hexafluorophosphate.
TRISUBSTITUTED OLEFINS: Hydridotris(triphenylphosphine)carbonylrhodium-(I). Hydrobromic acid. 3-Methoxyisoprene. Triethyl orthoacetate.
γ,β-UNSATURATED ACIDS: Zinc.
α,β-UNSATURATED ALDEHYDES: *n*-Butyllithium. 1,3-Bis(methylthio)allyllithium.
β,γ-UNSATURATED ALDEHYDES: 1,3-Dithiane.
γ,δ-UNSATURATED ESTERS: Ethoxycarbonylmethylcopper.
α,β-UNSATURATED KETONES: Dicyclohexylcarbodiimide.
γ,δ-UNSATURATED KETONES: Divinylcopperlithium–Tri-*n*-butylphosphine.
γ,δ-UNSATURATED NITRILES: Cyanomethylcopper.
UREAS: Selenium.
URETHANES: Diphenylphosphoryl azide. Methyl N-sulfonyl urethane triethylamine complex.
VALENCE ISOMERS OF ARENES: Methylene chloride–*n*-Butyllithium.
VINYL ETHERS: Phenylmethoxycarbenepentacarbonyltungsten(O).
VINYL ISOCYANIDES: Isocyanomethane-

phosphonic acid diethyl ester.
VINYLKETENE THIOACETALS: 2-Lithio-2-trimethylsilyl-1,3-dithiane.
VINYL TETRAHYDROPYRANYL ETHERS: Hexamethylphosphoric triamide.
VINYL TRIFLUOROMETHANESULFONATES: Trifluoromethanesulfonic acid.

TELOMERIZATION: Bis(triphenylphosphine)nickel(O). Carbon tetrachloride.
THORPE-ZIEGLER CYCLIZATION: Sodium N-methylanilide.
TOSYLATION: N-Methyl-N-tosylpyrrolidinium perchlorate.
TRANSETHERIFICATION: Silver oxide.
TRIFLUORACETOXYLATION: Lead tetra(trifluoroacetate).
TSCHUGAEFF REACTION: Sodium methylsulfinylmethylide.

ULLMANN REACTION: Copper bronze. Copper powder.

VINYLATION: Dimethylacetamide. Divinylcopperlithium.

WADSWORTH-EMMONS REAGENT: Diethyl 1-(methylthio)ethylphosphonate.
WAGNER-MEERWEIN REARRANGEMENT: *p*-Toluenesulfonic acid.
WITTIG REACTION, *which see.* Triphenylphosphine.
WITTIG REAGENT: Chlorofluoromethylenetriphenylphosphorane.
WURTZ DEHALOGENATION: Sodium.
WURTZ REACTION: Napthalene–Sodium.

# AUTHOR INDEX

602　Author index

614     Author index

# SUBJECT INDEX